AF903471

THE COMPLEX WORLD OF POLYSACCHARIDES

Edited by **Desiree Nedra Karunaratne**

The Complex World of Polysaccharides
http://dx.doi.org/10.5772/2947
Edited by Desiree Nedra Karunaratne

Contributors

Susana P. Miranda Castro, Eva G. Lizárraga Paulín, Stefan Kwiatkowski, Stefan Edgar Kwiatkowski, Rosa Eugenia Reyes, Carolina Romo González, Rafael Coria Jiménez, Maribel Ortiz Herrera, Alejandra Aquino Andrade, Tanya V. Ivashina, Vladimir N. Ksenzenko, Shauna L. Reckseidler-Zenteno, Natalya Nikolaevna Trofimova, Elena Nikolaevna Medvedeva, Nadezhda Viktorovna Ivanova, Yuriy Alekseevich Malkov, Vasiliy Anatolievich Babkin, Mona A. Esawy, Eman F. Ahmed, Wafaa A. Helmy, Nahla M. Mansour, Waled M. El-Senousy, Mounir M. El-Safty, Desiree Nedra Karunaratne, R.G.U. Jayalal, V. Karunaratne, Richard A. Cunha, Thereza A. Soares, Victor H. Rusu, Frederico J.S. Pontes, Eduardo F. Franca, Roberto D. Lins, Aleksandr N. Zimnitskii, V. Poinsot, M.A. Carpéné, F. Couderc, Ranieri Urbani, Paola Sist, Galja Pletikapić, Tea Mišić Radić, Vesna Svetličić, Vera Žutić, Pierre Lembre, Cécile Lorentz, Patrick Di Martino, Amit K. Ghosh, Prasun Bandyopadhyay, Alireza Alishahi, Rosa M. Raybaudi-Massilia, Jonathan Mosqueda-Melgar, Marina Dello Staffolo, Alicia E. Bevilacqua, María Susana Rodríguez, Liliana Albertengo, María Victoria Busi, Mariana Martín, Diego F. Gomez-Casati, Edmund M. K. Lui, Chike G. Azike, José A. Guerrero-Analco, Ahmad A. Romeh, Hua Pei, Sherif J. Kaldas, John T. Arnason, Paul A. Charpentier, María Josefina Carlucci, Cecilia Gabriela Mateu, María Carolina Artuso, Luis Alberto Scolaro, Mohit S. Verma, Frank X. Gu, A. V. Dushkin, T. G. Tolstikova, M. V. Khvostov, G. A. Tolstikov, Máira Regina Rodrigues, Alexandre de Souza e Silva, Fábio Vieira Lacerda, Dodi Safari, Ger Rijkers, Harm Snippe

Published by InTech

Janeza Trdine 9, 51000 Rijeka, Croatia

Copyright © 2012 InTech

All chapters are distributed under the Creative Commons Attribution 3.0 license, which allows users to download, copy and build upon published articles even for commercial purposes, as long as the author and publisher are properly credited, which ensures maximum dissemination and a wider impact of our publications. However, users who aim to disseminate and distribute copies of this book as a whole must not seek monetary compensation for such service (excluded InTech representatives and agreed collaborations). After this work has been published by InTech, authors have the right to republish it, in whole or part, in any publication of which they are the author, and to make other personal use of the work. Any republication, referencing or personal use of the work must explicitly identify the original source.

Notice

Statements and opinions expressed in the chapters are these of the individual contributors and not necessarily those of the editors or publisher. No responsibility is accepted for the accuracy of information contained in the published chapters. The publisher assumes no responsibility for any damage or injury to persons or property arising out of the use of any materials, instructions, methods or ideas contained in the book.

Publishing Process Manager Marina Jozipovic
Technical Editor InTech Prepress, Novi Sad
Cover InTech Design team

Edition 2014

Additional hard copies can be obtained from orders@intechopen.com

The Complex World of Polysaccharides, Edited by Desiree Nedra Karunaratne
p. cm.
ISBN 978-953-51-0819-1

We are IntechOpen, the first native scientific publisher of Open Access books

3,350+
Open access books available

108,000+
International authors and editors

115M+
Downloads

151
Countries delivered to

Our authors are among the
Top 1%
most cited scientists

12.2%
Contributors from top 500 universities

WEB OF SCIENCE™

Selection of our books indexed in the Book Citation Index in Web of Science™ Core Collection (BKCI)

Interested in publishing with us?
Contact book.department@intechopen.com

Numbers displayed above are based on latest data collected.
For more information visit www.intechopen.com

For EU product safety concerns:
IN TECH d.o.o., Prolaz Marije Krucifikse Kozulić 3, 51000 Rijeka, Croatia,
info@intechopen.com or visit our website at intechopen.com.

Meet the editor

Having obtained a B.Sc Chemistry Special (hons) degree from the University of Colombo, Sri Lanka, Desiree Nedra Karunaratne pursued a Ph.D in Bacterial polysaccharides under the tutelage of Prof. G.G.S. Dutton at the University of British Columbia (UBC) in Vancouver, Canada. Later, she researched on bacterial LPS and design of antimicrobial cationic peptides with Prof. R.E.W. Hancock (Dept. of Microbiology and Immunology, UBC). Presently, she is a professor at the University of Peradeniya, Sri Lanka, where she conducts research in the area of carbohydrate liquid crystals, and use of carbohydrate nanoparticles and liposomes for delivery of bioactive compounds. She has 22 publications in peer reviewed journals, 6 US patents and 3 US provisional patents to her credit. She has authored 8 books/chapters.

Contents

Preface

When I was invited by InTech Open Access Publisher to edit a book on polysaccharides, I accepted the challenge since to me polysaccharides constitute a wide variety of biological polymers with diverse composition, physical characteristics and biological activity and have been the focus of my research career. These naturally occurring entities have been studied for their chemical and physical properties and more recently for bioactivity. They have been used in the food industry for functions such as thickeners and protective coatings. Industrial uses of polysaccharides in cosmetics, textiles and medicines are based on rheological, emulsifying and stabilizing properties of polysaccharides. Even though carbohydrates have a long history of chemical and physical study, properties of polysaccharides with relation to structure activity/function has not been an area of in depth study. Polysaccharides of bacterial origin however gained interest in the 1980's due to their potential as vaccine formulations. Therefore the detailed chemical structures of capsular polysaccharides, LPS and exopolysaccharides were elucidated leading to discovery of new naturally occurring sugars. The plant derived polysaccharides such as the hemicelluloses and starch and other specific polysaccharides such as inulin, beta glucans, alginates and pectins are very well documented and have been studied over a longer period of time.

Other than the chemical and physical properties of the polysaccharides, the genetic involvements of the biosynthetic processes which imparts specificity to the structure and thereby its action, have also warranted much study leading to a better understanding of the structure activity relationship. Thus the areas of study of polysaccharides cover several disciplines. In compiling this book, the contributions on polysaccharides from diverse sources such as animals, plants and microorganisms were received and sectioned according to their properties and applications.

The first section deals with sources of polysaccharides and their biological properties. A wide range of polysaccharides from bacterial origin to plants and lichens are presented along with their biological applications. The many applications of chitosan, the most abundant polysaccharide of animal origin, in areas from food, medicine, agriculture, pharmacy and other industries is revisited in the first chapter. The second chapter focusses on glucan polysaccharides, abundantly produced by microorganisms, having properties valuable in food uses. Here the yeast cell wall derived glucans and their products with applications in the food and health industries are presented.

Biosynthesis of Bacterial polysaccharides occurs through elaborate mechanisms. The biosynthetic mechanisms leading to various structural changes in the O-antigens of bacterial polysaccharides are discussed in chapter 3. Variations in structure of the polysaccharides are shown to affect the biological properties and hence pathogenicity. Chapter 4 considers the genetic control of the biosynthesis of exopolysacchrides of Rhizobium leguminosarum. It is shown that diversity of the exopolysaccharides biosynthesized results from genetic rearrangements of the glycosyl transferase genes and other genes involved in translocation of the repeating units. The next chapter is devoted to the study of virulence and pathogenesis due to the capsular polysaccharides of *Burkholderia pseudomallei* . Several studies have been performed on bacterial polysaccharides as candidates for vaccines and it has been shown that virulence is due to changes in the capsular polysaccharide. Plant polysaccharides may be used for specific applications. However, the extraction of polysaccharides from plant waste products in timber industries with conversion of these polysaccharides into useful byproducts is a novel application. Chapter 6 addresses this showing that biomass obtained from large scale processing of Larch wood can be converted into valuable materials with many biological applications. The biological applications of the fructose rich levan polymer found abundantly in honey is discussed in the next chapter. The section ends with a lesser studied polysaccharide source: Lichens composed of a symbiotic relation between algae and fungi yielding polysaccharides which have been investigated for biological significance indicating antitumour, immunomodulatry and anti-inflammatory activities.

Chemical and physical characterizations are important aspects when dealing with understanding the uses of polysaccharides in relation to their properties. The second section discusses methods required for characterization and estimation of spatial arrangement and results obtained therefrom. The five chapters in this section deal with physical properties, methods of characterization and chemical analysis techniques useful for structure determination of a range of polysaccharides from the animal world (chitosan) to polysaccharides from microbes (diatoms). The structural dynamics of chitosan, its conformation and its interactions with biological materials starts off this section. The importance of conformation and molecular modelling is known with drug design studies. Likewise the structural dynamics of polysaccharides are useful for identifying interactions between polysaccharides and biological entities as well as nanoparticles. The use of Quantum chemical methods to explain the template synthesis of proteoglycans is described in the next chapter. Chapter 11 in this section deals with some methods which are essential for elucidating the structure of a polysaccharide. The focus is on the use of mass spectral analysis for determining structures at very low concentration. Other than the basic techniques for chemical characterization, the application of physico-chemical techniques such as laser light scattering and atomic force microscopy is described in the characterization of diatom polysaccharides. The final chapter on biofilm matrices deals with problems encountered in isolation of the biofilm polysaccharides and reviews the chemical and physical methods available.

The remaining two sections comprise applications of polysaccharides in the food industry and applications in the pharmaceutical industry. Various studies on polysaccharides as carriers of drugs, film formers in food protection applications exhibit the versatility of polysaccharides in many areas beneficial for human health. Section 3 deals with applications of polysaccharides in the food industry. Food consists of many components with proteins, fat and polysaccharides being the macro constituents. The composition and the interaction among the food components are important determinants for stability and organoleptic properties of the food item. The importance of polysaccharide-protein interaction and their relevance in food colloids is presented as a factor in emulsion stability determination in chapter 14. In the next chapter, the advantages of antibacterial and antioxidant activities of chitosan in bioactive coatings used for marine based foods, as well as exploiting the physical properties for gelling action and encapsulation are discussed. A whole gamut of functions of polysaccharides in films and coatings as carriers and protectors of bioactive additives and their role in improving food quality follows in chapter 16. Dietary fibre in food imparts health benefits. The polysaccharides cellulose and hemicelluloses are recognized as dietary fibres. The importance of dietary fibres for availability of nutrients is presented in the next chapter. The section ends with the advantages of starch as a source of energy. Modification of the properties of starch through biotechnological manipulation and production of high amylose starches is reviewed in this chapter.

The final section of the book is devoted to pharmaceutical applications involving polysaccharides. The first two chapters deal with the use of polysaccharides as therapeutic agents. The well- known medicinal properties of ginseng with emphasis on the activity of its polysaccharides, followed by the interaction of polysaccharides from red sea weed with virus, starts off this section. The next three chapters are devoted to the use of polysaccharides as carriers of drugs. The first of these, (chapter 21) deals with the use of beta glucans as drug delivery vehicles. Drug delivery vehicles using synthetic polymers as well as natural polymers have been in circulation for some time. With the advances in nanotechnology, nanoparticle use for drug delivery has taken the centre stage. This chapter looks into formation of glucan nanoparticles to enhance the property of the drug delivery vehicle. Chapter 22 on the other hand presents the formation of supramolecular complexes for efficient delivery of poorly water soluble drugs. On a different note, the effect of dietary fibre on availability of drugs with chitosan as the dietary fibre is evaluated in the next chapter. In conclusion, the long standing debate on the use of polysaccharides as vaccines and the future direction of carbohydrates as successful candidates is argued.

As evident from the diversity of the applications of polysaccharides presented in this book, study of carbohydrates brings us to a rare world where the abundance of sources and variety of structures is both mind boggling and intriguing. Carbohydrates have been explored since the beginning of chemical investigations and polysaccharides will continue to exert its sweet essence on researchers dabbling in the chemistry, physics and biology of this ubiquitous biopolymer. There is still a wealth of knowledge to be explored in the study of polysaccharides.

I wish to acknowledge with thanks, the assistance provided by the publishing team at InTech and their courteous service and prompt responses that made this a pleasant task. To the contributors who provided valuable insight into various aspects of polysaccharides, a big thank you. I wish them success in their future endeavours on polysaccharide research. Last but not least, the support and encouragement provided by my husband and family during this assignment which sustained me throughout the project is valued highly.

Professor Desiree Nedra Karunaratne
Department of Chemistry,
University of Peradeniya,
Peradeniya,
Sri Lanka

Section 1

Sources and Biological Properties of Polysaccharides

Chapter 1

Is Chitosan a New Panacea? Areas of Application

Susana P. Miranda Castro and Eva G. Lizárraga Paulín

Additional information is available at the end of the chapter

http://dx.doi.org/10.5772/51200

1. Introduction

Polysaccharides are extremely common in nature and cellulose is the most common organic compound on the planet. It is said that the second most common polysaccharide in the world after cellulose is chitin. "Chitin is to shellfish what cellulose is to trees".

It's been more than two centuries since chitin was discovered formally and considered very important from the scientific and industrial point of view, as it has many applications in many different areas.

The development of commercial applications for chitin and chitosan has progressed. The first known use of chitosan was a durable, flexible film used as a component in the varnish applied to Stradivarius violins, however new efforts are changing its vision in the market. The emphasis on environmentally friendly technology has stimulated interest in biopolymers, which are more versatile and far more biodegradable than their synthetic counterpart.

The purpose of this chapter is to highlight the basic concepts of chemistry and the application of this polysaccharide that is gaining much interest due to the properties it presents and the many applications in various fields. Thousands of scientific articles have been reported in the last 20 years where companies appeared engaging and exploiting this material worldwide. Through investigation many questions have arisen but have not yet answered, however, this polysaccharide has been very successful in many applications.

Furthermore, this chapter aims to convince young readers to further research on possible technology that tend to care for the environment and health.

2. The origin and discovery of chitin

The universe began about 15 million years ago. Materials with high temperature and density were expanded, released energy, then cooled and gave birth to stars, planets and all living beings. The sun was born 5 billion years and 0.4 million years later gave birth to Earth.

INTECH
open science | open minds

© 2012 The Author(s). Licensee InTech. This chapter is distributed under the terms of the Creative Commons Attribution License http://creativecommons.org/licenses/by/3.0), which permits unrestricted use, distribution, and reproduction in any medium, provided the original work is properly cited.

Why talk about the birth of the earth? This is because the chitin could be a constituent of the first living cell. It actually came into existence long before the dinosaurs. In the late Precambrian period, two billion years ago, living cells appeared with nuclei containing chitin around it. In the Silurian period 440 million years, land plants appeared containing cellulose. Fish appeared in the Carboniferous period and later, the arthropods in the Devonian period. The first dinosaurs lived two hundred million years ago and during the second half of the Jurassic period the crab rich in chitin appeared.

After dinosaurs occupied the Earth for 100 million years, from the Jurassic to Cretaceous, they were extinguished by a comet that crashed into the Yucatan Peninsula 65 million years ago, but crabs and small animals escaped this catastrophy.

Since living beings appeared, cellulose and chitin have been beneficial in general and both maintained an ecological balance. Chitin is the animal version of the cellulose and it is the second most abundant in nature, but Professor M Peter has challenged that assumption by saying that Chitin is certainly a very abundant material even if much of it is not readily accessible for industrial use and suggested that hemicelluloses, which occur in conjunction with cellulose in trees and other plants, are actually more abundant than chitin. The hemicellulose component averages about half of the cellulose component, whereas the normal estimate of chitin production is that it is one whole order of magnitude less than that of cellulose. Another possible contender is lignin, which again occurs in conjunction with cellulose in most plants and, like hemicelluloses, averages about half of the cellulose component. A fourth possible contender is starch which like cellulose it is a major component of vegetable matter where it acts as a reserve material rather than a structural component [1].

The English word "chitin" comes from the French word chitine, which first appeared in 1836. These words were derived from the Greek word chitōn, meaning mollusk that is influenced by the Greek word khitōn, meaning "tunic" or "frock". That word may come from the Central Semitic word *kittan, the Akkadian words kitû or kita'um, meaning flax or linen, and the Sumerian word gada or gida. A similar word, "chiton", refers to a marine animal with a protective shell (also known as a "sea cradle" [2].

It is normally accepted as a fact that chitin was first isolated from mushrooms and called "fungine" by the French chemist Henri Bracconot in 1811. Charles Jeuniaux suggested in a paper presented at the 1st International Conference of the European Chitin Society held in Brest in 1995, that chitin had previously been isolated from arthropod cuticle by the English scientist A Hachett in 1795. However, as pointed out by Professor Jeuniaux, Hachett only reported the presence in the cuticle of an organic material particularly resistant to the usual chemical reagents but did not investigate it further. Braconnot on the other hand carried out chemical analysis on his fungical culture, and reported the formation of acetic acid from it on treatment with hot acid, and concluded it was a new material. Braconnot may be considered the discoverer of chitin even though his name for the new material, 'fungine', was soon replaced by its current name "chitin"which was first proposed by Odier [3,4].

Chitin is a big molecule composed of –beta-1,4-N-acetylglucosamine (GlcNAc) monomers. There are three forms of chitin: α, β, and γ chitin. The α-form, is mainly obtained from crab and shrimp. Both α and β chitin/chitosan are commercially available [5].

3. Sources of chitin

In the book "Chitin" published by Muzarelli in 1977, we can find a complete list of organisms that contain chitin: Fungi, Algae, Cnidaria (jellyfish), Aschelminthes (round worm), Entoprocta, Bryozoa (Moss or lace animals, Phoronida (Horseshoe worms), Brachiopoda(Lamp shells), Echiruda, Annelida (Segmented worms), Mollusca, Arthropoda and Ponogophora [6]). Herring, P.J in 1979 wrote that chitin is the main component of arthropod exoskeletons, tendons, and the linings of their respiratory, excretory, and digestive systems, as well as insects external structure and some fungi. It is also found in the reflective material (iridophores) of both epidermis and eyes of arthropods and cephalopods (phylum: Mollusca) and the epidermal cuticle of the vertebrate Paralipophrys trigloides (fish) is also chitinous [7,8]. The main commercial sources of chitin are the shell wastes of shrimp, lobsters, crabs and krill. There are three forms of chitin: α, β, and γ chitin. The α-form, is composed of alternating antiparallel polysaccharide strands and is mainly obtained from crab and shrimp. α -Chitin is by far the most abundant; it occurs in fungal and yeast cell walls, krill, lobster and crab tendons and shells, shrimp shells, and insect cuticle. The rarer β -chitin is composed of parallel strands of polysaccharides, is found in association with proteins in squid pens [9,10] and in the tubes synthesized by pogonophoran and vestimetiferan worms [11,12]. It also occurs in aphrodite chaetae [13] as well as in the lorica, built by some seaweeds or protozoa [14,15,16]. And 2 parallel chains alternating with an antiparallel strand constitute gamma chitin and are found in fungi [15].

Figure 1. Chitinous structure of worm and insects

4. Chitin from crustacean

Currently most commercial production of chitin is based on extracting it from the exoskeleton of shrimp, prawn, crab and other crustaceans. This source contains a high percentage of inorganic material, primarily $CaCO_3$ and a rough calculation indicates that for every tonne of chitin produced, 0.8 tonne of CO_2 is released into the environment. In view of current concerns about global warming this cannot be considered to be a truly environmentally friendly process [3].

Another source of chitin that is more environmentally friendly, although much more limited in volume, is squid pen. This waste contains very little in the way of inorganic material and very little, if any CO_2 would be released in the extraction and purification process. Another and perhaps more sustainable source in the long run is vegetable chitin from fungal sources such as waste mycelia. There is extensive literature on the topic, but it is only recently that it has become commercially available [3].

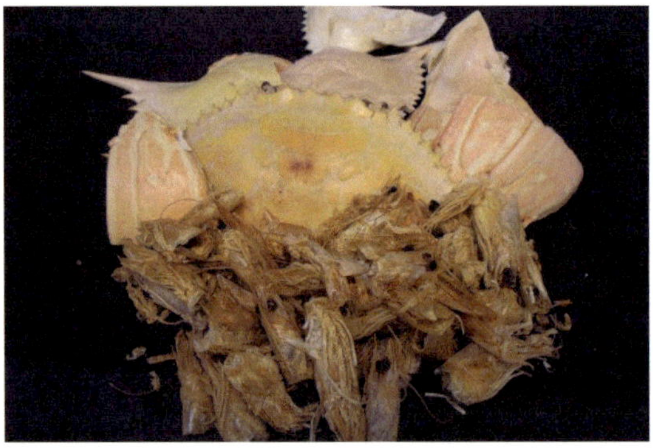

Figure 2. Exosqueleton of crustacea, this is the source of commercial chitin

5. Chitin from fungi

Chitin is widely distributed in fungi, occurring in *Basidiomycetes, Ascomycetes,* and *Phycomycetes,* where it is a component of the cell walls and structural membranes of mycelia, stalks, and spores. The amounts vary between traces and up to 45% of the organic fraction, the rest being mostly proteins, glucans and mannans. However, not all fungi contain chitin, and the polymer may be absent in one species that is closely related to another. Variations in the amounts of chitin may depend on physiological parameters in natural environments as well as on the fermentation conditions in biotechnological processing or in cultures of fungi [4].

The chitin in fungi possesses principally the same structure as the chitin occurring in other organisms. However, a major difference results from the fact that fungal chitin is associated with other polysaccharides which do not occur in the exoskeleton of arthropods. The molecular mass of chitin in fungi is not known. However, it was estimated that bakers' yeast synthesizes rather uniform chains containing 120 ± 170 GlcNAc monomer units which corresponds to 24,000 ± 34,500 Daltons [4].

6. Chemical methods to prepare chitin

Several procedures have been developed through the years to prepare chitin; they are at the basis of the chemical processes for industrial production of chitin and chitosan. Various methods are reported in Muzzarelli's book such as: Method of Rigby (1936 and 1937); Hackman (1954); Foster and Hackman (1957); Horowitz, Roseman and Blumenthal (1957);

Whistler and Be Miller (1962); Takeda and Abe (1962); Takeda and Katsuura (1964); Broussignac (1968); Lovell, Lafleur and Hoskins (1968); Madhavan and Ramachandran (1974) [6]. There is also a review that summarizes methods of preparation of various chitin and its conversion to chitosan [17].

7. Enzymatic methods to prepare chitin

A new process for deproteinization of chitin from shrimp head was studied [18]. Recovery of the protein fraction of the shrimp waste has been widely studied by enzymatic hydrolysis method [19,20].The enzymatic deproteinization process has limited value due to residual small peptides directly attached to chitin molecules ranging from 4.4% to 7.9% of total weight [21]. As these processes are costly because of the use of commercial enzymes, there is now a need to develop an efficient and economical method for extracting proteins from shellfish waste. One interesting new technology for extraction of chitin that offers an alternative to the more harsh chemical methods is fermentation by using microorganisms. Fermentation has been envisaged as one of the most ecofriendly, safe, technologically flexible, and economically viable alternative methods [22-28]. Fermentation of shrimp waste with lactic acid bacteria results in production of a solid portion of chitin and a liquor containing shrimp proteins, minerals, pigments, and nutrients [26,29]. Deproteinization of the biowaste occurs mainly by proteolytic enzyme produced by Lactobacillus [30]. Lactic acid produced by the process of breakdown of glucose, creating the low pH condition of ensilation; suppress the growth of microorganisms involved in spoilage of shrimp waste [31]. The lactic acid reacts with calcium carbonate component in the chitin fraction leading to the fermentation of calcium lactate, which gets precipitated and can be removed by washing. There is now a need to develop an efficient, simpler, eco- friendly, economical, and commercially viable method.

8. Chitosan

Despite the wide spread occurrence of chitin, up to now the main commercial sources of chitin have been crab and shrimp shells. In industrial processing, chitin is extracted from crustaceans by acid treatment to dissolve calcium carbonate followed by alkaline extraction to solubilized proteins. In addition a decolorization step is often added to remove leftover pigments and obtain a colorless product. These treatments must be adapted to each chitin source but by partial deacetylation under alkaline conditions, one obtains "chitosan" [16]. Chitosan is the most important derivate of this naturally occurring polymer being one of the most abundant polysaccharides after cellulose. Chitosan is a copolymer composed of N-acetyl-D-glucosamine and D-glucosamine units. It is obtained in three different ways, thermochemical deacetylation of chitin in the presence of alkali, by enzymatic hydrolysis in the presence of a chitin deacetylase, or naturally found in certain fungi as part of their structure. In chitosan part of the amino groups remain acetylated. It is generally accepted that N-acetylglucosamine residues are randomly distributed along the whole polymer chain. In an acid medium, amino groups are protonated and thus determine the positive charge of the chitosan molecule. Thus, chitosan behaves like a polycation in solution [32]. Properties of chitosan, such as the mean polymerization degree, the degree of N-deacetylation, the positive charge, and the nature of chemical modifications of its molecule, strongly influence its biological activity.

Chitin contains 6–7% nitrogen and in its deacetylated form, chitosan contains 7–9.5% nitrogen. In chitosan, between 60 to 80% of the acetyl groups available in chitin are removed [33]. The chain distribution is dependant on the processing method used to obtain biopolymer [34-36]. It is the N-deacetylated derivative of chitin, but the N-deacetylation is almost never complete [35]. Chitin and chitosan are names that do not strictly refer to a fixed stoichiometry. Chemically, chitin is known as poly-N-acetylglucosamine, and in accordance to this proposed name, the difference between chitin and chitosan is that the degree of deacetylation in chitin is very little, while deacetylation in chitosan occurs to an extent but still not enough to be called polyglucosamine [37].

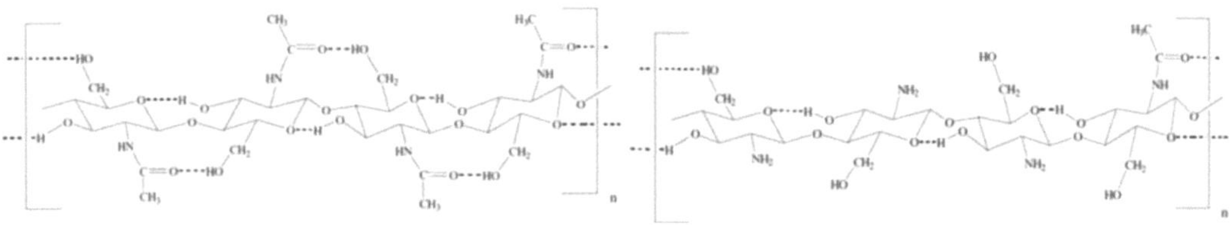

Figure 3. Chitin and chitosan chemical structure

9. Sources of chitosan

Chitosan is commercially produced from deacetylated chitin found in shrimp and crab shell. However, supplies of raw materials are variable and seasonal and the process is laborious and costly [38]. Furthermore, the chitosan derived from such process is heterogeneous with respect to its physiochemical properties [38]. Recent advances in fermentation technology provide an alternative source of chitosan. Fungal cell walls and septa of *Ascomycetes, Zygomycetes, Basidiomycetes* and *Deuteromycetes* contain mainly chitin, which is responsible for maintaining their shape, strength and integrity of cell structure [38]. The production of chitosan from fungal mycelia has a lot of advantages over crustacean chitosans such as the degree of acetylation, molecular weight, viscosity and charge distribution of the fungal chitosan. They are more stable than crustacean chitosans. The production of chitosan by fungus in a bioreactor at a technical scale offers also additional opportunities to obtain identical material throughout the year. The fungal chitosan is free of heavy metal contents such as nickel, copper [39-41]. Moreover the production of chitosan from fungal mycelia gives medium-low molecular weight chitosans (1–12 × 10^4 Da), whereas the molecular weight of chitosans obtained from crustacean sources is high (about 1.5 ×10^6 Da) [41]. Chitosan with a medium-low molecular weight has been used as a powder in cholesterol absorption [42] and as thread or membrane in many medical-technical applications. For these reasons, there is an increasing interest in the production of fungal chitosan.

There are some examples of chitosan extracted from fungi. Chitosans isolated from *Mucorales* typically show Mw in the range 4 x 10^5 to 1.2 x 10^6 Daltons and FA values between 0.2 and 0.09. Amino acid analysis of chitosan prepared from *Aspergillus niger* reveals covalently bound arginine, serine, and proline. Nadarajah et. al., 2001, studied chitosan production from mycelia of *Rhizopus* sp KNO1 and KNO2, *Mucor sp* KNO3 and *Asperigullus niger* with the highest amount of extractable chitosan obtained at the late exponential phase. *Mucor* sp KNO3, produced the highest amount of 557mg per 2.26 g of dry cell weight /250 ml of culture. Kishore et. al.(1993), examined the production of chitosan from mycelia of *Absidia coerulea, Mucor rouxii, Gongronella butieri, Phvcomyces blakesleeanus* and *Absidia blakesleeana*. Chitosan yields of *A. coerulea, M. rouxii, G. butieri, P. blakesleeanus* and *A. blakesleeana* were 47–50, 29–32, 21–25, 6 and 7 mg/100 mL of medium, respectively. The degree of acetylation of chitosan ranged from 6 to 15%; the lowest was from strains of *A. coerulea*. Viscosity average molecular weights of fungal chitosans were equivalent, approximately 4.5 x 10^5 Daltons. Wei-Ping Wang et.al., (2008) evaluated the physical properties of fungal chitosan from *Absidia coerulea* (AF 93105), *Mucor rouxii* (Ag 92033), and *Rhizopus oryzae* (Ag 92033). Their glucosamine contents and degrees of deacetylation (DD) were over 80%, differences had been observed in their molecular weight (Mw), ranging from 6.6 to 560 kDa. Chitosan was isolated and purified from the mycelia of *Rhizomucor miehei* and *Mucor racemosus* with a degree of deacetylation of 97 y 98 respectively [43-45].

Considerable research has been carried out on using mycelium waste from fermentation processes as a source of fungal chitin and chitosan. It is argued that this would offer a stable non-seasonal source of raw material that would be more consistent in character than shellfish waste, but so far this route does not appear to have been taken up by chitosan producing companies. Currently there is only one commercial source of fungal chitosan and is produced by the company Kitozyme. However their raw material is not mycelium waste from a fermentation process, which is what is normally envisaged when fungal chitosan is referred to, but actually conventional edible mushrooms grown under contract in France and shipped to Belgium for processing. So mycelium waste still remains a vast and as yet untapped potential source of chitosan.

10. Genetic engineering approach to produce chitin

It is difficult to obtain pure carbohydrates, especially chitin, through conventional techniques. Bacterial cells have been engineered in an effort to overcome this problem [46]. *E. coli* has been engineered to produce chitobiose. This method took advantage of NodC, which is a chito-oligosaccharide synthase, and genetically engineered chitinase to make a cell factory with the ability to produce chito-oligosaccharides [47]. Recombinant chito-oligosaccharides have also been obtained using *E coli* cells which expressed nodC or nodBC genes [48]. By expressing different combinations of nod genes in *E. coli*, O-acetylated and sulfated chito-oligosaccharide have been produced [49].

11. Parameters influencing the behavior of the biopolymer

The main parameters influencing the characteristics of chitosan are its degree of deacetylation (DD) and molecular weight (Mw), which affect the solubility, rheological and physical properties. Various grades of chitosan are available commercially, which differ primarily in the degree of deacetylation and molecular weight. Different conditions such as type and concentration of reagents, time and temperature employed throughout the processing can affect the physical characteristics and performance of the final chitosan product [50]. However, both DD and molecular weight can be further modified. For example, DD can be lowered by reacetylation [51-55] and molecular weight can be lowered by acidic or enzymatic depolymerisation [56-58].

12. Degree of Deacetylation (DD)

Deacetylation describes a reaction that removes an acetyl functional group. When the degree of deacetylation of chitin reaches about 50% (depending on the origin of the polymer), it becomes soluble in aqueous acidic media and is called chitosan. The solubilization occurs by protonation of the $-NH_2$ function on the C-2 position of the D-glucosamine repeat unit, whereby the polysaccharide is converted to a polyelectrolyte in acidic media. Chitosan is the only pseudonatural cationic polymer and thus, it finds many applications that follow from its unique character (flocculants for protein recovery, depollution, etc.). Being soluble in aqueous solutions, it is largely used in different applications as solutions, gels, or films and fibers.

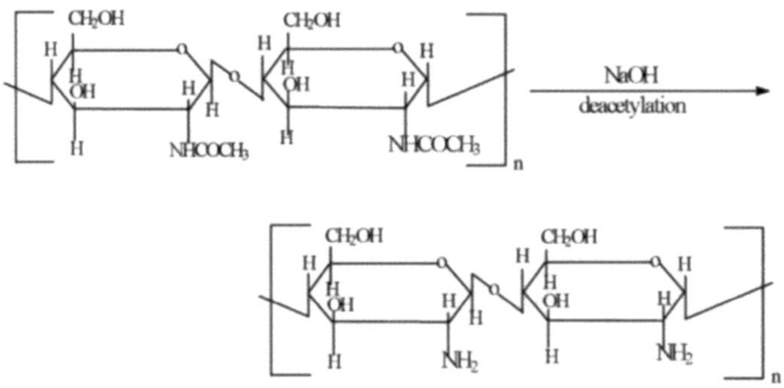

Figure 4. Chitin deacetylation

A highly deacetylated polymer has been used to explore methods of characterization [59]. The solution properties of a chitosan depend not only on its average DA but also on the distribution of the acetyl groups along the main chain in addition of the molecular weight [60-62]. Several methods have been proposed for alkaline deacetylation to obtain chitosan [6,17]. The conditions used in the deacetylation determines the polymer molecular weight and degree of deacetylation (DD).

Chitosan has been largely employed in many areas, such as photography, biotechnology, cosmetics, food processing, biomedical products (artificial skin, wound dressing, contact lens, etc.), system of controlled liberation of medicines (capsules and microcapsules), treatment of industrial effluents for removal of metallic and coloring ions. The amino groups are responsible for the distinct characteristics attributed to this basic polymer (compared to an acidic biopolymer). Therefore, the characterization of the polymer in either chitin or chitosan is extremely important according to the structure-properties relationship, defining a possible industrial application. Thus many techniques are available to determine the degree of deacetylation. Elson Santiago de Alvarenga (2011) published on line describing the most important parameters to be evaluated in chitosan as "deacetylation degree" (DD) [63].

The methods for carrying out the analysis of the degree of deacetylation are: Elemental analysis; Titration; HPLC; Infrared; ^{1}H nuclear magnetic resonance; CP-MAS ^{13}C NMR; CP-MAS ^{15}N NMR; steric exclusion; nitrous acid deamination; thermal analysis.

13. Molecular weight

Another important characteristic to consider for these polymers is the molecular weight and its distribution. The first difficulty encountered in this respect concerns the solubility of the samples and dissociation of aggregates often present in polysaccharide solutions [16, 57, 64, 65, 66]. As to choice of a solvent for chitosan characterization, various systems have been proposed, including an acid at a given concentration for protonation together with a salt to screen the electrostatic interaction. The solvent is important also when molecular weight has to be calculated from intrinsic viscosity using the Mark–Houwink relation.

14. Biological properties of chitosan

14.1. Biocompatibility

Biocompatibility of a biomaterial refers to the extent to which the material does not have toxic or injurious effects on biological systems [67, 68]. One of the present trends in biomedical research requires materials that are derived from nature as natural materials have been shown to exhibit better biocompatibility with humans and because chitosan's monomeric unit, N-acetylglucosamine, occurs in hyaluronic acid, an extracellular macromolecule that is important in wound repair. Additionally, the N-acetylglucosamine moiety in chitosan is structurally similar to glycosaminoglycans (GAGs), heparin, chondroitin sulphate and hyaluronic acid in which they are biocompatible, and hold the specific interactions with various growth factors, receptors and adhesion proteins besides being the biologically important mucopolysaccharides and in all mammals. Therefore, the analogous structure in chitosan may also exert similar bioactivity and biocompatibility [69, 70].

The potential of chitosan stems from its cationic nature and high charge density in solution. An effective approach for developing a clinically applicable chitosan is to modify the surface of the material that already has excellent biofunctionality and bulk properties [71]. Altering the physical and chemical properties of the chitosan in order to improve its medicinal quality will also influence its biocompatibility [69.70].

The excellent biological properties of chitosan can be potentially improved with a variety of additional chemicals such as polyethylene glycol and carboxymethyl N-acyl groups in order to produce biocompatible chitosan derivatives for use as wound dressings [72]. Chitosan's positive surface charge enables it to effectively support cell growth [73]. Chitosan-gelatin sponge wound dressing had demonstrated a superior antibacterial effect. Additionally, chitosan gelatin sponge allowed the wound site to contract markedly and shortened the wound healing time, as compared with sterile Vaseline gauze [74]. Widely used surface modification techniques include coating, oxidation by low temperature plasma for better printing and adhesion and surfactant addition for antistatic. Blends are often used to improve tensile properties and to provide a stronger structural component for separation media that supports the active polymer. The physical properties of a polymer can also be altered by introducing a second polymer that improves the properties of the original polymer in certain aspects, such as hydrophobility, lowered melt temperature, raised glass transition temperature, etc [75]. A thorough understanding of cell and proteins interactions with artificial surfaces is of importance to design suitable implant surfaces and substrates. The surface properties of newly synthesized biomedical grade chitosan derivatives, including surface composition, wettability, domain composition, surface oxidation, surface charge and morphology, may influence protein adsorption and subsequently, the cellular responses to biomaterial implants [76-81].

15. Biodegradability

The claim "biodegradable" is often associated with environmentally friendly products. It is defined as being able to be broken down by natural processes, into more basic components. Products are usually broken down by bacteria, fungi or other simple organisms [82].

An important aspect in the use of polymers as drug delivery systems is their metabolic fate in the body or biodegradation. In the case of the systemic absorption of hydrophilic polymers such as chitosan, they should have a suitable Mw for renal clearance. If the administered polymer's size is larger than this, then the polymer should undergo degradation. Biodegradation (chemical or enzymatic) provides fragments suitable for renal clearance. Chemical degradation in this case refers to acid catalysed degradation i.e. in the stomach. Enzymatically, chitosan can be degraded by enzymes able to hydrolyse glucosamine–glucosamine, glucosamine–N- acetyl-glucosamine and N-acetyl-glucosamine–N-acetyl- glucosamine linkages [83]. Even though depolymerisation through oxidation–reduction reaction [84] and free radical degradation [85] of chitosan have been reported these are unlikely to play a significant role in the in vivo degradation.

Chitosan is thought to be degraded in vertebrates predominantly by lysozyme and by bacterial enzymes in the colon [83, 86]. However, eight human chitinases (in the glycoside hydrolase 18 family) have been identified, three of which have shown enzymatic activity

[87]. A variety of microorganisms synthesises and/or degrades chitin, the biological precursor of chitosan. In general, chitinases in microorganisms hydrolyze N-acetyl-β-1,4-glucosaminide linkages randomly i.e. they are endo-chitinases (EC 3.2.1.14). Chitinases are also present in higher plants, even though they do not have chitin structural components.

Chemical characterisation assays determining the degradation of chitosan commonly use viscometry and/or gel permeation chromatography to evaluate a decrease in Mw [88]. Lysozyme has been found to efficiently degrade chitosan; 50% acetylated chitosan had 66% loss in viscosity after a 4 h incubation in vitro at pH 5.5 (0.1 M phosphate buffer, 0.2 M NaCl, 37 °C) [88]. This degradation appears to be dependent on the degree of acetylation with degradation of acetylated chitosan (more chitin like) showing the faster [89,90].

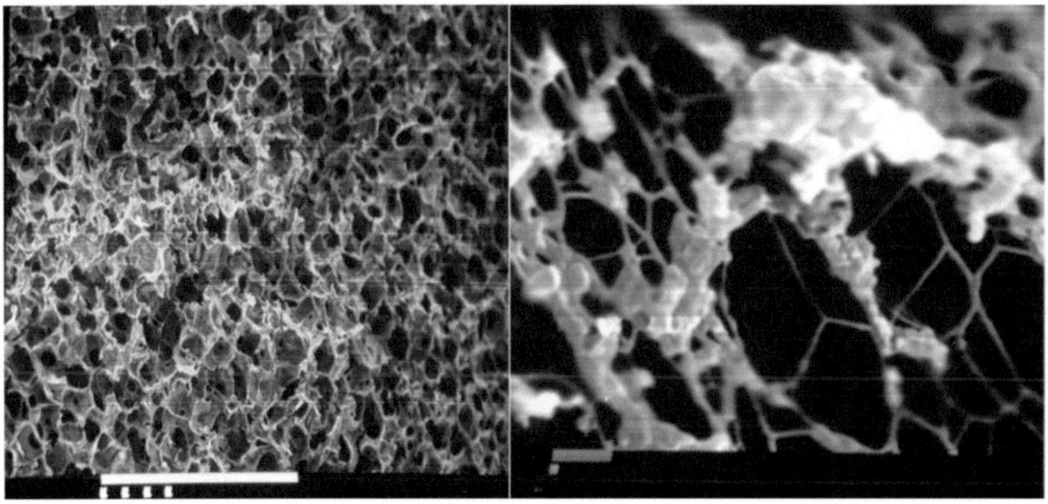

Figure 5. Biodegradation of chitosan thermosensible gel inside the rat's body after 5 days.

16. Safe biomaterial

Chitosan is a potentially biologically compatible material that is chemically versatile (–NH_2 groups and various Mw). These two basic properties have been used by drug delivery and tissue engineering to create a great amount of formulations and scaffolds that show promise in healthcare. It is approved for dietary applications in Japan, Italy and Finland [91] and it has been approved by the FDA for use in wound dressings [92] but is not approved for any product in drug delivery. The term "Chitosan" represents a large group of structurally different chemical entities that may show different biodistribution, biodegradation and toxicological profiles.

The formulation of chitosan with a drug may alter the pharmacokinetic and biodistribution profile.The balancing, or reduction, of the positive charges on the chitosan molecule has effects on its interaction with cells and the microenvironment, often leading to decreased uptake and a decrease in toxicity. The modifications made to chitosan could make it more or less toxic and any residual reactants should be carefully removed. In addition, the route of administration determines the uptake, concentration, contact time and cell types affected.

Chitosan details (DD, MW)	Modification	Assessment	IC50
95% DD, 18.7 kDa	Steric acid conjugation micelle	In vitro ,A549 cells	369±27 μg/ml
95% DD, 18.7 kDa	teric acid conjugation and entrapment in micelle	In vitro ,A549 cells	234±9 μg/ml
97% DD, 65 kDa	N-octyl-O-sulphate	Invitro, primary rat hepatocytes	>200 mg/ml
87% DD, 20, 45, 200, 460 kDa	None, aspartic acid salt	In vitro, Caco-2 cells, pH 6.2	0.67±0.24, 0.61±0.10, 0.65±0.20, 0.72±0.16 mg/ml
87% DD, 20, 45, 200, 460 kDa	None, glutamic acid salt		0.56±0.10, 0.48±0.07, 0.35±0.06, 0.46±0.06 mg/ml
87% DD, 20, 45, 200, 460 kDa	None, Lactic acid salt		0.38±0.13, 0.31±0.06, 0.34±0.04, 0.37±0.08 mg/ml
87% DD, 20, 45, 200, 460 kDa	None, hydrochloride salt		0.23±0.13, 0.22±0.06, 0.27±0.08, 0.23±0.08 mg/ml
78% DD, <50 kDa	None, lactic acid salt	In vitro B16F10 cells	2.50 mg/ml
82% DD, 150–170 kDa	None, lactic acid salt	In vitro B16F10 cells	2.00±0.18 mg/ml
>80% DD, 60–90 kDa	None, glutamic acid salt	In vitro B16F10 cells	2.47±0.14 mg/ml
77% DD, 180–230 kDa	None, lactic acid salt	In vitro B16F10 cells	1.73±1.39 mg/ml
85% DD, 60–90 kDa	None, hydrochloric acid salt	In vitro B16F10 cells	2.24±0.16 mg/ml
81% DD, 100–130 kDa	None, hydrochloric acid salt	In vitro B16F10 cells	0.21±0.04 mg/ml
100% DD, 152 kDa	Glycol chitosan	In vitro B16F10 cells	2.47±0.15 mg/ml
100% DD, 3–6 kDa	20, 44, 55% Trimethyl chitosan, chloride salt	In vitro, MCF7 and COS7 cells, 6 h & 24 h	>10 mg/ml
100% DD, 3–6 kDa	94% Trimethyl chitosan, chloride salt	In vitro, MCF7, 6 h	1.402±0.210 mg/ml
100% DD, 3–6 kDa	94% Trimethyl chitosan, chloride salt	In vitro, COS7, 6 h	2.207±0.381 mg/ml
100% DD, 100 kDa	36% Trimethyl chitosan, chloride sal	In vitro, MCF7, 6 h	0.823±0.324 mg/ml
100% DD, 100 kDa	36% Trimethyl chitosan, chloride sal	In vitro, COS7, 6 h	>10 mg/ml

Chitosan details (DD, MW)	Modification	Assessment	IC50
84.7% DD, 400, 100, 50, 25, 5 kDa	40% Trimethyl chitosan	In vitro, L929 cells, 3 h	30, 70, 90, 270, >1000 μg/ml
84.7% DD, 1.89 MDa	12% PEG modified 40% trimethyl chitosan	In vitro, L929 cells, 3 h	220 μg/ml
84.7% DD, 3.6 MDa	25.7% PEG modified 40% trimethyl chitosan	In vitro, L929 cells, 3 h	370 μg/ml
84.7% DD, 300 kDa	6.44% PEG modified 40% trimethyl chitosan (and all PEG modified TMC with lower Mw)	In vitro, L929 cells, 3 h	>500 μg/ml
97% DD, 65 kDa	N-octyl-O-sulphate	In vivo, IV, mice	102.59 mg/kg
97% DD, 65 kDa	N-octyl-O-sulphate	n vivo, IP, mice	130.53 mg/kg

Table 1. Toxicity of chitosan and chitosan derivatives
Table taken from [94]

In a series of articles are described the effects of chitosans with differing molecular weights and degree of deacetylation in vitro and in vivo. Toxicity was found to be degree of deacetylation and molecular weight dependent. At high DD the toxicity is related to the molecular weight and the concentration, at lower DD toxicity is less pronounced and less related to the molecular weight [93].

A summary of toxicities of chitosan and derivatives assessed through IC_{50} values is presented in the next table [94].

From this table it can be gathered that most chitosans (and derivatives) are not significantly toxic compared to a toxic polymer such as polyethylenimine [94].

It appears that the toxicity of chitosan is related to the charge density of the molecule, toxicity increases with increasing density. It appears that there is a threshold level below which there are too few contact points between a molecule and the cell components to produce a significantly toxic effect. This balance is between 40 and 60% DD, or degree of trimethylation, although any sufficiently small chitosan (<10 kDa) is not appreciably toxic. Modifications that do not increase the charge on the molecule seem to have little effect on the toxicity beyond that of the native molecule [94].

17. Antimicrobial activity

The exact mechanism by which chitosan exerts its antimicrobial activity is currently unknown, it has been suggested that the polycationic nature of this biopolymer that forms from acidic solutions below pH 6.5 is a crucial factor. Thus, it has been proposed that the positively charged amino groups of the glucosamine units interact with negatively charged components in microbial cell membranes, altering their barrier properties, and thereby preventing the entry of nutrients or causing the leakage of intracellular contents [95-100]. Another reported mechanism involves the penetration of low- molecular weight chitosan into the cell, binding to DNA and the subsequent inhibition of RNA and protein synthesis [101]. Chitosan has also been shown to activate several defence processes in plant tissues and it inhibits the production of toxins and microbial growth because of its ability to chelate metal ions [102,103].

EI-Ghaouth et. al.(1992) have proposed possible antibacterial actions of chitosan and its derivatives [104]. They asserted that chitosan reacted with the cell surface, altered cell permeability, and further prevented the entry of material or caused the leakage of material. However, no evidence has been provided to demonstrate the relationship between the antibacterial activity of chitosan and the surface characteristics of the bacterial cell wall. Antimicrobial activity of chitosan has been demonstrated against many bacteria, filamentous fungi and yeasts [105-108]. Chitosan has wide spectrum of activity and high killing rate against Gram-positive and Gram-negative bacteria, but lower toxicity toward mammalian cells [109,110]. Variations in chitosan's antimicrobial efficacy arise from various factors. These factors can be classified into four categories as follows: (1) microbial factors, related to microorganism species and cell age; (2) intrinsic factors of chitosan, including positive charge density, Mw, concentration, hydrophilic/hydrophobic characteristic and chelating capacity; (3) physical state, namely water-soluble and solid state of chitosan; (4) environmental factors, involving ionic strength in medium, pH, temperature and reactive time [111].

Although owning a broad spectrum of antimicrobial activity, chitosan exhibits different inhibitory efficiency against different fungi, Gram-positive and Gram-negative bacteria. Chitosan exerts an antifungal effect by suppressing sporulation and spore germination [112]. In contrast, the mode of antibacterial activity is a complicating process that differs between Gram positive and Gram-negative bacteria due to different cell surface characteristics.

Based on the available evidences, bacteria appear to be generally less sensitive to the antimicrobial action of chitosan than fungi. The antifungal activity of chitosan is greater at lower pH values [113]. For a given microbial species, age of the cell can influence antimicrobial efficiency.

Chitosan has a broad spectrum of unique biological activities, including its ability to induce resistance to viral infections in plants, inhibit viral infection in animal cells, and prevent the development of phage infection in infected microbial cultures. High polymeric chitosan, when added to a nutrient medium, prevents the accumulation of the infectious phage progeny in infected cultures of Gram-positive and Gram-negative organisms. The yield of infectious DNA containing phage can decrease by several orders of magnitude in the presence of chitosan [114, 115]. The observed effect also depends on the concentration, degree of polymerization, and molecular structure of chitosan. Thus, chitosans with a polymerization degree of 250 and higher were much more effective in inhibiting coliphage infection than their fragments with a polymerization degree of 15–19. On the other hand, chitosan oligomers were more effective than their polymeric precursor in inhibiting the replication of 1-97A phage in *Bacillus thuringiensis* cultures. Factors determining such strong differences are currently unclear. Anionic derivatives of chitosan, such as 6-O-sulfate and N-succinate-6-O-sulfate, caused no effect on phage infection [114]. This finding showed that the positive charge of a chitosan molecule is important for inhibition of phage infection.

It has been suggested that chitosan can inhibit the replication of bacteriophages by several mechanisms: it can (a) decrease the viability of cultured bacterial cells, (b) neutralize the infectivity of mature phage particles in the inoculum and/or daughter phage particles, and (c) block the replication of the virulent phage [114].

The condition of the phage culture is known to be of paramount importance for the development of phage infection because phages can reproduce only in viable cells. However, there is evidence that chitosan displays a bactericidal activity toward many microbial species including *Escherichia coli* [116, 117] and species of the genus Bacillus [118]. Chitosan, because of its polycationic nature, binds to the external membrane of Gram negative microorganisms by electrostatic forces, which is demonstrated in experiments with core phosphate groups of lipid A, thereby decreasing the potency of endotoxin. It was suggested that the antibacterial effects of chitosan and many other cationic agents are based on their ability to increase the permeability of the outer membrane of Gram negative microorganisms to an extent incompatible with their viability [116-119].

Great amount of literature support the essential importance of polycationic structure in antimicrobial activity. A higher positive charge density leads to strong electrostatic interaction. Therein, the positive charge is associated with DD or degree of substitution (DS) of chitosan or its derivatives, which affect positive charge density. Concerning chitosan derivatives, antimicrobial activity mostly depends on DS of the grafting groups.

There are several works regarding the influence of the molecular weight of chitosan on its antimicrobial properties [120-126]. Some of them have demonstrated that COS (chitooligosaccharides), which are soluble in water, were the least effective in terms of biocide properties [124-126]. Recent work carried out by Qin et al. 2006, on the evaluation of chitosan solutions against the growth of *Candida albicans, Escherichia coli* and *Staphylococcus aureus,* it has shown that only water insoluble chitosan in organic acidic solutions, i.e. chitosonium salts, exhibit efficient biocide properties. On the other hand, a research performed by Fernandes et al, 2008 showed that the growth of *E. coli* was markedly inhibited by COS, and this inhibition decreased slightly as molecular weight increased. In another work performed by Fernandez-Saiz et al. 2009, changes in molecular weight of the chitosan materials tested, i.e. 310–375 and 50–190 kDa, did not lead to significant variations in biocide properties [127, 128,129].

Concerning DD, there are several works that consider this feature, and there is no doubt that the antimicrobial properties of chitosan increased with this variable. The antimicrobial performance tends to increase upon an increase in the DD of chitosan, which is related to an increase in the positive charge of the polymer [130-133].

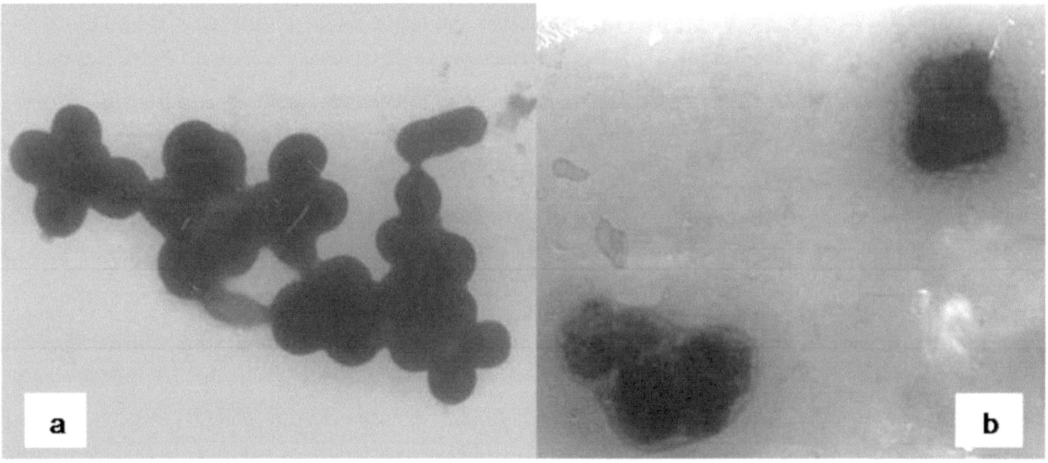

Figure 6. Staphylococcus bacteria (a), covered with chitosan (b)

18. Analgesic activity

Some investigators reported that chitin and chitosan induce analgesia. Allan in 1984 found that chitosan provide a cool and pleasant soothing effect when applied to open wounds. Ohshima et al in 1987, reported that chitin proved excellent pain relief in 83 out of 91 cases who received the agent topically over open wounds such as burns, skin abrasion, skin ulcers, skin graft areas and so on [134-135]. Minami in 1993 and Okamoto in 1993, reported that animals did not feel pain when their wounds were covered with chitin and chitosan [136-137].

Chitin and chitosan have been found to reduce the inflammatory pain due to intraperitoneal administration of acetic acid in rats [138]. When the chitosan suspension was mixed with acetic acid, the amino groups in C2 the position were protonated to NH_3 subsequently the particles resolved in the solution. Bradykinin is one of the main substances related to pain and the levels of this substance decrease in the presence of chitin and chitosan. Chitin absorbs Bradykinin more extensively than chitosan and this could be one of the main analgesic effect [138].

19. Antitumor activity

In some medical applications of chitin/chitosan, as antitumor compounds, for example, their degradation products are preferred, as they have a lower viscosity and a better solubility in water. The antitumor activity of chitin/chitosan is manifested by the stimulation of the immune system (production of lymphokines, including interleukins 1 and 2, stimulation of NK, etc.) [139-141]. Jeon and Kim in 2002, tested the antitumor activity of three kinds of COSs (high molecular weight ranging from 6.5 to 12 kDa – HMWCOS, medium molecular weight ranging from 1.5 to 5.5 kDa -MMWCOS, and low molecular weight ranging from 0.5 to 1.4 kDa – LMWCOS) against Sarcoma 180 solid (S180) and Uterine cervix carcinoma No.

14 (U14) [140]. The efficiency of tumor growth inhibition for both types of tumor cells in mice was best in the case of MMWCOS. There are many reports of S180 tumor cells being used for testing the antitumor activity of chitosan [141-142].

Maeda and Kimura 2004, investigated the antitumour effect of three water-soluble low molecular weight chitosans (21 kDa, 46 kDa, 130 kDa) and various doses of 650 kDa chitosan in sarcoma 180-bearing mice. They found that LMWC (21 and 46 kDa) and also smaller oligosaccharides could activate the intestinal immune system of animals, thus preventing tumor growth. But no antitumor effect was observed after the oral administration of chitosan samples, even of low molecular weight (46 kDa). The same authors confirmed that high molecular weight chitosan (650 kDa) prevents the adverse reactions of some cancer chemotherapeutic drugs [141]. Qin et al. in 2002 also tested the antitumor activity of LMWC against sarcoma 180, but they came to the opposite conclusions [143]. They noted that oral administration of LMW chitosan decreases the weight of the tumour [139, 142], although administration by intraperitoneal injection led to a higher inhibitory rate [142]. It was reported the higher the MW of LMWC, the better the tumor inhibitory effect [139]. The introduction of acidic groups as a result of chitosan oxidation has the opposite effect, and an increase in MW decreases antitumor activity. Qin Cai-qin et al in 2002, prepared low molecular weight chitosans by oxidative degradation with H_2O_2. They found that carboxyllc contents increased with decrease in molecular weight (M~). They also found that the introduction of carboxylic group is advantage to water-solubility of chitosan, but more acidic groups decrease the function of amino groups of chitosan against sarcoma 180 tumor. There is a correlation between the activity and the molecular weights of the oxidized chitosans, and a maximum of inhibition was found around 4 100 [143]. The influence of LMWC and COS (including the pentamer, hexamer, and higher oligomers) on the growth inhibition of Ehrlich ascites tumor (EAT) cells and tumor induced neovascularization was investigated [144]. Based on experimental results concerning the inhibition of angiogenesis and the induction of apoptosis, it was confirmed that COSs seem to be more potent angio inhibitory and antitumor compounds. Wang et al. reported that chitin oligosaccharides (DP 1-6) also reduced the number of K562 cells (human erythromyeloblastoid leukemia cell line) [145].

20. Chitosan applications

20.1. Biomedical

Potential applications of chitosan can only be exploited if its usable forms are properly developed and prepared. In solution and gel, it can be used as a bacteriostatic, fungistatic and coating agent. Gels and suspensions may play the role of carriers for slow release or controlled action of drugs, as an immobilising medium and an encapsulation material. Film and membranes are used in dialysis, contact lenses, dressings and the encapsulation of mammal cells, including cell cultures. Chitosan sponges are used in dressings, and to stop bleeding in mucous membranes. Chitosan fibres are used as resorbable sutures, non-wovens for dressings, and as drug carriers in the form of hollow fibres.

Figure 7. These are some ways you can shape the chitosan: films, gels, microspheres, tubes, sponges, powder

21. Artificial skin

The preparation of artificial skin by natural materials such as gelatin, pectin, starch, cellulose, alginate, chitin, collagen, polyamino acids, and dextran has been established to enhance the healing process. The structures of these natural materials are analogs of protein and growth factor structures in the human body that may be more relevant for stimulating the appropriate hysiological responses required for cellular regeneration and tissue reconstructing in wounds [146].

Dressing materials based on chitin, chitosan and derivatives are well-known on the market, and are produced mainly in Japan and the US. JEX KK Co produces composite dressings made of synthetic resins, chitosan and materials of collagen and acetylchitiosan. Eisai Co is manufacture of chitin dressings in sponge form (Chitopack C®) or a PET non-woven modified with chitin (Chitopack C®). The Japanese Unitika Co offers a dressing non-woven of chitosan fibres. The American 3M proposes a chitosan gel preparation (Tegasorb®) and a hydrocolloid (Tegaderm®) designed for the healing of extensive internal wounds.ChorioChit sponge is a biological dressing obtained by lyophilisation of human placenta blended with MCCh.

22. Scaffold for the regeneration of tissue

Chitin and its derivatives have been used as scaffolds for bone and other natural tissue regeneration [147] as well as structures by which three-dimensional formation of tissues are supported [148]. While looking for a good material for a good scaffold, there are at least four important factors that should be taken into account: (1) ability to form temporary matrix, (2) ability to form porous structure for tissue to grow, (3) biodegradability, and finally (4) non-toxic byproducts from the digestion [149,150]. Thus, neither the physical nor biological properties of such biomaterials should be ignored [147]. Chitin and its derivatives have been shown to possess these criteria.

23. Haemostasis and wound healing

Hemostasis through blood coagulation is an important step for wound healing. The main cellular components in blood coagulation are platelets. It has been shown that chitosan has a hemostatic effect [151]. Okamoto et al 2003, reported that chitin is an effective agent for hemostasis maintenance through aggregating platelets, and suggested that the effect of chitin and chitosan is due to both physical and chemical properties of these biopolymers, especially their amino groups [152].

Haemostatic dressings containing chitin and chitosan as bioactive agents are also well known, notably the Syvek patch, RDH (Marine Polymer Technologies), Clo-Sur PAD (Medtronic-Scion), Chito-Seal (Abbot), the M-Patch and Trauma DEX (Medafor).

24. Peripheral nerve prosthesis

Prosthesis is made from various forms of utility polysaccharides. The main objective of research is to develop replacement implants that will not be rejected by the body of the recipient and offering the ability to regenerate damaged nerve. Because chitin has high mechanical strength under physiological conditions (low for chitosan) chitin has the potential to be a good nerve guidance channel. Ferier et al, 2005, used this fact and made chitin tubes that could support nerve cell adhesion and neurite outgrowth [153]. In a research related to nerve regeneration, it was shown that rabbits with the crushed common peroneal nerve exhibit better improvement in peripheral nerve regeneration in the presence of chitooligosaccharide. As a result, chito-oligosaccharide can be used as neuroprotective material with an ability to improve injured peripheral nerve regeneration [154].

25. Immunology

The key property of chitin-derived products for application in various biomedical applications is the immuno-modulating effect [155,156]. Some mechanisms of immuno-enhancement activity of chitin and its derivatives have been reported, for example, chitosan exhibited the ability to boost NO production from macrophages in the presence of interferon-γ (IFN-γ) through the NF-κB signaling pathway [157]. Minami et al. in 1998 found that chitin and chitosan affected C3 and C5 components of complement system and concluded that complement system is activated by chitin and chitosan through the alternative pathway. After activating the complement, C5 is produced followed by an increase in migration of polymorphonuclear cells (PMN) to the injured tissue. This is a normal inflammatory reaction but in the presence of chitin and chitosan, there are no inflammatory symptoms, such as erythema, temperature elevation and abscess formation [158]. The intensity of complement [158] and macrophage [159] activation of chitin is less than chitosan; therefore, chitin is more immunomodulatory.

26. Blood cholesterol control

Chitin and chitosan are among the candidates to battle obesity and hypercholesterolemia. It has been reported that they can reduce the amount of cholesterol in rats [160]. Several mechanisms have been proposed to explain this phenomenon. One is through electrostatic interaction between lipids and aminopolysaccharides [161]. Chitin binds to lipid (cholesterol) micelles and inhibits their absorption. Another proposed mechanism is increasing the excretion of bile acid by which the amount of fecal fat increases [162]. The hypocholesterolemic effect of chitosan has also been found in humans. The proposed cholesterol lowering mechanism of chitosan was that it combines bile acids in the digestive tract, and excretes them into the feces, thus decreasing the resorption of bile acids, so that the cholesterol pool in the body was decreased and the level of serum cholesterol consequently decreased [163].

27. Drug delivery carriers

It is important for a drug delivery carrier to be efficiently removed after delivering drugs. In other words, it must not accumulate in the body nor must it be toxic [164].

Chitosan offers several advantages, and these include its ability to control the release of active agents and avoid the use of hazardous organic solvents while fabricating particles since it is soluble in aqueous acidic solution. Chitosan in the form of colloidal structures can entrap macromolecules by various mechanisms. These associated macromolecules have been shown to transport through mucosa and epithelia more efficiently [165]. Cationic chitosan in combination with other natural polymers has been shown to enhance the drug encapsulation efficiency of liposomes via the layer-by-layer (L-b-L) self-assembly technique [166]. Nanoparticles made of chitosan in association with polyethylene oxide have been used as protein carrier [167]. Moreover, an oral delivery system has been developed by using chitosan and tripolyphosphate. In this system, micro- and nano-particles were entrapped in beads made from chitosan in solution of tripolyphosphate [168].

28. Food

28.1. Chitosan films

Edible films and coatings have received considerable attention in recent years because of their advantages including use as edible packaging materials over synthetic films. This could contribute to the reduction of environmental pollution.

By functioning as barriers, such edible films and coatings can feasibly reduce the complexity and thus improve the recyclability of packaging materials, compared to the more traditional non-environmental friendly packaging materials, and may be able to substitute such synthetic polymer films [169].

Edible films are defined as a thin layer of material which can be consumed and provides a barrier to moisture, oxygen and solute movement for the food. The material can be a complete food coating or can be disposed as a continuous layer between food components [170]. Edible films can be formed as food coatings and free-standing films, and have the potential to be used with food as gas aroma barrier [171].

Chitosans are described in terms of the degree of deacetylation and average molecular weight and their importance resides in their antimicrobial properties in conjunction with their cationicity and their film forming properties [172]. Chitosan can form semi-permeable coatings, which can modify the internal atmosphere, thereby delaying ripening and decreasing transpiration rates in fruits and vegetables [173-176]. Films from aqueous chitosan are clear, tough, flexible and good oxygen barriers [177,178].

29. Bread

Applications of chitosan for extension of shelf life of bread by retarding starch retrogradation and/or by inhibiting microbial growth have been documented. Park and others in 2002 investigated the effect of chitosan (493 kDa) coating on shelf life of baguette [179].

Chitosan coating may offer a protective barrier for moisture transfer through the bread surface, thus reducing weight loss, retarding hardness, retrogradation, inhibiting microbial growth, retarding oxidation [179- 181].

30. Eggs

Several problems are encountered during storage of eggs, such as weight loss, interior quality deterioration, and microbial contamination [182-183]. The movement of carbon dioxide and moisture from the albumen through the shell governs quality changes in albumen and yolk, and weight loss of [184,185].

Chitosan coating may offer a protective barrier for moisture and gas transfer from the albumen through the egg shell, thus extending the shelf life of eggs [182, 186].

31. Fruits and vegetables

The major postharvest losses of fruits are due to fungal infection, physiological disorders, and physical injuries [102, 104, 187]. One of the potential approaches to extend the storability of these perishable commodities is to apply edible coatings on the surface, followed by a cold storage [188]. Edible coatings can be used as a protective barrier to reduce respiration and transpiration rates through fruit surfaces, retard microbial growth and color changes, and improve texture quality of fruits [171]. Coating fruits with semipermeable film has generally been shown to retard ripening by modifying the endogenous CO_2, O_2, and ethylene levels of fruits [102]. Chitosan coating is likely to modify the internal atmosphere without causing anaerobic respiration, since chitosan films are more selectively permeable to O_2 than to CO_2 [189]. Therefore, chitosan coating with its ability to modify internal atmosphere in the tissue and fungistatic property has a potential to prolong storage life and control decay of fruits.

32. Juice and beverages

Processing of clarified fruit juices commonly involves the use of clarifying aids, including gelatin, bentonite, silica sol, tannins, polyvinylpyrrolidone, or combinations of these compounds [190]. Chitosan with a partial positive charge has been shown to possess acid-binding properties [191] and to be effective in aiding the separation of colloidal and dispersed particles from food processing wastes [192,193]. These properties make chitosan an attractive processing aid in fruit juice production.

33. Mayonaise

Few studies have been conducted on the use of chitosan to enhance emulsification in mayonnaise preparation. Lee (1996) reported that addition of chitosan (1500 kDa, 0.1% based on egg yolk weight) increased emulsifying capacity of egg yolk by about 10% and enhanced emulsion stability of mayonnaise by 9.4% compared with those of the control [186]. Kim and Hur (2002) also suggested the use of chitosan as an emulsion stabilizer in commercial mayonnaise preparation [194].

Chitosan possesses a positive ionic charge and has both reactive amino and hydroxyl groups, which give it the ability to chemically bond with negatively charged protein. When pH is less than 6.5, chitosan solution carries a positive charge along its backbone. Because of its polar groups, chitosan also provides additional stabilization due to hydration forces [195]. According to Filar and Wirick in 1978, chitosan functions only in acid systems to show possible utility as a thickener and stabilizer [196].

34. Meat

Meat or meat products are highly susceptible to lipid oxidation, which leads to rapid development of rancid or warmed-over flavor. Chitosan possesses antioxidant and antibacterial capacity [126, 197], and may retard the lipid oxidation and inhibit the growth of spoilage bacteria in meat during storage.

Darmadji and Izumimoto in 1994 observed that addition of 1.0% chitosan to beef decreased the TBA value by about 70% compared to that of the control sample after 3 days of storage at 4 °C. Chitosan has a desirable effect on the development of the red color of beef during storage [198]. Sagoo et al in 2002, demonstrated that chitosan was an effective inhibitor of microbial growth in chilled comminuted pork products and that the effect of chitosan was concentration dependent [199].

35. Milk

A few attempts have been made to evaluate the possibility of using chitosan to improve the quality and shelf life of milk. Ha and Lee in 200, investigated the effectiveness of water-soluble chitosan (0.03%) to minimize the microbial (bacterial and yeast) spoilage of processed milk [200]. Complete inhibition of microbial growth was observed in the banana-flavored milk containing chitosan, in contrast to that observed in control milk (without chitosan), during storage for 15 days at 4 and 10 ◦C. The banana-flavored milk containing chitosan also maintained relatively higher pH than that of control milk during storage for 15 d at both temperatures [200].

36. Sausages

Sodium nitrite is generally used as a curing agent for color and flavor development as well as preservative effect in sausages {201]. However, nitrite reacts with amine in meat and may produce nitrosoamine, a strong toxicant detrimental to human health. Several workers [202] have investigated the possible role of chitosan, in lieu of sodium nitrite, as curing agent in sausage, and found that addition of chitosan could reduce or replace the use of nitrite without affecting preservative effect and color development.

37. Seafoods and seafood products

Seafood products are highly susceptible to quality deterioration due to lipid oxidation of unsaturated fatty acids, catalyzed by the presence of high concentrations of hematin compounds and metal ions in the fish muscle [203]. Furthermore, seafood quality is highly influenced by autolysis, contamination by and growth of microorganisms, and loss of protein functionality [204].

The oxidative stability of fish flesh with added chitosans was compared with those added with conventional antioxidants, butylated hydroxyanisole + butylated hydroxytoluene (BHA + BHT, 200 ppm) and tert butylhydroquinone (TBHQ, 200ppm), during storage at 4 ◦C. Chitosan was most effective in preventing lipid oxidation than the others. The antioxidant capacity of chitosan added to the fish muscle depended on the molecular weight and concentration of chitosan [204]. Similarly, Kim and Thomas in 2007 also observed that the antioxidative effects of chitosan in salmon depended on its molecular weight [205].

38. Chitosan in agriculture

Due to the antifungal, antibacterial and antiviral properties of chitosan, it has been used successfully in agriculture in recent years: in plant protection, like growth promoter, in soil correction, enhancer of secondary metabolites production, and activator of defense mechanisms to mention a few.

39. Seed coating

Chitosan application can be done by different ways: in the seed, in the soil or by foliar way. In seeds, it has been used as a coating material for cereals, nuts, fruits and vegetables [206-208]. It has been shown that this way of application alters permeability of the seed plasma membrane, increasing the concentrations of sugars and proline, and enhancing peroxidase (POD), catalase (CAT), phenylalanine ammonia-lyase (PAL) and tyrosine ammonia-lyase (TAL) activities [207,209]. By this way, germination rates increases significantly [210] and seedlings germinate quicker, better, and vigorously [211-214]. Chitosan is used not only in seed coatings, but also in fruits and vegetables, because it gives more firmness and it promotes diminution of the normal microbiological charge [215] increasing the product life.

40. Leaf coating

Chitosan foliar application increases stomatal conductance and reduces transpiration, without affecting plant height, root length, leaf area or plant biomass [216]. When chitosan is sprayed in leaves, abscisic acid (ABA) content increases [217]. It promotes the activation of defense mechanisms which allow plants to deal with stress and to defend against diseases due to the antiviral, antifungal and antibacterial nature of chitosan [218,219].

41. Fertilizer

By applying chitosan in soil, it has been demonstrated that it stops plant wilting because it acts as a potent fertilizer due to the high concentration of nitrogen content in its molecular structure [220,221]. Also, it has been used as a soil amendment, controlling diseases caused by fungal species like *Fusarium acuminatum, Fusarium sp, Cylindrocladium floridanum* and *Aspergillus flavus* [208, 218, 222].

42. Plants growth promoter

Chitosan acts as plant growth promoter in some crops like Faba bean plant, radish, passion fruit, potato, gerbera, cabbage, soybean and other crops when it is incorporated in solution, increasing plant production and protecting plants against pathogens too. Chitosan has a significant effect on growth rates of roots, shoots, flowering, and number of flowers [219, 223].

43. Plant self defense

Plants react naturally against most of biological and environmental adverse conditions, but sometimes defense has to be induced in order to fight against harder threats. It has been reported that chitosan is a great biopolymer used for this purpose, because it induces defense reactions in some plants, sensitizing them in order to increase their responses against pathogens attack. Some substances that get favored due to the presence of chitin and chitosan are phytoalexines, pathogenesis related proteins (PR), protein inhibitors, chitinases and glucanases, as well as Reactive Oxygen Species (ROS) and hydrogen peroxide generation [224]. This is because chitosan interacts with cellular DNA generating multiple biochemical reactions in the plant, generating a rapid response in the plant against pathogens attack. For this reason, chitosan has been considered as an elicitor, namely a defense mechanism activator in plants, generating a process at cellular level in which plant cells get and transduce biological signals in order to activate defense responses [225]. There are some specific elicitor-binding proteins which act like physiological receptors in signal transduction cascades, varying their specificity depending on the studied system, which allows researchers to find the molecular bases that origin the signal interchanges between host plants and microbial pathogens [225-227].

Not only at biochemical level but also at microbiological level, chitosan is effective on plant protection. It has been found that application of chitosan in plants by the ways mentioned in sections above reduces visibly the damages caused in the plants by pathogenic fungi because of the antibiotic nature of chitosan [215, 218]. Because of being a polysaccharide, chitosan acts as a bioremediator molecule that stimulates the activity of beneficial microorganisms in the soil such as Bacillus, fluorescent, Pseudomonas, Actinomycetes, Mycorrhiza and Rhizobacteria [228-233], which alter the microbial equilibrium in the rhizosphere disadvantaging plant pathogens, making them able to compete through mechanisms such as parasitism, antibiosis, and induced resistance [234,235].

44. Bioinsecticide

Chitosan research has been focused principally in controlling bacterial and fungal burden; nevertheless there are some investigations about the use of chitosan as bioinseciticide. One of the first findings was that chitosan is active against some insects like lepidopterous and homopterous, with a mortality of 80%, and this percentage increases when increasing oligo-chitosan concentration too [236].

Not only chitosan, but also its derivates (as N-acetyl (NAC) and N-benzyl (NBC) chitosan derivatives) had shown significant insecticidal activities superior to those of chitosan itself, particularly against species like Spodoptera littoralis, an important destructive pest of subtropical and tropical agriculture in northern Europe, affecting cotton, vegetable and ornamental crops [237]. Some other insects have been successfully attacked by chitosan derivates, like Helicoverpa armigera (H¨ubn), Plutella xylostella (L), Aphis gossypii (Glover), Metopolophium dirhodum (Walker), Hyalopterus pruni (Geoffroy), Rhopalosiphum padi L, Sitobium avenae (Fabricius) and Myzus persicae (Sulzer) [238].

Active chitinases from chitosan are relevant enzymes for biopesticide control mechanisms, being the hydrolysis of chitin-containing media a common practice to evaluate the efficiency of bioinsecticide organisms. It has been considered to add chitin derivatives to formulations containing these microorganisms to increase biopesticide effectiveness, to provide a favorable developmental environment and resistance against adverse conditions [239]. New chitosan derivatives with insecticidal or fungicidal properties may thus serve as good alternatives for broad-spectrum and highly persistent pesticides because they are non-toxic to vertebrates and humans, and have a biodegradable matrix.

45. Biopesticide

Tricoderma sp. and Bacilus sp. are microorganisms which often increase chitin and chitosan production, enhancing its efficiency to control pathogenic microorganisms and pests [238]. Native populations of biocontrol microorganisms became increased by adding chitin in soils infected with pathogenic agents. Thereafter, these endogenous control strains can be isolated, cultured and potentially used as biological controls. It has also been demonstrated a significant increase in chitinolytic microorganisms even in very infertile soils like in dunes, improving soil microbiota and its properties [239, 241].

46. Bionematicide

Nematodes proliferation can be controlled when chitosan is applied in soil, because chitinolytic microorganisms proliferate destroying nematode eggs and degrading the chitin-containing cuticle of young nematodes [240]. Because of the high content of nitrogen in chitosan and chitin molecules, concentrations of ammonia emissions increase turning toxic to nematodes which principally affect plant roots and shoots [239, 243].

Further research is still required to find more applications of chitosan in agriculture, but nowadays this polymer means to be a cheap and easy material to deal with crop problems pre-harvest, harvest and post-harvest level.

47. Conclusions

Scientific databases reveal thousands of articles and patents related to chitin, chitosan and its derivatives and increasingly opens up new possibilities to produce new derivatives as well as new applications.

The answer to the question if the chitosan is a "new panacea", is given by the multiple applications for this new biopolymer and its predecessor, the chitin. Two hundred years have passed since its discovery and this biopolymer has shown unique qualities that many other polymers do not have, as it can be applied in different areas like in the agricultural and medical field or in related areas such as pharmacy and biomedical.

As seen in this chapter, chitosan's behavior in different applications within diverse areas, is governed by its molecular weight, degree of deacetylation, degree of polymerization and source of obtention. Twenty years ago the articles published did not provide data on the characterization of material but today most papers focus on the properties of the polymer before the application.

Also, through the study of this biopolymer and due to the great demand of chitin and chitosan, it is very important to direct all efforts to seek methods of production through environment-friendly processes and on the other hand, through genetic engineering methods, finding the way to produce a more uniform material with characteristics previously designed, especially for medical or pharmaceutical items.

Acknowledgement

Gabriela Montaraz Miranda, Rosa Guadalupe Valadez Olguin and Ana María Sixto Berrocal for her technical assistance.

To PAPITT Program No. IT220411 Estudio fitopatológico, bioquímico y molecular de la respuesta contra estreses bióticos y abióticos en plántulas de maíz. For the financial support.

To PAPIME Program No. PE203211 Elaboración de material didáctico para las diferentes estrategias utilizadas en la enseñanza de la biotecnología de última generación. For the financial support.

Author details

Susana P. Miranda Castro* and Eva G. Lizárraga Paulín
Laboratorio de Biotecnología, Facultad de Estudios Superiores Cuautitlan, UNAM, Cuautitlan Izcalli, Estado de México, México

48. References

[1] Matsunaga A. (1998) Chitosan: The Ultimate Health Builder. Vantage Press, 134 p. ISBN 0533126290, 9780533126293.

[2] American Heritage dictionary of the English Language: Fourth Edition. 2000. http://www.specialist-online-dictionary.com/ Entry for chiton.

[3] Roberts George, European Chitin Society Newsletter, Editor: George A F Roberts, Truth or Myth? July 2006, N° 21.

[4] Peter M G (2005) Chitin and Chitosan in Fungi. Biopolymers. DOI: 10.1002/3527600035.bpol6005.

[5] Jang MK, Kong BG, Jeong YI, Lee CH, Nah JW (2004) Physicochemical characterization of α-chitin, β-chitin, and γ-chitin separated from natural resources. J. polym. sci. A. polym. chem. 42: 3423–3432. DOI: 10.1002/pola.20176.

[6] Muzzarelli, R.A.A. (1977) Enzymatic synthesis of chitin and chitosan. Occurrence of chitin. In: Chitin (Muzzarelli, R.A.A., ed.), pp. 5–44. Pregamon Press, New York, NY. ISBN 0080203671, 9780080203676

[7] Herring PJ(1979) Marine Ecology and natural products. Pure appl. chem. 51: 1901–1911. Available: http://pac.iupac.org/publications/pac/pdf/1979/pdf/5109x1901.pdf.

* Corresponding Author

[8] Wagner GP, Lo J, Laine R, Almeder M (1993) Chitin in the epidermal cuticle of a vertebrate (Paralipophrystrigloides, Blenniidae, Teleostei). Cell mol. life sci. 4: 317–319. DOI: 10.1007/BF01923410.

[9] Rudall KM, Kenchington W (1973) The chitin system. Biol. rev. 48:597–636. DOI: 10.1111/j.1469-185X.1973.tb01570.x.

[10] Rudall KM (1969) Chitin and its association with other molecules. J. polym. Sci Part C. 28: 83–102. DOI: 10.1002/polc.5070280110.

[11] Blackwell J, Parker KD, Rudall KM (1965) Chitin in pogonophore tubes. J. mar boil. Assoc UK. 45: 659–61. Available: http://sabella.mba.ac.uk/2313/01/Chitin_in_Pogonophore_tubes.pdf.

[12] Gaill F, Persson J, Sugiyama P, Vuong R, Chanzy H (1992) The chitin system in the tubes of deep sea hydrothermal vent worms. J. struct. Boil. 109: 116–28. Available: http://dx.doi.org/10.1016/1047-8477(92)90043-A.

[13] Lotmar W, Picken Ler (1950) A new crystallographic modification of chitin and its distribution. Experientia. 6: 58–59. DOI: 10.1007/BF02174818.

[14] Herth W, Kuppel A, Schnepf E (1977) Chitinous fibrils in the lorica of the flagellate chrysophyte Poterioochromonas stipitata (syn. Ochromonas malhamensis). J. cell biol. 73: 311–21. Available: http://www.ncbi.nlm.nih.gov/pmc/articles/PMC2109911/pdf/jc732311.pdf.

[15] Herth W, Mulisch M, Zugenmaier P (1986) Comparison of chitin fibril structure and assembly in three unicellular organisms. In: Muzzarelli R, Jeuniaux C, Gooday GW, editors. Chitin in nature and technology. New York: Plenum Publishing Corporation. p. 107–20. ISBN 0306422115, 9780306422119.

[16] Rinaudo M (2006) Chitin and chitosan: Properties and applications. Progress in Polymer Science (Oxford), 31 (7), pp. 603-632. http://dx.doi.org/10.1016/j.bbr.2011.03.031.

[17] No HK, Meyers SP (1995) Preparation and Characterization of Chitin and Chitosan—A Review. J. aquat. food prod. technol. 4: 27-52. DOI:10.1300/J030v04n02_03.

[18] Maryam MA and Mahmood A B (2007) A new process for deproteinization of chitin from shrimp head waste. Proceedings of European Congress of Chemical Engineering (ECCE-6) 16–20. Available: http://www.nt.ntnu.no/users/skoge/prost/proceedings/ecce6_sep07/upload/647.pdf

[19] Simpson BK and Haard NF (1985) The use of enzymes to extract carotene protein from shrimp waste. J appl. bioche. 7:212–222.

[20] Cano-Lopez A, Simpson BK, Haard NF (1987) Extraction of carotenoprotein from shrimp process wastes with the aid of trypsin from Atlantic cod. J. food sci. 52:503–504. DOI: 10.1111/j.1365-2621.1987.tb06656.x.

[21] Synowiecki J and Al-Khateeb NAAQ (2000) The recovery of protein hydrolysate during enzymatic isolation of chitin from shrimp Crangon crangon processing discards. Food chemistry, 68:147-152.ISSN: 03088146 http://dx.doi.org/10.1016/S0308-8146(99)00165-X.

[22] Rao MS, Munoz J, Stevens WF (2000) Critical factors in chitin production by fermentation of shrimp biowastes. Appl. microbiol. biotechnol. 54: 808–813. DOI: 10.1007/s002530000449.

[23] Shirai K, Guerrero I, Huerta S, Saucedo G, Castillo A, Gonzalez R (2001) Effect of initial glucose concentration and inoculation level of lactic acid bacteria in shrimp waste ensilation. Enz. microbiol. biotechnol. 28: 446–452. ISSN: 01410229. DOI: 10.1016/S0141-0229(00)00338-0.
[24] Healy M, Green M, Healy A (2003) Bioprocessing of marine crustacean shell waste. Acta biotechnol. 23: 151–160. ISSN: 01384988.
[25] Bhaskar N, Suresh PV, Sakhare PZ, Sachindra NM (2007) Shrimp biowaste fermentation with Pediococcus acidolactici CFR2182: optimization of fermentation conditions by response surface methodology and effect of optimized conditions on deproteination/ demineralization and carotenoids recovery. Enzym. microb. Tech. 40: 1427–1434. ISSN: 01410229. DOI: 10.1016/j.enzmictec.2006.10.019.
[26] Prameela K, Murali M CH, Hemalatha KPJ (2010a) Extraction of pharmaceutically important chitin and carotenoids from shrimp biowaste by microbial fermentation method. J. pharm. Res. 3: 2393–2395. ISSN: 0974-6943. Available: www.jpronline.info
[27] Prameela K, Murali MCH, Smitha PV, Hemalatha KPJ (2010b) Bioremediation of shrimp biowaste by using natural probiotic for chitin and carotenoid production an alternative method to hazardous chemical method. Int. j. appl. biol. pharm. technol. 1: 903–910. ISSN 0976-4550. Available: http://ijabpt.com/pdf/87022-Prameela-Gitam%5B1%5D.pdf.
[28] Prameela K, Murali M CH, Hemalatha KPJ (2010c) Bio- efficiency of Pediococcus acidilactici (ATCC 8042) for recovery of chitin and carotenoids in the fermentation of shrimp biowaste. Int. j. chem.tech. res. 2: 1924–1928. DOI: 10.1007/s00253-011-3651-2.
[29] Rao MS and Stevens WF (2005) Quality parameters of chitosan derived from fermentation of shrimp biomaterial using a drum reaction. J. chem. technol. Biotechnol. 80: 1080–1087. DOI: 10.1002/jctb.1286.
[30] Woods B (1998) Microbiology of fermented foods, vol 1. Blackie, NY. BN 978-0-7514-0216-2. ISBN 0-7514-0216-8
[31] Legarrenta GI, Zakaria Z, Hall GM (1996) Lactic fermentation of prawn waste: comparison of commercial and isolated starter culture. In: Domard A, Jeuniaux C, Muzzarelli RAA, Roberts GAF (eds) Advances in chitin science, vol 1. Jacques Andre, Lyon, pp 399–406 ISBN 2907922408, 9782907922401.
[32] Muzzarelli RAA and Peter MG, Eds.1997. Chitin Handbook,, Grottammare: Atec, ISBN 8886889011, 9788886889018
[33] Mathur NK and Narang CK (1990)Chitin and chitosan, versatile polysaccharides from marine animals. J. chem. educ. 67, 938-942. DOI: 10.1021/ed067p938.
[34] Dodane V and Vilivalam VD (1998) Pharmaceutical applications of chitosan. Pharm. sci. tech. today 1: 246–253. ISSN: 14615347. DOI: 10.1016/S1461-5347(98)00059-5.
[35] Kumar MNVR (2000) A review of chitin and chitosan applications. Reactive & Functional Polymers, 46: 1-27. http://dx.doi.org/10.1016/S1381-5148(00)00038-9.
[36] Khor E and Lim L (2003) Implantable applications of chitin and chitosan. Biomaterials. 24: 2339-2349. ISSN: 01429612. DOI: 10.1016/S0142-9612(03)00026-7.
[37] Muzzarelli RAA Natural (1973) Polymers, Pergamon Press, London. ISBN 0080172350, 9780080172354.

[38] Crestini C, Kovac B and Giovannozzi-Sermanni G (1996) Production and isolation of chitosan by submerged and solid-state fermentation from Lentinusedodes. Biotechnol. bioeng. 50: 207–210. DOI: 10.1002/bit.260500202.

[39] Tan SC, Tan TK, Wong SM, Khor E (1996) The chitosan yield of zygomycetes at their optimum harvesting time Carbohydrate Polymers. 30: 239-242. http://dx.doi.org/10.1016/S0144-8617(96)00052-5.

[40] Arcidiacono S and Kaplan D L (1992) Molecular weight distribution of chitosan isolated from Mucor rouxii under different culture and processing conditions. Biotechnol. bioeng. 39: 281–286. doi: 10.1002/bit.260390305.

[41] Nwe N, Stevens WF(2002) Chitosan isolation from the chitosan-glucan complex of fungal cell wall using amylolytic enzymes. Biotechnology Letters 24:1461-1464. DOI: 10.1023/A:1019898715518.

[42] Ikeda I, Sugano M, Yoshida K, Sasaki E, Iwamoto Y, and Hatano K (1993) Effects of chitosan hydrolyzates on lipid absorption and on serum and liver lipid concentration in rats. J. agric. food chem. 41: 431-435. DOI: 10.1021/jf00027a016.

[43] Nadarajah K. Kader J, Mazmira M and Paul DC (2001) Pakistan Journal of Biological Sciences 4 (3):263-265. Available: http://docsdrive.com/pdfs/ansinet/pjbs/2001/263-265.pdf.

[44] Kishore D. Raneab and Dallas G (1993) Hoove Production of Chitosan by fun. Food Biotechnology .7: 11-33 DOI:10.1080/08905439309549843.

[45] Wei-Ping W, Yu-Min D and Xiao-Ying W (2008) Physical properties of fungal chitosan. World J. Microb. Biot. 24: 2717-2720. DOI: 10.1007/s11274-008-9755-x.

[46] Endo T and Koizumi S. (2000) Large-scale production of oligosaccharides using engineered bacteria. Curr. Opin. Struct. Biol. 10: 536–541. ISSN: 0959440X. DOI: 10.1016/S0959-440X(00)00127-5.

[47] Cottaz S and Samain E (2005) Genetic engineering of Escherichia coli for the production of NI,NII- diacetylchitobiose (chitinbiose) and its utilization as a primer for the synthesis of complex carbohydrates. Metab. Eng. 7: 311–317. http://dx.doi.org/10.1016/j.ymben.2005.05.004.

[48] Samain E, Drouillard S, Heyraud A, Driguez H, Geremia RA (1997) Gram-scale synthesis of recombinant chitooligosaccharides in Escherichia coli. Carbohydr. res. 302: 35–42. ISSN: 00086215. DOI: 10.1016/S0008-6215(97)00107-9.

[49] Samain E, Chazalet V, Geremia RA (1999) Production of O-acetylated and sulfated chitooligosaccharides by recombinant Escherichia coli strains harboring different combinations of nod genes. J. biotechnol. 72: 33–47. ISSN: 01681656. DOI: 10.1016/S0168-1656(99)00048-6.

[50] Khan TA, Peh KK, Ching H (2002) Reporting degree of deacetylation values of chitosan: the influence of analytical methods. J. pharm sci. 5:205–12. Available: (www.ualberta.ca/~csps) 5(3):205-212, 2002.

[51] Shigemasa Y, Matsuura H, Sashiwa H, Saimoto H (1996) Evaluation of different absorbance ratios from infrared spectroscopy for analyzing the degree of deacetylation in chitin. Int. j. biol. macromol. 18: pp. 237-242. ISSN: 01418130. DOI: 10.1016/0141-8130(95)01079-3.

[52] Muzzarelli RAA and Muzzarelli BB (1998) Structural and functional versatility of chitins. In: Structural Diversity and Functional Versatility of Polysaccharides. S. Dumitriu (ed) Marcel Dekker, New York. ISBN 0824754808, 9780824754808
[53] Brugnerotto J, Lizardi J, Goycoolea F, Arguelles-Monal W, Desbrieres J, Rinaudo M (2001) An infrared investigation in relation with chitin and chitosan characterization. Polymer 42: 3569-3580. ISSN: 00323861. DOI: 10.1016/S0032-3861(00)00713-8.
[54] Duarte ML, Ferreira MC, Marvao MR, Rocha J. (2002) An optimised method to determine the degree of acetylation of chitin and chitosan by FTIR spectroscopy. Int. J. biol. macromol. 31: 1-8. ISSN: 01418130. DOI: 10.1016/S0141-8130(02)00039-9.
[55] Cao W, Jing D, Li J, Gong Y, Zhao N and Zhang X (2005) Effects of the Degree of Deacetylation on the Physicochemical Properties and Schwann Cell Affinity of Chitosan Films. J. biomater appl. 20: 157-177. doi:10.1177/0885328205049897
[56] Zhang H, Du Y, Yu X, Mitsutomi M and Aiba SI (1999) Preparation of chitooligosaccharides from chitosan by a complex enzyme. Carbohyd. res. 320: 257-260. DOI: http://dx.doi.org/10.1016/S0008-6215(99)00154-8.
[57] Jia Z and Shen D (2002) Effect of reaction temperature and reaction time on the preparation of low-molecular-weight chitosan using phosphoric acid. Carbohyd polym. 49 :393-396. ISSN: 01448617. DOI: 10.1016/S0144-8617(02)00026-7.
[58] Qin C, Du Y, Zong L, Zeng F, Liu Y, Zhou B (2003) Effect of hemicellulase on the molecular weight and structure of chitosan. Polym. degrad. stabil. 80 :435-441. ISSN: 01413910. DOI: 10.1016/S0141-3910(03)00027-2.
[59] Domard A and Rinaudo M (1983) Preparation and characterization of fully deacetylated chitosan. Int.j. biol macromol. 5: 49–52. http://dx.doi.org/10.1016/0141-8130(83)90078-8.
[60] Kubota N and Eguchi Y (1997) Facile preparation of water-soluble N-acetylated chitosan and molecular weight dependence of its water-solubility. Polym j. 29:123–127. ISSN: 00323896.
[61] Aiba SI (1991) Studies on chitosan: 3. Evidence for the presence of random and block copolymer structures in partially N-acetylated chitosans. Int. J. Biol. Macromol.13 : 40-44. Available: http://dx.doi.org/10.1016/0141-8130(91)90008-I.
[62] Rinaudo M, Domard A (1989) Solution properties of chitosan. In: Skjak-Braek G, Anthonsen T, Sandford P, editors. Chitin and chitosan. Sources, chemistry, biochemistry, physical properties and applications. London and New York: Elsevier; p. 71–86. ISBN 1851663959, 9781851663958.
[63] Elson Santiago de Alvarenga (2011) Characterization and Properties of Chitosan, Biotechnology of Biopolymers, Prof. MagdyElnashar (Ed.). ISBN: 978-953-307-179-4. InTech, Available from: http://www.intechopen.com/books/biotechnology-of-biopolymers/characterization-and-properties-of-chitosan.
[64] Allan G and Peyron M (1995) Molecular weight manipulation of chitosan I: Kinetics of depolymerization by nitrous acid. Carbohyd. res. 277: 257-272. Available: http://dx.doi.org/10.1016/0008-6215(95)00207-A.
[65] Rinaudo M, Pavlov G, Desbrieres J (1999) Influence of acetic acid concentration on the solubilization of chitosan. Polymer. 40:7029-7032 Available: http://dx.doi.org/10.1016/S0032-3861(99)00056-7.

[66] Kasaai M R, Arul J and Charlet G (2000) Intrinsic viscosity–molecular weight relationship for chitosan. J. polym. sci. B Polym. phys. 38: 2591–2598. doi: 10.1002/1099-0488(20001001)38:19<2591::AID-POLB110>3.0.CO;2-6.

[67] Dutta PK (2005) Chitin and chitosan opportunities and challenges. P.K. Dutta (Ed.) SSM International Publication. Contai, Midnapore , India p 6.

[68] Keong LC and Halim A S(2009) In Vitro Models in Biocompatibility Assessment for Biomedical-Grade Chitosan Derivatives in Wound Management. Int. j. mol. sci. 10:1300-1313. doi:10.3390/ijms10031300.

[69] Li Q, Dunn ET, Grandmaison EW, Goosen MFA (1992) Applications and properties of chitosan. J. compat. pol. 7:370–97. doi: 10.1177/088391159200700406.

[70] Suh FJK and Matthew HWT (2000) Applications of chitosan-based polysaccharide biomaterials in cartilage tissue engineering: A review. Biomater. 21: 2589-2598. http://dx.doi.org/10.1016/S0142-9612(00)00126-5.

[71] Ikada Y (1994) Surface modification of polymers for medical applications. Biomaterials.15:725–36. http://dx.doi.org/10.1016/0142-9612(94)90025-6.

[72] Zhang M, Li XH, Gong YD, Zhao NM and Zhang XF (2002) Properties and biocompatibility of chitosan films modified by blending with PEG. Biomaterials 23: 2641-2648. http://dx.doi.org/10.1016/S0142-9612(01)00403-3.

[73] Zileinski BA and Acbischer P (1994) Chitosan as a matrix for mammalian cell encapsulation. Biomaterials. 15: 1049-1056. http://dx.doi.org/10.1016/0142-9612(94)90090-6.

[74] Deng CM, He LZ, Zhao M, Yang D. Liu Y (2007) Biological properties of the chitosan-gelatin sponge wound dressing. Carbohyd. Polym. 69: 583-589. http://dx.doi.org/10.1016/j.carbpol.2007.01.014.

[75] Rathke TD and Hudson SM (1994) Review of chitin and chitosan as fiber and film formers. J Mater Sci: Rev Macromolecular ChemPhys. C34:375–437. DOI:10.1080/15321799408014163.

[76] Karel SJR, Jaromir L, Vera P, Jiřina B, Fu-Tong L, Jiří V, Hans-Joachim G. (1997) Effect of chemical structure of hydrogels on the adhesion and phenotypic characteristics of human monocytes such as expression ofgalectins and other carbohydrate-binding sites. Biomaterials. 18:1009–1014.Available: http://dx.doi.org/10.1016/S0142-9612(97)00037-9.

[77] Den Braber ET, De Ruijter JE, Ginsel LA, Von Recum AF, Jansen JA (1996) Quantitative analysis of fibroblast morphology on microgrooved surfaces with various groove and ridge dimensions.Biomaterials.17:2037-2044. http://dx.doi.org/10.1016/0142-9612(96)00032-4.

[78] Singhvi R, Stephanopoulos G, Daniel IC (1994) Effect of substrate morphology on cell physiology. Biotechnol. bioeng.43: 764–71. DOI: 10.1002/bit.260430811.

[79] Suzuki T and Mizushima Y (1997) Characteristics of silica-chitosan complex membrane and their relationships to the characteristics of growth and adhesiveness of L-929 cells cultured on the biomembrane. J. ferment. Bioeng. 84: 128-132. ISSN: 0922338X. DOI: 10.1016/S0922-338X(97)82541-X.

[80] Hallab NJ, Bundy KJ, O'Connor K, Clark R, Moses RL (1995) Cell adhesion to biomaterials: correlations between surface charge, surface roughness, adsorbed protein and cell morphology. J. long-term defects med impl. 5:209–31. PMID:10172729.
[81] Ruardy TG, Moorlag HE, Schakenraad JM, Van Der Mei HC, Busscher HJ (1997) Growth of fibroblasts and endothelial cells on wettability gradient surfaces. J. Colloid interf- sci. 188:209–217. http://dx.doi.org/10.1006/jcis.1997.4769.
[82] Kean T and Thanou M, (2009) Chitin and chitosan—sources, production and medical applications, in: P.A. Williams, R. Arshady (Eds.), Desk reference of Natural Polymers, their Sources, Chemistry and Applications, Kentus Books, London, pp. 327–361. ISBN: 978-1-84973-351-9. DOI:10.1039/9781849733519-00292
[83] Kean T and Thanou M. (2010) Biodegradation, biodistribution and toxicity of chitosan. Adv. Drug. Deliver. Rev. 62: 3-11. ISSN: 0169409X. DOI: 10.1016/j.addr.2009.09.004.
[84] Hsu SC, Don TM, Chiu WY (2002) Free radical degradation of chitosan with potassium persulfate. Polym. Degrade. Stabil. 75: 73-83. ISSN: 01413910. DOI: 10.1016/S0141-3910(01)00205-1.
[85] Zoldners J, Kiseleva T, Kaiminsh I (2005) Influence of ascorbic acid on the stability of chitosan Carbohyd. Polym. 60: 215-218. ISSN: 01448617. DOI: 10.1016/j.carbpol.2005.01.013.
[86] Saimoto H, Takamori Y, Morimoto M, Sashiwa H, Okamoto Y, Minami S, Matsuhashi A, Shigemasa Y (1977) Biodegradation of Chitin with Enzymes and Vital Components. Macromolecular Symposia. 120:11-18. ISSN: 10221360.
[87] Funkhouser JD, Aronson J NN (2007) Chitinase family GH18: Evolutionary insights from the genomic history of a diverse protein family BMC. Evol. biol. 7, art. no. 96. ISSN: 14712148. DOI: 10.1186/1471-2148-7-96. Available: http://www.biomedcentral.com/1471-2148/7/96.
[88] Onishi H and Machida Y (1999) Biodegradation and distribution of water-soluble chitosan in mice. Biomaterials. 20: 175-182. ISSN: 01429612. DOI: 10.1016/S0142-9612(98)00159-8.
[89] Senel S and McClure SJ (2004) Potential applications of chitosan in veterinary medicine. Adv. drug. deliver. rev.56: 1467-1480. ISSN: 0169409X. DOI: 10.1016/j.addr.2004.02.007.
[90] Zhang H and Neau SH (2002) In vitro degradation of chitosan by bacterial enzymes from rat cecal and colonic contents. Biomaterials, 23: 2761-2766. ISSN: 01429612. DOI: 10.1016/S0142-9612(02)00011-X.
[91] Illum L (1998) Chitosan and its use as a pharmaceutical excipient. Pharm. res. 15: 1326-1331. DOI: 10.1023/A:1011929016601.
[92] Wedmore I, McManus JG, Pusateri AE, Holcomb JB (2006) A special report on the chitosan-based hemostatic dressing: Experience in current combat operations. J. trauma. 60: 655-658. DOI: 10.1097/01.ta.0000199392.91772.44.
[93] Schipper NGM, Varum KM, Artursson P (1996) Chitosans as absorption enhancers for poorly absorbable drugs. 1: Influence of molecular weight and degree of acetylation on drug transport across human intestinal epithelial (Caco-2) cells. Pharm. res. 13:1686-1692. DOI: 10.1023/A:1016444808000.

[94] Bruno Sarmento, José das Neves (2012) Chitosan-Based Systems for Biopharmaceuticals: Delivery, Targeting and Polymer Therapeutics. Bruno Sarmento, José das Neves (eds). John Wiley & Sons. ISBN 111996296X, 9781119962960.

[95] Fernandez-Saiz P, Ocio MJ, Lagaron JM (2010) The use of chitosan in antimicrobial films for food protection, CAB Reviews: Perspectives in Agriculture, Veterinary Science, Nutrition and Natural Resources. Vol 5. ISSN: 17498848. DOI: 10.1079/PAVSNNR20105024.

[96] Ralston GB, Tracey MV, Wrench PM (1964) The inhibition of fermentation in baker's yeast by chitosan. BBA - General Subjects. 93:652-655. ISSN: 03044165. PubMed ID: 14263164.

[97] Helander IM, Nurmiaho-Lassila EL, Ahvenainen R, Rhoades J, Roller S (2001) Chitosan disrupts the barrier properties of the outer membrane of Gram-negative bacteria. Int. j. food. microbiol. 71: 235-244. ISSN: 01681605. DOI: 10.1016/S0168-1605(01)00609-2.

[98] Liu H, Du Y, Xiaohui Wang X, Sun L (2004) Chitosan kills bacteria through cell membrane damage. Int. j. food. Microbial. 95:147–55. PMID:15282127.

[99] Mayachiew P, Devahastin S, Mackey BM, Niranjan K.(2010). Effects of drying methods and conditions on antimicrobial activity of edible chitosan films enriched with galangal extract. Food res. Int. 43(1):125–32. Available: http://dx.doi.org/10.1016/j.foodres.2009.09.006,

[100] Ganan M, Carrascosa V, Martinez-Rodriguez AJ (2009) Antimicrobial activity of chitosan against Campylobacter spp. and other microorganism and its mechanism of action. Journal of Food Protection 8:1735–1738. PMID:19722411.

[101] Hadwiger LA, Kendra DF, Fristensky BW, Wagoner W (1985) Chitosan both activates genes in plants and inhibits RNA synthesis in fungi. In: Muzzarelli RAA, Jeuniaux C, Gooday GW, editors. Chitin in Nature and Technology. Plenum Press, New York; p. 209–22.

[102] El Ghaouth A, Arul J, Ponnampalam R, Boulet M (1991) Chitosan coating effect on storability and quality of fresh strawberries. J Food Sci 56:1618–20. DOI: 10.1111/j.1365-2621.1991.tb08655.x.

[103] Cuero RG, Osuji G, Washington A (1991) N-carboxymethyl chitosan inhibition of aflatoxin production: role of zinc. Biotechnol. lett.13:441–444. Available: http://www.springerlink.com/content/x63116370379vn54/fulltext.pdf.

[104] El Ghaouth A, Arul J, Grenier J, Asselin A (1992a) Antifungal activity of chitosan on two postharvest pathogens of strawberry fruits. Phytopathology 82:398-402. Available: http://www.apsnet.org/publications/phytopathology/backissues/Documents/1992Articles/Phyto82n04_398.pdf

[105] Hirano S and Nagao N, (1989) Effects of chitosan, pectic acid, lysozyme, and chitinase on the growth of several phytopathogens. Agricultural and Biological Chemistry 53, 3065–3066. DOI: http://dx.doi.org/10.1271/bbb1961.53.3065.

[106] Kong M, Chen G, Xing K, Park HJ (2010) Antimicrobial properties of chitosan and mode of action: A state of the art review. Int. j .food microbial. 144:51-63. ISSN: 01681605. DOI: 10.1016/j.ijfoodmicro.2010.09.012.

[107] Uchida Y, Izume M, Ohtakara A (1989) In: Skjak-Braek, G., Anthonsen, T., Sandford, P. (Eds.), Chitin and chitosan. Elsevier, London, UK, p. 373.
[108] Ueno K, Yamaguchi T, Sakairi N, Nishi N, Tokura S (1997) In:Domard, A., Roberts,G.A.F., Varum, K.M. (Eds.), Advances in chitin science. Jacques Andre, Lyon, p. 156.
[109] Franklin TJ and Snow GA (1981) Biochemistry of Antimicrobial Action, 3rd ed. Chapman and Hall, London, p. 175. ISBN 0412302608, 9780412302602.
[110] Kong M, Chen XG, Liu CS, Liu CG, Meng XH, Yu LJ (2008) Antibacterial mechanism of chitosan microspheres in a solid dispersing system against E. coli. Colloid. Surface. B. 65: pp. 197-202.DOI: http://dx.doi.org/10.1016/j.colsurfb.2008.04.003.
[111] Kong M, Chen XG, Xing K, Park HJ (2010)Antimicrobial properties of chitosan and mode of action: A state of the art review. Int. j .food microbial. 144: 51-63. PMID:20951455. Available: http://www.sciencedirect.com/science/article/pii/S0168160510005167.
[112] Hernandez-Lauzardo AN, Bautista-Banos S, Velazquez-del Valle MG, Mendez-Montealvo MG, Sanchez-Rivera MM, Bello-Perez LA (2008) Antifungal effects of chitosan with different molecular weights on in vitro development of Rhizopus stolonifer (Ehrenb.:Fr.) Vuill.Carbohyd. polym. 73: 541-547. ISSN: 01448617. DOI: 10.1016/j.carbpol.2007.12.020.
[113] Roller S, Covill N (1999) The antifungal properties of chitosan in laboratory media and apple juice. Int. j .food microbial. 47: 67-77. ISSN: 01681605. DOI: 10.1016/S0168-1605(99)00006-9.
[114] Kochkina Z M and Chirkov S N (2000a) Influence of chitosan derivatives on the development of phage infection in the Bacillus thuringiensis culture. Microbiology 69: 217-219. DOI: 10.1007/BF02756202.
[115] Kochkina ZM, Surgucheva NA, Chirkov SN (2000b) Influence of chitosan derivatives on the development of phage infection in the Bacillus thuringiensis culture. Microbiology. 69: 217-219. DOI: 10.1007/BF02756202.
[116] Sudarshan NR, Hoover DG, Knorr D(1992) Antibacterial action of chitosan. Food biotechnol. 6: 257-272. ISSN: 08905436
[117] Tsai GJ and Su W (1999) Antibacterial Activity of Shrimp Chitosan against Escherichia coli .J. Food Prot. 62:239–243.
[118] Barrette J, Champagne C, Goulet J (1999) Development of bacterial contamination during production of yeast extracts. J. appl. environ. microbiol. 65: 3261–3263. ISSN: 00992240.
[119] Vaara M (1992) Agents that increase the permeability of the outer membrane. Microbiol. Rev. 56: 395-411. ISSN: 01460749 PubMed ID: 1406489.
[120] Jeon Y-J, Park P-J, Kim S-K(2001) Antimicrobial effect of chitooligosaccharides produced by bioreactor Carbohyd. Polym. 44: 71-76. ISSN: 01448617. DOI: 10.1016/S0144-8617(00)00200-9
[121] Gerasimenko DV, Avdienko ID, Bannikova GE, Zueva OY, Varlamov VP (2004) Antibacterial effects of water-soluble lowmolecular- weight chitosan on different

microorganisms. Appl. Biochem. Micro. 40:253–7. DOI: 10.1023/B:ABIM.0000025947.84650.b4.

[122] Liu N, Chen X, Park H, Liu C, Liu C, Meng X, Yu L (2006) Effect of MW and concentration of chitosan on antibacterial activity of Escherichia coli Bioreactor. Carbohyd.polym.64:60–65. ISSN: 01448617 DOI: 10.1016/j.carbpol.2005.10.028.

[123] Liu X, Guan Y, Yang D, Li Z, Yao K (2001) Antibacterial action of chitosan and carboxymethylated chitosan. Journal of Applied Polymer Science. 79(7):1324–35. ISSN: 00218995. DOI: 10.1002/1097-4628(20010214)79:7<1324::AID-APP210>3.0.CO;2-LDocument

[124] Zivanovic S, Basurto C, Chi S, Davidson P, Weiss J (2004) Molecular weight of chitosan influences antimicrobial activity in oil-in-water emulsions. J. food protect. 67:952–959. ISSN: 0362028X.

[125] Uchida Y, Izume M, Ohtakara A (1989) Preparation of chitosan oligomers with purified chitosanase and its application. In: Skja˚ k-Brk G, Anthonsen T, Sandford P, editors. Chitin and Chitosan: Sources, Chemistry, Biochemistry, Physical Properties and Applications. Elsevier Applied Science, London. p. 373–82.

[126] No HK, Park NY, Lee SH, Meyers SP (2002) Antibacterial activity of chitosans and chitosan oligomers with different molecular weights. Int J FoodMicrobiol 74:65–72. DOI: http://dx.doi.org/10.1016/S0168-1605(01)00717-6.

[127] Qin C, Li H, Xiao Q, Liu Y, Zhu J, Du Y (2006) Water-solubility of chitosan and its antimicrobial activity. Carbohyd.polym. 63: 367-374. DOI: http://dx.doi.org/10.1016/j.carbpol.2005.09.023.

[128] Fernandes C, Tavaria F, Soares J, Ramos O, João Monteiro J, Pintado M, Xavier Malcata F (2008) Antimicrobial effects of chitosan and chitooligosaccharides, upon Staphylococcus aureus and Escherichia coli, in food model systems. Food microbial. 25:922–928. ISSN: 07400020 DOI: 10.1016/j.fm.2008.05.003.

[129] Fernandez-Saiz P, Lagaron J, Ocio M (2009) Optimization of the biocide properties of chitosan for its application in the design of active films of interest in the food area. Food hydrocolloid. 23(3):913–921. ISSN: 0268005X DOI: 10.1016/j.foodhyd.2008.06.001.

[130] Takahashi T, Imai M, Suzuki I, Sawai J (2008) Growth inhibitory effect on bacteria of chitosan membranes regulated with deacetylation degree. Biochem. Eng. j. 40:485–491. ISSN: 1369703X. DOI: 10.1016/j.bej.2008.02.009.

[131] Chen Yen-Men, Chung Ying-Chien, Woan Wang Li, Chen Kung-Tung, Li Shyh-Yuan(2002) Antibacterial properties of chitosan in waterborne pathogen. J. environ. sci. heal. A. 37: 1379-1390. ISSN: 1093-4529. DOI: DOI: 10.1081/ESE-120005993.

[132] Hongpattarakere T, Riyaphan O(2008) Effect of deacetylation conditions on antimicrobial activity of chitosans prepared from carapace of black tiger shrimp (Penaeusmonodon). Songklanakarin Journal of Science and Technology. 30:1–9. Available: http://rdo.psu.ac.th/sjstweb/journal/30-Suppl-1/0125-3395-30-S1-1-9.pdf.

[133] Tsai G-J, Su W-H, Chen H-C, Pan C-L(2002) Antimicrobial activity of shrimp chitin and chitosan from different treatments and applications of fish preservation. Fisheries sci. 68:170-177. ISSN: 09199268. DOI: 10.1046/j.1444-2906.2002.00404.x.

[134] Allan G (1984) Biochemical applications of chitin and chitosan. In Zikakis J, editor. Chitin, Chitosan and related enzymes. Orland, FL: Academic Press. pp 119-133. ISBN 0127809503, 9780127809502.

[135] Ohshima Y, Nishino K, Yonekura Y, Kishimoto S and Wakabayashy S (1987). Clinical application of chitin non-woven fabric as a wound dreesing. Eur. J. plast. surg. 10:66-69. DOI: 10.1007/BF00578375.

[136] Minami S, Okamoto Y, Tanioka S, Sashiwa H,Saimoto H, Matsuhashi A, Shigemasa Y (1993) Effects of chitosan on wound healing. In M. Yalpani, Carbohydrates and carbohydrate polymers: analysis, biotechnology, modification, antiviral, biomedical and other applications. pp. 141-152. Vol 1Mount Prospect: ATL Press. ISBN 1882360400, 9781882360406.

[137] Okamoto Y, Minami S, Matsuhashi A ,Sashiwa H, Saimoto H, Shigemasa Y, Taniagua T, Tanaka Y, and Tokura S (1993) Application of polymeric N-acetyl-D-glucosamine (chitin) to veterinary practice. J. vet. med. sci. 55, 743-747. (PMID:8286525).

[138] Okamoto Y, Kawakami K, Miyatake K, Morimoto M, Shigemasa Y, Minami S (2002) Analgesic effects of chitin and chitosan. Carbohyd. Polym. 49:249-252. ISSN: 01448617. DOI: http://dx.doi.org/10.1016/S0144-8617(01)00316-2,.

[139] Qin CQ, Du YM, Xiao L, Li Z and Gao XH (2002a) Enzymic preparation of water soluble chitosan and their antitumor activity. Int. j. biol. macromol. 31:111-117. ISSN 0141-8130. DOI http://dx.doi.org/10.1016/S0141-8130(02)00064-8.

[140] Jeon Y-J and Kim S-K (2002) Antitumor Activity of Chitosan Oligosaccharides Produced In Ultrafiltration Membrane Reactor System. J. microbiol. biotechnol. 12:503-507. ISSN 1017-7825.

[141] Maeda Y and Kimura Y (2004) Antitumor Effects of Various Low-Molecular-Weight Chitosans Are Due to Increased Natural Killer Activity of Intestinal Intraepithelia lymphocytes in sarcoma 180-bearing mice. J. nutr. 134:945-950.

[142] Qin CQ, Zhou B, Zeng L, Zhang Z, Liu Y.; Du, Y.M. & Xiao, L. (2004). The physicochemical properties and antitumor activity of cellulose-treated chitosan. Food Chem., 84, 107-115, ISSN 0308-8146. doi:10.1016/S0308-8146(03)00181-X.

[143] Qin CQ, Du YM, Xiao L, Gao XH, Zhou JL and Liu HL (2002b) Effect of Molecular Weight and Structure on Antitumor Activity of Oxidized Chitosan. Wuhan univ. j. nat. sci., 7, 231-236, ISSN 1007-1202. DOI: 10.1007/BF02830325.

[144] Harish Prashanth KV and Tharanathan RN (2005) Depolymerized products of chitosan as potent inhibitors of tumor-induced angiogenesis. Biochim. Biophys. Acta, 1722, 22-29, ISSN 0006-3002. DOI: 10.1016/j.bbagen.2004.11.009.

[145] Wang S-L, Lin T-Y, Yen Y-H, Liao H-F and Chen Y-J (2006) Bioconversion of shellfish chitin wastes for the production of Bacillus subtilis W-118 chitinase. Carbohydr. res., 341, 2507–2515, ISSN 0008-6215. DOI: 10.1016/j.carres.2006.06.027.

[146] Parvez S, Rahman MM, Khan MA, Khan MAH, Islam JMM, Ahmed M, Rahman MF, Ahmed B (2012) Preparation and characterization of artificial skin using chitosan and gelatin composites for potential biomedical application. Polymer Bulletin: 1-17. Article in Press. DOI: 10.1007/s00289-012-0761-7.

[147] Brandl F, Sommer F, Goepferich A (2007) Rational design of hydrogels for tissue engineering: Impact of physical factors on cell behavior. Biomaterials. 28:134–146. ISSN: 01429612. DOI: 10.1016/j.biomaterials.2006.09.017.

[148] Tsioptsias C, Tsivintzelis I, Papadopoulou L, Panayiotou C A.(2009) novel method for producing tissue engineering scaffolds from chitin, chitin-hydroxyapatite, and cellulose. Mater. Sci. Eng. C. 29: 159–164. ISSN: 09284931. DOI: 10.1016/j.msec.2008.06.003.

[149] Drury JL and Mooney DJ(2003) Hydrogels for tissue engineering: scaffold design variables and applications. Biomaterials 24, 4337–4351. ISSN: 01429612. DOI: 10.1016/S0142-9612(03)00340-5.

[150] Khor E and Lim LY (2003) Implantable applications of chitin and chitosan. Biomaterials 24: 2339–2349. ISSN: 01429612 DOI: 10.1016/S0142-9612(03)00026-7.

[151] Klokkevold PR, Fukayama H, Sung EC, Bertolami CN (1999) The effect of chitosan (poly-N-acetyl glucosamine) on lingual hemostasis in heparinized rabbits. J. oral maxillofac. surg. 57: 49–52. PII: S0278-2391(99)90632-8

[152] Okamoto Y, Yano R, Miyatake K, Tomohiro I, Shigemasa Y, Minami S (2003) Effects of chitin and chitosan on blood coagulation. Carbohyd. polym. 53: 337-342. ISSN: 01448617. DOI: 10.1016/S0144-8617(03)00076-6.

[153] Freier T, Montenegro R, Shan Koh H, Shoichet S (2005) Chitin-based tubes for tissue engineering in the nervous system. Biomaterials. 26: 4624–4632. ISSN: 01429612. DOI: 10.1016/j.biomaterials.2004.11.040.

[154] Gong Y, Gong L, Gu X, Ding F (2009) Chitooligosaccharides promote peripheral nerve regeneration in a rabbit common peroneal nerve crush injury model. Microsurgery. 29(8): 650–656. ISSN: 07381085 DOI: 10.1002/micr.20686

[155] Aam B, Heggset E, Norberg A, SØrlie M, Vårum K, Eijsink V (2010)Production of Chitooligosaccharides and Their Potential Applications in Medicine. Mar. Drugs. 8(5)1482–1517. ISSN: 16603397 DOI: 10.3390/md8051482

[156] Muzzarelli RAA(2010) Chitins and chitosans as immunoadjuvants and non-allergenic drug carriers. Mar. Drugs, 8, 292–312. ISSN: 16603397. DOI: 10.3390/md8020292.

[157] Jeong HJ, Koo HN, Oh EY, Chae HJ, Kim HR, Suh SB, Kim CH, Cho KH, Park BR, Park ST, Lee YM, Kim HM (2000)Nitric oxide production by high molecular weight water-soluble chitosan via nuclear factor-kappaB activation. Int. j. immunopharmacol. 22: 923–933. http://dx.doi.org/10.1016/S0192-0561(00)00055-2.

[158] Minami S, Suzuki H, Okamoto Y, Fujinaga T, Shigemasa Y (1998) Chitin and chitosan activate complement via the alternative pathway. Carbohyd. polym. 36: 151-155. ISSN: 01448617. http://dx.doi.org/10.1016/S0144-8617(98)00015-0.

[159] Freier T, Montenegro R, Shan Koh H, Shoichet MS (2005) Chitin-based tubes for tissue engineering in the nervous system. Biomaterials. 26: 4624–4632. ISSN: 01429612. DOI: 10.1016/j.biomaterials.2004.11.040.

[160] Razdan A and Pettersson D (1994) Effect of chitin and chitosan on nutrient digestibility and plasma lipid concentrations in broiler chickens. Br. j. nutr. 72: 277–288. ISSN: 00071145 . DOI: 10.1079/BJN19940029.

[161] Muzzarelli RAA and Muzzarelli C (2006) Chitosan, a dietary supplement and a food technology commodity. In Functional Food Carbohydrates; Biliaderis, C.G., Izydorczyk,

M.S., Eds.; Francis and Taylor: Orlando, FL, USA, pp. 215–248. eBook ISBN: 978-1-4200-0351-2. DOI: 10.1201/9781420003512.ch6.

[162] Gallaher CM, Munion J, Hesslink R Jr, Wise J, Gallaher DD (2000) Cholesterol Reduction by Glucomannan and Chitosan Is Mediated by Changes in Cholesterol Absorption and Bile Acid and Fat Excretion in Rats. J. nutr. 130, 2753–2759. ISSN: 00223166 . PubMed ID: 11053517.

[163] Maezaki Y, Tsuji K, Nakagawa Y, Kawai Y, Akimoto M, Tsugita T, Takekawa W, Terada A, Hara H, Mitsuoka T (1993) Hypocholesterolemic Effect of Chitosan in Adult Males. Biosci. biotech. biochem. 57, 1439–1444. Available: http://ci.nii.ac.jp/els/110002676529.pdf?id=ART0002948600&type=pdf&lang=en&host=cinii&order_no=&ppv_type=0&lang_sw=&no=1338944039&cp=.

[164] Dev A, Mohan JC, Sreeja V, Tamura H, Patzke GR, Hussain F, Weyeneth S, Nair SV, Jayakumar R (2010) Novel carboxymethyl chitin nanoparticles for cancer drug delivery applications. Carbohydr. polym. 79:1073–1079. ISSN: 01448617. DOI: 10.1016/j.carbpol.2009.10.038.

[165] Janes KA, Calvo P, Alonso MJ (2001) Polysaccharide colloidal particles as delivery systems for macromolecules. Adv. drug deliv. rev. 47: 83–97. ISSN: 0169409X. DOI: 10.1016/S0169-409X(00)00123-X.

[166] Haidar ZS, Hamdy RC, Tabrizian M (2008) Protein release kinetics for core-shell hybrid nanoparticles based on the layer-by-layer assembly of alginate and chitosan on liposomes. Biomaterials. 29: 1207–1215. ISSN: 01429612. DOI: 10.1016/j.biomaterials.2007.11.012.

[167] Calvo P, Remuñán-López C, Vila-Jato JL, Alonso M J (1997) Novel hydrophilic chitosan-polyethylene oxide nanoparticles as protein carriers. J. appl. polym. sci. 63, 125–132. ISSN: 00218995. DOI: 10.1002/(SICI)1097-4628(19970103)63:1<125::AID-APP13>3.0.CO;2-4.

[168] Bodmeier R, Chen H, Paeratakul O (1989) A Novel Approach to the Oral Delivery of Micro- or Nanoparticles. Pharm. Res. 6: 413–417. PMID:2748533.

[169] Bourtoom T (2008) Factor Affecting the Properties of Edible Film Prepared from Mung Bean Proteins, International Food Research Journal 15: 167-180. ISSN: 19854668.

[170] Guilbert S, Gontard N and Cuq B (1995) Technology and applications of edible protective films. Packag. technol. sci., 8: 339–346. ISSN: 08943214. doi: 10.1002/pts.2770080607.

[171] Kester J J and Fennema OR (1986) Edible films and coatings: A review in Food Technol. 40 : 47-59

[172] Muzzarelli R A A (1996) Chitosan-based dietary foods. Carbohydrate Polymer 29: 309-316. ISSN: 01448617. DOI: 10.1016/S0144-8617(96)00033-1

[173] Miranda SP, Garnica O, Lara-Sagahon V and Cárdenas G (2004) Water Vapor Permeability and Mechanical Properties of Chitosan Films. J. chil.chem. soc. 49:173-178. ISSN: 07179324. doi: 10.4067/S0717-97072004000200013

[174] Salvador L, Miranda S P, Aragón N and Lara V (1999) Recubrimiento de quitosán en aguacate. Journal of the Mexican Chemical Society 43: 18-23. Available : http://redalyc.uaemex.mx/redalyc/src/inicio/ArtPdfRed.jsp?iCve=47543204. ISSN 1870-249X.

[175] Trejo V, Aragón N, Miranda P. (2001) Estimación de la permeabilidad al vapor de agua en películas a base de quitosán. Journal of the Mexican Chemical Society .45:1-5. ISSN 1870-249X . Available:
http://redalyc.uaemex.mx/redalyc/src/inicio/ArtPdfRed.jsp?iCve=47545101>

[176] Cardenas G and Miranda S P (2004) FTIR and TGA studies of chitosan composite films. J. Chil. Chem. Soc. 49:291-295 .Available:
<http://www.scielo.cl/scielo.php?script=sci_arttext&pid=S0717-97072004000400005
&lng=es&nrm=iso>. ISSN 0717-9707. doi: 10.4067/S0717-97072004000400005.

[177] Sandford P A)1989) Chitosan: commercial uses and potential application. In Skjak-Braek, G., Anthosen, T. and Sandford, P. (Eds.). Chitin and Chitosan: Source, Chemistry, Biochemistry, Physical Properties, and Application, p. 51-69. New York: Elsevier Applied Science. ISBN 1851663959, 9781851663958.

[178] Xu J, McCarthy SP, Gross RA, Kaplan DL (1996) Chitosan film acylation and effects on biodegradability Macromolecules. 29: 3436-3440. DOI: 10.1021/ma951638b.

[179] Park IK, Lee YK, Kim MJ, Kim SD (2002) Effect of surface treatment with chitooligosaccharide on shelf-life of baguette. J. chitin chitosan. 7:214–8. ISSN 1229-4160.

[180] Butler BL, Vergano PJ, Testin RF, Bunn JM, Wiles JL (1996) Mechanical and barrier properties of edible chitosan films as affected by composition and storage. J. food sci 61:953–5, 961. ISSN 0022-1147. DOI: 10.1111/j.1365-2621.1996.tb10909.x.

[181] Nadarajah K, Prinyawiwatkul W, Hong KN, Sathivel S, Xu Z (2006) Sorption behavior of crawfish chitosan films as affected by chitosan extraction processes and solvent types .Journal of Food Science, 71 (2) ISSN: 00221147. DOI: 10.1111/j.1365-2621.2006.tb08894.x.

[182] Bhale S, No HK, Prinyawiwatkul W, Farr AJ, Nadarajah K, Meyers SP (2003) Chitosan coating improves shelf life of eggs. J. food s.ci 68:2378–83. ISSN: 00221147. DOI: 10.1111/j.1365-2621.2003.tb05776.x.

[183] De Reu K, Grijspeerdt K,Messens W, Heyndrickx M, Uyttendaele M, Debevere J, Herman L (2006) Eggshell factors influencing eggshell penetration and whole egg contamination by different bacteria, including Salmonella enteritidis. Int. j food microbial. 112:253–60. ISSN: 01681605. DOI: 10.1016/j.ijfoodmicro.2006.04.011 .

[184] Murray MW and Rutherford PP (1963) The relationship between the loss of water and carbon dioxide from eggs and the effect upon albumen quality. Poult. sci 42: 505–8. doi: 10.3382/ps.0420505.

[185] Stadelman WJ (1986) The preservation of quality in shell eggs. In: Stadelman WJ, Cotterill OJ, editors. Egg science and technology. Westport, Conn.: AVI Publishing.p 63–73. ISBN 1560220031, 9781560220039.

[186] Lee SH (1996) Effect of chitosan on emulsifying capacity of egg yolk. J. korean soc food nutr. 25:118–22. Available: http://210.101.116.28/W_kiss10/79900052_pv.pdf.

[187] El Ghaouth A, Ponnampalam R, Castaigne F, Arul J (1992b) Chitosan coating to extend the storage life of tomatoes. Hort. Science. 27:1016–1018. Available:
http://hortsci.ashspublications.org/content/27/9/1016.full.pdf

[188] Park SI, Stan SD, Daeschel MA, Zhao Y (2005) Antifungal coatings on fresh strawberries (Fragaria × ananassa) to control mold growth during cold storage. J. food sci. 70:M202–M207. DOI: 10.1111/j.1365-2621.2005.tb07189.x.

[189] Bai RK, Huang MY, Jiang YY (1988) Selective permeabilities of chitosan-acetic acid complex membrane and chitosan-polymer complex membranes for oxygen and carbon dioxide. Polym. bull. 20:83–88. DOI: 10.1007/BF00262253 .
[190] Soto-Peralta NV, Mller H, Knorr D (1989) Effects of chitosan treatments on the clarity and color of apple juice. J. food sci 54:495–496. DOI: 10.1111/j.1365-2621.1989.tb03119.x.
[191] Imeri AG and Knorr D (1988) Effect of chitosan on yield and compositional data of carrot and apple juice. J. food sci 53:1707–1709. DOI: 10.1111/j.1365-2621.1988.tb07821.x.
[192] Knorr D (1985) Utilization of chitinous polymers in food processing and biomass recovery. In: Colwell RR, Pariser ER, Sinskey AJ, editors. Biotechnology of marine polysaccharides. Washington, D.C.: Hemisphere Publishing Corp. p 313–332. ISBN 0891164332, 9780891164333.
[193] No HK and Meyers SP (2000) Application of chitosan for treatment of wastewaters. Rev. environ. contam. toxicol. 163:1–27. DOI: Version: za2963e q8za2 q8zb9 q8zc0 q8zdf q8ze3 q8zfe q8zg3.
[194] Kim JW and Hur JW. 2002. Improvement of functional properties of mayonnaise with egg-shell calcium and chitosan. Food eng. prog. 6:195–200.
[195] Del Blanco LF, Rodríguez MS, Schulz PC, Agulló E(1999) Influence of the deacetylation degree on chitosan emulsification properties. Colloid. polym sci. 277:1087–1092. ISSN: 0303402X. DOI: 10.1007/s003960050495.
[196] Filar LJ, Wirick MG (1978) Bulk and solution properties of chitosan. In: Muzzarelli RAA, Pariser ER, editors. Proceedings of the first international conference on chitin/chitosan. Mass.: MIT Sea Grant Program. p 169–81
[197] Kamil JYVA, Jeon YJ, Shahidi F (2002) Antioxidative activity of chitosans of different viscosity in cooked comminuted flesh of herring (Clupea harengus). Food chem. 79:69–77. http://dx.doi.org/10.1016/S0308-8146(02)00180-2.
[198] Darmadji P, Izumimoto M (1994) Effect of chitosan in meat preservation. Meat sci. 38:243–54. http://dx.doi.org/10.1016/0309-1740(94)90114-7.
[199] Sagoo S, Board R, Roller S (2002) Chitosan inhibits growth of spoilage microorganisms in chilled pork products. Food microbio.l 19:175–82. http://dx.doi.org/10.1006/fmic.2001.0474.
[200] Ha TJ and Lee SH (2001) Utilization of chitosan to improve the quality of processed milk. J. korean soc. food sci. nut. 30:630–634.
[201] Park SM, Youn SK, Kim HJ, Ahn DH (1999) Studies on the improvement of storage property in meat sausage using chitosan-I. J. Korean soc. food sci. nutr. 28:167–71.
[202] Youn SK, Park SM, Kim YJ, Ahn DH (1999) Effect on storage property and quality in meat sausage by added chitosan. J Chitin Chitosan 4:189–95.
[203] Decker EA and Hultin HO (1992) Lipid oxidation in muscle foods via redox iron. In: Angelo AJ, editor. Lipid oxidation in food.Washington, D.C.: American Chemical Society. P 33–54. ISBN13: 9780841224612eISBN: 9780841213562.
[204] Jeon YJ, Kamil JYVA, Shahidi F (2002) Chitosan as an edible invisible film for quality preservation of herring and Atlantic cod. J. agric. food chem. 50:5167–78. ISSN: 00218561. DOI: 10.1021/jf011693l.

[205] Kim KW, Thomas RL (2007) Antioxidative activity of chitosans with varying molecular weights. Food chem. 101:308–13. ISSN: 03088146. DOI: 10.1016/jfood chem.2006.01.038.

[206] Photchanacha S, Singkaew J and Thamthong J (2006) Effects of chitosan seed treatment on Colletotrichum sp. and seedling growth of chili cv. 'Jinda'. Acta Horticulturae. 712:585-590. ISSN: 05677572. ISBN: 9066055391;978-906605539-1

[207] Guan YJ, Hu J, Wang XJ and Shao CX (2009) Seed priming with chitosan improves maize germination and seedling growth in relation to physiological changes under low temperature stress. J. Zhejiang univ. sci. B, 10, 6, pp. 427–433. DOI: 10.1631/jzus.B0820373.

[208] Lizárraga-Paulín EG, Torres-Pacheco I, Moreno-Martínez E and Miranda-Castro S P (2011a) Chitosan application in maize (Zea mays) to counteract the effects of abiotic stress at seedling level. Afr. j. biotechnol. 10:6439-6446. Available online at http://www.academicjournals.org/AJB. ISSN 1684–5315.

[209] Uthairatanakij A, Teixeira JA and Obsuwan K (2007) Chitosan for Improving Orchid Production and Quality. Science. 1: 1-5. Available:
http://www.globalsciencebooks.info/JournalsSup/images/Sample/OSB_1(1)1-5.pdf.

[210] Zhou YG, Yang YD, Qi YG, Zhang ZM, Wang XJ and Hu XJ (2002) Effects of chitosan on some physiological activity in germinating seed of peanut. J. Peanut sci. 31: 22–25. DOI: cnki:ISSN:1002-4093.0.2002-01-004.

[211] Reddy MV, Arul J, Angers P and Couture L (1999) Chitosan treatment of wheat seeds induces resistance to Fusarium graminearun and improves seed quality. J. agric. food chem. 47: 1208–1216. DOI: 10.1021/jf981225k.

[212] Ruan SL and Xue QZ (2002) Effects of chitosan coating on seed germination and salt-tolerance of seedlings in hybrid rice (Oryza sativa L.). Acta agron. sinica, 28: 803–808. DOI: cnki:ISSN:0496-3490.0.2002-06-014.

[213] Shao CX, Hu J, Song WJ and Hu WM (2005) Effects of seed priming with chitosan solutions of different acidity on seed germination and physiological characteristics of maize seedling. J. Zhejiang univ. agric. life sci. 31: 705–708. DOI: CNKI:SUN:ZJNY.0.2005-06-007.

[214] Lizárraga-Paulín EG, Torres-Pacheco I, Moreno-Martínez E y Miranda-Castro S P. (2011b) Protección contra estrés biótico inducida por quitosán en plántulas de maíz (Zea mays). Rev. mex. cienc. agríc. 2: 813-827. ISSN 2007-0934. Available:
http://www.inifap.gob.mx/revistas/ciencia_agricola/vol2_num6_2011.pdf.

[215] Devlieghere F, Vermeulen A and Debevere J (2004) Chitosan: antimicrobial activity, interactions with food components and applicability as a coating on fruit and vegetables. Food microbial. 21, 6, pp. 703-71 . http://dx.doi.org/10.1016/j.fm.2004.02.008.

[216] Bittelli M, Flury M, Campbell GS and Nichols EJ (2001) Reduction of transpiration through foliar application of chitosan. Agric. forest meteorol. 107: 167–175. http://dx.doi.org/10.1016/S0168-1923(00)00242-2.

[217] Iriti M, Picchi V, Rossoni M, Gomarasca S, Ludwig N, Garganoand M and Faoro F (2009) Chitosan antitranspirant activity is due to abscisic acid-dependent stomatal closure. Env. exp. bot. 66: 493–500. Doi: 10.1016/j.envepbox.2009.01.004

[218] Rabea EI, El Badawy MT, Stevens CV, Smagghe G and Steurbaut W (2003) Chitosan as antimicrobial agent: Applications and mode of action. Biomacromolecules, 4: 57–1465. DOI: 10.1021/bm034130m.
[219] El Hadrami A, Lorne RA, El Hadrami I, Fouad D (2010) Chitosan in plant protection. Marine drugs. 8:968-87. doi: 10.3390/md8040968 .
[220] Utsunomiya N, Kinai H, Matsui Y, Takebayashi T (1998) The Effects of Chitosan Oligosaccharides Soil Conditioner and Nitrogen Fertilizer on the Flowering and Fruit Growth of Purple Passionfruit (Passiflora edulis Sims var. edulis). Journal of the Japanese Society for Horticultural Science. 67: 567-571. ISSN:0013-7626.
[221] Wu L and Liu M (2008) Preparation and properties of chitosan-coated NPK compound fertilizer with controlled-release and water-retention. Carbohyd. polym. 72: 240–247. http://dx.doi.org/10.1016/j.biortech.2006.12.027.
[222] Benhamou N, Lafontaine PJ and Nicole M (1994) Induction of systemic resistance to Fusarium crown and root rot in tomato plants by seed treatment with chitosan. Phytopathology 84: 1432–1444. Available: http://www.apsnet.org/publications/phytopathology/backissues/Documents/1994Articles/Phyto84n12_1432.pdf.
[223] Hirano S (1989) Production and application of Chitin and chitosan in Japan. In Chitin and chitosan, (Skjak-Braek G, Anthonsesn, T., Sandford P (Eds) pp 51-69, Elsevier Applied Science, London, UK. ISBN 1851663959, 9781851663958.
[224] R.A.A. Muzzarelli (2008) Aspects of chitin chemistry and enzymology, In Paoletti M and Musumeci S (Eds.). Binomium chitin-chitinase: emerging issues, auppauge. NY, Nova Science. ISBN: 978-1-60692-339-9.
[225] Hahn M (1996) Microbial elicitors and their receptors in plants. Annu. Rev. phytopathol. 34: 87-412. DOI: 10.1146/annurev.phyto.34.1.387.
[226] Radman R, Saez T, Bucke C and Keshavarz T (2003) Elicitation of plants and microbial cell systems. Biotechnol. Appl. Bioc. 37: 91-102. DOI: 10.1042/BA20020118.
[227] Angelova Z, Georgiev S and Ross W (2006) Elicitation of plants. Biotechnol. Biotechnol. 20: 72-83. Available: http://www.diagnosisp.com/dp/journals/view_pdf.php?journal_id=1&archive=1&issue_id=11&article_id=298.
[228] Utsunomiya N and Kinai H (1994) Effect of chitosan-oligosaccharides soil conditioner on the growth of passionfruit. Journal of the Japanese Society for Horticultural Science. 64: 176-177. ISSN:0013-7626.
[229] Wanichpongpan P, Suriyachan K and Chandrkrachang S (2001) Effect of chitosan on the growth of Gerbera flower plant (Gerbera jamesonii). Chitin and chitosan: Chitin and Chitosan in Life Science, Yamaguchi, Japan, pp 198-201. ISBN 906605087X, 9789066050877.
[230] Hadwiger L A, Klosterman S J and Choi J J (2002) The mode of action of chitosan and its oligomers in inducing pant promoters and developing disease resistance in plants. Advances in Chitin Science. 5: 452-457
[231] Asghari-Zakaria R, Maleki-Zanjani B, Sedghi E (2009) Effect of in vitro chitosan application on growth and minituber yield of Solanum tuberosum L. Plant soil environ. 55: 252-256. Available: http://www.agriculturejournals.cz/publicFiles/08422.pdf.

[232] Bell A A, Hubbard JC, Liu L, Davis R M and Subbarao K V (1998) Effects of chitin and chitosan on the incidence and severity of Fusarium yellows in celery. Plant dis. 82: 322–328. http://apsjournals.apsnet.org/doi/pdf/10.1094/PDIS.1998.82.3.322.

[233] Murphy J G, Rafferty S M and Cassells A C (2000) Stimulation of wild strawberry (Fragaria vesca) arbuscular mycorrhizas by addition of shellfish waste to the growth substrate: interaction between mycorrhization, substrate amendment and susceptibility to red core (Phytophthora fragariae). Appl. soil ecol .15: 153–158. http://dx.doi.org/10.1016/S0929-1393(00)00091-3.

[234] Daayf F, El Bellaj M, El Hassni M, J'aiti F and El Hadrami I (2003) Elicitation of soluble phenolics in date palm (Phoenix dactylifera L.) callus by Fusarium oxysporum f. sp. albedinis culture medium. Env. exp. bot. 49: 41–47. http://dx.doi.org/10.1016/S0098-8472(02)00048-5.

[235] Uppal A K, El Hadrami A, Adam LR, Tenuta M and Daayf F (2008) Biological control of potato Verticillium wilt under controlled and field conditions using selected bacterial antagonists and plant extracts. Biol. Control. 44: 90–100. http://dx.doi.org/10.1016/j.biocontrol.2007.10.020.

[236] Zhang M, Tang T, Yuan H and Rui C (2003) Insecticidal and Fungicidal Activities of Chitosan and Oligo-chitosan. J bioact. compat. pol. 18: 391-400. doi: 10.1177/0883911503039019.

[237] Rabea EI, Badawy MEI, Rogge TM, Stevens CV, Steurbaut W, Hofte M, Smagghe G (2006) Enhancement of fungicidal and insecticidal activity by reductive alkylation of chitosan. Pest manag. sci. 62: 890-897. DOI: 10.1002/ps.1263.

[238] Rabea EI, Badawy MEI, Rogge TM, Stevens CV, Hofte M, Steurbaut W, Smagghe G (2005) Insecticidal and fungicidal activity of new synthesized chitosan derivatives. Pest manag. sci. 61: 951-960. DOI: 10.1002/ps.1085.

[239] Ramírez MA, Rodríguez AT, Alfonso L and Peniche C (2010) Chitin and its derivatives as biopolymers with potential agricultural applications. Biotecnol. apl. 27: 270-276. ISSN: 08644551.

[240] Gohel V, Singh A, Vimal M, Ashwini P and Chhatpar HS (2006) Bioprospecting and antifungal potential of chitinolytic microorganisms. African j. biotechnol. 5: 54-72. ISSN: 16845315. Available: https://tspace.library.utoronto.ca/bitstream/1807/6640/1/jb06009.pdf.

[241] Gomes RC, Semêdo LT, Soares RM, Alviano CS, Linhares LF, Coelho RR (2000) Chitinolytic activity of actinomycetes from a cerrado soil and their potential in biocontrol. Lett, appl. microbiol. 30: 146-50. ISSN: 02668254. PubMed ID: 10736018.

[242] Rodríguez-Kabana R, Morgan-Jones G and Ownley-Gintis B (1984) Effects of chitin amendments to soil on Hetedodera glycines, microbial population and colonization of cyst by fungi. Nematropica, 14: 9-25.

[243] Bélair G and Tremblay N (1995) The influence of chitin-urea amendments applied to an organic soil on Meloidogyne hapla population and growth of green house tomatoes. Phytoprotection. 76: 75-80. ISSN: 0031-9511. Available: http://www.erudit.org/revue/phyto/1995/v76/n2/706087ar.pdf.

Chapter 2

Yeast (*Saccharomyces cerevisiae*) Glucan Polysaccharides – Occurrence, Separation and Application in Food, Feed and Health Industries

Stefan Kwiatkowski and Stefan Edgar Kwiatkowski

Additional information is available at the end of the chapter

http://dx.doi.org/10.5772/48100

1. Introduction

Energy is stored in plants in form of carbohydrates, including sucrose (i.e., saccharose, a disaccharide) and starch (a polysaccharide). Sucrose and starch are derived from glucose, the product of carbon dioxide and water in photosynthesis. These carbohydrates are not only essential as energy supplies in food and animal feed, but in recent years have also been tapped as a renewable fuel energy source by the fuel ethanol industry. Some plants, like sugar cane or sugar beet, produce juice rich in sucrose, a mixed glucose-fructose disaccharide. Brewer's yeast (*Saccharomyces cerevisiae*) can directly ferment sucrose into ethanol, whereas starch, like that found in corn or potatoes, must first be hydrolyzed, to glucose or fermentable glucose oligosaccharides by an enzyme (e.g., amylase) before fermenting into ethanol. Cellulose, structural polysaccharides in plants, is even more difficult to hydrolyze to glucose. To obtain fermentable sugars, cellulose-containing plant material is usually subjected to a physicochemical pretreatment followed by enzymatic hydrolysis with cellulases. Such pretreatment processes are expensive and not yet used commercially on a large scale.

Because sugar and starch are major components of human food, and cellulose is the principal component of ruminant feed, their large-scale use in fuel alcohol production is already affecting food prices in the world. Brazil began producing fuel ethanol from sugar cane 36 years ago and in 2010 supplied enough ethanol to sell gasoline containing 25% ethanol throughout the country. Similarly, a large-scale, government sponsored program of fuel ethanol production from starch, using maize as its source, was started in the United States in 2007. This industry has created vast quantities of yeast byproducts; including a commercial-scale source of yeast extract, yeast glucose and mannose polysaccharides. Sugar cane juice and sugar beet juice fermentations are especially convenient for yeast collection,

INTECH
open science | open minds

© 2012 The Author(s). Licensee InTech. This chapter is distributed under the terms of the Creative Commons Attribution License http://creativecommons.org/licenses/by/3.0), which permits unrestricted use, distribution, and reproduction in any medium, provided the original work is properly cited.

because the post-fermentation solution does not contain solids such as non-fermentable residues, like those produced during corn mash fermentation.

The yeast cell wall contains three major constituents: glucan (glucose polysaccharide), mannan (mannose polysaccharide) and a protein fraction. The separation of these natural polymers is simple and inexpensive. However, it is difficult and expensive to obtain more than 65% pure fractions and, therefore, these components are produced and sold at these low levels of purity. Their largest commercial application has been as animal feed nutritional supplements (see, http://www.Alltech.com).

There are three well known types of glucose polysaccharides in yeast: poly-(1→3)(1→6)-β-D-glucopyranose - commercial name: yeast β-D-glucan, yeast glucan; poly-(1→4)-α-D-glucopyranose (commercial name yeast glucogen) and poly-(1→6)(1→3)(1→4)-α-D-glucopyranose a recently "rediscovered" polysaccharide (Arvindecar & Patil, 2002) yet without a commercial name. The prefix "poly" informs about the polymeric nature of the material built of D-glucose cyclic monomers (D-glucopyranose rings) and the letter D informs that glucose belongs to the group of naturally occurring plant sugars with D stereochemistry. The letter α or β refers to the configuration of the glycosidic bonds between the C-1 of a nonreducing ring and the C-6 of a reducing ring or the C-1 of a nonreducing ring and the C-3 of a reducing ring (see Fig. 1, Lindhorst 2003). Figure 1 shows structures of β-D-glucopyranose isomer with a C-1 OH group in the β position (solid line) or a C-1 OH group in the (α) position (dotted line). Penta-(1→6)(1→3)-β-D-glucopyranose drawing shows the location of all OH groups, glycosidic bonds and glucopyranose rings within the structure. (The hydrogen atoms attached to carbon atoms are not shown.)

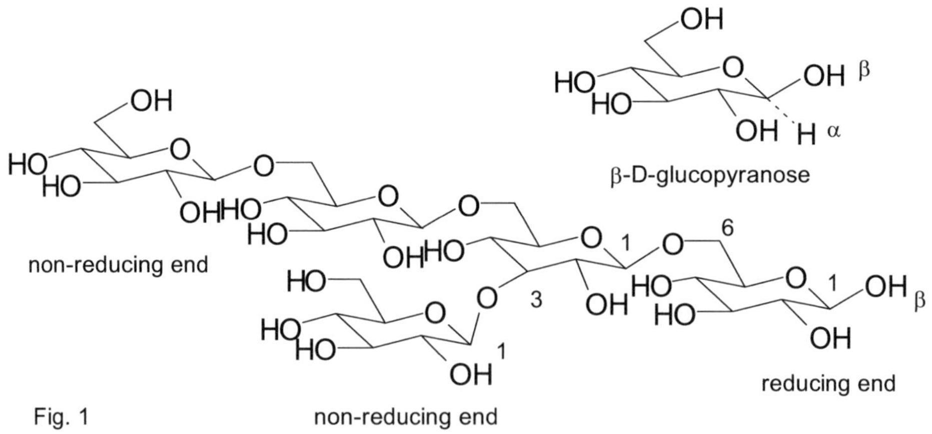

Figure 1.

Solving polysaccharide structure is not simple. At first, a sample needs to be homogenous and 98-99% pure. Different separation techniques (i.e., size exclusion chromatography, affinity (lectin) chromatography and in some cases high performance liquid chromatography (HPLC) are suitable for the purification of water or solvent soluble samples. Spectroscopic methods, such as ^1H and ^{13}C nuclear magnetic resonance (NMR) 1D/2D and mass spectrometry (MS) combined with chemical treatments of the sample, can provide further clues about the kind, sequence and ratios of the monosaccharides present, the type of bonds (α- or β-) between individual sugars, the branching points and reducing and non-reducing ends, and its molecular weight.

The experimental procedures that are most frequently used in the process of solving the polysaccharide structure are:

1. Sample purification

A solid, crude, non-homogenous, polysaccharide must be solubilized in an appropriate solvent (water) or, in the case of insoluble polysaccharides, it must be subjected to chemical or enzymatic degradation (with minimal damage to the original structure) to reduce the polymer size which then allows the sample to become soluble. Sample solubilization makes the use of various chromatographic techniques (i.e., ion exchange chromatography, size exclusion chromatography, affinity-chromatography, sorbent-liquid chromatography) for sample fractionation and purification, possible.

2. Composition analysis

Complete acid hydrolysis is used to cleave all glycosidic bonds in the polysaccharide and to release D-glucose as a single product (in the case of glucans), which is then identified by comparison to a D-glucose standard.

3. Methylation linkage analysis

All of the polysaccharide hydroxyl (OH) groups are methylated using an excess of methylation reagent (dimsyl sodium followed by methyl iodide). Obtained this way, the fully O-methylated derivative is then subjected to acidic hydrolysis of all glycosidic linkages. The released O-methylated derivatives of D-glucose: 2, 3, 4, 6-O-tetramethyl-D-glucose; 2, 3, 4-O-trimethyl-D-glucose; 2, 4-O-dimethyl-D-glucose are identified and quantified using HPLC/MS and appropriate standards. In the case of pentasaccharide (Fig.1) the ratios of the methylation products would be 2:2:1.

4. Spectroscopy

Finally, ^1H and ^{13}C NMR and M S are used to establish the sequence of glucopyranose rings, the position of branching points, the number of non-reducing ends and the nature of glycosidic bonds (α or β). If the polysaccharide contains repeating sequences, their ratios can be also established.

5. Enzymatic action

Glycosidases and lectins of known specificity are applied to confirm glucose oligosaccharide structures, from which the complete polysaccharide architecture can be deduced.

The Complex Carbohydrates Research Center located at The University of Georgia (http://www.ccrc.uga.edu) offers services covering all of the aspects of the polysaccharide structural assignments.

2. Fungal and yeast cell wall, D-glucose homo- and hetero-polysaccharides

Fungi constitute an independent evolutionary group of eukaryotic organisms equal in rank to that of plants or animals. They can excrete hydrolytic enzymes that break down biopolymers (polysaccharides and proteins) of plant or animal origin and use sugars and/or amino acids for their nutrition. Yeast is a single-cell fungus and the *Saccharomyces cerevisiae* species (baker's or brewer's yeast) has been utilized by people from early human civilization in food and beverage preparation. Mushrooms are fungal fruiting bodies; edible mushrooms are taste improving, "healthy" food ingredients that are very popular on European and Asian menus. Extracts and dried powders made by processing "medicinal mushrooms" are important ingredients of traditional Eastern medicinal remedies.

Fungal cell walls are made of polysaccharides. Chitin, a homopolymer of (1→4)-β-D-N-acetylglucosamine and various glucans (homopolymers of D-glucose, with α- or β-glycosidyl linkages between C-1 and C-2, C-3, C-4 and C-6 of glucopyranose rings) are the most frequently found building blocks used by nature in construction of the cell wall that separates the fungus from the environment. In yeast, α- and β-glucans and α-mannan are the major polysaccharides (chitin is present at 1-2%) that are utilized in cell wall construction. Their ratio changes with yeast strain and growth stage and is dependent upon growth conditions including oxygen and nutrient availability and the temperature and pH of the medium (Stewart & Russell, 1998). Table 1 shows the average abundance of yeast cell wall components.

Component	Cell wall mass (%, dry weight)
(1→3)-β-D-glucan	50-55
(1→6)-β-D-glucan	5-10
(1→4)-α-(1→3)-β-D-glucan	3-7*
Mannoprotein complex	35-40
Chitin	2

Mannan structure can be represented by the following formula: poly-(1→2) (1→3) (1→6)-α-D-mannopyranose. This polysaccharide is built out of a long backbone chain of (1→6)-α-D-mannopyranose rings with short mannopyranose chains (one to four rings long, connected through (1→2) and (1→3)-α-glucoside bonds (Ballou, 1980).

Table 1. The cell wall components of *Saccharomyces cerevisiae* (Klis et al. 2002; Kath & Kulicke, 1999; Lessage & Bussey, 2006; Kwiatkowski et al., 2009)

The yeast cell wall forms a border that defines the yeast cell's dimensions and separates its organelles from the negative influences of the environment. The individual constituents of the cell wall connect to each other by covalent bonds forming a single supra-molecular biopolymer. To find the nature of these connections, as well as the individual polysaccharides structures, all of the architectural elements must be dissected, solubilized, purified and analyzed. In this process it is almost impossible not to damage particular structural elements and to afterwards distinguish between the original components of the cell wall and soluble polysaccharides that could be present in the cytoplasm or might be trapped (physically absorbed) within cell wall. More than 50 years of research were required to establish the structure of the major, high molecular weight yeast homo-polysaccharides (Bacon & Farmer, 1969; Aimaniada et al., 2009).

The (1→3)-β-D-glucan and (1→6)-β-D-glucan form a single structure in which the (1→3)-β-D-glucan forms the backbone chain with (1→6)-β-D-glucan branches that are attached to branching glucopyranose rings at C-6 (Lessage & Bussey, 2006). The (1→3)-β-D-glucan chains form triple helix tridimensional structures with spring-like mechanical properties, responsible for the yeast cell wall's strength (Klis et al., 2002) and its ability to absorb toxins (Yiannikouris et al., 2004). The (1→6)-β-D-glucan is a linker between (1→3)-β-D-glucan, chitin and mannoproteins (Kaptein et al., 1999; Kollar et al., 1997) that stabilizes the whole structure and is the main cause of yeast cell wall insolubility. The properties and role of a mixed (1→4)-α-(1→3)-β-D-glucan from yeast cells will be discussed in section 3 of this chapter.

Mannoproteins are mostly located on the outside surface of the cell wall (Osumi, 1998). They play a sensory function for nutrients and chemical and bacterial/viral toxins, actively participate in the transport of nutrients and metabolites through the cell wall, and also participate in mating (Klis et al., 2002; Stewart & Russell, 1998). Some cell wall enzymes such as glucanase, mannanase, invertase, alkaline phosphatase and lipase are mannoproteins (Stewart & Russell, 1998) that hydrolyze nutrients and participate in the reconstruction of cell polysaccharides during cell growth and budding. Cell wall mannans are connected with cell wall glucans via covalent bonds, but they can be released under action of an alkaline medium in which they are perfectly soluble. The process of separating yeast cell wall glucans (which stay as an insoluble fraction) from yeast mannans is used on an industrial scale (Sedmak, 2006). The soluble, mannan-rich fraction, can be added back to the cell wall which contains only 10-17% mannan by weight, to increase the content of mannan to 30% by weight. Such a product is sold by Alltech Inc. under the name of Actigen™ to the animal nutrition industry.

The small fraction of chitin present in yeast cell wall is primarily located in the scar rings around buds in budding yeast and secures closure of the gaps in the mother cell and the departing daughter cell (Lessage & Bussey, 2006; Stewart & Russell, 1998).

3. α-D-glucans from baker's yeast (*Saccharomyces cerevisiae*)

The role of (1→3)-β-D-glucan in the maintenance of yeast cell wall shape and rigidity (Lessage & Bussey, 2006; Klis, et al. 2006) and (1→6)-β-D–glucan as a polysaccharide that links together all of the cell wall polysaccharides (Aimaniada et al., 2009; Kollar et al., 1997) is well documented and has been reviewed by Lessage & Bussey (2006) and Klis et al. (2002). The presence of starch-like, "alkali-soluble glycogen, "an "energy storage polysaccharide" in cell cytosol, and "difficult to dissolve acid-soluble," glycogen like (1→4)-α-D-glucan in the cell walls of *Saccharomyces cerevisiae* grown aerobically, was frequently mentioned in early yeast literature (Grba et al., 1976). The two forms of glycogen synthetase have also been identified (Rothman-Denes & Cabib, 1970; Lille & Pringle, 1980) but the yeast literature is still treating the "difficult to dissolve" α-glucan as physically adsorbed cytosol glycogen, trapped within yeast cell wall, and not as an independent component attached by a covalent bond to the other cell wall polysaccharides. In 1973, Manners et al. reported: "β-1, 6-glucan purified after acid extraction had to be exhaustively treated with α-amylase and still showed the presence of glucose and maltose in paper chromatographic analysis, along with gentiobiose and higher gentiooligosaccharides," which are (1→4)-α-D-glucan oligosaccharides formed during enzymatic hydrolysis of the mixed α, β-glucan from the yeast cell wall.

The α-glucan content in the yeast cell wall is reported to vary from as little as 1% (Lille & Pringle, 1980) to as much as 29% (Sedmak, 2006) of the dry weight, depending upon the nutritional status of the cells, the method of isolation, the method of analysis and the phase of growth during which the cells were harvested (Lille & Pringle, 1980). Spectrophotometric analysis of soluble glycogen can be run directly on a water extract, and if properly done, yields reliable results (Quain, 1981). Quantitative analysis of "insoluble" glycogen requires additional, enzymatic release of glucose from its mixed polysaccharide with β-glucan (see, http://www.megazyme.com). The industrially produced brewer's yeast described in Sedmak's US patent application (Sedmak, 2006) contains glucans (α + β) at 28.9% dry weight, which includes 12.4% α-glucan. Hydrolysis of these water-insoluble cells with alkaline protease solubilizes the mannoprotein complex and yields water-insoluble cell wall polysaccharides (including α-glucan) that contain 54.5% dry weight of glucans with more than half of the total weight (29.2%) as α-glucan.

Arvindekar and Patil (2002) proposed an explanation for the presence of α-glucan in the insoluble fraction from yeast cell walls, which has been described by others as "difficult to wash away" yeast glycogen (Gunia-Smith et al., 1977; Manners & Fleet, 1976). They found that the ratio of "soluble" to "insoluble" glycogen in different strains of *Saccharomyces cerevisiae* is in the range of 1:2.5–1:3. When the insoluble fraction was digested with purified lyticase, which possessed only (1→3)-β-D–glucan hydrolyzing activity and no (1→6)-β-D–glucan or (1→4)-α-D-glucan hydrolyzing activities, all of the material dissolved. The affinity chromatography of the solution on a Concanavalin A (ConA) column retained a fraction that contained mixed (1→4)-α-(1→6)-β-D–glucan polysaccharide. The eluent contained exclusively (1→3)-β-D–glucan oligosaccharides. When the (1→4)-α-D-glucan-rich fraction was released from its binding with ConA and treated with amyloglucosidase followed by dialysis to remove glucose, subjecting the resulting solution to affinity chromatography on a ConA column showed none of the material was retained. The material recovered from the solution was then proved to be pure (1→6)-β-D–glucan. These simple experiments showed that the "insoluble glycogen" from yeast cell wall is a mixed glucan in which (1→4)-α-D-glucan is connected to (1→3)-β-D–glucan through a (1→6)-β-D–glucan link. The structure of this mixed yeast cell wall (1→4)-α-(1→6)-β-(1→3)-β-D-glucan was confirmed by our ^{1}H NMR study (Kwiatkowski et al., 2009) and methylation analysis done at CCRC.

We used yeast cell wall α-glucan (prepared in our laboratory) in the development of the enzyme linked immunosorbent assay (ELISA) for yeast cell wall quantitative analysis in a complex polysaccharide matrix (Moran et al., 2011). The antigen was prepared in two stages.

At first a sample of cell wall (1→4)-α-D-glucan (fraction 50-100 kDa separated with the help of an ultra centrifugal filter (Aldrich Inc.) was oxidized mostly at the C-6 carbons using the Albright-Goldman oxidation reagent (Albright & Goldman, 1967; Zekovic et al., 2006) which converted the $-CH_2OH$ groups into aldehyde groups –CH=O in glucopyranose rings. The reaction produced a highly cross-linked polymer, with internal acetal bonds, but still well soluble in DMSO (dimethyl sulfoxide) and water, which was coupled to a bovine serum albumin (BSA), yielding the antigen (glucan-BSA conjugate). To restore the original α-glucan structure, but with $-CH_2$–NH-BSA groups instead of $-CH_2$-OH groups at the glucan's C-6 carbon, the conjugate C-6 imino (-CH=N-BSA) groups were reduced with sodium cyanoborohydride (Baxter & Reitz, 2002). Sodium dodecyl sulfate (SDS) polyacrylamide gel electrophoresis (PAGE) of the conjugate indicated the presence of significant quantities of free BSA (line at 66 kDa) together with the glucan-BSA conjugate of molecular size between 100 and 175 kDa. To remove the excess BSA, the concentrated solution of the conjugate was washed with water of pH 7.0, 8.5 and 4.5 using an ultra centrifugal filter with the 50-kDa membrane. This "purified" conjugate was used then in rabbit immunization. We found that the rabbit polyclonal antibodies were very sensitive to yeast cell wall α-glucan, but (as expected) they were also cross-reactive with BSA. To delete this activity we synthesized the BSA-coated affinity phase (Moran et al., 2011) and removed antibodies specific to BSA from the crude fraction of polyclonal rabbit anti-α-glucan antibodies. We also found that antibodies grown against this α-glucan polysaccharide did not recognize any commercially available starch or glycogen samples, including yeast glycogen. Such specificity confirms structural differences between the cell wall and the cytosol α-glucans from *Saccharomyces cerevisiae*. Because of the high specificity and very low cross reactivity of the "purified" rabbit antibodies with soy proteins and corn carbohydrates, the ELISA cell wall assay detects and quantifies samples of yeast cell wall down to 50 g/ton of feed (Moran et al., 2011). A similar assay was used for quantification of α-D-glucan-protein complex in mushrooms with immunomodulatory activity (Hirotaka et al., 2007).

What is the role of this polysaccharide in yeast cell life? Some remarks in the literature point to its role in yeast flocculation (Patel & Ingledew, 1975), acting as part of a possible killer toxin receptor (Hutchkins & Bussey, 1983) and as being sensitive to environmental conditions and changes in medium (Slaughter & Nomura, 1992; Dake et al., 2010; Jadhav et al., 2008). The cell wall bound α-glucan from fission yeast *Schizosaccharomyces pombe* has been studied in detail (Villar-Tajadura et al., 2008; Vos et al., 2007 ; Garcia et al., 2006; Grün et al., 2005), and systematic identification of the genes affecting glycogen storage in the yeast *Saccharomyces cerevisiae* was attempted (Wilson et al., 2002). Unfortunately, the presence of the two different pools of glycogen was not addressed. One of the possible functions of bound α-glucan might be the temporary formation of mixed α, β-glucan oligosaccharides that become soluble in cytoplasm and can be transported inside the cell wall where the β-glucan fragment is used in the construction of insoluble yeast cell wall β-glucan polysaccharide.

Is it possible that α-glucan polysaccharides from yeast or mushrooms can have medicinal properties? There is no literature concerning the medicinal properties of α-glucan as a separate fraction of the yeast cell wall, which is a consequence of acknowledging it as part of an energy storage cellular polysaccharide-yeast glycogen. Bioactive components containing α-glucan, mixed α, β-glucan or α-glucan-protein complexes are known to be present in extracts from the fruiting bodies of edible medicinal-mushrooms. Like in the case of many other traditional herbal remedies, it is difficult to separate the single bioactive component's (α-glucan) bioactivity from the activities of the other components of the extract, and in some cases, early assignments were incorrect.

Mushrooms with medicinal activities that contain extractable α-glucan in their fruiting bodies, belong to one of two phyla: Ascomycota or Basidiomycota. Hot-water extracts from *Peziza vesiculosa* (Ohno et al., 1985; Suzuki et al., 1982) and *Cordyceps chinensis* (Yan et al., 2011; Holliday & Cleaver, 2008; Khan et al., 2010; Liu et al., 2006; Kiho et al., 1993; Kiho et al., 1996; Li et al., 2006), which are both Ascomycota, have been shown to have antitumor and anti-diabetic activities. The linear α-(1→4)-D-glucan is less active than α-(1→4) (1→6)-D-glucan with 1 to 6 branches on the backbone 1→4 chain with a ratio of 1:8. The biological activity of glycogen-like polysaccharides was found to be molecular-size dependent with smaller ~9.5 kDa molecules being active, whereas 14-24 kDa molecules were not (Kakutani et al., 2007).

More is known about edible mushrooms belonging to Basidiomycota which contain both: α- and β-glucans in their fruiting bodies (Rop et al., 2009). Hot-water extracts from *Peziza vesiculosa* show antitumor activity (Ohno et al., 1985; Suzuki et al., 1982) and extracts from *Grifola frondosa* (Maitake) show glucose suppression (Tanaka et al., 2011) and antidiabetic (Lei et al., 2007) activities. Extracts (Shida & Matsuda, 1974) from *Lentinus edodes* (Shiitake) stimulate the immune system (Shah et al., 2011; Terakawa et al., 2008) and possess antibacterial, antiviral (Mach et al., 2008) and anticancer properties (Shida & Matsuda, 1974; Hyodo et al., 2005; Shah et al., 2011). Extracts from *Tricholoma matsutake* (Matsutake) show antitumor (Ebina et al., 2002) and immunomodulatory (Hirotaka et al., 2007; Hirotaka et al., 2005) activities. Extracts (Smiderle et al., 2010) from *Agaricus bisporus* (Portobello) inhibit breast cancer cell proliferation (Grube et al., 2001) and stimulate the immune system (Ren et al., 2008; Koppada et al., 2009). A glycogen-like polysaccharide from Portobello mushrooms potently activated macrophages, stimulating TNF-α production and phagocytosis of RAW264.7 cells (Kojima et al., 2010). Extracts from *Pleurotus ostreatus* (Oyster mushroom) show antiproliferative and proapoptotic activities on colon cancer cells (Lavi et al., 2006). *Pholiota nameko* (Butterscotch mushroom), which has been in use as a major component of miso soup in Japan, shows antinflammatory activity (Li et al., 2008). In Asian and European culinary traditions, meals containing these mushrooms are believed to be healthy and to heal a variety of ailments.

Whether α-glucan from yeast cell walls can contribute to the known medicinal activities of various extracts from whole yeast is still an open question. Newly developed, simple methods for producing large quantities of this material (Kwiatkowski et al., 2009; Moran et al., 2011) should stimulate new research in this area.

The abundant literature regarding bioactivity of the whole yeast cell wall and its extracts rich in α-mannoprotein or β-glucan, will be discussed in sections 4, 5 and 6 of this chapter.

4. Baker's yeast cell wall β-D-glucan/α-mannoprotein complex

Yeast cell wall biogenesis was studied by Smits (Smits et al, 2001) who found that "The yeast cell wall forms a border that defines the yeast cell's dimensions and separates its organelles from the negative impact of the environment. All of the individual constituents of a yeast cell wall connect to each other by covalent bonds forming a single supra-molecular biopolymer". The (1→3) (1→6)-β-D-glucan and the (1→2) (1→3) (1→6)-α-D-mannan/protein complexes (Vinogradov et al., 1998) are the major components of this supra-molecular biopolymer (Lessage & Bussey, 2006; Kath & Kulicke, 1999). They connect to each other by a covalent bond the nature of which is still under investigation (Kaptein et al., 1999; Kollar et al., 1997). In the process of yeast cell wall isolation the cell wall is first cracked open, with the help of physical, chemical and/or enzymatic treatments (Kath & Kulicke, 1999) and then separated from the soluble yeast cell components by centrifugation (Sedmak, 2006; Jamas et al., 1998). The concentrated yeast cell wall solids containing ~30% dry weight material are then spray-dried. The resulting product is a fine powder with a light tan color. The cell wall polysaccharides can be separated from each other by alkaline extraction, which solubilizes the α-mannoprotein fraction and leaves β-glucan particles in suspension. The β-glucan particulate can be separated from the soluble α-mannoprotein fraction by centrifugation and spray-dried to yield light, yellow colored, fine powder, free of any smell or taste, containing ~65% β-glucan. The α-mannoprotein solution can then be concentrated by using membrane ultrafiltration and spray-dried to produce a light brown, "mannan rich fraction" with ~40% by weight of α-mannan. Yeast cell walls as well as both fractions of polysaccharides are produced on a large scale and have practical applications as animal feed nutritional supplements. Furthermore, this yeast cell wall preparation, sold by Alltech Inc. under the name of Bio-Mos®, was proved (in more than 600 feeding trials) to have a positive impact on the immune system of livestock (Baurhoo et al., 2009; Yang et al., 2008; Rosen et al., 2007; Jacques & Newman, 1994; Morrison et al., 2010) fish (Dimitroglu et al., 2009), and companion animals (Swanson et al., 2002) when added to feed at the rate of 0.5-2.0 kg/ton. Bio-Mos® contains ~17% of α-mannan, which is possibly its bioactive component. An improved form of Bio-Mos® is marketed by Alltech Inc. under the name Actigen™. This improved product contains 30% α-mannan and is manufactured, by mixing yeast cell wall with a mannan-rich fraction. Actigen™ product is four times more active than Bio-Mos® (~13% of it is soluble and therefore, better available to interaction with microbes and animal gut) and its application rate of 250-500 g/ton approaches that of antibiotics. Alpha-D-mannans have a "brush like" structure built out of α-(1→2)- and α-(1→3)-D-mannopyranose branches, 1 to 5 rings long (brush hair), which are attached to a ~120-ring-long α-(1→6)-D-mannopyranose chain, the brush handle (Vinogradov et al., 1998). This structure creates a specific combination of various functionalities that also involve protein conjugates. It can fit with various receptors present on the walls of animal digestive tracts (Mansour & Levitz, 2003) and with the receptors on the membranes of pathogenic bacteria (Wellens et al., 2008). Alpha-D-mannan/protein-conjugates are involved in interactions with animal immune systems and as a result enhance immune system activity (Wismar et al., 2010) and contribute to animal antioxidant and antimutagenic defenses (Krizkova et al., 2006). Immune system strength, blockage of bacterial adhesion to the gut and modification of the gut structure contributes to improved survival and better growth in young animals. The second major yeast cell wall polysaccharide, (1→3)(1→6)-β-D-glucan, might also contribute to yeast cell wall biological properties (section 5 of this chapter), but its major role in animal nutrition is its ability to bind mycotoxins and detoxify animal feed. The (1→3)(1→6)-β-D-glucan is a part of the cell wall's "triple helix tridimensional structure, with spring-like mechanical properties, responsible for yeast cell wall strength and ability to absorb toxins" (Yannikouris et al., 2004) (see section 2, this chapter). Toxins occurring in plant-derived animal feed belong to one of two groups. Mycotoxins (the first group) are the by-products of the secondary metabolism of pathogenic fungi (Bennet & Klich, 2003), whereas chemical toxins (the second group) are incorporated into plant tissue as a result of plants metabolizing agrochemicals from the soil or water used for irrigation (McLean & Bledsoe, 1992). Some of the most problematic mycotoxins in causing human or animal diseases (i.e., aflatoxin, citrinin, ergot alkaloids, fumonisins, ochratoxin A, patulin, trichothecenes and zeralenone; Smith et al., 1995) can be absorbed by yeast cell wall β-glucans (Yiannikouris et al., 2004; Yannikouris et al., 2006). A mixture of cell wall β-glucan with clay (bentonite) sold as Mycosorb® by Alltech Inc. offers a spectrum of mycotoxin absorption superior to that of yeast cell wall glucans alone and also absorbs heavy metals (Brady et al., 1994). As in the case of Bio-Mos® a multitude of feeding trials have demonstrated the efficacy of Mycosorb® as an animal feed detoxifier using companion animals (Leung et al., 2006), horses (Raymond et al., 2003), pigs (Kogan & Kohler, 2007) and poultry (Dvorska et al., 2003; Karaman et al., 2005; Valarezo et al., 1998).

The whole yeast *Saccharomyces cerevisiae* grown in a medium containing inorganic selenium (Demirici & Pometto, 1999; Demirici et al., 1999; Mapelli et al., 2011) is used to produce yet another yeast-based human/animal nutritional supplement SelPlex® (Alltech Inc.). To maintain healthy metabolism, the human body requires 17 μg of selenium a day. Dietary selenium is used to synthesize selenocysteine (in liver) and is then incorporated into more than 25 Se-containing enzymes that play important roles in body's defense against free radical species (Tapiero et al., (2003) and in many other cellular processes including the generation of energy in mitochondria (Rayman, 2000). Because yeast does not possess genes that control selenium metabolism (Rodrigo, 2002), selenium, which has chemical properties extremely similar to sulfur, is metabolized in the same manner as sulfur and randomly incorporated into yeast cytosol, small-molecules like Se-glutathione, Se-adenosylhomocysteine (Uden et al. 2003) and proteins as Se-methionine (Tastet et al., 2008). Unlike yeast, mushrooms can metabolize selenium and can accumulate large quantities of it (Turlo et al., 2007) in the form of selenomethionine and selenocysteine. Animal feeding studies clearly showed that SelPlex® is not just a selenium source, but it also carries a variety of beneficial effects such as increased animal fertility (Rayman, 2000) or improved animal immune system activity (Rayman, 2000). Additionally, feeding SelPlex® to mice has been shown to delay the development of brain tumors from malignant human cancer cells implanted in mice brains (Toborek, 2011) and can significantly limit the deposition of Aβ amyloid plaques in APP/PS1 mouse brains that carry human Alzheimer's disease genes (Lovell et al., 2009).

Only some of these effects can be promoted by regular yeast cell wall preparations. However separating the bioactivities caused by the selenomethionine-containing yeast proteins from the activity caused by the polysaccharide parts of selenized yeast cell walls is difficult. Comparison of genomic activity in tissues taken from animals fed with SelPlex® to those fed with Bio-Mos® spiked with selenomethionine indicates very large differences in the regulation of multiple groups of genes for both treatments (Kwiatkowski et al., 2011), which may indicate that the bioactivity of Sel-Plex® involves cooperation between the selenoprotein and the polysaccharide components of selenized yeast cell walls.

Indeed, selected extracts from selenized yeast/yeast cell walls (Kwiatkowski et al., 2011) show potential as future, possible treatments of diseases such as type 2 diabetes, cancer and Alzheimer's.

5. β-(1→3)(1→6)-D-glucan as valuable by product from yeast fermentation

The commercial source from which the bulk yeast cell wall polysaccharides (including β-glucan) are produced uses the same strains of yeast as are used in fuel alcohol production. Le Saffre/ADM, Lallemand, Enzyme Development (New York) and DSM Life Sciences (Delft) are the largest suppliers of yeast for fuel ethanol producers in the United States and the European Union. Major factors that affect yeast cell wall composition include yeast strain (Hahn-Hagerdal et al., 2005), growth conditions (growth medium, temperature, osmotic pressure, toxic metabolites) and the time of harvesting (Aguilar-Uscanaga & Francois, 2003; Klis et al., 2002; Klis et al., 2006). Fuel alcohol fermentation is a high stress process (Devantier et al., 2005) and the cell walls of the yeast collected as its byproduct contain a high amount of β-glucans (Basso et al., 2008; Knauf & Kraus, 2006; Jones & Ingledew, 1994). In general, yeast strains of *Saccharomyces cerevisiae* that are used in baking (baker's yeast) have a higher β-glucan-to-α-mannan ratio than those that are used for alcohol fermentation (brewer's yeast), therefore it is advantageous to use pure baker's yeast for producing high quality (1→3)(1→6)-β-D-glucan for medicinal applications (Kim et al., 2007). The process of separating various yeast components has been heavily patented. However, the differences in the technologies are minor and in principle do not differ from the methodology described by Manners and Fleet (1976). The process starts from autolysing yeast cells at a temperature between 45°C and 65°C at slightly acidic pH, to release yeast cell walls that are insoluble and denser than the cytoplasmic contents and can be separated by centrifugation (Wheatcroft et al., 2002). These steps can be followed by incubating the yeast cell walls with alkaline protease at a pH of 9 to 10 to solubilize mannans and leave behind insoluble β-glucan (Zapata et al, 2008), which can then be physically separated from the liquid fraction by centrifugation and subsequent washing (Sedmak, 2006). Additional enzymes, like glucoamylase and lipase, can be used to hydrolyze α-glucan from α β-glucan, which is still present in the cell wall material and to solubilize the residue lipids from cell membranes. The final step of β-glucan production is a spray-drying that produces a white-to-maroon colored powder that does not carry any taste or aroma and is useful for feed and food applications. Further alkaline and acidic treatments of the food-grade β-glucan (Kelly,

2001) yields high purity (98.5% β-glucan, <0.1% mannan, 0.4% α-glucan, 0.3% protein, 0.2% chitin) microparticulate β-glucan with reduced molecular weight (from ~1-3 MDa to ~150 kDa) that is much more easily absorbed by the digestive tract and shows improved activity compared with food-grade products containing only ~65% β-glucan. Even further hydrolysis produces soluble yeast β-glucan (Jamas et al., 1998; Lee et al., 2001) that still retains most of the particular β-glucan bioactivities (Janusz et al., 1986; Wakshull et al., 1999).

Yeast *Saccharomyces cerevisiae*, its cell wall and products of its fractionation are generally recognized as safe (GRAS) by the US Food and Drug Administration (FDA, 1997), and they can be legally used as food ingredients but not as food additives. The European Food Safety Authority (EFSA) issued an opinion that yeast β-glucans are a "safe food ingredient" (EFSA, 2011) that can be used as a "food supplement" up to 375 mg/day and in foods for "particular nutritional uses" at dose levels up to 600 mg/day. (The uses of yeast cell wall as an animal feed ingredient were discussed in section 4 of this chapter).

Food-grade yeast β-glucans such as BetaRight® and WGP® (Biothera, Inc.) are used as ingredients in baked foods, beverages, ceral, yogurt, fruit juices, chocolate and as food thickeners in salad dressings, ice cream, mayonnaise, sauces and cheese. The majority of these applications have been patented (Zechner-Krpan et al., 2009; Thammakiti et al., 2004) and a critical review of 300 patented applications is available (Laroche & Michaud, 2007). Yeast β-glucans improve food rheological properties, gelling, water and oil-holding properties, without impacting its taste or odor (Petravic-Tominac et al., 2011). Beta-glucans also add health benefits (Laroche & Michaud, 2007) like antioxidative, bacteriostatic and immunostimmulating activities. Cosmetic products used in skin treatment contain yeast β-glucans as moisturizing and moisture-retaining components that also provide a proper moistening feeling. Because of its emulsion-stabilizing effects, pleasant texture and antioxidant activity yeast β-glucans can prevent skin injuries caused by solar radiation and therefore are used in sun-screens, oils and gels (Michiko & Yutaka, 2007). Deodorants containing yeast β-glucans have proved to be useful in oral preparations, mouthwashes and diapers (Michiko et al., 2005). Acid-treated cell walls (AYC) can be used as new binders in pharmaceutical formulations and, when mixed with traditional fillers like hydroxypropylcellulose or polyvinylpyrrolidone, yield harder pills with very short (~2 min) dissolution times (Yusa et al., 2002). Its adhesive and biological properties can be also utilized in producing coating for surgical instruments (Klein, 2003) and in the manufacture of packaging for the food industry (Cope, 1987). Its antibacterial and antiviral properties have found application in the control of plant pests (Kitagawa, 2007) and viral invasions (Slovakova et al., 1997).

6. Medicinal application of native and chemically modified forms of β-(1→ 6)(1→3)-D-glucan

Approximately 2000 research and review papers covering β-(1→6)(1→3)-D-glucan bioactivity and its medicinal applications have been published since the 1960's, and the majority of the literature has been reviewed (Bohn & BeMiller, 1995; Kogan, 2000; Zekovic et al., 2005; Vetvicka, 2011; http://www.betaglucan.org; and for patented applications see Laroche & Michaud, 2007). Many early assumptions have been confirmed by more rigorous studies, and many others have been disproven. Many of the studies that initially showed promise utilized poorly purified and not sufficiently standardized samples of yeast cell wall glucan (Jaehrig et al., 2007; Vetvicka & Vetvickova, 2007). Highly purified, insoluble, whole-glucan particles (**WGP**) with the size 2-4 μm (Yan et al., 2005), or soluble, yeast poly-(1→6)-β-D-glucopyranosyl-(1→3)-β-D-glucopyranose (**PGG**) with molecular weight ~150 kDa (Gawronski et al., 1999) is presently used in animal and human studies. Glucan samples used in earlier studies could contain significant quantities of (also) bioactive impurities like yeast mannoproteins or α-glucans. The existence of a correlation between the β-(1→6)(1→3)-D-glucan structure and its bioactivity is well established having been confirmed by multiple animal and human studies (Novak & Vetvicka, 2009). The WGP-glucan fate after oral administration and PGG-glucan after intravenous administration has been studied in animals (Vetvicka & Vetvickova, 2008). Mammalian cells lack β-glucan, which is present in the cell wall of many infectious microbes. Thus, mammalian macrophages, which are present "on guard" in the cells lining the digestive track, recognize and bind purified β-glucan, but not the whole yeast cells (which have mannoproteins on the cell surface), via the dectin-1 receptor, which is a glucan-sensing receptor expressed on dectin a transmembrane protein. The macrophages transport β-glucan molecules to the spleen, bone marrow and lymph nodes where they are disintegrated into smaller, soluble, β-glucan oligosaccharides that are released back into the bloodstream. The circulating monocytes, macrophages, neutrophils, natural killer and dendritic cells (Chan et al., 2009) which possesses β-glucan recognizing receptors, including: toll like receptor 2 (TLR-2, Underhill et al., 1999), dectin-1 (Brown et al., 2007), complement receptor 3 (CR-3) (Ross et al., 1987) and lactosyl ceramide receptor, bind to the β-glucan molecules triggering the non-specific-innate immune response, including phagocytosis and production of proinflammatory factors (Qi et al., 2011). This mechanism leads to the elimination of infectious agents (Chan et al., 2009). Cancer cells are recognized by the host immune system, but antibody response is too weak to destroy them. When humanized, anti-cancer monoclonal antibodies are used in cancer therapy the treatments are not uniformly effective. Combined antitumor monoclonal antibodies (mAb's) and β-glucan therapy yields much better results with fewer adverse side effects (Salvador et al., 2008; Liu et al., 2009). The mechanism of action is only partially understood with neutrophils and macrophages playing a role in the process of killing cancer cells (Liu et al., 2009). Strengthening immunity to various diseases (Hofer & Pospisil, 2011) and combating those that are already active in humans and animals are the main applications of yeast β-glucan. In Eastern culinary tradition, eating mushrooms (which contain large quantities of soluble β-glucan in their fruiting bodies) has long been recognized as healthy, and four pharmaceutical preparations, based on mushroom extracts, have been registered as drugs in Japan (Hyodo et al., 2005). In the Western world several clinical trials (Babineu et al., 1994; Weitberg, 2008; Spruijt et al., 2010) proved the beneficial properties of WGP and PGG glucans, which are still waiting to be registered as drugs (Vetvicka, 2011; Lehtovaara & Gu, 2011). The following therapeutic activities of yeast β-glucan are well-established in treating the following conditions: post-surgical infections, hospital pneumonia, acute renal failure (Koc et al., 2011), pressure ulcers (bed sores), wound healing (surgical and as result of injury, Spruijt et al., 2010) and burns caused by heat, UV or X-ray radiation. As an adjuvant yeast β-glucans alleviate stress and improves the mAb's plus PGG anticancer treatment for breast, colorectal, colon, leukemia, lung, ovarian and skin cancers, chemotherapy and radiotherapy; antireumatic drug therapy (Sener et al., 2006), and antifungal therapies. They stimulate bone marrow healing and bone repair (due to injuries), which increases red blood cell count and neutrophil production. Many of these activities were studied in the 1990s and 2000s and have been reviewed. Studies since 2010 have been referred to herein.

Some of the negative side effects that have been observed in response to prolonged or large doses of β-glucan preparations are the consequences of their mode of action as well as their physical and chemical properties. Their therapeutic activity is based upon immune system mobilization and the production of monocytes, macrophages, neutrophils, natural killer and dendritic cells which fight and destroy pathogens. However, prolonged pro-inflammatory changes in cells can cause the development of autoimmune diseases. The human body is missing enzymes that can hydrolyze β-glucan polysaccharides and for their degradation uses less efficient oxidative pathways. This insufficiency extends the cellular half-life of WGP and PPG glucans and can lead to the formation of granulomas in the liver, causing inflammation that can result in liver cirrhosis. Therefore the use of soluble and relatively low molecular weight PGG glucan is recommended. The well documented β-glucan abilities to lower blood sugar and blood pressure justify further study aimed at producing a glucan material with molecular weight lower than that of PGG glucan to preclude accumulation in granulomas, when used as a long-term food additive. The next stage of development of less toxic, soluble β-glucan preparations should include its size fractionation to produce more standardized material for medicinal applications. The majority of *in vivo* studies on β-glucan medicinal applications were done using rodent models, therefore it is easy to foresee veterinary applications of this food/feed ingredient to improve immune systems, facilitate burn and wound healing, stimulate post-trauma bone repair and help fight cancers.

Author details

Stefan Kwiatkowski
Alltech Inc. Nicholasville KY, USA

Stefan Edgar Kwiatkowski
University of Kentucky, USA

Acknowledgement

The author expresses deep gratitude to Dr. Pearse T. Lyons, President, and Dr. Karl Dawson, CSO, Alltech Inc. for their continuing support and inspiration.

7. References

Aimaniada, V., Clavaud, C., Simenel, C., Fontaine, T., Delepierre, M. & Latage, J-P. (2009). Cell wall (1→6)-β-D-glucan of *Saccharomyces cerevisiae* – structural characterization and in situ synthesis, *The Journal of Biological Chemistry* Vol. 284: 13401-13412.

Aguilar-Uscanaga, B., & Francois, J.M. (2003). A study of the yeast cell wall composition and structure in response to growth conditions and mode of cultivation, *Letters in Applied Microbiology* Vol. 37: 268-274.

Albright, J.D. & Goldman, L. (1967). Dimethyl sulfoxide-acid anhydride mixtures for oxidation of alcohols, *Journal of the American Chemical Society* Vol. 89: 2416.

Arvindekar, A.U. & Patil, N.B., (2002). Glycogen – a covalently linked component of the cell wall in Saccharomyces cerevisiae, *Yeast* Vol. 19(No. 2): 131-139.

Babineau, T.J., Marcello, P., Swails, W., Keuler, A., Bistrian, B. & Forse, A. (1994). Randomized Phase I/II Trial of a Macrophage-Specific Immunomodulator (PGG-Glucan) in High Risk Surgical Patients, *Annals of Surgery* Vol. 220(No. 5): 601-609.

Bacon, J.S.D., Farmer, V.C., Jones, D. & Taylor I.F. (1969). The glucan components of the cel wall of Baker's Yeast *(Saccharomyces cerevisiae)* considered in relation to its ultrastructure, *Biochemistry Journal* Vol. 114: 557-567.

Ballou, C.E. (1980). Genetics of mannoprotein biosynthesis" in *Fungal Polysaccharides*, Sandford, P., et al; ACS Symposium Series; American Chemical Society: Washington D.C. (1980) p 1-14.

Basso, L.C., de Amorim, H.V., de Oliveera, A.J. & Lopes, M.L. (2008). Yeast selction for fuel alcohol production in Brazil, *FEMS Yeast Research* Vol. 8(No. 7): 1155-1163.

Baxter, E.W. & Reitz, A.B. (2002). Reductive aminations of carbonyl compounds with borohydride and borane reducing agents, *Organic Reactions* Vol. 59: 1. Wiley: New York 2002.

Baurhoo, B., Ferket, P.R. & Zhao, X. (2009). Effects of diets containing different concentrations of mannooligosaccharides or antibiotics on growth performance, intestinal development, cecal and litter microbial populations, and carcass parameters in broilers, *Poultry Science* Vol. 88(No.11): 2262-2272.

Bennet, J.W. & Klich, M. (2003). Mycotoxins, *Clinical Microbiology Reviews*. Vol. 16(No. 3), 497-516.

Borchers, A.T., Keen, C.L. & Gershwin, M.E.et al. (2004). Mushrooms, tumors and immunity: an update, *Experimental Biology and Medicine* Vol. 229(No. 5): 393-406.

Bohn, J.A. & Be Miller, J.N. (1995). (1→3)(1→6)-β-D-Glucans as biological response modifiers: a review of structure-functional activity relationships, *Carbohydrate Polymers* Vol. 28(No. 1): 3-14.

Brady, D., Stoll, A.D. & Duncan, J.R. (1994). Chemical and enzymatic extraction of heavy metal binding polymers from isolated cell walls of *Saccharomyces cerevisiae, Biotechnol. Bioeng. (USA)* Vol. 44(No.3), 297-302.

Brown, J., O'Callaghan, Ch., Marshall, A.S.J., Gilbert, R.J.C., Siebold, Ch., Gordon, S., Brown, G.D. & Jones, Y.J. (2007). Structure of the fungal β-glucan binding receptor dectin-1: implications for function, *Protein Science* Vol. 16: 1042-1052.

Chan, G.Ch-F., Chan, W.K. & Sze, D.M-Y. (2009). The effects of β-glucan on human immune and cancer cells, *Journal of hematology & oncology* Vol. 2: 25.

Cope, Roy (1987) Adhesive substances from yeast cell wall, *Eur. Pat. Appl.* EP 230378 A2.

Dake, M.S., Jadhv, J.P. & Patil, N.P. (2010). Variations of two pools of glycogen and carbohydrate in *Saccharomyces cerevisiae* grown with various ethanol concentrations, *Journal of Industrial Microbiology & Biotechnology* Vol. 37(No. 7): 701-706.

Demirici, A. & Pometto III, A.L. (1999). Production of organically bound selenium yeast by continuous fermentation, *J. Agric. Food Chem.* Vol. 47(No. 6): 2491-2495.

Demirici, A., Pometto III, A.L. & Cox, D.J. (1999). Enhanced organically bound selenium yeast production by fed-batch fermentation, *J. Agric. Food Chem.* Vol. 47(No. 6): 2496-2500.

Devantier, R., Pedersen, S. & Olsson, L. (2005). Characterization of very high gravity ethanol fermentation of corn mash. Effect of glucoamylase dosage, pre-saccharification and yeast strain, *Appl Microbiol Biotechnol* Vol. 68: 622-629.

Dimitroglu, A., Merrifield, D.A., Moate, R., Davies, S.J., Spring, P., Sweetman, J. & Bradley, G. (2009). Dietary mannan oligosaccharide supplementation modulates intestinal microbial ecology and improves gut morphology of rainbow trout, *Oncorhynchus mykiss (Walbaum), Journal of Animal Science* Vol. 87(No. 10): 3226-3234.

Dvorska, J.E. & Surai, P.F. (2003). Protective effect of modified glucomannans against aurofusarin induced changes in quail egg and embryo" *Asian-Australian Journal of Animal Science* Vol.17(No.3): 434-440.

Ebina, T., Kubota., Ogama, N. & Matsunaga, K. (2002). Antitumoreffect of a peptide-glucan preparation extracted from a mycelium of *Tricholoma matsutake* (S. Ito and Imai) Sing, *Biotherapy* Vol. 16(No. 3): 255-259.

EFSA. (2011) Scientific opinion on the safety of 'yeast β-glucans' as novel food ingredient, *EFSA Journal* Vol. 9(No. 5): 2137-2159.

FDA. (1997) Substances generally recognized as safe, *Federal Register* 62 FR 18938, 17 April 1997.

Garcia, I., Tajadura, V., Martin, V., Toda, T. & Sanchez, Y. (2006). Synthesis of α-glucans in fission yeast spores is carried by three α-glucan synthase paralogs Mok12p, Mok13p and Mok14p, *Molecular Microbiology* Vol. 59(No. 3): 836-853.

Gawronski, M., Park, J.T., Magee, A.S. & Conrad, H. (1999). Microfibrillar structure of PGG glucan in aqueous solution as triple helix aggregates by small angle x-ray scattering, *Biopolymers* Vol. 50: 569-578.

Grba, S., Oura, E. & Soumalainen, H. (1976). Formation of glycogen and trehalose in baker's yeast, *European Journal of Applied Microbiology* Vol. 2(No. 1): 29-37.

Grube B.J., Eng, E.T., Kao, Y.C., Kwon, A. & Chen, S. (2001). White button mushroom phytochemicals inhibit aromatase activity and breast cancer cell proliferation, *J. Nutrition* Vol. 131(No. 12): 3288–93.

Grün, Ch.H., Hochstenbach, F., Humbel, B.M., Verkleij, A.J., Sietsma, J.H., Klis, F.M., Kamerling, J.P. & Vliegenthart, J.F.G. (2005). The structure of cell wall α-glucan from fission yeast, *Glycobiology* Vol. 15(No. 3): 245-257.

Gunia-Smith, Z., Patil, N.B. & Smith, E.E., (1977). Two pools of glycogen in *Saccharomyces, Journal of Bacteriology* Vol. 1977: 818-825.

Hahn-Hagerdal, B., Karhumaa, K., Larsson, Ch.U., Gorwa-Grauslund, M., Gorgens, J. & van Zyl, W.H. (2005). Role of cultivation media in the development of yeast strains for large scale industrial use, *Microbial Cell Factories* Vol. 4: 31-47.

Hirotaka, H., Yoko, Y., Kenichi, M., Yoko, I. & Tdeshi, Y. (2007). Development of enzyme-linked immunoassay to detect immunomodulatory α-D-glucan-protein complex, MPG-1, in basidiomycete Tricholoma matsutaka and related processed food, *J. Agric. Food Chem.* Vol. 55(No. 21): 8508-8515.

Hirotaka, H., Yoko, Y., Kenichi, M., Yoko, I. & Tdeshi, Y. (2005). Isolation and characterization of a novel immunomodulatory α-glucan-protein complex from the mycelium of *Tricholoma matsutake* in basidiomycetes, *J. Agric. Food Chem.* Vol. 53(No. 23): 8948-8956.

Hofer, M. & Pospisil, M. (2011). Modulation of animal and human hematopoiesis by β glucans: a review, *Molecules* Vol. 16: 7969-7979.

Holliday J., Cleaver, M. (2008). Medicinal Value of the Caterpillar Fungi Species of the Genus Cordyceps (Fr.) Link (Ascomycetes), *International Journal of Medicinal Mushrooms* Vol. 10(No. 3): 219

Huang, D., Farkas, I. & Roach, P.J. (1996). Pho85p, a cyclin-dependent protein kinase, and the Snf1p protein kinase act antagonistically to control glycogen accumulation in *Saccharomyces cerevisiae, Molecular and Cellular Biology* Vol. 16(No. 8): 4357-4365.

Hutchins, K. & Bussey, H. (1983). Cell wall receptor for yeast killer toxin: involvement of (1, 6)-β-D-glucan, *Journal of Bacteriology* Vol. 154: 161-169.

Hyodo I., Amano, N., Eguchi, K., Narabayashi, M., Imanishi, J., Hirai, M., Nakano, T. & Takashima S. (2005). Nationwide survey on complementary and alternative medicine in cancer patients in Japan, *Journal of Clinical Oncology* Vol. 23(No. 12): 2645–54.

Jacques, K.A. & Newman, K.E. (1994). Effect of oligosaccharide supplements on performance and health of Holstein calves pre- and post-weaning, *Journal of Animal Science* Vol. 72(Suppl. 1): 295.

Jadhav, J.P. Dake, M.S. & Patil, N.P. (2008). α, β-Glucan complex in *Saccharomyces carlsbergensis, Asian Journal of Chemistry* Vol. 20(No. 1): 55-65.

Jaehrig, S.C., Rohn, S., Kroh, L.W. Fleisher, L-G. & Kurtz, T. (2007). In vitro potential antioxidant activity of (1→3) (1→6)-β-D-glucan and protein fractions from *Saccharomyces cerevisiae* cell walls, *J. Agric. Food Chem.* Vol. 55(No. 12): 4710-4716.

Jamas, S., Easson Jr, D.D. & Ostroff, G.R. (1997). Glucan preparation, *US Patent* US 5622939, Apr 22, 1997.

Jamas, S., Easson Jr, D.D. & Ostroff, G.R. (1998). Underivatized, aqueous soluble β-(1, 3)-glucan, composition and making some, *US Patent* US 5817643, Oct 6, 1998.

Janusz, M.J., Austen, K.F. & Czop, J.K. (1986). Isolation of soluble yeast beta-glucan that inhibit human monocyte phagocytosis mediated by beta-glucan receptors, The Journal of Immunology Vol.137 (No. 10):3270-3276.

Jones, A.M. & Ingledew, W.M. (1994). Fuel alcohol production: optimization of temperature for very efficient very-high gravity fermentation, *Applied and Environmental Microbiology* Vol. 1994: 1048-1051.

Kakutani, R., Adachi, Y., Kajiura, H., Takata, H., Kuriki, T. & Ohno, N. (2007). Relationship between structure and immunostimulating activity of enzymatically synthesized glycogen, *Carbohydrate Research* Vol. 342: 2371-2379.

Kaptein, J.C., Van Den Ende, H. & Klis, F.M. (1999). The contribution of cell wall proteins to the organization of the yeast cell wall, *Biochimical Biophysical Acta-General Subjects* Vol. 1426(No. 2): 373-383.

Karaman, M., Basmocioglu, H., Ortatatli, M. & Oguz, H. (2005). Evaluation of detoxifying effect of yeast glucomannan on aflatoxicosis in broilers as assessed by gross examination and histopathology, *British Poultry Science* Vol. 46(No. 3): 394-400.

Kath, F. & Kulicke, W-M. (1999). Mild enzymatic isolation of mannan and glucan from yeast *Saccharomyces cerevisiae, Die Angewandte Makromoleculare Chemie* Vol. *268*: 59-68.

Kelly, G.E., (2001) Process for glucan preparation and therapeutic uses of glucan, *US Patent* US 6,242,594 B1.

Khan, M.A., Mousumi, T., Dian-Zheng Zhang & Han-Chun Chen (2010). *Cordyceps* Mushroom: A Potent Anticancer Nutraceutical, *The Open Nutraceutical Journal* Vol. 3: 179–183.

Kiho, T., Hui, J., Yamane A. & Ukai S. (1993). Polysaccharides in fungi. XXXII. Hypoglycemic activity and chemical properties of a polysaccharide from the cultural mycelium of Cordyceps sinensis, *Biol. Pharm. Bull.* Vol. 16(No. 12): 1291–1293.

Kiho, T., Hui, J., Yamane A. & Ukai S. (1996). Polysaccharides in fungi. XXXVI. Hypoglycemic activity of a polysaccharide (CS-F30) from the cultural mycelium of Cordyceps sinensis and its effect on glucose metabolism in mouse liver, *Biol. Pharm. Bull.* Vol. 19(No. 2): 294–296.

Kim, Y-H., Kang, S.W., Lee, J.H., Chang, H-I., Yun, Ch-W., Paik, H-D., Kang, Ch-W. & Seung, W. (2007). High density fermentation of *Saccharomyces cerevisiae* JUL3 in fed-batch culture for the production of β-glucan, *Journal of Industrial and Engineering Chemistry (Seoul)* Vol. 13(No. 1): 153-158.

Kitagawa, T. (2007). Yeast cell wall degradation product as plant pest control agent, *Jpn. Kokai Tokkyo Koho* JP 2007045708 A.

Klein, B (2003). Immunostimulating coating for surgical devices, *US Patent* US 6541678 B2.

Klis, FM., Mol, P., Hellingwerf, K. & Brul S. (2002) Dynamics of cell wall structure in *Saccharomyces cerevisiae, FEMS Microbiology Reviews* Vol. 26(No. 3): 239-256.

Klis, F.M., Boorsma, A. & DeGroot, P.W.J. (2006). Cell wall construction in *Saccharomyces cerevisiae, Yeast* Vol. 23: 185-202.

Knauf, M. & Kraus, K. (2006). Specific yeast developed for modern ethanol production, *Sugar Industry* Vol. 131: 753-758.

Koc, E., Reis, K.A., Ebinc, F.A., Pasaoglu, H., Demirtas, C., Omeroglu, S., Derici, U.B., Erten, Y., Bali M., Arinsov, T. & Sindel, S. (2011). Protective effect of beta-glucan on contrast induced-nephropathy [acute renal failure] and a comparison of beta-glucan with nebivolol and N-acetylcysteine in rats, *Clin. Exp. Nephrol.*Vol. 15(No. 5): 658-665.

Kogan, G. (2000). (1→3, 1→6)-β-D-Glucans of yeast and fungi and their biological activity, *Studies in Natural Products Chemistry* Vol. 23: 107-151/Atta-ur-Rahman (Ed.) Elsevier 2000.

Kogan, G. & Koher, A. (2007). Role of yeast cell wall polysaccharides in pig nutrition and health protection, *Livestock Sci.* Vol. 109: 161-165.

Kogan, G., Pajtinka, M., Babincova M., Miadokova, E., Ranko, P., Slamenova, D. &

Korolenko, T.A., (2008).Yeast Cell Wall Polysaccharides as antioxidants and antimutagens: can they fight cancer? *Neoplasma* Vol. 55: 5.

Kojima, H., Akaki, J., Nakajima, S., Kamei, K. & Tamesada, M. (2010). Structural analysis of glycogen like polysaccharides having macrophage-activating activity in extracts, *J. Nat. Med.* Vol. 64:16-23.

Kollar, R., Reinhold, B.B., Petrakova, E., Yeh, H.J.Ashwell, G., Drgonova, J., Kaptein, J.C.,

Klis, F.M. & Cabib E., (1997). Architecture of the yeast cell wall β-(1-6)-glucan interconnects mannoprotein, β-(1-3) glucan and chitin, *The Journal of Biological Chemistry* Vol. 272(No. 28): 17762-17775.

Koppada, R., Norozian, F. N., Torbati, D., Kalomiris Sophia, Ramachandran, Ch. & Totapally B.R. (2009). Physiological effects of a novel immune stimulator drug (1→4)-α-D-glucan in rats, *Basic and Clinical Pharmacology and Toxicology* Vol. 105(No. 4): 217-221.

Krizkova L., Zitnanova, I., Mislovicova, D., Masarova, J., Sasinkova, V., Durackova, Z. & Krajovic, J. (2006). Antioxidant and antimutagenic activity of mannan neoglycoconjugates: mannan-human serum albumin and mannan-penicilin G acylase, *Mutation Research/Genetic Toxycology and Environmental Mutagenesis* Vol. 606(No. 1-2): 72-79.

Kwiatkowski S., Thielen, U., Glenny, P. & Moran, C. (2009). A study of *Saccharomyces cerevisiae* cell wall glucans, *The Institute of Brewing & Distilling* Vol. 115(No. 2):151-158.

Kwiatkowski, S., Powers, R., Matney, C., Ghorghchian, P.P. & Ostertag, E.M. (2011). Compositions and methods for separating, characterizing and administering selenoglycoproteins, USPTO Application: 20110280948.

Laroche, C. & Michaud, P. (2007). New developments and perspective applications for β- (1, 3)-glucans, *Recent Patents on Biotechnology* Vol. 2007(No. 1): 59-73.

Lavi, I., Friesem, D., Geresh, S., Hadar, Y. & Schwartz, B. (2006). An aqueous polysaccharide extract from the edible mushroom *Pleurotus ostreaus* induces antiproliferative and pro-apoptotic effects on HT-29 colon cancer cells, *Cancer Letters (Amsterdam, Netherlands)* Vol. 244(No. 1): 61-70.

Lee, J.N., Lee, D.Y., Kim, G.E., Kim, H.N., Sohn, J., Kim, S. & Kim, C.W. (2001). Purification of soluble beta-glucan with immune-enhancing activity from cell wall of yeast, *Biosci. Biotechnol. Biochem.* Vol. 65(No. 4): 831-841.

Lei, H., Ma, X. & Wu, W. (2007). Anti-diabetic effect of α-glucan from fruiting body of maitake (*Grifola frondosa*) on KK-Ay mice, *Journal of Pharmacy and Pharmacology* Vol. 59(No. 4): 575-582.

Lehtovaara, B.C. & Gu, F.X. (2011). Pharmacological, Structural, and Drug Delivery Properties and Applications of 1, 3-β-Glucans, *J. Agric. Food Chem.* Vol. 59(No. 5): 6813-6828.

Lessage, G. & Bussey, H. (2006). Cell wall assembly in *Saccharomyces cerevisiae, Microbiology and Molecular Biology Reviews* Vol. 70(No. 2): 317-343.

Leung, M.C.K., Diaz-Llano, G. & Smith, T.K. (2006). Mycotoxins in pet food: a review on worldwide prevalence and preventive strategies, *J. Agric. Food Chem.* Vol. 54 (No. 26): 9623-9635.

Li, S.P., Zhang, G.H., Zeng, Q., Huang, Z.G., Wang, Y.T., Dong, T.T.X & Tsim, K.W.K. (2006). Hypoglycemic activity of polysaccharide, with antioxidation, isolated from cultured *Cordyceps mycelia, Phytomedicine* Vol. 13(No. 6): 428–433.

Li H., Lu, X., Zhang, S., Lu, M. & Liu, H. (2008). Anti-inflammatory activity of polysaccharide from *Pholiota nameko, Biochemistry (Moscow)* Vol. 73(No. 6): 669-675.

Lichtenthaler, F.W. & Peters, S. (1980). Carbohydrates as green raw materials for the chemical industry, *Comptes Rendus Chimie* Vol. 7(No. 2): 65-90.

Lille, S.H. & Pringle, J.R. (1980). Reserve carbohydrate metabolism in *Saccharomyces cerevisiae*: response to nutrition limitation, *Journal of Bacteriology* Vol. 143: 1384-1394.

Lindhorst, T.K. (2003). Essentials of carbohydrate chemistry and biochemistry, Wiley-VCH, Weinheim, ISBN 3-527-30664-1

Liu, J., Gunn, L., Hansen, R. & Yan, J. (2009). Combined yeast derived β-Glucan with antitumor monoclonal antibody for cancer immunotherapy, *Experimental and Molecular Pathology* Vol. 86(No. 3): 208-214.

Liu, W-Ch., Wang, S-Ch., Tsai, M-L., Chen, M-C., Wang, Y-C., Hong, J-H., McBride W. & Chiang, Ch-S. (2006). Protection against radiation-induced bone marrow and intestinal injuries by *Cordyceps sinensis*, a Chinese herbal medicine, *Radiation Research* Vol. 166(No. 6): 900–907.

Lovell, M.A., Xiong, S., Lyubartseva, G. & Markesbery, W.R. (2009). Organo-selenium (Sel Plex) decreases amyloid burden and RNA and DMA oxidative damage in APP/PS1 mice, *Free. Radic. Biol. Med.* Vol. 46(No. 11): 1527-1533.

Mach C.M., Fugii, H., Wakame, K. & Smith, J. (2008). Evaluation of active hexose correlated compound hepatic metabolism and potential for drug interactions with chemotherapy agents, *Journal of the Society for Integrative Oncology* Vol. 6(No. 3): 105–9.

Manners, D.J. & Fleet, G.H. (1976). Isolation and composition of alkali soluble glucan from cell walls of *Saccharomyces cerevisiae, J. Gen. Microbiol.* Vol. 94: 180-192.

Manners, D.J., Masson, A.J., Patterson, J.C., Bjorndal, H. & Lindberg L., (1973) The structure of β-(1→6)-D-glucan from Yeast cell wall, *Biochemical Journal* Vol. 135: 19-30.

Mansour, M.K. & Levitz S.M. (2003). Fungal mannoproteins: the sweet path to immunodominance, *ASM News* Vol. 69 (No. 12): 595-560.

Mapelli, V., Hillestrom, P.R., Kapolna, E., Larsen, E.H. & Olsson, L. (2011). Metabolic and bioprocess engineering for production of selenized yeast with increased content of seleno-methylselenocysteine" *Metabolic Engineering* Vol. 13(No. 3): 282-293.

McLean J.E. & Bledsoe B.E. (1992). Behavior of metals in soils, *EPA Ground Water Issue 1992*, EPA/540S-92/018.

Michiko, Y., Tomohiko, N. & Yutaka, K. (2005). Deodorants containing yeast cell wall fraction, *Jpn. Kokai Tokkyo Koho* JP 2005325065 A.

Michiko, Y. & Yutaka, K. (2007) Skin conditioning composition containing cell wall fractions, JP 2007045714 Feb 22, 2007.

Moran, C., Kwiatkowski, S., Yannikouris, A. & Thielen, U. (2011). Yeast cell wall components and detection thereof, *International Patent Application* WO 2011143613 (A1) 20-11-17.

Morrison S.J., Dawson, S. & Carson, A.F. (2010). The effects of mannan oligosaccharide and *Streptococcus faecium* addition to milk replacer on calf health and performance, *Livestock Science* Vol. 131(No. 2): 292-296.

Novak, M. & Vetvicka, V. (2009). Glucans as biological response modifiers, *Endocrine, Metabolic and Immune Disorders-Drug Targets* Vol. 9: 67-75.

Ohno, N., Mimura, H., Suzuki, I. & Yadomae, T. (1985). Antitumor activity and structural characterization of polysaccharide fractions extracted with cold alkali from a fungus, *Peziza vesiculosa, Chem. Pharm. Bull. (Tokyo)* Vol. 33(No. 6): 2564-8.

Osumi, M. (1998) The ultrastructure of yeast: cell wall structure and formation, *Micron* Vol. 29: 207-233.

Patel, G.B. & Ingledew, W.M. (1975). The relationship of acid soluble glycogen to yeast flocculation, *Can. J. Microbiol.* Vol. 21: 1608-1613.

Petravic-Tominac V., Zechner-Krpan, V., Berkovic, K., Golovic, P. Hercog, Z., Srecec, S. & Spoliaric, I. (2011). Rheological properties, water-holding and oil-binding capacities of particulate β-glucans isolated from spent Brewer's yeast by three different procedures, *Food Technol. Biotechnol.* Vol. 49(No. 1): 56-64.

Qi, C., Cai, Y., Ding, Li. B., Kloecker, G., Qian, K., Vasilakos, J., Saijo, S., Iwakura, Y., Yannelli, J.R. & Yan, J. (2011). Differential pathways regulating innate and adaptive antitumor immune responses by particulate β-glucan, *Blood* Vol. 117(No. 25): 6825-6836.

Quain, D.E. (1981). The determination of glycogen in yeast, *J.Inst.Brew.* Vol. 87: 289-291.

Rayman, M.P. (2000). The importance of selenium to human health, *Lancet* Vol. 356: 233-241.

Raymond, S.L., Smith, T.K. & Swamy H.V.L.N. (2003). Effects of feeding a blend of grains naturally contaminated with *Fusarium* mycotoxins on feed intake, serum chemistry and hematology of horses and the efficacy of a polymeric glucomannan mycotoxin adsorbent, *J. Anim. Sci.* Vol. 81: 2123-2130.

Ren Z., Guo, Z., Meydani, S.N. & Wu, D. (2008). White button mushroom enhances maturation of bone marrow-derived dendritic cells and their antigen presenting function in mice, *J.Nutrition* Vol. 138 (No. 3): 544–50.

Rodrigo, M-J., Moskovitz, J., Salamini, F. & Bartels, D. (2002). Reverse genetic approaches in plants and yeast suggest a role for novel, evolutionary conserved, selenoprotein-related genes in oxidative stress defense, *Mol. Genet. Genomicks* Vol. 267: 613-621.

Rop, O., Mlcek, J. & Jurikova, T. (2009) Beta-glucans in higher fungi and their health effects, *Nutrition Reviews* Vol. 67 (No. 11): 624–31.

Rosen, G.D. (2007). Holo-analysis of the efficacy of Bio-Mos in pig nutrition, *Animal Science* Vol. 82: 683 689.

Ross, G.D., Cain, J.A., Moynes, B.L., Newman, S.L. & Lachman, P.J. (1987). Role of complement receptor type three and serum opsonins in the neutrophil response to yeast, *Complemment Inflamm.* Vol. 4: 61-74.

Rothman-Denes, L.B. & Cabib, E. (1970). Two forms of glycogen synthetase and their role in glycogen accumulation, *P.N.A.S.* Vol. 66: 967-974.

Salvador, C., Li, B., Haqnsen, R., Cramer, D.E., Kong, M & Jun, Y. (2008). Yeast derived β glucan augments the therapeutic efficacy mediated by anti-vascular endothelial growth factor monoclonal antibody in human carcinoma xenograft models, *Clin. Cancer Res.*Vol. 14(No. 4): 1239-1247.

Sedmak, JJ. (2006). Production of β-glucans and Mannans, *US Patent Application* Nov 23, 2006 US 2006/0263415 A1.

Sener, G., Eksioglu-Demiraop, E., Cetiner, M., Ercan, F. & Yegen, B.C. (2006). Beta-glucan ameliorates methotrexate-induced oxidative organ injury via its antioxidant and immunomodulatory effects, *European J Pharmacology* Vol. 542(No. 1-3): 170-178.

Shah, S.K., Walker, P.A., Moore-Olufemi, S.D., Sundaresan, A., Kulkarini, A.D. & Andrassy, R.J. (2011). An evidence-based review of a *Lentinula edodes* mushroom extract as complementary therapy in the surgical oncology patient, *Journal of Parenteral and Enteral Nutrition* Vol. 35(No. 4): 449–58.

Shida, M. & Matsuda, K. (1974). Studies on polysaccharides from *Lentinus edodes* (Shii-ta-ke), *Mushroom Science* Vol. 9(pt1): 531-539.

Slaughter, J.C. & Nomura, T. (1992). Intracellular glycogen and trehalose contents as predictors of yeast viability, *Enzyme and Microbial Technology* Vol. 14(No. 1): 64-67.

Slovakova, L., Subikova, V., Sandula, J. & Huncikova, S. (1997) An agent for preventive protection of plants from viral diseases, SK 278413 Apr 9, 1997.

Smiderle, F.R., Sassaki, G.L., Van Arkel, J., Jacomini, M., Wickers, H.J. & Van Griensven, L.J.L.D. (2010). High molecular weight glucan of the culinary medicinal mushroom *Agaricus bisporus* in an α-glucan that forms complexes with low molecular weight galactan, *Molecules* Vol. 15(No. 8): 5818-5830.

Smith, J.E., Solomons, G., Lewis, C. & Anderson, J.G. (1995. Role of mycotoxins in human and animal nutrition and health, *Natural Toxins* Vol. 3(No. 4): 187-192.

Smits, G.J., van den Ende, H. & Klis, F.M. (2001). Differential regulation of cell wall biogenesis during growth and development in yeast, *Microbiology* Vol. 147: 781-794.

Spruijt, N.E., Visser, T. & Leenen, L.P.H. (2010). A systematic review of randomized controlled trials exploring the effect of immunomodulative interventions on infection, organ failure, and mortality in trauma patients, *Crit. Care* Vol. 14: R150.

Stewart G.G. & Russell I. (1998). Brewer's Yeast, *Brewing Science & Technology Series III* (1998): 28-46, The Institute of Brewing ISBN No. 0900489 13 8.

Suzuki, I., Yadomae, T., Yonekubo, H., Nishijima, M. & Miyazaki, T. (1982). Antitumor activity of an immunomodulating material extracted from a fungus, *Peziza vesiculosa, Chem. Pharm. Bull. (Tokyo)* Vol. 30(No. 3): 1066-8.

Swanson, K.S., Grieshop, C.M., Fuckinger, E.A., Bauer, U., Healy, H-P., Dawson, K.A., Merchen, N.R. & Fahey Jr., G.C. (2002). Supplemental fructooligosaccharides and mannooligosaccharides influence immune function, ileal and total tract nutrient digestibilities, microbial populations and concentrations of protein catabolites in the large bowel of dogs, *The Journal of Nutrition* Vol. 132(No. 5): 980-989.

Tanaka, A. (2011). Blood glucose suppression effect of Maitake mushroom extract, *BRAIN Techno News* Vol. 146: 30-34.

Tapiero, H., Townsend, D.M. & Tew, K.D. (2003). The antioxidant role of selenium and seleno-compounds, *Biomedicine & Pharmacotherapy* Vol. 57: 134-144.

Tastet, L., Schaumloffel, D., Bouisierre, B. & Lobinski, R. (2008). Identification of selenium containing proteins in selenium rich yeast aqueous extract by 2D gel electrophoresis, nanoHPLC-ICP MS and nanoHPLC-ESI MS/MS, *Talanta* Vol. 75: 1140-1145.

Terakawa, N., Matsui, Y. Satoi, S., Yanagimoto, H., Takahashi, K., Yamamoto, T., Yamao, Y.,

Takai, S., Kwon, A-Hon. & Kamiyama, Y. (2008). Immunological effect of active hexose correlated compound (AHCC) in healthy volunteers: a double-blind, placebo-controlled trial, *Nutrition and Cancer* Vol. 60(No. 5): 643–51.

Thammakiti, S., Suphantharika, M., Phaesuwan, T. & Verduyn, C. (2004). Preparation of brewer's yeast β-glucans for potential application in the food industry, *International Journal of Food Science and Technology* Vol. 39(No. 1): 21-29.

Toborek, M. (2011) University of Miami, USA, Private communication.

Turlo, J., Gutkowska, B. & Malinowska, E. (2007). Relationship between the selenium, selenomethionine and selenocysteine in submerged cultivated mycelium of *Lentinula edodes* (Berk.), *Acta Chromatographica* Vol. 18: 36-48.

Tzianabos, A.O. (2000). Polysaccharide immunomodulators as therapeutic agents: structural aspects and biologic function, *Clinical Microbiology Reviews* Vol. 2000: 523-533

Uden, P.C., Boakye, H.T., Kahakachchi, C., Hafezi, R., Nolibos, P., Block, E., Johnson, S Tyson, J.F. (2003). Element selective characterization of stability and selectivity of selenium species in selenized yeast, *J. Anal. At. Spectrom.* Vol. 19: 65-73.

Underhill, D.M., Ozinsky, A., Hajjar, A.M., Stevens, A., Wilson, C.B., Bassetti, M. & Andersen, A. (1999). The Toll like receptor 2 is recruited to macrophage phagosomes and discriminates between pathogens, *Nature* Vol. 401: 811-815.

Valarezo, S. (1998). Comparative effects of antibiotic, mannooligosaccharide and mycotoxin adsorbent on performance of commercial broilers fed pelleted diets, *Poultry Science* Vol. 77(Suppl. 1): 137.

Vetvicka, V. (2011). Glucan-immunostimulant, adjuvant, potential drug, *World Journal of Clinical Oncology* Vol. 2(No. 2): 115-119.

Vetvicka, V. & Vetvickova, J. (2007). Physiological effects of different types of β-glucan, *Biomed. Pap. Med. Fac. Univ. Palacky Olomouc. Czech Repub.* Vol. 151(No. 2): 225-231.

Vetvicka, V. & Vetvickova, J. (2008). A comparison of injected and orally administered β glucans, *JANA* Vol. 11(No. 1): 42-49.

Villar-Tajadura M., Madrid, M., Casando, J., Santos, B. & Perez, P. (2008). Rga2 is a Rho2 GAP that regulates morphogenesis and cell integrity in *Saccharomyces pombe*, *Molecular Microbiology* Vol. 70(No. 4): 867-881.

Vinogradov E., Petersen, B. & Bock, K. (1998). Structural analysis of intact polysaccharide mannan from *Saccharomyces cerevisiae* yeast using ^{1}H and ^{13}C NMR spectroscopy at 750MHz, *Carbohydrate Research* Vol. 307(No. 1-2): 177-183.

Vos, A., Dekker, N., Distel, B., Leunissen, J.A.M. & Hochstenbach, F. (2007). Role of the synthase domain of Ags1p in cell wall α-glucan boiosynthesis in Fission Yeast, *J. Biol. Chem.* Vol. 282(No. 26): 18969-18979.

Wakshull, E., Brunke-Reese, D., Lindermuth, J., Fisette, L., Nathans, R.S., Crowley, J.J., Tufs, J.C., Zimmerman, J., Mackin, W. & Adams, D.S. (1999). PGG-Glucan, a soluble β-(1, 3)-glucan enhances the oxidative burst response, microbial activity, and activities NF-κB-like factor in human PMN: evidence for glycosphingolipid β-(1, 3)-glucan receptor, *Immunopharmacology* Vol. 41: 89-107.

Weitberg, A.B. (2008). A phase I/II trial of β-(1→3) (1→6)-D-glucan in the treatment of patients with advanced malignances receiving chemotherapy, *J. Exp. Clin. Cancer Res.* Vol. 27: 40-44.

Wheatcroft, R., Kulandai, J., Gilbert, R.W., Sime, K.J., Smith, C.G. & Langeris, W.H., (2002). Production of β-glucan-mannan preparations by autolysis of cells under certain pH, temperature and time conditions, *US Patent* US 6444448 B1 Sep. 3, 2002.

Wellens, A., Garafalo, C., Nguyen, H., Van Gerven, N., Slattegard, R., Hernalsteens, J-P.,

Wyns, L., Oscarson, S., De Greve, H., Hultgren, S. & Bouckaert, J. (2008). Intervening with urinary tract infections using anti-adhesives based on the crystal structure of the FimH-oligomannose-3 complex, *PLoS ONE* Vol. 3(No. 4): e2040, 1-13.

Wilson, W.A., Wang, Z. & Roach, P.J. (2002). Systematic identification of the genes affecting glycogen storage in the yeast *Saccharomyces cerevisiae, Molecular & Cellular Proteomics* Vol. 1(No. 3): 232-242.

Wismar, R., Pedersen, S.B., Frokiaer, H. & Laerke, H.N. (2010). Dietary fibers as immunoregulatory compounds in health and disease, Annals *of the New York Academy of Sciences* Vol. 1190: 70-85.

Yan, J., Allendorf, D.J. & Bradley, B. (2005). Yeast whole glucan particle (WGP) β-glucan in conjunction with antitumor monoclonal antibodies to trat cancer, *Expert Opin. Biol. Ther.* Vol. 5: 691-702.

Yan, J-K., Wan, W-Q., Li, L. & Wu, J-Y. (2011). Physicochemical properties and antitumor activities of two α-glucans isolated from hot water extract of *Cordyceps* (CS-HKI) fungal mycelia, *Carbohydrate Polymers* Vol. 85(No. 4): 753-758.

Yang, Y., Iji, P., Kocher, A., Mikkelsen, L.L. & Choct, M. (2008). Effects of mannooligosaccharide and fructooligosaccharide on the response of broilers to pathogenic *Escherichia coli* challenge, *British Poultry Science* Vol. 49(No. 5): 550-559.

Yannikouris, A., Francois, J., Poughon, L., Dussap, C-G., Bertin, G., Jeminet, G. & Jouany, J. P. (2004). Alkali extraction of β-D-glucans from *Saccharomyces cerevisiae* cell wall and study of their absorptive properties toward Zeralenone, *J. Agric. Food Chem.* Vol. 52(No. 11): 3666-3673.

Yannikouris, A., Andre, G., Poughon, L., Dussap, C-G., Bertin, G., Jeminet, G. & Jouany, J-P. (2006). Chemical and conformational study of the interactions involved in mycotoxins complexation with β-Glucans, *Biomacromolecules* Vol. 7(No. 4): 1147-1155.

Yusa, H., Katsuyama, H., Ozeki, T., Takashima, Y., Okada, H., Kasai, T., Eguchi, T. & Kakiuchi, H. (2002). Application of acid treated yeast cell wall (AYC) as a novel binder, *Yakuzaigaku* Vol. 62(No. 4): 153-160.

Zapata, G.M., Meza, M. & Meza Mendoza, H.A. (2008). Method for obtaining yeast glucan by autolysis of *Saccharomyces cerevisiae* Baker's yeast cells, *PCT Int. Appl.*, WO2008032134 A1.

Zechner-Krpan, V., Petravic-Tominac, V., Panjkota-Krbavicic, I., Grba, S. & Berkovic, K. (2009). Potential application of yeast β-glucans in food industry, *Agriculturae Conspectus Scientificus* Vol. 74(No. 4): 277-282.

Zekovic, D.B., Kwiatkowski, S., Vrvic, M. M., Jakovlievic, D. & Moran, C. (2005). Natural and modified (1→3)-β-D-glucans in health promotion and disease alleviation, *Critical Reviews in Biotechnology* Vol. 25(No. 4): 205-230.

Zekovic, D.B., Radulovic, M., Nastasovic, A., Vrvic, M.M., Jakovlievic, D. & Kogan, G. (2006). Mild Pfitzner-Moffat oxidation of the (1→3)-β-D-glucan from *Saccharomyces cerevisiae, Chemical Papers* Vol. 60(No. 3): 24-53.

Chapter 3

Mechanisms of O-Antigen Structural Variation of Bacterial Lipopolysaccharide (LPS)

Rosa Eugenia Reyes, Carolina Romo González,
Rafael Coria Jiménez, Maribel Ortiz Herrera and Alejandra Aquino Andrade

Additional information is available at the end of the chapter

http://dx.doi.org/10.5772/48147

1. Introduction

One of the most complex topics within bacterial anatomy and physiology is that of exopolysaccharides. These molecules have various structures and functions and also provide different types of advantages to their producing microorganisms, including surface variability, resistance to innate and acquired immunity mechanisms, the ability to adhere to different surface and cell types and resistance to antibiotic activity.

These bacterial systems are closely linked to the different genera and species that contain them. The organisation and expression of the genes that code for these external structures, genetic expression mechanisms and biosynthesis routes are extremely heterogeneous.

Although numerous classic microbial physiology and biochemical studies have focused on analysis of the external structures of microorganisms, not until recently has the study of exopolysaccharides become important, due to the role of exopolysaccharides in bacterial pathogenicity and the ecology of microbial populations and their possible role in the colonisation, residence and adaptation mechanisms in various ecosystems. Among the most important exopolysaccharides are those described below.

2. The glycocalyx

Although it can not be considered a bacterial structure, the glycocalyx is a heterogeneous set of exopolymers that have diverse biochemical compositions [1]. The exopolymers are located immediately adjacent to the microorganism wall and are present as hydrophobic gels that are weakly associated with the external bacterial structures.

The production and presence of these exopolymers provide bacteria with a high degree of surface hydrophobicity that enables interaction between cellular and inert surfaces and subsequent bacterial colonisation through the development of microcolonies and biofilms.

INTECH
open science | open minds

© 2012 The Author(s). Licensee InTech. This chapter is distributed under the terms of the Creative Commons Attribution License http://creativecommons.org/licenses/by/3.0), which permits unrestricted use, distribution, and reproduction in any medium, provided the original work is properly cited.

Furthermore, the presence of certain biopolymers is related to the ability to resist antibiotic action by capturing these compounds through periplasmic glucans. An example of this is resistance to tobramycin, which is captured by cyclic-β (1,3)-glucan [2].

3. Capsules

In contrast to the glycocalyx, a bacterial capsule is a well-defined external structure with a characteristic composition for each bacterial genus and species that provides a number of advantages, which are primarily related virulence, to the producing microorganism.

One of the classic examples of the importance of the capsule is provided by *Streptococcus pneumoniae*, which is a gram-positive diplococci agent that causes pneumonia, meningitis and septicaemia. The virulent strains of this species produce a capsule that inhibits opsonisation and phagocytosis. The composition of the capsule is dependent on the producing bacteria. In the case of *Escherichia coli* K1, the capsule is formed by polysialic acid, in *S. pyogenes,* by hyaluronic acid and in *Streptococcus* group B by sialic acid. The biochemical composition of the capsules may be extremely diverse, which gives rise to a great amount of antigenic variability and presents a problem for the immunological mechanisms of the host in recognising these organisms [2].

It has been documented that the capsule participates in bacterial adhesion mechanisms and that its synthesis is stimulates by low stress conditions, such as the presence of serum, low Fe^{++} concentration and high CO_2 tensions. Although in certain microorganisms, the presence of a capsule is discreet, in others, such as *Cryptococcus neoformans,* the capsule plays an essential role in the mechanisms for aggression against the host.

4. Biofilms

Because of the production and exportation of bacterial exopolymers, the strains increase their degree of surface hydrophobicity, which facilitates interaction, adsorption and residence on a wide range of surfaces that, in principle, hinder bacterial colonisation. Regarding the production of biofilms, it has been documented that each bacterial genus and species responds to different signals from the environment and the host, as is the case of induction by tobramycin and the response capacity of the *quorum sensing* mechanisms. One of the most studied examples of the formation of bacterial biofilms is the case of *Pseudomonas aeruginosa,* which is an opportunistic pathogen that is associated with infections in immunocompromised hosts, such as cystic fibrosis (CF) patients. In CF patients colonised by *P. aeruginosa,* the bacteria exhibits two colonial phenotypes. The first phenotype is associated with the production of alginate (a polymer of mannuronic acid and glucuronic acid that forms a viscous gel around the bacteria), and the second phenotype is rough and is related to a lack of alginate. The production of alginate is a marker of virulence in which the producing strains are more aggressive than the non-producing strains. Additionally, alginate provides bacteria with the ability to form microcolonies and biofilms [3].

In the relationship between *P. aeruginosa* and the infected patient, the host exerts different types of selective pressure that favour the persistence of the mucoid strains with elevated alginate production. This biosynthesis is regulated by different bacterial signalling systems (quorum sensing) that detect changes in the host environment and modulate the metabolism to adapt. To date, it is not known precisely which signals in hosts with CF favour and allow for a predominance of mucoid strains compared to non-mucoid strains. *In vitro* studies, it has been demonstrated that both bacterial phenotypes exhibit similar behaviour and that successive passages in culture media inhibit the expression of the non-mucoid phenotype [4,5].

The physiology of bacteria that are found in biofilms is heterogeneous and depends on the specific site that the microorganism occupies in the microcolony. Nutrient gradients occur from the surface of the biofilm to the most internal parts, thereby influencing the bacterial physiology and consequently modifying the speed of growth, the generation time, the susceptibility to antibiotics (due to factors such as the presence of a diffusion barrier to antibiotics), the antigenic variability of individuals, the susceptibility to opsonisation and phagocytosis and even the alginate functions as a negative immunomodulator for the host [4].

Another example of the formation of biofilms is that produced by bacteria of the genus *Staphylococcus*, predominantly the coagulase-negative forms in which the production and excretion of biopolymers is capable of increasing cell surface hydrophobicity and, through hydrophobic interactions, adhering to surfaces. These mechanisms enable the bacteria to enter hosts, who subsequently require invasive procedures or therapeutic procedures, such as catheters. In the case of the bacterial populations that colonise the oral cavity, heterogeneous populations participate in the biofilms, of which *Streptococcus mutans* has a fundamental role in adherence because of its capacity to produce biofilms, thereby providing a high degree of surface hydrophobicity. This hydrophobicity allows the bacteria to adhere to dental enamel and begin the process of colonisation and the subsequent adherence to an increasingly heterogeneous bacterial population, which may ultimately cause harm and generate a cariogenic process.

5. Gram-negative lipopolysaccharide (LPS): Structure and function

LPS is essential in the structure and function of the external membrane of gram-negative bacterial cell walls. LPS intervenes in the transportation of hydrophobic molecules to the interior of bacterial cells and are an essential factor in host-microorganism interactions.

LPS is an amphipathic glycoconjugate that constitutes 10% to 15% of the total molecules in the external membrane and represents 75% of the total of bacterial surface [1]. There are three different LPS domains: a) Lipid A, which is the domain that is anchored to the membrane and the hydrophobic and endotoxic portions of the structure; b) The core oligosaccharide, which is the domain that connects lipid A to antigen O and is divided into the inner core and the outer core. The inner core is joined to lipid A and consists of unusual monosaccharides, including 2-keto-3-deoxy-octanoate (Kdo) and L-glycero-D-mannoheptose.

The outer core is joined to the O antigen and is made of common sugars such as hexoses and hexosamines [2]; and c) The O polysaccharide, which is the hydrophilic and immunodominant domain of LPS and is an oligosaccharide of repeated units that is projected from the core toward the exterior of the bacterial surface.

The O antigen has a polysaccharide chain that varies in length with up to 40 repeated units of dideoxyhexoses. At least 20 different sugar molecules may compose the O antigen, including molecules that are rarely found in nature, such as abequose, colitose, paratose and tyvelose. These components are strain-specific. The O antigen displays a large degree of inter-species and intra-species variation, which is related to the nature, order and union of the different sugars (**Figure 1**) [3].

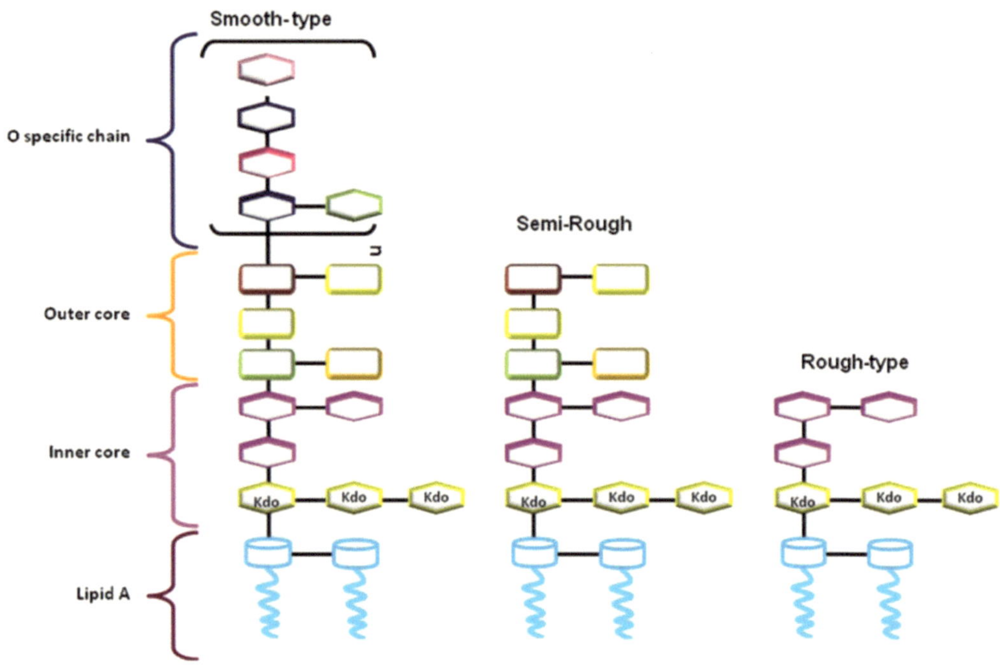

Figure 1. Schematic representation of lipopolysaccharide structure. Smooth-type (left), Semi-Rough type with only one O-chain subunit (center) and rough-type (right).

The O antigen is the immunodominant part of LPS and therefore is the easiest target for the humoral response of the host. For this reason, the O antigen is the basis for the serological classification of gram-negative bacteria. The O antigen is recognised by the innate immune response and participates in complement activation and in the inhibition of the formulation of the complex that attacks the membrane [6,7].

6. Biosynthesis of LPS

LPS is the primary component in the surface of gram-negative bacteria. The synthesis of LPS structures, which consist of lipid A, the core and antigen O, begins in the cytoplasm, where these structures are assembled. The structures are translocated to compartments such as the periplasm until the final destination is reached, which is the surface of the external membrane. The synthesis process has been widely studied in *E. coli* and *Salmonella*. The biosynthesis and exportation pathways of LPS are common among the majority of gram-negative bacteria, as is explained below (**Figure 2**). However, unique characteristics may exist in certain bacteria with respect to the types of enzymes and particular pathways.

The formation of lipid A is carried out in the internal face of the cytoplasmic membrane, and nine enzymes participate: LpxA, LpxC, LpxD, LpxH, LpxB, LpxK, KtdA, LpxL and LpxM. The biosynthesis of LPS begins with the formation of uridine diphosphate-diacyl- *D*-glucosamine (UDP-diacyl-GlcN) from uridine diphosphate *N*-acetylglucosamine (UDP-GlcNAc). This reaction is catalysed by the enzymes LpxA, LpxC and LpxD and results in two 3-OH fatty acid chains in the 2 and 3 position of UDP-glucosamine (UDP-GlcNAc) to form UDP-2,3-diacyl- glucosamine (UDP-diacyl-GlcN). Subsequently, this molecule is hydrolysed by LpxH to form lipid X, the enzyme LpxB condenses the lipid X and its precursor UDP-diacyl-GlcN to form disaccharide-1-P and the enzyme LpxK phosphorylates this molecule at the 4 position of the disaccharide-1-P molecule to form the lipid IV_A. Subsequently, KdtA incorporates the waste from 3-deoxy-D-manno-octulosonic acid (Kdo) in the 6′ position of lipid IV_A using the nucleotide cytidine monophosphate-3-deoxy-D-manno-octulosonic acid (CMP-Kdo) as a donor to produce Kdo_2 – lipid IV_A, which is exposed to more reactions catalysed by LpxL and LpxM to form Kdo_2 – lipid A. The enzyme LpxL adds a second lauryl, and LpxM adds a residue of myristoyl to the distal glucosamine unit.

Notably, the acyltransferases – LpxA, LpxD, LpxL and LpxM – selectively catalyse the different substrates and employ different acyl donors. For the first steps of the synthesis pathway of lipid A, the enzymes LpxA, LpxB and LpxD are required, with 3R-hydroxyacyl - Acyl Carrier Protein (3R-hydroxyacyl-ACP) serving as a donor. This compound is dehydrated by FabZ to form trans-2-acyl-ACP, which is also used as a donor of fatty acids in the biosynthesis of phospholipids. The synthesis of other LPSs in bacteria, such as *Neisseria meningitidis*, also involve trans-2-acyl-ACP[8]. Importantly, the structure of lipid A is the most conserved compared to the structure of the core oligosaccharides and antigen O [9,10].

6.1. The core oligosaccharides

The assembly of lipid A from the core oligosaccharides (Kdo_2 – lipid A) is the next step in the synthesis of LPS. This step is performed on the cytoplasmic surface of the internal membrane by glycosyltransferases, which are associated with the membrane and with nucleotide sugars as donors.

The core oligosaccharides normally contains 10 to 15 monosaccharides and may be divided into two structural regions, which are the inner core and the outer core, which are ultimately connected to lipid A and antigen O, respectively, in the final structure of LPS. The inner core contains residues of Kdo and Hep (L-glycero-D-manno-heptose). Kdo is the most conserved component in the nuclear region of the LPS. In contrast, the outer core is more variable, depend on the strain. However, the vertebral column of the oligosaccharide is typically composed of six units, and upon joining with other units, the column forms structures. The sugars commonly found in the core oligosaccharides are D-glucose, D-galactose, Kdo and Hep [9,10].

6.2. The O antigen

The majority of O antigens are heteropolymers, although a portion of O antigens may be composed of a single monosaccharide. The synthesis of the O antigen is performed in the same location as the core oligosaccharides, and this synthesis also uses nucleotide sugars as donors. In the majority of bacteria, a cluster of genes known as *rfb* codes the enzymes necessary for 1) the biosynthesis of the nucleotide sugars of antigen O, 2) the transfer of the sugars to form the polysaccharide chain (glycosyltransferases) and 3) the assembly and transfer of antigen O toward the periplasm. The synthesis routes of the nucleotide sugars are grouped according to the nucleotide that bonds to the sugar, which may be CDP, UDP, dTDP or GDP. Antigen O may be a homopolymer or a heteropolymers, and the sugars may be formed linearly or in a ramification. Glycosyltransferases may be grouped according to their function, and they carry undecaprenyl phosphate, which is also used for the synthesis of capsular polysaccharides and peptidoglycans.

The following hypotheses have been proposed regarding the assembly and transfer process of antigen O: a) a pathway dependent on Wzy, which is the prototype system; b) a pathway dependent on ABC transporters, which are typically used by linear polysaccharide structures; c) a pathway dependent on synthase, which involves glycosyltransferases capable of synthesising within a single polypeptide and is an uncommon pathway and finally d) seroconversion reactions, in which the addition of acetyl residues or glucose residues modifies antigen O. Within the prototype pathway dependent on Wzy, in bacteria such as *Salmonella enterica* and *E. coli*, a multi-step process occurs. When the lipid A-core and the O antigen are synthesised, they are transported to the periplasm. The protein MsbA, which displays homology with MDR (multi-drug resistant) eukaryote proteins, transports the lipid A-core, and Wzx transports the O antigen, which was previously polymerised by the proteins Wzy and Wzz. With the help of WaaL, the structures of the lipid A-core and the O antigen are assembled, finally producing the LPS [10].

6.3. LPS and its transportation to the external membrane

When an LPS is formed, it must pass through the periplasmic space to reach the external membrane [9,10], and this process is facilitated by protein LptA (periplasmic), LptB (cytosolic), LptC, LptF, LptG (internal membrane) and LptD and LptE (external membrane). Several of these proteins act in complexes. For example, in the case of the transporter ABC, LptBFG and LptA and LptC translocate the LPS to the internal side of the external membrane such that the proteins LptD and LptE place it on the surface of the membrane. It has been observed an absence of LptA or LptB or both causes the accumulation of LPS in the periplasm [11–17].

In the majority of bacteria, the genes that code for the enzymes involved in the biosynthesis of the O antigen are found in clusters. However, in the case of *Helicobacter pylori*, which is a pathogenic bacteria of the human stomach, these enzymes are found distributed throughout its chromosome, which probably contributes to the fact that the assembly pathway of its LPS has not been completely characterised. However, several enzymes that participate in the synthesis of *H. pylori* LPS has been identified and characterised, including several glycosyltransferases [18,19]. The glycosyltransferase WecA and the ligase WaaL also participate in the biosynthesis of *H. pylori* LPS. However, translocases are typically not involved in the translocation of the O antigen to the periplasm for its assembly with the lipid A-core structure [20]. Only the participation of a translocase named Wzk, which directs N-glycosylation in other bacteria, has been observed, and this fact suggests that the translocase Wzk of *H. pylori* could indicate an evolutionary connection between the biosynthesis pathways of LPS and glycoproteins [18]. Recently, it has been observed that there is an analogy and homology between the biosynthesis of LPS and the biosynthesis of glycoproteins in other bacteria, such as *S. enterica, P. aeruginosa* PAO1, *Neisseria* spp., *Paenibacillus alvei, Campylobacter jejuni and E. coli O8*. This homology could have enormous biotechnological potential. However, further studies are required to confirm this fact [21].

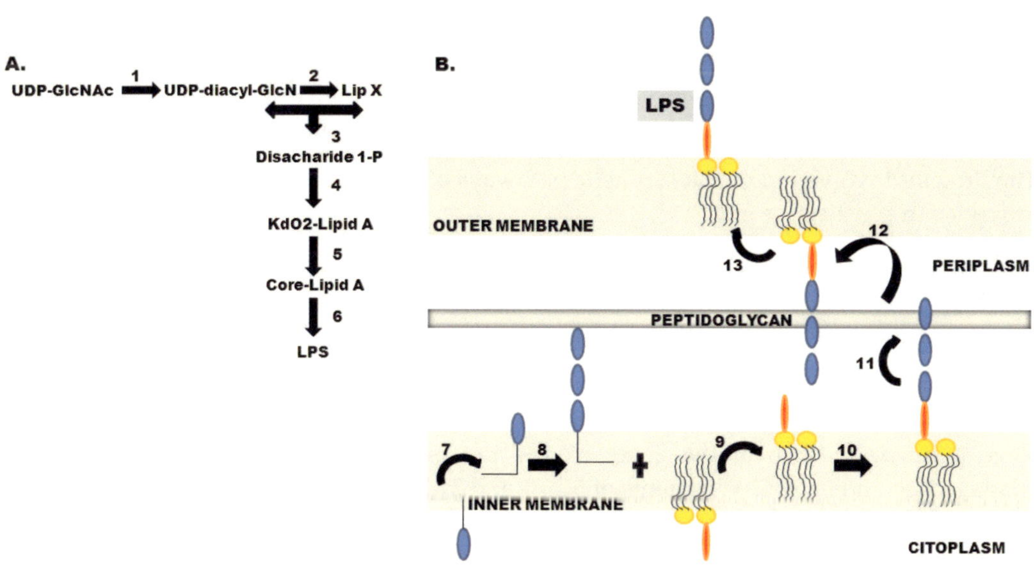

Figure 2. The biosynthetic pathway and transport of lipopolysaccharide. **A.** A representation of the biosynthetic pathway of the structures of LPS in *E. coli*. **B.** The assembly and transport of the LPS; the antigen O is assembled to the structure core-lipid A in the periplasm and is later transported toward the outer membrane. The names of the enzymes involved in these processes are: 1. LpxA, LpxC, LpxD; 2. LpxH; 3. LpxB; 4. LpxK, KdtA, LpxL, LpxM; 5. Glycosyltransferases; 6 and 10. WaaL; 7. Wzx; 8. Wzy, Wzz; 9. MsbA; 11. Lpt B C F G; 12. LptA; 13. Lpt D E. The lipid A (Yellow circles), the core (Red ovals) and antigen O (Blue ovals)

7. Regulation of the expression of LPS

The biosynthesis of LPS is performed through two separate pathways. One pathway involves the formation of lipid A and the core, and the other pathway involves the formation of the O antigen. In the synthesis of LPS, a large number of genes participate, many of which are part of clusters located in different regions of the bacterial chromosome and, in some organisms, in plasmids [22,23].

7.1. Regulation of the biosynthesis of lipid A

Lipid A and the core oligosaccharide are formed in a continuous process, which is separate from the synthesis of the O antigen. In the majority of *Enterobacteriaceae,* the genes involved in this synthesis are found in a single copy and share several characteristics with pathogenicity islands (PAI) [10,24].

The genes involved in the first steps of the biosynthesis of lipid A in *E. coli, S. enterica* serovar Typhimurium, *Yersiia enterocolitica, Haemophilus influenzae* and *Rickettsia rickettsii* are grouped into the *lpx*D-*fab*Z-*lpx*A-*lpx*B cluster[8]. The genes *lpx*A and *lpx*D code for N-acyltransferases, which add fatty acids to the glucosamine disaccharide. Both enzymes contain a conserved repeated structure, which is the hexapeptide $[(I,V,L)GXXXX]_n$. The gene *lpx*B is a co-transcript with *lpx*A and codes for the disaccharide synthase of lipid A, which catalyses the formation of the disaccharide of lipid A from UDP-2.3-diacylglucosamine and 2,3-diacylglucosamine-1-phosphate. The gene *fab*Z codes the enzyme that catalyses the dehydration of (3R)-hydroxyacyl-ACP to trans-2-acyl-ACP, which is used as a donor of fatty acid in the biosynthesis of phospholipids [10].

The proteins involved in the biosynthetic pathways of UDP-GlcNAc, UDP-Glc and UDP-Gal are coded in constitutive genes.

L-glycero-D-manno-heptose is added to its derivative ADP, which is synthesised from sedoheptulose 7-phosphate in four steps. The genes *gmhA* and *gmhD* code for enzymes in the first and last steps. The G+C content of the *gmhA, gmhD* and *waaE* genes are 51%, 51% and 52.7%, respectively.

Kdo is transferred from CMP-Kdo and synthesised from arabinose-5P and PEP by a three-stage pathway[9]. Two of the genes in this process, namely, *kdsA* and *kdsB,* are well-characterised and have G+C contents of 51.6% and 52.7%, respectively.

The gene *waaA,* which is located in the *waa* cluster, codes for a bifunctional Kdo transferase that adds residues from 2 Kdo. These genes have a G+C content of between 51% and 54%. Several of the modifications of lipid A are regulated by the concentration of Mg^{2+} through the regulon *phoP-phoQ* [25].

PhoP-PhoQ is a two-component system that regulates virulence through adaptation to limited magnesium environments and regulates numerous cellular activities in gram-negative bacteria. This regulon consists of an external membrane sensor, PhoQ, and a cytoplasmic regulator, PhoP, and is activated by the acidic pH and by certain antimicrobial peptides (APs). PhoP-PhoQ is repressed by millimolar concentrations of magnesium and calcium. PhoQ senses the concentration of magnesium and of APs throughout a periplasmic domain, which undergoes a conformational change when it is joined to these compounds and results in autophosphorylation. The activation of the PhoP-PhoQ system may allow for the activation or repression of 40 genes [26].

The regulon PrmAB of *Salmonella* is also a two-component system coded in the operon *pmrCAB,* which has protein products that include a phosphoethanolamine phosphotransferase (PmrC), a response regulator (PmrA) and a kinase sensor (PmrB). It has been confirmed that PmrAB regulates more than 20 genes in *Salmonella*. However, several studies suggest that more than 100 genes show activity[27]. One of the primary roles of the activation of PmrAB is the modification of LPS, such as the addition of Ara4N to lipid A, which, as explained below, impacts the susceptibility to some antimicrobial agents and the addition of phosphoethanolamine (pEtN) in the core of the LPS [26].

7.2. Regulation of the biosynthesis of the core

In *E. coli, Salmonella* and *K. pneumoniae,* the genes involved in the biosynthesis of the core are grouped on the chromosome. These loci code for the activities required in the assembly of the outer core and also code for the transferases necessary for the synthesis of the inner core. In *E. coli* and *Salmonella,* the *waa* locus (also called *rfa*) is formed by three operons and is located between the genes *cysE* and *pyrE*. The operons are defined by the name of the first gene in each transcriptional unit, such as *gmhD, waaQ* and *waaA* (**Figure 3**) [28].

Genes in the *waa* operon code for all of the transferases that assemble the core, including the gene that codes for the enzyme for the last step of the synthesis of ADP-L-Glycero-D-manno-heptose and the gene of the ligase of the O antigen, *waaL*. In addition, the *waa* operon contains the gene *waaA*, which codes for a bifunctional transferase Kdo.

The operon *gmhD,* which is located on the extreme 5′ end of the *waa* cluster, contains *gmhD, waaF, waaC* and *waaL*. The genes *gmhD-waaFC* are required for the biosynthesis and transfer of L, D-heptose. GmhD catalyses the last reaction of the synthesis of ADP-L-Glycero-D-manno-heptose. WaaC is the transferase for Hepl, and WaaF is a putative transferase of Hepll. WaaL is the ligase that bonds the O polysaccharide to the lipid A and core.

The transcription of the operon *gmhD* in *E. coli* K12 is regulated by heat shock promoters, indicating a requirement for the heptose domain of LPSs for growth at high temperatures [23,28,29].

The central operon *waaQ* contains 10 genes that are necessary for the biosynthesis of the outer core and for the modification of the core. WaaQ is the transferase for HepIII. The proteins WaaG, WaaO and WaaR are transferases for Glcl, Glcll and GlcIII, respectively, and WaaB is the transferase for the GaL residue. WaaP and WaaY are involved in the phosphorylation of the residue of heptose, whereas the functions of WaaU, WaaS and WaaZ are still not clear. In isolates of *E. coli* with the R1 and R4 cores, this operon contains the structural gene of ligase *waaL,* which must be produced for the union of the O polysaccharide with the complete core.

The *waaQ* operon is preceded by a JUMPStart sequence (Just Upstream of Many Polysaccharide-associated gene Starts), which includes a conserved region of 8 bp that is known as *ops* (operon polarity suppressor). The expression of the cluster for the biosynthesis of the core is regulated by the protein RfaH and also in response to thermal shock. RfaH is homologous to the factor NusG, which regulates the expression of the operon of hemolysin, genes of the polysaccharides of the capsule and genes for the transfer of the F plasmid. Regulation by RfaH occurs at the level of the polymerisation of the mRNA and depends on *ops* sequences that act in *cis*, as in the case of the operon *waaQGPSBIJYZK*, which includes 10 genes of the cluster of the core of LPS [22,30].

The transcript *waaA* contains the structural gene *waaA* (formally called *kdtA*), which codes for the transferase Kdo and a "non LPS" gene that codes for the adenylyltransferase pantetheine (*coaD*, formally *kdtB*) [31].

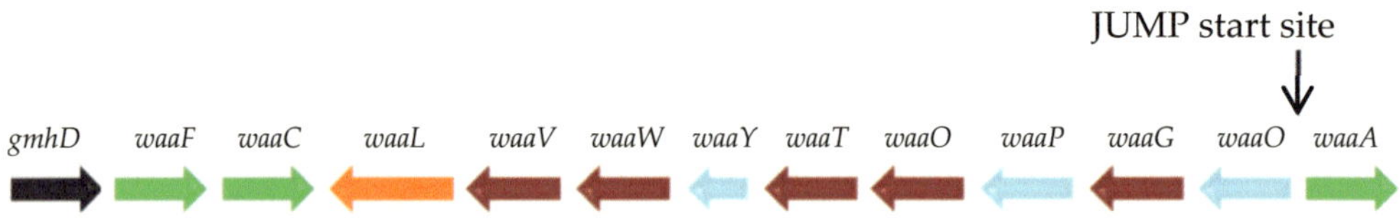

Figure 3. Structure of the genes involved in the synthesis of the core of *E. coli* R1. The genes of the glycosyltransferases that form the bonds of the inner core are shown in green. Those genes of the enzymes that modify the structure are in red, and the glycosyltransferases of the outer core are in blue. In orange are genes that modify the ligase enzyme.

7.3. Regulation of the biosynthesis of the O polysaccharide

The genes involved in the biosynthesis of the O antigen are generally found in the chromosome in the cluster of the O antigen or *rfb*. These genes have a lower GC content than the average for genomes, between 30% and 40%, which provides evidence that these genes have been acquired through lateral inter-species transference [24].

The genes that code for the proteins that participate in the synthesis of the O antigen form three main groups: a) proteins involved in the biosynthesis of the precursors of nucleotide sugars of the O antigen; b) protein glycosyltransferases, which sequentially transfer various precursor sugars to form an oligosaccharide of a lipid carrier, undecaprenyl phosphate (UndP), which is located in the cytoplasmic face of the internal membrane and c) genes for the processing of the O antigen, which are involved in the translocation through the membrane and polymerisation (**Figure 4**) [32,33]. A fraction of O antigens includes acetyl-O groups, and others include residues; therefore, in the corresponding clusters, the transferases for them are coded. The differences among the many forms of the O antigen are due to the genetic variation in the cluster of the O antigen. The genes for the initial steps, which are also involved in conserved functions, do not duplicate in the cluster of the O antigen [24].

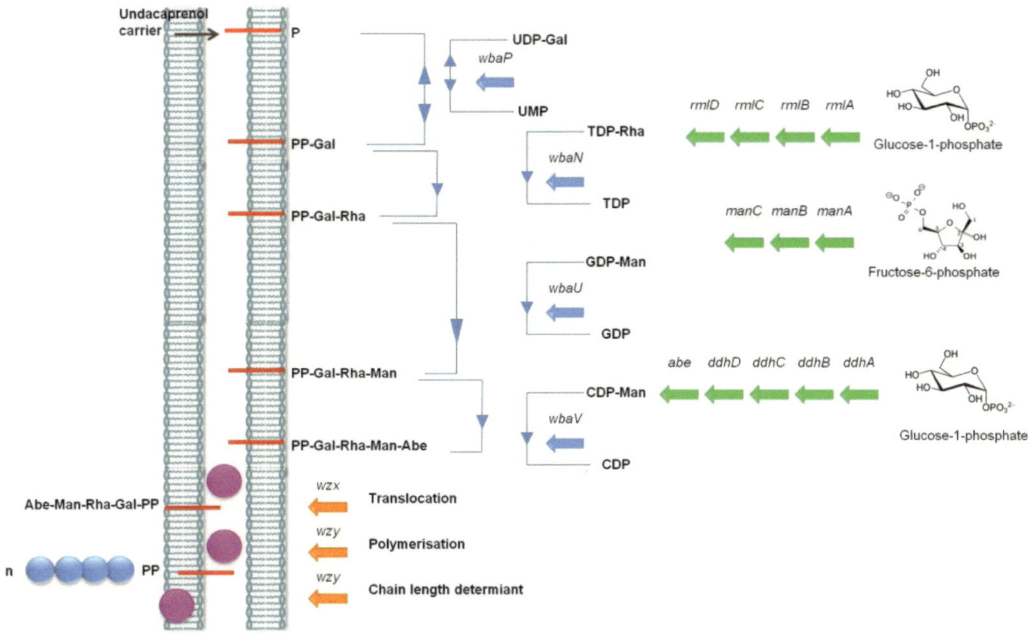

Figure 4. The synthesis of the O antigen involves three types of genes: a) genes related to the biosynthesis of precursors of sugar (green boxes), whose protein products perform their function in the cytoplasm; b) genes for glycosyltransferases (blue boxes), whose corresponding proteins transfer nucleotide sugars in the UndP lipid to form the O unit. This process occurs on the cytoplasmic side of the internal membrane and c) genes for the assembly of the O antigen and its exportation (orange box). Modified from [38].

In *E. coli* and *S. enterica,* the genes involved in the synthesis of the O antigen are typically found among the constituent genes *galF* and *gnd*. In *P. aeruginosa,* these genes are found among *himD* and *tyrB* [15]). In *V. cholerae,* O antigen synthesis genes are found among *gmhD* and *rjg* [34], and in *Yersinia* spp., these genes are found among *hemH* and *gsk.* Some exceptions, such as the polysaccharide O54 of *S. enterica,* which is a cluster that is coded in a plasmid [35] (**Figure 5A**).

The genes of the biosynthetic pathway of the three precursors of the nucleotides of sugar are grouped within the genetic cluster of the O antigen. *manB and manC* code for enzymes that convert mannose-6-P into GDP-mannose. The operon *rmlABCD* codes for enzymes that form TDP-rhamnose from glucose-1-P. *ddhABCD* and *abe* code for enzymes to make CDP-abequose from glucose 1-P. UDP-Gal is used in other pathways and is synthesised by conserved enzymes. The transferase galactose, which is coded by *wbaP*, initiates the synthesis of the O units by transferring galactose phosphate from UDP-Gal to UndP.

The transferases coded by *wbaZ*, *wbaW* and *wbaQ* are positioned over two residues of mannose and one of rhamnose before the residues of mannose, which is acetylated by WbaL. The residue of abequose is bonded to acetyl rhamnose by WbaR to form a complete O unit. The flippase of the bond of the O antigen is coded by *wzx* and is responsible for moving the O unit bonded to UndP through the internal membrane before the polymerase Wzy exerts its action over the long-chain O antigen.

The majority of the O antigen operons are constitutively expressed and are preceded by a sequence of 39 bp, known as JUMPStart **(Figure 5B)**.This sequence includes two elements known as *ops*, which are involved in the recruiting of elongation factors during the transcription. The JUMPStart region is controlled by RfaH, which acts as a positive regulator for the genes involved in the outer core of LPS and also increases the expression of the locus *rfb* [1,18,19.]

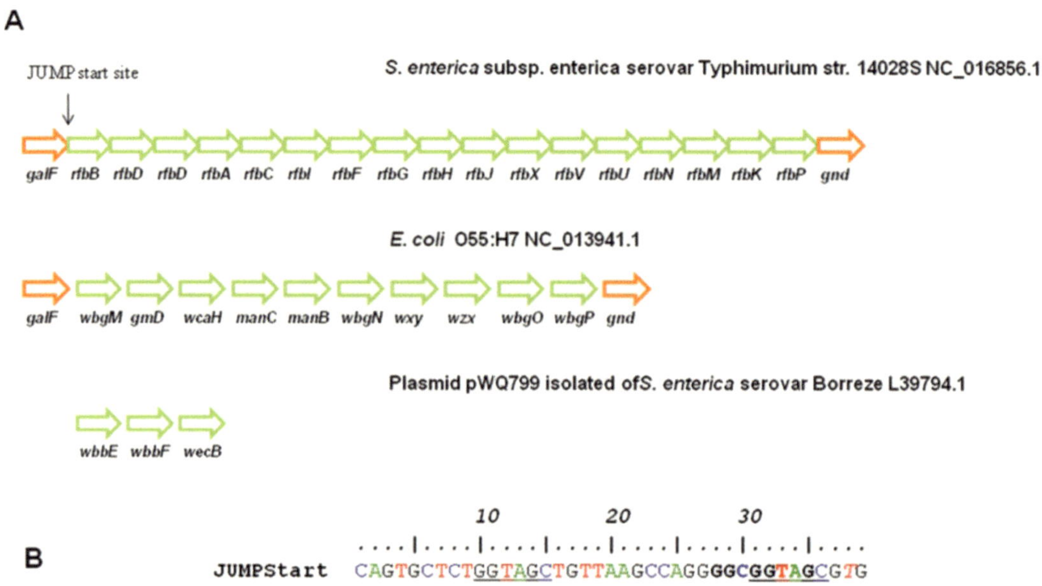

Figure 5. A. The location of the genes involved in the synthesis of the O antigen in several species of *Enterobacteriaceae*. **B.** The JUMPStart sequence of *S. flexneri* strain IB1833. The *ops* sequence is indicated in bold for eight base pairs. A residue of thymine that is highly conserved in the far 3' is indicated by cursive text, and repeated sequences are underlined. Accession number: U118168.1.

8. Variability of the LPS

In bacterial pathogens, the most variable structures are those that are expressed on the cell surface. Intra-species genetic individuality has inheritable traits that may be observed in a sample of individuals. The genetic variability is derived from the combination of a certain number of genes, which each exists as a family of alleles that differ in structure and function. An example is the family located in the biosynthetic locus of the O antigen, and an example of this phenomenon at the population level is the hypervariability in the structure of this biomolecule [39]. Among the components of the surface of the gram-negative bacteria, LPS is the primary constituents of the external membrane, which is a heterogeneous surface that is significantly involved in the process of the microorganism adapting to its environment.

LPS primarily consists of conserved segments, such as lipid A and the core, and secondarily consists of a hypervariable segment, which is the O antigen. The conserved domains of the LPS are shared regions among bacterial species, which intervene in the development and in the survival of the bacteria. The O antigen may display modifications such alterations in the length of the oligosaccharide chain and changes in the surface composition and in the chemical configuration, due to the addition of glycosyl or fucosyl groups or even non-hydrocarbonated substitutes, such as acetyl or methyl groups, which could affect the cellular structure [40].

8.1. Intra-species variability in the composition of the O antigen

The heterogeneity in the expression of LPS may provide a medium to discriminate among the bacterial species. This heterogeneity is responsible for the well-known ladder profile that can be detected in silver-stained SDS-PAGE gels (**Figure 6**). This method is used to determine the number and repeated units of oligosaccharides that constitute the O antigen, which has been useful in epidemiological studies [41]. The smooth strains contain the entire LPS, whereas the semi-rough strains have a subunit of the O antigen, and the rough strains lose the subunits of the O antigen.

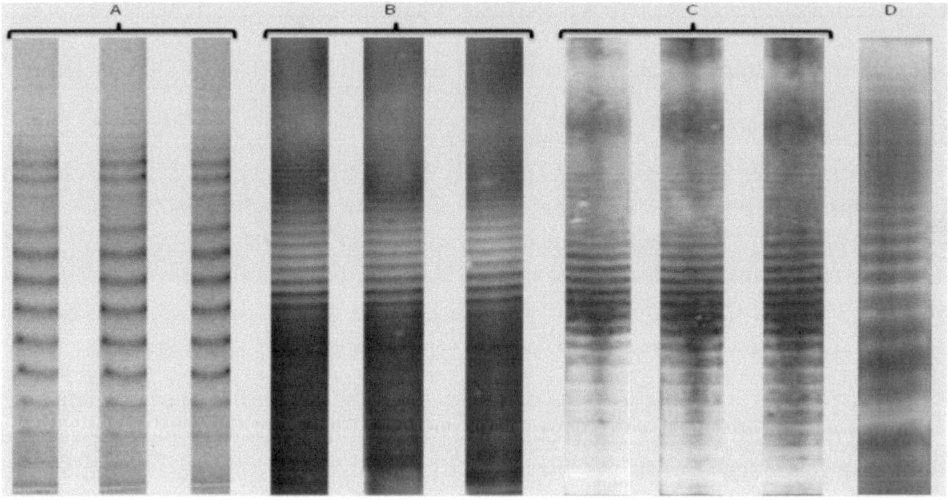

Figure 6. LPS profiles in SDS-PAGE. The smooth forms of LPS display a ladder profile that is strain-specific. **A)** *C. freundii* E9750; **B)** *S. senftenberg* 74210; **C)** *E. coli* O157:H7; **D)** *S. marcescens* biotype TC.

In **Table 1**, examples are presented of various sequences from different genera and species, which mark the variability of intra-species and inter-species LPS that impedes the immune response control of bacterial infections, among other issues.

Strain	Antigenic unit
E. coli O157 *S. enterica* O30 *C. freundii* F90	[D-PerNAc α 1,3 → L-Fuc α 1,4 → D-Glc β 1,3 → D-Gal2NAc] α 1,2 →
E. coli O55 *S. enterica* O50	[D-Gal α 1,3 → D-Gal2NAc β 1,6 → D-Glc2NAc] → 1,3 ↑ β 1,3 (on D-Glc2NAc) Col α 1,2 → D-Gal
E. coli O111	Col ↓ α 1,3 → 4 [D-Glc α 1,4 → D-Gal α 1,3 → D-GlcNAc] β 1 → ↑ α 1,6 Col
E. coli 113	→ 4 [D-GalA α 1,3 → D-Gal α 1,3 → D-GlcNAc β 1,4 → D-GalNAc] β 1 → ↑ β 1,3 (on D-GalNAc) D-Gal
S. enterica E1	→ 6 [L-Man β 1,4 → L-Rha α 1,3 → L-Gal] α 1 →

D-Col, D-colitose; L-Fuc, L-fucose; D-Gal, D-galactose; D-Gal2NAc, 2-*N*-acetylgalactosamine; D-Glc, D-glucose; D-Glc2NAc, D-2-*N*-acetylglucosamine; D-Per, D-perosamine; D-Per4NAc, D-4-*N*-acetylperosamine.

Table 1. Examples of different structures of the bacterial O antigen.

An increase in temperature during microbial growth causes changes in the concentration of carbohydrates in LPS, which modifies their composition [42]. In a study on *in vitro* passages of the *C. freundii* E9750 strain cultivated under different temperatures, variability was observed in the ladder profiles of the O antigen, in the concentration of carbohydrates and in the agglutination reactions with the anti-O serum of *C. freundii* E9750 and *Salmonella senftenberg* 74210 [43]. According to the ladder profiles of the isolates of *C. freundii* E9750, six profiles were distinguished: A (control strain), B, C, D, E (immunoreactive isolates) and F (isolates that had lost the immunoreactive chain). These profiles were generated according to their similarity. Each LPS profile displayed a typical ladder profiles (**Figure 7).**

The different profiles from the isolates obtained from the sub-cultures of *C. freundii* E9750 were out of phase with respect to the control. This result may be explained by variation in the number of oligosaccharide units present in the O antigen. However, when the length of the chain is increased, the differences are more marked, and the bands of the ladder profiles are out of phase, as Lawson et al., report with strains of *S. enterica* serovar Typhimurium [41]. There are other reports in the literature on the variability of the length of the O antigen chain as a response to change in temperature, as in strains of *P. aeruginosa* and enteric bacteria [44,45].

Several isolates of *C. freundii* E9750 displayed cross-reactivity in agglutination tests with the anti-O serum of *S. senftenberg* 74210, which suggests changes in the conformation of the epitopes that may be associated with the addition of glucosyl groups or residues derived from N-acetyl [46].

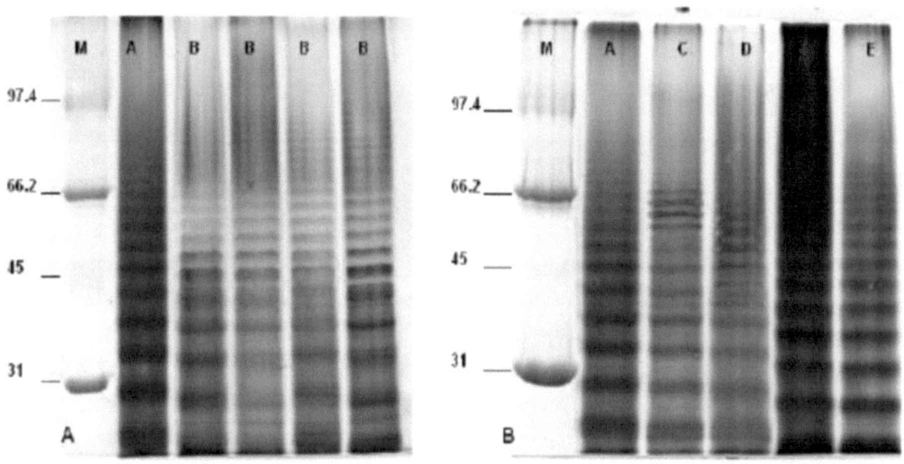

Figure 7. LPS profiles in SDS-PAGE stained with silver. Profiles of colonies of *C. freundii* E9750. A) isolates at a temperature of 42°C; B) isolates at a temperature of 37°C. M= weight markers.

The isolates of *C. freundii* E9750 that display variability suggest heterogeneity in the composition or conformation of the epitopes of the O antigen. The cross-reactivity of the isolates of *C. freundii* and *S. senftenberg* 74210 is associated with the specificity of the epitopes of the O polysaccharide that these species share. The addition at O antigen of glucosyl groups or derived N-acetyl residue could be involved in these cross-reactions.

8.2. Heterogeneity of populations and phase variation

Phase variation is used by various bacterial species to generate diversity within a population. Phase variation is a process of change in the expression of the epitopes of the cellular surface of the bacteria [19]. Bacterial cells may phenotypically vary even within a clone population, which allows them to adapt to their environment or even to evade the immune response of the host. Phase variation is a phenomenon that generates phenotypic heterogeneity within a population by means of gene regulation, which changes genes from a state of expression in which they are "turned on" to a state of non-expression in which they are "turned off". The state of expression is inheritable, reversible and affects the same phenotype.

Antigenic variation is referred to as the expression of an alternative form of an antigen of the cellular surface, such as polysaccharides, lipoproteins and type IV pili, which at the molecular level, share characteristics with the phase variation mechanisms. During this adaptation process, the bacteria display reversible phenotypic changes as a result of genetic changes or epigenetic alterations in a specific locus. The mechanisms which allow for phase variation are: genetics (Slipped-strand mispairing, recombination) and epigenetic (DNA methylation) [47,48].

a. Genetic mechanisms

Numerous studies have been performed to reveal the genetic basis of the variation of the O antigen. In certain cases, the variability in the expression of the genes is regulated by elements in *cis*, which cause changes in the composition of the structure of the antigens of the bacterial surface. Certain pathogens change the structure of the O antigen through the acquisition of phage genetic material followed by recombination processes [7].

i. Slipped-strand mispairing

One of the mechanisms that regulate phase variation at the molecular level is the slipping of one of the DNA strands, which causes a mispairing between the daughter strand and the parent strand during the replication of the DNA. This process is known as slipped-strand mispairing (SSM). The genomic sequences susceptible to SSM are those which contain short repetitions, microsatellites or a variable number of in tandem repetitions, which may cause a change in the expression of genes at the level of the transcription processes or translation according to the location of the repeated sequence in relation to the promoter and the codifying sequence [49]. At a transcriptional SSM, this may lead to the activation or deactivation of the promoter region of the target gene, as occurs in *H. influenzae (hif*A/B). At a translational level, SSM may affect the codifying region, as for example, with the genes involved in the biosynthesis of the LPS of *H. influenzae* and *Neisseria* spp. [50].

Within the genome of *H. pylori*, certain loci have been identified with repeated sequences of a single nucleotide or a pair of nucleotides. Several of these repetitions are within the open reading frames (ORFs) (**Figure 8)**. In the transcription process, the mispairing between nucleotides, when one of the strands of DNA slips over another chain, causes the "gain" or "loss" of a unit in the reading frame, which leads to the loss of the start codon or mutations in the proteins. Therefore, SSM increases the genetic variability of *H. pylori*. Similar repetitive sequences have been found in other microorganisms, such as *H. influenzae* [51].

One group of genes that generate phase variation are those that code for enzymes that intervene in the biosynthesis of LPS, which may cause variants of the gene product in the same bacterial population. The LPS of the majority of *H. pylori* strains contains complex carbohydrates known as the Lewis antigen. Type 1 (Le^a,Le^b) and type 2 (Le^x,Le^y) Lewis antigens are epitopes of fucosylated oligosaccharides. At least 80% of the strains of *H. pylori* express type 2. Some research on the antigenic determinants involved in the biosynthesis of the Lewis antigen have allowed for the identification of the fucosyltransferases (FucTs) that are involved in the formation of these antigens [19].

The genes that code for the FucTs have elements in *cis* that are differentiated by containing poliA and poliC sequences of different lengths that mediate SSM. The size of these sequences regulates the activation and deactivation of the genes of the FucTs. However, in some cases, such as in *H. pylori* UA948, the inhibition of the expression of *futB* is due to mutations outside of the hypervariable region (the elimination of 80 nucleotides in the promoter region) [52].

```
                                                            poly C tract
         10        20        30        40        50        60
     ....|....|....|....|....|....|....|....|....|....|....|....|
26695  ATACTTTGATGCTATATCCCCTTTAATCAAGCAAACCTTCACTCTACCCCCCCCCCCCCC
J99    ATATTTTGATGCTATATCCTCTTTAATCAAGCAAACCTTCACCCTACCCCCCCCCCCCCG
                 Imperfect TAA repeats                   HP0094 stop***

         70        80        90       100       110       120
     ....|....|....|....|....|....|....|....|....|....|....|....|
26695  GAAAATAATAAGAATAATAATAAAAAAGAGGAAGAATATCAGTGCAAGCTTTCTTTGATT
J99    AAA----ATGGAAATAATAAAAAAAAAAGAGGAAGAATACCACCGCAAGCTTTCTTTGATT

        130       140       150       160       170       180
     ....|....|....|....|....|....|....|....|....|....|....|....|
26695  TTAGCCGCTAAAAACAGCGTGTTTGTGCATATAAGAAGAGGGGATTATGTGGGGATTGGC
J99    TTAGCCGCTAAAAACAGCGTATTTGCGCATATAAGAAGAGGGGATTATGTGGGGATTGGC
                                                    ***Star of HP0093
        190       200       210       220       230       240
     ....|....|....|....|....|....|....|....|....|....|....|....|
26695  TGTCAGCTTGGTATTGACTATCAAAAAAAAGGCGCTTGAGTATATGGCAAAGCGCGTGCCA
J99    TGTCAGCTTGGTATTGACTATCAAAAAAAAGGCTGTTGAGTATATGGCAAAGCGCGTGCC
                                             ***Stop of -frame
```

Figure 8. The nucleotide sequence of the central region of the Hp *fuc*T2 gene. The sequences show the characteristics (simple repetitions) responsible for the ORF in the *H. pylori* J99 strain and the 26695 type variant. Due to the number of different repetitions of the residues of poli C, start sequences of an ORF of *fucT2* of the 26695 strain are found in the TGA stop codon (marked with asterisks) shortly after reading 1 (HP0093), which is the same as the marker of the *fuc*T J99 reading. The three supposed motifs X XXY YYZ are highlighted in bold and are underlined.

The α 1,2-fucosyltransferase (FutC) catalyses the addition of fucose in the conversion process of Le^X to Le^Y. Sanabria-Valentín et al., through *in vitro* and *in vivo* studies, confirmed the main function of *futC* slipped-strand mispairing in the variation of the Le antigen [53].

The *futC* gene includes an internal Shine-Dalgarno type sequence and a heptamer (AAAAAAG) followed by a loop structure. During translation, when the ribosomes are in the heptameric sequence of the mRNA, a phase shift occurs in the reading frame. The presence of Shine-Dalgarno type sequences and the loop structure accelerate the translation process by an interaction with the ribosome components [19].

ii. Phage recombination: seroconversion

The O antigen is a determinant of the virulence necessary for the pathogenicity of *S. flexneri*. The O antigen of *S. flexneri* is called the Y serotype and consists of repeated units of a tetrasaccharide of N-acetylglucosamine-rhamnose I-rhamnose II, rhamnose III, which forms the structure of the vertebral column of the polysaccharide unit of all the serotypes of *S. flexneri*, except the 6 and 6a serotypes. There are 13 serotypes, which are differentiated by the addition of glucosyl groups or acetyl residue to the different sugar molecules in the tetrasaccharide unit [54,55].

The temperate phages of *S. flexneri* play an important role in the processes of seroconversion (antigenic variation). The bacteriophages SfV, SfII, SfX, the cryptic prophages SfI and SfIV code for the *gtr* genes, which are the proteins involved in the glycosylation of the O antigen. When these phagic elements lysogenise, a conversion of Y serotype strains into 5a, 2a, X, 1a and 4a serotypes, respectively, occurs (**Figure 9**).

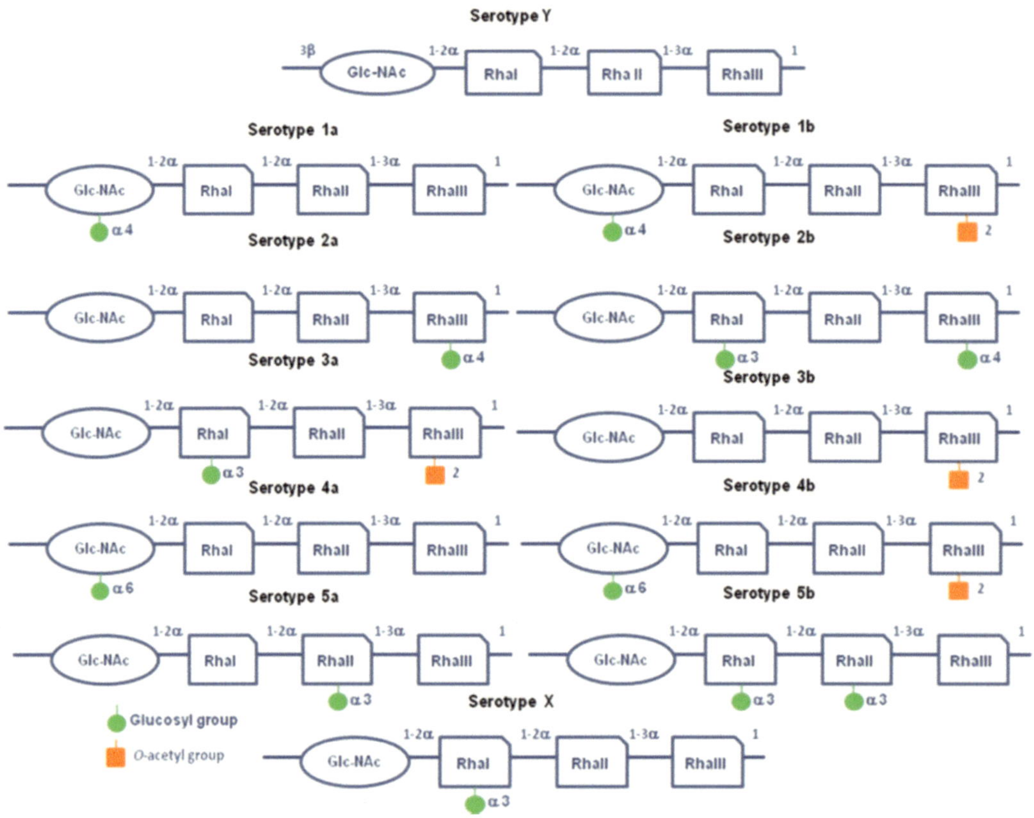

Figure 9. The chemical composition of the different serotypes of *S. flexneri*. The serotype Y formed by repeated units of tetrasaccharide N-acetylglucosamine-rhamnose I–rhamnose, II–rhamnose III. The serotypes are differentiated by the bonding of the glucosyl or the acetyl group. Adapted from [54].

The bacteriophage codes for an acetyltransferase and produces a conversion to the 3b serotype. The lysogenisation of the SfV bacteriophage produces modifications of the type V O antigen, which involves the addition of a glycosyl group through a bond of α 1,3 to the rhamnose II of the repeated tetrasaccharide unit. Similar to other phages that intervene in the glycosylation process, the genes involved in the conversion of serotypes are located immediate downstream from the *attP* locus, which is preceded by the genes *int* and *xis*. These phages are inserted into the *thrW* locus of the host [56].

The *gtr* genes of the temperate phages that code the glycosyltransferases are located in the genome of the phage, downstream from the *attP* locus. These genes are found in a cluster of three genes: *gtr*A, *gtr*B and *gtr* (type), which are cotranscribed. *grt*A and *grt*B are homologous and are interchangeable among the serotypes of *S. flexneri*. The Gtr (type) protein is specific for the formation of the glucosyl bond in a particular sugar molecule of the O antigen [54].

It has been suggested that GtrB catalyses the transfer of glucose from UDP-glucose to bactoprenol phosphate to form UndP-β-glucose in the cytoplasm. This molecule is subsequently translocated by GtrA in the periplasm before the glucosyl residue is joined by Gtr(type) for the growth of the O antigen unit [57].

The genes *gtr*V and *grt*X code for the glycosyltransferases GtrV and GtrX, respectively, which are membrane proteins that catalyse the transference of glucosyl residues through the bonding of the 1,3 rhamnose II and rhamnose I of the O antigen unit. This intervenes in the conversion of the serotype of *S. flexneri* from Y to the 5a serotype and the X serotype, respectively. GtrlV adds glucosyl residues to N-acetylglucosamine of the repeated unit of the O antigen through an α 1,6 bond, converting the Y serotype into the 4a serotype [57,58].

b. Epigenetic mechanisms: DNA methylation

The term epigenetic is defined as "inheritable changes in genetic expression that occur without alterations in the DNA nucleotide sequence". Thus, an epigenetic mechanism may be understood as a complex system to use the genetic information selectively by activating and deactivating various functional genes. Epigenetic modifications may imply methylation of cytosine residue in the DNA. DNA methylation has been observed in various bacterial species. In bacteria, methylation is part of a defence mechanism to reduce the amount of horizontal genetic transference among species. DNA methylation constitutes an epigenetic marker that identifies the template strand during the replication of the DNA. Generally, the methylation of the regulatory elements of genes, such as promoters, enhancers, insulators and repressors, suppresses this function [59].

The modifications of the O antigen that may affect the serotype are related to those that contain the operon that code for the glycosyltransferases (*gtr*).

Within a clone population of *S. enterica* serovar Typhimurium, the lysogenic phage P^{22} may lead to variability of the O antigen. The phase variation of the gtr (glycosylation of the O antigen) indirectly contributes to the diversity of the serotypes of *Salmonella*. The cluster that codes for the glycosyltransferases consists of three genes: *gtrA* codes for a membrane protein, *gtrB* codes for a glycosyl translocase and *gtrC* codes for the glycosyltransferase, which mediates the bonding of glucose to the O antigen [58].

Through studies based on the analysis of gene expression, the presence of mutations, the level of DNA methylation and the *in vitro* interaction of DNA-proteins, Broadbent et al., demonstrated that the Dam methyltransferase proteins together with OxyR regulate phase variation at the level of the *gtr* P22 promoter in *S.enterica* serovar Typhimurium. OxyR is an activator or a repressor of the *gtr* system, which depends on the alternative side (GATC sequences) to which OxyR bonds in the *gtr* P22 regulatory region (**Figure 10)**. The bonding of OxyR is inhibited by the methylation of the Dam target sequence, and the state of expression of the system is inheritable [60].

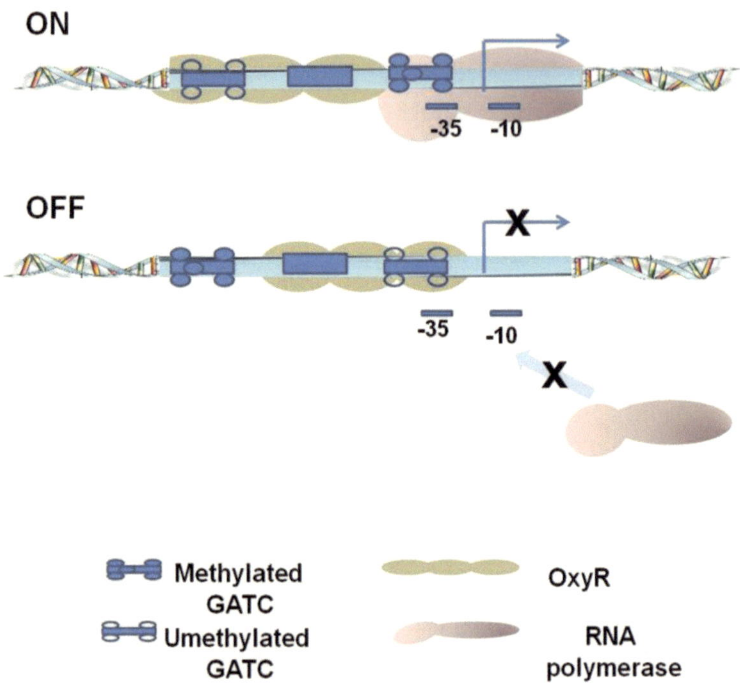

Figure 10. Model of the phase variation of the regulatory region of the *gtr* operon. Illustration of the interaction of DNA proteins in the regulatory region of the *grt* P22 operon, which consists of methylation and demethylation in the GATC sequence in the activated and deactivated phases. Adapted from [60].

Understanding the variation of the LPS structure is important because the composition and the length of the O antigen chain may be an indicator of the virulence, and this characteristic often differs within a single bacterial strain [7].

9. The importance of the variability of the O antigen of LPS

The modifications that are present in the O antigen and that cause its variability play an important role in infections by gram-negative bacteria, given that the modifications may influence adherence, colonisation and the ability to evade the host's defence mechanisms.

9.1. The role of the variation of LPS in the immune response.

LPS activates not only the innate immune response but also the adaptive response. The first contact that LPS has with the immune system is with lipid A, which is recognised by the receptors involved in the innate immune response, while the structure of the O antigen participates in the adaptive response (synthesis of antibodies). LPS is a potent stimulator of the cells of the immune system, given that it induces the production of pro-inflammatory cytokines, such as tumor necrosis factor α (TNF-α), interleukin 6 (IL-6), interleukin 1β (IL-1β) and acute phase proteins.

Although the variability is greater in the O antigen within LPS, there is also variability within lipid A. This variability is due to the length of the chains and the saturation of its acyl groups. Because these groups are strong immunostimulants, both the changes in the number of their chains and the presence of phosphorylations within the structure of the lipid A may influence its biological activity [61].

Lipid A is the structure of LPS that is recognised by the TLR4 receptors, which are part of the so-called Toll Like Receptors (TLRs), which are expressed by cells of the innate immune system and are stimulated by pathogen-associated molecular patterns (PAMPs). The stimulation of the LPS in certain cells of the monocyte-macrophage lineage, lymphoid cells and even cells that are not part of the immune system, such as epithelial cells, endothelial cells and vascular cells, occurs with the participation of other molecules, such as LPS binding protein (LBP), CD14 and MD-2. The transduction signals of TLR4 are divided into MyD88-dependent and MyD88-independent (also called TRI-dependent) groups. These signals may be regulated at various levels. For example, the RP105 and SIGIRR (Single immunoglobulin IL-IR related molecule) molecules inhibit the start of the signalling cascade [62].

Recognition through TLR4 is crucial for the control of infection, but changes in the signalling pathways may cause sepsis or evasion of the pathogen. The importance of signalling via TLR4-MD2 in response to gram-negative pathogens make this pathway an alternative to search for therapeutic targets not only for infectious diseases but also for other diseases with inflammatory aetiology, such as cancer, atherosclerosis, asthma and autoimmune conditions.

Antagonists of TLR4-MD2 have been identified, and several of these are based on the lipid A structures and other inhibitor molecules [63–65]. The intention is to use this type of antagonist therapy to treat septic shock. Additionally, several TLR4 antagonists primarily those that activate the TRAF (TNF receptor- associated factor) or TRIM (Tripartite motif) pathways have been proposed as adjuvants.

However, certain pathogens have the ability to modify the structure of lipid A and its detection by the host. For example, some isolates of *P. aeruginosa* are capable of modifying the structure of lipid A into a penta-acylated moiety, which does not activate the TLR4 and allows the immune response to be evaded. Other isolates of *P. aeruginosa* colonise respiratory pathways of patients with CF during its adaptation, producing hexa-acylated structures that are highly pro-inflammatory [61].

The large variability that the O antigen displays allows for the existence of various clones within a single species, which offers a selective advantage in the niche occupied by this clone and is precisely the interaction between the O antigen and the immune system that permits this advantage.

Many pathogens have the capability of varying the antigens that are attached to their surface and therefore can vary their antigenic composition. This variation is typically mediated by the regulation of the expression of genes. By varying their antigenicity, the pathogens have a greater ability to evade the immune response of the host, and this variability makes it more difficult to design vaccines for these pathogens [66].

The O antigen is considered to be highly immunogenic and induces the production of antibodies that may activate the complement pathway, either through the classic pathway or an alternate pathway, which leads to cellular death or phagocytosis. Certain modifications in the oligosaccharide chain of the O antigen may alter the interaction of the complement pathway. Several O antigens of pathogens are similar to host molecules and facilitate invasion through mimicking in the host; for example, O antigens of the LPS of *H. influenzae* and of *N. gonorrhoeae* mimic epitopes of glycosphingolipids [67].

The mimicking property may also serve to evade the immune system, as is the case of *H. pylori*. The chains of the O antigen that contain the surface of the LPS of *H. pylori* express Lewis antigens, mainly Le^x and Le^y, although some isolates may contain other antigens, such as Le^a, Le^b, Le^c, Sialyl-Le^x and H-1, in addition to type A and B blood groups [68].

The expression of Lewis antigens and their fucosylation have biological effects in the pathogenesis of this bacterium. The O antigen of *H. pylori* exhibits molecular mimicking with the Lewis antigens of the host within the gastric epithelium. The expression of the Lewis antigens is subject to phase variation, given that the regulation of the glycosyltransferase genes is regulated by SSM, which in the O antigen structure promotes variations among the strains. The antigenic mimicking is essentially involved in the evasion of the immune system and gastric adaptation. Several studies show that mimicking also plays a role in the colonisation and adhesion of Le^x of the bacteria with the galectin-3 of a gastric receptor.

Moreover, *H. pylori* is capable of evading the binding effect of surfactant protein D, which is expressed in the gastric mucosa and is a component of the innate immune response [69]. This microorganism impedes the bonding of the surfactant protein through the variation of its LPS. This phenomenon is associated with changes in the fucosylation of the O antigen chain. In addition, the expression of Lewis antigens affects both the inflammatory response and the polarisation of the T cells that are triggered after an infection. Because it is a chronic pathogen, several studies have shown that *H. pylori* may induce anti-Lewis auto-reactive antibodies, which enable the gastric mucosa to be recognised and contribute to the development of gastric atrophy [70].

9.2. Changes in the LPS related to resistance

The hydrophobic antibiotics that reach the interior of the cells due to the permeability of the external membrane are aminoglycosides, macrolides, rifamycins, novobiocin, fusidic acid and cationic peptides. The tetracycline and the quinolones use pathways that are mediated by lipids and porins. The central region of the LPS is important because it provides a barrier against hydrophobic antibiotics and other components; isolates that express a long LPS have intrinsic resistance to these factors [71].

The polymyxins, which include polymyxin B and colistin (polymyxin E), belong to a group of natural antimicrobials that are found in eukaryotic cells; this group is known as the cationic antimicrobial peptides. The polymyxins are active against gram-negative pathogens, such as *P. aeruginosa, Acinetobacter baumanii, Klebsiella* spp., *E. coli* and other *Enterobacteriaceae* [71–73].

The LPS has a negative charge and provides integrity and stability to the external membrane of the bacteria. Polymyxin has a positive charge, displacing the Mg^{+2} or Ca^{+2} and bonding to LPS, which as a consequence, destabilises and destroys the internal and external membrane [74].

Gram-negative bacteria may develop resistance to colistin and polymyxin B. The most important mechanisms involve modifications in the external membrane through changes of the LPS. The modification of the LPS occurs with the addition of 4-amino-4-deoxy-L-arabinose (Lara4N) to a phosphate group in lipid A. This addition causes an decreases in the negative charge of the lipid A, which decreases the affinity of the positively charged polymyxins [68–70].

The biosynthesis of LAra4N is mediated by the regulatory systems PmrA/PmrB and PhoP/PhoQ [26]. One of the primary roles of the activation of PmrAB is the modification of the LPS. These modifications include additions of Ara4N and pEtN to the lipid A and of pEtN to the core of LPS. The modifications mask phosphate groups with positive charges, thereby affecting the electrostatic interaction with certain cationic compounds. The biosynthesis of LAra4N depends on the genes of the operon of resistance to polymyxin, which is known as *arn*. This operon includes the genes *pmrHFIJKLM* [71–73].

10. Conclusion

In the bacterial pathogens, the most variable structures are those expressed in the cell surface. LPS is one of the principal antigenic structures of cell surface of gram-negative bacteria. A great variability in LPS has been demonstrated and principally in O antigen of gram-negative bacteria. This variability is present not only in the longitude of the oligosaccharide chains but also in the composition and structure of LPS.

Many of the functions of O antigen are associated to the longitude of the chain and to the variability of its structural features. This variability could affect the function, physical and chemical properties as well as the target site of LPS and determines the changes in the virulence of the microorganism that favor its adaptation to fluctuating enviroment which in many occasions are hostile to the microorganism and permit its evasion of the immune response of the host. The variation in O antigen structure has demonstrated that its composition and the longitude of its chain could be biological markers of virulence and this characteristic could differ within the same bacterial strain. The variability of LPS could derive from adaptations that involve associated changes to the synthesis of this molecule. The antigenic variability could occur by means of genetic and epigenetic mechanisms. The lost or gain of genes associated to variability of LPS is due to the events of genetic material interchange produced by lateral transference of genes which leads to strain selection with new characteristics and the evolution of the bacteria by modification of this structure.

One of the most important aspects of LPS function is its participation as immunogenic molecule and its role in bacterial classification based on O antigen and its variability. In general, it is seen that the modifications of O antigen play an important role in the process of infection including the adherence, the colonization, and the ability to evade defensive mechanisms of the host especially the innate resistance.

The study of the events of variation of LPS and its effects on pathogenecity and virulence represents a field of study of great interest to understand bacterial physiology and its mechanisms of adaptation and evolution.

The immunogenicity and variability of O antigen confer to gram-negative bacteria an important characteristic for its serological typification. The 0-antigen is subject to an intense selection on the part of immune system, which could be the principal factor for the different forms in which it is presented. For this reason, the variability of O antigen has been an area of intense research.

Author details

Rosa Eugenia Reyes, Alejandra Aquino Andrade, Rafael Coria Jiménez,
Carolina Romo González and Maribel Ortiz Herrera
Departamento de Medicina Experimental, Instituto, Nacional de Pediatría, México D.F., México

Rosa Eugenia Reyes
Departamento de Sistemas Biológicos, División CBS, Universidad Autónoma Metropolitana Unidad, Xochimilco, Coyoacán, México D.F., México

11. References

[1] Erridge C, Bennett-Guerrero E, Poxton IR. (2002) Structure and function of lipopolysaccharides. Microbes Infect 4:837–51.
[2] Silhavy TJ, Kahne D, Walker S. (2010) The bacterial cell envelope. Cold Spring Harb Perspect Biol 2:a000414.
[3] Flemming H-C, Neu TR, Wozniak DJ. (2007) The EPS matrix: the «house of biofilm cells». J. Bacteriol 189:7945–7.
[4] Pier GB. (2007) *Pseudomonas aeruginosa* lipopolysaccharide: a major virulence factor, initiator of inflammation and target for effective immunity. Int. J. Med. Microbiol 297:277–95.
[5] Franklin MJ, Nivens DE, Weadge JT, Howell PL. (2011) Biosynthesis of the *Pseudomonas aeruginosa* extracellular polysaccharides, Alginate, Pel, and Psl. Front Microbiol 2:167.
[6] Erridge C, Stewart J, Bennett-Guerrero E, McIntosh TJ, Poxton IR. (2002) The biological activity of a liposomal complete core lipopolysaccharide vaccine. J. Endotoxin Res 8:39–46.
[7] Lerouge I, Vanderleyden J. (2002) O-antigen structural variation: mechanisms and possible roles in animal/plant-microbe interactions. FEMS Microbiol. Rev 26:17–47.

[8] Steeghs L, Jennings MP, Poolman JT, van der Ley P. (1997) Isolation and characterization of the *Neisseria meningitidis lpx*D-*fab*Z-*lpx*A gene cluster involved in lipid A biosynthesis. Gene 190: 263–70.

[9] Raetz CRH, Whitfield C. (2002) Lipopolysaccharide endotoxins. Annu. Rev. Biochem 71: 635–700.

[10] Wang X, Quinn PJ. (2010) Lipopolysaccharide: Biosynthetic pathway and structure modification. Prog. Lipid Res 49:97–107.

[11] Sperandeo P, Cescutti R, Villa R, Di Benedetto C, Candia D, Dehò G, Polissi A.(2007) Characterization of *lpt*A and *lpt*B, two essential genes implicated in lipopolysaccharide transport to the outer membrane of *Escherichia coli*. J. Bacteriol 189:244–53.

[12] Sperandeo P, Lau FK, Carpentieri A, De Castro C, Molinaro A, Dehò G, Silhavy TJ, Polissi A. (2008) Functional analysis of the protein machinery required for transport of lipopolysaccharide to the outer membrane of *Escherichia coli*. J. Bacteriol 190:4460–9.

[13] Sperandeo P, Dehò G, Polissi A. (2009) The lipopolysaccharide transport system of Gram-negative bacteria. Biochim. Biophys. Acta 1791:594–602.

[14] Ruiz N, Gronenberg LS, Kahne D, Silhavy TJ. (2008) Identification of two inner-membrane proteins required for the transport of lipopolysaccharide to the outer membrane of *Escherichia coli*. Proc. Natl. Acad. Sci. U.S.A. 105:5537–42.

[15] Ma B, Reynolds CM, Raetz CRH. (2008) Periplasmic orientation of nascent lipid A in the inner membrane of an *Escherichia coli* LptA mutant. Proc. Natl. Acad. Sci. U.S.A. 105:13823–8.

[16] Wu T, McCandlish AC, Gronenberg LS, Chng S-S, Silhavy TJ, Kahne D. (2006) Identification of a protein complex that assembles lipopolysaccharide in the outer membrane of *Escherichia coli*. Proc. Natl. Acad. Sci. U.S.A. 103:11754–9.

[17] Bos MP, Tefsen B, Geurtsen J, Tommassen J. (2004) Identification of an outer membrane protein required for the transport of lipopolysaccharide to the bacterial cell surface. Proc. Natl. Acad. Sci. U.S.A. 101:9417–22.

[18] Hug I, Couturier MR, Rooker MM, Taylor DE, Stein M, Feldman MF. (2010) *Helicobacter pylori* lipopolysaccharide is synthesized via a novel pathway with an evolutionary connection to protein N-glycosylation. PLoS Pathog 6:e1000819.

[19] Wang G, Ge Z, Rasko DA, Taylor DE. (2000) Lewis antigens in *Helicobacter pylori*: biosynthesis and phase variation. Mol. Microbiol 36:1187–96.

[20] Raetz CRH, Reynolds CM, Trent MS, Bishop RE. (2007) Lipid A modification systems in gram-negative bacteria. Annu. Rev. Biochem.76:295–329.

[21] Hug I, Feldman MF. (2011) Analogies and homologies in lipopolysaccharide and glycoprotein biosynthesis in bacteria. Glycobiology 21:138–51.

[22] Marolda CL, Valvano MA. (1998) Promoter region of the *Escherichia coli* O7 specific lipopolysaccharide gene cluster: structural and functional characterization of an upstream untranslated mRNA sequence. J. Bacteriol. 180:3070–9.

[23] Schnaitman CA, Klena JD. (1993) Genetics of lipopolysaccharide biosynthesis in enteric bacteria. Microbiol. Rev. 57:655–82.

[24] Reeves PP, Wang L. (2002) Genomic organization of LPS-specific loci. Curr. Top. Microbiol. Immunol. 264:109–35.

[25] Guo L, Lim KB, Gunn JS, Bainbridge B, Darveau RP, Hackett M, Miller SI. (1997) Regulation of lipid A modifications by *Salmonella typhimurium* virulence genes phoP-phoQ. Science 276:250–3.

[26] Gunn JS. (2008) The *Salmonella* PmrAB regulon: lipopolysaccharide modifications, antimicrobial peptide resistance and more. Trends Microbiol. 16:284–90.
[27] Marchal K, De Keersmaecker S, Monsieurs P, van Boxel N, Lemmens K, Thijs G, Vanderleyden J, De Moor B. (2004) In silico identification and experimental validation of PmrAB targets in *Salmonella typhimurium* by regulatory motif detection. Genome Biol. 5:R9.
[28] Heinrichs DE, Yethon JA, Whitfield C. (1998) Molecular basis for structural diversity in the core regions of the lipopolysaccharides of *Escherichia coli* and *Salmonella enterica*. Mol. Microbiol. 30:221–32.
[29] Rietschel ET, Kirikae T, Schade FU, Mamat U, Schmidt G, Loppnow H, Ulmer AJ, Zähringer U, Seydel U, Di Padova F. (1994) Bacterial endotoxin: molecular relationships of structure to activity and function. FASEB J. 8:217–25.
[30] Bailey MJ, Hughes C, Koronakis V. (1997) RfaH and the ops element, components of a novel system controlling bacterial transcription elongation. Mol. Microbiol. 26:845–51.
[31] Geerlof A, Lewendon A, Shaw WV. (1999) Purification and characterization of phosphopantetheine adenylyltransferase from *Escherichia coli*. J. Biol. Chem. 274:27105–11.
[32] Feng L, Han W, Wang Q, Bastin DA, Wang L. (2005) Characterization of *Escherichia coli* O86 O-antigen gene cluster and identification of O86-specific genes. Vet. Microbiol. 106:241–8.
[33] Clark CG, Kropinski AM, Parolis H, Grant CCR, Trout-Yakel KM, Franklin K, Ng L-K, Paramonov NA, Parolis LAS, Rahn K, Tabor H. (2009) *Escherichia coli* O123 O antigen genes and polysaccharide structure are conserved in some *Salmonella enterica* serogroups. J. Med. Microbiol. 58:884–94.
[34] Sozhamannan S, Deng YK, Li M, Sulakvelidze A, Kaper JB, Johnson JA, Nair GB, Morris JG Jr. (1999) Cloning and sequencing of the genes downstream of the *wbf* gene cluster of *Vibrio cholerae* serogroup O139 and analysis of the junction genes in other serogroups. Infect. Immun. 67:5033–40.
[35] Keenleyside WJ, Whitfield C. (1996) A Novel Pathway for O-Polysaccharide Biosynthesis in *Salmonella enterica* Serovar Borreze. J. Biol. Chem. 271:28581–92.
[36] Wang L, Jensen S, Hallman R, Reeves PR. (1998) Expression of the O antigen gene cluster is regulated by RfaH through the JUMPstart sequence. FEMS Microbiol. Lett. 165:201–6.
[37] Liu B, Knirel YA, Feng L, Perepelov AV, Senchenkova SN, Wang Q, Reeves PR, Wang L. (2008) Structure and genetics of *Shigella* O antigens. FEMS Microbiol. Rev. 32:627–53.
[38] Samuel G, Reeves P. (2003) Biosynthesis of O-antigens: genes and pathways involved in nucleotide sugar precursor synthesis and O-antigen assembly. Carbohydr. Res. 338:2503–19.
[39] Olson MV, Kas A, Bubb K, Qui R, Smith EE, Raymond CK, Kaul R. (2004) Hypervariability, suppressed recombination and the genetics of individuality. Philos. Trans. R. Soc. Lond., B, Biol. Sci .359:129–40.
[40] Dixon DR, Darveau RP. (2005) Lipopolysaccharide heterogeneity: innate host responses to bacterial modification of lipid a structure. J. Dent. Res. 84:584–95.
[41] Lawson AJ, Chart H, Dassama MU, Threlfall EJ. (2002) Heterogeneity in expression of lipopolysaccharide by strains of *Salmonella enterica* serotype Typhimurium definitive phage type 104 and related phage types. Lett. Appl. Microbiol. 34:428–32.
[42] Tso MD, Dooley JS. (1995) Temperature-dependent protein and lipopolysaccharide expression in clinical *Aeromonas* isolates. J. Med. Microbiol. 42:32–8.

[43] Reyes RE, Ramírez Saad H, Pérez J, Meléndez E, Navarro A, Murata Ch, Cravioto A. (2010) Intra-strain variation of lipopolysaccharides from *Citrobacter freundii* E9750 by *in vitro* passages. Afr. J. Microbiol. Res. 4:1324–31.
[44] McGroarty EJ, Rivera M. (1990) Growth-dependent alterations in production of serotype-specific and common antigen lipopolysaccharides in *Pseudomonas aeruginosa* PAO1. Infect. Immun. 58:1030–7.
[45] Guard-Petter J, Lakshmi B, Carlson R, Ingram K. (1995) Characterization of lipopolysaccharide heterogeneity in *Salmonella enteritidis* by an improved gel electrophoresis method. Appl. Environ. Microbiol. 61:2845–51.
[46] Nishiuchi Y, Doe M, Hotta H, Kobayashi K. (2000) Structure and serologic properties of O-specific polysaccharide from *Citrobacter freundii* possessing cross-reactivity with *Escherichia coli* O157:H7. FEMS Immunol. Med. Microbiol 28:163–71.
[47] van der Woude MW, Bäumler AJ. (2004) Phase and antigenic variation in bacteria. Clin. Microbiol. Rev. 17:581–611.
[48] Wisniewski-Dyé F, Vial L. (2008) Phase and antigenic variation mediated by genome modifications. Antonie Van Leeuwenhoek 94:493–515.
[49] Torres-Cruz J, van der Woude MW. (2003) Slipped-strand mispairing can function as a phase variation mechanism in *Escherichia coli*. J. Bacteriol. 185:6990–4.
[50] van Belkum A, Scherer S, van Leeuwen W, Willemse D, van Alphen L, Verbrugh H. (1997) Variable number of tandem repeats in clinical strains of *Haemophilus influenzae*. Infect. Immun. 65:5017–27.
[51] Olivares D, Gisbert JP. (2006) Factors involved in the pathogenesis of *Helicobacter pylori* infection. Rev Esp Enferm Dig 98:374–86.
[52] Wang G, Rasko DA, Sherburne R, Taylor DE. (1999) Molecular genetic basis for the variable expression of Lewis Y antigen in *Helicobacter pylori*: analysis of the alpha (1,2) fucosyltransferase gene. Mol. Microbiol. 31:1265–74.
[53] Sanabria-Valentín E, Colbert M-TC, Blaser MJ. (2007) Role of *fut*C slipped strand mispairing in *Helicobacter pylori* Lewisy phase variation. Microbes Infect. 9:1553–60.
[54] Allison GE, Verma NK. (2000) Serotype-converting bacteriophages and O-antigen modification in *Shigella flexneri*. Trends Microbiol. 8:17–23.
[55] Perepelov AV, L'vov VL, Liu B, Senchenkova SN, Shekht ME, Shashkov AS, Feng L, Aparin PG, Wang L, Knirel YA. (2009) A similarity in the O-acetylation pattern of the O-antigens of *Shigella flexneri* types 1a, 1b, and 2a. Carbohydr. Res. 344:687–92.
[56] Allison GE, Angeles D, Tran-Dinh N, Verma NK. (2002) Complete genomic sequence of SfV, a serotype-converting temperate bacteriophage of *Shigella flexneri*. J. Bacteriol. 184:1974–87.
[57] Nair A, Korres H, Verma NK. (2011) Topological characterisation and identification of critical domains within glucosyltransferase IV (GtrIV) of Shigella flexneri. BMC Biochem. 12:67.
[58] Korres H, Verma NK. (2006) Identification of essential loops and residues of glucosyltransferase V (GtrV) of Shigella flexneri. Mol. Membr. Biol. 23:407–19.
[59] Rodríguez-Dorantes M, Téllez-Ascencio N, Cerbón MA, López M, Cervantes A. (2004) DNA methylation: an epigenetic process of medical importance. Rev. Invest. Clin. 2004;56:56–71.

[60] Broadbent SE, Davies MR, van der Woude MW. (2010) Phase variation controls expression of *Salmonella* lipopolysaccharide modification genes by a DNA methylation-dependent mechanism. Mol. Microbiol. 77:337–53.
[61] Bryant CE, Spring DR, Gangloff M, Gay NJ. (2010) The molecular basis of the host response to lipopolysaccharide. Nat. Rev. Microbiol. 8:8–14.
[62] Lu Y-C, Yeh W-C, Ohashi PS. (2008) LPS/TLR4 signal transduction pathway. Cytokine 42:145–51.
[63] Kawamoto T, Ii M, Kitazaki T, Iizawa Y, Kimura H. (2008) TAK-242 selectively suppresses Toll-like receptor 4-signaling mediated by the intracellular domain. Eur. J. Pharmacol. 584:40–8.
[64] Toshchakov VY, Fenton MJ, Vogel SN. (2007) Cutting Edge: Differential inhibition of TLR signaling pathways by cell-permeable peptides representing BB loops of TLRs. J. Immunol. 178:2655–60.
[65] Lee H-K, Brown SJ, Rosen H, Tobias PS. (2007) Application of beta-lactamase enzyme complementation to the high-throughput screening of toll-like receptor signaling inhibitors. Mol. Pharmacol. 72:868–75.
[66] Maskell D, Frankel G, Dougan G. (1993) Phase and antigenic variation the impact on strategies for bacterial vaccine design. Trends Biotechnol. 11:506–10.
[67] Reeves P. (1995) Role of O-antigen variation in the immune response. Trends Microbiol. 3:381–6.
[68] Nilsson C, Skoglund A, Moran AP, Annuk H, Engstrand L, Normark S. (2008) Lipopolysaccharide diversity evolving in *Helicobacter pylori* communities through genetic modifications in fucosyltransferases. PLoS ONE 3:e3811.
[69] Khamri W, Moran AP, Worku ML, Karim QN, Walker MM, Annuk H, Ferris JA, Appelmelk BJ, Eggleton P, Reid KBM, Thursz MR. (2005) Variations in *Helicobacter pylori* lipopolysaccharide to evade the innate immune component surfactant protein D. Infect. Immun. 73:7677–86.
[70] Moran AP. (2008) Relevance of fucosylation and Lewis antigen expression in the bacterial gastroduodenal pathogen *Helicobacter pylori*. Carbohydr. Res. 343:1952–65.
[71] Delcour AH. (2009) Outer membrane permeability and antibiotic resistance. Biochim. Biophys. Acta 1794:808–16.
[72] Gatzeva-Topalova PZ, May AP, Sousa MC. (2005) Structure and mechanism of ArnA: conformational change implies ordered dehydrogenase mechanism in key enzyme for polymyxin resistance. Structure 13:929–42.
[73] Falagas ME, Rafailidis PI, Matthaiou DK. (2010) Resistance to polymyxins: Mechanisms, frequency and treatment options. Drug Resist. Updat. 13:132–8.
[74] Clausell A, Garcia-Subirats M, Pujol M, Busquets MA, Rabanal F, Cajal Y. (2007) Gram-negative outer and inner membrane models: insertion of cyclic cationic lipopeptides. J Phys Chem B 111:551–63.

Chapter 4

Exopolysaccharide Biosynthesis in *Rhizobium leguminosarum*: From Genes to Functions

Tanya V. Ivashina and Vladimir N. Ksenzenko

Additional information is available at the end of the chapter

http://dx.doi.org/10.5772/51202

1. Introduction

Gram-negative soil α-proteobacteria belonging to the genera *Allorhizobium, Azorhizobium, Bradyrhizobium, Mesorhizobium, Rhizobium,* and *Sinorhizobium* (*Ensifer*) are able to infect the roots of their legume hosts in a host-specific way and induce the formation of specialized new plant organs - nodules, in which endosymbiotic bacteria reduce atmospheric nitrogen to ammonia. *Rhizobium leguminosarum* comprises two biovars, namely *trifolii* (nodulating *Trifolium*) and *viciae* (nodulating *Pisum, Vicia, Lathyrus,* and *Lens*) [1]. Closely related to the *R. leguminosarum* is *R. etli* (formely *R. leguminosarum* bv. *phaseoli*) nodulating *Phaseolus* beans. To avoid confusions, we will use former names for *R. leguminosarum* bv. *phaseoli* strains as they were described in original papers.

The *R. leguminosarum* strains investigated so far synthesize different types of polysaccharides, including acidic exopolysaccharide (EPS), capsule polysaccharide (CPS), gel-forming polysaccharide (GPS), cellulose fibrils, galactomannan, lipopolysaccharide, and cyclic glycans [2]. The cyclic neutral β-(1,2)-glucans are predominantly accumulated within the periplasmic space and play an essential role during hypoosmic adaptation as well as during plant infection [3]. Glucomannan was shown to be important for lectin-mediated polar attachment to *Vicia sativa* and *Pisum sativum* root hairs and competitive nodulation [4,5]. The CPS is tightly associated with the cell surface of bacteria forming a polysaccharide matrix surrounding the bacteria [6]. Differences in noncarbohydrate substitutions, such as *O*-acetyl, pyruvate, and 3-hydroxybutyrate, may distinguish anionic capsule-bound polysaccharides from secreted EPS. In late-stationary-phase cultures, CPS was replaced by a polysaccharide with strong gel-forming properties having an unknown function [7]. The LPS present at the outer leaflet of the outer membrane and consists of three parts: lipid A, the core polysaccharide and the O-antigen polysaccharide [8]. More and more data appear indicating that LPS plays a specific role in the later stages of establishment of symbiosis

INTECH
open science | open minds

© 2012 The Author(s). Licensee InTech. This chapter is distributed under the terms of the Creative Commons Attribution License http://creativecommons.org/licenses/by/3.0), which permits unrestricted use, distribution, and reproduction in any medium, provided the original work is properly cited.

(reviewed by [9]). EPS refers to extracellular polysaccharides with slight or no cell association [9-11].

Among the above mentioned polysaccharides, acidic EPSs attract attention due to their diverse functions both in free-living rhizobia and during the establishment of nitrogen-fixing symbiosis with host legume plants. EPS forms a biofilm layer on the cell surface which is thought to contribute to the following processes: cellular protection against environmental stresses, attachment to surfaces, nutrient gathering, and the preferential absorption of flavonoids secreted by plants along the membrane surface [9]. In addition, EPS biosynthesis is required for the effective nodulation of legumes such as *Medicago, Pisum, Trifolium, Leucaena,* and *Vicia* spp., which form indeterminate-type nodules (otherwise called meristematic or cylindrical) [12]. EPS-deficient mutants of *Rhizobium leguminosarum* bv. *viciae* (hereafter *Rlv*), *R. leguminosarum* bv. *trifolii* (*Rlt*), and *Sinorhizobium meliloti* induce symbiotically defective phenotypes which include delayed root hair curling, nodules devoid of bacteria due to infection threads that abort within peripheral cells of the developing nodule, and small, partially infected, non-nitrogen-fixing nodules [12]. The precise function(s) of the EPS molecules in these associations is still unclear; however, recent studies suggest that these extracellular polymers may function as symbiotic signaling molecules which regulate plant responses in the infection process [13-17].

The current knowledge of EPS biosynthesis in rhizobia was based almost exclusively on the research with succinoglycan (EPS I) produced by *S. meliloti* [18]. More than 26 genes needed for the synthesis, modification, polymerization, export, and processing of *S. meliloti* EPS I were identified, and a pathway for the polysaccharide biosynthesis was proposed [19-21]. Recent discoveries on succinoglycan biosynthesis and its functioning in symbiosis are summarized in the exhaustive reviews [22,23].

The EPSs produced by *R. leguminosarum* strains investigated to date structurally differ from those of *S. meliloti.* At present there is fragmentary information about the *R. leguminosarum* EPS assembly, modification, and processing, as well as the role of the individual gene products in these processes. These data were reviewed in [9-11,24]. Here we report some recent and complementary data obtained in our group. On the basis of these data as well as the analysis of the sequenced *R. leguminosarum* and *R. etli* genomes we tried to outline main regularities in the EPS synthesis.

2. Structural characteristics of exopolysaccharides

Structures of known *R. leguminosarum* and *R. etli* (*Re*) EPSs are presented in Table 1. It should be noted that EPSs secreted by the most of the *Rlv* and *Rlt* strains were shown to have repeating units of identical structure [25-35]. These EPSs are branched heteropolysaccharides of the octasaccharide repeating unit consisting of a backbone of two glucose and two glucuronic acid residues, a side chain of three glucose and one galactose residues. The backbone contains β1-4 and α1-4 linkages only while the branch point is linked β1-6. The side chain contains β1-4 and β1-3 linkages. In general, the octasaccharide is modified by two pyruvyl, one or two non-stoichiometric O-acetyl, and one non-stoichiometric hydroxybutanoyl groups. The distribution pattern of O-acetyl and 3-hydroxybutanoyl groups may vary for some *R. leguminosarum* strains and was shown to be dependent on the growth phase of bacteria and culture medium [27,32].

	Repeating unit structure	Strains	References
Backbone chain	→4)-α-**Glc**-(1→4)-α-**GlcA**-(1→4)-β-**GlcA**-(1→4)-β-**Glc**-(1 6 ↑	all strains	[25-41]
Side chains	β ↑ β-**Gal**-(1→3)-β-**Glc**-(1→4)-β-**Glc**-(1→4)-β-**Glc** 6 4 6 4 **Pyr** **Pyr**	*Rlv*: VF39, 3841, LPR1, 128c53, 128c63; *Rlt*: TA1, LPR5, ANU843, NA30, 0403; *Re* CFN42; *Rlp*: LPR49, 127K36; *Rhizobium* sp. GRH2	[25-35]
	β ↑ β-**Glc**-(1→3)-β-**Glc**-(1→4)-β-**Glc** 6 4 **Pyr**	*Rlt* 4S	[37]
	β ↑ β-**Gal**-(1→3)-β-**Glc**-(1→3)-β-**Glc**-(1→3)-β-**GlcA**-(1→4)-β-**Glc** 6 4 6 4 **Pyr** **Pyr**	*Rlv* 248	[38]
	β ↑ α-**Gal**-(1→4)-β-**GlcA**-(1→3)-β-**Glc**-(1→4)-β-**Glc**-(1→4)-β-**Glc** 6 4 **Pyr**	*Rlp* 127K44	[39]
	β ↑ β-**Glc**-(1→6)-α-**Gal**-(1→4)-β-**GlcA**-(1→4)-β-**Glc**-(1→4)-β-**Glc**-(1→4)-β-**Glc** 6 4 **Pyr**	*Rlp* 127K38	[40]
	β ↑ β-**Gal**-(1→6)-β-**Gal**-(1→6)-β-**Glc**-(1→6)-α-**Gal**-(1→4)-β-**GlcA**-(1→4)-β-**Glc**-(1→4)-β-**Glc** 6 4 **Pyr**	*Rlp* 127K87	[41]

Table 1. Structure of *R. leguminosarum* and *R. etli* EPS repeating units. Abbreviations: Glc, glucose; GlcA, glucuronic acid; Gal, galactose and Pyr, ketal pyruvate group.

However, several *R. leguminosarum* strains produce EPS with divergent side chains though with identical backbones and the same β1-6-linked glucosyl residue branching the side chain. Side chains of these EPS may consist of three to seven sugar residues. Up to three Gal residues can be present as in the *R. leguminosarum* bv. *phaseoli* (*Rlp*) 127K87 EPS, or terminal Gal residue can be absent as in the *Rlt* 4S EPS. In addition, in some cases side chains of EPS contain GlcA residues (*Rlp* 127K44, *Rlp* 127K38, *Rlp* 127K87, *Rlv* 248). Besides β1-4 and β1-3 linkages, sugar residues can be attached by α1-6 or β1-6 glycosidic bonds (*Rlp* 127K38, *Rlp* 127K87).

The acidic nature of EPS is explained by the presence of uronic acids and negatively charged pyruvyl groups. Similar to other representatives of *Rhizobiaceae*, the *R. leguminosarum* strains synthesize EPS in high-molecular-weight (HMW) and low-molecular-weight (LMW) forms [13,42]. The latter were proposed to act as signaling factors during the development of symbiosis [14,16-17,42].

3. Organization of exopolysaccharide biosynthesis genes

According to the modern conception, the synthesis of heteropolysaccharides requires a complex pathway starting with the synthesis of sugar nucleotide precursors as well as of the non-carbohydrate donors followed by sequential assembly of the repeating unit on polyprenyl lipid carries, their modification, polymerization, and export outside of the cell [20,21,43].

We started the study of the genetic control of the biosynthesis of acidic exopolysaccharide with isolation of non-mucoid Tn*5*-derived mutants in *Rlv* VF39. As a result, five non-slimy mutants (GL1-5) were obtained. The mutations were mapped within four separate chromosomal loci. The open reading frames (*orfs*) interrupted by insertion of the Tn*5* transposon were named as *pss* (**p**oly**s**accharide **s**ynthesis) according to Borthakur and co-workers [44]. The Tn*5* insertion in the GL4 mutant was localized within the *pssA* gene [45], the ortholog of which was previously identified in *Rlp* 8002 [44]. The mutations in GL2 and GL6 were mapped within the *pssE* and *pssD* genes, respectively. Their orthologs were found earlier in *Rlt* LPR5 [46]. Chromosomal walking around these genes in *Rlv* VF39 allowed us to identify a 15.5–kb multi-cistronic operon which included a core set of genes needed for the assembly of the repeating unit of the EPS (*pssEDCFGHIJS*), its modification (*pssKMR*), polymerization (*pssL*) and processing (*pssW*) (Fig. 1) [47]. It should be mentioned that the *pssV-E* operon was found in all *R. leguminosarum* and *R. etli* genomes, whose complete or partial sequences are available now. Moreover, nine out of the fifteen genes from this operon have orthologs in all these genomes. At the same time, certain *Rlv* VF39 genes are absent in some other genomes, certain genes are substituted for non-orthologous genes, and some additional genes are also present (Fig. 1). We will discuss the functioning of all these genes bellow. Here we would like only to consider the problem with their names.

All fifteen genes from the *pssV-E* operon were named as *pss* genes. In addition, the same gene name abbreviation was assigned to six genes (*pssA, pssB, pssN, pssO, pssP* and *pssT*) localized in other operons. It is easy to count up that only five letters of the alphabet left that can be used with the "*pss*" body in the names of new genes involved in EPS biosynthesis. Meanwhile, in our opinion even at present new names for eight genes from *Rlt* WSM2304, *Re* CFN42, *Re* CNPAF512 and *Re* CIAT 652 have to be assigned. Therefore, we propose (i) to retain the existing names for all orthologous genes, and (ii) to introduce a new set of genes with the body name "*psa*" (**p**oly**s**accharide repeating unit **a**ssembly). Our propositions concerning new names to be assigned for certain genes involved in EPS biosynthesis are summarized in Table 2.

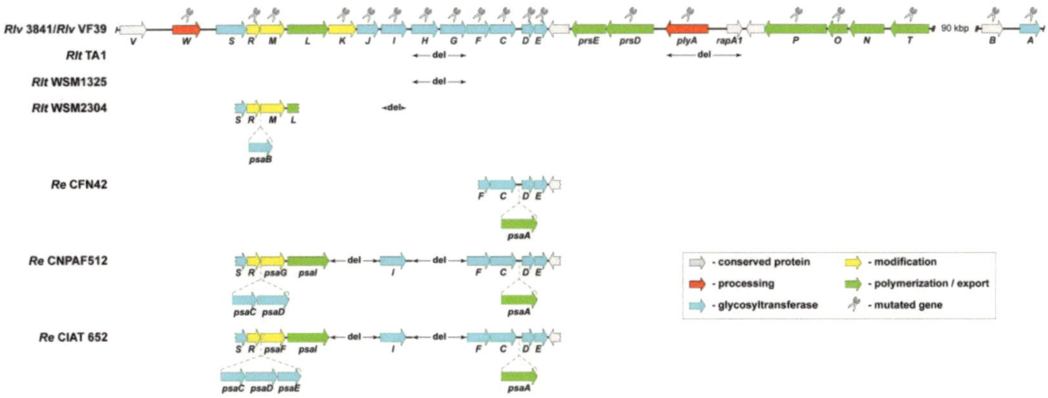

Figure 1. Gene arrangements in PssI-cluster of *R. leguminosarum* and *R. etli* strains.

Proposed gene name	Existing gene names in *R. leguminosarum* and *R. etli* genomes*				Predicted function
	Rlt WSM2304	*Re* CFN42	*Re* CNPAF512	*Re* CIAT 652	
psaA	Rleg2_0274	RHE_CH03224	RHECNPAF_430080	RHECIAT_CH0003464	O-antigen ligase
psaB	Rleg2_2965	none	none	none	GT, family 2
psaC	none	none	RHECNPAF_430069	RHECIAT_CH0003456	GT, family 2
psaD	none	none	RHECNPAF_430070	RHECIAT_CH0003457	GT, family 1
psaE	none	none	none	RHECIAT_CH0003458	GT, family 2
psaF	none	none	none	RHECIAT_CH0003459	Polysaccharide pyruvyl transferase
psaG	none	none	RHECNPAF_430071	none	Polysaccharide pyruvyl transferase
psaI	none	none	RHECNPAF_430072	RHECIAT_CH0003459	Polysaccharide biosynthesis protein

* EMBL/GenBank/DDBJ accession numbers: *Rlt* WSM2304 (CP001191), *Re* CFN42 (CP000133), *Re* CNPAF512 (AEYZ01000266) and *Re* CIAT 652 (CP001074)

Table 2. Proposed names for genes controlling EPS biosynthesis in some *R. leguminosarum* and *R. etli* strains.

The *pssV-E* operon is neighboring with the region comprising several operons containing genes for the Type I secretion system (*prsED*), EPS processing (*plyA*), and EPS polymerization/export (*pssTNOP*). This whole chromosomal region is known now as the Pss-I gene cluster [48].

The *pssA* gene controlling the first step in the repeating unit assembly is localized approximately at 90 kb from the Pss-I cluster. The gene was shown to be transcribed as a monocistronic mRNA [45]. Upstream of the *pssA* is the *pssB* gene encoding inositol monophosphatase [49-51]. Effects of mutations within the *pssB* gene on the synthesis of EPS and symbiotic behavior have been analyzed in the *Rlv* VF39 and *Rlt* TA1 backgrounds and have been shown to be contradictory. In *Rlv* VF39 the *pssB* mutants retained the ability to produce EPS in amounts equal to those of the wild-type strain. In *Rlt* TA1 the *pssB* inactivation displayed an increased overall production of EPS versus the wild-type strain, and alterations in the LPS PAGE-banding pattern and the O-antigen sugar composition [51,52]. Nevertheless, *pssB* mutants of both strains elicited non-effective nodules on the *Vicia faba* or *Trifolium pratense* roots, respectively [49,52].

In the case of GL1 and GL3 mutants we have not carried out the extended chromosomal walking around the Tn*5* insertions, but only short genome sequences flanking Tn*5* were determined. We localized the Tn*5* insertion in the GL3 mutant within some small *orf* (213 bp only). The ortholog of this *orf* (RL2260) was found in the chromosome sequence of *Rlv* 3841 located far away from the Pss-I cluster. It encodes a 7.2 kDa positively charged protein (pI 10.8), which is conserved in numerous bacteria. As for the GL1 mutation, we have not been able to map it up to now. Probably, this mutation targets a gene located in one of the *Rlv* VF39 strain-specific plasmids.

Taking together these data, one can conclude that the core set of EPS synthesizing genes are clustered in the chromosomal Pss-I region. Clustering of genes involved in EPS biosynthesis is not unique for *R. leguminosarum*, but is widespread in different polysaccharide producing bacteria [53,54]. Such a type of genes organization could reflect their coordinated expression and tightly regulated control.

Evidently, EPS biosynthesis is linked with other metabolic pathways in the cell. Therefore, localization of mutations affecting EPS production distant from the Pss-I region can reflect this linkage. For example, the *pssA* gene is located in another chromosomal region probably due to involvement in the initiation of not only EPS synthesis, but the synthesis of the other polysaccharides. Recently, several regulatory genes influencing EPS production (*psiA, psrA, exoR, expR, rosR, praR*) were found to be localized either in different regions of the chromosome or at the endogenious plasmids (reviewed by [24]). We can not exclude that mutations in GL3 and GL1 also target the regulatory genes.

4. Synthesis of nucleotide-sugar precursors

According to the sugar composition of acidic EPS, biosynthesis of its repeating units requires nucleotide sugars: UDP-glucose, UDP-glucuronic acid and UDP-galactose that are formed by central carbon metabolism. Several genes involved in synthesis of nucleotide-sugar precursors were identified in *R. leguminosarum* genomes. The *exoB* encodes UDP-glucose 4-epimerase, responsible for UDP-galactose production [55]. Mutations in this gene have a pleiotropic effect and influence the synthesis of different classes of galactose containing polysaccharides; namely acidic EPS, glucomannan, lipopolysaccharide, and probably the galactose-rich gel-forming polysaccharide [5,55,56].

The *exo5* gene encodes a UDP-glucose dehydrogenase responsible for oxidation of UDP-glucose to UDP-glucuronic acid. A mutation in *exo5* affects the production of extracellular acidic polysaccharide and capsular polysaccharide both of which contain glucuronic acid residues in the backbone chain [57,58].

5. Assembly of the EPS repeating unit

As mentioned above, heteropolysaccharides are polymers consisting of identical repeating units which can vary only by the distribution of modifying groups per monomer. Obviously, their assembly has to be stringently controlled. The biosynthetic pathways for the number of heteropolysaccharides have been elucidated [54,59]. It is evident from the obtained data that the unique structure of the repeating unit is governed by the specificity of nonprocessive glycosyltransferase (GT) catalyzing a certain step of the biosynthesis. This specificity is likely based on the ability of GT to recognize the sugar residue to be transferred, the acceptor, and the linkage to be formed. At present the complete pathway of EPS biosynthesis has not been determined for any of *R. leguminosarum* biovars, and only some individual steps have been characterized. Nevertheless, a detailed consideration even of these fragmentary data together with a comparative analysis of the available *R. leguminosarum* genome sequences allowed us to predict the genetic control of all steps in the repeating unit assembly at least for *Rlv* VF39 and closely related strains.

It was shown previously that assembly of a repeating unit of *R. leguminosarum* EPS starts with the addition of a glucose residue to the lipid carrier [60]. Biochemical studies and complementation analysis provided strong evidence that this reaction is conducted by the *pssA* gene product [36,61]. The *pssA* gene encodes the integral inner membrane protein UDP-glucose:polyprenyl-phosphate glucosephosphotransferase, belonging to a family of diverse bacterial sugar transferases. Members of this family catalyze the formation of a phosphodiester bond between polyprenol phosphate and hexoso-1-phosphate, which is donated by nucleotide sugars. The *pssA* gene is highly conserved in *R. leguminosarum* biovars and *R. etli* [44-46,50]. The *pssA* mutants do not produce EPS and as a consequence impaired the normal development of the nitrogen-fixing nodules on the appropriate plant hosts and the formation of biofilms both *in vitro* and on root hairs [5,45,46,50]. Some contradictory results exist concerning the influence of *pssA* mutations on the synthesis of CPS which displays similar structure to EPS in terms of glycosyl composition [5,30,62]. In the *Rlt* 5599 genetic background, *pssA* mutants still produce CPS at the level similar to that of a wild-type strain [46]. In contrast, both EPS and CPS are absent in the *pssA* mutant of *Rlv* 3841 [5], which indicates that the PssA protein might be involved also in the initiation of the CPS synthesis. It has been shown that expression of *pssA* depends on the environmental factors such as phosphate and ammonium concentrations and also on root exudates [63].

Bossio and co-workers have demonstrated that subsequent to addition of glucose to isoprenylpyrophosphate, two glucuronic acid residues are attached [64]. The attachment of the first GlcA is catalyzed by the *pssE* and *pssD* gene products [47,61]. This conclusion is based on the results of *in vivo* reciprocal complementation between the *pssED* and *spsK* genes of the *Sphingomonas* strain S88 and the *in vitro* sugar incorporation studies. Recently we confirmed directly the function of PssED. The corresponding mRNAs were translated in a wheat-germ cell-free system in the presence of liposomes obtained from *Rlv* VF39 phospholipids. The resulting proteoliposomes were used as an enzyme source in experiments on PssDE specificity (Ivashina et al., unpublished).

Both, PssD and PssE display similarity to GT family 28 (CAZy database, http://www.cazy.org/). Notably, the amino acid sequences of PssD and PssE are similar to the N-terminal and C-terminal halves of the glucuronosyl-(ß1,4)-glucosyltransferase SpsK, respectively. This observation has led us to the conclusion that PssD and PssE represent two subunits of the same enzyme. The proposed catalytic domain is localized in the peripheric inner membrane of the *pssE* subunit, in contrast to PssD, which was shown to be an integral inner membrane protein. We have also found that the integration of PssE into a membrane or liposomes strongly depended on the presence of PssD and *vice versa* (Ivashina et al, unpublished). It is interesting to note that the same was observed in the case of yeast proteins Alg13 and Alg14. It was found that Alg14 was needed for the correct positioning of Alg13 on the cytosolic face of the endoplasmic reticulum membrane mediating the formation of the active UDP-N-acetylglucosamine transferase complex [65,66]. Mutations in the *pssE* and *pssD* genes fully abolished EPS production and as a consequence resulted in defects of nodule infection [67-69].

PssC belongs to the GT family 2 containing a variety of inverting glycosyltransferases (enzymes that form glycosidic bonds with stereochemistry opposite to that of the glycosyl donor) that utilize a diverse range of nucleotide-sugar donors and participate in the synthesis of various types of polysaccharides [70]. This GT was assigned by Pollock and co-workers to glucuronosyl-(β-1,4)-glucuronosyltransferase, which catalyzed the attachment of the third sugar residue (GlcA) to the disaccharide (GlcA-β-1,4-Glc) lipid-linked intermediate with the formation of the β,1-4 glycosidic bond [61]. This conclusion was based on the comparative data on the genetic control of the first three steps of *R. leguminosarum* EPS and the *Sphingomonas* strain S88 sphingan assembly. Obviously, these data can not be considered as direct evidence, and conclusion on the PssC assignment needs additional experimental proofs.

Several *pssC* mutations have been characterized in various *R. leguminosarum* biovars backgrounds to date [46,47]. All of them were mapped at the N-terminus of PssC and resulted in the decreased amount (27-38%) of EPS in culture supernatants. However, structural analysis of EPSs secreted by these mutants showed them to be identical to that of the wild-type strains [46,47]. We proposed that the initiation of translation of *pssC* could be realized from the second potential start codon GTG located downstream from the spots of mutations and therefore leading to the synthesis of protein retaining enzymatic activity. In fact, Western blot analysis with antibodies against PssC demonstrated the synthesis of a truncated protein in the *pssC* mutant (Ivashina et al., unpublished). Attempts to introduce mutations into the central part of *pssC* in the *Rlv* VF39 background were unsuccessful. However, it was easy to homogenote *pssC* in the same strain carrying a mutation in the *pssD* gene, which failed to produce EPS. These results pointed to the detrimental effect of such mutations most probably due to the accumulation of lipid-linked intermediates in the cytoplasmic membrane and as a result to inability of the lipid carrier to be released for other essential cellular functions. The data obtained with the use of different genetic systems led to the same conclusions [61].

The PssJ protein is the last glycosyltransferase for which its biochemical function was ascertained experimentally. The *Rlt* RBL5515 strain carrying mutation in the *pssJ* gene (known as *exo*344::Tn*5*), synthesizes residual amounts of EPS, the repeating unit of which lacks the terminal galactose of the side chain. On the basis of the structural features of the polysaccharides synthesized and the results of an analysis of the enzyme activities involved, it was hypothesized that the galactosyltransferase catalyzing formation of the β1-3 linkage between sub-terminal (Glc) and terminal (Gal) sugar residues in the octasaccharide unit is affected in this strain [6]. PssJ did not reveal any homologs in protein databases and therefore it could be referred to a family of "not-classified glycosyltransferases" (CAZy database).

It can be seen from the EPS repeating unit structure that the third (GlcA) and the fourth (Glc) sugar residues in the backbone chain are linked by the α1-4 glycosidic bond (Table 1). PssS is the only enzyme which can be responsible for this reaction. According to homology search data, the PssS was referred to the GT family 1 (CAZy database), which integrates the retaining glycosyltransferases forming glycosidic bonds with stereochemistry identical to that of the glycosyl donor. These enzymes were shown to be involved in exopolysaccharide, lipopolysachharide, and slime polysaccharide colanic acid biosynthesis. We were unable to disrupt the *pssS* gene in the wild-type strain *Rlv* VF39, but easily inactivated it in the *pssD* mutant (Eps$^-$). It is likely that in this case the inactivation of *pssS* also leads to the accumulation of toxic lipid-linked intermediates as it was proposed for *pssC* mutants.

Summarizing, specific GTs were assigned to the assembly of the backbone chain of the octasaccharide unit, as well as in the attachment of the terminal Gal in the side chain. It should be emphasized that in all known *R. leguminosarum* EPS structures the backbone chains are identical (Table 1). At the same time, in genomes of *R. leguminosarum* strains, which synthesize similar EPSs the orthologs of *pssAEDCS* genes are present (Fig. 1). Therefore, we can conclude that the prediction has been made correctly. The same statement is true for PssJ catalyzing the attachment of the terminal Gal residue: in the *R. leguminosarum* genomes, where *pssJ* has not been found, the *pssK*, which has been predicted to modify this Gal residue (see below), is also missing.

As mentioned above, *R. leguminosarum* and *R. etli* can produce EPSs with side chains varying in their length, sugar composition and type of glycosidic linkages. It should be noted, that the set of GT genes in Pss-I clusters can also vary. Studies on the genetic control of EPS biosynthesis are impeded to a considerable degree due to the fact that when genome sequences are available, nothing is known about the structure of synthesized EPS and *vice versa*. The *Rlv* VF39 and *Rlt* TA1 represent the only pair when (i) the structure of EPSs and sequences of the Pss-I clusters are determined; (ii) both strains produce structurally identical EPSs but differ in the sets of GT genes; (iii) data on mutational analysis of *Rlv* VF39 GT genes are obtained. Taking into account all these considerations, we have picked the *Rlv* VF39/*Rlt* TA1 pair for prediction of the pathway of side chain biosynthesis. In this case the question arises, which glycosyltransferase initiates branching by attachment of the Glc residue via the β1-6 bond, and which GT(s) is (are) responsible for the attachment of two subsequent Glc residues by formation of the β1-4 glycosidic linkage. It is obvious, that in *Rlt* TA1 only two GTs (PssF and PssI) can perform these functions. In addition to PssF and PssI, two other GTs (PssH and PssG) can participate in the side chain assembly in the *Rlv* VF39. We introduced mutations into all four genes (*pssFGHI*) in *Rlv* VF39 and found that the structures of EPSs of mutant strains were identical to that of the parental strain, and only the level of acidic EPS production decreased. Based on these results, we can conclude that the action of each GT considered in this system can be interchangeable.

In our opinion, PssF is the best candidate to play the role of GT which catalyzes the attachment of the Glc residue by formation of the β1-6 bond. Firstly, in all *Rhizobium leguminosarum* strains the EPS side chain starts with the Glc residue attached to the backbone chain via the β1-6 bond. At the same time, PssF is present in all PssI-clusters sequenced up to now. Secondly, PssI, PssH and PssG reveal a rather high level of similarity with each other especially in the N-terminal parts of their amino acid sequences where catalytic domains are located. In contrast, PssF is practically non-homologous to that of three GTs but shows although weak but yet reliable homology of its N-terminal half with GTs attaching the Glc residue via the β1-6 bond (e.g. ExoO from *S. meliloti* [20]).

If our prediction on the PssF function is correct, the attachment of two subsequent Glc residues could be achieved by single GT (PssI) in the *Rlt* TA1, and as many as three GTs (PssI, PssH and PssG) could participate in this process in the *Rlv* VF39. Apparently, PssI in the *Rlt* TA1 strain is to a certain extent tolerant to the acceptor structure and the identity of EPS repeating units probably is attained at the expense of high specificity of PssJ, which catalyzes the last step of the EPS assembly.

It seems that in the *Rlv* VF39 the subsequent attachment of Glc residues is achieved by two separate GTs, namely PssI and PssH. This assumption is based on a comparison of the amino acid sequences of these homologous GTs. A rather low level of similarity of their C-terminal parts, containing the putative acceptor recognition domain was observed. In contrast, PssG reveals a very high level of homology to PssI over its entire amino acid sequence (more than 80% similarity). It is plausible to assume that PssI and PssG are isoenzymes, which handle the same step in the EPS assembly. Thereby, genetic control of the repeating unit biosynthesis in the *Rlv* VF39 resembles that of *S. meliloti*, where the attachment of the sugar residue at each step of biosynthesis is catalyzed by specific GT, and even two GTs can participate in catalysis at some steps of the pathway.

A presumptive circuit of the EPS repeating unit assembly in *Rlt* TA1 and *Rlv* VF39 is presented in Figure 2. It should be noted that PssA, PssDE, PssC, PssS, PssF and PssI/PssG apparently comprise the basic set of GTs involved in biosynthesis of all EPS under consideration. Taking into account the mentioned variability of side chains of these EPS, one can expect that the basic set of GTs in corresponding *R. leguminosarum* strains is augmented. Indeed, PssJ is present in *Rlv* 3841, *Rlv* VF39, *Rlt* TA1, *Rlt* WSM2304, *Re* CFN42, but is absent in *Re* CNPAF512 and *Re* CIAT 652. At the same time, genes for family 2 GT PsaC and family 1 GT PsaD were found in the PssI-clusters of the latter strains. In addition, genes encoding family 2 GTs PsaB and PsaE, amino acid sequences of which are similar to GTs attaching the Glc residue via the β1-6 bond, were localized in *Rlt* WSM2304 and *Re* CIAT 652, respectively.

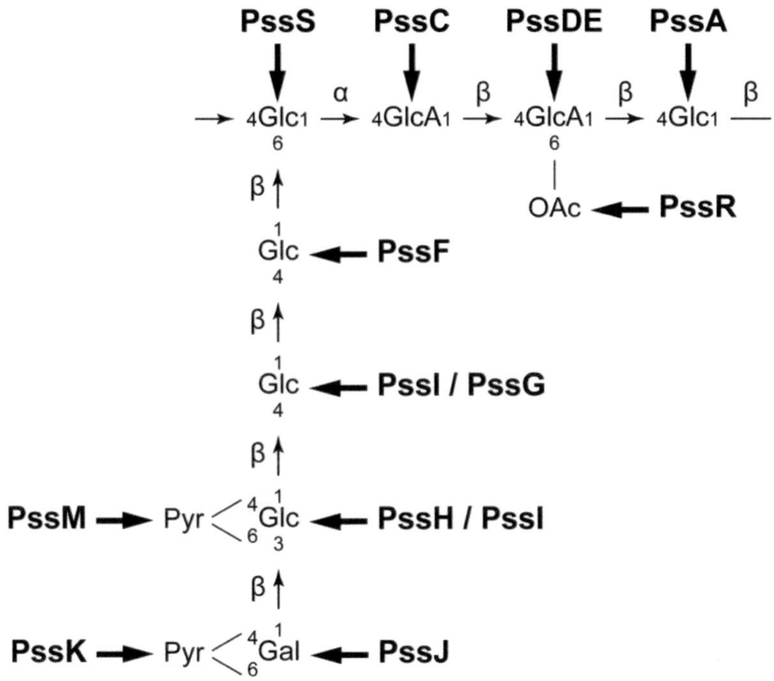

Figure 2. Model for the EPS repeating unit assembly in *Rlv* VF39 and *Rlt* TA1 strains. Abbreviations: Glc, glucose; GlcA, glucuronic acid; Gal, galactose and Pyr, ketal pyruvate group.

6. Genes involved in modification of EPS

Three genes, *pssR*, *pssM*, and *pssK*, were identified within the Pss-I cluster that may be involved in the modification of EPS in all *R. leguminosarum* strains as well as in *R. etli* CFN42 (Fig. 1). A homology search for the *pssR* gene product revealed similarity of its C-terminal region (amino acids 87-136) with the corresponding parts of a large number of acetyltransferases, including well characterized rhizobial NodL [71-73], *E. coli* LacA and CysE [74,75], *S. aureus* Cap1G [76]. The hexapeptide repeat [LIV]-G-x(4) (IPR001451, bacterial transferase hexapeptide repeat) is present in the homologous regions of all these proteins including the PssR. It should be noted that PssR shares similarity over nearly its entire length with conservative putative acetyltransferases from different representatives of *Bacillaceae*. Proteins from this family use acetyl coenzyme A as the acetyl donor and acetylate different substrates including capsular and extracellular polysaccharides, lipooligosaccharides, chitin fragments, N-acetylglucosamine and antibiotics [64,77,78].

Notably *pssR* orthologs were found in all PssI-clusters (Fig. 1). This observation is in agreement with the data concerning the major site of O-acetylation localized at the second GlcA residue in the backbone chains which are identical in all EPS with known structure.

Insertional inactivation of *pssR* in the *Rlv* VF39 genome does not result in a complete absence of acetyl groups in EPS. This suggests the existence of other gene(s) elsewhere in the *Rlv* VF39 genome needed for the EPS acetylation. Decreasing of the level of acetylation has no effect on nodule development and nitrogen fixation. Similar data were obtained for *S. meliloti* ExoZ mutants, which failed to acetylate succinoglycan. It was shown that the acetyl decoration of succinoglycan is not absolutely required for a nodule formation; however it increased the efficiency of infection threads initiation [79,80].

The amino acid sequence of PssM shares homology with several known and putative ketal pyruvate transferases, including ExoV from *S. meliloti* and GumL from *Xanthomonas campestris*. Knock-out of the *pssM* gene does not result in the loss of ability to produce HMW EPS, but leads to the absence of the pyruvic acid ketal group at subterminal glucose in the repeating unit of EPS as it was shown by ^{13}C and ^{1}H NMR analyses. Complementation *in trans* restored the EPS modification in the *pssM* mutant [81]. Disruption of the *pssM* gene led to essential disturbances in symbiosis. Thus, the *pssM* mutation resulted in the formation of aberrant non-nitrogen-fixing nodules on peas. Ultrastructural studies of mutant nodules indicated that the infection thread formation, release of bacteria into the plant cell cytoplasm and early steps of differentiation of bacteroids were not affected. However, further stages in the symbiosome development and maintenance were arrested. We proposed that the induction of early senescence of symbiosomes depends on the failure in recognition mechanisms and, what is essential, that recognition of a micro-symbiont by the host plant is important not only at early stages of symbiosis, but also during its intracellular period of life [81]. Moreover, an accumulation of very large starch granules observed in infected and non-infected cells, suggests that the plant-derived photosynthates, which serve as an energy source for nitrogen fixation [82] are not fully consumed in *pssM* induced nodules. The mechanisms which modify the "symbiotic" nodule to starch accumulation may include alteration in the starch phosphorylase activity and (or) its expression [83].

Our finding that mutation in *pssM* abolishes pyruvylation of only one of the two sugar residues in *Rlv* VF39 EPS permits to propose that pyruvylation of the terminal galactose may be controlled by the *pssK* gene localized within the Pss-I cluster. The PssK amino acid sequence was similar to proteins containing the pyruvyltransferase domain IPR007345, including Pvg1p from *Shizosaccharomyces pombe*, YveS, YvfF and YxaB from *Bacillus subtilis* [84], and EpsL from *Streptococcus thermophilus* [85]. It was shown that Pvg1 catalysed the transfer of the pyruvyl group to Galβ1,3-residues in N-linked galactomannan chains [86]. Interestingly, no sequence homology was observed between the PssM and PssK proteins that can reflect different substrate specificities of these enzymes. No direct evidence for the *pssK* function was obtained in any of the *R. leguminosarum* strains. Our preliminary data indicate that knock-out of *pssK* abolishes the EPS synthesis and results in a non-slimy phenotype of colonies (Ivashina et al., unpublished). It is possible that pyruvyl modification of the terminal sugar residue may be necessary for the efficient polymerization or export of EPS as it was proposed for *S. meliloti exoV* which is involved in pyruvylation of succinoglycan [19,21].

As seen from Figure 1, in the *Re* CNPAF512 and *Re* CIAT 652 the *pssK* gene is absent and *pssM* is replaced by non-orthologous *psaG* and *psaF* genes, respectively. The latter genes presumably can also encode ketal pyruvate transferases since IPROO7345 domain (Polysacch_pyruvyl_Trfase) was found in their amino acid sequences. Unfortunately, EPS structure of both *R. elti* strains remains unknown, but one can suppose that at least in their side chains it differs from that of *Rlv* VF39.

This assumption is based on the observation that the sets of GTs in their Pss-I clusters differ from that of *Rlv* VF39. One can see that genes of ketal pyruvyl transferases are different also. Therefore, this finding additionally argues towards high substrate specificity of these modifying enzymes.

7. Polymerization and secretion of EPS

At present three pathways are known for the export of carbohydrate polymers in bacteria: (i) Wzx/Wzy-dependent; (iii) ATP-binding cassette (ABC) transporter-dependent; and (iii) synthase-dependent (reviewed in detail by [87,88). In the Wzx/Wzy dependent mode individual undecaprenol diphosphate-linked polysaccharide repeating units are assembled and translocated across the cytoplasmic membrane by a transport process requiring a Wzx protein (putative translocase or "flippase") followed by their polymerization at the periplasmic space by the Wzy protein [87,89]. Further export of polysaccharides from the periplasm to the cell surface has been shown to be dependent upon additional protein(s) assigned to the polysaccharide co-polymerase (PCP) and the outer membrane polysaccharide export (OPX; formerly OMA) families [87,89]. The best characterized member of the OPX family is an *E. coli* K30 outer membrane lipoprotein Wza which forms a multimeric 'secretin-like' structure mediating translocation of the group 1 capsular polysaccharide across the outer membrane. The high-resolution crystal structure of Wza has been determined and this shed light on the CPS traffic across the outer membrane [90]. It has been postulated that the Wza protein together with co-polymerase Wzc form a molecular scaffold that spans the cell envelope and promotes the export of CPS (reviewed by [87]).

Current data indicate that polymerization and secretion of acidic EPS in *R. leguminosarum* biovars might be realized in a Wzx/Wzy-dependent manner. This supposition is based on the structural similarity of *R. leguminosarum* proteins PssTNOP and PssL to enzymes involved in CPC/EPS biosynthesis. The main data concerning the elucidation of the role of the mentioned proteins in EPS biosynthesis were obtained in the group of A. Skorupska.

The precise function of PssL has not been determined due to the inability to knock-out the *pssL* gene in the *Rlt* TA1 strain. However, the amino acid sequence similarity and hypothetical protein secondary structure allow placing the PssL protein within Wzx-like translocases that belong to the polysaccharide specific transport (PST) family. The predicted secondary structure of the *Rlt* TA1 PssL inner membrane protein has been supported experimentally with a series of PssL-PhoA and PssL-LacZ translational fusions. The obtained results clearly show that PssL displays characteristic features of members of the PST protein family comprising transporters with 12 membrane spanning segments, a large cytoplasmic domain, located between the sixth and seventh transmembrane segments, and amino and carboxyl termini located in the cytoplasm [91].

In addition to *pssL*, four closely linked *pssTNOP* genes were identified in the Pss-I cluster of various representatives of *R. leguminosarum* (Fig. 1) and assigned to be involved in polymerization and export of EPS [42,92,93]. The PssT protein has been predicted to be a Wzy-like protein that together with PssL might be responsible for Wzx/Wzy-like-dependent EPS polymerization and translocation. This conclusion is based on structural homology of PssT with inner membrane proteins belonging to the PST family of proteins that are involved in transport of complex polysaccharides [42]. The PssT consists of 12 transmembrane helices, a large periplasmic loop between the ninth and tenth transmembrane segments, and cytoplasmic N- and C-termini. The predicted topology of PssT has been confirmed with the use of a series of PssT-PhoA fusion proteins and a complementary set of PssT-LacZ fusions. The role of PssT in EPS biosynthesis has been investigated further by plasmid integration mutagenesis. The *Rlt* TA1 *pssT* mutant lacking the C-terminal part of PssT (starting after the 363-rd amino acid located in the periplasmic loop) produced increased amounts of total EPS with an altered distribution of high- and low-molecular-weight forms in comparison to the wild-type strain [42]. The PssT was structurally and functionally homologous to *S. meliloti* ExoT, which together with ExoP and ExoQ proteins is involved in the final stages of succinoglycan biosynthesis [21,94].

The PssP protein displays significant structural features with members of the copolymerase (PCP2a) family that are involved in the synthesis of high-molecular-weight CPS/EPS including the well characterized ExoP protein from *S. meliloti* and a Wzc protein from *E. coli* [95-98]. Membrane topology of the PssP protein resembles that of ExoP. Both proteins consist of a periplasmic hydrophilic N-terminal domain flanked by two potential transmembrane helices and a cytoplasmic C-terminal domain. The C-terminus contains the conserved Walker motifs A and B for ATP binding. Coiled-coil regions characteristic of PCP2a members were found both in periplasmic and cytoplasmic C-terminal domains of PssP [97]. ExoP has been shown to be an autophosphorylating protein tyrosine kinase. Site-directed mutagenesis of specific tyrosine residues in the cytoplasmic domain of ExoP has been demonstrated to result in an altered ratio of LMW succinoglycan to HMW succinoglycan [98]. It has been hypothesized that the phosphorylation state of ExoP might regulate the degree of succinoglycan polymerization by controlling polymerization activities of other proteins, e.g., ExoQ and ExoT [94]. A putative site for tyrosine phosphorylation has been found in the PssP protein, however, the functional significance of this site for phosphorylation of PssP is still unknown. Unlike ExoP, no tyrosine-rich region is found at the C-terminus of PssP.

Several mutations have been introduced into the *Rlt* TA1 *pssP* gene and shown to display different effects. The *Rlt* TA1 mutant with the deletion of the entire coding region of *pssP* is deficient in EPS production. A mutant that synthesizes a functional N-terminal periplasmic domain but lacks the C-terminal part of PssP produces significantly reduced amounts of EPS with a slightly changed low- to high-molecular form ratio. A *pssP* mutant with the disrupted 5'-end of the gene synthesizes exclusively low-molecular-weight EPS suggesting the importance of the functional N-terminal domain in the degree of polymerization [99].

The *pssN* gene encodes a protein which is homologous to the outer membrane polysaccharide export OPX protein family involved in CPC/EPS export [96,99]. Like other members of the OPX family, PssN contains a conserved signal peptidase II cleavage site in the lipobox. With the use of *pssN-phoA* and *pssN-lacZ* gene fusions and *in vivo* acylation with [^{3}H]-palmitate it has been shown that PssN is a lipoprotein associated with the outer membrane and with the N-terminal signal sequence directed to the periplasm. Several experimental approaches (indirect immunofluorescence with anti-PssN and fluorescein isothiocyanate-conjugated antibodies and protease digestion of spheroplasts and intact cells of *Rlt* TA1) indicated that PssN is not exposed to the surface, but oriented towards the periplasmic space. Investigation of the secondary structure of the purified PssN-His$_6$ protein by Fourier transform infrared spectroscopy (FTIR) revealed the predominant presence of beta-structure; however, alpha-helices, which could be involved in association with murein and/or other proteins, were also detected. Similar to OPX proteins, PssN has been shown to exist in a homo-oligomeric form of at least two monomers suggesting that together with PssP it might be involved in the formation of efflux channels for EPS export. No *pssN* mutants have been obtained so far. However, the increased amount of the PssN protein in *Rlt* TA1 correlated with a moderate enhancement of EPS production [92,99].

It was hypothesized by Mazur and co-workers [42], that the PssT protein, acting in complex with PssP and PssN, could be involved in controlling the rate of polymerization of repeating units and export of EPS to the cell surface. The PssN could interact with the periplasmic loop of the PssP protein, whereas the transmembrane regions of PssP could associate with the corresponding PST transporter, facilitating polymer export across the bilayer sructure.

The *pssO* gene product reveals no homology with known bacterial proteins. However, its participation in EPS biosynthesis has been confirmed by mutagenesis analysis: deletion of *pssO* in *Rlt* TA1 abolished EPS production and overproduction of PssO increased EPS secretion. Subcellular fractionation, *pssO–phoA* and *pssO-LacZ* translational fusion analyses and immunolocalisation of PssO on the *Rlt* TA1 cell surface by electron microscopy demonstrated that PssO is secreted to the extracellular medium and remains attached to the cell. The secondary structure of PssO–His$_6$, as determined by FTIR spectroscopy, is rich in α-helices (32%) [100]. It was speculated by Marczak and co-workers, that PssO may function as a periplasmic "chaperon" coating the EPS polymer and protecting it from the action of glycanases and/or be co-transported with the polysaccharide through a channel formed in the outer membrane. However, the authors can not exclude that PssO forms some kind of a cell surface structure essential for the assembly of the EPS transporter complex and its stability [100].

Using plasmid-borne transcriptional fusions of promoters of *pss* genes with the reporter gene *lacZ,* the effect of root exudate, phosphate, and ammonia on expression of *pssT, pssN, pssO,* and *pssP* genes in the wild-type *Rlt* TA1 background was examined. A stimulating effect of these environmental factors on *pssO* and *pssP* was observed. Interestingly, within the putative *pssO* promoter the divergent *nod*-box element was found. The *pssO* promoter was slightly inducible in a flavonoid-dependent manner in wild-type strains *Rlt* TA1 and *Rlt* 843 and very weakly in a mutant of *Rlt* 843 that lacks the regulatory *nodD* gene. The regulation of EPS production by NodD might be an important finding that connects EPS synthesis to the symbiosis of *R. leguminosarum* with clovers [101].

pssTNOP genes from *Rlt* TA1 have corresponding orthologs in genomes of *R. leguminosarum* and *R. etli* (Fig. 1) suggesting that there is a common mechanism of their action at least in these strains. The *pssL* gene is not so conserved: in *Re* CNPAF512 and *Re* CIAT 652 it was replaced by the non-homologous gene designated as *psaI*. However, the PsaI protein can be assigned to the same family IPROO2797 (Polysacc_synth.) as PssL. As mentioned above, these are just the strains for which EPS side chains were predicted to have different structure. We suggest that namely PssL/PsaI can be specific for the structure of EPS to be translocated across the inner membrane to the periplasmic space. Probably the main function of PssL-like proteins consists in the stringent control of the identity of repeating units which are further polymerized by the action of PssT.

Noteworthy, another gene designated as *psaA* was found in the *pssV-E* operons of *R. etli* strains described here. It encodes the protein, which can be assigned to the O-antigen ligase-like protein family PF13425. The function of this protein in EPS biosynthesis still remains unclear, however its participation in translocation or polymerization of the EPS repeating units can not be excluded. It should be noted that *psaA* homologs were found in all *R. leguminosarum* genomes under consideration, but they localize in different chromosomal regions far from the Pss-I cluster.

8. Processing of EPS

It was well documented that several rhizobial species (e.g., *S. meliloti, R. leguminosarum, Rhizobium sp.* NGR234, *Bradyrhizobium*) produce two distinct EPS classes that differ in size: HMW forms symbiotically active LMW forms [13-17,27,102]. LMW forms of succinoglycan can be produced in *S. meliloti* either by direct export of the low-polymerized octasaccharide repeating units [95,103] or by depolymerization of HMW polysaccharide by specific glycanases [104,105]. These observations are also true for *R. leguminosarum* model. As discussed above, mutations in the genes controlling the polymerization and transport of EPS in the *Rlt* TA1 strain could contribute to the production of LMW EPS [93].

Moreover, three gene products have been shown to participate in degradation of HMW EPS in *R. leguminosarum*: PlyA, PlyB, and PssW (formely PssT in *Rlv* VF39 and PssT1 in *Re* CFN42). PlyA and PlyB, which are similar to each other display homology to bacterial and fungal polysaccharide lyases [106,107]. Ten copies of a novel heptapeptide repeat motif were found in the sequences of these proteins which may constitute a fold similar to that found in the family of extracellular pectate lyases. These proteins are secreted via the PrsDE system of the Type I secretion system which is conserved in different *Rhizobium* species [108]. PlyA appears to remain attached to the cells, while the PlyB diffuses beyond the edge of the colony [109]. It has been proposed that the presence of extra 50 amino acids near the C-terminal domain of PlyA could be responsible for maintaining the protein attachment to the cell surface. Both proteins were inactive in the EPS-defective mutants and did not degrade mature EPS. They may be only active in association with the rhizobial cell surface suggesting the activation of PlyB by an EPS-related (nascent EPS or an intermediate in EPS biosynthesis) component [109]. The PlyA and PlyB glycanases are not specific for EPS but can also degrade carboxymethyl cellulose (CMC). In cultured bacteria the *plyA* gene is expressed at a very low level, while a *plyB* mutant has a very large reduction in degradation of EPS and CMC [106]. Cultures of *plyB* mutants contained an increased ratio of EPS repeating units to the reducing ends indicating that EPS was present in a longer-chain form, and this correlated with a significant increase in the culture viscosity. A double *plyAB* mutant retained residual CMC degradation, indicating the existence of additional activities in the cell. Recently, a third gene named *plyC* has been found in the genome of *Rlv* 3841. PlyC is secreted via the PrsDE secretion system and displays common structural features with PlyA and PlyB proteins [107], indicating that it may perform a similar function. The analysis of the symbiotic properties of a *plyAB* double mutant revealed that genes involved are not required for symbiotic nitrogen fixation and that nodulation was not significantly affected [109]. In *Rlt* TA1 the *plyA* gene is missing.

Recently we have characterized another glycosylhydrolase encoded by the *pssW* gene and provided experimental evidence for its participation in the EPS processing (Kanapina et al, unpublished). The *pssW* gene has its counterparts in *R. leguminosarum* and *R. etli* genomes, and is located within the *pssV-E* operon. The PssW protein was referred to the family 10 of glycosylhydrolases of the GH-A clan, which are retaining glycoside hydrolases displaying endo-1,3-β-xylanase and endo-1,4-β-xylanase activities (CAZy database). We have shown that PssW is synthesized as a precursor of 42.4 kDa followed by its translocation across the cytoplasmic membrane and by cleavage of the 46 amino acid signal peptide. The periplasmic localization of the PssW indicates that it might be active toward the nascent EPS before its secretion outside the cell. The deletion of the *pssW* gene resulted in approximately a 3-fold increase in the ratio of HMW to LMW EPS and as a result in the increase of viscosity of the culture supernatant. Complementation of the *pssW* mutation restored the wild-type phenotype, and even increased the level of secreted LMW EPS. The PssW purified from the periplasmic space did not degrade CMC and succinoglycan, and revealed a 2-fold decrease in hydrolysis of EPS in the *Rlv* VF39 *pssM* mutant which lacks one of the two pyruvyl groups. The latter suggests the importance of pyruvyl modification on the degradation activity of the PssW. The knock-out of *pssW* did not significantly affect nodulation of peas probably due to the incomplete block in the LMW EPS synthesis.

It can be concluded that *R. leguminosarum* strains like *S. meliloti* were able to realize different strategies for production of LMW forms of polysaccharide: regulation of the degree of polymerization of EPS and hydrolase-mediating cleaving of EPS. Complex mechanisms directing the synthesis of LMW EPS can reflect the evolutionary benefit of rhizobia possessing different pathways of LMW EPS production that can be considered as important molecules for cell to cell communications during the development of nitrogen-fixing nodules.

9. EPS-deficient mutants as a possible model for studying bacterial gene expression in symbiosis

A decade ago in collaboration with the group of Prof. B. Rolfe we demonstrated that mutations in the *pssA* gene in *Rlv* VF39 and *Rlt* ANU794 strains led not only to abolishing the capacity of these strains to synthesize EPS but also to induction or up-regulation of at minimum 22 proteins. The differences identified in the *pssE* and *pssD* mutants of the *Rlv* VF39 strain were a distinct subset of the same protein synthesis changes that occurred in the *pssA* mutant (9 out of 22 changes) [110]. Genetic complementation of *pssA* restored wild-type protein synthesis levels. We concluded that the observed alterations in protein synthesis are caused either by dysfunction of the PssA protein itself or are a response to the absence of EPS.

The N-terminal sequence analysis of 15 members of the *pssA* mutant stimulon was performed and the unique amino acid sequences were determined for 11 proteins [110]. However, our attempts to identify these proteins were unsuccessful, and it was clear why. None of *R. leguminosarum* and *R. etli* genomes was sequenced at that time. We have repeated our attempts now and assigned 9 proteins to their known orthologs in *Rhizobiaceae* species. The results of this analysis are summarized in Table 3. The functions of all of these orthologs are predicted. Five proteins belong to ABC transporter systems; two proteins are assigned as NADH-dependent FMN reductases, one as taurine dioxygenase, and the last as disulfide isomerase. Most genes encoding the proteins were mapped on indigenous plasmids.

It is interesting why mutation in a single gene resulted in induction or up-regulation of genes encoding proteins with such different functions. We propose here the following hypothesis. It is based on the data that the expression of the *pssA* gene was not observed in bacteroids [63,111]. In our opinion, members of the *pssA* mutant stimulon might be just those proteins which are expressed at the symbiotic state of bacteria. In this case, PssA, directly or indirectly, could play the role of a negative regulator of their gene expression in free-living bacteria. EPS itself could play the same role.

Obviously, the hypothesis can be easily verified. It is sufficient to examine whether the genes listed in Table 3 are expressed in bacteroids. If our assumption turns out to be correct, EPS-deficient mutants of *R. leguminosarum* could be considered as an attracting model for studying bacterial gene expression during symbiosis.

Rlt ANU437 mutant-responsive proteins			Characteristics of *Rlt* ANU437 orthologs from the other *Rhizobiaceae* species					
Spot/status[a]	M_r/pI[*]	N-terminal sequence (position in the ortholog)[**]	Protein (accession number)/organism	M_r/pI[***]	Signal peptide position	Domain/motif	Predicted function	Gene location
n2/up-reg. (+)	37.436/6.31	**QITDAPLLSKV** 2[illegible] 36	PRL100398 (YP_770674)/*Rlv* 3841	34,376/5.65	1-26	IPRO15168, NMT1/TH15-like	ABC transporter protein	pRL10
n4/induced n6/induced	22,147/5.90 22,147/6.23	**SNDLIVGFSG** 2 11 **SNDLIVGFSGN**L 2 13	ATCR1_07864 (EHH06861)/*Agrobacterium tumefaciens* CCNWWGS0386	19,791/5.35	-	IPRO05025, FMN_red	NADH-dependent FMN reductase	Not known
n5/induced n7/induced	22,880/6.09 22,880/6.38	**SAPRLVGLAGSF** 2 13 **SAPRLVGLAGSF** 2 13	PRL90304 (YP_765590)/*Rlv* 3841	20,211/5.46	-	IPRO05025, FMN_red	NADH-dependent FMN reductase	pRL9
n8/up-reg. (++)	34,000/7.50	**SNPVLVNQIIPES**RVV(G/T)**LT**(A/D)S**V** 2 23	Rleg2_5751 (YP_002278025)/*Rlt* WSM2304	33,910/6.48	-	IPRO03819, Taurine_dOase	Taurine dioxygenase	pRLG202
n15/up-reg. (++) n21/up-reg. (+++)	34,321/5.15 34,197/5.02	**ADKKVVVAYQTDALP** 21 35 **ADKKVVVAYQTDALP** 21 35	PRL100239 (YP_770519)/*Rlv* 3841	34,400/4.94	1-20	IPRO15168, NMT1/TH15-like	ABC transporter protein	pRL10
n16/induced	30,566/5.11	**D**XX**QQKKEFGEFIKQYL** 25 42	RL2091 (YP_767689)/*Rlv* 3841	24,820/5.04	1-25	IPRO01853, DSB-like_Thioredoxin_dom	Disulfide isomerase	chromosome
n17/up-reg. (+) n22/up-reg. (+++)	30,456/4.97 114,862/4.99	**EDKSIKVGIMAGEEE** 40 55 **EDKSIKVGIMAGEEEDIXRVVASEAAK** 40 66	PRL90303 (YP_765589)/*Rlv* 3841	26,417/4.86	1-39	IPRO04872, Lipoprotein NlpA	Lipoprotein, component of ATP-driven transport system	pRL9
n18/up-reg. (+++)	45,109/4.78	**DGLSGAPAPFDKGGV** 25 39	PRL90245 (YP_765531)/*Rlv* 3841	34,271/4.69	1-24	PF13407, Peripla_BP_4	Periplasmic solute-binding component of ABC transporter	pRL9
n19/up-reg. (++)	34,633/4.94	**TDTVKLRYLASQGNL** 25 39	Rleg_6388 (YP_002979383)/*Rlt* WSM1325	32,191/4.98	1-25	IPRO15168, NMT1/TH15-like	ABC transporter protein	pR132505

[*] Spot number: arbitrary numbers were assigned to proteins which exhibited a change in their relative synthesis levels in response to a Tn*5* insertion in the *pssA* gene of *Rlt* ANU794. Status: induced, when protein is absent in the wt strain and synthesized in the mutant strain; up-reg (+, ++, or +++), when 2-, 3- or 5-fold (or more) level of individual protein synthesis was observed in the mutant strain. Observed M_r and pI values of the protein based on the migration of proteins in gels. Data were taken from Table 2 [110].

[**] Data on N-terminal amino acid sequences were taken from Table 3 [110]. Identical amino acids in the matched sequences in the ortholog proteins are shown in bold letters, and similar amino acids are underlined. Positions of these sequences in the orthologs are indicated bellow the sequences.

[***] M_r and pI of the ortholog proteins were calculated from their amino acid sequences.

Table 3. Identification of proteins from the *pssA* mutant stimulon in *Rhizobiaceae* species.

10. Conclusion

The EPSs described in this review have common structure in that their repeating units possess identical backbone chains and the same β1-6-linked glucosyl residue starting the side chain. The diversity of EPSs is specified by the structure of their side chains. Presumably, structural information contained in side chains determines participation of EPS in symbiosis as signaling factors. This assumption follows from at least the observation that the absence of a single pyruvyl group in the side chain dramatically disturbed symbiotic properties of bacteria.

Most of the *R. leguminosarum* and *R. etli* genes involved in EPS biosynthesis are localized within the single cluster Pss-I. Gene arrangement in this cluster is similar. This implies that Pss-I clusters of these bacterial species have been evolved from a common ancestor. The differences mostly include genetic rearrangements of GTs genes, genes for modification of certain glycosyl residues in the side chain, and genes *pssL/psaI* thought to be involved in controlling translocation of the repeating units. It seems that each variant of Pss-I cluster originated as a result of co-evolution of all these genes.

Author details

Tanya V. Ivashina*
Skryabin Institute of Biochemistry and Physiology of Microorganisms, Russian Academy of Sciences, Pushchino, Moscow Region, Russia

Vladimir N. Ksenzenko
Institute of Protein Research, Russian Academy of Sciences, Pushchino, Moscow Region, Russia

Acknowledgement

We are thankful to Yu.A. Knirel for providing an information dealing with rhizobia EPS structures and A.S. Glukhov, T.B. Kuvshinkina and L.A. Kulakov for the help in the manuscript preparation. This work was supported by grant 11-04-02120a from the Russian Foundation for Basic Research.

11. References

[1] Weir BS (2011) The Current Taxonomy of Rhizobia. New Zealand Rhizobia Website. Available: http://www.rhizobia.co.nz/taxonomy/rhizobia.html.

[2] Zevenhuizen LP, Scholten-Koerselman HJ (1979) Surface Carbohydrates of *Rhizobium*. I. Beta-1, 2-glucans. Antonie Van Leeuwenhoek. 45: 165-175.

[3] Breedveld M, Miller KJ (1994) Cyclic β-Glucans of Members of the Family *Rhizobiaceae*. Microbiol. Rev. 58: 145-161.

* Corresponding Author

[4] Laus MC, Logman TJ, Lamers GE, Van Brussel AA, Carlson RW, Kijne JW (2006) A Novel Polar Surface Polysaccharide from *Rhizobium leguminosarum* Binds Host Plant Lectin. Mol. Microbiol. 59: 1704-1713.

[5] Williams A, Wilkinson A, Krehenbrink M, Russo DM, Zorreguieta A, Downie JA (2008) Glucomannan-Mediated Attachment of *Rhizobium leguminosarum* to Pea Root Hairs is Required for Competitive Nodule Infection. J. Bacteriol. 190: 4706-4715.

[6] Breedveld MW, Cremers HC, Batley M, Posthumus MA, Zevenhuizen LP, Wijffelman CA, Zehnder AJ (1993) Polysaccharide Synthesis in Relation to Nodulation Behavior of *Rhizobium leguminosarum*. J. Bacteriol. 175: 750-757.

[7] Zevenhuizen LPTM, van Neerven ARW (1983) Surface Carbohydrates of *Rhizobium*. IV. Gel-Forming Capsular Polysaccharide of *Rhizobium leguminosarum* and *Rhizobium trifolii*. Carbohydr. Res. 124: 166-171.

[8] Campbell GR, Reuhs BL, Walker GC (2002) Chronic Intracellular infection of Alfalfa Nodules by *Sinorhizobium meliloti* Requires Correct Lipopolysaccharide Core. Proc. Natl. Acad. Sci. USA. 99: 3938-3943.

[9] Fraysse N, Couderc F, Poinsot V (2003) Surface Polysaccharide Involvement in Establishing the *Rhizobium*-Legume Symbiosis. Eur. J. Biochem. 270: 1365-1380.

[10] Lepek VC, D'Antuono AL (2005) Bacterial Surface Polysaccharides and Their Role in the Rhizobia-Legume Association. Lotus Newsletter. 35: 93-105.

[11] Skorupska A, Janczarek M, Marczak M, Mazur A, Król J (2006) Rhizobial Exopolysaccharides: Genetic Control and Symbiotic Functions. Microb. Cell Factories. 5:

[12] Gage DJ (2004) Infection and Invasion of Roots by Symbiotic, Nitrogen-Fixing Rhizobia During Nodulation of Temperate Legumes. Microbiol. Mol. Biol. Rev. 68: 280-300.

[13] Djordjevic SP, Chen H, Batley, M, Redmond JW, Rolfe BG (1987) Nitrogen Fixation Ability of Exopolysaccharide Synthesis Mutants of *Rhizobium* sp. Strain NGR234 and *Rhizobium trifolii* is Restored by the Edition of Homologous Exopolysaccharides. J. Bacteriol. 169: 53-60.

[14] Battisti L, Lara JC, Leigh JA (1992) Specific Oligosaccharide Form of the *Rhizobium meliloti* Exopolysaccharide Promotes Nodule Invasion in Alfalfa. Proc. Natl. Acad. Sci. USA. 89: 5625-5629.

[15] Urzainqui A, Walker GC (1992) Exogenous Suppression of the Symbiotic Deficiencies of *Rhizobium meliloti exo* Mutants. J. Bacteriol. 174: 3403-3406.

[16] Wang L-X, Wang Y, Pellock B, Walker GC (1999) Structural Characterization of the Symbiotically Important Low-molecular Weight Succinoglycan of *Sinorhizobium meliloti*. J. Bacteriol. 181: 6788-6796.

[17] Staehelin C, Forsberg LS, D'Haeze W, Gao MY, Carlson RW, Xie ZP, Pellock BJ, Jones KM, Walker GC, Streit WR, Broughton WJ (2006) Exo-oligosaccharides of *Rhizobium* sp. Strain NGR234 are Required for Symbiosis with Various Legumes. J. Bacteriol. 188: 6168-6178.

[18] Becker A, Puhler A (1998) Production of Exopolysaccharides. In: Spaink HP, Kondorosi A, Hooykaas JJ, editors. The *Rhizobiaceae*. Kluwer Academic Publishers, Dordrecht, the Netherlands. pp. 97–118.

[19] Becker A, Kleickmann A, Kuster H, Keller M, Arnold W, Pühler A (1993) Analysis of the *Rhizobium meliloti* Genes *exoU, exoV, exoW, exoT* and *exoI* Involved in Exopolysaccharide Biosynthesis and Nodule Invasion: *exoU* and *exoW* Probably Encode Glucosyltransferases. Mol. Plant-Microbe Interact. 6: 735-744.

[20] Glucksmann MA, Reuber TL, Walker GC (1993) Family of Glycosyl Transferases needed for the Synthesis of Succinoglycan by *Rhizobium meliloti*. J. Bacteriol. 175: 7033-7044.

[21] Glucksmann MA, Reuber TL, Walker GC (1993) Genes Needed for the Modification, Polymerization, Export, and Processing of Succinoglycan by *Rhizobium meliloti*: a Model for Succinoglycan Biosynthesis. J. Bacteriol. 175: 7045-7055.

[22] Jones K.M., Kabayashi H, Davies BW, Taga ME, Walker GC (2007) How Rhizobial Symbionts Invade Plants: the *Sinorhizobium-Medicago* Model. Nat. Rev. Microbiol. 5: 619–633.

[23] Long SR (2001) Genes and Signals in the *Rhizobium*-Legume Symbiosis. Plant Physiol. 125: 69-72.

[24] Janczarek M (2011) Environmental Signals and Regulatory Pathways that Influence Exopolysaccharide Production in Rhizobia. Int. J. Mol. Sci. 12: 7898-7933.

[25] Robertsen BK, Aman P, Darvill AG, McNeil M, Albersheim P (1981) Host-Symbiont Interactions. V. The Structure of Acidic Extracellular Polysaccharides Secreted by *Rhizobium leguminosarum* and *Rhizobium trifolii*. Plant Physiol. 67: 389-400.

[26] Dudman WF, Franzén L-E, McNeil M, Darvill AG, Albersheim P (1983) The Structure of the Acidic Polysaccharide Secreted by *Rhizobium phaseoli* strain 127 K87. Carbohydr. Res. 117: 169–183.

[27] McNeil M, Darvill J, Darvill A, Albersheim P, van Veen R, Hooykaas P, Schilperoort R, Dell A (1986) The Discernible Structural Features of the Acidic Exopolysaccharides Secreted by Different *Rhizobium* Species are the Same. Carbohydr. Res. 146: 307-326.

[28] Philip-Hollingsworth S, Hollingsworth RI, Dazzo FB (1989) Host-Range Related Structural Features of the Acidic Extracellular Polysaccharides of *Rhizobium trifolii* and *Rhizobium leguminosarum*. J. Biol. Chem. 264: 1461-1466.

[29] Philip-Hollingsworth S, Hollingsworth RI, Dazzo FB, Djordjevic MA, Rolfe BG (1989) The Effect of Interspecies Transfer of *Rhizobium* Host-Specific Nodulation Genes on Acidic Polysaccharide Structure and *In Situ* Binding by Host Lectin. J. Biol. Chem. 264: 5710-5714.

[30] Carlson RW, Lee RP (1983) A Comparison of the Surface Polysaccharides from *Rhizobium leguminosarum* 128C53 Sm^r Rif^r with the Surface Polysaccharides from Its exo^{-1} Mutant. Plant Physiol. 71: 223-228.

[31] Orgambide G, Philip-Hollingsworth S, Cargill L, Dazzo F (1992) Evaluation of Acidic Heteropolysaccharide Structures in *Rhizobium leguminosarum* Biovars Altered in Nodulation Genes and Host range. Mol. Plant-Microbe Interact. 5: 484-488.

[32] Canter-Cremers HC, Batley M, Redmond JW, Wijfjes AH, Lugtenberg BJ, Wijffelman CA (1991) Distribution of O-acetyl Groups in the Exopolysaccharide Synthesized by *Rhizobium leguminosarum* Strains is not Determined by the Sym Plasmid. J. Biol. Chem. 266: 9556-9564.

[33] O'Neill MA, Darvill AG, Albersheim P (1991) The Degree of Esterification and Point of Substitution by *O*-acetyl and *O*-(3-hydroxybutanoyl) Groups in the Acidic Extracellular Polysaccharides Secreted by *Rhizobium leguminosarum* Biovars *viciae, trifolii,* and *phaseoli* are not Related to Host-Range. J. Biol. Chem. 266: 9549-9555.

[34] Gil-Serrano A, Sanchez del Junco A, Tejero-Mateo P, Megias M, Caviedes MA (1990) Structure of the Extracellular Polysaccharide Secreted by *Rhizobium leguminosarum* var. *phaseoli* CIAT 899. Carbohydr Res. 204:103-107.

[35] Lopez-Lara IM, Orgambide G, Dazzo FB, Olivares J, Toro N (1993) Characterization and Symbiotic Importance of Acidic Extracellular Polysaccharides of *Rhizobium* sp. Strain GRH2 Isolated from Acacia Nodules. J. Bacteriol. 175: 2826-2832.

[36] Ksenzenko VN, Ivashina TV, Dubeikovskaya ZA, Ivanov SG, Nanazashvili MB, Druzhinina TN, Kalinchuk, NA, Shibaev VN (2007) The *pssA* Gene Encodes UDP-glucose:polyprenyl Phosphate-glucosyl Phosphotransferase Initiating Biosynthesis of *Rhizobium leguminosarum* Exopolysaccharide. Rus. J. Bioorg. Chem. 33: 150-155.

[37] Amemura A, Tokuya H, Abe M, Shiro H (1983) Structural Studies on the Extracellular Acidic Polysaccharide from *Rhizobium trifolii* 4S. Carbohydr. Res. 115: 165–174

[38] Canter Cremers HC, Stevens K, Lugtenberg BJ, Wijffelman CA, Batley M, Redmond JW, Breedveld MW, Zevenhuizen LP (1991) Unusual Structure of the Exopolysaccharide of *Rhizobium leguminosarum* bv. *viciae* Strain 248. Carbohydr. Res. 218: 185-200.

[39] Franzén L-E, Dudman WF, McNeil M, Darvill AG, Albersheim P (1983) The Structure of the Acidic Polysaccharide Secreted by *Rhizobium phaseoli* Strain 127 K44. Carbohydr. Res. 117: 157.

[40] Aman P, Franzen L-E, Darvill JE, McNeil M, Darvill AG, Albersheim P (1982) The Structure of the Acidic Polysaccharide Secreted by *Rhizobium phaseoli* strain 127 K38. Carbohydr. Res. 103: 77-100.

[41] Ziegler T, Eckhardt E, Keller D (1997) Rhizobial Saccharides.3. Unexpected Alpha-Selective Glycosylation During Synthesis of Pyruvated Tetrasaccharide Fragments Related to Exopolysaccharides of *Rhizobium leguminosarum.* J. Carbohydr. Chem. 16: 719-738

[42] Mazur A, Król JE, Marczak M, Skorupska A (2003) Membrane Topology of PssT, the Transmembrane Protein Component of the Type I Exopolysaccharide Transport System in *Rhizobium leguminosarum* bv. *trifolii* strain TA1. J. Bacteriol. 185: 2503-2511.

[43] Whitfield C (2006) Biosynthesis and Assembly of Capsular Polysaccharides in *Escherichia coli.* Ann. Rev. Biochem. 75: 39-68.

[44] Borthakur D, Barker RF, Latchford JW, Rossen L, Johnston AW (1988) Analysis of *pss* Genes of *Rhizobium leguminosarum* Required for Exopolysaccharide Synthesis and Nodulation of Peas: Their Primary Structure and Their Interaction with *psi* and Other Nodulation Genes. Mol. Gen. Genet. 213: 155-162.

[45] Ivashina TV, Khmelnitsky MA, Shlyapnikov MG, Kanapin AA, Ksenzenko VN (1994) The *pssA* Gene from *Rhizobium leguminosarum* bv. *viciae* VF39: Cloning, Sequence and the Possible Role in Polysaccharide Production and Nodule Formation. Gene 150: 111-116.

[46] Van Workum WAT, Canter Cremers HCJ, Wijfjes AHM, van der Kolk C, Wijffelman CA, Kijne JW (1997) Cloning and Characterization of Four Genes of *Rhizobium leguminosarum* bv. *trifolii* Involved in Exopolysaccharide Production and Nodulation. Mol. Plant-Microbe Interact. 10: 290–301.

[47] Sadykov MR, Ivashina TV, Kanapin AA, Shlyapnikov MG, Ksenzenko VN (1998) Structure-functional Organization of Exopolysaccharide Biosynthetic Genes in *Rhizobium leguminosarum* bv. *viciae* VF39. Mol. Biol. (Mosk). 32: 665-671.

[48] Król J, Mazur A, Marczak M, Skorupska A (2007) Syntenic Arrangements of the Surface Polysaccharide Biosynthesis Genes in *Rhizobium leguminosarum*. Genomics. 89: 237-247.

[49] Ivashina TV, Sadykov MR, Chatuev BM, Kanapin AA, Shlyapnikov MG, Ksenzenko VN (1996) The *Rhizobium leguminosarum bv. viciae* VF39 *pssB* Gene Product Participates in Symbiotic Fixation of Molecular Nitrogen. Dokl. Akad. Nauk (Russian). 350: 712-715.

[50] Janczarek M, Krol J, Skorupska A (1999) The *pssB* Gene Product of *Rhizobium leguminosarum* bv. *trifolii* is Homologous to a Family of Inositol Monophosphatases. FEMS Microbiol. Lett. 173: 319–325.

[51] Janczarek M, Skorupska A (2001) The *Rhizobium leguminosarum* bv. *trifolii pssB* Gene Product is an Inositol Monophosphatase that Influences Exopolysaccharide Synthesis. Arch. Microbiol. 175: 143-151.

[52] Kutkowska J, Janczarek M, Kopcinska J, Urbanik-Sypniewska T, Skorupska A (2007) Effects of *pssB* Mutation on Surface Polysaccharides and Symbiotic Phenotype of *Rhizobium leguminosarum* bv. *trifolii*. Acta Biologica Cracoviensia/Botanica. 49: 81–89.

[53] Bernhard F, Coplin DL, Geider K (1993) A Gene Cluster for Amylovoran Synthesis in *Erwinia amylovora*: Characterization and Relationship to *cps* Genes in *Erwinia stewartii*. Mol. Gen. Genet. 239: 158-168.

[54] Franklin M, Nivens DE, Weadge JT, Howell PL (2011) Biosynthesis of the *Pseudomonas aeruginosa* Extracellular Polysaccharides, Alginate, Pel, and Psl. Front. Microbiol. 2: 167.

[55] Canter Cremers HCJ, Batley M, Redmond JW, Eydems L, Breedveld MW, Zevenhuizen LPTM, Pees E, Wijffelman CA, Lugtenberg BJJ (1990) *Rhizobium leguminosarum exoB* Mutants are Deficient in the Synthesis of UDP-glucose 4′-epimerase. J. Biol. Chem. 265: 21122-21127.

[56] Sanchez-Andujar B, Coronado C, Philip-Hollingsworth S, Dazzo FB, Palomares AJ (1997) Structure and the Role in Symbiosis of the *exoB* Gene of *Rhizobium leguminosarum* bv. *trifolii*. Mol. Gen. Genet. 255: 131-140.

[57] Laus MC, Logman TJ, van Brussel AAN, Carlson RW, Azadi P, Gao MY, Kijne JW (2004) Involvement of *exo5* in Production of Surface Polysaccharides in *Rhizobium leguminosarum* and its Role in Nodulation of *Vicia sativa* subsp. Nigra. J. Bacteriol. 186: 6617–6625.

[58] Muszyński A, Laus M, Kijne JW, Carlson RW (2011) Structures of the Lipopolysaccharides from *Rhizobium leguminosarum* RBL5523 and its UDP-Glucose Dehydrogenase Mutant (*exo5*). Glycobiology. 21: 55-68.

[59] Reuber TL, Walker GC (1993) Biosynthesis of Succinoglycan, a Symbiotically Important Exopolysaccharide of *Rhizobium meliloti*. Cell. 74: 269-280.

[60] Bossio JC, De Lannino NI, Dankert MA (1986) *In vitro* Synthesis of a Lipid-linked Acetylated and Pyruvylated Oligosaccharides in *Rhizobium trifolii.* Biochem. Biophys. Res. Commun. 134: 205-21.

[61] Pollock TJ, van Workum WA, Thorne L, Mikolajczak MJ, Yamazaki M, Kijne JW, Armentrout RW (1998) Assignment of Biochemical Functions to Glycosyl Transferase Genes which are Essential for Biosynthesis of Exopolysaccharides in *Sphingomonas* Strain S88 and *Rhizobium leguminosarum*. J. Bacteriol. 180: 586-593.

[62] Abe M, Sherwood JE, Hollingsworth RI, Dazzo FB (1984) Stimulation of Clover Root Hair Infection by Lectin-Binding Oligosaccharides from the Capsular and Extracellular Polysaccharides of *Rhizobium trifolii*. J. Bacteriol. 160: 517-520.

[63] Janczarek M, Skorupska A (2004) Regulation of *pssA* and *pssB* Gene Expression in *Rhizobium leguminosarum bv. trifolii* in Response to Environmental Factors. Antonie Van Leewenhoek. 85: 217-227.

[64] Bossio JC, Semino CE, De Lannino NI, Dankert MA (1996) The *in vitro* Biosynthesis of the Exopolysaccharide Produced by *Rhizobium leguminosarum* bv. *trifolii* strain NA30. Cell. Mol. Biol. 42: 737-758.

[65] Gao XD, Moriyama S, Miura N, Dean N, Nishimura S (2008) Interaction Between the C termini of Alg13 and Alg14 Mediates Formation of the Active UDP-N-acetylglucosamine Transferase Complex. J. Biol. Chem. 283: 32534-32541.

[66] Lu J, Takahashi T, Ohoka A, Nakajima K-I, Hashimoto R, Miura N, Tachikawa H, Gao X-D (2012) Alg14 Organizes the Formation of a Multiglycosyltransferase Complex Involved in Initiation of Lipid-linked Oligosaccharide Biosynthesis. Glycobiology. 22: 504-516.

[67] van Workum WAT, van Slageren S, van Brussel AAN, Kijne JW (1998) Role of Exopolysaccharides of *Rhizobium leguminosarum* bv. *viciae* as Host Plant-Specific Molecules Required for Infection Thread Formation During Nodulation of *Vicia sativa.* Mol. Plant-Microbe Interact. 11: 1233-1241.

[68] van Workum WAT, van Slageren S, van Brussel AAN, Kijne JW (1998) Role of Exopolysaccharides of *Rhizobium leguminosarum* bv. *viciae* as Host Plant-Specific Molecules Required for Infection Thread Formation During Nodulation of *Vicia sativa.* Mol. Plant-Microbe Interact. 11: 1233-1241.

[69] Król J, Wielbo J, Mazur A, Kopcinska J, Lotocka B, Golinowski W, Skorupska A (1998) Molecular Characterization of *pssCDE* Genes of *Rhizobium leguminosarum* bv. *trifolii* Strain TA1: *pssD* Mutant is Affected in Exopolysaccharide Synthesis and Endocytosis of Bacteria. Mol. Plant-Microbe Interact. 11: 1142-1148.

[70] Saxena IM, Brown RMJr, Fevre M, Geremia RA, Henrissat B (1995) Multidomain Architecture of β-Glycosyl Transferases: Implications for Mechanism of Action. J. Bacteriol. 177: 1419–1424.

[71] Downie A (1989) The *nod*L Gene from *Rhizobium leguminosarum* is Homologous to the Acetyl Transferases Encoded by *lacA* and *cysE*. Mol. Microbiol. 3: 1649-1651.

[72] Bloemberg GV, Thomas-Oates JE, Lugtenberg, BJJ Spaink HP (1994) Nodulation Protein NodL of *Rhizobium leguminosarum* O-Acetylates Lipo-Oligosaccharides, Chitin Fragments and N-Acetylglucosamine *in vitro*. Mol. Microbiol. 11: 793-804.

[73] Bloemberg GV, Lagas RM, van Leeuwen S, van der Marel GA, van Boom JH, Lugtenberg BJ, Spaink HP (1995) Substrate Specificity and Kinetic Studies of Nodulation Protein NodL of *Rhizobium leguminosarum*. Biochemistry. 34: 12712-12720.

[74] Pye VE, Tingey AP, Robson RL, Moody PC (2004) The Structure and Mechanism of Serine Acetyltransferase from *Escherichia coli*. J. Biol. Chem. 279: 40729-40736.

[75] Wang XG, Olsen LR, Roderick SL (2002) Structure of the *lac* Operon Galactoside Acetyltransferase. Structure. 10: 581-588.

[76] Luong TT, Ouyang S, Bush K, Lee CY (2002) Type 1 Capsule Genes of *Staphylococcus aureus* are Carried in a Staphylococcal Cassette Chromosome Genetic Element. J. Bacteriol. 184: 3623-3629.

[77] Claus H, Borrow R, Achtman M, Morelli G, Kantelberg C, Longworth E, Frosch M, Vogel U (2004) Genetics of Capsule O-Acetylation in Serogroup C, W-135 and Y Meningococci. Mol. Microbiol. 51: 227-239.

[78] Rende-Fournier R, Leclereq R, Galimand M, Duval J, Courvalin P (1993) Identification of the *sat*A Gene Encoding a Streptagramin A Acetyltransferase in *Enterococcus faecium* BM4145. Antimicrob. Agents Chemother. 10: 2119-2125.

[79] Buendia A M, Enenkel B, Koplin R, Niehaus K, Arnold W, Pühler A (1991) The *Rhizobium meliloti exoZ1 exoB* Fragment of Megaplasmid 2: ExoB Functions as a UDP-glucose 4-epimerase and ExoZ Shows Homology to NodX of *Rhizobium leguminosarum* Biovar *viciae* Strain TOM. Mol. Microbiol. 5: 1519-1530.

[80] Reuber TL, Walker GC (1993) The Acetyl Substituent of Succinoglycan is not Necessary for Alfalfa Nodule Invasion by *Rhizobium meliloti* Rm1021. J. Bacteriol. 175: 3653-3655.

[81] Ivashina TV, Fedorova EE, Ashina NP, Kalinchuk NA, Druzhinina TN, Shashkov AS, Shibaev VN, Ksenzenko VN (2010) Mutation in the *pssM* Gene Encoding Ketal Pyruvate Transferase Leads to Disruption of *Rhizobium leguminosarum* bv. *viciae-Pisum sativum* Symbiosis. J. Appl. Microbiol. 109: 731-742.

[82] Postma JG, Jager D, Jacobsen E, Feenstra WJ (1990) Studies on a Non-Fixing Mutant of Pea (*Pisum sativum* L.): Phenotypical Description and Bacteroid Activity. Plant Sci. 68: 151-161.

[83] Colebatch G, Desbrosses G, Ott T, Krusell L, Montanari O, Kloska S, Kopka J, Udvardi MK (2004) Global Changes in Transcription Orchestrate Metabolic Differentiation During Symbiotic Nitrogen Fixation in *Lotus japonicus*. Plant J. 39: 487–512.

[84] Fabret C, Quentin Y, Chapal N, Guiseppi A, Haiech J, Denizot F (1996) Integrated Mapping and Sequencing of a 115 kb DNA Fragment from *Bacillus subtilis*: Sequence Analysis of a 21 kb Segment Containing the *sigL* Locus. Microbiol. 142: 3089-3096.

[85] Stingele F, Neeser JR, Mollet B (1996) Identification and Characterization of the *eps* (Exopolysaccharide) Gene Cluster from *Streptococcus thermophilus* Sfi6. J. Bacteriol. 178: 1680-1690.

[86] Andreishcheva EN, Kunkel JP, Gemmill TR, Trimble RB (2004) Five Genes Involved in Biosynthesis of the Pyruvylated Galβ1,3-Epitope in *Schizosaccharomyces pombe* N-Linked Glycans. J. Biol. Chem. 279: 35644-35655.

[87] Cuthbertson L, Mainprize IL, Naismith JH, Whitfield C (2009) Pivotal Roles of the Outer Membrane Polysaccharide Export and Polysaccharide Copolymerase Protein Families

in Export of Extracellular Polysaccharides in Gram-Negative Bacteria. Microbiol. Mol. Biol. Rev. 73: 55-77.

[88] Cuthbertson L, Kos V, Whitfield C (2010) ABC Transporters Involved in Export of Cell Surface Glycoconjugates. Microbiol. Mol. Biol. Rev. 74: 341-362.

[89] Morona R, Purins L, Tocilj A, Matte A, Cygler M (2009) Sequence-Structure Relationships in Polysaccharide Co-polymerase (PCP) Proteins. Trends Biochem. Sci. 34: 78-84.

[90] Dong C, Beis K, Nesper J, Brunkan-LaMontagne AL, Clarke BR, Whitfield C, Naismith JH (2006) Wza the Translocon for *E. coli* Capsular Polysaccharides Defines a New Class of Membrane Protein. Nature. 444: 226-229.

[91] Mazur A, Marczak M, Król JE, Skorupska A (2005) Topological and Transcriptional Analysis of *pssL* Gene Product: a Putative Wzx-like Exopolysaccaride Translocase in *Rhizobium leguminosarum* bv. *trifolii* TA1. Arch. Microbiol. 184: 1-10.

[92] Mazur A, Król JE, Skorupska A (2001) Isolation and Sequencing of *Rhizobium leguminosarum* bv. *trifolii pssN*, *pssO* and *pssP* Genes Encoding the Proteins Involved in Polymerization and Translocation of Exopolysaccharide. DNA Seq. 12: 1-12.

[93] Mazur A, Król JE, Wielbo J, Urbanik-Sypniewska T, Skorupska A (2002) *Rhizobium leguminosarum* bv. *trifolii* PssP Protein is Required for Exopolysaccharide Biosynthesis and Polymerization. Mol. Plant-Microbe Interact. 15: 388-397.

[94] Gonzalez JE, Semino CE, Wang L-X, Castellano-Torres LE, Walker GC (1998) Biosynthetic Control of Molecular Weight in the Polymerization of the Octasaccharide Subunits of Succinoglycan, a Symbiotically Important Exopolysaccharide of *Rhizobium meliloti*. Proc. Natl. Acad. Sci. USA. 95: 13477-13482.

[95] Becker A, Niehaus K, Pühler A (1995) Low-molecular-weight Succinoglycan is Predominantly Produced by *Rhizobium meliloti* Strains Carrying a Mutated ExoP Protein Characterized by a Periplasmic N-terminal Domain and a Missing C-terminal Domain. Mol. Microbiol. 16: 191-203.

[96] Paulsen IT, Beness AM, Saier MH Jr (1997) Computer-based Analyses of the Protein Constituents of Transport Systems Catalysing Export of Complex Carbohydrates in Bacteria. Microbiol. 143: 2685-2699.

[97] Morona R, Van Den Bosch L, Daniels C (2000) Evaluation of Wzz/MPA1/MPA2 Proteins Based on the Presence of Coiled-coil Regions. Microbiol. 146: 1-4.

[98] Niemeyer D, Becker A (2001) The Molecular Weight Distribution of Succinoglycan Produced by *Sinorhizobium meliloti* is Influenced by Specific Tyrosine Phosphorylation and ATPase Activity of the Cytoplasmic Domain of the ExoP Protein. J. Bacteriol. 183: 5163-5170.

[99] Marczak M, Mazur A, Król JE, Gruszecki WI, Skorupska A (2006) Lipoprotein PssN of *Rhizobium leguminosarum* bv. *trifolii*: Subcellular Localization and Possible Involvement in Exopolysaccharide Export. J. Bacteriol. 188: 6943-6952.

[100] Marczak M, Mazur A, Gruszecki WI, Skorupska A (2008) PssO, a Unique Extracellular Protein Important for Exopolysaccharide Synthesis in *Rhizobium leguminosarum* bv. *trifolii*. Biochimie. 90: 1781–1790.

[101] Wielbo J, Mazur A, Król JE, Marczak M, Skorupska A (2004) Environmental Modulation of the *pssTNOP* Gene Expression in *Rhizobium leguminosarum bv. trifolii.* Can. J. Microbiol. 50: 201-211.

[102] Louch HA, Miller KJ (2001) Synthesis of a Low-molecular-weight Form of Exopolysaccharide by *Bradyrhizobium japonicum* USDA 110. Appl. Environm. Microbiol. 67: 1011-1014.

[103] Gonzalez JE, Semino CE, Wang L-X, Castellano-Torres LE, Walker GC (1998) Biosynthetic Control of Molecular Weight in the Polymerization of the Octasaccharide Subunits of Succinoglycan, a Symbiotically Important Exopolysaccharide of *Rhizobium meliloti.* Proc. Natl. Acad. Sci. USA. 95: 13477-13482.

[104] York GM, Walker GC (1997) The *Rhizobium meliloti exoK* Gene and *prsD/prsE/exsH* Genes are Components of Independent Degradative Pathways which Contribute to Production of Low-molecular-weight Succinoglycan. Mol. Microbiol. 25: 117-134.

[105] York GM, Walker GC (1998) The Succinyl and Acetyl Modifications of Succinoglycan Influence Susceptibility of Succinoglycan to Cleavage by the *Rhizobium meliloti* Glycanases ExoK and ExsH. J. Bacteriol. 180: 4184-4191.

[106] Finnie C, Zorreguieta A, Hartley NM, Downie JA (1998) Characterization of *Rhizobium leguminosarum* Exopolysaccharide Glycanases that are Secreted via a Type I Exporter and Have a Novel Heptapeptide Repeat Motif. J. Bacteriol. 180: 1691-1699.

[107] Krehenbrink M, Downie JA (2008) Identification of Protein Secretion Systems and Novel Secreted Proteins in *Rhizobium leguminosarum bv. viciae.* BMC Genomics. 9: 55.

[108] Finnie C, Hartley NM, Findlay KC, Downie JA (1997) The *Rhizobium leguminosarum prsDE* Genes are Required for Secretion of Several Proteins, Some of Which Influence Nodulation, Symbiotic Nitrogen Fixation and Exopolysaccharide Modification. Mol. Microbiol. 25: 135-46.

[109] Zorreguieta A, Finnie C, Downie JA (2000) Extracellular Glycanases of *Rhizobium leguminosarum* are Activated on the Cell Surface by an Exopolysaccharide-related Component. J. Bacteriol. 182: 1304-1312.

[110] Guerreirro N, Ksenzenko VN, Djordjevic MA, Ivashina TV, Rolfe BG (2000) Elevated Levels of Synthesis of Over 20 Proteins After Mutation of the *Rhizobium leguminosarum* Exopolysaccharide Synthesis Gene *pssA.* J. Bacteriol. 185: 4521-4532.

[111] Latchford JW, Borthakur D, Johnston AW (1991) The Products of *Rhizobium* Genes, *psi* and *pss,* which Affect Exopolysaccharide Production, are Associated with the Bacterial Cell Surface. Mol. Microbiol. 5: 2107-2114.

Chapter 5

Capsular Polysaccharides Produced by the Bacterial Pathogen *Burkholderia pseudomallei*

Shauna L. Reckseidler-Zenteno

Additional information is available at the end of the chapter

http://dx.doi.org/10.5772/50116

1. Introduction

Polysaccharide capsules are structures found on the cell surface of a broad range of bacterial species. The polysaccharide capsule often constitutes the outermost layer of the cell, and therefore is often involved in mediating direct interactions between the bacteria and its environment. It is due to these interactions that polysaccharide capsules have been implicated as important virulence factors for many bacterial pathogens.

Bacterial extracellular polysaccharides (EPS) may be classified as either capsular polysaccharides (CPS), where the polysaccharide is intimately associated with the cell surface, or as slime polysaccharides, where the polysaccharide is loosely associated with the cell [1]. Differentiation between these forms is difficult since CPS may be released from the cell, giving the appearance of a slime polysaccharide [1]. In turn, distinguishing between CPS and other cell surface polysaccharides, such as O-antigenic moieties of lipopolysaccharide (LPS), may also be difficult, since CPS may be found associated with LPS [1].

Capsular polysaccharides are highly hydrated molecules that are over 95% water [2]. They are often linked to the cell surface of the bacterium via covalent attachments to either phospholipid or lipid-A molecules, although some CPS may be associated with the cell in the absence of a membrane anchor [1, 3]. Capsular polysaccharides can be either homo- or heteropolymers composed of repeating monosaccharides joined by glycosidic linkages [4]. The multiple hydroxyl groups present within each monosaccharide may be involved in the formation of the glycosidic bond, therefore, any two monosaccharides may be joined in a number of configurations, which leads to large structural diversity among CPS types. In the case of human pathogens, a large number of different capsule serotypes have been identified, and certain CPS or K-antigens have been associated with specific infections [4]. For example, the *Escherichia coli* K1 antigen, a homopolymer of α2,8-linked N-acetylneuraminic acid (NeuNAc), is the major cause of neonatal meningitis [5]. While

INTECH
open science | open minds

© 2012 The Author(s). Licensee InTech. This chapter is distributed under the terms of the Creative Commons Attribution License http://creativecommons.org/licenses/by/3.0), which permits unrestricted use, distribution, and reproduction in any medium, provided the original work is properly cited.

bacterial species may demonstrate great structural diversity in synthesizing capsules, chemically identical capsular polysaccharides may also be synthesized by different bacterial species. The *Neisseria meningitidis* group B capsular polysaccharide is identical to the K1 polymer of *E. coli,* and the *E. coli* K18, K22, and K100 antigens have the same constituents and structure as the *Haemophilus influenza* serotype b capsule [6, 7]. The conservation of CPS types between bacterial species raises interesting questions regarding the evolution of capsules and the transmission and acquisition of capsule biosynthesis genes [4].

The genetic loci necessary for the production of bacterial capsules are primarily clustered at a single chromosomal locus, which allows for the coordinate regulation of a large number of genes that may be involved in both the biosynthesis and export of capsular polysaccharides [4, 1]. In most bacterial species, the capsule gene clusters demonstrate conserved genetic organization. The capsules of *E. coli* have been classified into four groups based on genetic organization and biosynthetic criteria and capsules of other bacteria may resemble these prototypes. Group 1 capsules include the *E. coli* K30 capsule and the capsules of *Klebsiella* sp. and *Erwinia* sp. [8]. Group 2 capsules include the *E. coli* K1 and K5 capsules, as well as the capsules produced by *Neisseria* sp. and *Haemophilus* sp. Group 3 capsules include the *E. coli* K10 capsule. Group 4 capsules include the capsules of *E. coli* K40 and 0111. Both group 2 and group 3 capsule gene clusters are organized into 3 regions. Regions 1 and 3 are involved in the export and modification of the capsular polysaccharides and are conserved between members of the group, while region 2 contains the genes responsible for the biosynthesis of the capsule and is usually serotype specific [4, 8]. Generally these regions are organized into one transcriptional unit and the regions within the capsule locus are divergently transcribed [4, 8]. In addition, some genes within a region may be translationally coupled, such as the *kpsU* and *kpsC* genes in group 2 *E. coli* capsules, and the *kpsM* and *kpsT* genes in group 3 *E. coli* capsules, which allows for balanced expression of two proteins [4].

The A+T composition of capsule gene clusters is often significantly higher than the rest of the chromosome, suggesting a common ancestry of capsule genes in gram-negative bacteria [9]. It is likely that these A+T rich regions have been horizontally transferred between bacterial species. In addition, the A+T ratio of region 2 DNA of group 2 *E. coli* capsule gene clusters compared to regions 1 and 3 confirms that capsule diversity has been achieved in part through the acquisition of different region 2 sequences [4].

The production of a polysaccharide capsule is widespread in pathogenic bacteria. A number of functions have been assigned to bacterial capsules including: prevention of dessication, adherence, resistance to non-specific host immunity, resistance to specific host immunity, and mediating the diffusion of molecules through to the cell surface [4, 1].

Capsules may form a hydrated gel around the surface of the bacterial cell, which may protect the bacteria from the harmful effects of dessication [10]. This may increase the survival of encapsulated bacteria outside of the host, promoting the transmission of pathogenic bacteria from one host to another [4]. Mucoid isolates of *E. coli, Acinetobacter calcoaceticus,* and *Erwinia stewartii* are more resistant to drying than isogenic nonmucoid strains [11]. Studies with *E. coli* have shown that the expression of genes encoding for the colanic acid capsule is increased by dessication [11]. In addition, alginate production by *Pseudomonas aeruginosa* is triggered by high osmolarity, which may be a consequence of dessication [12].

Capsular polysaccharides may promote adherence of bacteria to both surfaces and other bacterial cells, which may facilitate colonization of a particular niche and may lead to the formation of biofilms [13]. Cell-surface polysaccharides have been shown to mediate the attachment of bacterial cells to one another, leading to biofilm formation and persistence of the organisms during colonization [1, 14].

Capsular polysaccharides are one of the components responsible for resistance to the non-specific immunity of the host. The presence of a capsule is thought to confer resistance to non-specific host defense mechanisms such as complement and complement-mediated opsonophagocytosis [4]. Bacterial capsules may resist complement-mediated killing by providing a permeability barrier to complement components, which masks the underlying cell surface structures that activate complement [15]. The capsule may also act in concert with O-antigens to confer resistance to complement-mediated killing [16]. As a result, a combination of cell surface structures is responsible for conferring resistance to killing by the complement cascade [4]. Finally, capsules are responsible for resistance to complement-mediated opsonophagocytosis. This resistance may be due to steric effects, which results in the capsule acting as a barrier between the C3b deposited on the bacterial surface and the C3b receptors present on phagocytes [4]. Alternatively, the resistance to opsonophagocytosis may be due to the net negative charge of the polysaccharide capsule [17].

Capsules may also confer resistance to the specific immune response of the host. Although most capsular polysaccharides can elicit an immune response, some capsules are poorly immunogenic [4]. Examples of such capsules include those containing NeuNAc, such as the *E. coli* K1 capsule or the capsule of *Neisseria meningitidis* serogroup B, and the *E. coli* K5 capsule, which is similar to desulfoheparin [18, 19]. Because these capsules are structurally similar to polysaccharides encountered on host tissue, these capsules are poorly immunogenic, and elicit a poor antibody response in the host [20].

Burkholderia pseudomallei, the causative agent of melioidosis, is a gram-negative, facultatively anaerobic, motile bacillus that is commonly found in the soil and stagnant waters in a number of regions around the world, particularly in areas that fall between 20° north and 20° south of the equator [21, 22]. Infection by *B. pseudomallei* is often due to either direct inoculation into wounds and skin abrasions or to inhalation of contaminated material [22, 23, 24,]. This would explain the prevalence of the disease among rice farmers as well as helicopter pilots in the Vietnam War who developed melioidosis due to inhalation of contaminated dust [24]. Melioidosis may present as an acute pneumonia or an acute septicemia, which is the most severe form of the disease. The disease may also manifest as a chronic infection involving long-lasting suppurative abscesses in numerous sites in the body. Infection with *B. pseudomallei* may even result in a subclinical infection and remain undetected for a number of years. Both the chronic and subclinical forms generally remain undiagnosed until activated by a traumatic event or a decrease in immunocompetence [25]. *B. pseudomallei* is inherently resistant to a number of antibiotics, and even with aggressive antibiotic therapy, the mortality rate remains high, and the incidence of relapse is common [26, 27].

At the time our studies were initiated some cell-associated antigens had been identified and characterized in *B. pseudomallei*. Cell-associated antigens include exopolysaccharide (EPS) and lipopolysaccharide (LPS) [28, 29, 30]. The EPS produced by *B. pseudomallei* was determined to be an unbranched polymer of repeating tetrasaccharide units with the structure -3)-2-*O*-acetyl-β-D-Gal*p*-(1-4)-α-D-Gal*p*-(1-3)-β-D-Gal*p*-(1-5)-β-D-KDO*p*-(2- [31, 32]. The role of EPS in virulence was not known, but sera from patients with melioidosis had been shown to contain antibodies against EPS [30]. Two other EPS structures were also identified; a branched 1,4-linked glucan polymer ((CP-1a) and a triple-branched heptasaccharide repeating unit composed of rhamnose, mannose, galactose, glucose, and glucuronic acid (CP-2) [33]. The genes involved in the synthesis of these capsules, and the role of these capsules in virulence had not been identified. The LPS of *B. pseudomallei* was structurally characterized and reported to contain two types of O-polysaccharide moieties termed type I O-PS and type II O-PS [34, 35]. Type II O-PS was found to be an unbranched heteropolymer with repeating D-glucose and L-talose residues with the structure -3)-β-D-glucopyranose-(1-3)-6-deoxy-α-L-talopyranose-(1-. Type II O-PS had been shown to be involved in serum resistance [36]. Type II O-PS mutants also demonstrated reduced virulence in three animal models of *B. pseudomallei* infection [36]. Type I O-PS was determined to be an unbranched homopolymer with the structure -3)-2-*O*-acetyl-6-deoxy-β-D-manno-heptopyranose-(1-, however, the role for this polysaccharide in infection had not been defined, nor the genes responsible for its biosynthesis been identified.

B. thailandensis is a nonpathogenic soil organism originally isolated in Thailand [37]. Based on biochemical, immunological, and genetic data, *B. pseudomallei* and *B. thailandensis* are closely related species. However, these two organisms differ in a number of ways and have been classified into two different species [38]. The rRNA sequence of *B. thailandensis* differs from that of *B. pseudomallei* by 15 nucleotides, and there are significant differences in genomic macrorestriction patterns between these organisms [39]. The biochemical profiles of these two species differ in that *B. thailandensis* can utilize L-arabinose whereas *B. pseudomallei* does not [38, 40]. The most distinct difference between these two species, however, is their relative virulence. The 50% lethal dose (LD_{50}) for *B. pseudomallei* in the Syrian hamster model of acute melioidosis is <10 organisms, whereas the LD_{50} for *B. thailandensis* is approximately 10^6 organisms [38]. It has also been shown that the two species can be differentiated based on their propensity to cause disease in humans. Environmental strains isolated in Thailand that are able to assimilate L-arabinose are not associated with human infection, whereas clinical isolates are not able to utilize L-arabinose [41].

2. Subtractive hybridization to identify Capsular Polysaccharide 1 (CPS I)

To identify the genetic determinants that confer enhanced virulence in *B. pseudomallei*, a method combining subtractive hybridization, insertional mutagenesis, and animal virulence studies was developed [42]. Subtractive hybridization was carried out between the virulent *B. pseudomallei* and the weakly virulent *B. thailandensis* in order to isolate DNA sequences encoding for virulence determinants unique to *B. pseudomallei*. The genomic DNA sample from *B. pseudomallei* containing the sequences of interest was known as the tester DNA, and genomic DNA from *B. thailandensis*, the reference sample, was called the driver DNA. Tester and driver DNAs were digested and subjected to two rounds of hybridization. The remaining unhybridized sequences were considered tester-specific sequences. To enrich for tester-specific sequences, excess driver DNA was added in the hybridizations. The tester-specific sequences were then amplified by PCR and cloned into the plasmids pPCR or pZErO-2.1. Screening of the subtraction library revealed a number of DNA sequences unique to *B. pseudomallei*. Fifteen distinct plasmid inserts from the library were sequenced. The DNA inserts ranged from 100 to 800 bp in length and were found to contain an average G+C content of approximately 44 to 52%, which is considerably lower than the 68% G+C content of the *B. pseudomallei* chromosome. The DNA sequences were analyzed using the NCBI BLASTX program. One of the plasmid inserts, pDD1015, was found to share limited homology with WbpX, a glycosyltransferase, from *Pseudomonas aeruginosa* [43].

2.1. Demonstration that CPS I is important for the virulence of *B. pseudomallei*

The 373-bp DNA insert from pDD1015 was cloned into a mobilizable suicide vector, pSKM11 [44]. The resulting plasmid, pSR1015, was mobilized into wildtype *B. pseudomallei* 1026b to create the mutant strain SR1015. Since the insert from pDD1015 was found to demonstrate homology to a glycosyltransferase from *P. aeruginosa*, it was postulated that it might encode a protein involved in carbohydrate synthesis. Since three carbohydrate structures had been previously purified and characterized, antibodies to each of these polysaccharides were available. To define the phenotype of SR1015, an ELISA was performed with the EPS-specific monoclonal antibody 3015, and *B. pseudomallei* 1026b and SR1015 were both found to contain EPS [45]. SR1015 was also shown to contain type II O-PS and to be serum resistant. Immunogold electron microscopy studies using rabbit polyclonal sera specific for a type I O-PS–flagellin conjugate was performed on the parent strain, 1026b, and SR1015 (Figure 1). *B. pseudomallei* 1026b reacted with antibodies to both flagellin and type I O-PS, as was evident by the distribution of gold particles around the bacterial surface and extending out along the flagella (Figure 1A). Unlike *B. pseudomallei* 1026b, SR1015 reacted only with the antibodies to flagellin, as the gold particles were found associated only with the flagella (Figure 1B). *B. thailandensis*, the negative control, did not react with the antibodies either to flagellin or to type I O-PS (Figure 1C). *B. stabilis* LMG7000 was also shown to react to the antibodies to type I O-PS, indicating this organism may produce a similar capsule (Figure 1D). Western blot analysis of proteinase K-digested whole cells from *B. pseudomallei* 1026b, *B. thailandensis* E264, and *B. pseudomallei* SR1015 using rabbit polyclonal sera raised to O-PS–flagellin protein conjugate confirmed the lack of type I O-PS in SR1015. Type I and type II O-PS were stained in *B. pseudomallei* 1026b, while only type II O-PS was stained in the lanes corresponding to *B. pseudomallei* SR1015 and *B. thailandensis*. These results indicated that we had identified and insertionally inactivated a gene involved in the synthesis of the type I O-PS of *B. pseudomallei* [42]. SR1015 was tested for virulence in the Syrian hamster model of acute septicemic melioidosis. The LD_{50} for SR1015 after 48 h was 3.5 x 10^5 CFU, while the LD_{50} of the parent strain, 1026b, was <10 CFU. The LD_{50} for SR1015 was similar to that for the weakly virulent *B. thailandensis* (6.8 x 10^5 CFU) [42]. This demonstrated that SR1015 is severely attenuated for virulence in this animal model of melioidosis and that type I O-PS is a major virulence determinant of *B. pseudomallei*. We later determined that the type I O-PS was a capsular polysaccharide (CPS I), not an O-PS moiety, which will be discussed below.

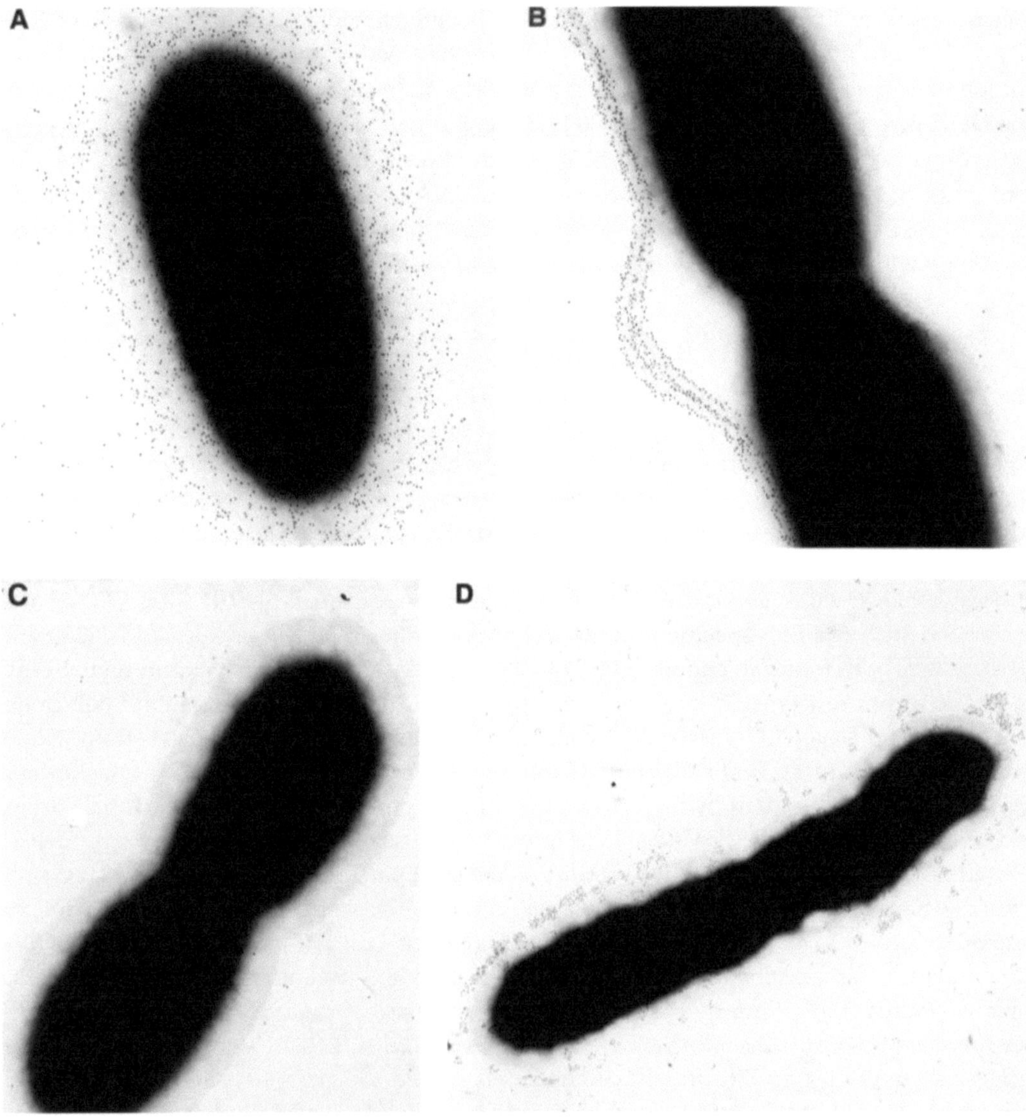

Figure 1. Immunogold electron microscopy of *B. pseudomallei* 1026b (A) and SR1015 (B), *B. thailandensis* E264 (C), and *B. stabilis* LMG7000 (D). Bacteria were reacted with polyclonal rabbit antiserum directed against an O-PS–flagellin protein conjugate absorbed with *B. thailandensis* E264 to remove the antibodies directed against type II O-PS, washed, and reacted with a goat anti-rabbit IgG-gold (5 nm) conjugate. Original magnification x330,000.

2.2. Cloning and sequencing of the genetic loci required for CPS I production and export

Two methods were used to clone the genes involved in the production and export of type I O-PS. The DNA flanking the insertion of pSR1015 was cloned from SR1015 and sequenced. We also used transposon mutagenesis to clone the genes involved in production of the polysaccharide; this was done to obtain any unlinked genes that may be involved in polysaccharide production. Approximately 1,300 transposon mutants were screened for loss of type I O-PS by ELISA. Six mutants were identified, and the DNA flanking the transposon insertion was cloned and sequenced. The Tn5-OT182 mutants SLR5, SLR8, SLR13, SLR18, and SLR19 mapped to the same region of the chromosome. Sequence analysis of the cloned fragments revealed the presence of 26 potential open reading frames involved in the synthesis and export of type I O-PS [42]. The open reading frames that predicted proteins involved in polysaccharide biosynthesis were found to demonstrate homology to proteins involved in the synthesis of a polysaccharide structure composed primarily of mannose. The other reading frames in the locus predicted proteins involved in the transport of capsular polysaccharides in a variety of bacteria, particularly those that produce group 2 and group 3 capsular polysaccharides [8]. The genes responsible for the production of type I O-PS were found to be similar to other loci encoding for capsular polysaccharides in that they are divergently transcribed [4]. The gene cluster involved in the production of this polysaccharide is also similar to group 3 capsule gene clusters in that there are no genes encoding KpsF and KpsU, which are present in group 2 capsule gene clusters [8]. However, the organization of the *B. pseudomallei* type I O-PS gene cluster differs in that it does not contain two export regions flanking a single biosynthetic region as seen in other group 3 capsule polysaccharide clusters [46]. The biosynthetic genes identified are not organized into one continuous transcriptional unit; instead, *wcbB*, *manC*, and *wcbP* are separated from the rest of the biosynthetic genes. The overall G+C content of this region is about 58%, lower than the G+C content of the rest of the chromosome (68%). The low G+C content in these clusters suggests that polysaccharide genes have a common origin and may have been transferred horizontally between species [9]. The genes involved in the production of this polysaccharide were named according to the bacterial polysaccharide gene nomenclature scheme [47]. The gene products associated with this cluster are shown in Figure 4. Mutations constructed in a number of these genes confirmed their role in the production of this polysaccharide [42].

2.3. CPS I was originally identified as a Type I O-polysaccharride moiety

The polysaccharide with the structure -3)-2-*O*-acetyl-6-deoxy-β-D-manno-heptopyranose-(1- was originally isolated and characterized as an O-PS component of LPS in *B. pseudomallei* and was designated type I O-PS [35]. However, our results suggested that this polysaccharide was a capsule rather than an O-PS moiety. The genes involved in the production of this capsule demonstrated strong homology to the genes involved in the production of capsular polysaccharides in many organisms, including *N. meningitidis*, *H. influenzae*, and *E. coli*. In addition, the export genes associated with this cluster are not associated with the previously characterized O-PS gene cluster [36]. Western blot analysis of proteinase K cell extracts and silver staining demonstrated that this polysaccharide has a high molecular mass (200 kDa) and lacks the banding pattern seen with O-PS moieties. Studies by our laboratory indicated that mutants in the production of the core oligosaccharide of the LPS are still capable of producing this polysaccharide [48]. Based on the above criteria and the genetic similarity to group 3 capsules, we proposed that this polysaccharide is a group 3 capsule and designated this capsule CPS I. This conclusion was further supported by Isshiki *et al* who separated this polysaccharide from a smooth lipopolysaccharide preparation of *B. pseudomallei* [49].

3. Role of the *B. pseudomallei* capsule in virulence

The role of CPS I in the pathogenicity of *B. pseudomallei* was investigated by performing further animal studies, serum bactericidal assays, complement protein C3b deposition assays, and radio-labelled phagocytic assays [50]. These experiments were facilitated by constructing a deletion strain harbouring a mutation in one of the CPS I genes and by complementation of this strain. An in-frame deletion was constructed in *wcbB*, a gene which encodes a glycosyltransferase, resulting in the capsule-minus strain SZ210. To confirm the role of *wcbB* in the biosynthesis of capsule, SZ210 was complemented by the introduction of a wild-type copy of the *wcbB* gene cloned into the mobilizable broad-host-range plasmid pBHR1 (MoBiTec). Western blot analysis of proteinase K-digested whole cells was performed using mouse monoclonal antibody directed to *B. pseudomallei* capsule to assess capsule production by these strains. Similar to the capsule minus strain SR1015 and *B. thailandensis* E264, which is known to lack this capsule, SZ210 was found to be negative for CPS I production, as indicated by the absence of a 200 kDa band that is present for wild-type 1026b. Complementation of SZ210 by providing the wild type *wcbB* gene in *trans* restored capsule production. Whole-cell extracts from the complemented strain SZ210(pSZ219) reacted to the capsule antibody producing the 200-kDa band corresponding to the *B. pseudomallei* capsule.

3.1. Production of CPS I correlates with clinical infection

To establish a correlation between CPS I production and clinical infection a number of strains of *B. pseudomallei* isolated from a variety of clinical specimens were tested for capsule production by western blot analysis with polyclonal rabbit antisera to *B. pseudomallei* CPS I. Out of the 55 clinical strains tested for capsule production, 52 were found to produce this capsule. Three strains, 420a, 415c, and 375a were found to be negative for capsule production, similar to *B. thailandensis* E264. However, one of the capsule genes, *wzt2*, was successfully amplified from these three strains and following inoculation in the animal model, all three of these strains were found to produce capsule by western blot analysis. This indicated that CPS I production may be regulated in some strains and its expression may be induced *in vivo*. Therefore all of the 55 clinical strains of *B. pseudomallei* tested were found to produce capsule, establishing a 100% correlation between CPS I production and clinical infection [51].

3.2. CPS I promotes survival of *B. pseudomallei in vivo*

Syrian golden hamsters were inoculated intraperitoneally with 10^1 to 10^5 cells of either wild type *B. pseudomallei* 1026b, capsule mutants SR1015 and SZ210, or the complemented strain SZ210(pSZ219). One group of animals inoculated with SR1015 also received 100 μg of purified *B. pseudomallei* capsule. After 48 h, the LD_{50} values were calculated, and the blood of the infected animals was diluted and plated for bacterial quantitation. The addition of purified capsule significantly increased the virulence of the capsule mutant strain SR1015. The LD_{50} value was calculated to be 34 CFU, similar to the LD_{50} value of wild-type *B. pseudomallei* 1026b (<10 CFU). In contrast, the LD_{50} value for SR1015 without the addition of purified capsule was calculated to be 3.5 x 10^5 CFU, 10,000- fold higher than when capsule was added to the inoculum. In addition, purified capsule enhanced the survival of SR1015 in the blood. Bacteria could not be detected in the blood of hamsters inoculated with SR1015 alone. However, the number of SR1015 CFU recovered from the blood of infected animals was 9.0 x 10^2 CFU/ml when capsule was added to the inoculum, an almost-1,000-fold increase. This number was comparable to the number of wild-type *B. pseudomallei* 1026b bacteria recovered from the blood. The addition of capsule was not toxic to the hamsters, as hamsters inoculated with 100 μg of purified capsule alone survived for the duration of the experiment without any ill effects. The LD_{50} value for the capsule mutant strain SZ210 containing an in frame deletion of the *wcbB* gene was calculated to be 9.6 x 10^4 CFU, and the number of bacteria in the blood was determined to be 10 CFU/ml. Complementation of this strain restored virulence in the animal model, resulting in an LD_{50} value of 12 CFU, comparable to that of wild type *B. pseudomallei* 1026b. Furthermore, the number of bacteria in the blood of animals infected with the complemented strain, SZ210(pSZ219), was determined to be 4.9 x 10^5 CFU/ml, similar to the number of bacteria recovered from animals infected with 1026b [50].

To further demonstrate the role of the capsule in infection by *B. pseudomallei*, an experiment was designed to investigate differences in tissue distribution between the capsule mutant strain and the wild type in infected hamsters. Animals were inoculated with 10^2 CFU of either wild-type *B. pseudomallei* 1026b or the capsule mutant SR1015. At different time points, the animals were sacrificed, and the numbers of bacteria in the blood, liver, lungs, and spleen of each animal were determined. As seen in Figure 2, the numbers of *B. pseudomallei* 1026b and SR1015 bacteria were nearly undetectable at 12 h (Figure 2A). By 24 h, the numbers of 1026b bacteria recovered from the blood, lung, liver, and spleen increased, while SR1015 was detected only in the spleen (Figure 2B). By 48 h, very high numbers of 1026b bacteria were recovered from all of the organs taken, representing a dramatic increase compared to the inoculum (Figure 2C). In contrast, all of the organs taken from hamsters infected with SR1015 contained fewer bacteria (Figure 2C). Of particular interest was the fact that the number of SR1015 bacteria recovered from the blood at 48 h was lower than in the inoculum, suggesting that the capsule mutant was cleared from the blood more effectively than the wild type. The number of SR1015 bacteria recovered from the spleen was higher than the number of SR1015 bacteria in the blood, suggesting that SR1015 was being cleared from the blood and sequestered in the spleen. The difference in virulence between the two strains can be attributed to capsule production, since the CPS I mutant strain was found to have a growth rate similar to that of the wild-type strain 1026b [50].

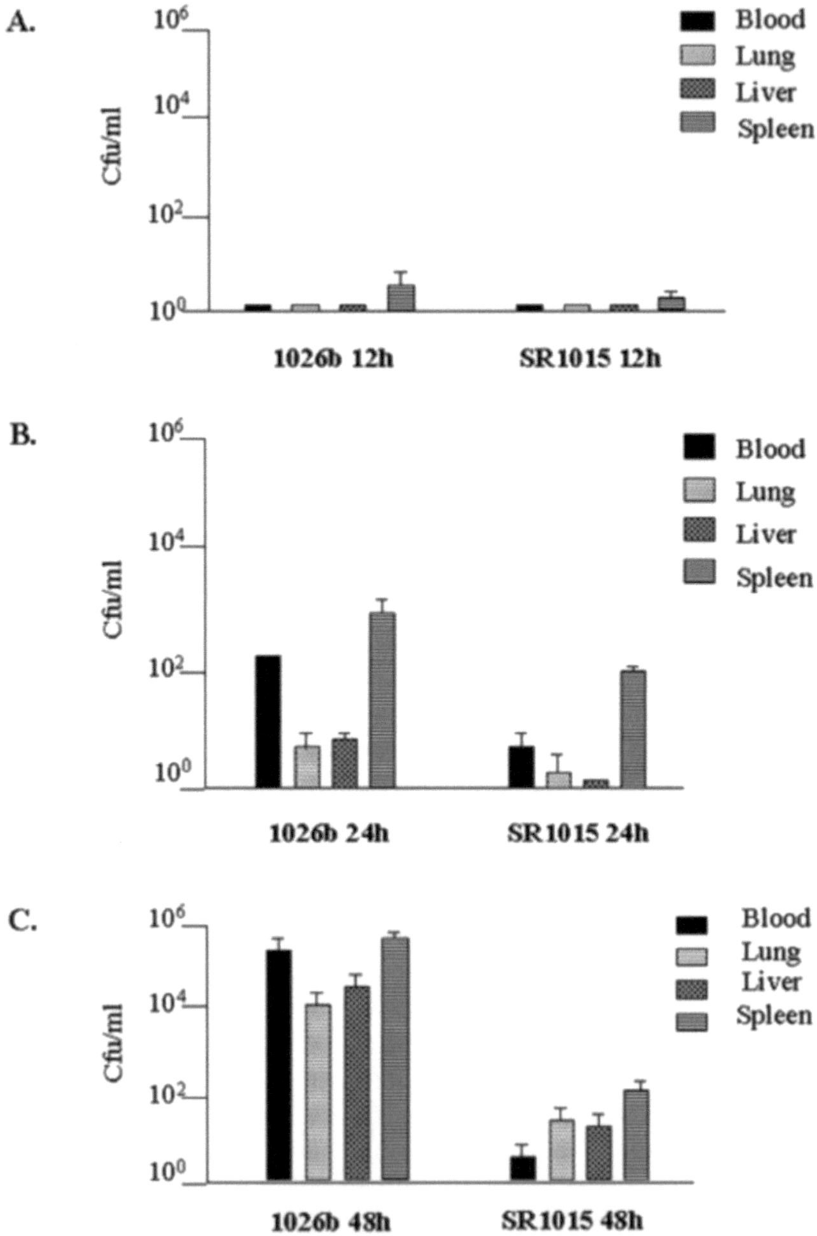

Figure 2. Differences in tissue distribution between *B. pseudomallei* strains 1026b and SR1015 in the Syrian hamster model of acute melioidosis. Female Syrian hamsters (three per group) were inoculated intraperitoneally with 10^2 CFU of either strain, and at 12, 24, and 48 h, two groups of animals were sacrificed and bacterial quantitation of the tissues was determined. The data represent the average number of bacteria found in each tissue and the standard deviation for a given time point.

3.3. CPS I production by *B. pseudomallei* is responsible for persistence in the blood by inhibiting complement

To define the role of the capsule for persistence in the blood, serum bactericidal assays were performed with the addition of purified capsule to determine if capsule had an effect on the survival of serum-sensitive strains of *B. pseudomallei*. For these experiments, we utilized a double mutant that we constructed in the laboratory, SLR5, which lacks both capsule and O-polysaccharide, since the capsule mutant SR1015 was previously found to be serum resistant [42]. The survival of SLR5 was extremely poor when incubated in the presence of 30% normal human serum (NHS). However, the addition of purified capsule increased the survival of SLR5 in NHS. The addition of 50 μg of capsule to the reaction increased the numbers of SLR5 to 5.9 x 10^1 CFU/ml, and the addition of 100 μg of capsule increased the survival of SLR5 by nearly 1,000-fold to 1.9 x 10^3 CFU/ml. Furthermore, pre-incubation of 30% NHS with 100 μg of capsule (PI-CPS) before the addition of bacteria increased the survival of SLR5 100,000-fold to 4.4 x 10^6 CFU/ml. This was similar to the survival of SLR5 when incubated with serum that was heat-inactivated (HI-NHS). These effects were found to be specific to capsule, since the addition of 50 or 100 μg of purified *B. pseudomallei* O-PS or preincubation of the serum with O-PS did not increase the survival of serum-sensitive SLR5 [50].

Since capsule mutants of *B. pseudomallei* are serum resistant in that they are not susceptible to lysis by the membrane attack complex (MAC) because they still produce O-PS, we postulated that the ability of the capsule to enhance survival in the blood could be due to its ability to inhibit C3b deposition and opsonization. To investigate the effect of capsule on C3b deposition, the amount of C3b deposited on the surfaces of wild-type *B. pseudomallei* 1026b and the capsule mutant, SR1015, in the presence of serum was determined by Western blot analysis using a mouse monoclonal antibody specific to human complement factor C3b. The deposition of C3b was found to be more pronounced in the capsule mutant SR1015 than in the wild type in both 10 and 30% NHS. Similar results were observed with the capsule mutant SZ210, a strain containing an in-frame deletion of the *wcbB* gene. More C3b was detected when SZ210 was incubated in both 10 and 30% NHS than with 1026b. Optical densitometry measurements were performed in order to quantitate the difference in C3b deposition between the strains. The average amount of C3b deposited on the surfaces of SR1015 and SZ210 bacteria was 3.5-fold higher than for 1026b in 10% NHS and 2.5-fold higher in 30% NHS. In addition, there was a shift in the molecular mass of C3b, which normally runs at 185 kDa, indicating a covalent attachment of the molecule to the bacterial surface. The nature of this attachment was not investigated; however, C3b is thought to covalently attach to the bacterial surface through an ester or amide linkage [50, 52].

Immunofluorescence microscopy analysis was also performed to demonstrate the difference in C3b deposition between the capsule mutant and the wild type. The same experiment described above was performed, and samples were reacted with the mouse monoclonal antibody to human complement factor C3b, except that the samples were reacted with a secondary antibody conjugated to Cy3 and stained with DAPI for visualization of bacterial cells. As shown in Figure 3, the *B. pseudomallei* capsule mutant SR1015 demonstrated more reactivity to the antibody to human C3b in the presence of serum than the wild-type 1026b.

This is evident from the red fluorescence that corresponds to the C3b bound to the bacterial surface surrounding the blue DAPI-stained cells seen when the capsule mutant was incubated in the presence of 10% NHS (Figure 3D to F). In contrast, the amount of red fluorescence surrounding the DAPI-stained wild-type cells was minimal in the presence of 10% NHS (Figure 3A to C). There was a dramatic difference in the amount of C3b deposited on the surface of the capsule mutant compared to the wild type, which was detectable after only 15 min of incubation of the bacteria with human serum (Figure 3B and E). By 60 min, there was some C3b deposition on wild-type *B. pseudomallei*; however, there was still more C3b deposited on the surface of the capsule mutant (Figure 3C and F) [50]. This experiment was not performed with 30% NHS due to excessive clumping of the samples during the fixation process, which resulted in inconsistent and poor staining of the cells. Western blot analysis was also performed to determine the amount of complement factor C3b deposition on the surface of *B. thailandensis* E264, a related nonpathogenic organism. The amount of C3b deposition in *B. thailandensis* E264 was more pronounced than with *B. pseudomallei* 1026b and was similar to the amount of C3b deposited on the surface of the capsule mutant, *B. pseudomallei* SR1015, in the presence of human serum. The amount of C3b deposition that occurred on the surface of *B. thailandensis* was expected, since the organism is known to lack this capsule [42, 50].

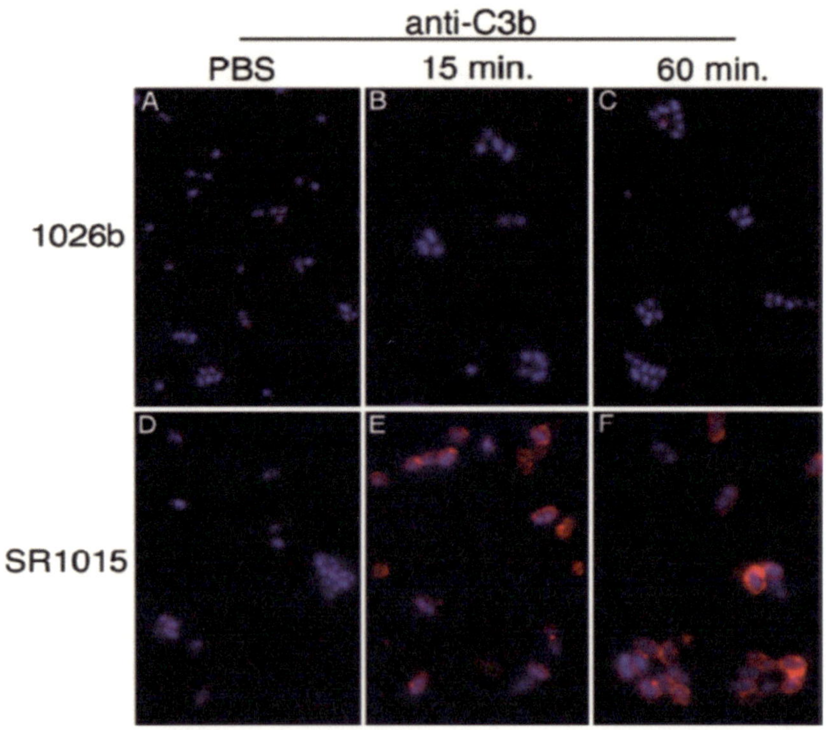

Figure 3. Immunofluorescence microscopy analysis of decreased complement factor C3b deposition in 10% normal human serum by *B. pseudomallei* capsule. *B. pseudomallei* 1026b and SR1015 were incubated in 10% normal human serum (NHS), reacted with a mouse monoclonal antibody to human complement factor C3b, reacted with a rabbit anti-mouse IgG conjugated to Cy3 (Jackson Laboratories), and stained

with DAPI for visualization of whole bacterial cells (Sigma). (A) *B. pseudomallei* 1026b incubated in PBS; (B) 1026b incubated in 10% NHS for 15 min; (C) 1026b incubated in 10% NHS for 60 min; (D) *B. pseudomallei* SR1015 incubated in PBS; (E) SR1015 incubated in 10% NHS for 15 min; (F) SR1015 incubated in 10% NHS for 60 min. The blue fluorescence indicates the DAPI stained bacteria, and the red fluorescence indicates the binding of complement factor C3b to the bacterial surface.

3.4. The *B. pseudomallei* capsule CPS I reduces phagocytosis

The capsule mutant SR1015 was phagocytosed more significantly by PMNL than the wild-type strain. The proportion of wild-type *B. pseudomallei* 1026b phagocytosed in the presence of 10% NHS was 35.9%, while the proportion of the capsule mutant SR1015 phagocytosed was 51.7% ($P < 0.001$). When each strain was incubated in the presence of 30% NHS, 59.3% of the wild-type strain 1026b was phagocytosed by the PMNL after 30 min compared to 82.3% for the capsule mutant ($P < 0.001$) [50].

3.5. *B. pseudomallei* CPS I expression is elevated in the presence of 30% normal human serum

The *lux* reporter strain *B. pseudomallei* SZ211 was constructed by cloning an internal fragment of the *wcbB* gene into pGSV3-*lux*, a suicide vector containing the *lux* operon from *Photorhabdus luminescens* [53]. Absorbance (OD540) and luminescence (in relative light units) measurements were taken every 2 h. Capsule expression was higher in the presence of M9 plus 1% glucose plus 30% normal human serum (NHS) and M9 plus 1% glucose plus 30% heat-inactivated serum (HI-NHS) than in M9 plus 1% glucose alone. The increase in light production of SZ211 in the presence of serum supports the requirement for capsule for survival in serum. The strain *B. pseudomallei* SZ213 was constructed by cloning an internal region of the *wbiA* gene, which encodes an *O*-acetyltransferase required for *O*-acetylation of the O-PS component of *B. pseudomallei* LPS [36]. Light production of this strain was measured under the same conditions to determine whether LPS expression was induced in the presence of serum. Similar to the capsule, LPS expression was elevated in the presence of both 30% NHS and 30% HI-NHS. The levels of expression of both capsule and LPS were not significantly different in NHS and HI-NHS, suggesting that the environment of the serum may be required for induction of gene expression rather than complement [50].

4. Other capsules produced by *B.pseudomallei*

Sequence analysis of the completed genome of *B. pseudomallei* revealed four operons with the predicted function of capsular polysaccharide biosynthesis and export [54]. One of these operons, with the gene identifiers BPSL2786-2810, corresponds to the previously characterized mannoheptose capsule designated CPS I [42, 50]. Three other operons were identified. These three capsule operons in the genome of *B. pseudomallei* were further analyzed using the BLAST program and Artemis. The operons are illustrated in Figure 4. The operon consisting of the genes BPSS0417-0429, was designated CPS II (Figure 4B). Another operon, BPSS1825-1835, was designated CPS III and the predicted homologues were investigated further (Figure 4C). A fourth operon, CPS IV, was found to contain genes that may be involved in the synthesis of a capsule with the gene identifiers BPSL2769-2785 (Figure 4D) [54, 55].

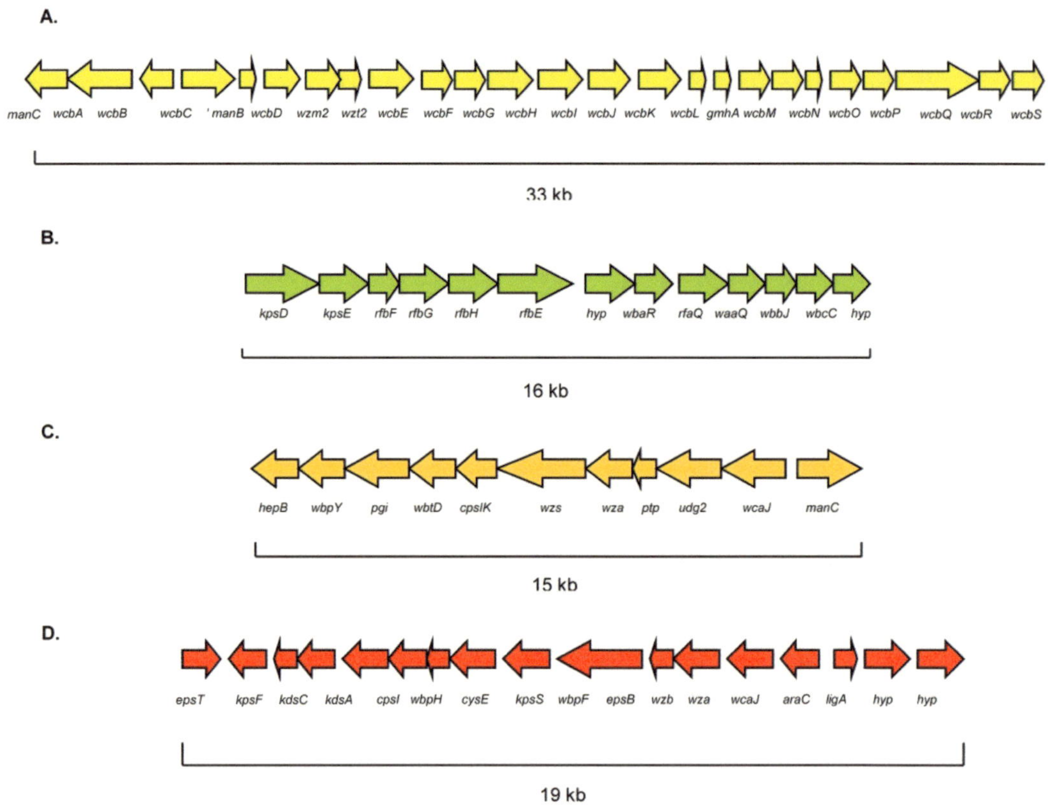

Figure 4. Organization of the chromosomal regions containing the genes comprising the *B. pseudomallei* capsule operons. The direction of transcription is represented by arrows, and the gene names demonstrating the highest degree of homology to the *B. pseudomallei* open reading frames are indicated. The relative sizes of each locus are indicated. (A) *B. pseudomallei* capsule cluster I (CPS I). (B) *B. pseudomallei* capsule cluster II (CPS II). (C) *B. pseudomallei* capsule cluster III (CPS III). (D) *B. pseudomallei* capsule cluster IV (CPS IV).

4.1. Distribution of capsule loci between three *Burkholderia* species

Comparative analysis of the genomes of three *Burkholderia* species, *B. pseudomallei, B. mallei* and *B. thailandensis,* was performed to determine whether all of the predicted *B. pseudomallei* capsule operons were present in *B. mallei* and *B. thailandensis* as well. CPS II, III, and IV were found to be present in *B. pseudomallei* and *B. thailandensis,* but not *B. mallei*. This is in contrast to CPS I, which is present in *B. pseudomallei* and *B. mallei,* but not *B. thailandensis* [42, 50, 54, 55, 56, 57]. CPS II was found to be identical between *B. pseudomallei* and *B. thailandensis,* but *B. thailandensis* was found to contain two flanking hypothetical genes not present in *B. pseudomallei*. The CPS II genes were found to be deleted entirely from *B. mallei*. A large chromosomal region ranging from open reading frames BPSS0404 to BPSS0491 including CPS II was shown to be deleted in *B. mallei* compared to *B. pseudomallei* and replaced with a large chromosomal region containing open reading frames BMAA0555 (IS407A *orfB*) to BMAA1784, a unique hypothetical protein not found in *B. pseudomallei*. The two genomes align with the presence of the alkyl hydroperoxidase reductase genes *ahpC* and *ahpF*, but these are organized in the opposite orientation in *B. mallei* compared to *B. pseudomallei*. The entire CPS III operon and flanking genes were shown to be the same in both *B. pseudomallei* and *B. thailandensis*. In contrast, the majority of the CPS III cluster was deleted from *B. mallei* with the exception of the *wcaJ* and *manC* genes, as well as the two flanking hypothetical genes on one side, and another hypothetical gene on the other side of the deleted region. The deletion of the CPS III genes in *B. mallei* was found to be replaced with the IS407A *orfA* and *orfB* genes. The entire CPS IV region was found to be replaced in *B. mallei* and flanked by two IS407A elements. The open reading frame BPSL2785, which encodes a hypothetical protein, is present in *B. mallei* (BMA2284.1) ATCC 23344 as well as a number of other *B. mallei* strains, but is organized in the opposite orientation. The genomes of *B. pseudomallei*, *B. thailandensis*, and *B. mallei* all diverge upstream of the CPS IV region, but all three organisms were found to align at the location of the *ompA* and hypothetical genes [55].

4.2. CPS III does not contribute to the virulence of *B. pseudomallei*

In order to assess the role of CPS III in virulence a mutant in the CPS III operon was tested for virulence compared to wild type *B. pseudomallei* in the Syrian hamster model of melioidosis. Syrian golden hamsters were inoculated intraperitoneally with 10^1 to 10^3 cells of either wild-type *B. pseudomallei* 1026b or the capsule mutant SZ1829. After 48 h, the LD_{50} values were determined. SZ1829 had LD_{50} values of <10 CFU, identical to that of wild type *B. pseudomallei*, indicating that this capsule is not required for virulence. In addition, the bacterial load in the blood of the infected hamsters was similar to that of wild-type and significantly higher than that of the non-pathogenic *B. thailandensis* E264 and the CPS I mutant, *B. pseudomallei* SR1015, both of which are incapable of establishing bacteremia [37, 50]. This indicates that CPS III does not contribute to persistence in the blood. Similar results were obtained for CPS II and CPS IV mutants, but we went on to further characterize CPS III.

4.3. Expression of CPS III in host and environmental conditions

A *lux* reporter strain was constructed in the CPS III operon by cloning an internal fragment of one the genes into pGSV3-*lux*, a suicide vector containing a promoterless *lux* operon from *Photorhabdus luminescens* [53]. Regulation of this capsule in an environment similar to that encountered in the host was determined by growing the *lux* reporter strain, SZ1829, in the presence of M9 plus 1% glucose versus M9 plus 1% glucose plus 30% normal human serum (NHS). Absorbance (OD_{540}) and luminescence (in relative light units) measurements were taken every hour. The expression of CPS III (SZ1829) was higher in M9 plus 1% glucose alone compared to M9 plus 1% glucose plus 30% NHS. The expression of SZ1829 was 3-4 fold lower in 30% NHS. This was in contrast to the expression of CPS I (SZ211) (see section 3.5), which was significantly more highly expressed in 30% NHS at a level of 3-4 fold compared to growth in M9 plus 1% glucose alone [55]. Although CPS III demonstrated higher expression initially in 30% NHS, this may have been due to the fact that the addition of NHS caused precipitation in the media which affected the optical density of the cultures.

The expression of the *lux* operon in reporter strains SZ211 (CPS I$^-$) and SZ1829 (CPS III$^-$) was also measured in water to determine whether CPS III was induced in this environment. Overnight cultures of SZ211 and SZ1829 were inoculated into sterile water and incubated at 37°C without shaking. Capsule expression was determined as described above, but the luminescence/absorbance calculations for water were compared to the values for these strains when grown in LB. CPS III was found to be induced in water compared to LB. The expression of SZ1829 was found to be significant with an increase of 2-3 fold over the course of the experiment. The expression of CPS I was found to be greater than 4 fold higher in LB compared to water [55].

Microarray analysis of capsule expression was performed using a low-density DNA microarray. RNA was isolated from the livers and lungs of hamsters infected with *B. pseudomallei* and from *B. pseudomallei* grown in LB. The results of the microarray experiment are shown in Table 1 [55]. The level of gene expression, or fold change, is represented as the ratio of gene expression in the hamster compared to growth in LB. As shown in Table 1, CPS III genes were not found to be significantly expressed *in vivo* since most of the fold changes were determined to be less than 2-fold. Many of the genes had negative fold change values, indicating that these genes are suppressed in the host environment. The highest fold change result was 2.337866 for BPSS1827, a predicted glucose-6-phosphate isomerase, which is still much lower than the fold changes observed for CPS I genes, which were significantly higher [58].

Gene ID	**Predicted function**	**Fold Change (*in vivo* vs. *in vitro*)**
BPSS1825	Glycosyltransferase	1.757155
BPSS1826	Glycosyltransferase	0.093565
BPSS1827	Glucose-6-phosphate isomerase	2.337866
BPSS1828	Glycosyltransferase	-0.66607
BPSS1829	Glycosyltransferase	-0.18061
BPSS1830	Capsule export, tyrosine-protein kinase	-0.30655
BPSS 1831	Capsule export, outer membrane protein	0.725849
BPSS1832	Transport, tyrosine-protein phosphatase	1.52665
BPSS1833	UDP-glucose-6-dehydrogenase	-0.71411
BPSS1834	Sugar transferase	-0.07963
BPSS 1835	Mannose-1-phosphate guanyltransferase	-0.75075

*Note: Gene expression representative of the liver and lungs of infected hamsters.

Table 1. Microarray analysis of *B. pseudomallei* CPS III expression following intraperitoneal inoculation in the hamster model of melioidosis.

4.4. Carbohydrate composition of CPS III

Glycosyl composition analysis was performed on the purified capsule by combined gas chromatography/mass spectrometry (GC/MS). GC/MS results indicated that CPS III is composed of galactose, glucose, mannose, xylose, and rhamnose residues, with the highest proportion of carbohydrate being galactose and glucose. Glycosyl linkage analysis was also performed [55]. The predominant glycosyl residue detected was a terminally-linked heptopyranosyl (t-Hep) at a percentage of 23.2. Other residues detected were a terminally-linked and a 4-linked glucopyranosyl (t-Glcp) (4-Glcp) at percentages of 14.6 and 10.8, respectively [55].

5. Conclusion

Although significant advances have been made in the field, melioidosis continues to be a public health concern in many regions of the world [59]. Completion of the sequencing of the *B. pseudomallei* genome has revealed potential virulence determinants and comparative genomics between the genomes of *B. pseudomallei*, *B. thailandensis*, and *B. mallei* species has contributed to a better understanding of the organism. Further studies are ongoing to define the pathogenesis of *B. pseudomallei* and to identify effective vaccine candidates and diagnostic targets [54, 59].

To obtain virulence determinants unique to *B. pseudomallei*, we used subtractive hybridization between this organism and a related nonpathogenic organism, *B. thailandensis*. Analysis of the subtractive hybridization library revealed that *B. pseudomallei* contains a number of DNA sequences that are not found in *B. thailandensis*. One of the subtraction clones, pDD1015, demonstrated weak homology to a glycosyltransferase, WbpX, from *P. aeruginosa* [43]. The insert from pDD1015 was cloned into a mobilizable suicide vector for insertional inactivation of the glycosyltransferase gene in wild-type *B. pseudomallei*. The resulting strain, SR1015, was markedly less virulent than the parent strain in an animal model. We determined that SR1015 harbored a mutation in a glycosyltransferase gene involved in the production of a capsular polysaccharide which we subsequently designated as CPS I. We then identified the operon involved in the biosynthesis and transport of this capsular polysaccharide (CPS I) [42]. The genes identified encode for proteins that are similar to proteins involved in the biosynthesis and export of capsular polysaccharides, particularly those involved in the production of group 3 capsular polysaccharides. Group 3 capsules include the *E. coli* K10 capsule and may also include the *H. influenzae* group b capsule and the capsule produced by *N. meningitidis* serogroup B [8]. Group 3 capsules are always coexpressed with O serogroups, are not thermoregulated, are transported by an ABC-2 exporter system, and do not contain the *kpsU* and *kpsF* genes, and usually the gene clusters map near the *serA* locus [8]. Thus far, no *serA* locus that is associated with the type I O-PS cluster was identified, but this polysaccharide is coexpressed with O antigen and lacks the *kpsU* and *kpsF* genes, and genes encoding for a putative ABC-2 transporter have been identified. The genes involved in the production of group 3 capsules are organized into regions and are divergently transcribed. Regions 1 and 3 are generally conserved and contain genes involved in export of the polysaccharide. These regions flank region 2, which contains the biosynthetic genes and is not conserved between serotypes [4]. The genetic organization of the CPS I is also similar to that of other capsule gene clusters in that the genes are organized into more than one transcriptional unit and appear to be divergently transcribed [42].

The polysaccharide with the structure -3)-2-*O*-acetyl-6-deoxy-β-D-manno-heptopyranose-(1- was originally isolated and characterized as an O-PS component of LPS in *B. pseudomallei* and was designated type I O-PS [35]. However, our results suggested that this polysaccharide is a capsule rather than an O-PS moiety. The genes involved in the production of this capsule demonstrated strong homology to the genes involved in the production of capsular polysaccharides in many organisms, including *N. meningitidis, H. influenzae,* and *E. coli.* In addition, the export genes associated with this cluster are not associated with the previously characterized O-PS gene cluster [36]. Western blot analysis of proteinase K cell extracts and silver staining showed that this polysaccharide has a high molecular mass (200 kDa) and lacks the banding pattern seen with O-PS moieties. This conclusion was further supported by another group of researchers that demonstrated this polysaccharide is a capsule rather than an O-PS component of LPS because it lacks a lipid A moiety and was not capable of macrophage activation [49]. Studies by our laboratory have indicated that mutants in the production of the core oligosaccharide of the LPS are still capable of producing this polysaccharide [48]. Based on the above criteria and the genetic similarity to group 3 capsules, we proposed that this polysaccharide is a capsule.

Virulence genes of a number of pathogenic bacteria are located on pathogenicity islands (PAIs), regions on the bacterial chromosome that are present in the genome of pathogenic strains but rarely present in those of nonpathogenic strains. The PAIs may range in size from about 30 kb to 200 kb and often differ in G+C content from the remaining bacterial genome; the PAIs are often associated with the carriage of many virulence genes. These genetic units are often flanked by direct repeats and may be associated with tRNA genes or insertion sequence (IS) elements at their boundaries. They may also be associated with the presence of mobility genes, such as IS elements, integrases, transposases, and origins of plasmid replication. These DNA regions are considered to be unstable in that they may be subject to deletion with high frequency or undergo duplications and amplifications [7]. A number of PAIs have been described for both gram-positive and gram-negative bacteria, and the application of subtraction hybridization has been used to successfully identify such genetic elements [7]. The subtractive hybridization that was carried out between *B. pseudomallei* and *B. thailandensis* led to the identification of a number of sequences that were found to be A-T rich compared to the rest of the *B. pseudomallei* chromosome. This, combined with the fact that insertional mutagenesis of the glycosyltransferase gene identified by this method resulted in an avirulent strain, suggests that we may have identified DNA sequences from a putative PAI and that the capsular polysaccharide gene cluster may be located on this island. It is possible that *B. pseudomallei, B. mallei,* and *B. stabilis* acquired DNA encoding for capsule as well as other potential, yet unidentified virulence factors by horizontal transfer recently in evolution. *B. pseudomallei* and *B. mallei* are known to contain IS elements that are present in *B. cepacia* but not in *B. thailandensis* [56, 60].

Capsule production has been correlated with virulence in many bacteria, particularly those causing serious invasive infections of humans [61]. Our studies demonstrated that CPS I is critical for the virulence of *B. pseudomallei* [42, 50]. A number of functions have been suggested for polysaccharide capsules: prevention of desiccation for transmission and survival, adherence for colonization, resistance to complement-mediated phagocytosis and complement-mediated killing, and resistance to specific host immunity due to a poor antibody response to the capsule [4].

To establish a correlation between capsule production and clinical infection a number of *B. pseudomallei* strains isolated from clinical specimens were tested for CPS I production. All 55 strains tested were found to produce CPS I by western blot analysis [51]. In addition 10 strains of *B. thailandensis* were tested and found negative for CPS I production, confirming the importance of CPS I in virulence as well as clinical infection.

CPS I production by *B. pseudomallei* was shown to contribute to the persistence of the organism in the blood of the host. All CPS I mutants tested in the animal model could not be isolated from the blood following infection. The addition of purified capsule was shown to increase the virulence of the CPS I mutant strains SR1015 and SZ210 in the animal model. Differences in tissue distribution between wild type *B. pseudomallei* and SR1015 in infected hamsters indicated that SR1015 was cleared from the blood because the numbers of SR1015 in the blood of infected hamsters was 10,000-fold lower than that of wild type 1026b and lower than the initial inoculum of 100 cfu/ml [50].

CPS I production was shown to be responsible for persistence in the blood by evasion of the complement cascade and the mechanism for this was determined to be through the reduction of C3b deposition and opsonophagocytosis. The addition of purified CPS I to serum bactericidal assays showed that the capsule contributes to increased resistance of serum sensitive strains lacking the O-polysaccharide moiety (O-PS) of LPS to the bactericidal effects of normal human serum. However, CPS I mutants themselves were not found to be serum sensitive because they still produced O-PS, which was previously shown to be responsible for serum resistance, because it prevents lysis by the MAC complex [36]. This led us to postulate that CPS I was affecting the complement cascade through some other mechanism and it was found that this mechanism was through the reduction of C3b deposition and opsonization [50]. Both Western blot analysis and immunofluorescence microscopy experiments using a mouse monoclonal antibody to human C3b demonstrated the inhibition of C3b deposition by CPS I. In both experiments C3b deposition was more pronounced on the surface of the CPS I mutant compared to wild type. Also evident was that some C3b deposition occurred in the wild type, but this was expected since bacterial capsules are known to allow the diffusion of some C3b to the bacterial surface and *B. pseudomallei* is capable of activating the alternative pathway of complement culminating in the formation of the MAC complex [36, 62]. The accumulation of C3b affects the amplification step of the complement cascade and therefore, the less C3b deposited the less C5a is generated for phagocyte recruitment [63]. This explains the increased clearance of CPS I mutants from the blood. This conclusion was supported by the fact that *B. thailandensis*, the non-pathogenic organism which lacks CPS I, has been shown to be serum resistant, but is not capable of establishing a bacteremia in the Syrian hamster model of acute melioidosis [36, 37, 42]. Effective opsonization of invading bacteria results in enhanced phagocytosis and clearance of organisms form the blood of an infected host [52]. Quantitative radiolabelled phagocytic assays were also performed to establish a correlation between opsonization of the bacteria and phagocytosis by polymorphonuclear leukocytes. In the presence of serum, the CPS I mutant was more readily phagocytosed than wild type [50].

The expression of CPS I in the presence of normal human serum was found to be significantly elevated, also confirming that this capsule contributes to survival in the host. The presence of CPS I enables *B. pseudomallei* to survive in the blood through the inhibition of complement factor C3b deposition and phagocytosis [50]. The presence of this capsule facilitates survival as well as spreading to other organs, which can explain the overwhelming septicemia that is common in culture-positive melioidosis patients [64]. Therefore CPS I production is critical to the virulence of *B. pseudomallei* and further research will enhance the development of preventative strategies for melioidosis since this polysaccharide is one of the components of a *B. pseudomallei* subunit vaccine [28, 65].

Sequence analysis of the genome of *B. pseudomallei* revealed the presence of four operons possibly involved in polysaccharide capsule biosynthesis. One of these operons, (CPS I), corresponded to the previously identified and characterized mannoheptose capsule that was shown to be responsible for virulence and comprises one of the currently proposed melioidosis and glanders subunit conjugate vaccine [28, 66, 42, 50, 67]. The CPS I capsule cluster is present in the genome of *B. mallei* as well, but the complete cluster is not found in the genome of *B. thailandensis* [56, 57, 68]. This correlates with previous studies that have shown that this capsule is produced by *B. mallei*, but not by *B. thailandensis* [36, 37, 42, 56].

Three other putative capsule operons were identified by sequence analysis and all of these operons were found to be present in *B. pseudomallei* and *B. thailandensis*, but not *B. mallei*. Since these capsules are found in *B. thailandensis* and *B. pseudomallei*, they may be required for either survival in the host or in the environment; however, further studies are required to determine the roles of CPS II and CPS IV.

CPS III, located on chromosome 2, was found to contain 11 genes involved in the biosynthesis of a polysaccharide and was shown to be present in the genomes of *B. pseudomallei* and *B. thailandensis*, but not *B. mallei*. A mutation in the CPS III cluster did not affect production of CPS I and so it can be concluded that this operon encodes for gene products responsible for the biosynthesis of a separate capsule. CPS III was not found to contribute to the pathogenesis of *B. pseudomallei*. This capsule was not shown to be highly expressed *in vivo* by microarray analysis and was not required for virulence in the animal model. The CPS III mutant, SZ1829, which contains a mutation in the BPSS1829 gene as a result of insertional inactivation, was found to be as virulent in the animal model as wild type *B. pseudomallei*. The expression of this capsule was shown to be elevated when incubated in water, but suppressed in the presence of normal human serum [55]. The presence of the CPS III cluster in *B. pseudomallei* and *B. thailandensis*, both of which can survive for long periods in the environment compared to *B. mallei*, the increased expression of this capsule in water, and the low level of expression of this capsule *in vivo*, suggests that this capsule may contribute to the survival of *B. pseudomallei* in the environment [69].

Previous studies have demonstrated that *B. pseudomallei* produces three other capsular polysaccharides in addition to CPS I and these have been structurally characterized. One is an acidic polysaccharide with the structure, -3)-2-*O*-acetyl-β-D-Gal*p*-(1-4)-α-D-Gal*p*-(1-3)-β-D-Gal*p*-(1-5)-β-D-KDO*p*-(2-, which is recognized by patient sera [32]. The other two are: a branched 1,4-linked glucan polymer ((CP-1a) and a triple-branched heptasaccharide repeating unit composed of rhamnose, mannose, galatose, glucose, and glucoronic acid (CP-2) [49]. Combined GC/MS analysis of CPS III revealed that the composition of this capsule demonstrates some similarity to the composition of the previously described capsule CP-2 composed of rhamnose, mannose, galactose, glucose, and glucoronic acid; however, the proportions of carbohydrate residues were not similar, and the CPS III capsule was also found to contain xylose and not glucoronic acid. In addition, CPS III was determined to be composed primarily of heptose [55]. Therefore it is evident that the capsule identified in this study is not one of the previously described capsule structures. Some of the previously characterized capsules produced by *B. pseudomallei* have been shown to be produced under unique conditions [32, 49]. Strain variation, differences in expression of the capsules, and discrepancies between purification strategies may also explain why a number of capsules have been shown to be produced by this organism. Nevertheless, the genes BPSS 1825-1835 appear to be involved in the biosynthesis of a capsule with this composition. Further analysis by 2D NMR would be required to definitively establish a connection between CPS III and one of the other published structures.

Studies by another laboratory have also focused on the presence of these capsule clusters in *B. pseudomallei*. Sarkar-Tyson *et al.* identified two polysaccharide clusters, one of which corresponds to the CPS III presented in this paper, but the authors identified this cluster as type IV O-PS (2007). The type IV O-PS was found to be involved in virulence in a mouse model [70]. However, a mutant in this polysaccharide did not demonstrate any difference in hydrophobicity compared to wild-type, indicating that this polysaccharide does not contribute to making the cell surface more hydrophobic, which is an advantageous characteristic for some pathogenic bacteria. The differences in virulence compared to the current work can be attributed to the use of different animal models; however, all other data seem to indicate that this capsule is not required for virulence.

A study was recently published which outlines the identification of another capsule produced by *B. pseudomallei* [71]. This capsule was determined to be composed of 1,3-linked α-D-mannose residues. This capsular polysaccharide was also found to be produced by *B. mallei*. The genes involved in the synthesis of this polysaccharide have not yet been identified and work is also underway to determine the role for this novel capsule in the pathogenesis of melioidosis and glanders.

B. pseudomallei is an environmental saprophyte often found in soil and stagnant water and incidence of the disease is high in rice farmers in Southeast Asia [22, 69]. This organism harbors a large genome which explains its ability to survive for long periods of time in the environment as well exist as a significant pathogen in both humans and animals. The presence of multiple polysaccharide clusters in the genome and the production of multiple capsule structures under differing conditions may contribute to the ability of this organism to adapt to a variety of conditions. As demonstrated in this study, capsule expression is dependent on the particular environment, which indicates that *B. pseudomallei* produces these capsules to promote a survival advantage either in the host or in the environment. Further studies aimed at characterizing the capsules of *B. pseudomallei* will be beneficial to understand the pathogenesis of this organism and to advance further vaccine development.

Author details

Shauna L. Reckseidler-Zenteno
Athabasca University, Athabasca, Alberta, Canada

Acknowledgement

The author would like to thank Dr. Donald Woods and all colleagues in the *Burkholderia pseudomallei* research community for their support and discussions over the years. The work described in this chapter was funded by the following sources: Department of Defense contract DAMD 17-98-C-8003, the Medical Research Council of Canada, the Canadian Bacterial Diseases Network of Centers of Excellence, Canadian Institutes for Health Research MOP 36343, and a Research Incentive Grant from Athabasca University. Carbohydrate analysis was conducted by the Complex Carbohydrate Research Center at the University of Georgia and this was supported in part by the Department of Energy-funded (DE-FG09-93ER-20097) Center for Plant and Microbial Complex Carbohydrates. Microarray analysis was performed at the Southern Alberta Microarray Facility.

6. References

[1] Whitfield C (1988) Bacterial Extracellular Polysaccharides. Can. J. Microbiol. 34: 415-420.
[2] Costerton J W, Irvin R T, Cheng K-J (1981) The Bacterial Glycocalyx in Nature and Disease. Annu. Rev. Microbiol. 35: 299-324.
[3] Whitfield C, and Valvano M (1993) Biosynthesis and Expression of Cell-surface Polysaccharides in Gram-negative Bacteria. Adv. Microbiol. Phys. 35: 135-146.
[4] Roberts I S (1996) The Biochemistry and Genetics of Capsular Polysaccharide Production in Bacteria. Ann. Rev. Microbiol. 50: 285-315.
[5] Robbins J B, McCracken G H, Gotschlich E C, Ørskov I, Hanson L A (1974). *Escherichia coli* K1 Capsular Polysaccharide Associated with Neonatal Meningitis. N. Engl. J. Med. 90: 267-271.
[6] Grados O, Ewing V M (1970) Antigenic Relationships between *Escherichia coli* and *Neisseria meningitidis* group B. J. Infect. Dis. 122: 100-103.

[7] Hacker J, Blum-Oehler G, Mühldorfer I, Tschäpe H (1997) Pathogenicity Islands of Virulent Bacteria: Structure, Function and Impact on Microbial Evolution. Mol. Microbiol. 23: 1089-1097.

[8] Whitfield C, Roberts I S (1999) Structure, Assembly, and Regulation of Expression of Capsules in *Escherichia coli*. Mol. Microbiol. 31: 1307-1319.

[9] Frosch M, Müller D, Bousset K, Müller A (1992) Conserved Outer Membrane Protein of *Neisseria meningitidis* Involved in Capsule Expression. Infect. Immun. 60: 798-803.

[10] Roberson E, Firestone M (1992) Relationship between Dessication and Exopolysaccharide Production in Soil *Pseudomonas* sp. Appl. Environ. Microbiol.58: 1284-1291.

[11] Ophir T, Gutnick D (1994) A Role for Exopolysaccharide in the Protection of Microorganisms from Dessication. Appl. Environ. Microbiol. 60: 740-745.

[12] Berry A, DeVault J, Chakrabarty A (1989) High Osmolarity is a Signal for Enhanced *algD* Transcription in Mucoid and Nonmucoid *Pseudomonas aeruginosa* strains. J. Bacteriol. 171: 2312-2317.

[13] Costerton J W, Cheng K-J, Geesy G G, Ladd T I, Nickel J C et al. (1987) Bacterial Biofilms in Nature and Disease. Annu. Rev. Microbiol. 41: 435-464.

[14] Jenkinson F F (1994) Adherence and Accumulation of Oral Streptococci. Trends. Microbiol. 2: 209-212.

[15] Howard C J, Glynn A A (1971) The Virulence for Mice of Strains of *Escherichia coli* Related to the Effects of K antigens on their Resistance to Phagocytosis and Killing by Complement. Immunology 20: 767-777.

[16] Kim K, Kang, J, Cross A (1986) The Role of Capsular Antigens in Serum Resistance and In Vivo Virulence of *Escherichia coli*. FEMS Microbiol. Lett. 35: 275-278.

[17] Moxon E, Kroll J (1990) The Role of Bacterial Polysaccharide Capsules as Virulence Factors. Curr. Top. Microbiol. Immunol. 150: 65-85.

[18] Bhattacharjee A, Jennings H, Kenny C, Martin A, Smith I (1975) Structural Determination of the Sialic Acid Polysaccharide Antigens of *Neisseria meningitidis* Serogroups B and C with Carbon 13 Nuclear Magnetic Resonance. J. Biol. Chem. 250: 1926-1932.

[19] Vann W F, Schmidt M A, Jann B, Jann K (1981) The Structure of the Capsular Polysaccharide (K5 antigen) of Urinary Tract Infective Escherichia coli O10:K5:H4. A Polymer Similar to Desulfo-heparin. Eur. J. Biochem. 116: 359-364.

[20] Wyle F, Artenstein M, Brandt B L, Tramont E C, Kapser D L et al. (1972) Immunologic Response of Man to Group B Meningococcal Polysaccharide Vaccines. J. Infect. Dis. 126: 514-521.

[21] Dance, D A B (1991) Melioidosis: the Tip of the Iceberg? Clin. Microbiol. Rev. 4: 52–60.

[22] Leelarasamee A, Bovornkitti S (1989). Melioidosis: review and update. Rev. Infect. Dis. 11: 413–425.

[23] Chaowagul, W, White N J, Dance D A, Wattanagoon Y, Naigowit P, Davis T M, Looareesuwan S, Pitakwatchara N (1989). Melioidosis: a Major Cause of Community Acquired Septicemia in northeastern Thailand. J. Infect. Dis. 159: 890–899.

[24] Howe C, Sampath A, Spotnitz M (1971) The Pseudomallei Group: a Review. J. Infect. Dis. 124: 598–606.

[25] Ip, M, Osterberg L G, Chau P Y, and Raffin T A (1995) Pulmonary Melioidosis. Chest 108: 1420–1424.

[26] Currie B J, Fisher D A, Anstey N M, Jacups S P (2000) Melioidosis: Acute and Chronic Disease, Relapse and Re-activation. Trans. R. Soc. Trop. Med. Hyg. 94: 301-304.

[27] White N J, Dance D A B, Chaowagul, W, Wattanagoon Y, Wuthiekanun V, Pitakwatchara N (1989) Halving of Mortality of Severe Melioidosis by Ceftazadime. Lancet 2: 697-701.

[28] Brett P J, Woods D E (1996) Structural and Immunologic Characterization of *Burkholderia pseudomallei* O-polysaccharide–flagellin Protein Conjugates. Infect. Immun. 64: 2824–2828.

[29] Bryan L E, Wong S, Woods D E, Dance D A, Chaowagul W (1994). Passive protection of diabetic rats with antisera specific for the polysaccharide portion of the lipopolysaccharide isolated from *Pseudomonas pseudomallei*. Can. J. Infect. Dis. 5: 170–178.

[30] Steinmetz I, Rohde M, Brenneke B (1995) Purification and Characterization of an Exopolysaccharide of *Burkholderia* (*Pseudomonas*) *pseudomallei*. Infect. Immun. 63: 3959–3965.

[31] Masoud, H, Ho M, Schollaardt T, Perry M B (1997) Characterization of the Capsular Polysaccharide of *Burkholderia pseudomallei* 304b. J. Bacteriol. 179: 5663–5669.

[32] Nimtz M, Wray V, Domke T, Brenneke B, Haussler S, Steinmetz I (1997). Structure of an Acidic Exopolysaccharide of *Burkholderia pseudomallei*. Eur. J. Biochem. 250: 608–616.

[33] Karawara K, Dejsirilert S, Ezaki T (1998) Characterization of Three Capsular Polysacharides Produced by *Burkholderia pseudomallei*. *FEMS Microbiol Lett* 169: 283-287.

[34] Knirel Y A, Paramonov N A, Shashkov A S, Kochetkov N K, Yarullin R G, Farber S M, Efremenko V I (1992) Structure of the Polysaccharide Chains of *Pseudomonas pseudomallei* Lipopolysaccharides. Carbohydr. Res. 233: 185–193.

[35] Perry M B, MacLean L L, Schollaardt T, Bryan L E, Ho M (1995) Structural Characterization of the Lipopolysaccharide O antigens of *Burkholderia pseudomallei*. Infect. Immun. 63: 3348–3352.

[36] DeShazer D, Brett P J, Woods D E (1998). The Type II O-antigenic Polysaccharide Moiety of *Burkholderia pseudomallei* Lipopolysaccharide is Required for Serum Resistance and Virulence. Mol. Microbiol. 30: 1081–1100.

[37] Brett P J, DeShazer D, Woods D E (1997) Characterization of *Burkholderia pseudomallei* and *Burkholderia pseudomallei*-like strains. Epidemiol. Infect. 118:137–148.

[38] Brett P J, DeShazer D, Woods D E (1998) *Burkholderia thailandensis* sp. nov., description of a *Burkholderia pseudomallei*-like species. Int. J. Syst. Bacteriol. 48: 317–320.

[39] Chaiyaroj S C, Kotrnon K, Koonpaew S, Anantagool N, White N J, Sirisinha S (1999) Differences in Genomic Macrorestriction Patterns of Arabinose-positive (*Burkholderia thailandensis*) and Arabinose-negative *Burkholderia pseudomallei*. Microbiol. Immunol. 43: 625–630.

[40] Wuthiekanun V, Smith M D, Dance D A, Walsh A L, Pitt T L, White N J (1996) Biochemical Characteristics of Clinical and Environmental Isolates of *Burkholderia pseudomallei*. J. Med. Microbiol. 45: 408–412.

[41] Trakulsomboon S, Dance D A B, Smith M D, White N J, Pitt T L (1997) Ribotype Differences between Clinical and Environmental Isolates of *Burkholderia pseudomallei*. J. Med. Microbiol. 46: 565–570.

[42] Reckseidler S L, DeShazer D, Sokol P A, Woods D E (2001) Detection of Bacterial Virulence Genes by Subtractive Hybridization: Identification of Capsular Polysaccharide of *Burkholderia pseudomallei* as a Major Virulence Determinant. Infect. Immun. 69: 34–44.

[43] Rocchetta H L, Burrows L L, Pacan J C, Lam J S (1998) Three Rhamnosyltransferases Responsible for Assembly of the A-band D-rhamnan Polysaccharide in *Pseudomonas aeruginosa*: a Fourth Transferase, WbpL, is Required for the Initiation of both A band and B-band Lipopolysaccharide Synthesis. Mol. Microbiol. 28: 1103–1119.

[44] Mongkolsuk S, Rabibhadana S, Vattanaviboon P, Loprasert S (1994) Generalized and Mobilizable Positive-selective Cloning Vectors. Gene 143: 145–146.

[45] Steinmetz I, Reganzerowski A, Brenneke B, Haussler S, Simpson A, White N J (1999) Rapid Identification of *Burkholderia pseudomallei* by Latex Agglutination Based on an Exopolysaccharide-specific Monoclonal Antibody. J. Clin. Microbiol. 37: 225–228.

[46] Clarke B R, Pearce R, Roberts I S (1999) Genetic Organization of the *Escherichia coli* K10 Capsule Gene Cluster: Identification and Characterization of Two Conserved Regions in Group III Capsule Gene Clusters Encoding PolysaccharideTransport Functions. J. Bacteriol. 181: 2279–2285.

[47] Reeves P R, Hobbs M, Valvano M A, Skurnik M, Whitfield C, Coplin D, Kido N, Klena J, Maskell D, Raetz C R, Rick P D (1996) Bacterial Polysaccharide Synthesis and Gene Nomenclature. Trends Microbiol. 498: 495–503.

[48] Burtnick M N, Woods D E (1999) Isolation of Polymyxin B-susceptible Mutants of *Burkholderia pseudomallei* and Molecular Characterization of Genetic Loci Involved in Polymyxin B Resistance. Antimicrob. Agents Chemother. 43: 2648–2656.

[49] Isshiki Y, Matsuura M, Dejsirilert S, Ezaki T, Kawahara K (2001) Separation of 6-deoxy-heptan from a Smooth-type Lipopolysaccharide Preparation of *Burkholderia pseudomallei*. FEMS Microbiol. Lett. 199: 21-25.

[50] Reckseidler-Zenteno S L, DeVinney R, Woods D E (2005). The Capsular Polysaccharide of *Burkholderia pseudomallei* Contributes to Survival in Serum by Reducing Complement Factor C3b Deposition. Infect. Immun. 73: 1106-1115.

[51] Reckseidler-Zenteno S L (2002) Role of the *Burkholderia pseudomallei* Capsule. University of Calgary. 228 p.

[52] Joiner K A, Brown E J, Frank M M (1984) Complement and Bacteria: Chemistry and Biology in Host Defense. Annu. Rev. Microbiol. 2: 461–491.

[53] Moore R A, Reckseidler-Zenteno S, Kim H, Nierman W, Yu Y, Tuanyok A, Warawa J, DeShazer D, Woods D E (2004) The Contribution of Gene Loss to the Pathogenic Evolution of *Burkholderia pseudomallei* and *Burkholderia mallei*. Infect. Immun. 72: 4172–4187.

[54] Holden M T, Titball R W, Peacock S J, Cerdeño-Tárraga A M, Atkins T, Crossman L C, Pitt T, Churcher C, Mungall K, et al. (2004) Genomic Plasticity of the Causative Agent of Melioidosis, *Burkholderia pseudomallei*. Proc. Natl. Acad. Sci. USA 101: 14240-14245.

[55] Reckseidler-Zenteno S L, Viteri D F, Moore R, Wong E, Tuanyok A, Woods D E (2010) Characterization of the Type III Capsular Polysaccharide Produced by *Burkholderia pseudomallei*. J. Med. Microbiol. 59(12): 1403-1414.

[56] DeShazer D, Waag D M, Fritz D L, Woods D E (2001) Identification of a *Burkholderia mallei* Polysaccharide Gene Cluster by Subtractive Hybridization and Demonstration that the Encoded Capsule is an Essential Virulence Determinant. Microb. Pathog. 30: 253-269.

[57] Yu Y, Kim H S, Chua H H, Lin C H, Sim S H, Lin D, Derr A, Engels R, DeShazer D et al. (2006) Genomic Patterns of Pathogen Evolution Revealed by Comparison of *Burkholderia pseudomallei,* the Causative Agent of Melioidosis, to Avirulent *Burkholderia thailandensis*. BMC Microbiol. 6: 46.

[58] Tuanyok A, Tom M, Dunbar J, Woods D E (2006) Genome-wide Expression Analysis of *Burkholderia pseudomallei* Infection in a Hamster Model of Acute Melioidosis. Infect. Immun. 74: 5465-5476.

[59] Cheng A C, Currie B J (2005) Melioidosis: Epidemiology, Pathophysiology, and Management. Clin. Microbio.l Rev. 18: 383-416.

[60] Mack K, Titball R W (1998) The Detection of Insertion Sequences within the Human Pathogen *Burkholderia pseudomallei* which have been Identified Previously in *Burkholderia cepacia*. FEMS Microbiol. Lett. 162: 69–74.

[61] Boulnois G J, Roberts I S (1990) Genetics of Capsular Polysaccharide Production in Bacteria. Curr. Top. Microbiol. Immunol. 150: 1–18.

[62] Egan, A M, Gordon, D L (1996) *Burkholderia pseudomallei* Activates Complement and is Ingested but not Killed by Polymorphonuclear Leukocytes. Infect. Immun. 64: 4952-4959.

[63] Densen P (1995) Complement. In: Mandell G L, Bennett J E, Dolin R, editors. Principles and Practice of Infectious Diseases. New York: Churchill Livingstone. pp. 58-78.

[64] Supputamongkol Y, Rajchanuwong A, Chaowagul W, Dance D A B, Smith M D, Wuthiekanun V, Walsh A L, Pukrittayakamee S, White N J (1994) Ceftazadime vs. Amoxicillin/Clavulanate in the Treatment of Severe Melioidosis. Clin. Infect. Dis. 19: 846-853.

[65] Patel N, Conejero L, De Reynal M, Easton A, Bancroft GJ, Titball RW (2011) Development of Vaccines against *Burkholderia pseudomallei*. Front. Microbiol. 2: 198.

[66] Brett P J, Woods D E (2000) Pathogenesis of and Immunity to Melioidosis. Acta. Trop. 74: 201-210.

[67] Warawa J, Woods D E (2002) Melioidosis Vaccines. Expert Rev. Vaccines 1: 477-482.

[68] Nierman W C, DeShazer D, Kim H S, Tettelin H, Nelson K E, Feldblyum T, Ulrich R L, Ronning C M, Brinkac L M, et al (2004) Structural Flexibility in the *Burkholderia mallei* Genome. Proc. Natl. Acad. Sci. USA 101: 14246-14251.

[69] Aldous P (2005) Tropical Medicine: Melioidosis? Never heard of it… Nature 434: 692-693.

[70] Sarkar-Tyson M, Thwaite J E, Harding S V, Smither S J, Oyston P C F, Atkins T P, Titball, R W (2007) Polysaccharides and Virulence of *Burkholderia pseudomallei.* J. Med. Microbiol. 56: 1005-1010.

[71] Heiss C, Burtnick M N, Wang Z, Azadi P, Brett P J (2012) Structural Analysis of Polysaccharides Expressed by *Burkholderia mallei* and *Burkholderia pseudomallei.* Carbohydrate Res. 349: 90-94.

Chapter 6

Polysaccharides from Larch Biomass

Natalya Nikolaevna Trofimova, Elena Nikolaevna Medvedeva,
Nadezhda Viktorovna Ivanova, Yuriy Alekseevich Malkov
and Vasiliy Anatolievich Babkin

Additional information is available at the end of the chapter

http://dx.doi.org/10.5772/53809

1. Introduction

There are two species of the genus *Larix* Mill, *Larix sibirica* Ledeb. (Figure 1) and *Larix gmelinii* (Rupr.) Rupr., considered to be the most abundant trees in the Russian Federation: the total stock of their wood exceeds 26 billion m^3. Traditionally, the main economic value of larch wood consists of the manufacture of roundwood (timber), the value of which is connected with the high quality of lumber from this breed of tree. Larch wood can sometimes be used in insignificant quantities in pulp-and-paper manufacture to obtain pulp. About 40% of this valuable breed (as bark, sawdust) enters waste through existing lumber manufacturing processes. Such an irrational approach to the development of larch wood does not allow exploitation of the richest potential of the given renewed source. Meanwhile, biologically active compounds contained in larch biomass can be used for the manufacture of products for medical, food and agricultural purposes with maximal benefit. The development of complex technology for chemical processing of larch biomass and waste timber will considerably raise the economic value of this biological resource.

This chapter is devoted to polysaccharides contained in the wood and bark of larch, i.e. heteropolysaccharide arabinogalactan (AG) and pectin polysaccharides (pectin substances, PS). The chapter also deals with the development of technologies for the preparation of these polysaccharides and the study of their useful properties. Wood hemicelluloses and cellulose represent a potential source of valuable monosaccharides, namely glucose, which can be readily obtained by hydrolysis transformation of these polysaccharides.

Larch wood contains a high quantity of arabinogalactan, a considerable part of which can be found in the butt of a tree, which frequently enters waste. Arabinogalactan possesses a wide spectrum of biological activity. It shows immunomodulating activity, gastroprotective, membranotropic and prebiotic properties, and can be used in the medical, veterinary, food

INTECH
open science | open minds

© 2012 The Author(s). Licensee InTech. This chapter is distributed under the terms of the Creative Commons Attribution License http://creativecommons.org/licenses/by/3.0), which permits unrestricted use, distribution, and reproduction in any medium, provided the original work is properly cited.

and cosmetic industries. Arabinogalactan is a perspective matrix for obtaining on its basis metal-, sulpho-, amino- and other derivatives due to the reactive hydroxyl and aldehyde groups contained in its molecule. The ability of arabinogalactan to form water-soluble stable substances with inorganic nanoparticles and low-molecular medical substances (MS) provides serious prospects for the development of materials with unique properties.

Figure 1. *Larix sibirica* Ledeb.

In the USA, arabinogalactan has been extracted from the wood of *L. occidentalis* Nutt. and *L. laricina* (Du Roi) by K. Koch for more than 40 years as a commercial product. Effective immunity-modulating and prebiotic biologically active food additives have been developed to improve quality of human life. Application of arabinogalactan in agriculture as fodder additive allows the greater efficiency of animal industries. Arabinogalactan is not currently manufactured in Russia.

The bark of larch does not however have industrial application. Annually, wood-processing industries and pulp-and-paper enterprises waste more than 30 million m^3 in volume. It has become a serious environmental problem because the bark is badly exposed to biodegradation. At the same time, the chemical compounds of the bark can be a source of valuable biologically active substances, including polysaccharides. The creation of medical, food and other useful products on the basis of polysaccharides is also possible.

Larch bark contains about 7–12% of pectin polysaccharides, based on the weight of absolutely dry raw material. Pectin is acid polysaccharide–glycogalacturonane, and is contained in practically all plants. It is obtained from diverse sources that differ in their chemical structure due to distinctions in the qualitative structure of carbohydrates and their quantitative parities. Pectin substances promote digestive processes, and help organisms to resist many diseases such as atherosclerosis, diabetes, cancer, etc. There are two basic sources of pectin production in Germany and Denmark. However, there is no industrial production of pectin in Russia.

As one of the possible sources of polysaccharides, cellulose–lignin residue can be formed by the extraction of polyphenolic substances from larch wood in the scheme for complex processing of larch biomass. Its chemical processing allows carbohydrate products to be produced, mainly crystalline glucose. Glucose is contained, mainly in the bound form, in considerable amounts of natural products as a constituent of various glycosides and polysaccharides. Taking into consideration that about 40% of the dry substance of plant mass is accounted for by cellulose formed by photosynthesis at a rate of about 70 kg/day per each inhabitant of our the planet, glucose is the most abundant sugar in nature.

All developed technologies are environmentally friendly. They are focused on processing of timber and industrial wood-processing waste, and possess high technical and economic parameters.

2. Larch Arabinogalactan

2.1. Physicochemical and biological properties of Arabinogalactan

Larch wood is distinctive for its high content of water soluble polysaccharide arabinogalactan, reaching up to 35% by weight of dry wood [1]. This valuable substance has been studied since the 1950s [1-6]. The physicochemical and biological properties of AG from the wood of *L. occidentalis* Nutt. and Siberian larch species *L. sibirica* Ledeb. and *L. gmelinii* (Rupr.) Rupr. are the most explored. The wood of Siberian larches contains up to 10–15% of AG [2,7] and is a reliable source of industrial raw material for AG production.

The AG macromolecule from larch wood has a highly branched structure. Its main chain consists of β-(1→3) linked galactose residues (Figure 2). Approximately one half of the side chains (in *L. occidentalis* Nutt., *L. sibirica* Ledeb. and *L. gmelinii* (Rupr.) Rupr.) is formed of β-(1→6)-linked dimers of galactopyranose; galactopyranose monomers comprise about a quarter; and the remainder contains the major part of the polysaccharide's arabinose in aggregates of two or more monomers [1,8,9]. Arabinose fragments mainly occur as side chains consisting of 3-O-substituted β-L-arabinofuranose residues and terminal residues of β-L-arabinopyranose, β-D-arabinofuranose and α-L-arabinofuranose [8,9]. However, arabinose fragments have also been found in the main chain [1]. Glucuronic acid fragments in AG from wood of various larch species is low in content. As for AG from the above species, no glucuronic acid fragments have been detected in the purified samples [6].

Monosaccharide composition and molecular mass (MM) of AG macromolecules differ among the species and also varies within single species. It has been established that the composition of AG macromolecules is dependent on the conditions under which it is isolated from larch wood and on the purification procedure [9,10] as well as on molecular weight [1,11]. AG macromolecules have low molecular weights (13–20 kDa, according to HPLC data) and a narrow molecular weight distribution (degree of polydispersity 1.1–2.3) [10].

Biological activity of AG in higher plants is directly linked to their structural characteristics, such as length of galactan chain, structure of side chains, molecular weight and ability to form intermolecular associates [12,13]. Larch AG is characterized by low toxicity, showing neither acute poisoning for doses of 5 g/kg nor chronic poisoning for doses of 500 mg/kg per day [14]. Diverse biological activity of AG includes immunomodulatory, prebiotic, hypolipidemic, gastro- and hepatoprotective, mitogenic, antimutagenic and antiviral effects, etc. There are reports on the inhibitory and destructive action of AG against certain types of malignant tumours [6,15-22]. Moreover, it has a good solubility in cold water, uniquely low viscosity of concentrated aqueous solutions, an ability to bind fat and retain liquid and dispersive capacity, etc. All these benefits are in high demand in medicine and veterinary science as well as in food and cosmetic industries [5,6]. There are a number of biologically active food supplements which incorporate AG [23-25]. In medicine, an ophthalmic composition (eye drops/contact lens care solution) has been developed [26]. Membranotropicity caused by galactose fragments and realized through receptor mediated endocytosis makes AG a promising drug carrier to increase absorbability and selectivity of medical substances that are characterized by low bioavailability [27-36]. Applications of AG in photodynamic diagnostics, in oncological disease therapy and in gene therapy (targeted delivery of functional genes) are currently being explored [37-39]. The unique properties of AG are prominent among the known polysaccharide carriers of medical substances [40,41].

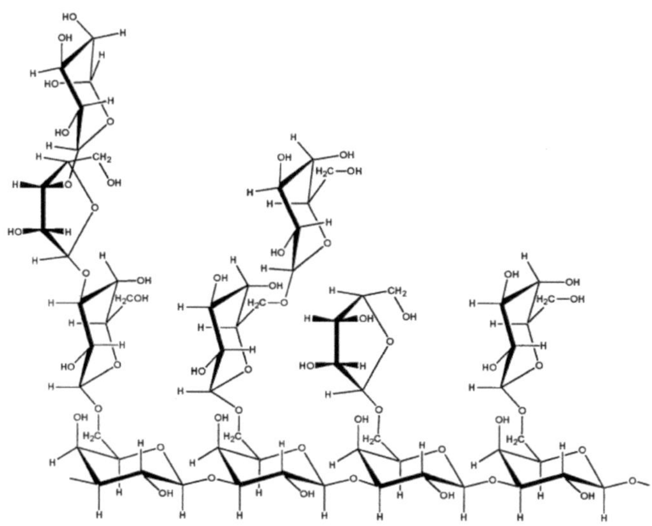

Figure 2. Structural fragment of AG macromolecule

Significant interest for medicine is raised by the products of AG modification [6]. Introducing diverse functional groups into the AG macromolecule makes it a manifold synthone to obtain a wide range of new biologically active substances. Oxidation is a promising way to functionalise AG. Methods of selective oxidation developed in classical carbohydrate chemistry [42] have been used to develop functionalized AG products [43,44]. Among others, oxidative destruction with simultaneous introduction of carboxylic groups into the macromolecule under the action of hydrogen peroxide in aqueous medium has been carried out [45]. It has been revealed that oligomeric products show anti-inflammatory and antiulcer activities. Reactions of AG and oxidized AG with some known MS give intermolecular complexes [46-48]. Such complexes of AG with 5-aminosalicylic acid show high antiulcer activity, while complexes incorporating 4-aminosalicylic or isonicotinic acid hydrazide demonstrate an antituberculosis effect. It has been established that conjugates of AG and products of its modification increase the physiological effect of MS and decrease their toxicity [28,46-48]; for instance, a conjugate of AG (9kDa) 9-β-D-arabinofuranosyladenine-5'-monophosphate was 25-fold more active than the parent compound 9-β-D-arabinofuranosyladenine (araA) in decreasing the amount of hepatitis B virus, and the toxicity of the complex preparation is much lower than araA [28]. To increase reactivity of AG with MS, the synthesis of conjugates proceeds by bromination, phosphorylation, amination, formation of hydrazides or reaction with $NaBH_4$ [27,33,34].

Mechanochemical activation is another promising method of AG modification, in which the target products are obtained in one stage without the use of solvents. Mechanochemical treatment can give rise to numerous physicochemical transformations of AG macromolecules, which are associated, first, with the breaking and formation of valence bonds and, second, with the disturbance and origination of weak intermolecular interactions (disordering, conformational rearrangements, etc.). As a result, the polysaccharide can change its biological activity, toxicity and pharmacological properties. HPLC and quantitative ^{13}C NMR spectroscopy has established [49] that mechanochemical treatment of AG isolated from Siberian larch wood changes its molecular weight distribution, monosaccharide composition and degree of branching due to partial destruction of the macromolecules and subsequent recombination of their fragments. The extent of these changes depends on the activation conditions. The IR and ^{13}C NMR spectra have not shown any functionalization of AG macromolecules in the conditions studied. Toxic-pharmacological study has established that a mechanochemically activated AG sample has the same LD_{50} (more 5000 mg/kg) than the starting AG. As for the effect of mechanochemically activated AG upon the central nervous system, it has demonstrated anxiolytic activity similar to sibason in a dose of 20 mg/kg tested in laboratory animals. Meanwhile, single intravenous injection of the substance, in a dose of 3.5 mg/kg, slightly but statistically significantly decreases arterial pressure (by 6%) in normotensive rats without affecting electrocardiogram parameters and heart rate. Thus, mechanochemically activated AG is a promising drug carrier [49].

Combined mechanochemical activation of MS with AG (pharmacon clathration) is even more effective in improving safety and bioavailability of drugs [50-53]. Essentially (up to 50 times) increased solubility of MS and dramatically decreased therapeutic doses of the same efficiency are reported for clathrates of poorly soluble anti-inflammatory, psychotropic and hypotensive drugs [51,52]; for instance, clathrates of AG with nifedipine containing a 10 times lower dose than the starting MS show pronounced hypotensive and antiarrhythmic effects [52]. Additionally, the side effects of MS in clathrates are decreased, for instance in the ulcerogenity of nonsteroidal anti-inflammatory drugs. HPLC, ^{13}C NMR and IR spectroscopic studies have shown the absence of any chemical reaction between AG and MS at pharmacon clathration [51,53]. The X-ray phase study and thermal analysis prove the destruction of the crystal structure of MS and its dispersion within the AG matrix. The polysaccharide macromolecules are cleaved similarly to the mechanochemical activation of AG alone [49].

L. sibirica Ledeb. and *L. gmelinii* (Rupr.) Rupr. contain much bioflavonoid dihydroquercetin (DHQ, taxifolin) and their diverse biological activity is well studied. It is an officinal drug with a wide range of therapeutic action and is also the basis for a number of efficient medical preparations and food supplements [54]. Complexes of AG and DHQ combining these unique properties are very promising. Such complexes have been obtained by pharmacon clathration [54] and have shown essentially improved solubility (up to 38 times) in comparison with starting DHQ and untreated AG/DHQ mixture. No chemical reaction between AG and DHQ takes place, as in clathrates [51]. According to HPLC data, AG in AG/DHQ clathrates has a narrower molecular weight distribution in comparison with pure mechanochemically activated AG, due to a decrease in both high- and low-molecular fractions, and thus DHQ stabilizes the polysaccharide macromolecules in mechanochemical treatment.

The most recent application of AG as a stabilizing polymer matrix in hybrid nanosized materials is based on iron oxides, cobalt, copper, nickel, ferrites and zero-valence metals such as silver, palladium and platinum [18,55-57]. Metal content in nanocomposite samples depends on synthetic conditions and on the type of metal ion used, varying in the range of 0.1–21.0%. In the case of metal oxide nanocomposites, AG shows properties of a nanostabilizing matrix, while in the composites of noble metals it reduces metal to a zero-valence state and stabilizes the metal nanoparticles formed. Nanocomposites based on AG retain high biological activity. Ferroarabinogalactans show synergy between the pronounced antianaemic activity of the ferric core and the unique membranotropic and immunomodulatory properties of AG. Parenteral administration of ferroarabinogalactan normalizes quantitative and qualitative characteristics of the erythrocyte system and iron depot level in animals (white rats) [18]. The original synthetic method of ferroarabinogalactan retains both membranotropic and immunomodulatory properties of AG. Studies of the natural effects of immunomodulators, together with the investigation of specific immunity to plague, have revealed that ferroarabinogalactan activates peritoneal macrophages in guinea pigs, in comparison with the animals' cells being immunized only with vital plague vaccine.

Antibiotic resistance of microorganisms has led to a new interest in silver preparations. The most efficient are preparations of ultradispersed silver. Highly dispersed (nanosized) particles increase bactericidal activity. It has been established that silver-containing nanocomposites with AG possess high antimicrobial activity against gram-negative enterobacteria (*Escherichia coli, Salmonella typhimurium, Candida albigans, Bacillus subtilis* and *Staphyllococcus aureus*) [57].

Thus, the method for synthesis of nanocomposites with available polysaccharide AG is an easy way to synthesize universal materials. The AG-based nanobiocomposites synergistically combine the properties of the stabilizing natural polysaccharide matrix and the nanocore materials. They are applicable as nanosized water-soluble enantioselective catalysts, magnet-controlled medical substances, materials for coherent and nonlinear optics, high-sensitivity optical markers, universal antimicrobial preparations, etc. The use of AG as a bioactive polysaccharide matrix participating in the processes of receptor-induced endocytosis leads to new approaches to therapy for metal deficiency states and to the development of new biomaterials of target action, which are in high demand in medicine and biology, both as controlled composite materials and as new water-soluble biodegradable metal-containing drugs.

At present, healthy diet is a question of public policy in all developed countries due to the undisputed role of food in public health, working capacity of people, adaptation, child growth and longevity. The increasing popularity of healthy diets has made manufacturers pay more and more attention to functional food, i.e. medically fortified food products. Food supplements and enriched products are becoming increasingly popular. Such products are functionalized by both natural substances (pectin, inulin, gum arabic, etc.) and semi-synthetic compounds (lactulose, polydextrose, resistant starches, chitosan, etc.).

Larch arabinogalactan adds nutrition and function to beverages, snack foods, nutrition bars, and more [6,58]. Not only does AG function as a prebiotic fibre and immunity enhancer, it also retains moisture, enhances mouthfeel and bulk, and improves shelf stability. Because of AG's low-viscosity profile and emulsification-enhancing properties, the most immediate applications include refrigerated and non-refrigerated beverages, and beverage mixes. Interest has also been expressed in snack foods, bars, ready-to-eat cereals, yogurt/dairy products and baked goods. Commercially available larch AG-containing products include beverages and nutrition bars.

Not only does AG add nutrition, it also provides functional benefits. An independent food laboratory confirmed that the inclusion of AG improved white pan bread make-up, external symmetry and internal grain scores. Fat-free flour tortillas with AG showed better handling, taste and aroma than the control. AG is a low-calorie additive for artificial sweeteners. It delivers mouthfeel, taste and bulking attributes that are most like sugar.

In confectionery and baked goods, AG lowers water activity and aids flavour and oil retention. AG can be used in browning compositions for uncooked foods, in seasoning powders to improve flow and reduce hygroscopicity, and in starch-containing foods to inhibit swelling.

Recent clinical investigations of AG have demonstrated not only benefits to gastrointestinal health and immunity, but also a significant reduction in serum cholesterol, glucose and insulin levels [58,59]. This opens the door to potential heart health label claims and provides an option for consumers looking for foods that are beneficial in terms of body weight, blood glucose or blood insulin control.

AG benefits, as determined by human and animal clinical trials, have been observed as low as 1.5 g/day, or more specifically 20 mg/kg of body weight. AG has been found to be totally safe as a food ingredient. On average, finished products containing a minimum of 60 mg/kg of body weight or about 4.5 g/day is recommended for foods [58]. Using arabinogalactan additives isolated from Siberian larch, the authors of one study [60] examined the soft wheat flour quality and quantity of gluten, physical properties of the dough, and quality of finished bread, depending on the quantity of the added polysaccharide. The addition of 1% of arabinogalactan to flour causes a significant improvement in the qualitative indices of bread. In this case, AG is totally consumed in the course of bread making because it is utilized by yeast. It is recommended that bread quality can be improved when the flour incorporates 1% mass of AG. When 2–3% of AG is added to flour, the AG content decreases. An excess of AG inhibits yeast growth, which leads to a decrease in bread quality.

The optimum compositions for AG-enriched bakery goods and pastry have been proposed [61,62]. It has been shown that when AG is added to flour in proportions of 1–5% by weight, bakery products are rich in dietary fibre of prebiotic and immunostimulating action, while their energy density is lower due to the decreased amount of sugar in the recipes. The AG-enriched bakery and pastry products have a medical effect [59]. Production technology for AG-enriched prebiotic cultured milk products has been developed [63]. The efficiency of AG in veterinary medicine has been proven [64-67].

2.2. Technology of AG production

The only regular production of AG is realized in the USA where the raw materials are derived from *L. occidentalis* Nutt. [5,6]. The known AG production techniques are all based on extraction of the polysaccharide from larch wood particles (chips, shavings, sawdust) and differ in pre-treatment of the raw material, conditions of extraction and methods of purifying the extracts and the final product [10]. The extraction process is highly dependent on the wood reduction range. Sawdust (specific surface of 164 cm^2/g) yields almost 90% of AG during the first 10 min of extraction while chips (specific surface of 17 cm^2/g) yield only 10% [68]. To intensify extraction, mechanochemical activation of larch wood and its further treatment with superheated water steam ("autohydrolysis explosion") is proposed [69]. The AG yields increase significantly (more than twice) if larch sawdust extraction is conducted with microwave or pulse ultrasound activation. Microwave or shock-and-acoustic processing of sawdust allows 90% of AG extraction to be completed in 30–60 seconds [70]. The products of larch wood water extraction, together with AG, are various phenolic compounds (low-molecular and oligomeric flavonoids, lignans, lignine substances and tannins), making production of high purity AG a complicated problem [71,72]. Purification of AG water extracts from those impurities is a current issue.

We propose a beneficial, economically and ecologically sound method to obtain dry 95–97% AG [17,73]. There are two main stages: first, extraction and purification of the extract, and second, dry product isolation. Every stage has been carefully studied and theoretically rationalized.

Study of extraction process kinetics is primarily aimed at determining duration of contact between the phases to give the target degree of isolation. Extraction kinetics data determine the geometry of the apparatus. AG water extraction proceeds in two stages, the first of which is fast and the second is slow (Figure 3).

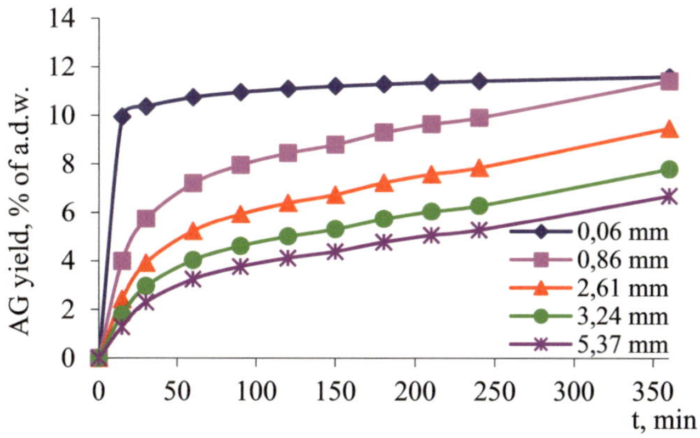

Figure 3. AG yield (% of absolutely dry wood, a.d.w.), showing dependence on particle size

The extraction process is determined by AG diffusion and penetration of extracting solution into wood pores as well as by hydrodynamic conditions. Thirty minutes is sufficient to isolate almost the total amount of AG from sawdust, while, in the same conditions, wood chips need several days to yield the major fraction of AG. The quality of AG and other water-soluble components is not affected by wood reduction range.

On the basis of the experimental data, we calculated diffusion constants, mass-transfer coefficients and the Biot diffusion criterion (Bi), characterizing the influence of hydrodynamic conditions on AG isolation rate. For the extract (Y_p), extraction was conducted until the equilibrium of the AG solution concentrations in wood pores reached equilibrium. As $\tau \rightarrow \infty$, Fourier diffusion criterion $Fo_g \rightarrow \infty$. Thus, the kinetic equation of the extraction process can be described by the equation [74]:

$$\frac{X - Y^*}{X_s - Y_s} = \sum_{n=1}^{\infty} A_n e^{-\mu_n^2 Fo_g} \tag{1}$$

where X is the average AG solution concentration in wood pores, at the moment of time τ, g/cm^3; Y^* is the equilibrium concentration of the substance isolated in the solution, g/cm^3; X_s is the initial AG solution concentration in the wood pores, g/cm^3; and Y_s - is the initial concentration of the substance isolated in the solution, g/cm^3;

$$A_n = \frac{6Bi^2}{\mu_n^2 \left(\mu_n^2 + Bi^2 - Bi\right)} \tag{2}$$

where μ_n represents characteristic equation roots.

According to substance balance data,

$$Y^* - Y_i = b\left(X - Y_p\right) \tag{3}$$

where Y_i is the concentration of the substance isolated in the solution at the moment of time τ, g/cm³; and

$$b = \frac{G\varepsilon}{\rho V} \tag{4}$$

where G is the mass of the solid, g; V is the volume of the liquid phase, cm³; ρ is the density of the solid, g/cm³; ε is the specific volume of wood pores occupied by solution, cm³/cm³.

The right side of the equation (3) determines the fraction of the substance transferred from the solid into the solution during the time interval from the moment under consideration until the end of the experiment. The left side determines the increase of the solution concentration during the interval mentioned. Substituting (X – Y*) into equation (1) according to (2) leads to:

$$\frac{Y^* - Y_i}{X_s - Y_s} = \sum_{n=1}^{\infty} B_n e^{-\mu_n^2 Fo_g} \tag{5}$$

where $B_n = bA_n$.

For a regular mode of extraction, the first member of the series in equation (5) is sufficient.

Figure 4 graphs the dependence of $\ln\frac{Y^* - Y_i}{X_s - Y_s}$ on τ according to equation (5), on the basis of experimental data.

Extrapolating the straight line (Figure 4) $f(\tau) = \ln\frac{Y^* - Y_i}{X_s - Y_s}$ to τ = 0 gives a diffusion constant, according to the equations:

$$tg(\alpha) = -\mu_1^2 \frac{D\tau}{l^2} \tag{6}$$

and

$$tg\mu_1 = \frac{\mu_1}{b} \tag{7}$$

where α is the angle of inclination of the straight line to the time axis.

Solving the characteristic equation (6) regarding μ_1 gives, according to equation (7), the diffusion constant D. On the basis of the experimental data, the diffusion constant at temperatures of 20–25 °C is $1.55–2.67 \cdot 10^{-10}$ m²/s. For particles of average size of 5 mm, $Bi>>1$, which demonstrates the insignificant affect of hydrodynamic conditions upon AG isolation rate.

These data on mass transfer were used to calculate the AG extraction process from larch wood particles in similar hydrodynamic conditions. The value of the diffusion constant D was used to calculate the processes conducted at a given temperature.

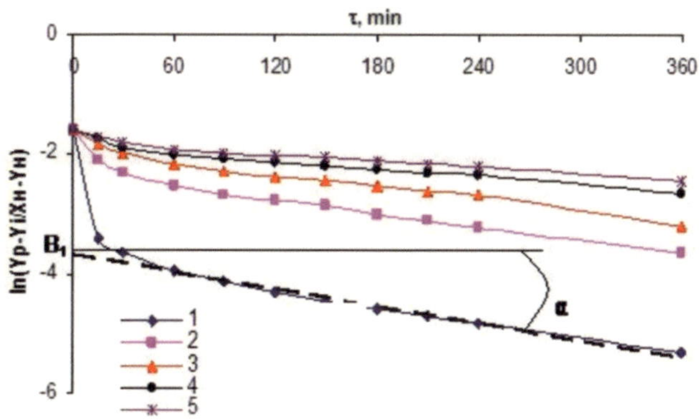

Figure 4. Graph of regular mode of AG extraction. Particle sizes, mm: 1 - 0.056; 2 - 0.86; 3 - 2.61; 4 - 3.24; 5 - 5.37

Based on the experimental data, a mathematical model was developed, material balance was calculated, and the optimal parameters of the extraction process were determined [75]. The model was used to optimize the technological process and production algorithm.

According to the method developed, firstly DHQ and other phenolic extractive substances were isolated from larch wood particles by organic solvent, and then were exposed to extraction by circulating water at 60–80 ºC for 2–3 h. The extract obtained was treated with a cationic flocculant solution to remove mechanical and colloidal impurities. Decolourized extract was ready for being concentrated and further purified by ultrafiltration.

Concentrations of AG water extracts were decolourized by flocculation, which was achieved by ultrafiltration using UAM-150P cellulose acetate membranes (Russia) [76]. Filtration rate decreased with time due to an increase in solution concentrations, viscosities and sedimentation of high-molecular particles on the membrane surface. Initial productivity decreased as initial AG solution concentration increased. However, initial productivity increased with increasing pressure gradient. At the final stage of ultrafiltration, when the process rate approaches a constant level, the higher the pressure gradient the lower the productivity, due to the higher rate of concentration at high pressure during equal time intervals.

Studies have been made of the influence of pressure upon the ultrafiltration process. The maximum degree of concentration is reached at the pressure gradient ΔP = 0.4 MPa. However, the optimal ratio between productivity and degree of concentration is at ΔP = 0.2 MPa.

Ultrafiltration results in simultaneous concentration of AG extracts and their purification by almost entirely filtering out low-molecular phenolic impurities. Purification efficiency depends on composition of the extract, membrane characteristics and conditions of filtration as well as a degree of concentration.

According to the IR spectroscopy and HPLC data, filtrate contains, together with phenolic substances, an oligomeric fraction of AG. Atomic absorption and X-ray fluorescence analyses have shown that ultrafiltration purifies AG from metal cations [76]. The total content of dry substances in filtrates is not more than 1–2.5%.

To increase productivity of the ultrafiltration module, we also tested the UAM-500P membrane (Russia). Ultrafiltration dynamics of decolourized AG extracts have shown that filtration rate is in an inverse ratio to initial extract concentration. The use of a macroporous membrane allows ultrafiltration without pre-treatment of AG extracts by a flocculating agent. It has been proven experimentally that productivity of this process is comparable to that of extract decolourized by flocculation. Thus, for the UAM-500P membrane, pore blocking at the initial stage is not a limiting factor, unlike the case for UAM-150P. The optimal conditions of ultrafiltration have been determined to make the technology profitable.

After ultrafiltration, the concentrate was dried in a drying unit. The known methods of dry product isolation, by precipitation in alcohol or acetone [6], are disadvantageous for industrial use from a technological, economical and ecological point of view. The filtrate, without additional treatment, was mixed with fresh water and reused for DHQ extraction.

The method proposed, as compared to known methods, enables the following improvements:

- AG extraction from larch wood is realized after the isolation of DHQ and resin substances, giving a rather high purity of the extract
- the process is simple, energy-efficient and economically viable
- no expensive sorbents are needed and no toxic or combustible organic solvents are involved
- concentrates of dry substance content of up to 40% can be produced
- the closed water cycle allows water consumption and the amount of waste water to be decreased.

Additional AG purification from high-molecular phenolic impurities was realized by treatment of the water extracts with ecologically harmless oxidant (hydrogen peroxide) [17]. The optimal conditions were found to oxidize impurities without affecting the polysaccharide macromolecule.

For final product isolation from the concentrate, spray drying, lyophilic drying or fluid bed drying can be used.

The experiments showed that spray drying is technically and economically optimal. In a manufacturing pilot, different modes of AG spray drying were tested by varying the starting concentration of AG solution, air temperature at the drier input, air temperature at the drier output and pressure of compressed air at spraying. The temperature of the drying gas (air) was the most technologically relevant. The required humidity of the final product (less than 7%) was reached at an air temperature higher than 100 ºC. The optimal process conditions produced AG of high quality.

On the basis of our study, a technological scheme for isolating high purity AG was developed, involving:

1. Flocculation of AG solution (**FR**)
2. Oxidation of impurities in the AG solution (**OR**)
3. Microfiltration of the AG solution (**MF**)
4. Concentration of the AG solution by ultrafiltration (**UF**)
5. Spray drying of the concentrate (**SD**).

The manufacturing pilot revealed drawbacks to the proposed scheme: low productivity and high time consumption per product unit. Thus, the scheme was optimized.

The most reasonable sequence of steps was determined using a decision tree [77-79] with limited operation sequence combinations: spray drying was always the last step and, thus, was excluded from the decision tree; microfiltration and oxidation always followed flocculation. The decision tree, taking into account the limitations mentioned, is shown in Figure 5.

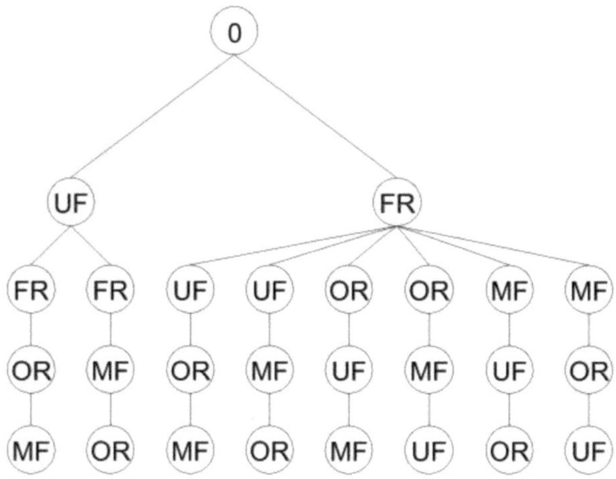

Figure 5. Decision tree

The variants were ranged from 1 to 7 according to these criteria:

- productivity
- material and reagent consumption
- steel intensity (investment)
- laboriousness (timetable)
- manufacturability.

The optimal variant is UF-FR-OR-MF (see Table 1), which has a four times higher productivity and a two times lower investment in comparison with the initial scheme.

An optimized technological scheme of arabinogalactan production was implemented at an experimental–industrial scale.

Criteria	Possible algorithms							
	FR-UF-OR-MF	FR-UF-MF-OR	FR-OR-UF-MF	FR-OR-MF-UF	FR-MF-UF-OR	FR-MF-OR-UF	UF-FR-OR-MF	UF-FR-MF-OR
Productivity	1	1	1	1	1	1	7	7
Material and reagent consumption	3	3	3	3	3	3	1	1
Steel intensity	1	1	1	1	1	1	7	7
Laboriousness	3	5	3	5	7	7	2	1
Manufacturability	2	1	2	5	7	6	6	4
Total	**10**	**11**	**10**	**15**	**19**	**18**	**23**	**20**

Table 1. Criteria of variant estimation

3. Larch bark pectins

3.1. Physicochemical and biological properties of pectins

Pectin substances are present in the majority of land and water plants, and in some freshwater algae [80]. Being an important component of cell walls, they are involved in ion exchange, water metabolism and cell wall structure formation. They stimulate seed germination and germ growth, provide turgor, etc.

The unique physicochemical properties of pectin make it indispensable in medical, food and cosmetic industries as a gelling agent, thickener, stabilizer and dietary fibre. Recently, it has become widely used as a matrix carrier for biologically active components in drugs. Pectins have physiological activities of their own (immunomodulating, hepatoprotective, anticarcinogenic, antimetastatic, etc.) making them applicable as medical preparations and biologically active food supplements.

Industrial demand for pectins in Russia is estimated at 2000 t/year, of which 10% is for the fragrance and cosmetic industries, 15% goes to medicine and pharmaceuticals, and 75% is for the food industry [81]. However, this demand is generally met by imported production. There are recent innovative Russian developments that are ecologically harmless and economically viable (there is no need to utilize aggressive acid media and to support treatment facilities), therefore having a low cost price. The raw material for pectin is the marc of citrus fruit, apple, sugar beet and sunflower head pith. There are proposals for using other plants as raw materials, such as amaranth, small mallow, duckweed, silene, coffee beans, etc. [82-86].

The bark of *L. sibirica* Ledeb. and *L. gmelinii* (Rupr.) Rupr., having a 12% pectin content, is a promising alternative raw material. At our laboratory, we are conducting systematic studies of the structure and properties of pectin from these larch species to determine a suitable technology for its industrial production.

3.2. Isolation of pectin substances from larch bark

There are a number of methods to isolate pectin polysaccharides from plant tissues, including hydrolysis extraction of dry raw material particles of certain sizes [87] using hot water, organic and inorganic acid solutions as well as salts, alkali or their mixtures as extracting solutions. Basic parameters of the pectin isolation process, such as raw material pre-processing, hydromodulus, temperature, extraction duration, medium pH and precipitator used, can all be varied depending on characteristics of the raw material [88]. We studied the influence of the following combinations of the basic parameters upon yield and product quality:

Experiment 1: 0.5% ammonium oxalate solution (hydromodulus 1:5)
Experiment 2: 0.5% oxalic acid solution (1:5)
Experiment 3: equimolar mixture of 0.5% oxalic acid and 0.5% ammonium oxalate solutions (1: 5)
Experiment 4: 0.25% sodium hydroxide solution (1:5)
Experiment 5: similar to experiment 3 (1:7)
Experiment 6: similar to experiment 3 (1:10)

In all the experiments, extraction process were the same (at 80 °C for 2 h of constant stirring).

The pectin samples obtained were white or light cream-coloured powders, tasteless and with no smell (Figure 6).

Figure 6. A laboratory sample of larch bark PS

Table 2 sets out the yield (% of weight of absolutely dry bark, a.d.b.) and composition data of pectin substances obtained in experiments 1–6.

The highest yields were observed with a weak alkali solution, but the ash content was too high (16.66%), which affected gelling ability [89]. The lowest ash content was found in the preparations isolated using an equimolar mixture of ammonium oxalate and oxalic acid (5–5.6%).

Experiment	Yield, % of a.d.b.	Medium pH	Composition, %		
			C	H	Ash
1	0.77	6.95	30.28	6.66	7.89
2	0.97	1.90	31.17	5.75	8.34
3	1.64	3.90	32.05	5.42	5.58
4	5.81	11.97	36.55	4.51	16.66
5	2.71	2.86	34.24	6.30	5.20
6	2.93	2.84	31.48	6.82	5.08

Table 2. Yield and elemental composition of pectin substances in larch bark

The optimal hydromodulus was observed in experiments 1–4 (hydromodulus 1:5) with yield increasing by 1.5 times. Hydromodulus provides insufficient penetration of extracting agent, lower than 1: 5. A 1:10 rise of hydromodulus (experiment 6) had no essential effect upon the yield and qualities of the product. Thus, the equimolar mixture of ammonium oxalate and oxalic acid used as an extracting solution at hydromodulus 1:7 (experiment 5) was the most effective in isolating pectin substances from larch bark, leading to a 2.7% yield of absolutely dry bark mass with ash content of 5.2%.

Raw material pre-processing by solvents of increasing polarity (hexane, ethyl acetate and water) resulted in both enzyme deactivation and elimination of the impurities, therefore increasing the extracting solution's ability to access the plant cell walls. Notably, the pre-extracted substances are valuable for medicine [90] and the leather industry [91].

We experimentally compared the yields of pectins isolated with and without raw material pre-processing in the conditions described above (see Figure 7). It was shown that prior elimination of impurities leads to higher yield of the product (about 1.5 times), clearly due to higher availability of pectin substances.

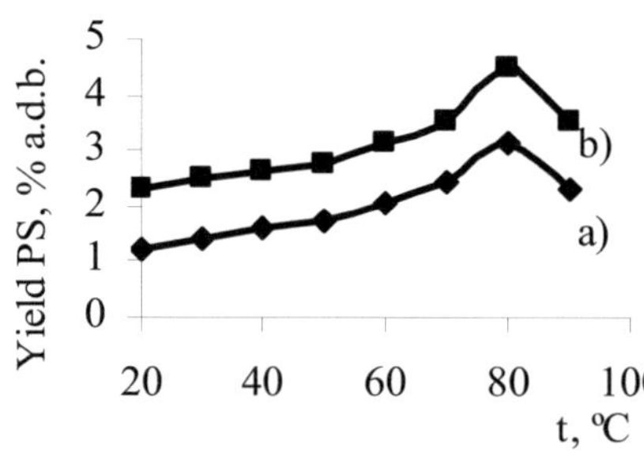

Figure 7. PS yield dependence on: a) raw material pre-processing, and b) extraction temperature

The PS yield increased as the extraction temperature rose, reaching a maximum at 80 °C. The data obtained were in good correspondence with the literature on classic pectin isolation [89]: raising temperature causes partial hydrolysis of protopectin. Thus, pectin yield increases while at temperatures higher than 80 °C the superstructure of pectin substances is broken. This is also confirmed by the dependence of the molecular weights of the resulting pectins on extraction temperatures (see Figure 8.).

Kinetic studies, particularly those concerning the pectin hydrolysis extraction process, have a particular interest. Pectin yields vs. extraction times are charted in Figure 9. A major part of PS is transferred into the extract within 1 h of extraction, after which there is no significant increase of yield.

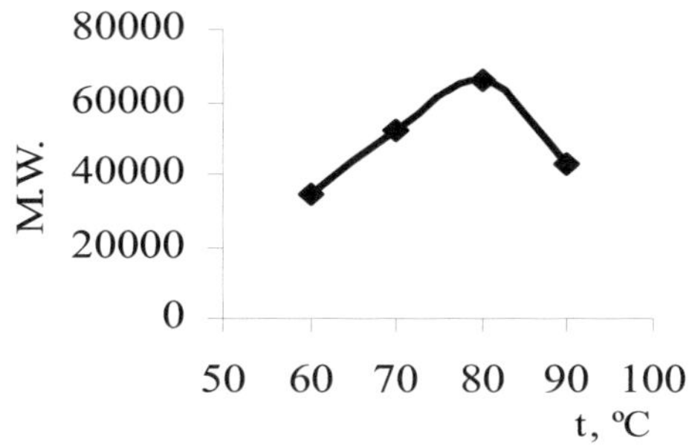

Figure 8. PS molecular weight (M.W.) dependence on extraction temperature

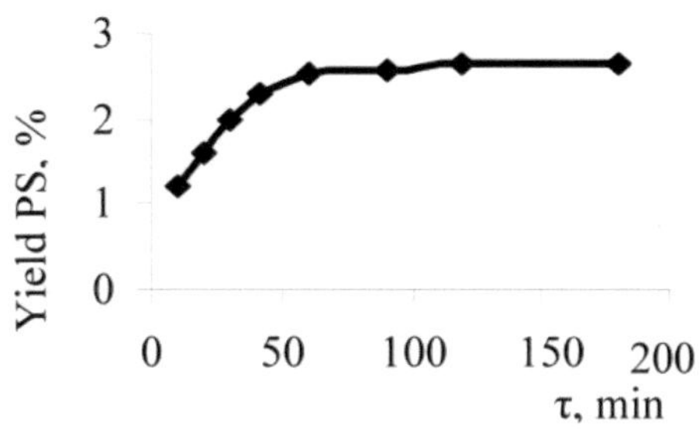

Figure 9. Dependence of PS yield on extraction time

We also studied the influence of type of precipitator used upon the yield and qualities of the product. For this purpose, pectin extract was prepared from larch bark by treating it with an equimolar mixture of 0.5% ammonium oxalate and 0.5% oxalic acid (hydromodulus 1:7) at 80 °C for 2 h. The extract was concentrated in a circulation vacuum evaporator until it reached one third of its original volume. One half of the concentrate was precipitated with acetone and the other half with ethanol. Precipitators were added in equal quantity, dropwise while continuous stirring was applied. The precipitate was vacuum-filtered, dissolved in 100 ml of distilled water by heating to 40–50 °C when necessary, and then again precipitated and filtered. The final precipitates were washed with the same precipitator and then with diethyl ether, dried in the air and then in a drier at 50 °C, cooled to room temperature in a desiccator and measured to determine yield. It is noteworthy that the dropwise addition of precipitator into the extract increased yield by 0.5% compared to the usual precipitation procedure. We established that acetone is less selective, and ethanol therefore gives a purer product. Purity of the pectin preparation obtained can also be estimated based on galacturonic acid content [92]: for larch bark pectins precipitated by acetone and ethanol, the result was 69.77 and 78.12%, respectively.

Thus, the optimal procedure for isolating pectin substances from larch bark involves pre-treatment by hexane, ethyl acetate and water, extraction by an equimolar mixture of 0.5% oxalic acid and ammonium oxalate solutions at hydromodulus 1:7 and an extraction temperature of 80 °C for 1 h, and precipitation by ethanol. The method has been patented [93] and used for preparing the samples for physicochemical and application studies.

3.3. Characterization of pectin substances

Pectinase enzyme hydrolysis of PS samples isolated from larch bark by the above method, and further analysis of hydrolysis products by paper chromatography (PC), have shown an essential destruction of PS with formation of free D-galacturonic acid.

Table 3 sets out the main maxima of absorption bands in the IR spectra of PS and their assignment, proving the PS pectin nature of the samples [94].

Frequency (ν, cm^{-1})	Assignment
3460	ν(OH), ν(H_2O)
3260	ν(NH)
2962, 2872	ν(CH_3)
2573	ν(OH),
1730	ν(C=O) в COOH
1640	δ(OH)$_A$
1540	δ(NH)
1380–1450	δ(C-CH_3), ν(C-O) pyranose rings
1331	δ(OH) in pyranose rings
1265	ν(C-O) in esters
1150	ν(C-O-C)
1095	ν(C-C)
1027	ν(C-OH)
890	δ(C1-H) in glucopyranose ring
766, 629, 528	pulse vibrations of pyranose ring

Table 3. Absorption band maxima in IR spectra of PS and their assignments

Thus, enzyme hydrolysis and IR spectroscopy data prove that the polysaccharide isolated from larch bark refers to the pectin group.

The monosaccharide composition of PS was determined by total acid hydrolysis with trifluoroacetic acid (TFA). Monosaccharide identification of PS was performed using gas–liquid chromatography (GLC) and the sample was shown to consist of galacturonic acid, protein compounds and monosaccharides of arabinose, galactose, rhamnose, glucose, mannose and (in minor quantities) xylose. Dominant monosaccharides were galactose and arabinose, in a ratio of 2.7: 1.

The degree of homogeneity for PS was determined by ion exchange chromatography on DEAE cellulose with sodium chloride aqueous solutions. Four fractions were detected (Table 4). In the fractions PS-1 and PS-2, arabinose and galactose were predominant (18.26/52.96% and 11.65/30.83%, respectively); thus, they refer to acidic arabinogalactans. The acidic nature of PS was developed with D-galacturonic acid residues with the PS-1 proportion five times less than in PS-2, while in PS-3 and PS-4 it was a major monosaccharide, and thus they refer to pectins. The content of neutral monosaccharides in PS-4 was minimal compared to other fractions (3% mass). All the fractions contained protein compounds that were not eliminated by gel filtration. It seems likely that the protein and polysaccharide compounds were strongly aggregated, or that their molecular weights were close to each other.

Sample*	Yield, %	Content, %							
		GalpA	Protein	Monosaccharides					
				Rha	Ara	Xyl	Man	Glu	Gal
PS-1	12.1	5.67	6.9	*traces*	18.26	1.54	2.53	5.95	52.92
PS-2	5.9	29.12	7.3	0.53	11.65	1.02	2.71	8.81	30.83
PS-3	17.0	65.93	5.7	1.91	4.45	0.75	1.18	1.12	9.42
PS-4	37.0	79.87	3.6	0.35	0.93	0.18	0.24	0.21	1.06

* PS-1 isolated with use of 0.01M NaCl solution, PS-2 – 0.1M NaCl solution, PS-3 and PS-4 – 0.2M NaCl solution

Table 4. Chemical characterization of PS sample after DEAE-cellulose fractioning

The amino acid composition of PS proteins was studied. The major components of PS were glutamic acid (6%) and aspartic acid (2.8%), while total content of amino acids with aliphatic side chains (glycine, alanine, valine, isoleucine, leucine) was 9% (Figure 10).

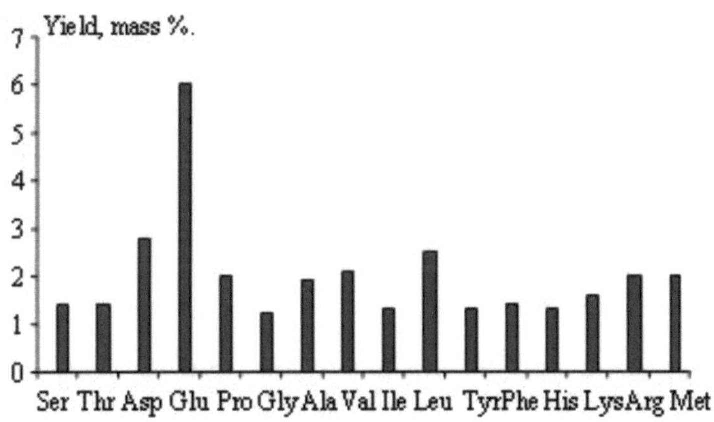

Figure 10. Amino acid composition of PS proteins

Thus, we isolated PS from the bark of *L. sibirica* Ledeb. and *L. gmelini* (Rupr.) Rupr., pre-treated with ethyl acetate and hot water. Enzyme hydrolysis with pectinase and IR spectroscopy were employed to prove the presence of pectin polysaccharides in the samples isolated. DEAE-cellulose chromatography revealed that PS includes four fractions, two of which are acidic arabinogalactans and the other two belong to the pectin group.

3.4. Structural study of main chain of larch bark pectin

Acid hydrolysis of PS by 2M TFA results in galacturonan PVG-1. The high value and positive sign of the rotation angle of +245.3º (*c* 0.1, H_2O) suggest α-D-configuration of D-galactopyranosyluronic acid residues.

Values of chemical shifts (CS) of carbon atoms in the ^{13}C NMR spectrum of PVG-1 (Table 5), compared to other data, [95] corresponded to those for carbon atoms in D-galacturonic acid residues in pyranose form which compose the linear fragment of pectin molecules (pectin core). The presence of an anomeric carbon atom signal at 101.9 ppm indicated both (1→4)-bonding between D-galacturonic acid residues and α-configuration of C-1 anomeric atoms. Signals at 176.2 ppm were assigned to the C-6 atom and indicated a free carboxyl group in D-galacturonic acid residue. Additionally, there were galacturonic acid residues esterified by methoxyl in the PVG-1 molecule, according to signals with CS at 172.2 ppm ($\underline{C}$-6-OCH_3) and 54.4 ppm (-$O\underline{C}H_3$). The ratio of the integrated signal intensity of carbon atoms observed in methoxyl and carboxyl groups suggests a high degree of galacturonan methoxylation. The ^{13}C NMR spectrum also showed signals at δ 76.1 and 74.9 referring to the C-3 carbon atom substitute in (...→4)-α-D-Gal*p*A-(1→...) galacturonic acid residue in the galacturonan molecule (the non-substituted atom has CS at 72.1 ppm).

Residue	C-1	C-2	C-3	C-4	C-5	C-6	C-6-(OCH_3)	-$O\underline{C}H_3$
→4)-α-D-Gal*p*A-(1→	100.9	68.9	70.1 76.1 74.9	79.2	73.4	176.2	172.2	54.4

Table 5. Chemical shifts of signals of galacturonic acid carbon atoms in ^{13}C NMR spectrum of PVG-1

Thus, according to spectral and chromatographic data, linear polysaccharide from larch bark has a structure of homogalacturonan consisting of (...→4)-α-D-Gal*p*A-(1→...)-linked fragments D-galacturonic acid has partially etherified by methoxyl groups with branching points at C-3 atom of galacturonopyranosyl residue.

3.5. Structural study of side branches of larch bark pectin

Partial acid hydrolysis of PS with 0.01M TFA for 3 h resulted in galacturonan PVG-2. According to ^{13}C NMR data, it was a pectin polysaccharide. The spectrum contained both typical signals of galacturonic acid residues, namely pronounced signals of anomeric carbon atoms at 100.4 and 104.4 ppm, and signals of carboxyl carbon atoms at 171.4, 166.5 and 53.7

ppm, the latter two being signals of carbon atoms in uronic acid residues methoxylated by the C-2 and/or C-3 atoms (Table 6).. Intensities and spectral positions of signals at 68.9, 70.8, 78.9 and 72.2 ppm corresponded to data in the literature for α-D-Gal*p*A residues connected by 1→4 bonds. There is a ratio of 1:5 between integral signal intensities of carboxyl and methoxyl carbon atoms, which suggests a high degree of PS methoxylation.

Residue	C1	C2	C3	C4	C5	C6	-O$\underline{C}H_3$ ($\underline{C}H_3$-)
→4)-α-D-Gal*p*A-(1→	100.4	68.9	70.8	78.9	72.2	171.4	-
2-MeO-α-D-Gal*p*A-(1→	100.9	166.5	69.6	78.9	73.8	171.4	53.7
3-MeO-α-D-Gal*p*A-(1→	100.9	68.9	166.5	78.9	73.8	171.4	53.7
β-D-Gal*p*-(1→	104.64	71.7	74.1	69.6	76.1	62.0	-
→6)-β-D-Gal*p*-(1→	104.38	71.7	73.8	69.6	74.3	70.8	-
→3,6)-β-D-Gal*p*-(1→	104.64	71.7	82.5	69.57	74.1	71.4	-
α-L-Ara*f*-(1→	108.6	80.7	78.9	84.9	62.0	-	-
β-L-Ara*p*-(1→	101.1	69.6	-	-	-	-	-
→3,5)-α-L-Ara*f*-(1→	108.6	80.7	84.9	83.2	67.8	-	-
→2,5)-α-L-Ara*f*-(1→	108.0	84.9	77.6	83.2	67.8	-	-

Table 6. Chemical shifts in signals of carbon atoms in the ^{13}C NMR spectrum of PVG-2

In the ^{13}C NMR spectrum of PVG-2 samples there were upfield signals at 17.9 and 18.13 ppm belonging to C-6 atoms in terminal rhamnose residues and in polysaccharide chains, respectively. The integral intensities of these signals and those of C-2 and/or C-3 and C-6 carbon atoms for galacturonan residues at 166.5 and 171.4 ppm were found to have a ratio 1:5. The total integral intensity of signals for anomeric C-1 atoms for rhamnose and the total integral intensity of signals of anomeric atoms of galacturonan residues were equal to each other, *i.e.* they had the same ratio for rhamnose and galacturonan residue content in the chain. According to data in the literature, signals at δ 99.7, 77.6, 70.8, 82.5, 68.9 and 17.9 ppm are assigned to C-1, C-2, C-3, C-4, C-5 and $\underline{C}H_3$ carbon atoms in →2,4)- α-L-Rha*p*-(1 → residues.

Thus, according to ^{13}C NMR spectral data, linear fragments of pectin polysaccharide isolated from larch bark are rhamnogalacturonans where D-galacturonic acid residues in pyranose form with an α-configuration of their anomeric centre are connected 1–4 by glycosidic bonds. One fifth of galacturonan residues associated with the C-6 atom were esterified by methoxyl groups. The ratio between 2,4-substituted rhamnopyranosyl and galacturonosyl residues (1:5), thus, the main chain structure of the pectin polysaccharide was highly branched at the C-4 atoms of rhamnopyranosyl residues.

Further ^{13}C NMR spectrum analysis of the PVG-2 sample showed that arabinogalactan fragments are present in rhamnogalacturonan as side chains. Concerning signals of anomeric carbon atoms, the ^{13}C NMR spectrum of the PVG-2 sample showed that there are signals at 101.1, 104.38, 104.64 and 108.6 ppm, as well as signals of anomeric carbon atoms in galacturonopyranosyl residues of the galactan core. According to [9], intensities and values of CS can be assigned to signals of anomeric carbon atoms in β-L-Ara$_p$, α-L-Ara$_f$ and β-D-Gal$_p$ residues. The most upfield of the signals mentioned (δ 101.1 ppm) belong to terminal β-L-Ara$_p$ residues. Signals at 104.38 and 104.64 ppm belong to C-1 in β-D-Gal$_p$ residues while CS values of C-2, C-3, C-4, C-5 and C-6 atom signals are calculated according to the official data for β-D-galactopyranosyl residues. Bonding at the C-3 and C-6 positions of β-D-galactopyranose was proven by downfield shifts of these signals at 8.7 and 8.8-9.4 ppm, respectively, due to glycosylation of these atoms as compared to their positions in non-substituted 1→3,6 linked β-D-Gal$_p$ residues. Signals at δ 108.6 ppm, like those at 80.7, 78.9, 84.9 and 62.0 ppm, are terminal α-L-arabinofuranose. The anomeric atoms of arabinose and galactose are monosaccharides integrated at a ratio of 1:2.

Hence, according to spectral data for the PVG-2 fragment of the pectin polysaccharide from larch bark, highly branched arabinogalactan was detected as side chains consisting of linear chains with →3,6)-β-D-Gal$_p$-(1→ residues with branching at C-6 atoms. Side chains of arabinogalactan fragments contain terminal arabinose, both in pyranose and in furanose form, as well as →2,5)-α-L-Ara$_f$-(1→ and →3,5)-α-L-Ara$_f$-(1→ residues as intermediate fragments.

3.6. Larch bark pectin peculiarities and implementation fields

It has been determined that larch bark pectin substances possess immunomodulatory, antineoplastic, gastroprotective and antitoxic action [96-98]. In order to understand larch bark pectin's physiological and pharmacological action, we have started research focussed on examining its membrane-acting action. The vacuoles of isolated cell plants and their membranes were found to be an appropriate object for our research. The influence of pectin on membranes and the peculiarities of their barriers were estimated according to the change of destruction dynamics in isolated vacuoles in comparison with the control. The results are depicted in Figure 11. It has been established that implementation of pectin aqueous solutions leads to their protective action on vacuolar membranes, exceeding the control threefold. Thereby, the experiments proved that larch bark pectin possesses a membrane stabilizing activity.

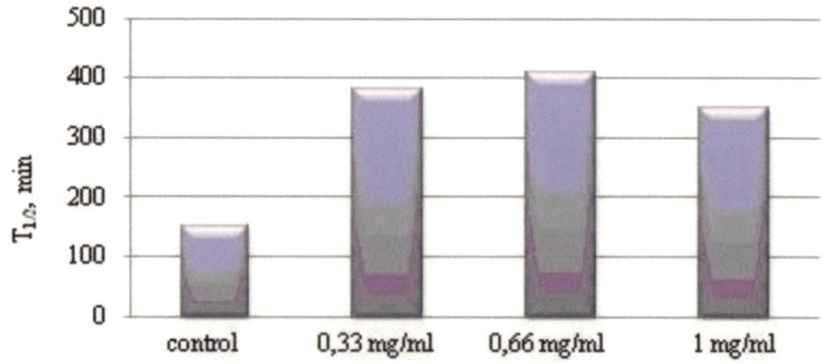

Figure 11. Influence of pectin upon isolated vacuole half-lives ($T_{1/2}$)

In order to broaden larch bark pectin implementation fields, we carried out research into its implementation as a reducer and stabilizer of noble metal particles in nanosized state. Supramolecular structure peculiarities, optical activity, carboxyl and abundance of hydroxyl groups, and polymeric pectin molecule stabilizing effect provided significant potential in nanobiocomposite formation processes in metals with a polysaccharide matrix ("pectin – metal (0)").

Synthesis of nanobiocomposites was carried out using the redox reactions of PS with silver nitrate. Nanobiocomposite samples 0.5 "pectin – Ag(0)" up to 72% content of silver were obtained in different reaction conditions. It was discovered that the effectiveness of the reaction to create a silver nanoparticle flow depends on medium spectrum pH. The spectra of the mixtures of pectin and silver nitrate water solutions versus time reaction are depicted in Figure 12a. It was determined that, with a reduction of pH to 3.5, the Ag(I) reaction proceeds very slowly. This is demonstrated by the appearance of a link in the absorption spectrum in the range of λ 280–470 nm only 24 h after the beginning of the reaction (Fig. 12a). The wide maximum low intensity link was indicated by the formation of silver metal primary centres. Despite this, the reaction speed of the reduction was so slow that even 96 h was not enough to create fully recovered Ag(0) centres. With pectin and Ag(I) interactions in reaction mixtures beginning at a pH of 7, a symmetric bond at λ_{max} 420 nm can be observed in the electron spectra at the start of reaction by proving the formation of Ag(0) nanoparticles (Fig.12b, line 2). Even so, it takes about 24 h for the full silver cation conversion which was experimentally evaluated according to the absorption bond intensity growth. Ag(I) reduction with pectin at pH 11–12 proceeded swiftly immediately after mixing of the components (Figure 12b, line 4 and Figure 12c line 1) and finished within 30 min. Reduction under these conditions was also accompanied by variations in the particle size of Ag(0), as shown by the shift in the Plasmon pick position into the short-wave region at 10 nm (Figure 12c).

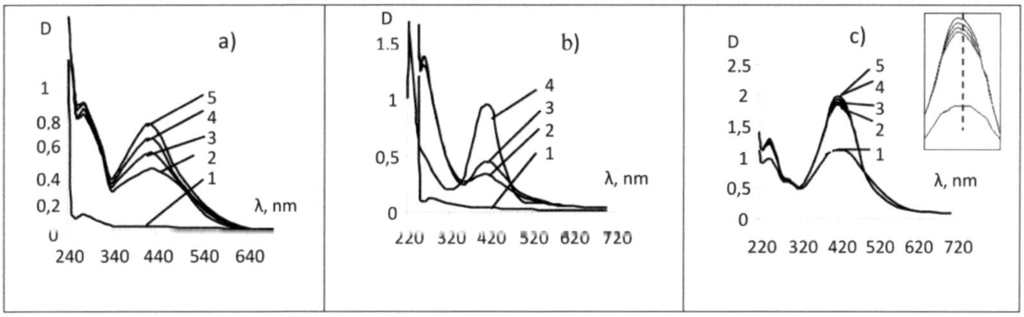

Figure 12. Absorption spectra of mixtures of aqueous solutions of pectin (0.5%) and silver nitrate (0.1%) in a ratio of 1:1 depending on: a) reaction duration: 1 min (1), 24 h (2), 48 h (3), 72 h (4), 96 h (5); b) medium pH: 3.5 (1), 7 (2), 9.7 (3), 11.5 (4); c) reaction duration at pH 11.5: 1 min (1), 30 min (2), 60 min (3), 180 min (4), 24 h (5)

Radiographic phase analysis of obtained nanobiocomposites of "pectin-Ag(0)" demonstrated it to be a mixture of radioamorphous and crystalline phases. There was a wide halo with maximum intensity at d ~ 0.46 nm in 2θ angle intervals from 8 to 60E (Figure13a) in a radioamorphous phase diffraction pattern typical of a pectin source. There were quite intensive but broadened lines typical of metallic silver (Figure 13b) during silver loading in the diffraction patterns of reaction products against a background of pectin reflection. The calculation of silver unit-cell parameters showed that in their quantity in the provided samples was lower than for massive silver and changed from 0.4036 to 0.4050 nm (∀ 0.0008 nm). Moreover, the average size of the coherent-scattering region (CSR) was calculated according to Selyakov–Sherarar's formula [99] to be in the range of 3 nm. The data obtained demonstrated that, in the samples of Ag (0), the persistence of nanosized particles was stabilized by an amorphous phase with pectin.

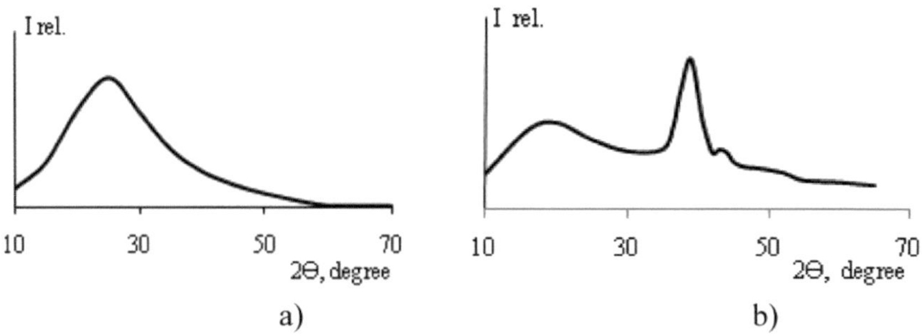

Figure 13. Diffraction patterns of pectin sample (a) and of the "pectin-Ag(0)" nanobiocomposite sample (b)

"Pectin – Ag (0)" nanobiocomposite scanning electron microscopy (Figure 14) showed that the analysed samples contain particles considerably smaller than 100 μm.

Figure 14. Electron microphotography of "pectin – Ag(0)" sample

Microphotography analysis of nanobiocomposites, obtained by the use of transmission electron microscopy, demonstrated that there are isolated silver particles of null valency in globular form (Figure 15a), of a size within the range from 4 to 17 nm (predominance (up to 80%) at 6–7 nm, Figure 15b).

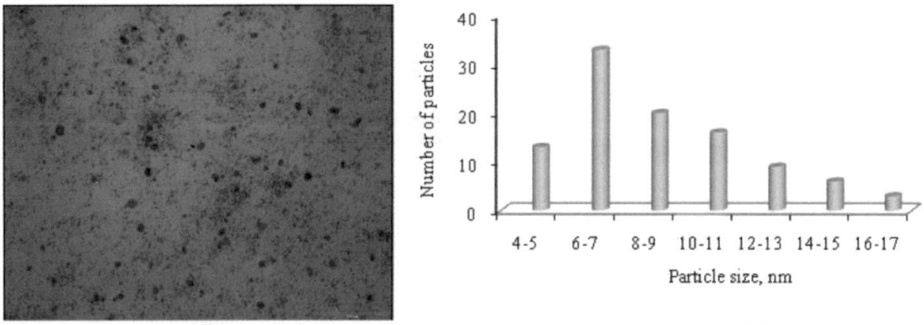

a) b)

Figure 15. Transmission electron microphotography, a), and size distribution graph of "pectin-Ag(0)" nanobiocomposite sample, b)

Thereby, "pectin-Ag(0)" nanobiocomposite formation takes place as a result of the interaction of pectin water solutions with Ag(I). Process speed increases significantly with variation within the alkaline pH range of the medium. The initial component proportion influences the results of the reaction: the more Ag(I) that falls per 1g of pectin, the less the quantity of Ag(0) particles that is created in a nanosized condition. Using pectin implements reduction and stabilizing functions, and also adjusts the sizes of obtained Ag(0) nanoparticles.

4. Cellolignin residue of larch wood as raw material for crystalline glucose production

The cellolignin residue is formed during the chemical processing of larch wood using the technology for obtaining dihydroquercetin and arabinogalactan [100]. CR represents larch wood chips initially extracted by ethyl acetate and hot water. The larch wood chip basically consists of cellulose, hemicellulose, and lignin. The polysaccharide content in larch wood chips is 65–75% of the mass of its absolutely dry wood (a.d.w.) [101], and the content of water-soluble substances is 10–16% (in some samples up to 30%) [102]. The content of holocellulose in CR (without water-soluble substances) is about 54% of its a.d.w. weight, whereas the content of holocellulose in the original larch wood is about 40% of its a.d.w. (also without water-soluble substances). The gain in the relative content of polysaccharides per mass of a.d.w. is about 13%, which permits one to consider the CR as a polysaccharide-"enriched" raw material from which it is possible to obtain sugar and other products by hydrolysis. Hemicellulose polysaccharides in the CR of larch wood are mainly represented by the water-soluble polysaccharide arabinogalactan and 4-O-methylglucuronoaraboxylans and galactoglucomannans [103], which are associated to differing degrees with cellulose. The yield of arabinogalactan from larch wood using dihydroquercetin production technology in which it is isolated as a by-product (without special optimization) accounts for 67% of its total content in the original raw material; therefore, the content of water-soluble substances in CR remains rather high, at 8.9%.

The data on the group composition of the components of CR were reported in [104]. The acid hydrolysis of hemicelluloses results in the formation of mono- and oligosaccharides, the presence of which substantially impairs the quality of hydrolysates and hinders the crystallization of glucose from these syrups. Therefore, it is necessary to purify CR from hemicelluloses. Hemicelluloses are commonly removed (to a particular limit) from the raw material, either by hot water extraction or by hydrolysis with diluted acid at elevated temperature.

In experiments with water hydrolysis of larch wood chips in laboratory autoclaves, the parameters of hydrolysis, i.e., temperature, hydromodulus, duration and the number of hydrolysis steps, were varied. The maximum total yield of water-soluble substances (18%) was attained based on four-step hydrolysis. However, it is economically more attractive to perform one-step hydrolysis, with the conditions specified for obtaining a maximum yield of sugars with this hydrolysis method being as follows: a gradual increase in temperature from 25 °C for 1 h and 160 °C for 1 h with a hydromodulus of 1:6. The yield of substances was 15%. As indicated by paper chromatography (PC) and thin layer chromatography (TLC), the pre-hydrolysate contained arabinose, galactose, xylose, mannose and trace amounts of glucose [105].

Mild hydrolysis of hemicelluloses of CR was performed using 1–5% sulphuric or hydrochloric acids at the boiling temperature of the solution (100–105 °C). By choosing the optimal hydrolysis conditions, a maximum yield of reducing substances (RS) in hydrolysate, 1.1%, was achieved with the use of 5% sulphuric acid, which corresponded to a 23% content of hemicelluloses or noncellulose polysaccharides in CR. The hexose content in the prehydrolysate was 33.9% of total sugars, as determined from the content of RS, and that of pentoses was 62.1%; arabinose, galactose, xylose, mannose, and glucose were identified qualitatively. Significantly less amounts of polysaccharides were hydrolyzed with the use of 2% hydrochloric acid under the same hydrolysis conditions - 9.6% .

Aqueous and acidic sugar solutions obtained during the pre-hydrolysis of CR (technological sugar solutions) contained greater amounts of fermentable sugars compared to other conifers and can be used in the production of feed for animals [106, 107] and other products.

To obtain pure glucose syrups, two alternative variants of acid hydrolysis of cellulose were considered: hydrolysis of CR and of cellulose itself after delignification of CR. Using the first variant, we studied the hydrolysis of CR by diluted (1.0–5.0%) hydrochloric acid at high temperatures (160–170 °C) and by concentrated (50–85%) sulphuric acid at room temperature to compare the crystallization properties of glucose syrups obtained by the two methods. A high temperature hydrolysis of the raw material by diluted acid was carried out in laboratory autoclaves in four steps. As the acid concentration was increased from 1 to 5%, a gain in the sugar content of the hydrolysates was observed. Thus, the mass portion of RS relative to the mass of a.d.w. during hydrolysis of sawdust was 5.4% using 1% H_2SO_4, 7.9% with 2% H_2SO_4, 9.2% with 3% H_2SO_4, and 11.6% with 5% H_2SO_4. However, simultaneously to an increase in the acid concentration, the quality of the hydrolysate was markedly impaired; an intensive dark colour appeared due to the formation of sugar degradation products, and the content of colloidal impurities in syrups at later stages caused severe problems. Conversely, the low temperature hydrolysis of CR (of the same raw material) by concentrated sulphuric acid made it possible to obtain a high yield of sugar in hydrolysates with a minimum content of degradation products. The maximum yield of RS was 64% using 70% sulphuric acid at the hydromodulus (1:5) and a hydrolysis time of 5 h, and 62.6% using 80% sulphuric acid at the hydromodulus (1:5) and hydrolysis time of 2 h. Hydrolysates contained no pentoses. These characteristics meet the requirements imposed upon hydrolysates from which crystalline glucose is isolated.

Thus, the use of concentrated sulphuric acid in the concentration range of 65–80% makes it possible to obtain hydrolysates with a maximum content of RS (up to 64%) [108]. After additional hydrolysis (the inversion stage), the hydrolysate has a pH value close to 2. On further evaporation of this hydrolysate, the acid is concentrated, which leads to further degradation of sugar; as a result, the yield of glucose decreases, and the syrup is contaminated with stained products.

Sulphuric acid was neutralized using barium acetate, sodium hydroxide and calcium hydroxide, and the contribution of each compound to formation of the mineral ash component in syrups was determined. The lowest ash content in the hydrolysate (0.2%) was achieved by applying barium acetate. However, because of the toxicity of barium and its salts, the use of barium was abandoned. Upon neutralization of sulphuric acid by calcium hydroxide the ash content in the hydrolysate was initially as high as 10%; nevertheless, preference was given to this compound alone since it is nontoxic, readily available, and convenient in operation. According to the "Glucose, crystalline hydrated" GOST 97588 Regulations [109], the ash content in the final product must not exceed 0.06–0.07%, calculated for dry substance; therefore, it was necessary to provide neutralization conditions to decrease the ash content in hydrolysates. For neutralizing sulphuric acid, pH value was brought to 4–4.5 at a temperature no higher than 80 °C. The amount of calcium hydroxide was calculated according to the sulphuric acid neutralization reaction so as to exclude the over-alkalization of the solution [110]. The resulting dihydrate gypsum crystals were filtered off. The ash content in hydrolysate decreased by up to 0.5%. In addition to glucose and mineral contaminations, the hydrolysate contained, dependent on hydrolysis conditions, the products of partial hydrolysis of the lignocarbohydrate complex of wood—mono-, di-, tri-, and oligosaccharides—as well as impurities belonging to different classes of organic compounds: acid-soluble lignin, furfurol, oxymethylfurfurol, a lignohumic complex, colloids, levulinic acid, and other organic acids [106]. At the next stage, we assessed the nature of substances by determining the colour of hydrolysates and selected how to remove them from sugar solutions [111]. It was possible to assign some impurities, that by their nature are associated with lignin, to substances based on their colouring. First of all, this was acid-soluble lignin. According to the published data, 2–3% of total lignins are dissolved in the hydrolysis of coniferous wood by a solution of 72% sulphuric acid [103].

In addition, coloured substances are formed during sugar degradation: hexosans form high molecular weight substances of brown colour, which partially precipitate from solution, and pentosans form furfurol, which imparts a yellow colour to hydrolysate. Under acidic conditions, lignin and sugar degradation products, i.e., furfurol and oxymethylfurfurol, form condensed products in small amounts, and insoluble humic compounds; and the products of incomplete hydrolysis of polysaccharides, oligosaccharides, can be partially adsorbed by acid-insoluble lignin [106]. Hydrolysates were clarified using activated carbon BAU (Russia). The UV spectrum of a neutralisate from CR shows an absorption band at 280 nm, which disappears after treatment of the neutralisate by activated carbon. Treating the hydrolysate with dichloroethane followed by IR analysis of the concentrated extract made it possible to identify it as acid-soluble lignin. Thus, the treatment of hydrolysates with activated carbon significantly reduces the content of acid-soluble lignin in hydrolysates (from 1.4 to 0.3%).

The scheme of the acid–hydrolytic transformation of cellulose to glucose with preliminary delignification of lignocellulose raw material by industrial methods [112] is the second variant, which also makes it possible to obtain high purity glucose syrups. Its main advantage is that it enables glucose syrups to be obtained with a factor of merit of no less than 85%, which are not contaminated with colouring impurities of ligno-carbohydrate origin and ash components. It is known that, during the low temperature hydrolysis of cellulose by concentrated acids, partial destruction of cellulose with the formation of water-soluble products occurs [113]. Under optimized conditions, concentrated sulphuric acid almost completely dissolves cellulose, and the cleavage of glycoside bonds proceeds in a homogeneous medium. As a result of hydrolysis, a mixture of products differing by the polymerization degree (PD) is formed: from comparatively high molecular weight cellulose and cellodextrins (PD from 7 to 50–60) to oligosaccharides (mainly di- and trisaccharides) and glucose. The composition of the mixture and the ratio of the products in the hydrolysate depend on the hydrolysis conditions. These products also vary in water solubility. Thus, cellodextrins, oligosaccharides, cellobioses and monosaccharides are water-soluble, and part of the cellulose itself, mainly its crystalline moiety, and hydrocellulose do not dissolve in water.

Thus, the hydrolysis of cellulose enables firstly the isolatation of intermediate water-soluble hydrolysis products with simultaneous removal of sulphuric acid without its chemical neutralization, which in turns prevents the entry of mineral impurities into syrups; and, secondly, obtaining of the required monomeric sugar, i.e., glucose, in one stage, by subsequent additional hydrolysis of the intermediate product. The hydrolysis of industrial cellulose was carried out by 72% sulphuric acid at room temperature for 1 h with regular stirring of the hydrolysate mass; in this case, cellulose had completely dissolved within the first 15 min. Increasing the hydrolysis duration up to 2 and 3 h did not significantly affect the final yield of the product, which was 80–90% of the weight of absolutely dry cellulose (a.d.c.) [114].

We arbitrarily called this product the inverted polysaccharide (IPS) since this name reflects its position in the technological scheme. Dried IPS is a white or pale cream powder. The product is partially soluble in water (the insoluble fraction accounts for 43% of the weight of IPS), is soluble in aqueous alkaline solutions, and exhibits a lower PD than the starting cellulose (150; PD for starting cellulose, 573). The content of cellulose in an aqueous IPS solution was estimated, using HPLC, to be 2% of the a.d.c. A comparative analysis of IR spectra of the starting cellulose and IPS indicated that IPS is cellulose with a high degree of amorphism [94]. In particular, this is evidenced by strong changes in the IR spectrum of IPS in the region of 600–1500 cm^{-1}, which accompanies changes in the polysaccharide hypomolecular structure, and smoothing of the intensity of so-called crystallinity bands at 1100, 1140, 1190, 1250, 1360, and 1420 cm^{-1} [115]. The IR spectrum of the product contained no absorption bands at 1112 and 1162 cm^{-1}, which are typical of the spectra of a highly ordered cellulose structure. The residual sulphuric acid content in an aqueous IPS solution was 0.2%, indicating that 98% of the acid taken for hydrolysis is removed simultaneously with the isolation of IPS (without chemical neutralization).

In order to convert IPS to the monomeric form of sugar (glucose), an inversion was carried out at high temperature using a diluted acid. We studied the kinetics of the IPS inversion using diluted (0.075–1.5%) hydrochloric acid [116]. The choice of this acid was primarily dictated by the fact that sodium chloride formed during the neutralization of the acid by NaOH is a part of the complex composite (CC) of glucose with the formula $(C_6H_{12}O_6)_2{\cdot}NaCl{\cdot}H_2O$, the decomposition of which results in the release of crystalline glucose. Considering that the potential yield of glucose, on inversion by 5% sulphuric acid at 100 °C for 5 h, is 1.5% in the hydrolysate, which corresponds to a glucose yield equal to 82% of the weight of a.d.c., the acid concentration of 0.125% and temperature of 170 °C represent the optimum inversion conditions under which the yield of RS in the invertion is at its maximum; the time taken for attaining the maximum yield in these conditions is minimal. Thus, during the hydrolysis of cellulose and the subsequent inversion of IPS, the main glucose content in the inverted solution is about 70% of the mass of RS; i.e., the real yield of glucose is 35–45% of a.d.c.

It should be noted that these glucose syrups are transparent, of a light yellow colour, and are distinguished by a high factor of merit (85–90% and more) (Figure 16). In comparison, the yield of glucose from CR (the first variant of hydrolysis) is 23–25%. CG is isolated from glucose solutions either by direct or salt crystallization [117]. We studied the crystallization properties of glucose syrups obtained by hydrolysis of CR and cellulose using both the direct and salt methods. As mentioned above, the factor of merit of starting syrups must be no less than 85% for the successful crystallization of glucose using the direct method. Glucose syrups obtained by hydrolysis of cellulose completely meet this requirement. The application of activated carbon increased the quality of the syrup since direct crystallization occurred only in clarified syrups.

In order to perform direct crystallization, cellulose hydrolysis was used to obtain a hydrolysate with a RS content of 1%, pH 4.4, and a factor of merit of 94%, which was allowed to stand at room temperature for spontaneous crystallization. After two weeks, the onset of crystallization was visually observed. Crystallization by itself, without the creation of special temperature conditions, progresses slowly (taking a month and more). The method of salt crystallization of glucose has some advantages over direct crystallization. It does not require a deep purification of hydrolysates, the crystallization process is shorter and simpler (there is no need for a multiple recrystallization), and the yield of glucose increases.

Figure 16. Glucose syrup from larch wood CR

We studied the crystallization conditions of glucose using its CC with sodium chloride $(C_6H_{12}O_6)_2 \cdot NaCl \cdot H_2O$ and examined CC crystals obtained from model mixtures of glucose, sodium chloride and water, as well as from experimental glucose syrups obtained by the hydrolysis of CR and cellulose [118].

The crystallization of CC was studied using artificial mixtures in which the NaCl content varied from 15% to a twofold excess relative to the glucose content. In all cases, a crystalline phase formed. The composition of the crystalline phase was determined by elemental analysis. We determined that the range of NaCl:glucose ratios from 0.2:1 to 0.7:1 (parts by weight) is optimal for the formation of CC. Similarly, we determined the crystallization conditions for CC in hydrolysates of CR and cellulose. CC crystals were isolated from hydrolysates; the artificial mixtures were colourless and transparent and had well-defined facets. According to the chemical analysis data, they have a composition close to being stoichiometric: C – 32.5%, H – 5.9% and Cl – 9.7%. Theoretically, CC with the general formula $(C_6H_{12}O_6)_2 \cdot NaCl \cdot H_2O$ contains C – 33%, H – 6%, and O – 18.13%.

According to X-ray phase analysis, CC monocrystals synthesized from pure solutions have the unit cell parameters a = b = 16.8, c = 17.0 Å and β = 120°, and represent a hexagonal prism. Based on the symmetry of lauegrams and weissenbergograms, they belong to the diffraction class $P\overline{3}m$ with regular extinctions at $1 \neq 3n$. Therefore, the spatial group of CC crystals was determined as $P3_1 12$ (151) and $P3_2 12$ (153).

The set of diffraction maxima obtained by X-ray phase analysis of CC crystals and the reference indicates that, under the experimental conditions used, glucose in the presence of sodium chloride crystallizes as CC with the formula $(C_6H_{12}O_6)_2 \cdot NaCl \cdot H_2O$.

Thus, when studying the crystallization properties of glucose syrups produced by acid hydrolysis of the crystalline glucose of larch wood, we obtained CC glucose crystals with sodium chloride, upon decomposition of which D-glucose is released in crystalline form. In addition, glucose can be directly crystallized from glucose syrups produced by the hydrolysis of cellulose with a high factor of merit (more than 85%).

The ways in which the efficiency of hydrolysis of polysaccharides from wood CR can be increased are of a great interest. One method is the steam explosion hydrolysis of cellulose containing raw material, which makes possible the efficient and completely ecologically safe decomposition of lignocellulose material into its constituents: lignin, hemicellulose, and cellulose.

We studied the steam explosion hydrolysis of CR from larch wood, and showed that this method can be used for effective prehydrolysis processing of larch wood CR [119].

Hence, a laboratory scheme has been developed which will be used as the technological basis for obtaining crystalline glucose from the CR in larch wood [120].

5. Experimental section

Arabinogalactan was extracted using technology from *L. sibirica* Ledeb. [76] at the experimental–industrial plant and purified with methods described in [73, 17]. Molecular masses of arabinogalactan were defined with a high-performance liquid chromatography (HPLC) method, assisted by the Agilent Technologies 1260 Infinity chromatographic system using 0.1 M sample solutions of $LiNO_3$ on PL aquagel-OH-40.8 mm, 300x7.5 mm column, with a PL aquagel-OH Guard 8 mm 50x7.5 mm precolumn, standardized according to dextrans with 25, 12 and 5 kDa molecular mass standard solutions and monosaccharides.

IR spectra were registered in KBr tablets on a "Specord 75IR" spectrophotometer with a 500–4000 cm^{-1} interval. UV spectra were registered with a "Specord UV-vis" spectrophotometer (10 mm layer thickness). NMR ^{13}C spectra of AG samples were registered with a "Varian VXR 500S" spectrometer with a 125.1 Hz operating frequency; D_2O was used as solvent. Deuteroaceton was used as an internal standard. The correlation of galactose and arabinose chains, composed of AG macromolecules, were calculated according to the correlation between integral intensities of carbon galactose anomeral atomic signals and arabinose. The ratio of galactose to arabinose units in AG macromolecules was calculated from the ratio of the intensities of signals from anomeric carbon atoms of glactose and arabinose [10].

Pectin polysaccharides (PS) were extracted from the bark of *L. sibirica* Ledeb. and from *L. gmelini* (Rupr.) Rupr. according to the scheme depicted in Figure 17. Larch air-dried bark (500 g), which was initially ground and treated with ethyl acetate, was extracted using distilled water at 70 °C over 3 h. Raw material residue was poured with a mixture (1:1, v/v) 0.5% of ammonium oxalate water solution and 0.5% oxalic acid water solution and heated at 80 °C for 2 h. The extract was concentrated. Polysaccharides were precipitated with a triple volume of ethyl alcohol or acetone and dried with lyophilization. As a result, we obtained PS. The PS was dissolved in water and aminoacids were detected on AAA339M automatic analyser.

PS (50 mg) was dissolved in 20 ml of water. Pektinaza (2 mg; Sigma, USA) water solution was added. The mixture was temperature-controlled at 37 ° C for 3 h. Then, a reaction mixture was heated for 5 min in a water bath at 100 °C. Coagulated protein was separated by centrifugation. The obtained supernatant was concentrated and up to 5 ml 96% ethyl alcohol was added (4 volumes). Deposition was separated with centrifugation. Alcoholic supernatant was concentrated and analysed with the help of PC.

Galacturonic acid content in PS was defined according to the reaction with 3,5-dimethyl phenol in the presence of concentrated H_2SO_4, protein using the Lowry method [121] and based on the calibrating schedule for a bovine serum albumin 80000 Da. Paper chromatography was carried out on "Filtrak FN-13" paper with a descending method in a n-butanol-pyridine-water system (volume correlations 6:4:3, respectively). To define carbohydrates, aniline phthalate was poured on the paper and heated at 105 °C. Gas–liquid chromatography was carried out using a Hewlett-Packard 4890A (USA) chromatograph equipped with a flame-ionization detector, RTX-1 (0.25 mm x 30 m) capillary column, argon carrier gas, and 1:60 dumping. Temperature rate: 175 °C (1 min)–250 °C (2 min), Δ 3°/min.

A full acid hydrolysis PS (5 mg) was carried out for the implementation of 2M trifluoroacetic acid (2 ml) which contained *myo*-inositol (1 mg/ml). The mixture was heated in a soldered ampoule for 5 h at 100 °C, and the acid was removed with a repeated dry evaporation with added methanol. As a result, we obtained PVG-1.

Ion-exchange chromatography PS (100 mg) was carried out on a DEAE-cellulose (25x2 cm) column. NaCl solutions were used as an eluent with increasing concentrations (0.01M–1M, 60 ml/h elution speed, fractions selection by 12 ml). Pick correspondent fractions at the output bents were combined, dialysed and lyophilized. As a result, we obtained PS 1-4 fractions. The monosaccharide composition of each fraction was defined with GLC in acetate polyol after preliminary hydrolysis.

In order to obtain acetate polyol, each PS 1-4 fraction was dissolved in a 1M ammonia solution (1 ml) and 5 mg of $NaBH_4$ was added. The mixture was kept for one day at a room temperature. Then, $NaBH_4$ was eliminated by adding 2–3 drops of concentrated acetic acid; 0.2 ml of dry pyridine and acetic anhydride were added to the dry residue. The mixture was acetilized at 100 °C for 1 h. The solution was dry-evaporated until pyridine and acetic anhydride were removed, first by adding 1 ml of toluene and then 1 ml of methanol. The obtained acetate mixture of PS 1-4 polyol fractions was dissolved in 0.2 ml of dry chloroform and moved quantitatively to Appendorf tubes, concentrated up to 0.1–0.2 ml and analysed with the GLC method.

PS (5 mg) partial acidic hydrolysis was carried out using 0.01M TFA (2 ml), which contained *myo*-inositol (1 mg/ml). The mixture was heated in a soldered ampoule at 100 °C for 3 h. The acid was removed using a repeated dry evaporation with added methanol. As a result, we obtained PVG-2.

After extraction of dihydroquercetin, arabinogalactan and resin, the larch chip presented as a cellolignin residue. The chip had the following dimensions: 25x15x5mm, and sawdust fraction, 1x2x2 mm.

Bleached pulp from Baikal Pulp Mill was used for hydrolysis: polymerization degree 573, ash 1.1%, humidity 3%. Cellulose hydrolysis was carried out using 72% sulphuric acid and water in a ratio of 1:3 at room temperature for 1 h. Hydrolysis products—inverted polysaccharides—were precipitated with a five-fold ethanol volume. The precipitate was filtered and washed with alcohol the last washed portion achieved a neutral reaction. The product was dried in the air at up to 6% humidity. Acid content of inverted polysaccharides was defined using 1N HCl titration. Inversion of IPS was carried out in 0.75–1.50% solutions, hydromodulus 1:30, at 100–170 °C, for 0.25–3.0 h inversion duration. The potential content of reducing substances in hydrolyzates was defined by inversion of water-soluble polysaccharides with 5% sulphuric acid. 20% NaOH was used to neutralize the hydrochloric acid. Glucose quantitative content in neutralisate (pH 4–5) was defined by HPLC methodology.

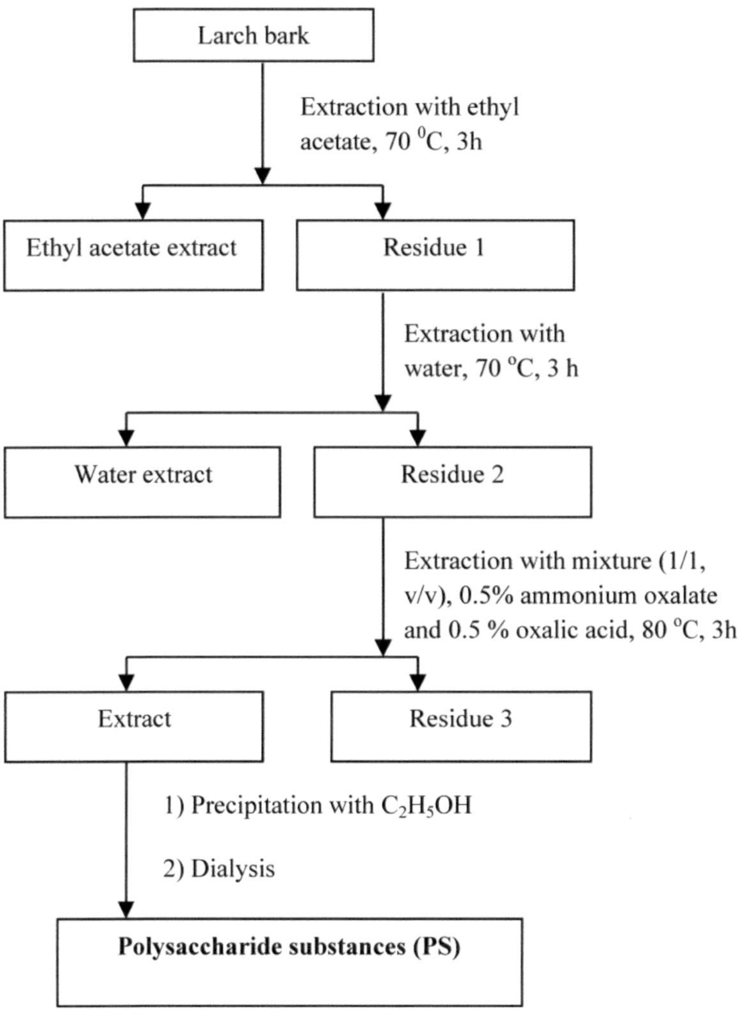

Figure 17. Extraction scheme of pectin substances from larch bark

Larch cellolignin timber residue with particles having dimensions of 25x15x5 mm was used for explosive autohydrolysis. Autoexplosive hydrolysis was carried out in a special 200 ml capacity autoclave, which allowed us to conduct a quick decompression of the reactor (steam explosion). Hydrolysis conditions were: 200 and 220 °C, duration 2 and 5min.

The laboratory scheme for obtaining crystalline glucose is depicted in Figure 18.

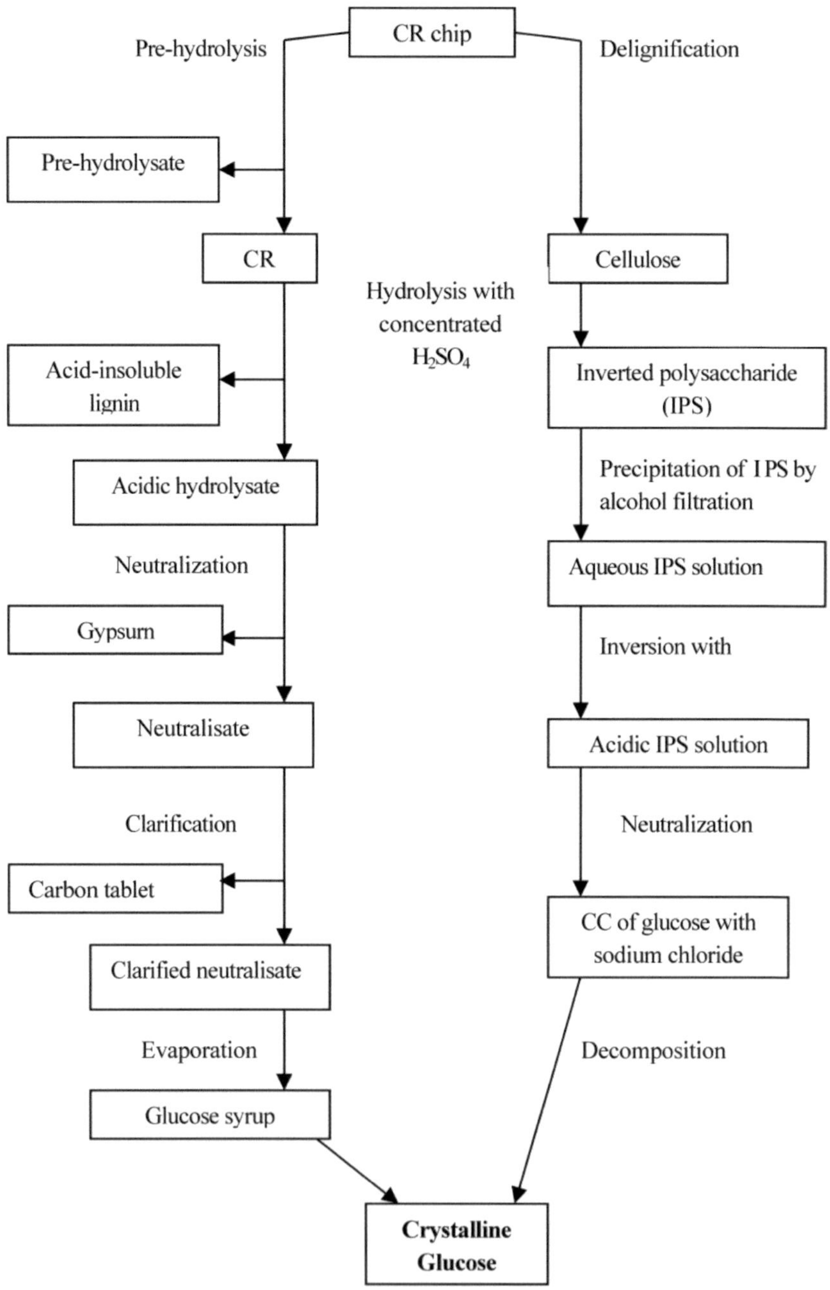

Figure 18. Laboratory scheme for obtaining crystalline glucose

6. Conclusion

This chapter therefore summarizes studies on polysaccharides in the context of the development of technology for 100% processing of larch wood and bark as forestry waste in order to provide new medicines, veterinary drugs, dietary supplements and valuable materials for the cosmetics and agricultural industries. There are data on larch wood and bark extraction by the two-phase solvent system, namely the kinetic study of extraction processes, diffusion constants, mass-transfer coefficients, mechanisms and physicochemical characterization of the transfer process, its mathematical modelling and structural characteristics of the samples isolated. This work aims to support the development of economically and ecologically viable production technology for high-demand products on the basis of renewable raw materials with a 15–20% increase of forestry efficiency due to waste processing. The technology will provide new medicines and food supplements, as well as cheaper, by 40–50%, analogues, to those currently known.

Author details

Natalya Nikolaevna Trofimova, Elena Nikolaevna Medvedeva, Nadezhda Viktorovna Ivanova, Yuriy Alekseevich Malkov and Vasiliy Anatolievich Babkin
Laboratory of Wood Chemistry, Federal Research Budget Institution, A.E. Favorsky Institute of Chemistry, Siberian Branch, Russian Academy of Sciences, Irkutsk, Russia

Acknowledgement

This project has been supported by a State Contract with the Department of Industry, Science and Technology of the Russian Federation №43.044.1.1.26.38(2002–2004); by a State Contract with the Agricultural Department of the Government of Irkutsk Oblast №02-66 (2010–2011); and by RAS and SB RAS Grants (2005–2012). The authors are grateful to all their colleagues at the Laboratory of Wood Chemistry of A.E. Favorsky Irkutsk Institute of Chemistry who have contributed to this project.

7. References

[1] Clarke A, Anderson R, Stone B (1979) Form and function of arabinogalactans and arabinogalactan-proteins. Phytochemistry. 18(4): 521-540.
[2] Antonova GF, Tukavkina NA (1983) Water soluble substances of larch and their possible applications. Khimia Drevesiny. 2: 89-96.
[3] Adams MF, Douglas C (1963) Arabinogalactan—a review of literature. TAPPI. 46: 544-548.
[4] Aspinall GO (1964) Some recent developments in the chemistry of arabinogalactans. In: Chimie et Biochimie de la Lignine, de la Cellulose et des Hemicelluloses. Grenoble: Actes du Symposium International de Grenoble. pp89-97.
[5] Medvedeva SA, Aleksandrova GP (2003) Modification strategy and biopotential of natural polysaccharide arabinogalactan. In: Panorama Sovremennoi Khimii Rossii [Polymer Synthesis and Modification]. Moscow: Khimia. pp328-356.

[6] Medvedeva EN, Babkin VA, Ostroukhova LA (2011) Larch arabinogalactan: properties and application prospects. Khimia rastitel'nogo syr'ia. 1: 27-37.

[7] Chochieva MM, Tsvetayeva IP, Yurieva MK, Zaytseva AF (1958) Arabinogalactan distribution in the Dahurian larch wood. Works of the USSR AS Forest Institute. 45: 31-49.

[8] Ponder G, Richards G (1997) Arabinogalactan from Western larch. Part III: Alkaline degradation revisited, with novel conclusions on molecular structure. Carbohydrate Polymers. 34: 251-261.

[9] Karacsonyi S, Kovacik V, Alfoldi J, Kubackova M (1984) Chemical and ^{13}C NMR studies of arabinogalactan from *Larix sibirica* L. Carbohydrate Research.134: 265-274.

[10] Medvedeva EN, Fedorova TE, Vanina AS, Rokhin AV, Es'kova LA, Babkin VA (2006) Arabinogalactan structure and properties vs. methods of its isolation from larch wood and purification techniques. Khimia rastitel'nogo syr'ia. 1: 25-32.

[11] Prescott J, Groman E, Gyongyi G (1997) New molecular weight forms of arabinogalactan from *Larix occidentalis.* Carbohydrate Research. 301: 89-93.

[12] Ovodov YS (1998) Polysaccharides of phanerogams: their structure and physiological activity (review article). Russian Journal of Bioorganic Chemistry (Bioorganicheskaya Khimiya). 24: 483-501.

[13] Arifkhodzhaev AO (2000) Galactans and galactan-containing polysaccharides of higher plants. Khimiya prirodnykh soedinenii. 3: 185-197.

[14] Groman E, Enriquez P, Jung Chu, Josephson L (1994) Arabinogalactan for hepatic drug delivery. Bioconjugate Chemistry. 5: 547-556.

[15] Grieshop CM, Flickinger EA, Fahey GC Jr (2002) Oral administration of arabinogalactan affects immune status and fecal microbial populations in dogs. The Journal of Nutrition. 132: 478-482.

[16] Robinson RR, Feirtag J, Slavin JL (2001) Effects of dietary arabinogalactan on gastrointestinal and blood parameters in healthy human subjects. Journal of the American College of Nutrition. 20: 279-285.

[17] Medvedeva EN, Babkin VA, Makarenko OA, Nikolaev SM, Khobrakova VB, Shulunova AM, Federova TE, Es'kova LA (2004) Preparation of high purity larch arabinogalactan and study of its immunomodulating properties. Khimiya rastitel'nogo syr'ia. 4: 17-23.

[18] Dubrovina VI, Medvedeva SA, Vityazeva SA, Kolesnikova OB, Aleksandrova GP, Gutsol LO, Grishenko LA, Chetverikova TD (2007) Structure and Immunomodulating Action of Arabinogalactan of Siberian Larch and its Metal Derivatives. Irkutsk: Asprint. 145 p.

[19] Robinson RR, Causey J, Slavin JL (2001) Nutritional benefits of larch arabinogalactan. In: McCleary BV, Prosky L. (eds.) Advanced Dietary Fiber Technology. Oxford: Blackwell Science Ltd. p443-451.

[20] Causey J, Robinson R, Feirtag J, Fulcher R, Slavin J (1999) Long term supplementation of arabinogalactan in humans and its effects on blood lipids, glucose and insulin. The FASEB Journal. Part I:3.

[21] Zarubaev VV, Ostrouhova LA, Medvedeva EN, Babkin VA, Kiselyov OI (2010) Antiviral preparations based on larch wood biologically active substances. Bulletin of Eastern-Siberian Science Centre SB RAMS. 1(71): 76-81.

[22] European Patent Office. Arabinogalactan as Immune Enhancer, EP 2 327 411 A1, Bulletin 2011/22. European Patent Office; 2011.
[23] Lonza. Nutrition. http://www.lonza.com/products-services/nutrition.
[24] Gaines Nutrition. Ara-Plus (90 capsules) (Mfr: D'Adamo Personalized Nutrition/North American Pharmacal) (Professional quality product). http://www.gaines.com/store/NorthAmerPharmacal/EM-ARAPLinfo.html (accessed Nov.1994).
[25] Kiselev OI, Babkin VA, Tikhonov VP (2010) Drug for Prophylaxis of Influenza and Upper Respiratory Tract Infections. RF Patent 2401117.
[26] Burke S, Scheuer C, Mullard K (2009) Ophthalmic compositions comprising a terpene and a natural polymer. WO 2009018060 (A1). European Patent Office. http://worldwide.espacenet.com/publicationDetails/biblio?FT=D&date=20090205&DB=EPODOC&locale=ru_RU&CC=WO&NR=2009018060A1&KC=A1. Accessed on April 2005.
[27] Jung C, Enriquez P, Palmacci S, Josephson L, Lewis JM (1995) Arabinogalactan derivatives and uses thereof. US Patent Number 5478576.
[28] Cui L, Faraj A, el Alaoui A (1997) Arabinogalactan (9kDa)-9-β-D-arabinofuranosyladenine-5'-monophosphate, a novel liver-targeted conjugate that selectively inhibits hepatitis B virus replication in vitro. Antiviral Chemistry & Chemotherapy. 8: 529-536.
[29] Josephson L, Rutkowski JV, Paul K, Frigo T, Korba BE, Tennant B, Groman EV (1996) Antiviral activity of a conjugate of adenine-9-β-D-arabinofuranoside - 5'- monophosphate and a 9 kDa fragment of arabinogalactan. Antiviral Therapy. 1(3): 147-156.
[30] Groman EV, Menz ET, Enriquez PM, Jung C (1996) Delivery of Therapeutic Agents to Receptors using Polysaccharides. US Patent Number 5544386.
[31] Josephson L, Groman EV, Jung C, Lewis JM. (1994) Targeting of therapeutic agents using polysaccharides. US Patent Number 5336506.
[32] Nogawa M, Akaike T, Maruyama A (2001) Polyethylenimine/arabinogalactan conjugate as a hepatocyte specific gene carrier. S.T.P. Pharmaceutical Sciences. 11: 91-102.
[33] Ehrenfreund T, Golenser J, Frankenburg S, Domb AJ (2000) Water soluble amphotericin B - arabinogalactan conjugates for the treatment of experimental leishmaniasis. In: Proceedings of the 27th International Symposium on the Controlled Release of Bioactive Materials, 7-13 July 2000, Porte Maillot, Paris, France. pp578-579.
[34] Ehrenfreund-Kleinman T, Azzam T, Falk R, Polacheck I, Golenser J, Domb AJ (2002) Synthesis and characterization of novel water soluble amphotericin B - arabinogalactan conjugates. Biomaterials. 23: 1327-1335.
[35] Pozharitskaya ON, Urakova IN, Shikov AN (2008) Solid Nanocomposition for Delivering Biologically Active Substances. WIPO Patent Application WO/2008/123798.
[36] Reis RL (2008) Natural-based Polymers for Biomedical Application. Cambridge: Woodhead. 802 p.
[37] Enriquez P, Jung C, Groman E (1996) Directed delivery of radioprotectants using a receptor - specific carrier. US Patent Number 5490991.
[38] Domb AA (2001) Biodegradable Polycation Composition for Delivery of an Anionic Macromolecule in Gene Therapy. PCT Int. Application WO 2001007486 A1. See URL http://www.patentstorm.us/patents/7001891.html.

[39] Azzam T, Eliyahu H, Shapira L, Linial M, Barenholz Y, Domb AJ (2002) Polysaccharide-oligoamine based conjugates for gene delivery. Journal of Medicinal Chemistry. 45: 1817-1824.
[40] Dutta RC (2007) Drug carriers in pharmaceutical design: promises and progress. Current Pharmaceutical Design. 13: 761-769.
[41] Jain A, Gupta Y, Jain SK (2007) Perspectives of biodegradable natural polysaccharides for site-specific drug delivery to the colon. Journal of Pharmacy and Pharmaceutical Sciences. 10: 86-128.
[42] Kochetkov NK (1967) Carbohydrate Chemistry Methods. Moscow: Mir. 512 p.
[43] Domb A, Benita S, Polachek I, Linden G (2000) Drug conjugates with oxidized arabinogalactan or dextran. US Patent Number 6011008.
[44] Ponedelkina IY, Saitgalina EA, Tolstikov GA, Dzemilev UM, Tolstikova TG, Odinokov VN, Lukina ES (2009) Modified arabinogalactans and a method for the production thereof. Patent RU 2359975, WO 2008/123799.
[45] Mudarisova RKh, Shirokova EN, Badykova LA, Borisov IM, Tolstikova TG, Sorokina IV, Dolgikh MP, Monakova YB (2005) Synthesis and pharmacological activity of the oxidized fractions of arabinogalactan from siberian larch (*Larix sibirica L.*). Pharmaceutical Chemistry Journal. 39: 418-421.
[46] Badykova LA, Mudarisova RKh, Tolstikova TG, Borisov IM, Monakov YuB (2005) Arabinogalactan poly- and oligosaccharides modified with 5-aminosalicylic acid. Chemistry of Natural Compounds. 41(3): 272-275.
[47] Badykova LA, Mudarisova RKh , Khamidullina GS, Aminev KhK, Borisov IM, Monakova YB (2006) Modification of arabinogalactan and its oxidized forms with *p*-aminosalicylic acid. Russian Journal of Applied Chemistry. 79: 1626-1628.
[48] Badykova LA (2007) Reaction of arabinogalactan from Siberian larch with 5-aminosalicylic acid, 4-aminosalicylic acid and isonicotinic acid hydrazide. Abstract of candidate's PhD thesis. Ufa: Institution of the Russian Academy of Sciens Institute of Organic Chemistry of Ufa Scientific Centre of the Russian Academy of Sciences (IOChem USC RAS), Russia. 26 p.
[49] Medvedeva EN, Neverova NA, Fedorova TE, Babkin VA, Meteleva ES, Dushkin AV, Tolstikova TG, Khvostov MV, Dolgikh MP (2010) Structural transformations of arabinogalactan from the Siberian larch during mechanochemical processing and the biological properties of the products. Russian Journal of Bioorganic Chemistry. 36: 853-859.
[50] Tolstikova TG, Tolstikov GA, Dushkin AV, Meteleva ES (2008) Water Soluble Medical Composition and Synthetic Method Thereof. RU Patent 2 337 710.
[51] Dushkin AV, Meteleva ES, Tolstikova TG, Tolstikov GA, Polyakov NE, Neverova NA, Medvedeva EN, Babkin VA (2008) Mechanochemical preparation and pharmacological activity of water-soluble intermolecular complexes of arabinogalactan and pharmaceutical drugs. Russian Chemical Bulletin. 57: 1299-1307.
[52] Tolstikova TG, Khvostov MV, Bryzgalov AO, Dushkin AV, Tolstikov GA (2010) Plant polysaccharide arabinogalactan as new pharmacon clathrating agent. RAS Reports 433: 713-714.
[53] Babkin VA, Medvedeva EN, Neverova NA, Chipanina NN, Aksamentova TN, Meteleva ES, Dushkin AV (2009) Study of mechanocomposites on the basis of arabinogalactan and medical substances. In: Materials of the IV Russian Scientific Conference: New

Achievements in Chemistry and Chemical Technology of Plant Raw Material. Barnaul. 21-23 April 2009. pp176-178.
[54] Medvedeva EN, Neverova NA, Ostroukhova LA, Babkin VA, Gus'kov SA, Meteleva ES, Dushkin AV (2010) Mechanical composites based on biologically active compounds from larch wood. Chemistry of Natural Compounds. 42: 1-4.
[55] Trofimov BA, Sukhov BG, Aleksandova GP, Medvedeva SA, Grishchenko LA, Mal'kina AG, Feoktistova LP, Sapozhnikov AN, Dubrovina VI, Martynovich EF, Tirskii VV, Semenov AL (2003) Nanocomposites with magnetic, optical, catalytic and biologically active properties based on arabinogalactan. Doklady Chemistry. 393(4): 287-288.
[56] Sukhov BG, Aleksandrova GP, Grishchenko LA, Feoktistova LP, Sapozhnikov AN, Proidakova OA, T'kov AV, Medvedeva SA, Trofimov BA (2007) Nanobiocomposites of noble metals based on arabinogalactan: preparation and properties. Journal of Structural Chemistry. 48(5): 922-927.
[57] Aleksandova GP, Grishchenko LA, Fadeeva TV, Medvedeva SA, Sukhov BG, Trofimov BA (2006) Antimicrobial Preparation. RU Patent 2 278 969.
[58] Ohr L (2001) Arabinogalactan Adds More than Health Benefits. Prepared Foods. 170: 55.
[59] Babkin VA, Medvedeva EN, Sambueva ZG, Khobrakova VB, Neverova NA, Suyuncheva BO, Tatsyi AA (2012) Medical and biological properties of bakery and pastry products enriched with larch arabinogalactan. Khimia rastitel'nogo syr'ia. 2: 185-193.
[60] Ermakova MF, Chistyakova AK, Shchukina LV, Pshenichnikova TA, Medvedeva EN, Neverova NA, Belovezhets LA, Babkin VA (2010) Effect of arabinogalactan isolated from Siberian Larch on the baking value of soft wheat flour and bread quality. Russian Journal of Bioorganic Chemistry. 36: 951–956.
[61] Neretina OV, Medvedeva EN, Babkin VA (2011) Production Technology of Prophylactic Bakery and Compositions of Prophylactic Biscuit, Cheese and Sugar Cookies. RU Patent 2430614.
[62] Nesteruk VV, Vershchako VA, Likhova NV, Babkin VA, Medvedeva EN (2011) Production Technology of Oatmeal Cookies Containing Prophylactic Arabinogalactan Supplement and Composition Thereof. Patent application 2011113945.
[63] Utochkina EA (2011) Study and technology development of arabinogalactan-enriched cultured milk product. Abstract of PhD thesis. Blagoveshchensk: Far Eastern State Agrarian University; Russia.
[64] Galochkin VA, Galochkina VP, Maksimenko SV (2008) Non-specific resistance and meat productivity in calves at arabinogalactan-enriched feeding. Sel'skokhozyaistvennaya biologia. 4: 89-95.
[65] Maksimenko SV (2008) Influence of arabinogalactan upon productivity and non specific resistance of calves. Problemy biologii produktivnykh zhivotnykh. 1: 82-86.
[66] Torshkov AA, Fomichev YuP (2010) Application of arabinogalactan in raising broiler chickens. Orenburg Agrarian University Reports. 1: 172-175.
[67] Torshkov AA, Gerasimenko VV (2010) Changes in chemical composition of broiler chicken meat at the use of arabinogalactan. Orenburg Agrarian University Reports. 2: 167-169.
[68] Moskovtsev NG, Belova TP, Valetov TA (1988) Isolation of water extracted substances from larch wood. Khimiya Drevesiny (Wood Chemistry). 1: 112-116.

[69] Kuznetsova SA, Kuznetsov BN, Aleksandrova NB, Danilov VG, Zhizhaev AM (2005) Obtaining arabinogalactan, dihydroquercetin and microcrystalline cellulose using mechanochemical activation. Chemistry for Sustainable Development. 13: 261-268.

[70] Kuznetsova SA, Mikhailov AG, Skvortsova GP, Aleksandrova NB, Lebedeva AB (2005) Intensifying the process of water extraction of arabinogalactan from larch wood. Khimia rastitel'nogo syr'ia. 1: 53-58.

[71] Antonova GF (1977) Study of fraction analysis of polysaccharide arabinogalactan from Siberian larch wood. Khimiya Drevesiny (Wood Chemistry). 4: 97-100.

[72] Medvedeva EN, Ostroukhova LA, Neverova NA, Onuchina NA, Babkin VA (2011) Phenolic admixtures in larch wood arabinogalactan. Khimiya rastitel'nogo syr'ia. 1: 45-48.

[73] Babkin VA, Kolzunova LG, Medvedeva EN, Malkov YuA, Ostroukhova LA (2005) Production method of arabinogalactan. RU Patent 2 256 668.

[74] Akselrud GA, Altshuler MA (1983) Introduction to Capillary-Chemical Technology. Moscow: Khimiya. 264 p.

[75] Malkov YuA, Ostroukhova LA, Babkin VA (2002) Application of mathematical modeling method for technology development of isolating extracted substances from larch wood. Khimiya rastitel'nogo syr'ia. 2: 133-138.

[76] Babkin VA, Malkov YuA, Medvedeva EN, Babkin DV, Ostroukhova LA, Kolzunova LG, Zolotar' RN, Nikolaev SM, Khobrakova VB (2005) Technology of arabinogalactan isolation from larch wood and research of its biological activity. In: Phytopharm 2005: proceedings of the Ninth International Congress, St. Peterburg. 22-25 June 2005. pp165-168.

[77] Kafarov VV, Vetokhin VN (1987) Fundamentals of Automatic Design of Chemical Production. Moscow: Nauka. 624 p.

[78] Mukhlenov IP (1986) Chemical-technological Systems. Leningrad: Khimiya. 425 p.

[79] Kafarov VV, Meshalkiin VP, Perov VL (1974) Principles of Mathematical Modeling of Chemical-technological Systems. Moscow: Khimiya. 624 p.

[80] Ovodov YuS, Golovchenko VV, Gunter EA, Kutchin AV (2009) Pectic Substances of Plants of the European North of Russia. Ekaterinburg: Uro RAN. 112 p.

[81] Tehnokonsalt Management. The organisation of production technology of pectin (in Russian). http://www.tc-m.ru/index.php/155.html. Accessed 2004 March.

[82] Minzanova ST, Mironov VF, Sosnina NA, Vyshtakalyuk AB, Lapin AA, Smolentsev AV, Mironova LG, Khirug SS, Konovalov AI (2007) Technological aspects of the manufacturing of amaranth fodder additives. Doklady Chemistry. 413(1): 75-78.

[83] Bubenchikov VN, Drozdova IL, Kochkarov VI (2002) Method of Isolation of the Polysaccharides Possessing Anti-inflammatory Activity. RU Patent 2177793.

[84] Ovodova RG, Golovchenko VV, Shazhkov AS, Popov SV, Ovodov YuS (2000) Structural investigation of pectin and physiological activity of lemnan and pectin from *Lemna minor L.* Bioorganicheskaya khimiya. 26:743-751.

[85] Ovodova RG, Bushneva OA, Golovchenko VV, Popov SV, Ovodov YuS (2000) Method of Isolation from Vegetative Raw Materials of the Polysaccharides Possessing Immunostimulating Action. RU Patent 2149642.

[86] Gelgay MK, Donchenko LV, Reshetnyak AI (2008) Innovative technology of pectin from secondary sources of raw material after processing coffee. New Technologies. 6:15-18.

[87] Graves F (1994) Process for Producing Pectin Through Ion Exchange. US Patent 5364851.

[88] Aleinikov IN, Sergeev VN, Rusakov AV, Merkulov YuG (2000) New approach to pectin production. Pishchevaya promyshlennost'. 1: 59-61.
[89] Aimukhomedova GB, Shelukhina NP (1964) Pectin Substances and Their Determination Methods. Frunze: Ilim. 119 p.
[90] Babkin VA, Ostroukhova LA, Ivanova NV, Malkov YuA, Ivanova SZ, Onuchina NA (2002) Phytocomplex of Antioxidant Activity and Production Method Thereof. RU Patent 2188031.
[91] Shelukhina NP, Abayeva RSh, Aimukhomedova GB (1987) Pectin and Parameters of its Production. Frunze: Ilim. 91 p.
[92] Kaisheva NSh, Shcherbak SN, Kompantsev VA, Krikova NI, Ivanova LI (1994) Analysis of protective action of pectins. Zhurnal analititcheskoi khimii. 49: 1158–1162.
[93] Babkin VA, Ivanova NV, Trofimova NN, Es'kova LA, Salyaev RK, Nurminski VN, Korzun AM, Feoktistova LP, Sapozhnikov AN, Lykhoshvai EV, Arsent'ev KYu (2010) Isolation of Pectin, Possessing Membrane Stabilizing Activity and Ability to Reduce Silver Ions, from Larch Wood. Silver Nanocomposites Stabilized by Pectin. RU Patent 2403263. http://www.nti.lin.irk.ru/bibl/rese.asp?n=2156&s=10 Accessed 29 June 2009.
[94] Nakanisi K (1965) Infrared Spectra and Structure of Organic Compounds. Moscow: Mir. p40.
[95] Odonomažig P, Balga D, Ebrigerova A (1992) Structures of pectin polysaccharides isolated from the Siberian apricot (*Armaniaca sibirica* Lam.). Carbohydrate Research. 226: 353-358.
[96] Grigoriuk GP (1998) Pectins from a bark of coniferous trees as biologically active components of food stuffs in the medical and practical purposes. In: International Conference on Natural Products and Physiologically Active Substances, 30 November-6 December 1998, Novosibirsk; Russia. p79.
[97] Kondratyuk TA (2009) Production and quality estimation of food products from Siberian larch bark. Abstract of PhD thesis; Krasnoyarsk: Krasnoyarsk State Trade and Economic institute, Russia; Kemerovo, 2009.
[98] Ministry of Agriculture of the Russian Federation. Manual on pectosorbitol application in veterinary. Registration PSP-2-2.0/00344. :Council on Veterinary Drugs of the Ministry of Agriculture of the Russian Federation.
[99] Lipson G, Still G (1972) Powder radiographinterpretation. Moscow: Mir. 384 p.
[100] Babkin VA, Ostroumova LA, D'yachkova SG, Svyatkin YuK, Babkin DV, Onuchina NA (1997) The complex wasteless Siberian and Dahurian larch biomass processing echnology. Khimiya v Interesah Ustoich. Razvit. (Chemistry for Sustainable Development) 5: 105–115.
[101] Levin ED, Denisov OB, Pen RZ (1978) Kompleksnaya pererabotka listvennitsy Complex Processing of Larch). Moscow: Lesnaya promyshlennost'. 224 p.
[102] Tsvetaeva IP, Yur'eva MK (1960) About water soluble larch wood substances. Izvestia Vuzov. Lesnoi Zhurnal. 1: 148–151.
[103] Fengel D, Vegener G (1988) Drevesina (Khimiya, ul'trastruktura, reaktsii) (Wood: Chemistry, Ultrastructure, Reactions)). Moscow: Lesnaya promyshlennost'. 512 p.
[104] Trofimova NN, Bichevina OB, Babkin VA (2004) The carbohydrates of the larch cellolignin. Khimiya rastitel'nogo syr'ya. 3: 11–14.

[105] Bichevina OB, Trofimova NN, Babkin VA (2001) Study of the aqueous prehydrolysis of cellolignin from Larch wood. In: Problems of Sustainable Regional Development, Abstracts of the Second School Workshop of Young Scientists of Russia, September 17-21, 2001, Ulan-Ude, Russia. pp24-25.
[106] Korol'kov II (1978) Percolation Hydrolysis of Row Plant Material. Moscow: Lesnaya promyshlennost'. 263 p.
[107] Fisher NP. Use of pentoses for manufacture of fodder yeast. In book: Problems of Using Pentose Containing Row Materials, Proceedings of the All Union Conference. Riga: Izdatel'stvo Akademii nauk Latviiskoi SSR, 1958. pp91-99.
[108] Trofimova NN, Gordienko II, Babkin VA (2005) Dependence of redusing substances yield on acid hydrolysis parameters of larch cellolignin. Khimiya rastitelnogo syr'ya. 4: 25-28.
[109] Glyukoza kristallicheskaya gidratnaya: Tekhnicheskie usloviya (Crystalline Hydrate Glucose: Technical Specification) GOST 97588.
[110] Sharkov VI, Sapotnitskii SA, Dmitrieva OA, Tumanov IF (1973) Technology of Hydrolytic Productions. Moscow: Lesnaya promyshlennost'. 480 p.
[111] Babkin VA, Ivanova SZ, Fedorova TE, Medvedeva EN, Malkov YuA, Ostroukhova LA, Trofimova NN, Ivanova NV (2007) Scientific bases of technology of complex processing of a biomass of a larch. Khimiya Rastiytel'nogo syr'ya 3: 9-21.
[112] Nikitin VM, Obolenskaya AV, Shchegolev VP (1978) Wood and Cellulose Chemistry. Moscow: Lesnaya promyshlennost'. 368 p.
[113] Rogovin ZA (1972) Khimiya tsellulosy. Moscow: Lesnaya promyshlennost'. 520 p.
[114] Trofimova NN, Babkin VA. (2003) The cellolignin residue of a larch wood as a drug for obtaining of a crystalline glucose. Khvoinye Boreal'noi Zony. 1: 116-122.
[115] Gert EV, Shishonok MV, Torgashov VI, Kaputskii FN (1988) Effect of dinitrogen etroxide on the structure and properties of cellulose. 8. Recrystallization of amorphous cellulose in the medium of nitrogen oxide (IV) and nitric acid resulting in the formation of the knecht compound. Khimiya Drevesiny (Wood Chemistry). 5: 52-56.
[116] Trofimova NN, Malkov YuA, Bichevina OB, Kurets IZ, Babkin VA (2003) The nvestigation of the sellulose hydrolysis pulp inversion process. Khimiya Rastitel'nogo syr'ya. 4: 11-15.
[117] Glazkova EL, Lebedev NV (1960) Sbornik trudov VNII, no. 8. 235 p;.
[118] Kurets IZ, Trofimova NN, Bichevina OB, Belonogova LN, Kashaev AA, Babkin VA 2003) The characteristics of glucose complex with sodium chloride crystals by X-Ray power diffraction and X-Ray stuctural analysis methods. Khimiya Rastitel'nogo syr'ya. 2: 15-18.
[119] Trofimova NN, Babkin VA, Chemeris MM (2002) Catalyzed steam hydrolysis of the cellolignin residue of a larch wood. Khimiya Rastitel'nogo syr'ya. 2: 53–56.
[120] Trofimova NN, Babkin VA (2010) Study of the acid hydrolysis of polysaccharides from arch wood for obtaining crystalline glucose. Russian Journal of Bioorganic Chemistry. 36(7): 847-852.
[121] Lowry OH, Roosebrough NJ, Farr AL, Randall RG (1951) Protein measurement with the Folin phenol reagent. Journal of Biological Chemistry. 193:265-275.

Chapter 7

Antiviral Levans from *Bacillus spp.* Isolated from Honey

Mona A. Esawy, Eman F. Ahmed, Wafaa A. Helmy,
Nahla M. Mansour, Waled M. El-Senousy and Mounir M. El-Safty

Additional information is available at the end of the chapter

http://dx.doi.org/10.5772/52483

1. Introduction

In recent years, significant progress has been made in discovering and developing new bacterial polysaccharides that possess novel and highly functional properties (Baird et al 1983). Although their ubiquitous role in biological processes and their versatility as biocompatible, environmentally friendly materials are beyond doubt, polysaccharides are still considered to be the "sleeping giant" of biotechnology.

Honey contained a great variety of dominant spores and in consequence their dominant spores are expected to be new expolysaccarides sources which could be isolated. This expectation comes from the honey constituents which is mainly fructose (about 38.5%) and glucose (about 31.0%) (Crosby and Alfred 2004). Aerobic spore forming Bacillus were the most frequently encountered microbes on the external surface, crop and intestine of the honey bees and consequently honey (Root, 1993, Esawy et al., 2011).

Most of the researches in the honey field focused on its antimicrobial, antioxidant and anticancer activities, also the identification of the dormant endospore inside it (Sabate, et al., 2009). None till now paid attention to the enzymatic products of these dormant endospores (Esawy et al., 2011). Osmophilic microorganisms survive environmental extremes of desiccation, pressure and acidity, it is expected that their biopolymers will also have some unique properties to adapt to such extreme conditions. This investigation concerned the question of whether honey collect bacteria that are good producers of levansucrase and levan yield. Recently, screening of 16 bacterial honey isolates for levansucrase production showed that all the tested isolates were levansucrase producers despite variations in the degree of activity (data not published yet). Levansucrase, one of the fructosyltransferases or glycansucrases, is produced by various microorganisms

INTECH
open science | open minds

© 2012 The Author(s). Licensee InTech. This chapter is distributed under the terms of the Creative Commons Attribution License http://creativecommons.org/licenses/by/3.0), which permits unrestricted use, distribution, and reproduction in any medium, provided the original work is properly cited.

(Iizuka et al.,1991; Hernandez, et al 1995; Kojima et al.,1993; Ben Ammar et al., 2002, Esawy et al., 2008). Bacterial levansucrases catalyze at least three different reactions: hydrolysis of sucrose, polymerization of fructose derived from sucrose and hydrolysis of levan. It is reported that levansucrase activity is involved in a variety of processes including survival of bacteria in soil (*B. subtilis*), phytopathogenesis (*Erwinia* and *Pseudomonas* species) and symbiosis (*Paenibacillus polymyxa*) of plant interactive bacteria (Hettwer et al.,1995). *Bacillus subtilis,* known as the hay bacillus or grass bacillus, is a Gram-positive, catalase positive bacterium commonly found in soil (Madigan & Martinko, 2005). Recently, Esawy et al 2011, Esawy [a] et al 2012 and Esawy [b] et al 2012 reported in novel *Bacillus subtilis* honey isolates as new sources of very important enzymes such as levansucrase, dextranase and.lipase.

Levan is one of two main types of fructans, which are natural homopolymers of fructose (Arvidson et al 2006). It is a naturally occurring polymer of β-D-fructofuranose with β (2→6) linkages between repeating five-member fructofuranosyl rings and branching at C-1 (Arvidson et al 2006, Barone and Medynets., 2007). Levans produced by different organisms differ in their molecular weight and degree of branching. Levans from plants generally have molecular weights about 2000 - 33.000Da (Rhee et al., 2002). The molecular weight of levan, and the fraction of residues incorporated in side chains, depends on both the source and the growth conditions, with plant levan and microbially-produced levan having very different characteristics (Arvidson et al 2006; Kasapis, and Morris 1994; Kasapis et al., 1994; Newbrun 1971; Stivala, and Bahary 1978; Huber et al., 1994). Recently it was reported in the *B. subtilis* NRC1aza levansucrase, the unique feature of this isolate its ability to produce two types of levan with different molecular weights (El Fattah et al, 2012). Bacterial levans are much larger than those produced by plants, with multiple branches and molecular weights (2-100 million Da) (Pontis and Del Campillo 1985 ; Keith et al., 1991). Levan is non-gelling, non-swelling in water, (Kasapis et al., 1994; Stivala, and Bahary 1978; Huber et al., 1994) and an unusual polysaccharide due to its relatively low intrinsic viscosity compared to other molecules of similarly high molecular weights. Levan can be used as food or a feed additive with prebiotic and hypocholesterolemic effects (Sanders, et al., 2003). Subsequently, there are a variety of potential industrial applications for levan such as a surfactant for household use due to its excellent surface-active properties, a glycol/levan aqueous two-phase system for the partitioning of proteins, etc. In addition, in vitro anti-tumor activity of levan produced from *Microbacterium laevaniformans, Rahnella aquatilis* and *Zymomonas mobilis,* has been shown against eight different tumor cell lines (Urdaci, et al 2004;Yoo,et al 2004; Yoon et al.,2004; Liu et al., 2012; El Fattah et al, 2012). Recently, Liu et al., (2012) and El Fattah et al, 2012 reported in the antioxidant activity of native levan and their derivatives. Dahech et al., (2011) reported that polysaccharide levan is efficient in inhibiting hyperglycemia and oxidative stress induced by diabetes and suggests that levan supplemented to diet may be helpful in preventing diabetic complications in adult rats. The market for levan will gradually increase in the various fields (Kang et al., 2009).

2. Experimental

2.1. Sources of honey and microorganisms

Three different honey samples were purchased; local honey bee collecting nectar from clover flower; Kashmiry honey, honey bee collecting nectar from desert flower (Saudi Arabian); and Gably honey, a honey bee collecting nectar from desert flower (Libya).

2.2. Isolation of bacterial strains from honey samples

One hundred micro liters of honey samples was spread on nutrient agar plates (g/L): beef ext., 1.0; yeast ext., 2.0; peptone, 5.0 and agar, 25.0. After drying for 20 min in a laminar flow hood, the plates were incubated at 50 °C to avoid the growth of any pathogenic spores for 24 h or until the colonies size was sufficient (approximately larger than 3–5 mm in diameter). The bacterial isolates were streaked onto agar plates and preserved at 4 °C. The purity of the isolates was assessed by colony morphology and microscopy.

2.3. Chromosomal DNA and plasmid extraction

Chromosomal DNA was prepared from overnight culture in LB,using AxyGEN Biosciences DNA extraction kit, according to manufacturer's instructions. Plasmid extraction was performed using Wizard mini prep. extraction kit (Promega) according to manufacturer's instructions with slight modification, where 50 L of lysozyme (200 mg/mL) were added to the resuspended buffer and incubated at 37 °C for 1 h then the protocol was carried on as described in the kit (O'Sullivan & Klaenhammer, 1993; Sambrook, et al., 1989).

2.4. PCR amplification for molecular identification

To amplify the 16S rRNA gene, a primer pair hybridizing to two conserved regions in 16S rRNA genes from *Bacillus spp*. was used: (bac-F and bac-R) (Ash et al., 1991; Kwon et al., 2009). For the amplification of the 16–23S intergenic region, a primer pair was used: L516SF and L523SR. While for recA gene, a primer pair corresponding to conserved regions in recA genes from *Bacillus sp*. was used: recA-F and recA-R. Speciesspecific primer set for *B. subtilis* corresponding for ytcP gene was used: ytcP-F and ytcP-R. All polymerase chain reaction amplifications were performed with the Taq DNA polymerase kit (Promega). Reaction mixtures consisted of 20 mm Tris–HCl (pH 8.4), 50 mm KCl, 3 mm $MgCl_2$, 50 mm of each of the four deoxynucleoside triphosphates (dNTP), 1 U Taq polymerase, 5 pmol of each primer and 1 L of template DNA in a final volume of 50 L. Samples were amplified in a GeneAmp polymerase chain reaction system 2700 (Applied Biosystems) programmed as follows: initial denaturation of DNA for 5 min at 94 °C, 30 cycles of 30 s at 94 °C, 30 s at 50 °C and 30 s at 72 °C. Polymerase chain reaction products were quantified by electrophoresis on a 1% (w/v) agarose gel containing ethidium bromide. Polymerase chain reaction products obtained from the selected isolates were purified, using QIA quick polymerase chain reaction purification KIT (Qiagen) and then sequenced commercially by Sigma-Egypt. The sequencings were performed and manually aligned, using DNAMAN software (version 4.0). Sequence homologies were examined by comparing the obtained sequence with those in the NCBI database and the NEB cutter V 2.0 database. NEB cutter V 2.0 is an on-line DNA sequence tool used to find large, non-overlapping, open-reading frames and works for all restriction enzymes. It provides a website, which allows users to check nucleotide sequences for restriction enzyme sites. The sequences were submitted and AluI enzyme was chosen for digestion. Finally gel photograph using 2% agarose was viewed (Ash et al., 1991).

2.5. Bacterial strains and growth conditions

Bacillus strains used in this work were cultivated in Luria-Bertani (LB) broth or agar at 37 °C. Defined medium was used for cellular production of levansucrase (Yanase et al., 1992). It had the following composition (g/L): yeast extract, 2.5; commercial sucrose, 80; $MgSO_4$, 0.2 and K_2HPO_4, 5.5. The medium was completed by the addition of 1 L distilled water and the pH was adjusted to 7.0 before autoclaving. The parameters included initial incubation temperature (25–45 °C); different concentrations of sucrose (80–160 g); incubation time (16–48 h); (50–150) rpm; pH (5-9) and NaCl (1–4% w/v) were studied. The sucrose was substituted with fructose, glucose, lactose (80 g/L) and beet molasses (equivalent to 80 g sucrose) to study their effects on enzyme production.

2.6. Cellular production

Cultivation was carried out in 250 mL Erlenmeyer flasks. Each flask contained 50 mL production medium and was autoclaved for 15 min. The flasks were then inoculated with 2.0 ml inoculum and incubated for 24 h at 30 °C. The culture broth was then centrifuged in a cooling centrifuge (K70; Janektzki, Germany) at 10,397×g to separate the bacterial cells from the supernatant.

2.7. Assay of levansucrase

Levansucrase assay was performed according to the method of Yanase et al. (1992) with some modification. 0.5 ml of culture filtrate was incubated with 1 ml 20% (w/v) sucrose and 0.5 mL 0.1 M acetate buffer at pH 5.2 and incubated at 37°C for min. The decreasing amounts of sugars produced were measured by glucose oxidase kits. One unit of enzyme activity was defined as the amount of enzyme that produced decreasing sugars equivalent to 1 _mol of glucose/min.

2.8. Separation of levan polymer

The levan producing organisms were cultivated on a defined medium as described above. After growth, the culture was centrifuged to remove bacterial cells; the levan was precipitated with two volumes of absolute ethanol. The precipitate was collected and dried under vacuum.

2.9. Chromatography

Paper chromatography was performed according to Block et al. (1995). Hydrolysate of products of levan were analyzed by either paper chromatography on whatman No. 1. The mixtures at the end of incubation time were boiled for 3 min to stop the reaction. Chromatographic development was carried out with a solvent system of n-butanol:acetone:water (4:5:1) and detected by spraying with aniline hydrogen phthalate. The acid hydrolysate of the polysaccharide produced by the six isolates was analyzed using high-pressure liquid chromatography (HPLC). A 7.8 mm ×300 mm PL-HI-PLEXPB column was linked to a differential refractometer. The column temperature was maintained at 80 °C. The aqueous mobile phase was delivered at a flow rate of 0.6 ml/min.

2.10. Determination of molecular weight

Different concentrations of levan and oligosaccharide were prepared and the flow time of equal volume for each concentration at 30°C was determined in a U-shaped Ostwald viscometer. Flow time of the same volume of distilled water was also determined as control. Thus, specific viscosity/C (gsp) was estimated. A plot of levan and oligosaccharide concentration (C) against intrinsic viscosity (C) (gsp/C) therefore yielded a straight line.

2.11. Antivirus detection

Two types of viruses were used, highly pathogenic avian influenza H5N1 virus Egyptian isolate, was used at titre of 106 EID50/mL (embryo infective dose per mL) and adenovirus type 40 with different doses $20^{1}\times10^{4}$, 1×10^{5}, and 1×10^{6} infectious particles/mL obtained from the Holding Company for Biological Products & Vaccines (VACSERA).

2.12. Specific pathogen free (SPF) eggs

SPF embryonated chicken eggs were used at nine days old and inoculated via the allantoic sac route. SPF eggs (Brown et al., 2007) were obtained from Nile SPF Eggs, Koomoshiem, Fayoum, Egypt.

2.13. Cytotoxicity test

It was done according to Simoes et al., (1999) and Walum et al., (1990). Briefly, All samples (100 mg) were dissolved in 500 _L of water or ethanol. Samples A, E, M, and K were dissolved in ethanol while samples C and G were dissolved in water. Decontamination of samples was done by adding 12 _L of 100x of antibiotic–antimycotic mixture to 500 _L of each sample. Then, bi-fold dilutions were done to 100 _L of original dissolved samples and 100 _L of each dilutions were inoculated in Hep-2 cell line (obtained from the Holding Company for Biological Products & Vaccines VACSERA, Egypt) previously cultured in 96 multi well plates (Greiner-Bio one, Germany) to estimate the non toxic dose of the tested samples. Cytotoxicity assay was done using cell morphology evaluation by inverted light microscope and cell viability test applying trypan blue dye exclusion method.

2.14. Cell morphology evaluation by inverted light microscopy

Hep-2 cell cultures (2× 10^5 cells/mL) were prepared in 96-well tissue culture plates (Greiner-Bio one, Germany). After 24 h incubation at 37 °C in a humidified 5% (v/v) CO2 atmosphere cell monolayers were confluent, the medium was removed from each well and replenished with 100 _L of bi-fold dilutions of different samples tested prepared in DMEM (GIBCO BRL). For cell controls 100 _L of DMEM without samples was added. All cultures were incubated at 37 °C in a humidified 5% (v/v) CO2 atmosphere for 72 h. Cell morphology was observed daily for microscopically detectable morphological alterations, such as loss of confluence, cell rounding and shrinking, and cytoplasm granulation and vacuolization. Morphological changes were scored (Simoes et al., 1999).

2.15. Cell viability assay

It was done by trypan blue dye exclusion method (Walum et al., 1990). Hep-2 cell cultures (2×10^5 cells/mL) were grown in 12-well tissue culture plates (Greiner-Bio one, Germany). After 24 h incubation, the same assay described above for tested samples cytotoxicity was followed by applying 100 L of tested samples dilutions (bifold dilutions) per well. After 72 h the medium was removed, cells were trypsinized and an equal volume of 0.4% (w/v). Trypan blue dye aqueous solution was added to cell suspension. Viable cells were counted under the phase contrast microscope.

2.16. Haemagglutinating activity assay

This was applied for the allantoic fluids of the inoculated eggs and measured by micro technique of haemagglutination (HA) test (Takatsy, 1955).

2.17. Evaluation for antiviral activity

Three experiments were conducted.

2.17.1. Experiment 1

One hundred and five emberyonated chicken eggs (ECEs) were examined; equal volumes of HPAI H5N1 virus and original extracts were separately used at three levels:

Level 1: Equal volumes of HPAI H5N1 virus and the original undiluted samples were mixed and incubated at room temperature for 1 h then inoculated into the allantoic sac of five ECEs for each product sample at dose 0.2 mL/ECE.

Level 2: Equal volumes of HPAI H5N1 virus and the 1/5 dilution of each sample were mixed and preceded as level 1.

Level 3: Equal volumes of the virus and the 1/10 dilution of each sample were mixed and preceded as level 1. In addition, five ECEs were inoculated with the virus that mixed with equal volume of saline at a dose of 0.2 mL/ECE of saline alone (negative control). The ECEs are inoculated at 37 °C and candled every 2 h till all the positive control ECEs died.

2.17.2. *Experiment 2*

One hundred and five SPF ECEs were used in this experiment; equal volumes of HPAI H5N1 virus and the original samples were mixed with equal volume of the original samples and inoculated directly into the allontoic sac of five ECEs for each product sample at a dose of 0.20 mL/ECE for each product sample at a dose of 0.2 mL/ECE. Five ECEs were inoculated with equal volume of the HPAI H5N1 virus and saline at dose of 0.2 mL/ECE (positive control). Another five ECEs were inoculated with 0.20 mL/ECE of saline alone (negative control). All the ECEs were incubated at 37 °C and controlled every 2 h till the ECEs of the positive control died

2.17.3. *Experiments 3*

One hundred and five SPF ECEs of nine days old were used in this experiment. 0.10 mL of the HPAI H5N1 virus was inoculated via the allontoic sac of each ECE into 100 ECEs and then the inoculated ECEs were incubated for 1 h at 37 °C. The original samples were inoculated into five ECEs, which previously inoculated with the virus at a dose of 0.1 ml. Another five ECEs were inoculated with 0.2 ml/ECE of the mixed virus and saline. Other five ECEs were inoculated with 0.2 ml/ECE of saline alone. The ECEs were inoculated at 37 °C and candled every 2 h till ECEs of the positive control died.

2.18. Antiviral effect of tested samples on adenovirus type 40

Seventy five microliters of non toxic dilutions were mixed with 75 L of different doses 1×10^4, 1×10^5, and 1×10^6 infectious viral particles/mL of adenovirus type 40 provided by American Type Culture Collection (ATCC). Then the mixture was incubated overnight at 4 °C. Inoculation of 100 L of 10 fold dilutions of treated and untreated adenovirus was done into Hep-2 cell line in 12 multi well-plates. After 1 h incubation for adsorption at 37 °C, 1 mL medium (DMEM) was added to each well. The cell line was observed daily for one week then, three times freezing and thawing for tested plates were done. Nested PCR was done for confirmation of adenovirus (presence/absence) in each well (Puig et al., 1994).

3. Results

3.1. Molecular identification of the levansucrase producers strains

3.1.1. *16S rRNA sequences and their analogical electrophoresis*

Six levansucrase producers' bacteria were isolated from different honey sources. The isolates resembled each other in cell morphology where cells were rod-shaped, Gram-positive, motile and spore-forming. Colonies were circular, creamy, and no pigment was formed. They were identified as *Bacillus* spp. based on morphological examination. The identification was confirmed by molecular biological analysis, using 16S rRNA sequencing method. The results showed that the 6 isolates were identified as *Bacillus* spp. (99%), or *B. licheniformis* (99%), or *B. amyloliquefaciens* (99%) (Fig. 1).

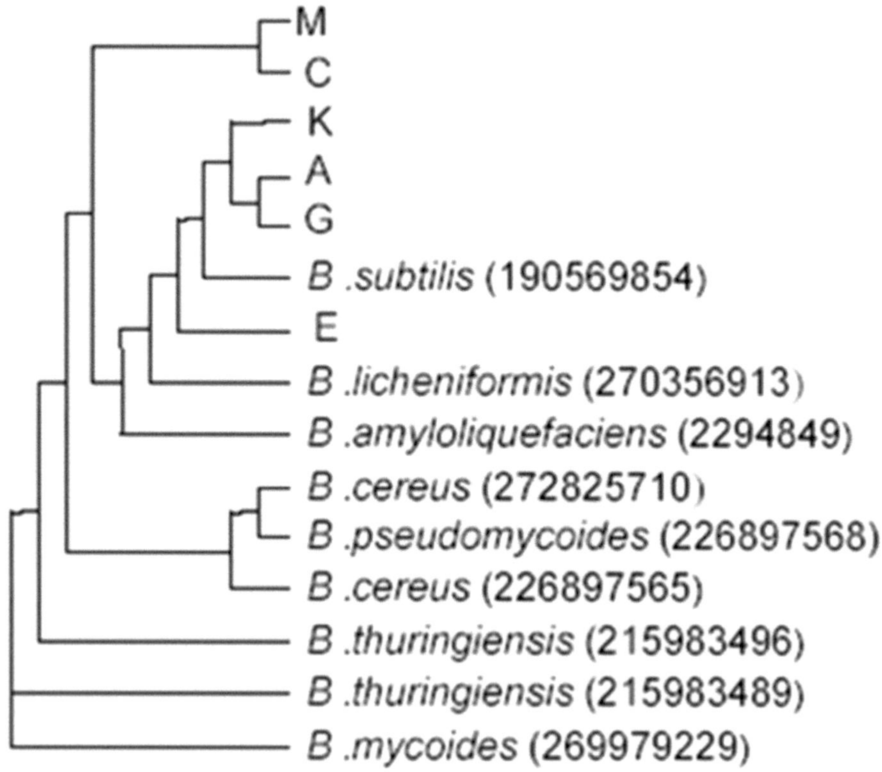

Figure 1. Phylogenetic neighbor-joining tree obtained by 16S rRNA sequence analysis of the tested isolates and other *Bacillus* spp. present in the gene bank database (accession numbers in parentheses).

The DNA of the isolates was extracted as described in Section 2 and the 1.5 kb 16S rRNA gene was amplified for each DNA by PCR using primers bac-F and bac-R. The PCR amplification, purification and sequencing were performed as described previously. The 1.5 kb obtained sequences were aligned and clustered with sequences from the NCBI database. 16S rRNA gene sequence analysis indicated that the six isolates (K, M, A, C, E, and G) were *Bacillus* spp. with 99% identity any of these three species *B. subtilis, or B. licheniformis, or B. amyloliquefaciens* and they clustered into a monophyletic line in a phylogenetic tree. To distinguish and clear identification of these strains on the species level the analogical electrophoresis, using NEB cutter was applied to identify the 16S rRNA results, which have been sequenced, as the strains of the same species expected to have almost the same sites when digested with AluI. Fig. 2 showed that the isolates M and G have the same size fragments as *B. subtilis* gi 269313996 while the other isolates A, C, E; K showed different AluIfragments which differ to the AluI fragments generated from 16S rRNA sequence of *B. subtilis, or B. licheniformis or B. amyloliquefaciens.* It was clear that 16S rRNA gene alone could not distinguish these three closely related Bacillus species.

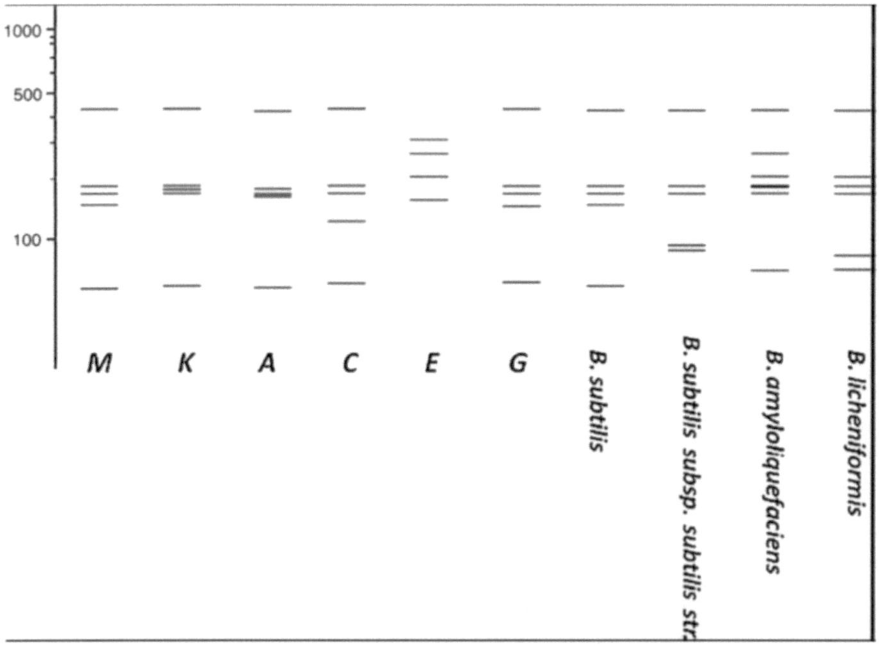

Figure 2. The analogical electrophoresis of Bacillus isolates compared to Bacillus strains (from the gene bank) by "AluI",using NEBcutter 2.0. The accession no. of the control Bacillus strains: *B. subtilis* gi|269313996|; *B. subtilis* subsp. subtilis str. AL009126.3; *B. amyloliquefaciens* gi|229484923|; and *B. licheniformis* gi|270356913|.

3.2. Identification by 16–23S intergenic region

To distinguish between these three closely related strains the 16–23S intergenic region was amplified by primers L516SF × L523SR, and then the sequence was determined for the six isolates. The homology results for the 16–23S intergenic region showed that: K and M strains showed 100% similarity to *B. subtilis* but strains A, C, E, G could be either *B. subtilis (99%) or B. amyloliquefaciens* (85%) **(Table. 1.).**

Target	Primer name	Oligonucleotide	Reference
16–23S intergenic region	L516SF L523SR	5′-TCGCTAGTAATCGCGGATCGGC-3′ 5′-GCATATCGGTGTTAGTCCCGTCC-3′	Yoon et al., 2001
recA gene	recA-F recA-R	5′-TGAGTGATCGTCAGGCAGCCTTAG-3′ 5′-CYTBRGATAAGARTACCAWGMACCGC-3′	Gun-Hee Kwon et al., (2009)
hypotheticl gene	ytcP-F ytcP-R′	5′-GCTTACGGGTTATCCCGC-3′ 5′CCGACCCCATTTCAGACATATC-3′	Gun-Hee Kwon et al., (2009)

Table 1. Primers used for gene amplification

3.3. Identification by recA sequence

Hence the recA gene has been used as a molecular chronometer in addition to rRNA genes. The 1.2 kb band was gel isolated and subjected to sequencing results and showed that the four strains A, C, E, and G were *B. subtilis* rather than *B. amyloliquefaciens*. Still these results need more confirmation.

3.4. Identification by specific-PCR for *B. subtilis*

To solve this problem, identification using specific-PCR for *B. subtilis* was described. Based on ytcP gene encoding a hypothetical protein, a PCR primer pair ytcF and ytcR were designed for *B. subtilis* species for specific amplification purpose. Using this primer pair, a 0.46 kb fragment was amplified only from *B. subtilis* strains, whereas no similar band was detected from *B. licheniformis or B. amyloliquefaciens* strains. These primers were subjected to the six isolates and PCR resulted in the 0.46 kb band as shown in Fig. 3 and this confirmed that all isolates A, C, E, G, K, M surely belong to *B. subtilis not B. amyloliquefaciens*.

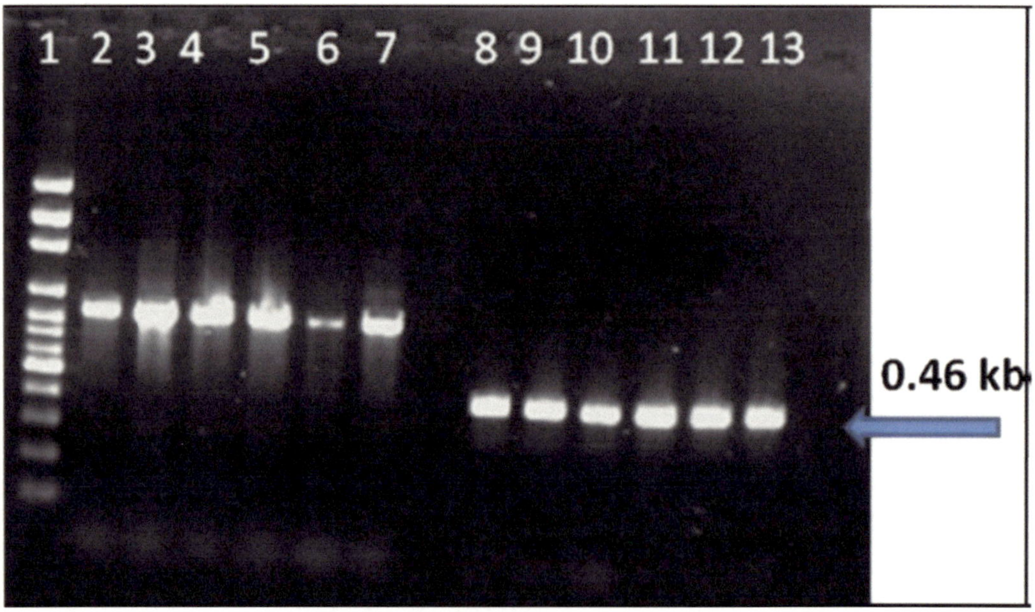

Figure 3. PCR using the chromosomal DNA of the 6 Bacillus isolates. Lane 1, 100 bp ladder; lane 2–7, using primers recR × recF; lane 8–13, using B. subtilis specific primers ytcF × ytcR.

3.5. Levan and levansucrase production

The six honey isolates were tested for production of levansucrase, the optimized conditions for the isolates ranged from 8 to 12% (w/v) commercial sucrose, 37–40 ºC, 24–28 h, 50–100 rpm and pH 6–7.0 (data not shown). Among all the tested isolates, M and K isolates showed the highest levansucrase activities (62 and 59 U/mL). The presence of

NaCl (1–4%, w/v) showed great influence in enzyme activity, the enzyme production increased from 2 to 3 folds according to the strain (Fig. 4). Paper chromatography of the product hydrolysate revealed the presence of only fructose and tiny traces of glucose, pointing to the levan nature of the product. Furthermore, the acid hydrolysate of the polysaccharide produced by the isolates was exclusively fructose, as revealed by HPLC. Levan was harvested by precipitation from the culture broth by addition of ethanol. The yield and consistency of the product varied according to the isolate. The final products were a brownish-white gummy material, which could be freeze-dried or vacuum-dried. The highest amount of levan was produced on the medium containing commercial sucrose, followed by beet molasses (Fig. 5). While a small amount of microbial polysaccharide (alcohol precipitate) was also produced when the organism was grown on lactose and glucose, it was worthy to record that no polysaccharides were produced on fructose. The amounts of levan decreased 40–50% in the presence of NaCl. Isolates M, K, A, C, E, and G produced 11, 16.25, 6.60, 1.81, 1.74, 6.6 g/L levan, respectively under the optimized conditions. The levan products consisted of one fraction characterized by high and low molecular masses (40.938, 71.887, 43.487, 154.638, 77.753 and 14.200 kDa for isolates K, M, A, C, E, and G, respectively) (Fig. 6).

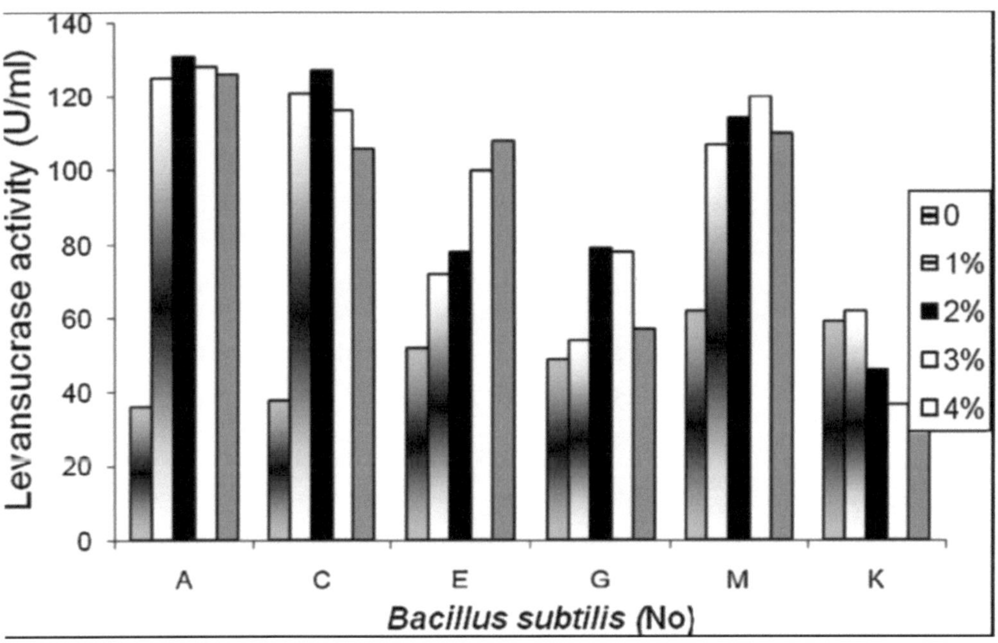

Figure 4. The effect of absence and presence of different concentrations of NaCl on levansucrase production from honey isolate. Note: NaCl conc (0–4%) represented from left to right.

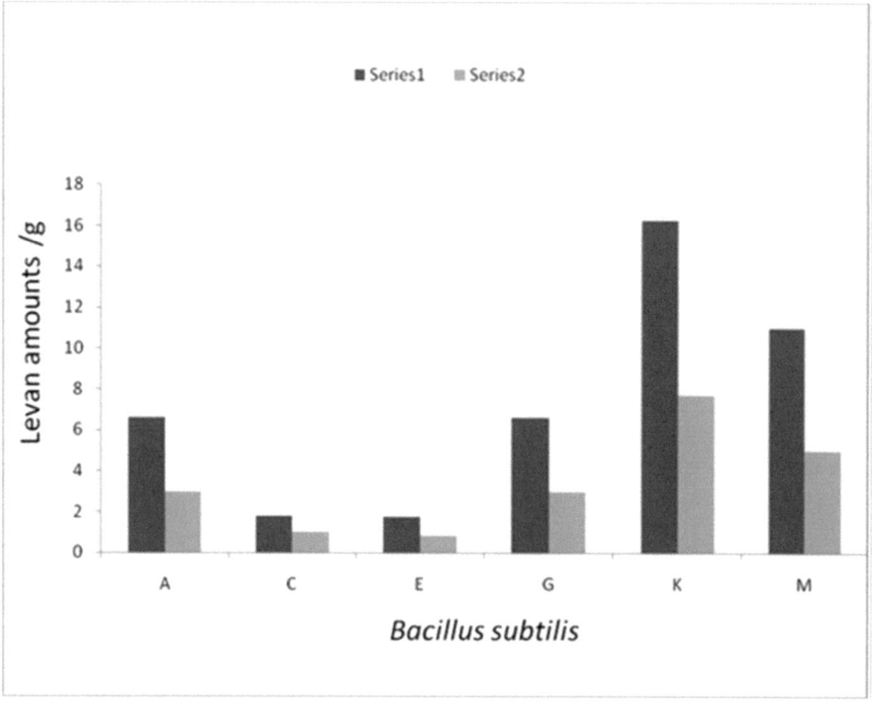

Figure 5. The difference between amounts of levan produced in sucrose medium (series 1) and in molasses medium (series 2)

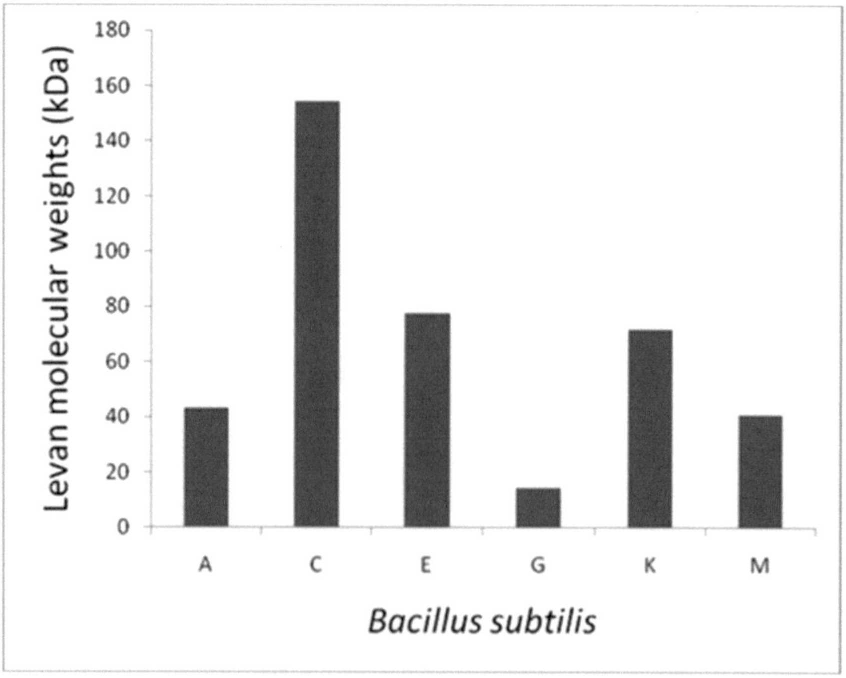

Figure 6. The levan molecular weights from *Bacillus subtilis* isolates

3.6. Cytotoxicity test

The non toxic doses for samples C and G were 0.5 mg/mL in water. On the other hand, the non toxic doses for samples A, E, and M were 0.2 mg/mL in ethanol while the nontoxic dose of sample K was 0.1 mg/ml in ethanol.

3.7. Anti-adenovirus type 40 assay

The samples (C, E, G and K) had weak effect on adenovirus 40 which did not exceed 10%. The two samples A and M revealed antiviral effect on adenovirus type 40 ranged from 50 to 60% as shown in Table 2.

Sample	Initial viral doses	Final viral doses	Percentage of reduction
A	$1x10^4$	$4x10^3$	60%
	$1x10^5$	$5x10^4$	50%
	$1x10^6$	$5x10^5$	50%
M	$1x10^4$	$4x10^3$	60%
	$1x10^5$	$4x10^4$	60%
	$1x10^6$	$5x10^5$	50%

Table 2. Effect of levan A and levan M on the infectivity of enteric adenovirus type 40 (DNA virus). viral infectivity.

3.8. Anti-H5N1 virus assay

All the embryos of the positive controls died and the allantoic fluid of each was positive for haemagglutination assay (HA), while all the embryo of negative control were not died and the allantoic fluid of each was negative for HA. Three levans of code K, M, E showed antiviral against HPAI H5N1. Each of these samples showed antiviral effect when inoculated with H5N1 virus 1 h before inoculation into nine days old ECEs, while they had no effect on the virus when inoculated simultaneously with the virus just after mixing or after the virus inoculation for 1 h (Table 3). These results also revealed that simultaneous inoculation of the levans product or even after infection was of no value.

Code of sample	Experiment 1						Experiment 2		Experiment 3	
	Level 1		Level 2		Level 3		NDE	+HA	NDE	+HA
	NDE	+HA	NDE	+HA	NDE	+HA				
con	5/5	5.0	5/5	5.0	5/5	5.0	5/5	5.0	5/5	5.0
M	0/5	0.0	5/5	5.0	5/5	5.0	5/5	5.0	5/5	5.0
C	5/5	5.0	5/5	5.0	5/5	5.0	5/5	5.0	5/5	5.0
E	0/5	0.0	5/5	5.0	5/5	5.0	5/5	5.0	5/5	5.0
K	0/5	0.0	5/5	5.0	5/5	5.0	5/5	5.0	5/5	5.0

Table 3. Showed that three levans of code K, M, E had antiviral against HPAI- H5N1 NDE: Number dead Haemagglutination assay

4. Discussion

The public health and the discovery of new drugs is a main objective of many research activities, however, sometime this type of research activity cost a lot of money. Although, in this proposed research we will be aiming towards the protection of the public health and introduce a new drugs contribute in solve the problem of serious diseases through products that will save a lot of money to our economy. Within this context, six mobile spore-forming, and Gram-positive facultative aerobic bacilli were isolated from different honey samples and identified as *Bacillus* spp. On the base of morphological, and molecular identification, using 16S rRNA sequence method. *B. subtilis* isolates are biologically and commercially important as producers of a great variety of secondary metabolites such as antibiotics, and enzymes (Desai & Banat 1997; Roberts et al., 1996). The16S rRNA sequence method could not identify the Bacillus isolates at the species level where revealed identical to any of three strains *B. subtilis or B. licheniformis or B. amyloliquefaciens.* In fact DNA-based identification methods such as 16S rRNA gene sequencing and 16S–23S intergenic region sequencing have been used widely for the purpose of identification and typing of microorganisms isolated from natural environments including fermented foods (Hansen et al., 2001; Levine, et al., 2005). But identification based on rRNA gene sequences fails to distinguish one species from the other if they share highly similar rRNA genes. This is true for some Bacillus species. It is difficult to distinguish *B. subtilis* from closely related *B. licheniformis* or *B. amyloliquefaciens* by rRNA gene sequences because of no significant differences in their rRNA sequences (Nakamura, 1989; Nakamura et al., 1999). Other genes such as recA (Rodriguez et al., 2007) and dnaJ (Shah et al., 2007) have been employed instead of rRNA genes. It is necessary to compare results from different identification methods as a whole before to reach a conclusion (Bourque et al., 1995).

To solve the ambiguity in differentiating them based solely on the 16S rRNA gene, it was turned to 16–23S intergenic region, recA gene, and ended with *B. subtilis* specific primers. It is specifically useful to distinguish organisms with highly similar rRNA genes recA gene was amplified from the chromosomal DNA of bacilli isolates by PCR, using primers (recF and recR) (Payne et al., 2005). When rRNA and recA gene sequences were considered together, it was possible to conclude that the G, M isolates are belonging to *B. subtilis* but the other isolates A, C, E needed further identification. For the accurate distinguish between the 6 isolates, the specific-PCR for *B. subtilis* based on ytcP gene was used and the results revealed that the six strains are belonging to *B. subtilis*. The present results proved high phenotypic and genotypic variability among *B. subtilis* isolates, where they showed different morphological and biological properties suggesting them as new different species of *B. subtilis* with valuable impact in the industry. Many authors reported in the production of levansucrase from *B. subtilis* (Euzenat et al., 1997; Le Gorrec et al., 2002). The various sugars, initial pH, fermentation temperature, and agitation speed affected the levansucrase production by *B. subtilis* (Abdel-Fattah et al., 2005; Shih et al., 2005). The result ensure the halophilc feature of *B. subtilis* levansucrases, this comes from its osmophilic character. Enhancements of levansucrase production in the presence of NaCl were reported (Euzenat et al., 2006; Poli et al., 2009). As far as we are aware no studies were reported on the effect of NaCl on the enzyme production. It seemed that levan production by the isolates was dependent mainly on commercial sucrose media, where the use of beet molasses, glucose, and lactose led to noticeable reduction in levan synthesis. Beet molasses was used as low-cost substitutes for sucrose in commercial levan yield (Han & Watson, 1992). The decrease in levan yield in the molasses medium (2.533 g/L) when compared to the commercial sucrose (21.685 g/L) was also reported (De Oliveira et al., 2007). Although higher Halomonas sp. AAD6 biomass concentrations were observed when glucose, maltose, fructose and galactose were used as carbon sources, levan levels were very low comparing with sucrose (Poli et al., 2009). The six isolates produced different levan weights, with wide range of molecular mass. On the other hand, it was reported that halophilic Halomonas sp. AAD6 cells grown in the presence of sucrose afforded the highest levan production levels (1.073 g/L) (Poli et al., 2009). Also, *B. polymyxa* produced about 40 g/L extracellular polysaccharide per liter in sucrose medium, which was about three times that produced by familiar levan producers (Han, 1989). Levan antitumor activity was reported by many authors (Yoon, Yoo, Cha, & Lee, 2004,) but as far as we are aware nothing was reported on antiviral activity of this fructose polymer. The present findings showed antiviral effects of K, M, and E levan products on H5N1 virus, While, A and M levan products showed antiviral effects on adenovirus type-40. It was obvious that the product M was entirely effective against both respiratory RNA virus (H5N1) and enteric adenovirus type 40 (DNA virus). It was apparent that each of effective levan showed antiviral effect when inoculated with H5N1 virus 1 h before inoculation into nine days old ECEs, while they had no effect on the virus when inoculated simultaneously with the virus just after mixing or after 1 h of inoculation.

The outcome of this study is the probable suitability of some types of levan as a safe and cheap natural product in antiviral treatments with applying the known roles concerning the use of these compounds. In addition, this article affords honey micro flora as a new and important sources of levansucrase enzymes, could be have biotechnological applications in pharmaceutical industries.

Author details

Mona A. Esawy*, Eman F. Ahmed and Wafaa A. Helmy
Department of Chemistry of Natural and Microbial Products,
National Research Centre, Dokki, Giza, Egypt

Nahla M. Mansour
Gut Microbiology and Immunology Lab,
Central of Excellence for Advanced Sciences (CEAS), National Research Centre, Dokki, Giza, Egypt

Waled M. El-Senousy
Water Pollution Research Department, National Research Centre, Dokki, Giza, Egypt

Mounir M. El-Safty
Central Laboratory for Evaluation of Veterinary Biologics (CLVB), Abbassia, Cairo, Egypt

---

* Corresponding Author

5. References

Abdel-Fattah, A. F., Mahmoud, D. A. R., & Esawy, M. A. (2005). Production of levansucrase from *Bacillus subtilis* NRC 33a and enzymic synthesis of levan and fructo-oligosaccharides. Current Microbiology, 51, 402–407.

Abdel-Fattah, A.M., Gamal-Eldeen, A.M., Helmy, W.A., & Esawy, M.A. Antitumor and antioxidant activities of levan and its derivative from the isolate *Bacillus subtilis NRC1aza. Carbohydrate Polymers xx (2012)89, 314-322*

Ash, C., Farrow, J. A. E., Dorsch, M. E., Stackebrandt, E., & Collins, M. D. (1991). Comparative analysis of Bacillus anthracis, Bacillus cereus and related species on the basis of reverse transcriptase sequencing of 16S rRNA. International Journal of Systemic Bacteriology, 141, 343–346.

Arvidson, S.A., Rinehart, B.T., & Gadala-Maria F. (2006). Concentration regimes of solutions of levan polysaccharide from *Bacillus* sp. Carbohydrate Polymers 65 144–149.

Baird[1], P. A. Sandford[1] & I. W. Cottrell. Industrial Applications of Some New Microbial Polysaccharides *Nature Biotechnology* 1, 778 - 783 (1983)

Barone, J.R., & Medynets, M., Thermally processed levan polymers, Carbohydrate Polymers 69 (2007) 554–561

Beine, R., Moraru, R., Nimtz, M., Na'amnieh, S., Pawlowski, A., & Buchholz, K. (2008). Synthesis of novel fructooligosaccharides by substrate and enzyme engineering. Journal of Biotechnology, 138, 33–41.

Ben Ammar, B. Y., Matsubara, T., Ito, K., Iizuka, M., Limpaseni, T., & Pongsawasdi, P., et al. (2002). Characterization of a thermostable levansucrase from Bacillus sp. TH4-2 capable of producing high molecular weight levan at high temperature. Journal of Biotechnology, 99, 111–119.

Block, R. J., Durrum, E. L., & Zweig, G. (1995). A manual of paper chromatography and paper electrophoresis. New York, NY: Academic Press Inc., pp. 127

Bourque, S. N., Valero, J. R., Lavoie, M. C., & Levesque, R. C. (1995). Comparative analysis of the 16 to 23S ribosomal intergenic spacer region sequences of *Bacillus thuringiensis* strains and subspecies and of closely related species. Applied Microbiology and Biotechnology, 61, 1623–1626.

Crosby, Alfred W. (2004). Ecological Imperialism: The Biological Expansion of Europe, 900–1900. Cambridge University Press, 188. ISBN 0-521-54618- 4. "There are many kinds of bees and other insects producing honey all round the world"

Dahech I, Belghith, K.S, Hamden, K, Feki, A., Belghith H, & Mejdoub, H. (2011). Antidiabetic activity of levan polysaccharide in alloxan-induced diabetic rats. International Journal of Biological Macromolecules. 2011 49:742-6.

De Oliveira, M. R., Da Silva, R. S. S. F., Buzato, J. B., & Celligoi, M. A. P. C. (2007, November). Study of levan production by *Zymomonas mobilis* using regional low-cost carbohydrate sources. Biochemical Engineering Journal, 37: 177–183.

Desai, J. D., & Banat, I. M. (1997). Microbial production of surfactant and their commercial potential. Microbiology and Molecular Biology Review, 61, 47–64.

EsawyI, M. A.; MahmoudI; D. A. R. & Abd- Fattah A. F. (2008). Immobilisation of *Bacillus subtilis* NRC33a levansucrase and some studies on its properties. Brazilian Journal of Chemical Engineering 25: 237-246

Esawy, M.A., Ahmed, E.F., Helmy, W. A., Mansour, N.M., El-Senousy,W.M., & El-Safty , M.M. (2011). Production of levansucrase from novel honey *Bacillus subtilis* isolates capable of producing antiviral levans. Carbohydrate Polymer, 36, 823-830

Esawy[a], M. A.; Mansour, S. H; Ahmed, E. F., Hassanein, N.M. & El Enshasy, H.A. (2012). Characterization of Extracellular Dextranase from a Novel Halophilic *Bacillus subtilis* NRC-B233b a Mutagenic Honey Isolate under Solid State Fermentation E-Journal of Chemistry 9: 1494-1510

Esawy[b], M.A.; Awad G.E.A., Mansour, S. H; Ahmed, E. F., & Danial, E.N. Evaluation of honey as new reservoir for probiotic bacteria. Advances in food science. V: 43 (2) in press.

Euzenat, O., Guibert, A., & Combes, D. (1997). Production of fructo-oligosaccharides by levansucrase from *Bacillus subtilis* C4. Process Biochemistry, 32, 237–243.

Euzenat, O., Guibert, A., & Combes, D. (2006). Production and purification of *Bacillus subtilis* C4 levansucrase kinetic characterization of the enzyme. Annals of the New York Academy of Sciences, 864, 288–294.

Han, Y. W. (1989). Levan production by *Bacillus polymyxa*. Journal of Industrial Microbiology, 4, 447–452.

Han, Y. W., & Watson, M. A. (1992). Production of microbial levan from sucrose, sugarcane juice and beet molasses. Journal of Industrial Microbiology, 9, 257–260.

Hansen, B. M, Leser, T. D., & Hendriksen, N. B. (2001). Polymerase chain reaction assay for the detection of Bacillus cereus group cells. FEMS Microbiology Letters, 202, 209–213.

Hernandez, L., Arrieta, J., Menendez, C., Vazquez, R., Coego, A., Suarez, V., et al. (1995). Isolation and enzymatic properties of levansucrase secreted by *Acetobacter diazotrophicus* SRT4, a bacterium associated with sugar cane. Biochemistry journal, 309, 113–118.

Hettwer, U., Gross, M., and Rudolph, K., (1995): Purification and characterization of an extracellular levansucrase from *Pseudomonas syringae* pv Phaseolicola. Journal of Bacteriology, pp. 2834–2839.

Huber, A. E., Stayton, P. S., Viney, C., & Kaplan, D. L. (1994). Liquid crystallinity of a biological polysaccharide: the levan/water phase diagram. Macromolecules, 27, 953–957.

Kang, (2009): "Levan: Applications and Perspectives". Microbial Production of Biopolymers and Polymer Precursors. Caister Academic Press. ISBN 978-1-904455-36-3.

Kasapis, S. & Morris, E., (1994) Conformation and physical properties of two unusual microbial polysaccharides: *Rhizobium trifolii* CPS and Levan. In Food Hydrocolloids: Structures, Properties and Functions, ed by K. Nishinari, Plenum Press, New York, pp 97-103.

Kasapis, S., Morris, E., Gross, M., & Rudolph, K., (1994): Solution properties of levan polysaccharide from *Pseudomonas syringae* pv. phaseolicola, and its possible primary role as a blocker of recognition during pathogenesis. Carbohydrate Polymers. 23: 55–64.

Keith, J., Wiley, B., Ball, D., Arcidiacono, S., Zorfass, D., Mayer, J., & Kaplan, D. (1991): Continuous culture system for production of biopolymer levan using *Erwinia herbicola.* Biotechnology and Bioengineering, 38: 557– 560.

Kojima, I., Saito, T., Iizuka, M., Minamiura, N., & Ono, S. (1993). Characterization of levan produced by Serratia sp. Journal of Fermentation and Bioengineering, 75, 9–12.

Kwon, H. K., Lee, H. A., Park, J. Y., Kim, J. S., Lim, J., & Park, C. S., et al. (2009). Development of a RAPD-PCR method for identification of Bacillus species isolated from Cheonggukjang. International Journal of Food Microbiology, 129, 282–287.

Le Gorrec, K., Connes, C., & Guibert, A., et al. (2002). Identification of three inducible and extracellular enzymatic activities working on sucrose in Bacillus subtilis NCIMB 11871 and 11872 supernatant. Enzyme and Microbial Technology, 31, 44–52.

Levine, S. M., Tang, Y. W., & Pei, Z. (2005). Recent advances in the rapid detection of *Bacillus anthracis*. Reviews in Medical Microbiology, 16, 125–133.

Lizuka, M., Tanaka, T., Yamamoto, S., et al. (1991). Synthesis of fructan and oligosaccharides by microbial and plant fructosyltransferases. Denpun Kagaku, 38, 217–222.

Liu, J., Luo, J., Ye, H., & Zeng, X. (2012). Preparation, antioxidant and antitumor activities in vitro of different derivatives of levan from endophytic bacterium *Paenibacillus polymyxa* EJS-3. Food and Chemical Toxicology, 50, 767–772.

Madigan, M. T., & Martinko, J. M. (2005). Brock Biology of Microorganisms (11th ed.). Pearson Prentice Hall.

Nakamura, L. K. (1989). Taxonomic relationship of black-pigmented Bacillus subtilis strains and a proposal for *Bacillus atrophaeus* sp. nov. International Journal of Systematic Bacteriology, 39, 295–300.

Nakamura, L. K., Roberts, M. S., & Cohan, F. M. (1999). Relationship of Bacillus subtilis clades associated with strains 168 and W23: A proposal for Bacillus subtilis subsp. subtilis subsp. nov. and *Bacillus subtilis* subsp. spizizenii subsp. nov. International Journal of Systematic Bacteriology, 49, 1211–1215.

Newbrun, E., and Baker, S., (1968): Physico-chemical characteristics of the levan produced by *Streptococcus salivarius*. Carbohydrate Research, 6: 165–170.

O'Sullivan, D. J., & Klaenhammer, T. R. (1993). Rapid mini-prep isolation of high quality plasmid DNA from Lactococcus and Lactobacillus spp. Applied and Environment Microbiology, 59, 2730–2733.

Payne, G. W., Vandamme, P., Morgan, S. H., et al. (2005). Development of a recA gene-based identification approach for the entire Burkholderia genus. Applied and Environmental Microbiology, 71, 3917–3927.

Poli, A., Kazak, H., & Gürleyenda˘g, B., et al. (2009). High level synthesis of levan by a novel Halomonas species growing on defined media. Carbohydrate Polymers, 78, 651–657.

Pontis, H. G., & Del Campillo, E. (1985). Fructans. In M. Dey, & R. A. Dixon (Eds.), Biochemistry of storage carbohydrates in green plants (pp. 205–277). Academic Press Inc.

Puig, M., Jofre, J., Lucena, F., Allard, A., Wadell, G., & Girones, R. (1994). Detection of adenovirus and enterovirus in polluted waters by nested PCR amplification. Applied and Environmental Microbiology, 60, 2963–2970.

Roberts, M.S., Nakamura, L.K., & Cohan, F.M. (1996). *Bacillus vallismortis* sp. nov., a close relative of *Bacillus subtilis*, isolated from soil in Death Valley, CA. Interational Journal of Systemic Bacteriology, 46, 470–475.

Rodriguez, H., de las Rivas, B., & Muňoz, R. (2007). Efficacy of recA gene sequence analysis in the identification and discrimination of Lactobacillus hilgardii strains isolated from stuck wine fermentations. International Journal of Food Microbiology, 115:70–78

Root, A.L., The ABC and XYZ of bee culture. London: Edward Arnold publication Ltd.; 1993. pp. 5–28.

Rhee, S., Song, K., Kim, C., Park, B., Jang, E., & Jang, K., (2002): Levan. In A.Steinbu¨ chel (Ed.), Biopolymers; polysaccharides , pages 351–377.

Sabaté, D. C., Carrillo, L., & Audisio, M. C. (2009). Inhibition of Paenibacillus honey samples. Research Microbiology, 160, 193–199.

Sambrook, J., Fritsch, E. F., & Maniattis, T. (1989). Molecular cloning: A laboratory manual. Cold Spring Harbor, NY: Cold Spring Harbor Laboratory.

Sanders, M. E., Morelli, L., & Tompkins, T. A. (2003). Sporeformers as human probiotics: Bacillus, Sporolactobacillus, and Brevibacillus. Comprehensive Reviews inFood Science and Food Safety (CRFSFS), 2, 101–110.

Shah, M. M., Lihara, H., Noda, M., Song, S. X., Nhung, P. H., & Ghkusu, K., et al. (2007). dnaJ gene sequence-based assay for species identification and phylogene grouping in the genus Staphylococcus. International Journal of Systematic and Evolutionary Microbiology, 57, 25–30.

Shih, I., Shieh, Y., Yu, C., & Hsieh, C. (2005). Selective production and characterization of levan by Bacillus subtilis (Natto) Takahashi. Journal of Agriculture Food Chemistry, 53, 8211–8215.

Simões, C., Amoros, M., & Girre, L. (1999). Mechanism of antiviral activity of triterpenoid saponins. Phytotherapy Research, 21, 317–325.

Stivala, S., and Bahary, W., (1978): Some dilute-solution parameters of the levan of *Streptococcus salivarius* in various solvents. Carbohydrate Research. 67: 17–21

Takatsy, G. (1955). The use of spiral loops in serological and virological micromethods. Acta Microbiologica Academic Science Hungarica, 3, 191.

Urdaci, M. C., Bressollier, P., & Pinchuk, I. (2004). *Bacillus clausii* antimicrobial and immunomodulatory activities. Journal of Clinical Gastroenterology, 38, 586–590.

Walum, E., Strenberg, K., & Jenssen, D. (1990). Understanding cell toxicology: Principlesand practice. NewYork: Ellis Howood., pp. 97–111.

Yanase, H., Iwata, M., Nakahigashi, R., Kita, K., Kato, N., & Tonomura, K. (1992). Purification, crystallization, and properties of extracellular levansucrase from *Zymomonas mobilis*. Bioscience, Biotechnology and Biochemistry, 56, 1335–1337.

Yoo, S. H., Yoon, E. J., Cha, J., & Lee, H. G. (2004). Antitumor activity of levan 536 polysaccharides from selected microorganisms. International Journal of Biological Macromolecules, 34, 37–41.

Yoon, E. J., Yoo, S. H., Cha, J., & Lee, H. G. (2004). Effect of levan's branching structure on antitumor activity. International Journal of Biological Macromolecules, 34, 191–194.

Chapter 8

Lichen Polysaccharides

Desiree Nedra Karunaratne, R.G.U. Jayalal and V. Karunaratne

Additional information is available at the end of the chapter

http://dx.doi.org/10.5772/51021

1. Introduction

Lichens are symbiotic organisms composed of a fungal partner, the mycobiont, and one or more photosynthetic partners, the photobiont, which may be either a green alga or a cyanobacterium (Nash III, 1996; Wolseley and Aguirre-Hudson, 1994 and Yoshimura, 2004). The mycobiont in this combination is an ascomycete or a basidiomycete and the photobiont a green alga or a cyanobacterium. Some lichen species can contain more than one green algal species as photobionts (Friedl and Gärtner, 1988; Friedl, 1989; Ihda *et al.*, 1993). At present, about 13,500 fungal species have been recognized to be involved in lichen symbiosis (Hawksworth *et al.* 1995; Kirk *et al.*, 2001). However, Sipman and Aptroot (2001) stated that this number could be as high as 20,000 after including "orphaned" species; according to Lumbsch *et al.*, (2011), about 10,000 of lichenized fungi have been estimated as undescribed species.

The hyphae of mycobionts are septate, branched, thin or thick walled, and the walls are colourless or variously coloured. It is necessary that the photobiont, which is essentially aquatic in nature, remains protected from desiccation in a lichenized terrestrial condition. The protection is provided by the mycobiont which forms the bulk of the thallus, by the development of specialized hyphal tissues in the form of a cover or cortex over the stratum of the photobiont. The development of the cortex is assumed to be stimulated by the photobiont (Ahmadjian, 1987). In some gelatinous lichens with cyanobacteria, the polysaccharidic sheath produced by the photobiont (cyanobacteria) contributes to water retention (Prieto *et al.*, 2008). Lichens have been used for ecological, medicinal and other economic purposes for over 100 years and these beneficial effects have been correlated to some extent with their polysaccharide content. Amongst the identified lichens so far, about 100 species have been studied for their polysaccharides and their composition (Cordeiro *et al.*, 2005).

2. Separation and characterization

The cell wall of fungi is composed mainly of polysaccharides such as lichenan, isolichenan, and galactomannan (fig.1) (Elix and Stocker-Wörgötter, 2008). Bernard and Latgé (2001)

INTECH
open science | open minds

© 2012 The Author(s). Licensee InTech. This chapter is distributed under the terms of the Creative Commons Attribution License http://creativecommons.org/licenses/by/3.0), which permits unrestricted use, distribution, and reproduction in any medium, provided the original work is properly cited.

described that the fungal wall is a complex structure composed typically of chitin, 1,3-β- and 1,6-β-glucan, mannan and proteins, although wall composition frequently varies markedly between the species of fungi.

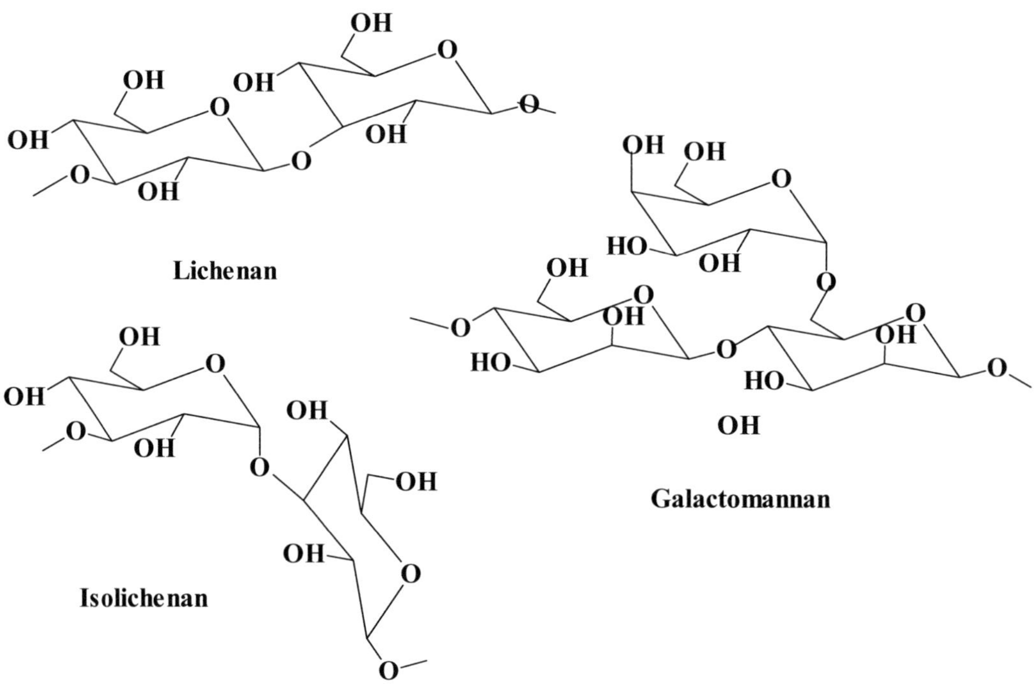

Figure 1. Structures of some polysaccharides contained in fungal cell wall

For the separation and isolation of lichen polysaccharides, traditional methods and modern techniques are used. Traditional methods basically involved freezing and thawing of the material originally extracted with hot water. Dialysis and ethanol precipitation has been employed for further purification. Also, alkali solutions have been used to extract these compounds. The polysaccharides present in the thalli of *Umbilicaria mammulata* were isolated by an exhaustive method starting with successive extraction with hot water at 100°C followed by extraction of the residue with hot 2% KOH at 100 °C. The alkaline extract was further treated with Fehlings solution to yield the 1,3-β- glucan from the supernantant and a 1,6-β- glucan was obtained from the precipitated Fehlings complex (Carbonero *et al.*, 2006). In our laboratory, we have used a similar alkaline extraction procedure to isolate a polysaccharide from *Usnea* cf. *cornuta* Körb. Chromatographic techniques and filtration devices were utilized to separate lichen polysaccharides (Paulsen *et al.*, 2002). However, different column chromatographic methods, including GP-HPLC, HPLC and ion exchange chromatography have been used recently to separate polysaccharides from lichens. Olafsdottir *et.al* (1999), fractionated the polysaccharides from *Cetraria islandica* on DEAE Sepharose CL-6B anion exchange columns with a 0 -1 M NaCl gradient. Further purification was performed by Sephacryl S-400 gel filtration. To determine homogeneity and M_r of these compounds, electrophoretic methods and gel permeation chromatography are used. For the determination of monosaccharides, TLC and GC methods are used. The monosaccharides could be derivatized into alditol acetates or into trimethylsilyl (TMS) ethers for analysis by GC. The linkage analysis is performed by using NMR spectroscopy (1D proton, 2D-COSY, NOESY, 2D-TOCSY, 1H 13C-HSQC, HMBC, H2BC and HSQC-NOESY) (Paulsen *et al.*, 2002 and Jensen *et al.*, 2010) and methylation analysis (Omarsdottir *et.al.*, 2005; Olafsdottir *et. al.*, 1999; Ruthes *et al.* 2010).

3. Biological source of the polysaccharide fraction

Whether these polysaccharides are produced by the mycobiont or photobiont separately or in symbiosis has been debated for a long time. Takahashi *et al.* (1979) showed that aqueous extracts from cultivated myco and photobiont had different monosaccharide composition and physical properties. It was also found that while the extracts of the mycobiont had a similar composition to that of the parent intact lichen, the photobiont fractions were different from those of the symbiotic thalli and its mycobiont. Complete structural analysis by Cordeiro *et al.* (2004) confirmed Takahashi's results, and showed that the nigeran, laminaran and the galactomannan, found previously in the symbiotic thalli of *Ramalina peruviana* (Cordeiro *et al.* 2003) were also generated by the aposymbiotically cultured fungal partner.

Three different polysaccharide structural types: β-glucans, α-glucans (linear or lightly substituted), and galactomannans (branched) (Carbonero *et al*, 2005a) are present in the fungal cell wall. According to Olafsdottir and Ingólfsdóttir (2001), all the polymers present in lichen thalli are categorized into glucan type [β-(1→3)(1→4)], lichenan type [α-(1→3)(1→4)] and pustulan [β-(1→6)]. Some of these types are depicted in Table 1. However, the recent discovery of a few additional complex heteroglycans, such as rhamnogalactofuranan (Olafsdottir, 1999), galactomannoglucans (Woranovicz-Barreira, 1999 and Carbonero *et al.*, 2002) and thamnolan (Carbonero *et al.*, 2005a) necessitated the closer examination of the origins of the polysaccharides. Studies carried out by Cordeiro *et al.*, (2005), revealed that the algal partner (*Trebouxia* sp.) also consists of some carbohydrates such as β-galactofuranan heteropolysaccharide. Later in 2007, new polysaccharides xylorhamnogalactofuranan and Xylan in the photobiont *Asterochloris* sp. were recorded by Cordeiro *et al.* Thereafter, Jensen *et al.* (2010), isolated another heteropolysaccharide called colleman from a cyanobacteria present in the lichen *Collema flaccidum*.

In the past, lichen polysaccharides have been extracted from the whole thallus without giving consideration to the origin of components such as the fungal partner or the photobiont (Gorin and Iacomini, 1984, 1985; Gorin *et al.*, 1993; Teixeira *et al.*, 1995). Later Cordeiro *et al.* (2005, 2007, 2008), found that, dissimilarities between the polysaccharides extracted from the cultivated photobionts *Trebouxia* and *Asterochloris* with those from their respective lichens could indicate that the photobiont or mycobiont was responsible for these differences. Galactoglucomannans were isolated from the mycobiont of *Parmotrema* species including *Parmotrema austrosinense, Parmotrema delicatulum, Parmotrema mantiqueirense,*

Polysaccharide	**Main unit**	**Side chain**	**Lichen species**	**Mycobiont/ Photobiont**	**reference**
Lichenan	Homoglucan with β(1→3) (1→4) linkage		*Cetraria ilandica, C. nivaris, C. richardsonii, Usnea barbata, U. lingissima, U. bayleyi, Parmelia tinctorum, P. conspersa, P. hypotrypella, P. nikkoensis, Alectoria sulcata, A. sarmentosa*	Whole thallus	Shibata, 1973
Isolichenan	Homoglucan with α(1→3) (1→4) linkage		-do-	-do-	Shibata, 1973
Pustulan	Glucan with β(1→6) linkage		*Lasallia pustulata, L. papulosa, Umbilicaria hirsuta, U. angulata, U. caroliniana, U. polyphylla,Gyrophora esculenta*	-do-	Shibata, 1973
Everniin	Glucan with α(1→3) (1→4) linkage		*Evernia prunastri*	-do-	Shibata, 1973
Acroscyphan	Homoglucan with α(1→3) (1→4) (1→6) linkage		*Acroscyphus sphaerophoroides*	-do-	Shibata, 1973
Galacto-glucomannans	a (1→6)-linked main chain of ◎-Manp	O-2 and O-4 by α-Galp and β-Galp nonreducing end-units	*Parmotrema austrosinense, P. delicatulum, P. mantiqueirense, P. schindlerii, P. tinctorum and Rimelia (R. cetrata* and *R. reticulata)*	mycobiont	Carbonero *et al.*, 2005b
Xylorhamno-galactofuranan	(1→3)-linked galactofuranosyl	galactofuranosyl units 5-*O* and 6-*O*-substituted, rhamnopyranosyl units 2-*O*, 3-*O* and 2,3-di-*O*-substituted in position 6.	*Cladina confusa*	Photobiont	Cordeiro *et al.*, 2007

Polysaccharide	Main unit	Side chain	Lichen species	Mycobiont/ Photobiont	reference
laminaran and pustulan and galacto-furanomannan	(1→3)-and (1→6)-linked β-glucans	(1→6)-linked α-mannopyranosyl	*Umbilicaria mammulata*	Mycobiont	Carbonero *et al.*, 2006
Hetero-polysaccharide	(1→5)-linked galactofuranosyl units	Complex	*Ramalina gracilis*	Photobiont	Cordeiro *et al.*, 2008
O-methylated mannogalactan	(1→6)-linked β-galactopyranose	at O-3 by β-Gal*p*, 3-OMe-α-Manp or α-Man*p* units	*Peltigera aphthosa*	Photobiont	Cordeiro *et al.*, 2010
Colleman like	Complex heteroglycan	2-OMe Man*p*, 2-OMe-Arap, Xyl*p* and Glc*pA*	*Collema flaccidum*	Whole thallus (cyanobacteri a)	Jensen *et al.*, 2010
	(1→4)-linked β-D-xylan		*Peltigera canina*	Mycobiont	Ruthes *et al.*, 2010

Table 1. Summarized data of lichen polysaccharides and their origin

Parmotrema schindlerii, Parmotrema tinctorum and *Rimelia cetrata* and *Rimelia reticulata* by Carbonero *et al.*, in 2005b. These galactoglucomannans consisted of (1→6)-linked main chain of α-Manp units, which were substituted preferentially at O-2 and O-4 by α-Gal*p* and β-Gal*p* nonreducing end-units, respectively. Further, they also isolated two galactomannan fractions from the lichen, *Roccella decipiens*. One galactomannan fraction had a main chain with (1→4)-linked α-D-Man*p* units, substituted at O-2 with side chains containing a nonreducing end, 2-O- and 6-O-substituted α-Man*p* units. The other fraction had a similar α-D-Man*p* core structure, but with side chains containing nonreducing end, 5-O-, 6-O-, and 5,6-di-O-substituted β-D-Gal*f* units (Cordeiro *et al.*, 2005). Another heteropolysaccharide xylorhamnogalactofuranan was isolated from the lichen *Cladina confusa.* It consisted of (1→3)-linked galactofuranosyl units with side chains of galactofuranosyl units 5-*O* and 6-*O*-substituted, as well rhamnopyranosyl units 2-*O*, 3-*O* and 2,3-di-*O*-substituted at position 6. Nonreducing end units were composed of Xylose (Cordeiro *et al.*, 2007). In 2006, (1→3)-and (1→6)-linked β-glucans, namely laminaran and pustulan and galactofuranomannan which have a main chain of (1→6)-linked α-mannopyranosyl residues, partially substituted at O-2, O-4 were isolated from another lichen *Umbilicaria mammulata* (Carbonero *et al.*, 2006). Later in 2008, Cordeiro *et al.* found Galactofuranose-rich heteropolysaccharide from the lichen *Ramalina gracilis.* This polysaccharide has (1→5)-linked galactofuranosyl units at the main chain with very complex branched structures of side chains in position 6. They found that this polysaccharide arose from the algal symbiont *Trebouxia sp.* of the lichen *Ramalina gracilis.* An O-methylated mannogalactan was isolated from *Peltigera aphthosa* by Cordeiro *et al.* in 2010. This consisted of (1→6)-linked β-galactopyranose main chain partially substituted at O-3 by β-Gal*p*, 3-OMe-α-Manp or α-Man*p* units. The algal symbiont *Coccomyxa mucigena* of the lichen *Peltigera aphthosa* was thought to be the origin of this heteropolysaccharide since the lichen thallus yielded a polysaccharide of different structure. A colleman like heteropolysaccharide was isolated from the cyano lichen *Collema flaccidum.*

Colleman is a complex heteroglycan containing the unusual monosaccharides 2-OMe Man*p* and 2-OMe-Arap as well as Xyl*p* and Glc*pA* (Jensen *et al.*, 2010). The presence of uronic acids has been reported previously from cyanobacterial polysaccharides. Since the structural features and sugar content of colleman is representative of polysaccharides of cyanobacterial origin, it is proposed that colleman originates from the cyanobacterial partner. Ruthes *et al.* (2010) were able to isolate (1→4)-linked β-D-xylan (an EPS) and heteropolysaccharide with a complex structure of β-L-Ara*p* and β-D-Xyl*p*-(1→4)-linked units from *Peltigera canina.* Again it was opined that the photobiont, *Nostoc muscorum* was the source of this heteropolysaccharide.

4. Polysaccharides as a taxonomic tool

The identification and classification is generally based on morphology of the organism. Taxonomy of lichen species have been corroborated by phylogenetic applications with the advances in DNA technology. Lichen polysaccharides have been used as a taxonomic tool and chemotaxonomic classification has resulted in clarification of conflicting taxonomic data.

The lichen-forming ascomycete order Lichinales comprises around 250 species and is distributed among 52 genera and four families (Eriksson, 2006). Earlier molecular studies (Wedin *et al.*, 2005) did not confirm its phylogenetic relationships, although the order was treated as a separate class, *Lichinomycetes* (Hibbett *et al.*, 2007). Since alkali and water-soluble polysaccharides from *Lichina pygmaea* and *L. confinis* reflect phylogeny in other ascomycetes (Prieto *et al.* 2008), an isolated polysaccharide was purified to investigate whether such polysaccharide compounds in the Lichinomycetes are distinctive. Results support molecular studies showing that lichen species are remote from Lecanoromycetes as the galactofuranose residues are in the α-configuration. That the Lichinomycetes were part of an ancestral lichenized group cannot be established from the present data because the extracted polysaccharide does not have the galactofuranose residue in the β configuration; however, the data suggests that an ancestor of the Lichinomycetes contained a mannan and was part of an early radiation in the ascomycetes. Polysaccharides present support the molecular data obtained recently that the Lichinales are distinct from other ascomycete groups and should be treated as a separate order in the separate class Lichenomycetes (Reeb *et al.*, 2004). However, more representatives in the order must be subjected to molecular studies and more polysaccharides be investigated before confirming this hypothesis. Interestingly, the relative basal position of *Lichina* with respect to the water soluble polysaccharides agrees with the suggestion that ancestral lichens contained cyanobacteria as the photosynthetic partner (Hawksworth, 1982).

Investigation of mannose containing polysaccharides as a taxonomic tool centers around the structural diversity of the galactomannans isolated from several lichenized fungi. The taxonomic value of these galactomannans depends on the side-chain substituents on (1 →6)-linked α-D-mannopyranosyl main chains (Gorin and Lacomini, 1985).

Although classical taxonomy regarded *Cladina* as sub-species of *Cladonia,* lichenologists considered them to be distinct species. It was shown that galactomannans are important chemotypes in determining the taxonomy of *Clodonia* spp. and other related genera

(Woranovicz-Barreira *et al.*, 1999). However, based on molecular phylogentic results *Cladina* and *Cladoni*a were confirmed as synonyms (Ahti and Depriest, 2001).

Lichen polysaccharides, can also be used as a taxonomic tool to differentiate some lichen species, since some of the heteropolysaccharides and their chemical characters are unique to certain groups of lichens (Carbonero *et al.*, 2006 and Cordeiro *et al.*, 2007). Further, the polysaccharides content of the lichen photobiont may be used as a marker in algal symbiont taxonomy (Cordeiro *et al.*, 2007).

5. Biological activities of lichen polysaccharides

Many lichens are known to have immunomodulating properties, potent antibiotic, antitumour, antiviral as well as antioxidant properties which are mostly attributed to the secondary metabolites (Malhotra *et al.*,2008, Behera, *et al.*, 2007). According to Yanaki *et al.*, (1986) and Bohn and BeMiller, (1995), functional activity of polysaccharides mainly depends on molecular weight, degree of branching, water solubility, structure and configuration. Hence they have different uses in different fields. The biological activities of lichen polysaccharides reported have been limited to anti-tumor, anti-inflamatory or immunomodulatory activity (Omarsdottir *et al.*, 2007).

Cordeiro *et al.*, (2008) reported that one of the β-galactofuranan polysaccharides isolated from *Trebouxia* sp., the algal symbiont of the lichen *Ramalina gracilis* expressed *in vitro* activity on peritoneal macrophages. Further studies carried out on 4 lichen polysaccharides by Omarsdottir, *et al.*, (2006) showed that, three heteroglycans namely Ths-4, Ths-5 and thamnolan and a β-glucan (Th-2) isolated from the lichen *Thamnolia vermicularis* var. subuliformis showed an effect on the human immune system. Thamnolan, a galactofuranorhamnan had less mitogenic effect than Ths-5 and Ths-2 indicating that its unusual galactofuranorhamnan structure may be responsible for its different immunomodulatory activity. In a study on the immunomodulatory activities of an aqueous lichen extract from *Cetraria islandica* , Freysdottir *et al.*, (2007) discovered that the extract was able to upregulate IL-10 secretion. Interestingly, when the individual components of this extract (lichenan and isolichenan and secondary metabolites protolichesterinic and fumarprotocetraric acids) were subjected to the same assay, only lichenan displayed antiinflammatory effects (Freysdottir, *et al.*, 2008).

The cytotoxic activity, phagocytic activity and antitumor activity of an α-D glucan from *Ramalina celastri* has been reported (Leão *et al.*, 1997 and Stuelp-Campelo *et al.*, 2002). Based on this, De Araújo *et al.*, (2011) found that a sulphated α-D glucan lichen polysaccharide extracted from *Ramalina celastri* exhibited antischistosomal activity. These α-D glucans with α(1–3) and α(1–4) linkages are linear and water soluble and are known for their ability to stimulate the mononuclear phagocyte system and improve host resistance to viral, bacterial and parasitic infections.

According to Nishikawa *et al.*, (1970), O-acetylated pustulan isolated from three species of *Umbilicaria* showed a significant antitumor effect against the implanted Sarcoma-180 in mice. The polysaccharide fractions isolated from six species of lichens were studied for antitumor activity (Nishikawa *et al.*, 1974). The active polysaccharide present in *Lasallia pensylvanica,* was the partially O-acetylated (1-6) β-glucan, pustulan while that of *Usnea rubescens,* was identified as a lichenan type polysaccharide. The other 4 polysaccharides were from the *Cladonia* sp.; *Cladonia crispata, Cladonia rangiferina* subsp. *grisea, Cladonia mitis* and *Cladonia squamosa* all of which showed significant antitumor activity. These *Cladonia* polysaccharides which were complex heteroglycans of the galactomannan type were found to exert moderate antitumor activity. However, previous studies have shown that the isolichenin type polysaccharides were not as effective in their antitumour effect. In another study, isolichenin isolated from the lichen *Usnea fasciata* showed moderate activity against Ehrlich tumor cells (Periera *et al.*, 1994). In 1989, Hirabayashi *et al.*, showed inhibitory effect of a lichen polysaccharide sulfate (GE-3-S), isolated from the lichen *Umbilicaria esculenta* on the replication of human immunodeficiency virus (HIV) in vitro.

Behera *et al.*, (2007) showed correlation between lichen protein/polysaccharide ratio and their antioxidant properties. Since cultured lichen extracts were used in the study, the effect of secondary metabolites as well as polyphenols present in the extract have not been evaluated. Thus the antioxidant activity cannot be solely attributed to the polysaccharide specially since polyphenols are known antioxidants. In addition, the mechanism of scavenging activity of polysaccharides on free radicals is not fully understood yet.

6. Conclusion

Polysaccharides from lichens unlike bacterial polysaccharides do not show a wide range of variation in sugar content. The predominant sugars are limited to glucose, galactose and mannose, with arabinose and xylose present in minor proportions in addition to others such as rhamnose. These polysaccharides however have been useful in chemotaxonomic studies due to this conservation of sugar structures and regular structural patterns. As far as biological activity is concerned, very few studies have been reported and it would be worthwhile further investigating the immunomodulating effects of the polysaccharides.

Author details

Desiree Nedra Karunaratne, R.G.U. Jayalal and Veranja Karunaratne
University of Peradeniya, Peradeniya, Sri Lanka

7. References

Ahmadjian, V. (1987). Coevolution in lichens. *Annals of the New york Academi of Sciences* 503, 307–315.

Ahti, T. and Depriest, P.T., (2001). New combination of *Cladina* epithets in *Caldonia* (Ascomycotina, Cladonaceae). *Mycotaxon,* 78, 499-502.

Behera, B.C., Verma, N., Sonone, A. and Makhija, U. (2007). Tissue culture of some lichens and screening of their antioxidant, antityrosinase and antibacterial properties. *Phytotherapy Research* 21, 1159–1170.

Bernard, M. and Latgé, J.P. (2001). *Aspergillus fumigatus* cell wall: composition and biosynthesis. *Med Mycol* 39 (1), 9–18.

Bohn, J.A. and BeMiller, J.N. (1995). (1–3)-β-Glucans as biological response modifiers: a review of structure-functional activity relationships. *Carbohydr Polym* 28, 3–14.

Carbonero, E.R., Cordeiro,L.M.C., Mellinger,C.G., Sassaki,G.L., Wörgötter,E.S., Gorin, P.A.J. and Iacomini, M. (2005a). Galactomannans with novel structures from the lichen *Roccella decipiens* Darb. *Carbohydrate Research* 340, 1699–1705.

Carbonero, E. R., Mellinger, C. G., Eliasaro, S., Gorin, P. A. J. and Lacomini, M. (2005b). Chemotypes significance of lichenized fungi by structural characterization of heteropolysaccharides from the genera *Parmotrema and Rimelia*. *FEMS Microbiology Letters* 246, 273–278.

Carbonero, E. R., Montai, A. V., Woranovicz-Barreira, S. M., Gorin, P. A. J. and Iacomini, M. (2002). Polysaccharides of lichenized fungi of three *Cladina* spp.: significance as chemotypes. *Phytochemistry* 61, 681–686.

Carbonero, E.R., Smiderle, F.R., Gracher, A.H.P., Mellinger, C.G., Torri, G., Ahti, T., Gorin, P.A.J. and Iacomini, M. (2006). Structure of two glucans and a galactofuranomannan from the lichen *Umbilicaria mammulata*. *Carbohydrate Polymers* 63, 13–18.

Cordeiro, L.M.C., Wörgötter, E.S., Gorin, P.A.J. and Iacomini, M. (2003). Comparative studies of the polysaccharides from species of the genus *Ramalina*-lichenized fungi of three distinct habitats. *Phytochemistry* 60, 967–975.

Cordeiro, L.M.C., Wörgötter, E.S., Gorin, P.A.J. and Iacomini, M. (2004). Elucidation of polysaccharide origin in *Ramalina peruviana* symbiosis. *FEMS Microbiol. Lett.* 238 (2004) 79–84.

Cordeiro, L.M.C., Carbonero, E.R., Sassaki, G.L., Reis, R.A., Wörgötter, E.S., Gorin, P.A.J. and Iacomini, M. (2005). A fungus-type β-galactofuranan in the cultivated *Trebouxia* photobiont of the lichen *Ramalina gracilis*. *FEMS Microbiol. Lett.* 244, 193–198.

Cordeiro, L.M.C., de Oliveira, S.M., Buchi, D.F. and Iacomini, M. (2008). Galactofuranose-rich heteropolysaccharide from *Trebouxia* sp., photobiont of the lichen *Ramalina gracilis* and its effect on macrophage activation. *International Journal of Biological Macromolecules* 42, 436–440.

Cordeiro, L.M.C., Sassaki, G.L., Gorin, P.A.J. and Iacomini, M. (2010). O-Methylated mannogalactan from the microalga *Coccomyxa mucigena,* symbiotic partner of the lichenized fungus *Peltigera aphthosa*. *Phytochemistry* 71, 1162–1167.

Cordeiro, L.M.C., Sassaki, G.L. and Iacomini, M. (2007). First report on polysaccharides of *Asterochloris* and their potential role in the lichen symbiosis. *International Journal of Biological Macromolecules* 41, 193–197.

De Araújo, I.W.F., Vanderlei, E.S.O., Rodrigues, A.R.G., Coura, C.O., Quinderé, A.L.G., Fontes, B.P., Queiroz, I.N.L., Jorge, R.J.B., Bezerra, M.M., Silva, A.A.R., Chaves, H.V., Monteiro, H.S.A., Paula, R.C.M. and Benevides, N.M.B. (2011). Effects of a sulfated polysaccharide isolated from the red seaweed *Solieria filiformis* on models of nociception and inflammation. *Carbohydr. Polym.*, 86: 1207-1215.

Elix, J. A. and Stocker-Wörgötter, E. (2008). Biochemistry and secondary metabolites. In Lichen Biology 2nd Edn. Vol. 1. ed. T.H. Nash III. Cambridge University Press, Chapter 7, 104-133.

Eriksson, O.E. (2006). Outline of Ascomycota - 2006. *Myconet* 12, 1–82.

Freysdottir,J., Omarsdottir, S., Ingólfsdóttir, K., Vikingsson, A. and Olafsdottir, E.S. (2007). Abstracts, Scandinavian Society for Immunology. 37th Annual Meeting Turku, Finland. Pp. 596.

Freysdottir,J., Omarsdottir, S., Ingólfsdóttir, K., Vikingsson, A. and Olafsdottir, E.S. (2008). In vitro and in vivo immunomodulating effects of traditionally prepared extract and purified compounds from *Cetraria islandica. International Immunopharmacology* 8: 423–430.

Friedl, T. (1989). Systematik und Biologie von *Trebouxia* (Microthamniales, Chlorophyta) als Phycobiont der Parmeliaceae (lichenisierte Ascomyceten). Ph.D. thesis, University Bayreuth, Bayreuth, 218.

Friedl, T. and Gärtner, G. (1988). *Trebouxia* (Pleurastrales, Chlorophyta) as a phycobiont in the lichen genus *Diploschistes. Arch. Protistenk.*143, 153–61.

Gorin, P.A.J., Baron, M., Silva, M.L.C., Teixeira, A.Z.A. and Iacomini, M. (1993). Lichen carbohydrates. *Ciência e Cultura* 45, 27–36.

Gorin, P.A.J. and Iacomini, M. (1984). Polysaccharides of the lichens *Cetraria islandica* and *Ramalina usnea. Carbohydr. Res.* 128, 119–132.

Gorin, P.A.J. and Iacomini, M. (1985). Structural diversity of D-galacto-D-mannan components isolated from lichens having ascomycetous mycosymbionts. *Carbohydr. Res.* 142, 253–267.

Hawksworth, D.L. (1982). Co-evolution and the detection of ancestry of lichens, *J. Hattori Botanical Laboratory* 52, 323-329.

Hawksworth, D.L., Kirk, P.M., Sutton, B.C. and Pegler, D.N. (1995). Ainsworth and Bisby's dictionary of the fungi. 8th edn. CAB International, Wallingford

Hibbett, D.S., Binder, M., Bischoff, J.F., Blackwell, M., Cannon, P.F., Eriksson, O.E., Huhndorf, S., James, T., Kirk, P.M., Lucking, R., Lumbsch, T., Lutzon, F., Matheny, P.B., McLaughlin, D.J., Powell, M.J., Redhead, S., Schoch, C.L., Spatafora, J.W., Stalpers, J.A., Vilgalys, R., Aime, M.C., Aptroot, A., Bauer R., Begerow, D., Benny, G.L., Castlebury, L.A., Crous, P.W., Dai, Y.C., Gams, W., Geiser, D.M., Griffith, G.W., Gueidan, C., Hawksworth, D.L., Hestmark, G., Hosaka, K., Humber, R.A., Hyde, K., Ironside, J.E., Koljalg, U., Kurtzman, C.P., Larsson, K.H., Lichtwardt, R., Longcore, J., Miadlikowska, J., Miller, A., Moncalvo, J.M., Mozley-Standridge, S., Oberwinkler, F., Parmasto, E., Reeb, V., Rogers, J.D., Roux, C., Ryvarden, L., Sampaio, J.P., Schussler, A., Sugiyama, J., Thorn, R.G., Tibell, L., Untereiner, W.A., Walker, C., Wang, Z., Weir, A., Weiss, M., White, M.M., Winka, K., Yao, Y.J. and Zhang, N. (2007). A higher-level phylogenetic classification of the Fungi. *Mycological Research* 111: 509–547.

Hirabayashi, K., Iwata, S., Ito, M., Shigeta, S., Narui, T., Mori, T. and Shibata, S. (1989). Inhibitory effect of a lichen polysaccharide sulfate, GE-3-S, on the replication of human immunodeficiency virus (HIV) in vitro. *Chem Pharm Bull (Tokyo)* 37(9), 2410-2412.

Ihda, T. A., Nakano, T., Yoshimura, I. and Iwatsuki, Z. (1993). Phycobionts isolated from Japanese species of *Anzia* (lichenes). *Archiv für Protistenkunde* 143, 163–172.

Jensen, J.S.R.E., Petersen, B.O., Veselinovic, T., Olafsdottir, E.S., Duus, J.Ø. and Omarsdottir, S. (2010). Structural characterisation of a new O-methylated heteroglycan, colleman, from the cyanolichen *Collema flaccidum. Carbohydrate Polymers* 80, 799 –807.

Kirk, P.M., Cannon, P.F., David, J.C. and Stalpers, J.A. (2001). *Ainsworth & Bisby's dictionary of the fungi*, 9th edition. CAB International, Wallingford.

Leão, A.M., Buchi, D.F., Iacomini, M., Gorin, P.A. and Oliveira, M.B. (1997). Cytotoxic effect against HeLa cells of polysaccharides from the lichen *Ramalina celastri*. *J Submicrosc Cytol Pathol.* 29(4), 503-9.

Lumbsch H. T., Ahti T., Altermann S., De Paz G.A., Aptroot A., Arup U., Peña A.B., Bawingan P.A., Benatti M.N., Betancourt L., Björk C.R., Boonpragob K., Brand M., Bungartz F., Cáceres M.E.S., Candan M., ChavesJ. L., Clerc P., Common R., Coppins B. J., Crespo A., Dal-Forno M., Divakar P.K., Duya M.V., Elix J.A., Elvebakk A., Fankhauser J.D., Farkas E., Ferraro L.I., Fischer E., Galloway D.J., Gaya E., Giralt M., Goward T., Grube M., Hafellner J., J.E. Hernández M., M. De L. A. H. Campos, Kalb K., Kärnefelt I., Kantvilas G., Killmann D., Kirika P., Knudsen K., Komposch H., Kondratyuk S., Lawrey J. D., Mangold A., Marcelli M. P., Mccune B., Messuti M. I., Michlig A., González R. M., Moncada B., Naikatini A., Nelsen M. P., Øvstedal D. O., Palice Z., Papong K., Parnmen S., Pérez-Ortega S., Printzen C., Rico V. J., Plata E. R., Robayo J., Rosabal D., Ruprecht U., Allen N. S., Sancho L., De Jesus L. S., Vieira T. S., Schultz M., Seaward M. R. D., Sérusiaux E., Schmitt I., Sipman H. J. M., Sohrabi M., Søchting U., Søgaard M. Z., Sparrius L. B., Spielmann A., Spribille T., Sutjaritturakan J., Thammathaworn A., Thell A., Thor G., Thüs H., Timdal E., Truong C., Türk R., Tenorio L. U., Upreti D. K., Van Den Boom P., Rebuelta M. V., Wedin M., Will-Wolf S., Wirth V., Wirtz N., Yahr R., Yeshitela K., Ziemmeck F., Wheeler T. and Lücking R.. (2011). One hundred new species of lichenized fungi: a signature of undiscovered global diversity. *Phytotaxa* 18, 1–127.

Malhotra, S., Subban, R. and Singh, A. (2008). Lichens- Role in Traditional Medicine and Drug Discovery. *The Internet Journal of Alternative Medicine* 5(2).

Nash. III, T. H. (1996). Photosynthesis, respiration, productivity and growth. *Lichen Biology* (ed. by T.H.Nash), Cambridge University Press, Cambridge, 88–120,

Nishikawa, Y., Tanaka, M., Shibata, S. and Fukuoka, F. (1970). Polysaccharides of lichens and fungi. IV. Antitumor active O-acetylated pustulan type glucans from the lichens of *Umbilicaria* species. *Chem. Pharm. Bull.* 18 (7), 1431-1434.

Nishikawa, Y., Ohki, K., Takahashi, K., Kurono, G., Fukuoka, F. and Emori, M. (1974). Studies on the water soluble constituents of lichens. II. Antitumor polysaccharides of *Lasallia, Usnea* and *Cladonia* species. *Chem. Pharm. Bull.* 22 (11), 2692-2702.

Olafsdottir, E.S. and Ingólfsdóttir, K. (2001). Polysaccharides from lichens: structural characteristics and biological activity. *Planta Medica* 67, 199-208.

Olafsdottir, E.S., Omarsdottir, S., Paulsen, B.S., Jurcic, K. and Wagner, H., (1999). Rhamnopyranosylgalactofuranan, a new immunologically active polysaccharide from *Thamnolia subuliformis*. *Phytomedicine* 6, 273-279.

Omarsdottir, S., Petersen, B.O., Barsett, H., Paulsen, B.S., Duus, J.Ø. and Olafsdottir, E.S. (2006). Structural characterisation of a highly branched galactomannan from the lichen *Peltigera canina* by methylation analysis and NMR-spectroscopy. *Carbohydrate Polymers* 63, 54–60

Omarsdottir, S., Freysdottir, J. and Olafsdottir, E.S. (2007). Immunomodulating polysaccharides from the lichen *Thamnolia vermicularis* var. *subuliformis*. *Phytomedicine* 14, 179–184.

Paulsen, B.S., Olafsdóttir, E.S. and Ingólfsdóttir, K. (2002). Chromatography and electrophoresis in separation and characterization of polysaccharides from lichens. *Journal of Chromatography* 967, 163–171

Periera, E. C, Nascimento, S. C. and Lima, R. C. (1994). Analysis of *Usnea fasciata* crude extracts with antineoplastic activity. *The Tokai journal of experimental and clinical medicine* 19 (1-2), 47-52.

Prieto, A., Leal, J.A., Bernabe, M. and Hawksworth, D.L. (2008). A polysaccharide from *Lichina pygmaea* and *L. confinis* supports the recognition of Lichinomycetes. *Mycological Research* 112, 381 – 388.

Reeb, V., Lutzoni, F. and Roux, C., (2004), Contribution of RPB2 to multilocus phylogenetic studies of the Euascomycetes (Pezizomycotina, Fungi) with special emphasis on the lichen forming acaromsporaceae and evolution of polyspory. *Molecular Phylogenetics and Evolution* 32, 1036-1060.

Ruthes, A.C., Komura, D.L., Carbonero, E.R., Sassaki, G.L., Gorin, P.A.J. and Iacomini, M. (2010). Structural characterization of the uncommon polysaccharides obtained from *Peltigera canina* photobiont *Nostoc muscorum*. *Carbohydrate Polymers* 81, 29–34.

Shibata, S. (1973). Polysaccharides of lichens. *Journal of the National Science Council of Sri Lanka* 1, 183-188.

Sipman, H.J.M. and Aptroot, A. (2001). Where are the missing lichens? *Mycological Research* 105, 1433–1439.

Stuelp-Campelo, P.A., Oliveira, M.B.M., Carneiro-Leão, A.M.A., Carbonero, E.R., Gorin, P.A.J. and Iacomini, M. (2002). Effect of a soluble α-D-glucan from the lichenized fungus *Ramalina celastri* on macrophage activity. *Int J Immunopharmacol* 2, 691-698.

Takahashi, K., Takeda, T. and Shibata, S. (1979). Polysaccharides of lichen symbionts. *Chem. Pharm. Bull.* 27, 238–241.

Teixeira, A.Z.A., Iacomini, M. and Gorin, P.A.J. (1995). Chemotypes of mannose containing polysaccharides of lichen mycobionts: a possible aid in classification and identification. *Carbohydr. Res.* 266, 309–314.

Wedin, M., Wiklund, E., Crewe, A., Döring, H., Ekman, S., Nyberg, Å., Schmitt, I. and Lumbsch, H.T. (2005). Phylogenetic relationships of Lecanoromycetes (Ascomycota) as revealed by analyses of mtSSU and nLSU rDNA sequence data. *Mycological Research* 109, 159–172.

Wolseley, P.A. and Aguirre-Hudson, B. (1994). Key to lichen genera in Thailand with special reference to epiphytic taxa, part I, macrolichens. *Natural History Bulletin*. Siam. Soc. 43, 303-335

Woranovicz-Barreira, S.M., Gorin, P.A.J., Sassaki, P.L., Marcelli, M.P. and Iacomini,M.(1999). Galactomannoglucans of lichenized fungi of *Cladonia* spp.: significance as chemotypes. *FEMS Microbiology Letters* 181(2), 313–317.

Yanaki, T., Itoh, W. and Tabata, K. (1986). Correlation between the antitumor activity of Schizophyllan and its triple helix. *Agric Biol Chem* 50, 2415–2416.

Yoshimura, I. (2004). Mysterious Lichens - An Introduction to their taxonomy, morphology, physiology, and chemistry. Proceedings of the 1st KoLRI International Symposium of Lichenology.

Section 2

Physical and Chemical Characteristics of Polysaccharides

Chapter 9

The Molecular Structure and Conformational Dynamics of Chitosan Polymers: An Integrated Perspective from Experiments and Computational Simulations

Richard A. Cunha, Thereza A. Soares, Victor H. Rusu,
Frederico J.S. Pontes, Eduardo F. Franca and Roberto D. Lins

Additional information is available at the end of the chapter

http://dx.doi.org/10.5772/51803

1. Introduction

Chitosan is a semi-synthetic linear copolymer composed of a variable number of β-(1-4) linked units of 2-acetamide-2-deoxy-β-d-glucopyranose (GlcNAc) and 2-amino-2-deoxy-β-d-glycopyranose (GlcN) [1]. The two monomers differ with respect to the C2-substituent in the sugar ring, which is either an amino or acetamide group (Figure 1). Chitosan is obtained via the alkaline deacetylation of chitin. However, the deacylation reaction hardly proceeds completely in a normal heterogeneous reaction, leading to a random distribution of GlcNAc and GlcN residues in the chitosan polymer [2, 3]. The degree of acetylation of a polymer is a measure of the average number of GlcNAc per 100 chitosan monomers in percentile unit. The degree of acetylation governs important physical-chemical properties of the chitosan polymer such as solubility and conformation, being critical for the effectiveness of various technological applications [4-6]. The threshold for the conversion of chitin into chitosan depends on the solubility of the oligosaccharide in a slightly acid solution (0,1 mol/L of acetic acid). Conventionally, chitin polymers with a level of acetylation below 50% is considered as chitosan. The level of protonated amino groups in the glucosamine monomers dictates the solubility of chitosan, conferring the cationic nature to the polymer [7].

Chitin is the most abundant amino polysaccharide, being produced in the amount of one hundred billion tons per year in nature [8]. Its main source is the exoskeleton of crabs and shrimps, whose availability in nature makes chitin a renewable source of chitosan. In the last decades chitosan has emerged as a biomaterial with unique properties for advanced applications in green chemistry, biomedical, pharmaceutical, food [9] and agriculture [10].

INTECH
open science | open minds

© 2012 The Author(s). Licensee InTech. This chapter is distributed under the terms of the Creative Commons Attribution License http://creativecommons.org/licenses/by/3.0), which permits unrestricted use, distribution, and reproduction in any medium, provided the original work is properly cited.

The variety of applications of chitosan is determined by its chemical structure, which varies with respect to size (average molecular weight; MW), degree of acetylation (DA) and numerous chemical modifications [6, 11-15]. An increasing number of chitosan chemical modifications have been described in the literature [14, 15]. Chitosan is also a highly absorptive material used as heavy-metal chelators in water [16, 17]. The chelation of metals occurs via electrostatic interactions with chitosan reactive groups (hydroxyl, acetamide and primary amino groups). These functional groups are also responsible for properties such as high hydrophilicity, reactivity, and structural flexibility that make chitosan soluble in near-neutral acid solution [17]. The soluble adsorptive properties of chitosan enable its use for removal of pesticide and dyes from water, for adsorption of proteins, as flocculant agent, and even as a catalyst support for biodiesel production [18-24]. Besides being an abundant renewable resource, chitosan stands out due to some unique properties such as exceptional biocompatibility, biodegradability, non-toxicity and the easiness of production of chemically modified forms [14, 25, 26]. Furthermore, chitosan exhibits antimicrobial and antifungal activity [27-30]. These singular properties makes chitosan welll-suited for a wide range of biomedical applications such as drug delivery, platform for neural stem cell growth, tissue engineering (bone, cartilage, nerve, skin), immunoprophylaxis, gene therapy, wound healing and treatment of infections [31-44].

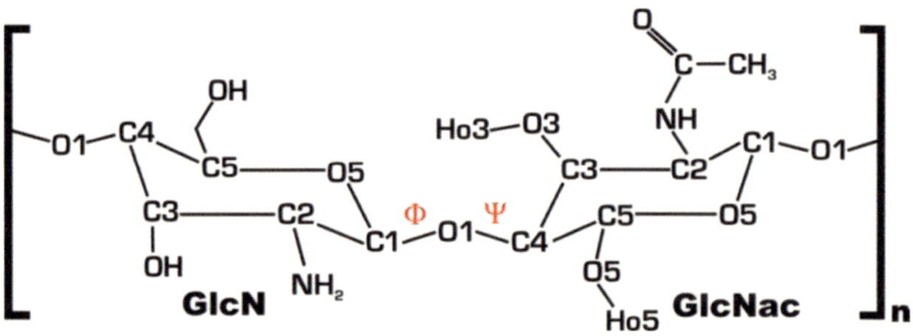

Figure 1. Schematic representation of a disaccharide formed by 2-amino-2-deoxy-*β*-d-glycopyranose (GlcN) bound to 2-acetamide-2-deoxy-*β*-d-glucopyranose (GlcNac). Φ and Ψ dihedral angles are shown in the scheme in red and here are represented by atoms: O5-C1-O1-C4′ and C1-O1-C4′-C5′, respectively. GlcN and GlcNac are the two units forming chitosan. Solubility is typically achieved when GlcN > 50%.

This short review will cover the structural dynamics of chitosan from a microscopic perspective, focusing on the interplay between its conformational variability and macroscopic properties such as solubility and aggregation. It is not the goal of the review to provide a detailed summary of the extensive literature on carbohydrate structure and its characterization. Excellent reviews on the subject can be found in the literature [45-47]. The text is organized in four main sections. First, we present an overview of chemical interactions between chitosan and biological materials, in particular lipid bilayers. In what follows, we describe the advantages and limitations of experimental and computational techniques used for the structural characterization of oligosaccharides, emphasizing the necessity of combining different approaches in order to obtain high-resolution structural data on chitosan. In this section we also introduce the theoretical principles underlying molecular dynamics (MD) simulations, which has been widely used to study the structural dynamics of carbohydrates in solution. The third section, we review the types of secondary structure observed for chitin and chitosan in the crystalline state. In the final section, we offer a detailed account of the structure and conformational dynamics of chitosan in solution as unveiled by computational simulations carried out in our group.

2. The molecular interactions underlying chitosan bioactivity

Chitosan is a very promising material with wide range of biomedical applications. This oligosaccharide incorporates highly sought properties for biomedical applications (biocompatibility, biodegradability and bioresorbability) with the easy processing into gels, membranes, nanofibers, beads, microparticles, nanoparticles, scaffolds and sponges forms [48-52]. Yet, chitosan has a flexible, hydrophilic helical structure with reactive amine groups, which offers a multitude of possible inter- and intra-molecular interactions. A detailed understanding of the effects of different materials and environmental conditions on such interactions can enable the design of novel chitosan-based technologies.

Chitosan amino groups are the major players in metal chelating processes. However, it has been previously shown that electrostatic interactions involving the protonated amino groups *per se* in low pH are not sufficient to explain the biological behavior of chitosan in presence of biological membranes [53]. A comparative study between chitosan and a fully cationic polymer has shown that hydrophobic interactions play an important role in the polysaccharide action. A study on the effect of the pH and the molecular weight of chitosan in multilamellar vesicles of dipalmitoylphosphatidylcholine (DPPC) has shown that increasing the biopolymer molecular weight (213 kDa) and decreasing of pH can lead to disruption of the membrane [54]. In contrast, another study has shown that the interactions between chitosan and DPPC lipids in liposomes led to an increase in thermodynamic stability of the composite. It has been proposed that this stabilization results from a shielding mechanism based on the electrostatic interactions between the chitosan chains and phospholipids polar heads [55]. In contrast, anti-fungal and bactericidal activity of chitosan has been attributed to the ability of chitosan to disrupt the inner and outer membranes of cells [56, 57]. The contribution of electrostatic, hydrophobic and hydrogen bond interactions between chitosan and three different lipids have been evaluated using Langmuir films to mimic the interaction of chitosan and bacterial membranes [29]. It has been shown that chitosan had a negligible effect on DPPC monolayers but it distinctly affected dipalmitoylphosphatidylglycerol (DPPG) and cholesterol monolayers [29]. The effect on DPPG was found to decrease with increasing pH, ascribed to the charge-mediating action of chitosan, whereas the pH did not affect the cholesterol monolayers where interactions occurs mainly via hydrogen bonding. A recent study has suggested that sensitivity of fungi to chitosan depends of the membrane fluidity and dynamics [58]. The same group has also suggested in a previous work that chitosan kills fungal cells by an unknown mechanism that does not involve endocytosis [59]. Although chitosan has been used in a variety of biologically relevant applications involving interactions with lipids, proteins, inorganic and organic compounds, a microscopic picture of these interactions remains lacking. The conformational flexibility of chitosan has hampered the acquisition of high-resolution structural data through of X-ray crystallography and NMR spectroscopy [46, 47]. However, current molecular modeling techniques can be used to bridge the gap of experimental resolution, thus providing complementary information to measurements.

3. Structural characterization of polysacchrides

Several analytical techniques have been applied to the characterization of a variety of oligosaccharide properties [60-63]. Among them, the molecular geometry is one of the most important properties that experimental data can provide on carbohydrates. Its characterization is critical for the understanding of the function and recognition mechanisms of carbohydrates in living organisms. However, sugars are inherently flexible, undergoing conformational changes in response to chemical modifications, complexation to biomolecules, changes in the pH, ionic strength and solvent type [2]. In solution, oligosaccharides tend to adopt a coiled conformation, which fluctuates between local and overall conformations, adopting an enormous variety of spatial arrangements around glycosidic linkages.

As a first approach to the complexity of the conformational flexibility of polysaccharides, let us assume that its monosaccharide units have a rigid ring structure. Thus, the determination of the conformation of oligosaccharide structures is reduced to the characterization of the glycosidic linkages between rigid monosaccharide monomers, *i.e.*, the description of two or three torsion angles for each glycosidic linkage would suffice to characterize the conformations of the oligosaccharide chain. However, the description of these torsion angles faces two major issues [46, 64]. First, the motions associated to each glycosidic linkage range across large-scale vibrations of a single well-defined conformation to transitions between several different conformations. Therefore, the accurate characterization of a given glycosidic linkage requires information on the number of conformations adopted, the time spent in each conformation and the flexibility of each conformation [46]. An additional difficulty is given by the fact that the conformational transitions in different linkages of an oligosaccharide chain are coupled. Second, the two experimental techniques most effective in providing atomic level structural information on biomolecules, namely X-ray diffraction and nuclear magnetic resonance (NMR) spectroscopy, have appreciable limitations when applied to oligosaccharides. In this section, we briefly overview the strengths and limitations of the most representative experimental techniques used for structural characterization of chitin and chitosan.

Mass spectrometry (MS) can be used to determine the total mass of a carbohydrate or differentiate distinct oligosaccharide as function of the respective weight [65]. Although MS cannot offer a detailed description of the oligosaccharide structure, it can identify the location of branch points [66-69]. Further fragmentation will not result in additional information, because the fragments can be alike. Despite the inadequacy of MS to generate information on the molecular geometry of oligosaccharides, MS can be coupled with separation techniques such as high performance liquid chromatography (HPLC) to differentiate between solutions containing different types of carbohydrates [65, 70-73].

The techniques of X-ray diffraction and NMR spectroscopy determine time and spatial averages of molecular properties in atomic coordinates measured from an ensemble of molecules corresponding the Avogadro's number [74-76]. Yet, the two techniques differ significantly with respect to the spatial distribution of molecules and time scale accessible to each one [74-76]. X-ray data represents an average over molecules arranged in a periodic crystal lattice over the second to hour timescale whereas NMR data represents an average over semi-randomly oriented molecules in solution over the nanosecond to second timescale. Despite the robustness of single X-ray crystallography in protein structure determination, the technique is not easily applicable to oligosaccharides. The difficulty to obtain highly crystalline samples for oligosaccharides imposes limits on the quality of the diffraction pattern. X-ray diffraction of carbohydrate structures larger than tetramers are rare and only seen when co-crystallized with proteins [18]. Due to the difficulty to obtain single crystals from oligosaccharides, oriented fibers have often been used for X-ray diffraction studies. Fibers exhibit helical symmetry rather than the three-dimensional symmetry seen in single crystals. Analysis of the diffraction pattern from orientated fibers allows deducing the helical symmetry of the molecule, in some cases also the structure. This is possible by constructing a model of the fiber and calculating the expected diffraction pattern. By comparing the calculated and observed diffraction patterns one eventually arrives at a better model. However, oligosaccharides in crystalline fibers can be affected by intra-molecular and crystal lattice packing, which may lock the structure in a conformation not representative of the conformational ensemble in solution.

On the other hand, NMR spectroscopy can provide information on the covalent structure and the complex conformational equilibria of oligosaccharides in solution [46, 47, 64, 77-79]. Moreover, it is the only available technique that can determine both the anomericities and linkages of a novel glycan. Another practical advantage of NMR spectroscopy is the possibility of measuring relative dilute solutions of oligosaccharides. Sample requirement amounts to as little as 1 mg, which is within the limits of enzyme-assisted synthesis [47]. NMR spectroscopy is probably the most used experimental tool to characterize the atomic structure of carbohydrates. For this reason, it has been the subject of numerous reviews [46, 47, 64, 78]. Biomolecular NMR spectroscopy has progressed appreciably in the last decades. Developments in the instrumentation, pulse sequences and spectral interpretation associated to molecular modeling techniques led to great advances in the determination of primary and three-dimensional structures of biomolecules in solution [17]. Such progress has been more noticeable in the structural characterization of proteins and nucleic acids. Notwithstanding, carbohydrates are not too far behind despite difficulties in proton assignment of each atom due to the structural similarity of the monosaccharide units [19-29].

The NMR spectroscopy data reflect primarily short-range through-bond interactions (J-coupling constants), short-range through-space interactions via the nuclear Overhauser effect (NOEs) or local perturbations to electronic shielding (chemical shifts). NOE is the main source of conformational information on carbohydrates. The strength of the NOE signal between two nuclei is proportional to the inverse sixth power of the distance between the atoms. However, a given distance between two protons is often consistent with a range of distinct conformations that will be represented by a set of NOE-derived distance constraints. The larger the number of available NOE constraints, the more consistent a single structure will be with this collection of spatial constraints. Nonetheless, the use of NOE constraints in structure determination of oligosaccharides is beset by a few issues (reviewed in [46]). For instance, the number of NOE constraints may not suffice to unambiguously define a conformation, particularly around the glycosidic linkage. In addition, NOE is sensitive only to short-distance nuclei (< 5-6 Å). For that reason, NOE constraints are obtained only between nuclei within a monosaccharide unit or across a glycosidic linkage. Due to the lack of long-range information, the accurate determination of the whole structure of oligosaccharides depends on combining the local conformations of the individual linkages. Such procedure leads to the cumulative addition of any uncertainties or errors in the local structure and its dissemination to the whole oligosaccharide structure (except if in the presence of sufficient sequential NOEs). A last issue concerns space-time ensemble averaging effects. Accordingly, different NMR parameters are averaged over time scales ranging from 50 ms to 1 s. In the case of oligosaccharides transitioning between several conformations, NOE constraints will represent average values that cannot be easily decomposed into each of the single conformation contributing to the average constraints. The conformational uncertainty ensuing from these issues can be minimized to some degree by the use of additional conformational constraints such as scalar coupling constants (J values), which are simple linear averages over the ensemble of individual conformers. ^{1}H-^{1}H J-coupling constants can be used to define ring conformations and dihedral angles. This information can also be obtained via NMR residual dipolar couplings measurements. Non-anomeric protons can be assigned through 2D homonuclear correlation COSY and/or TOCSY experiments. NOESY experiments can be used to provide monosaccharide sequence information due to the absence of coupling over the glycosidic linkage of the COSY and TOCSY spectra. ^{1}H-^{13}C HSQC and HMQC experiments provide important correlations in the determination of repeating units of polysaccharides [2]. Yet, the identification of distinct carbohydrate conformations requires combining complementary techniques. These techniques vary from other experimental methods to atomistic molecular dynamics simulations [46, 64, 80-82].

Classical molecular dynamics (MD) simulations can be used to complement incomplete experimental data and to provide detailed conformational distributions in time and space that experimental measurements can only obtain as averages [80, 82, 83]. It can also be used to predict properties under environmental conditions that may not be accessible to experimental measurements. As such, MD is an indispensable tool to interpret experimental data. However, the accuracy of MD simulations is intrinsically dependent on the quality of the empirical potential energy functions and the force-field parameters used. Robust force fields for MD simulations of carbohydrate-based systems are available. Some of the most used are CHARMM[84-87], GLYCAM/AMBER[88, 89], GROMOS[90] and OPLS-AA[91]. These force fields offer a realistic description of the structural dynamics of oligosaccharides within the limitations of the experimental data available, making MD simulations a reliable procedure for the prediction of molecular interactions [92-95]. Therefore, the importance of accurate measurements of the spatial arrangements of carbohydrates can never be overstated as they are the principal component in the development of physical chemical parameters (force fields) governing molecular simulations. The availability of high-quality experimental data is critical for biomolecular modeling [80, 81]. Classical force fields used for simulations of biomolecules are built from quantum chemistry calculations and/or experimental measurements. Without experimental measurements, the development of classical force fields would be extremely difficult as the expensive costs of quantum-chemical theoretical models limit their use in force-field construction [80, 90, 96-100]. In addition, quantum-mechanical data is not an ideal validation target as it only yields gas-phase quantities. Model validation and comparisons of biomolecular simulations are often best done against condensed-phase experimental data [82, 98, 101]. The availability of structural data on carbohydrates has made possible the creation of several databases like the SUGABASE, CarbBank, EUROCarbDB, Glycoconjugate DB, GLYCOSCIENCES.de, GlycoSuiteDB, JCGGDB, KEGG-Glycan, CFG-Glycan Database and GlycoBase. These databases represent a convenient tool for the building of molecular models as well as for comparison of atomistic simulations of carbohydrates against experimental data.

3.1. Theorical foundations of molecular dynamics simulations

The MD method has its foundations in the laws of classical mechanics [102, 103]. It allows the simulation of the time-dependent behavior of molecular systems according to Newton's laws of motion. The atom nuclei are treated classically as spheres connected to each other through a set of springs emulating chemical bonds. The forces acting on each atom, necessary to simulate their motion, are derived from a set of force field parameters, and the set of coordinates and velocities that mapped during the whole process comprise the phase space. In a simulation, the force *F* on each atom is expressed as a function of time, and is equal to the negative gradient of the potential energy *V* with respect to the position r_i of each atom, in a distinct expression of the more common form of the equation $F = ma$:

$$-\frac{dV}{dr_i} = m\frac{d^2 r_i}{dt^2}$$

The MD method integrates iteratively and numerically the classical equations of motion for every atom in the system at time increments (*Δt – time step*) defined by the user. A number of algorithms exist for this purpose and are implemented in different computational codes [104]. There are several algorithms available for performing the numerical integration of the equations of motion. The Verlet-type algorithms (Verlet, velocity-Verlet and leap-frog) are widely used because it requires a minimum amount of computer memory and CPU time [105, 106]. The velocity Verlet, for instance, uses positions, velocities and accelerations at the current time step, which gives a more accurate integration than the original Verlet algorithm. Other algorithms, as the Beeman gives better energy conservation at the expense of computer memory and CPU time [107]. The Gear predictor-corrector algorithm predicts the next set of positions and accelerations, and then compares the accelerations to the predicted ones to compute a correction for the step [108]. Each step can thus be refined iteratively. Predictor-corrector algorithms give an accurate integration but are seldom used due to their large computational cost. In the classical Verlet algorithm, the positions in the next time step (Δt) are calculated from a given set of particles with coordinates r_i using a Taylor expansion:

$$r_{i+1} = r_i + \frac{\partial r}{\partial t}(\Delta t) + \frac{1}{2}\frac{\partial^2 r}{\partial t^2}(\Delta t)^2 + \frac{1}{6}\frac{\partial^3 r}{\partial t^3}(\Delta t)^3 + ...$$

$$r_{i+1} = r_i + v_i(\Delta t) + \frac{1}{2}a_i(\Delta t)^2 + \frac{1}{6}b_i(\Delta t)^3 + ...$$

where the last equation links the spatial coordinates with the velocities v_i (the first derivative of the positions in respect to time (dr_i/dt) at time t_i. The accelerations a_i (the second derivatives (d^2r/dt^2) at time t_i and so on. If the goal is to determine the positions for a small time step (Δt) earlier, the equation becomes:

$$r_{i-1} = r_i - v_i(\Delta t) + \frac{1}{2}a_i(\Delta t)^2 - \frac{1}{6}b_i(\Delta t)^3 + ...$$

Adding the last two equations makes it is possible to find a new equation that predicts the position at a chosen time step using the current and previous atom positions and current acceleration. The latter can be calculated from the force or potential.

$$r_{i+1} = (2r_i - r_{i-1}) + a_i(\Delta t)^2 + ...$$

$$a_i = \frac{F_i}{m_i} = -\frac{1}{m_i}\frac{dV}{dr_i}$$

At the beginning of the simulation, when the previous positions are not available, this quantity can be estimated from the following approximation:

$$r_{-1} = r_0 - v_0 \Delta t$$

The time increment in MD simulation should be sufficiently small that errors in the integration equations keep small, preserving the conservation of the energy. Normally Δt is on the order of femtoseconds (10^{-15} s). This time order is one order of magnitude smaller than the fastest molecular process. Furthermore, because the forces F_i should be recalculated for every step, MD is a computation intensive task. Currently achieved timescale of MD simulations is on the order of multi-nanoseconds to a few microseconds. This time is shorter than many relevant phenomena, for this reason MD results should be analyzed under the point of view of the sample of the phase space close to the starting condition, in spite of this capacity of sampling different configurations. One strategy is to increase the time step value and to allow longer simulation times is to freeze the bond lengths related to hydrogen atoms. The fastest processes in molecules are stretching vibrations, especially involving hydrogen atoms. As these degrees of freedom have little influence on many properties, some algorithms were developed to keep frozen these chemical bonds as the SHAKE [109], RATTLE [110] and LINCS [111]. Alternative MD-based methodologies have also been recently developed aiming to partly overcome this limitation. The so-called enhanced sampling techniques artificially drive a given system according a set of pre-defined rules that result in a larger sampling of the configurational phase space within the same simulation time (e.g., simulated annealing, replica-exchange, parallel-tempering, local elevation search, metadynamics) [112, 113].

An accurate description of the aqueous medium that shapes the structure, dynamics and function of biological molecules is essential for the realistic reproduction of its kinetics and thermodynamics properties. It is known that simulations of a small arrangement of atoms do not reproduce satisfactorily the properties of bulk liquids due to surface effects suffered by a large fraction of the molecules. The obvious solution for this problem, which would be to increase the number of solvent molecules, can lead to issues in the evaluation of the force between the atoms. An alternative solution to treat explicit solvent molecules in MD simulations is the use of periodic boundary conditions [104]. In this approach the simulation box is replicated throughout the space to form an infinite lattice, where the number of molecules entering or leaving the simulation box is kept constant during the simulation and as a consequence, surface effects are canceled. There are currently numerous water models used in MD simulations. Some of the models currently implemented in major classical MD softwares are the SPC model[114] and the TIP3P, TIP4P, and TIP5P models [115, 116]. These models were parameterized assuming that a cut-off is applied to nonbonded interactions and treat water as a rigid molecule. Although bond stretching and bond-angle bending[117], or polarization effects and many-body interactions [118], have been introduced into water models, they involve a large increase of computational expense, which has limited their use as widely as the SPC or TIP models. The water models are usually parameterized at a single temperature (ca. 298 K) and therefore may not capture correctly the temperature dependence of properties such as the solvent density or diffusion coefficients [119].

The basic principle underlying the MD theory is that if one allows the system to evolve in time indefinitely, it will eventually pass through all possible states. Thus, MD simulations should cover time scales sufficiently long to generate enough representative conformations to satisfy this principle. In other words, the simulations must sample a sufficient amount of the phase space corresponding to the system in consideration. In that case, experimentally relevant information concerning structural, dynamic and thermodynamic properties can be calculated using a feasible amount of computational resources. The connection between theoretical results and experiments is made through the use of the Ergodic hypothesis. This fundamental axiom of statistical mechanics states that the average obtained by following a small number of particles over a long time is equivalent to averaging over a large number of particles for a short time. Exploring the limit of a sufficient large time scale, the Ergodic hypothesis implies that the time average over a single particle is equivalent to the average over a large number of particles at any given time. This theoretical justification in the scope of a MD simulation validates the calculation of thermodynamic averages for molecular systems if finite molecular dynamics trajectories are "long enough" in the ergodic sense.

$$\lim_{\tau \to \infty} \langle A(r,p) \rangle_{\tau} = \langle A(r,p) \rangle_{Z}$$

4. Chitosan molecular structure

In the solid state, chitosan is characterized by an ordered fibrillar structure with a high degree of crystallinity, and polymorphism [120, 121]. X-ray measurements of the chitosan polymer have shown an extended two-fold helix in a zigzag structure [122, 123]. The crystal packing is mainly formed by chitosan chains arranged in an antiparallel fashion (Figure 2A), and similar to the anhydrous form of the α-chitin structure. The structure of the α and β forms differ only in the arrangement of the piles of chains, which is alternately antiparallel in α-chitin and all parallel in β-chitin [92, 124]. The crystallographic structure of chitin and chitosan have also revealed that although both biopolymers exhibit a hydrated and anhydrous forms, chitin occurs exclusively in the conventional extended 2-fold helical conformation (Figure 2A) [123, 125-128]. The presence of free amino groups in the structure of chitosan gives rise to different types of helical conformations in acid (Figure 2) [128]. These structures can be classified in four main types: type I (anhydrous), type II (hydrated), type IIa (hydrated) and type III (anhydrous), which adopt a helical conformations in a two-fold helix, relaxed two-fold helix, a 4/1 helix and a five-fold helix, respectively (Figure 2) [128, 129]. The diversity of chitosan structural types depend on the experimental conditions (kind and concentration of acid, temperature and salt preparation) used for the conversion of chitin into chitosan [128]. The helical structure propensities can be determined according to the repeating unit and helical symmetry as observed in chitosan crystal structures [121, 123, 128, 130, 131]. A less common motif, classified as 3-fold, has also been identified (Figure 2B).

The type I salts are the anhydrous form of the unreacted chitosan crystal. The polysaccharide chains in these crystals have a 2/1 helical symmetry with a repeating pattern of 1.0 nm. This conformation is similar to that of chitin, and characterizes the two-fold helix (Figure 2A) [92, 132, 133]. Type II chitosan exhibits a hydrated crystal with a fiber repeat of about 4.08 nm long and an asymmetric repeating units consisting of tetrasaccharides. In this type, the helical conformation is composed of eight glucosamine residues with repeating units related by a 2/1 helical symmetry. This pattern suggests a two-fold helix even though the corresponding asymmetric unit is rather distinct from that of type I where the asymmetric unit has only one glucosamine residue. The main difference between the type I and type II conformations is that the latter is almost four times longer than chitosan, and originated the designation of relaxed two-fold helix (Figure 2E) [92, 134-136]. A type II salt variant, called Type IIa, has a similar fiber repeat (4.05 nm), but with an asymmetric unit consisting of a glucosamine dimer in a 4/1 helical symmetry. This right-handed helix comprised of four asymmetric subunits is classified as 4/1-helix conformation, being also called four-fold helix (Figure 2C) [121, 129]. The most recently discovered type III form has a chain repeat of 2.55 nm, a 5/3 helical symmetry, and an asymmetric unit of a single glucosamine residue. Type III helical conformation is classified as five-fold helix, and displays a less symmetric helicoidal conformation (Figure 2D) [129, 137].

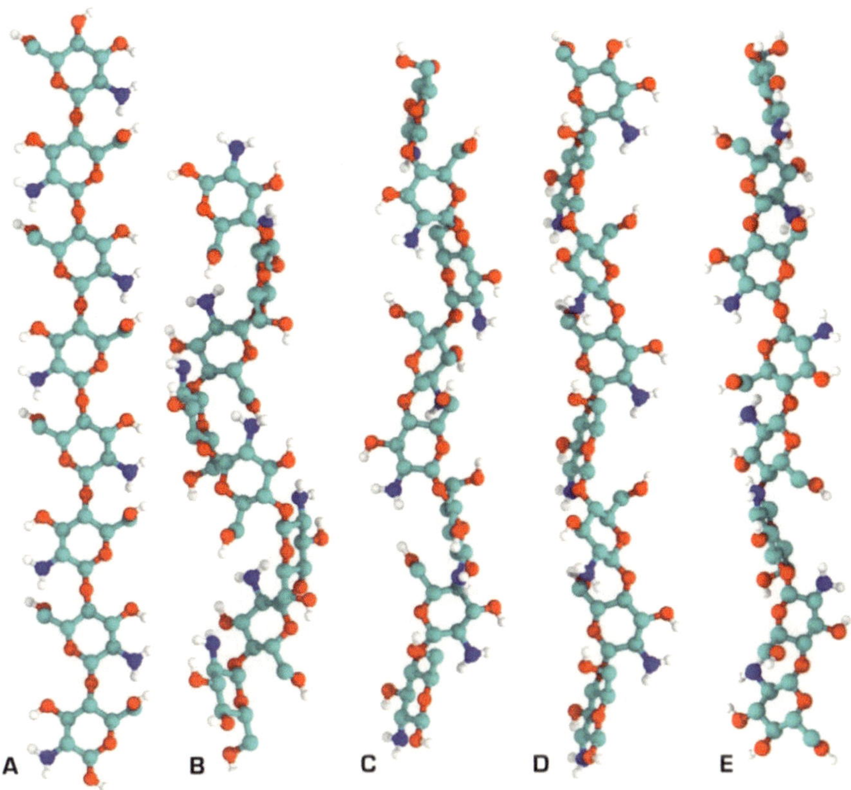

Figure 2. Chitosan secondary structures as determined by solid X-ray crystallography. A) two-fold; B) 3-fold; C) 4-fold; D) 5-fold; and, E) two-relaxed-fold.

In solid state, the two-fold helix pattern is stabilized by O3-HO3••• O5′ intra-chain hydrogen bonds across the glycosidic linkages [120]. In order to verify these helical properties in aqueous solution, MD simulations were carried out for chitin and chitosan [92, 93]. These simulations have shown that chitin chains assume exclusively a two-fold helix conformation which indeed is stabilized by the O3-HO3••• O5′ intra-chain hydrogen bonds [92]. However, chitosan chains can adopt several distinct conformations, including all the helical conformation observed in solid state. Helical preferences and conformational interchangeability were shown to be affected by the level of acetylation of the chitosan chains [92].

5. Structural dynamics of chitin and chitosan biopolymers

Structural characterization of chitin and chitosan conformations and their underlying interactions (intra- or inter-chain) has been largely determined by X-ray crystallography. The high flexibility of these oligosaccharides in solution has limited the acquisition of high-resolution structural data almost exclusively to X-ray diffraction of solid states (fiber, powder and tablet) (see section 3). Although NMR techniques are more suitable for structural characterization in solution, the flexibility of oligosaccharides makes NMR-derived geometrical constraints scant and limits the application of NMR spectroscopy to the determination of chitosan tridimensional structure [138]. Experimental data describing dynamic processes such as solvation, particle formation and aggregation remain limited to a macroscopic view, which is based on the measurement of chain stiffness and intrinsic viscosity [139]. Transmission electron microscopy has also been used as a complementary technique. Combining the latter with uranyl staining, electrostatic interactions involving chitosan protonated amino groups were attributed a major role on chitin and chitosan agglomeration in solution [140]. Therefore, the role of intra- and inter-chain hydrogen bonds, ionic strength and temperature on the structural dynamics of chitosan cannot be addressed exclusively by the means of experimental techniques [141]. Towards this end, MD simulations can be used to obtain information on the time-evolution of carbohydrate conformations at the atomic level and under varied environmental conditions that can be complementary to experimental measurements [92-95].

Chitosan conformational diversity influences its solubility/physical state (soluble, gel, aggregate), porosity, particle size and shape (fiber, nanoparticle, hollow fiber), ability to chelate metal ions and organic compounds, biodegradability and consequently its biological activity. The transition between these distinct conformational states is modulated by the percentage and distribution of acetyl groups. The level of chitosan acetylation and the distribution of N-acetyl groups along the chain have been shown to influence properties such as solubility [142, 143], biodegradability [144] and apparent pK_a values [145, 146]. Therefore, the percentage and distribution of acetyl groups are key parameters for determining if chitosan can effectively interact with biological systems [147]. The degree of acetylation can be experimentally determined by infra-red spectroscopy [148, 149], enzymatic reaction [150], ultra-violet spectroscopy [151], ^{1}H liquid-state NMR [152], and solid-state ^{13}C NMR [63, 153]. However, the interplay between chitosan acetylation and conformational transitions in solution cannot be characterized at high-resolution by experimental techniques. In these cases, atomistic MD simulation is a more suitable approach.

MD simulations in explicit solvent have been carried out for chitosan single chains and nanoparticle aggregates with varied percentage and distribution of acetyl groups [92, 93]. Four degrees of acetylation were considered: 0% (fully deacetylated chains), 40% (60% of the sites having a N-acyl group uniformily distributed), 40%-block (60% of the sites having a N-acyl group in two spatially located well-defined regions of the particle), 60% (40% of N-acyl uniformly distributed), 60%-block (40% of N-acyl groups spatially located in two well-defined regions of the particle), and 100% (fully N-acetylated nanoparticle), i.e., a chitin nanoparticle. Snapshots of molecular dynamics simulations after 40 ns for a chithin (100%) and fully deacetylated chitosan nanoparticles (0%) are shown in Figure 3. Both simulations started from aggregate crystal-like particles. It can be seen that chitin remain insoluble (in an aggregate form, Figure 3A), while chitosan chains separate apart one from another until each chain become fully hydrated (Figure 3B). Water molecules are not display in Figure 3 for clarity. These simulations have also shown a strong dependence of chitosan conformation and solubility with pH and degree of acetylation. An increase in the level of acetylation was shown to cause a progressive loss of flexibility and conformational interchangeability (Figure 4). Thus, acetylation promotes a shift from more flexible structural motifs such as 5-fold and relaxed 2-fold towards a 2-fold conformation (Figure 4). It was also shown that the spatial location of the N-acetyl groups influences significantly chitosan conformational preferences, and therefore its solubility (Figure 4). Analyses of the MD trajectories have also shown that a high degree of acetylation and/or an increase in pH leads to a 3-fold increase of the lifetime of O3-HO3•••O5′ intra-chain hydrogen bonds across the glycosidic linkages. The increase in the lifetime of this hydrogen bond was associated to a decrease in chitosan solubility. Chitosan with a high degree of acetylation favored the 2-fold conformation, but higher pH values did not affect significantly the secondary structure pattern of this oligosaccharide. In addition, we have also addressed the influence of spatial distribution of *N*-acetyl groups along the chitosan chain on swelling and the relative solubility of chitosan nanoparticles [93]. Simulations of chitosan with a uniform and block-wise distribution of *N*-acetyl groups along chains of a nanoparticle have shown that the latter displayed lower solubility [93]. The mechanism was attributed to the fact that 2-fold crystalline-like regions are created by the block distribution of acetyl groups, which is responsible to keep a more stable aggregate than its uniformly distributed counterpart.

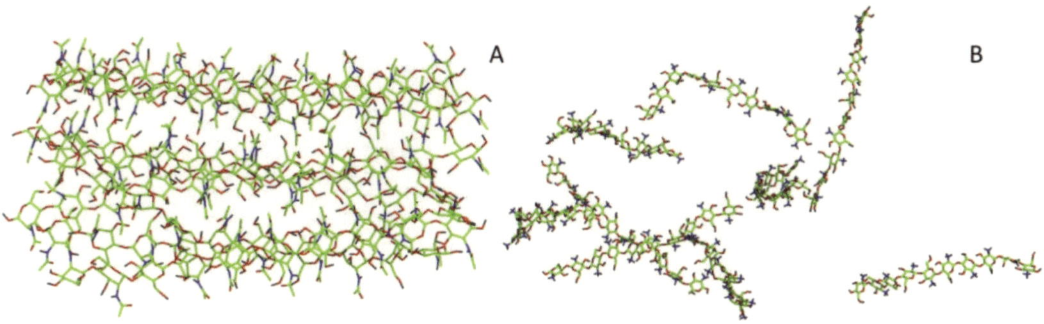

Figure 3. Molecular structure representation of nanoparticles of chitin (A) and chitosan in circumneutral pH (B) at the end of 40-ns molecular dynamics simulations in explicit water. Water molecules and hydrogen atoms have been removed for clarity. Structures are represented in stick model and atoms color coded as: green: carbon; red: oxygen, blue: nitrogen.

Analysis of the cumulative average water content around each chain in the nanoparticles illustrates the relative solubility of each system (Figure 5a). On average, there is 0.26 water molecule per monosaccharide a 0.5 nm radial distance from each chitin chain. That corresponds to one water molecule for roughly every four monosaccharides. The number of water molecules increases to one water molecule per monosaccharide within the same radial distance for fully *N*-deacetylated chitosan. As expected, nanoparticle swelling is directly proportional to its solubility. The relative swelling can be expressed as the average radius of gyration of each chain in a particle as a function of the degree of acetylation (Figure 5b). Chitosan particles with a degree of acetylation ≥ 60% did not display any significant swelling in water. At this level of acetylation, only a small increase in the relative solvation content of chitosan with a uniform distribution (ca. 0.13 water molecules per monosaccharide) than its counterpart with a block distribution was observed. This small difference in solvation did not affect the overall solubility of the particles, supporting the empirical observation of a solubility threshold around a level of 50% N-acetylation. Unexpectedly, water molecules within the insoluble chitosan particles were identified contributing for the maintenance of the regions in a 2-fold motif. The N-acetyl-glucosamine residues trapped water molecules between the chitosan chains, creating a hydrogen bond network between water molecules and the different chains without direct interaction between sheets. This finding substantiates a mechanism previously postulated by Ogawa and coworkers outlining the role of water molecules in chitin [123].

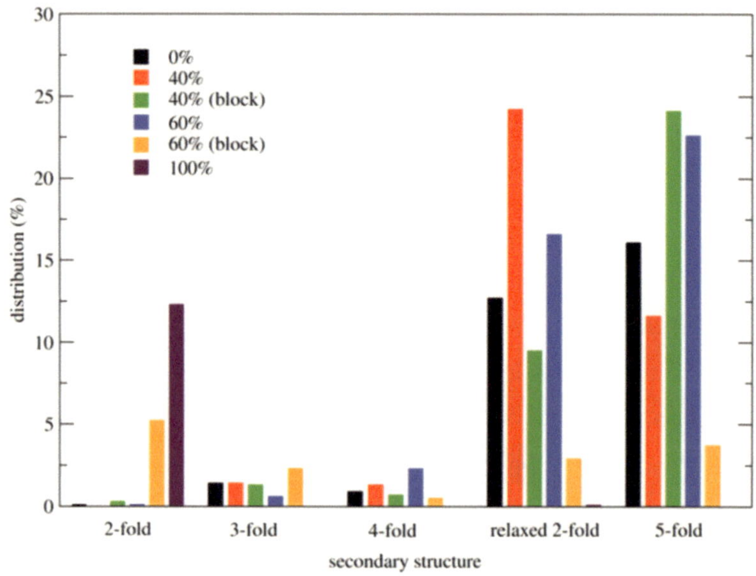

Figure 4. Secondary structure preferences for chitin and chitosan particles in water as a function of the degree of acetylation and spatial location of the N-acetyl groups. Results are the average of all chains in each nanoparticle, which are averaged over the last 5 ns of a 30-ns molecular dynamics simulation. The sum is each case does not reach 100% due to the increased flexibility and conformational diversity of these polymers in water. To be classified within a given secondary structure, as determined by solid X-ray crystallography, all dihedral angles in a chain should not present a deviation larger than 15% of the experimentally determined value. Data representation has been modified from [93].

Chitosan is a polyelectrolyte in acid medium. Its structure, physical state and conformational dynamics are greatly influenced by pH. The net charge of this cationic polyelectrolyte can be altered by its degree of acetylation [154]. Moreover, its apparent pK_a is directly related to the level of acetylation, varying from 6.1 to 7.32 units accordingly to proton concentration in the milieu [145, 155-158]. Based on these observations, it was proposed that aggregation occurs upon high levels of acetylation due to reduction of the biopolymer net charge [145], implying in a predictable behavior of chitosan chains depending on its electric charge distribution in aqueous solution [145, 146, 159]. It was also proposed that the low tendency of fully deacetylated chitosan to form aggregates is due to electrostatic repulsion among protonated amino groups. As result, chitosan electrostatic behavior was divided in three distinct patterns: i. DA < 20%, where it displays a polyelectrolyte behavior; ii. 20% < DA < 50%, where it is characterized by a counterbalance between hydrophilic and hydrophobic interactions;

and iii. DA > 50%, where associations of chitosan chains lead to the formation of stable aggregates. The results from atomistic molecular dynamics simulations in explicit water offer support to this hypothesis based on the accurate molecular description of the effect of the degree and distribution of *N*-acetyl groups on the swelling and aggregation stability of chitosan. Calculations of the hydrophobic and electrostatic contributions to the solvation free energy of the central chain in different particles as a function of acetylation are also consistent with the hypothesis (Figure 6) [129].

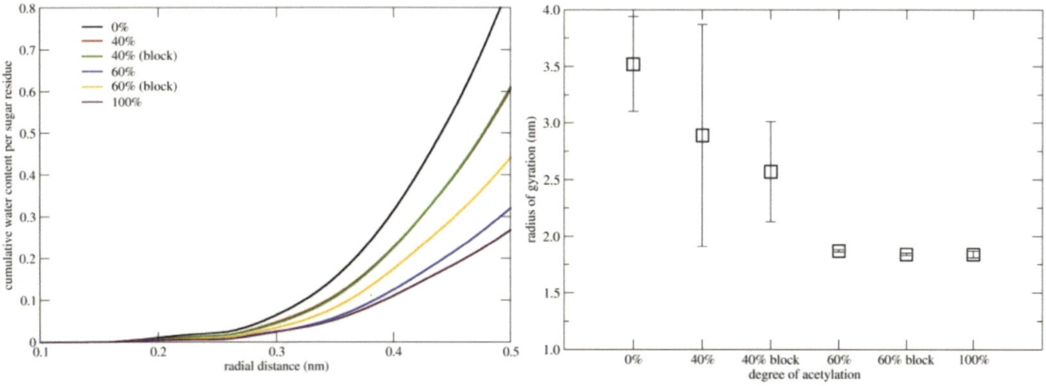

Figure 5. Relative solvation of chitin and chitosan nanoparticles as a function of the degree of acetylation and spatial location of the N-acetyl groups. A) Average coordination number of water molecules per residue as a function of distance (from 0.1 to 0.5 nm); B) Radius of gyration (and its variance, represented in bars) averaged per chain, over the last 5 ns of a 30-ns molecular dynamics simulation. Data representation has been modified from [93].

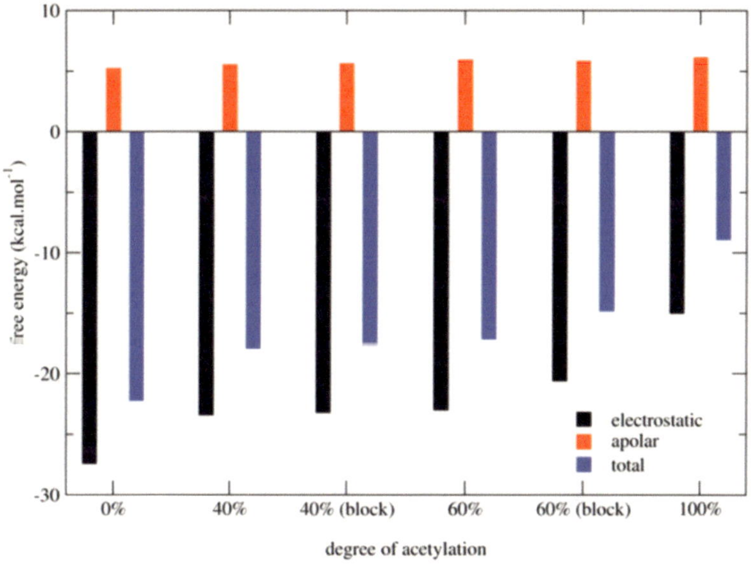

Figure 6. Average apolar and electrostatic contribution for the solvation free energy, per sugar residue, in chitin and chitosan nanoparticles as a function of the degree of acetylation and spatial location of the N-acetyl groups. Data representation has been modified from [92] and [93].

These contributions should be examined only as relative values as there are no experimental data for calibration or comparison of the calculated values. The apolar contribution remained nearly unaffected by the presence of water, while the electrostatic contribution is dominant even for insoluble chitin (100% acetylation). This finding suggests that hydrogen bond interactions, either intra-chains or between polymer chains and water molecules, play far a more important role in the solubility of chitin and chitosan than hydrophobic interactions. These results have further shown that fine tuning the electrostatic contributions in chitosan can be used to promote remodeling of its the physical state. Additional simulations have shown that the overall net charge and solubility of chitosan can be altered by changes in the pH. Comparison of the electrostatic response of a chitosan and chitin chains to pH changes shows a rather distinct surface charge profile for the two polymers. The electrostatic similarity between chitin and chitosan in basic pH aids to explain the loss of solubility of chitosan at high pH values (Figure 7). The positively charged character of chitosan chains in acid pH is shown by patches in blue (Figure 7D). On the other hand, chitin (Figure 7A) and chitosan chains in basic medium (Figure 7B) show a similar electrostatic potential at their molecular surfaces.

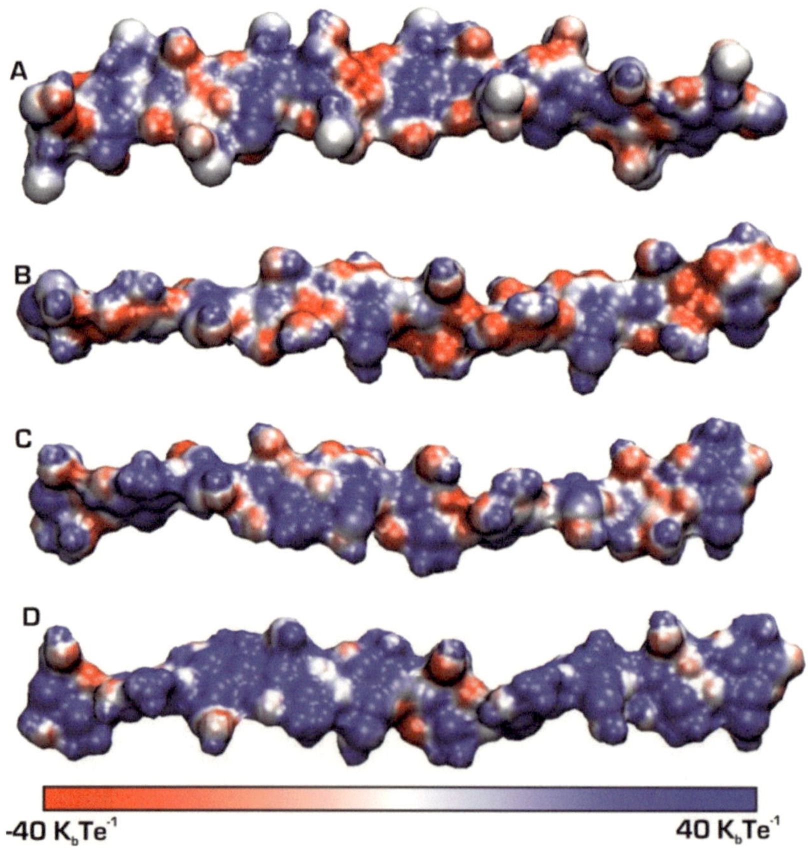

Figure 7. Electrostatic potential represented at the molecular surface of chitin (A) and chitosan (B-D) chains. Molecular structures were obtained from 20-ns molecular dynamics simulations. The different

chitosan chains were simulated at different pH values: B) basic (pH > 10); C circumneutral (pH = 6.5); and, D acid (pH < 3.5). Data representation has been modified from [92].

6. Final remarks and perspectives

Chitosan-based materials are involved in a plethora of medical, industrial and bioengineering applications such as bioremediation, radionuclide tissue decontamination and bone replacement to name a few. Due to the intrinsic flexibility and conformational variability of chitosan, the development of novel materials has been conducted mostly empirically. In this review, we have summarized the potential of using computer modeling to characterize in details the conformational behavior of chitin and chitosan. Understanding of the molecular properties of a given material allows for a more efficient/rational design. Therefore, this approach can be used to tailor these properties for specific needs. In this case, a systematic use of concerted experimental-theoretical information can provide a much clearer picture of the structural dynamics of polysaccharides and consequently can aid in such endeavor. This is still an emerging field that will benefit in the few years to come from the development of more accurate/extension of parameters for carbohydrate simulations, as well as novel models capable of better bridging the micro- and macroscopic scales.

Author details

Richard A. Cunha and Eduardo F. Franca*
Institute of Chemistry, Federal University of Uberlândia, Uberlândia, MG, Brazil

Thereza A. Soares, Victor H. Rusu, Frederico J.S. Pontes and Roberto D. Lins*
Department of Fundamental Chemistry, Federal University of Pernambuco, Recife, PE, Brazil

Acknowledgement

This work and the data herein contained was supported by the following sponsoring agencies: CNPq, FACEPE, FAPEMIG, INCT-INAMI (Brazilian National Science and Technology Institute for Integrated Markers), NanoBiotec-BR, CAPES, nBioNet and the Swedish Foundation for International Cooperation in Research and Higher Education (STINT). Part of the computational resources were provided by the Environmental Molecular Sciences Laboratory, an U.S. national scientific user facility sponsored by the U.S. Department of Energy located at the Pacific Northwest National Laboratory.

7. References

[1] Muzzarelli R A A (1973) Natural Chelating Polymers. Pergamon Press. 83.
[2] Toffey A, Samaranayake G, Frazier C E, Glasser W G (1996) Chitin derivatives. I. Kinetics of the heat-induced conversion of chitosan to chitin. Journal of Applied Polymer Science. 60: 75-85.

* Corresponding Authors

[3] Chang K L B, Tsai G, Lee J, Fu W-R (1997) Heterogeneous N-deacetylation of chitin in alkaline solution. Carbohydrate Research. 303: 327-332.

[4] Kasaai M R (2010) Determination of the degree of N-acetylation for chitin and chitosan by various NMR spectroscopy techniques: A review. Carbohydrate Polymers. 79: 801-810.

[5] Knaul J Z, Kasaai M R, Bui V T, Creber K A M (1998) Characterization of deacetylated chitosan and chitosan molecular weight review. Canadian Journal of Chemistry-Revue Canadienne De Chimie. 76: 1699-1706.

[6] Wang Q Z, Chen X G, Liu N, Wang S X, Liu C S, Meng X H, Liu C G (2006) Protonation constants of chitosan with different molecular weight and degree of deacetylation. Carbohydrate Polymers. 65: 194-201.

[7] Wang W, Du Y, Qiu Y, Wang X, Hu Y, Yang J, Cai J, Kennedy J F (2008) A new green technology for direct production of low molecular weight chitosan. Carbohydrate Polymers. 74: 127-132.

[8] Kurita K (2006) Chitin and Chitosan: Functional Biopolymers from Marine Crustaceans. Marine Biotechnology. 8: 203-226.

[9] Wei Z, Wang C, Zou S, Liu H, Tong Z (2012) Chitosan nanoparticles as particular emulsifier for preparation of novel pH-responsive Pickering emulsions and PLGA microcapsules. Polymer. 53: 1229-1235.

[10] Hernandez-Lauzardo A N, Velazquez-del Valle M G, Guerra-Sanchez M G (2011) Current status of action mode and effect of chitosan against phytopathogens fungi. African Journal of Microbiology Research. 5: 4243-4247.

[11] Abd-Elmohdy F A, El Sayed Z, Essam S, Hebeish A (2010) Controlling chitosan molecular weight via bio-chitosanolysis. Carbohydrate Polymers. 82: 539-542.

[12] Bagheri-Khoulenjani S, Taghizadeh S M, Mirzadeh H (2009) An investigation on the short-term biodegradability of chitosan with various molecular weights and degrees of deacetylation. Carbohydrate Polymers. 78: 773-778.

[13] Guo Z, Xing R, Liu S, Zhong Z, Ji X, Wang L, Li P (2008) The influence of molecular weight of quaternized chitosan on antifungal activity. Carbohydrate Polymers. 71: 694-697.

[14] Kean T, Thanou M (2010) Biodegradation, biodistribution and toxicity of chitosan. Advanced Drug Delivery Reviews. 62: 3-11.

[15] Vinsova J, Vavrikova E (2011) Chitosan Derivatives with Antimicrobial, Antitumour and Antioxidant Activities - a Review. Current Pharmaceutical Design. 17: 3596-3607.

[16] Muzzarelli R A A, Isolati A (1971) Methyl mercury acetate removal from waters by chromatography on chelating polymers. Water, Air, and Soil Pollution. 1: 65-71.

[17] Miretzky P, Cirelli A F (2009) Hg(II) removal from water by chitosan and chitosan derivatives: A review. Journal of Hazardous Materials. 167: 10-23.

[18] Chatterjee S, Lee M W, Woo S H (2010) Adsorption of congo red by chitosan hydrogel beads impregnated with carbon nanotubes. Bioresource Technology. 101: 1800-1806.

[19] Lu L C, Wang C I, Sye W F (2011) Applications of chitosan beads and porous crab shell powder for the removal of 17 organochlorine pesticides (OCPs) in water solution. Carbohydrate Polymers. 83: 1984-1989.

[20] Wan Ngah W S, Teong L C, Hanafiah M A K M (2011) Adsorption of dyes and heavy metal ions by chitosan composites: A review. Carbohydrate Polymers. 83: 1446-1456.

[21] Zeng D, Wu J, Kennedy J F (2008) Application of a chitosan flocculant to water treatment. Carbohydrate Polymers. 71: 135-139.

[22] Hoven V P, Tangpasuthadol V, Angkitpaiboon Y, Vallapa N, Kiatkamjornwong S (2007) Surface-charged chitosan: Preparation and protein adsorption. Carbohydrate Polymers. 68: 44-53.

[23] Krajewska B (2005) Membrane-based processes performed with use of chitin/chitosan materials. Separation and Purification Technology. 41: 305-312.

[24] da Silva R B, Lima Neto A F, Soares dos Santos L S, de Oliveira Lima J R, Chaves M H, dos Santos Jr J R, de Lima G M, de Moura E M, de Moura C V R (2008) Catalysts of Cu(II) and Co(II) ions adsorbed in chitosan used in transesterification of soy bean and babassu oils – A new route for biodiesel syntheses. Bioresource Technology. 99: 6793-6798.

[25] Muzzarelli R A A, Tanfani F (1982) N-(Ortho-Carboxybenzyl) Chitosan, N-Carboxymethyl Chitosan and Dithiocarbamate Chitosan - New Chelating Derivatives of Chitosan. Pure and Applied Chemistry. 54: 2141-2150.

[26] Pavinatto F J, Caseli L, Oliveira O N (2010) Chitosan in Nanostructured Thin Films. Biomacromolecules. 11: 1897-1908.

[27] Rabea E I, Badawy M E T, Stevens C V, Smagghe G, Steurbaut W (2003) Chitosan as antimicrobial agent: Applications and mode of action. Biomacromolecules. 4: 1457-1465.

[28] Ren J M, Liu J L, Li R C, Dong F, Guo Z Y (2012) Antifungal properties of chitosan salts in laboratory media. Journal of Applied Polymer Science. 124: 2501-2507.

[29] Krajewska B, Wydro P, Janczyk A (2011) Probing the Modes of Antibacterial Activity of Chitosan. Effects of pH and Molecular Weight on Chitosan Interactions with Membrane Lipids in Langmuir Films. Biomacromolecules. 12: 4144-4152.

[30] Kong M, Chen X G, Xing K, Park H J (2010) Antimicrobial properties of chitosan and mode of action: A state of the art review. International Journal of Food Microbiology. 144: 51-63.

[31] Shi W, Nie D, Jin G, Chen W, Xia L, Wu X, Su X, Xu X, Ni L, Zhang X, Chen J (2012) BDNF blended chitosan scaffolds for human umbilical cord MSC transplants in traumatic brain injury therapy. Biomaterials. 33: 3119-3126.

[32] Yannas I V, Burke J F, Orgill D P, Skrabut E M (1982) Wound Tissue Can Utilize a Polymeric Template to Synthesize a Functional Extension of Skin. Science. 215: 174-176.

[33] Muzzarelli R A A (2009) Chitins and chitosans for the repair of wounded skin, nerve, cartilage and bone. Carbohydrate Polymers. 76: 167-182.

[34] Leffler C C, Muller B W (2000) Chitosan-gelatin sponges for controlled drug delivery: the use of ionic and non-ionic plasticizers. Stp Pharma Sciences. 10: 105-111.

[35] Panos I, Acosta N, Heras A (2008) New drug delivery systems based on chitosan. Curr Drug Discov Technol. 5: 333-41.

[36] Nimesh S, Saxena A, Kumar A, Chandra R (2012) Improved transfection efficiency of chitosan-DNA complexes employing reverse transfection. Journal of Applied Polymer Science. 124: 1771-1777.

[37] Tikhonov V E, Stepnova E A, Babak V G, Yamskov I A, Palma-Guerrero J, Jansson H B, Lopez-Llorca L V, Salinas J, Gerasimenko D V, Avdienko I D, Varlamov V P (2006) Bactericidal and antifungal activities of a low molecular weight chitosan and its N-/2(3)-(dodec-2-enyl)succinoyl/-derivatives. Carbohydrate Polymers. 64: 66-72.

[38] Martinez-Camacho A P, Cortez-Rocha M O, Castillo-Ortega M M, Burgos-Hernandez A, Ezquerra-Brauer J M, Plascencia-Jatomea M (2011) Antimicrobial activity of chitosan nanofibers obtained by electrospinning. Polymer International. 60: 1663-1669.

[39] Liu H, Du Y M, Wang X H, Sun L P (2004) Chitosan kills bacteria through cell membrane damage. International Journal of Food Microbiology. 95: 147-155.

[40] Cho C H, Eliason J F, Matthew H W (2008) Application of porous glycosaminoglycan-based scaffolds for expansion of human cord blood stem cells in perfusion culture. Journal of Biomedical Materials Research Part A. 86: 98-107.

[41] Shi Z L, Neoh K G, Kang E T, Wang W (2006) Antibacterial and mechanical properties of bone cement impregnated with chitosan nanoparticles. Biomaterials. 27: 2440-2449.

[42] Suh J K F, Matthew H W T (2000) Application of chitosan-based polysaccharide biomaterials in cartilage tissue engineering: a review. Biomaterials. 21: 2589-2598.

[43] Cunha C, Panseri S, Antonini S (2011) Emerging nanotechnology approaches in tissue engineering for peripheral nerve regeneration. Nanomedicine-Nanotechnology Biology and Medicine. 7: 50-59.

[44] Tchemtchoua V T, Atanasova G, Aqil A, Filee P, Garbacki N, Vanhooteghem O, Deroanne C, Noel A, Jerome C, Nusgens B, Poumay Y, Colige A (2011) Development of a Chitosan Nanofibrillar Scaffold for Skin Repair and Regeneration. Biomacromolecules. 12: 3194-3204.

[45] Woods R J (1998) Computational carbohydrate chemistry: what theoretical methods can tell us. Glycoconjugate Journal. 15: 209-216.

[46] Wormald M R, Petrescu A J, Pao Y L, Glithero A, Elliott T, Dwek R A (2002) Conformational studies of oligosaccharides and glycopeptides: complementarity of NMR, X-ray crystallography, and molecular modelling. Chemical Reviews. 102: 371-86.

[47] Duus J, Gotfredsen C H, Bock K (2000) Carbohydrate structural determination by NMR spectroscopy: modern methods and limitations. Chemical Reviews. 100: 4589-614.

[48] Jayakumar R, Prabaharan M, Nair S V, Tamura H (2010) Novel chitin and chitosan nanofibers in biomedical applications. Biotechnol Advances. 28: 142-50.

[49] Jayakumar R, Menon D, Manzoor K, Nair S V, Tamura H (2010) Biomedical applications of chitin and chitosan based nanomaterials—A short review. Carbohydrate Polymers. 82: 227-232.

[50] Muzzarelli R A A, Muzzarelli C (2005) Chitosan chemistry: Relevance to the biomedical sciences. Polysaccharides 1: Structure, Characterization and Use. 186: 151-209.

[51] Cardenas G, Anaya P, von Plessing C, Rojas C, Sepulveda J (2008) Chitosan composite films. Biomedical applications. Journal of Materials Science-Materials in Medicine. 19: 2397-2405.

[52] Araujo J V, Lopes-Da-Silva J A, Almeida M M, Costa M E V (2006) Preparation and characterization of chitosan-based scaffolds for biomedical applications. Advanced Materials Forum Iii, Pts 1 and 2. 514-516: 1005-1009.

[53] Pavinatto A, Pavinatto F J, Barros-Timmons A, Oliveira O N (2010) Electrostatic Interactions Are Not Sufficient to Account for Chitosan Bioactivity. Acs Applied Materials & Interfaces. 2: 246-251.

[54] Fang N, Chan V, Mao H-Q, Leong K W (2001) Interactions of Phospholipid Bilayer with Chitosan: Effect of Molecular Weight and pH. Biomacromolecules. 2: 1161-1168.

[55] Mertins O, Lionzo M I Z, Micheletto Y M S, Pohlmann A R, da Silveira N P (2009) Chitosan effect on the mesophase behavior of phosphatidylcholine supramolecular systems. Materials Science & Engineering C-Biomimetic and Supramolecular Systems. 29: 463-469.

[56] Je J Y, Kim S K (2006) Chitosan derivatives killed bacteria by disrupting the outer and inner membrane. Journal of Agricultural and Food Chemistry. 54: 6629-6633.

[57] Helander I M, Nurmiaho-Lassila E L, Ahvenainen R, Rhoades J, Roller S (2001) Chitosan disrupts the barrier properties of the outer membrane of Gram-negative bacteria. International Journal of Food Microbiology. 71: 235-244.

[58] Palma-Guerrero J, Lopez-Jimenez J A, Perez-Berna A J, Huang I C, Jansson H B, Salinas J, Villalain J, Read N D, Lopez-Llorca L V (2010) Membrane fluidity determines sensitivity of filamentous fungi to chitosan. Molecular Microbiology. 75: 1021-1032.

[59] Palma-Guerrero J, Huang I C, Jansson H B, Salinas J, Lopez-Llorca L V, Read N D (2009) Chitosan permeabilizes the plasma membrane and kills cells of Neurospora crassa in an energy dependent manner. Fungal Genetics and Biology. 46: 585-594.

[60] Van de Velde K, Kiekens P (2004) Structure analysis and degree of substitution of chitin, chitosan and dibutyrylchitin by FT-IR spectroscopy and solid state C-13 NMR. Carbohydrate Polymers. 58: 409-416.

[61] Cardenas G, Cabrera G, Taboada E, Miranda S P (2004) Chitin characterization by SEM, FTIR, XRD, and (13)C cross polarization/mass angle spinning NMR. Journal of Applied Polymer Science. 93: 1876-1885.

[62] Jeffrey G A (1990) Crystallographic Studies of Carbohydrates. Acta Crystallographica Section B-Structural Science. 46: 89-103.

[63] Heux L, Brugnerotto J, Desbrieres J, Versali M F, Rinaudo M (2000) Solid state NMR for determination of degree of acetylation of chitin and chitosan. Biomacromolecules. 1: 746-751.

[64] Bubb W A (2003) NMR spectroscopy in the study of carbohydrates: Characterizing the structural complexity. Concepts in Magnetic Resonance Part A. 19A: 1-19.

[65] Zaia J (2004) Mass spectrometry of oligosaccharides. Mass Spectrom Rev. 23: 161-227.

[66] van der Hoeven R A M, Hofte A J P, Tjaden U R, van der Greef J, Torto N, Gorton L, Marko-Varga G, Bruggink C (1998) Sensitivity improvement in the analysis of

oligosaccharides by on-line high-performance anion-exchange chromatography/ion spray mass spectrometry. Rapid Communications in Mass Spectrometry. 12: 69–74.
[67] Ahn Y H, Yoo J S (1999) Effect of the labeling group in structural analyses of malononitrile-labeled oligosaccharides by matrix-assisted laser desorption/ionization time-of-flight mass spectrometry and electrospray ionization mass spectrometry. Rapid Commun Mass Spectrom. 13: 1985-90.
[68] Sato Y, Suzuki M, Nirasawa T, Suzuki A, Endo T (2000) Microsequencing of glycans using 2-aminobenzamide and MALDI-TOF mass spectrometry: occurrence of unique linkage-dependent fragmentation. Anal Chem. 72: 1207-16.
[69] Yamagaki T (1999) A new analytical method for oligosaccharides and glycoconjugates using post-source decay fragmentation by matrix-assisted laser desorption/ionization time-of-flight mass spectrometry. Trends Glycoscience Glycotechnology. 11: 227-232.
[70] Mechref Y, Novotny M V, Krishnan C (2003) Structural characterization of oligosaccharides using MALDI-TOF/TOF tandem mass spectrometry. Anal Chem. 75: 4895-903.
[71] Li M, Kinzer J A (2003) Structural analysis of oligosaccharides by a combination of electrospray mass spectrometry and bromine isotope tagging of reducing-end sugars with 2-amino-5-bromopyridine. Rapid Communications in Mass Spectrometry. 17: 1462-6.
[72] Cheng H L, Her G R (2002) Determination of linkages of linear and branched oligosaccharides using closed-ring chromophore labeling and negative ion trap mass spectrometry. J Am Soc Mass Spectrom. 13: 1322-30.
[73] Sturiale L, Naggi A, Torri G (2001) MALDI mass spectrometry as a tool for characterizing glycosaminoglycan oligosaccharides and their interaction with proteins. Seminars in Thrombosis and Hemostasis. 27: 465-72.
[74] Brunger A T (1997) X-ray crystallography and NMR reveal complementary views of structure and dynamics. Nature Structural Biology. 4 Suppl: 862-5.
[75] Wagner G, Hyberts S G, Havel T F (1992) NMR structure determination in solution: a critique and comparison with X-ray crystallography. Annu Rev Biophys Biomol Struct. 21: 167-98.
[76] Billeter M (1992) Comparison of protein structures determined by NMR in solution and by X-ray diffraction in single crystals. Quaterly Reviews in Biophysics. 25: 325-77.
[77] Dwek R A, Edge C J, Harvey D J, Wormald M R, Parekh R B (1993) Analysis of glycoprotein-associated oligosaccharides. Annu Rev Biochem. 62: 65-100.
[78] Bush C A, Martin-Pastor M, Imberty A (1999) Structure and conformation of complex carbohydrates of glycoproteins, glycolipids, and bacterial polysaccharides. Annual Reviews in Biophysics and Biomolecular Structure. 28: 269-93.
[79] Wang Y, Hollingsworth R I (1996) An NMR spectroscopy and molecular mechanics study of the molecular basis for the supramolecular structure of lipopolysaccharides. Biochemistry. 35: 5647-5654.
[80] van Gunsteren W F, Bakowies D, Baron R, Chandrasekhar I, Christen M, Daura X, Gee P, Geerke D P, Glattli A, Hunenberger P H, Kastenholz M A, Ostenbrink C, Schenk M,

Trzesniak D, van der Vegt N F A (2006) Biomolecular modeling: Goals, problems, perspectives. Angewandte Chemie - International Edition. 45: 4064-4092.

[81] van Gunsteren W F, Bakowies D, Burgi R, Chandrasekhar I, Christen M, Daura X, Gee P, Glattli A, Hansson T, Oostenbrink C, Peter C, Pitera J, Schuler L, Soares T A, Yu H B (2001) Molecular dynamics simulation of biomolecular systems. Chimia. 55: 856-860.

[82] Gulin V I, Samoilova R I, Dikanov S A, Tsvetkov Y D (1991) 2-mm Band EPR Spectroscopy of Semiquinone Radicals in Rigid Media. Applied Magnetic Resonance. 2: 425-434.

[83] Adcock S A, McCammon J A (2006) Molecular dynamics: Survey of methods for simulating the activity of proteins. Chemical Reviews. 106: 1589-1615.

[84] Guvench O, Hatcher E, Venable R M, Pastor R W, MacKerell A D (2009) CHARMM Additive All-Atom Force Field for Glycosidic Linkages between Hexopyranoses. Journal of Chemical Theory and Computation. 5: 2353-2370.

[85] Guvench O, Mallajosyula S S, Raman E P, Hatcher E, Vanommeslaeghe K, Foster T J, Jamison F W, MacKerell A D (2011) CHARMM Additive All-Atom Force Field for Carbohydrate Derivatives and Its Utility in Polysaccharide and Carbohydrate-Protein Modeling. Journal of Chemical Theory and Computation. 7: 3162-3180.

[86] Hatcher E, Guvench O, MacKerell A D (2009) CHARMM Additive All-Atom Force Field for Aldopentofuranoses, Methyl-aldopentofuranosides, and Fructofuranose. Journal of Physical Chemistry B. 113: 12466-12476.

[87] Hatcher E R, Guvench O, MacKerell A D (2009) CHARMM Additive All-Atom Force Field for Acyclic Polyalcohols, Acyclic Carbohydrates, and Inositol. Journal of Chemical Theory and Computation. 5: 1315-1327.

[88] Yongye A B, Tessier M B, Kirschner K, Tschampel S M, Gonzalez-Outeirino J, Daniels C R, Foley B L, Woods R J (2008) CARB 43-GLYCAM06, a generalizable biomolecular force field: Carbohydrates, lipids, lipid bilayers and glycolipids. Abstracts of Papers of the American Chemical Society. 235.

[89] Kirschner K N, Yongye A B, Tschampel S M, Gonzalez-Outeirino J, Daniels C R, Foley B L, Woods R J (2008) GLYCAM06: A generalizable Biomolecular force field. Carbohydrates. Journal of Computational Chemistry. 29: 622-655.

[90] Lins R D, Hunenberger P H (2005) A new GROMOS force field for hexopyranose-based carbohydrates. Journal of Computational Chemistry. 26: 1400-1412.

[91] Kony D, Damm W, Stoll S, van Gunsteren W F (2002) An improved OPLS-AA force field for carbohydrates. Journal of Computational Chemistry. 23: 1416-1429.

[92] Franca E F, Lins R D, Freitas L C G, Straatsma T P (2009) Characterization of Chitin and Chitosan Molecular Structure in Aqueous Solution. Journal of Chemical Theory and Computation. 4: 2141-2149.

[93] Franca E F, Lins R D, Freitas L C G (2011) Chitosan Molecular Structure as a Function of N-Acetylation. Biopolymers. 95: 448-460.

[94] Skovstrup S, Hansen S G, Skrydstrup T, Schiott B (2010) Conformational Flexibility of Chitosan: A Molecular Modeling Study. Biomacromolecules. 11: 3196-3207.

[95] Prathab B, Aminabhavi T M (2007) Molecular modeling study on surface, thermal, mechanical and gas diffusion properties of chitosan. Journal of Polymer Science Part B-Polymer Physics. 45: 1260-1270.

[96] Brooks B R, Brooks C L, Mackerell A D, Nilsson L, Petrella R J, Roux B, Won Y, Archontis G, Bartels C, Boresch S, Caflisch A, Caves L, Cui Q, Dinner A R, Feig M, Fischer S, Gao J, Hodoscek M, Im W, Kuczera K, Lazaridis T, Ma J, Ovchinnikov V, Paci E, Pastor R W, Post C B, Pu J Z, Schaefer M, Tidor B, Venable R M, Woodcock H L, Wu X, Yang W, York D M,Karplus M (2009) CHARMM: The Biomolecular Simulation Program. Journal of Computational Chemistry. 30: 1545-1614.

[97] Oostenbrink C, Soares T A, van der Vegt N F A, van Gunsteren W F (2005) Validation of the 53A6 GROMOS force field. European Biophysics Journal with Biophysics Letters. 34: 273-284.

[98] Soares T A, Daura X, Oostenbrink C, Smith L J, van Gunsteren W F (2004) Validation of the GROMOS force-field parameter set 45A3 against nuclear magnetic resonance data of hen egg lysozyme. Journal of Biomolecular Nmr. 30: 407-422.

[99] Soares T A, Hunenberger P H, Kastenholz M A, Krautler V, Lenz T, Lins R D, Oostenbrink C, Van Gunsteren W F (2005) An improved nucleic acid parameter set for the GROMOS force field. Journal of Computational Chemistry. 26: 725-737.

[100] Wang J M, Wolf R M, Caldwell J W, Kollman P A, Case D A (2004) Development and testing of a general amber force field. Journal of Computational Chemistry. 25: 1157-1174.

[101] van Gunsteren W F, Mark A E (1998) Validation of molecular dynamics simulation. Journal of Chemical Physics. 108: 6109-6116.

[102] Brooks C L, Karplus M, Pettitt B M (1988) Proteins: A Theoretical Perspective of Dynamics, Structure, and Thermodynamics. New York: John Wiley & Sons. p.

[103] McCammon J A, Harvey D J (1988) *Dynamics of Proteins and Nucleic Acids*. In. Cambridge: Cambridge University Press. pp. 248.

[104] Frenkel D, Smit B (1996) Understanding Molecular Simulation: From Algorithms to Applications. Orlando, FL, USA: Academic Press Inc. p.

[105] Verlet L (1967) Computer Experiments on Classical Fluids .I. Thermodynamical Properties of Lennard-Jones Molecules. Physical Review. 159: 98-&.

[106] Hockney R W (1970) The potential calculation and some applications. Methods Comput Phys. 9: 136-211.

[107] Beeman D (1976) Some Multistep Methods for Use in Molecular-Dynamics Calculations. Journal of Computational Physics. 20: 130-139.

[108] Gear C W (1971) Numerical Initial Value Problems in Ordinary Differencial Equations. New York: Prentice-Hall Englewoodcliffs. p.

[109] Ryckaert J P, Ciccotti G, Berendsen H J C (1977) Numerical integration of the cartesian equations of motion of a system with constraints: Molecular dynamics of *n*-alkanes. Journal of Computational Physics. 23: 327–341.

[110] Andersen A M, Norby P (2000) Ab initio structure determination and Rietveld refinement of a high-temperature phase of zirconium hydrogen phosphate and a new

polymorph of zirconium pyrophosphate from in situ temperature-resolved powder diffraction data. Acta Crystallogr B. 56: 618-25.
[111] Hess B, Bekker H, Berendsen H J C, Fraaije J (1997) LINCS: A linear constraint solver for molecular simulations. Journal of Computational Chemistry. 18: 1463-1472.
[112] Liwo A, Czaplewski C, Oldziej S, Scheraga H A (2008) Computational techniques for efficient conformational sampling of proteins. Current Opinion in Structural Biology. 18: 134-9.
[113] Trzesniak D, Lins R D, van Gunsteren W F (2006) Protein under pressure: Molecular dynamics simulation of the arc repressor. Proteins-Structure Function and Bioinformatics. 65: 136-144.
[114] Berendsen H J C, Grigera J R, Straatsma T P (1987) The Missing Term in Effective Pair Potentials. Journal of Physical Chemistry. 91: 6269-6271.
[115] Jorgensen W L, Chandrasekhar J, Madura J D, Impey R W, Klein M L (1983) Comparison of simple potential functions for simulating liquid water. Journal of Chemica Physics. 79: 926-935.
[116] Mahoney M W, Jorgensen W L (2000) A five-site model liquid water and the reproduction of the density anomaly by rigid, non-polarizable models. Journal Chemical Physics. 112: 8910-8922.
[117] Cornell W D, Cieplak P, Bayly C I, Gould I R, Merz K M, Ferguson D M, Spellmeyer D C, Fox T, Caldwell J W, Kollman P A (1995) A 2nd Generation Force-Field for the Simulation of Proteins, Nucleic-Acids, and Organic-Molecules. Journal of the American Chemical Society. 117: 5179-5197.
[118] Sprik M, Klein M L (1988) A polarizable model for water using distributed charge sites. Journal of Chemical Physics. 89: 7556-7561.
[119] Horn H W, Swope W C, Pitera J W, Madura J D, Dick T J, Hura G L, Head-Gordon T (2004) Development of an improved four-site water model for biomolecular simulations: TIP4P-Ew. Journal of Chemical Physics. 120: 9665-9678.
[120] Rinaudo M (2006) Chitin and chitosan: Properties and applications. Progress in Polymer Science. 31: 603-632.
[121] Mogilevskaya E L, Akopova T A, Zelenetskii A N, Ozerin A N (2006) The crystal structure of chitin and chitosan. Polymer Science Series A. 48: 116-123.
[122] Ogawa K (1991) Effect of Heating an Aqueous Suspension of Chitosan on the Crystallinity and Polymorphs. Agricultural and Biological Chemistry. 55: 2375-2379.
[123] Okuyama K, Noguchi K, Miyazawa T, Yui T, Ogawa K (1997) Molecular and crystal structure of hydrated chitosan. Macromolecules. 30: 5849-5855.
[124] Sugiyama J, Boisset C, Hashimoto M, Watanabe T (1999) Molecular directionality of beta-chitin biosynthesis. Journal of Molecular Biology. 286: 247-255.
[125] Blackwel J (1969) Structure of beta-chitin or parallel chain systems of poly-beta-(1-4)-N-acetyl-d-glucosamine. Biopolymers. 7: 281-298.
[126] Gardner K H, Blackwell J (1975) Refinement of Structure of Beta-Chitin. Biopolymers. 14: 1581-1595.

[127] Minke R, Blackwell J (1978) Structure of Alpha-Chitin. Journal of Molecular Biology. 120: 167-181.

[128] Ogawa K, Yui T, Okuyama K (2004) Three D structures of chitosan. International Journal of Biological Macromolecules. 34: 1-8.

[129] Franca E F, Leite F L, Cunha R A, Oliveira O N, Freitas L C G (2011) Designing an enzyme-based nanobiosensor using molecular modeling techniques. Physical Chemistry Chemical Physics. 13: 8894-8899.

[130] Ogawa K (1988) Chain Conformation of Chitin and Chitosan. Nippon Nogeikagaku Kaishi-Journal of the Japan Society for Bioscience Biotechnology and Agrochemistry. 62: 1225-1228.

[131] Ogawa K, Inukai S (1987) X-Ray-Diffraction Study of Sulfuric, Nitric, and Halogen Acid Salts of Chitosan. Carbohydrate Research. 160: 425-433.

[132] Yui T, Imada K, Okuyama K, Obata Y, Suzuki K, Ogawa K (1994) Molecular and Crystal-Structure of the Anhydrous Form of Chitosan. Macromolecules. 27: 7601-7605.

[133] Lertworasirikul A, Yokoyama S, Noguchi K, Ogawa K, Okuyama K (2004) Molecular and crystal structures of chitosan/HI type I salt determined by X-ray fiber diffraction. Carbohydrate Research. 339: 825-833.

[134] Okuyama K, Noguchi K, Kanenari M, Egawa T, Osawa K, Ogawa K (2000) Structural diversity of chitosan and its complexes. Carbohydrate Polymers. 41: 237-247.

[135] Okuyama K, Osawa K, Hanafusa Y, Noguchi K, Ogawa K (2000) Relaxed 2/1-helical conformation of type II chitosan has a tetrasaccharide motif. Journal of Carbohydrate Chemistry. 19: 789-794.

[136] Lertworasirikul A, Tsue S, Noguchi K, Okuyama K, Ogawa K (2003) Two different molecular conformations found in chitosan type II salts. Carbohydrate Research. 338: 1229-1233.

[137] Kawahara M, Yui T, Oka K, Zugenmaier P, Suzuki S, Kitamura S, Okuyama K, Ogawa K (2003) Fourth 3D structure of the chitosan molecule: Conformation of chitosan in its salts with medical organic acids having a phenyl group. Bioscience Biotechnology and Biochemistry. 67: 1545-1550.

[138] Brugnerotto J, Desbrieres J, Heux L, Mazeau K, Rinaudo M (2001) Overview on structural characterization of chitosan molecules in relation with their behavior in solution. Macromolecular Symposia. 168: 1-20.

[139] Weinhold M X, Thöming J (2011) On conformational analysis of chitosan. Carbohydrate Polymers. 84: 1237-1243.

[140] Pedroni V I, Schulz P C, Gschaider M E, Andreucetti N (2003) Chitosan structure in aqueous solution. Colloid and Polymer Science. 282: 100-102.

[141] DeMarco M L, Woods R J (2008) Structural glycobiology: A game of snakes and ladders. Glycobiology. 18: 426-440.

[142] Cho Y W, Jang J, Park C R, Ko S W (2000) Preparation and solubility in acid and water of partially deacetylated chitins. Biomacromolecules. 1: 609-614.

[143] Kubota N (1997) Permeation properties of chitosan transition metal complex membranes. Journal of Applied Polymer Science. 64: 819-822.

[144] Kristiansen A, Varum K M, Grasdalen H (1998) Quantitative studies of the non-productive binding of lysozyme to partially N-acetylated chitosans. Binding of large ligands to a one-dimensional binary lattice studied by a modified McGhee and von Hippel model. Biochimica Biophysica Acta. 1425: 137-50.

[145] Sorlier P, Denuziere A, Viton C, Domard A (2001) Relation between the degree of acetylation and the electrostatic properties of chitin and chitosan. Biomacromolecules. 2: 765-772.

[146] Sorlier P, Viton C, Domard A (2002) Relation between solution properties and degree of acetylation of chitosan: Role of aging. Biomacromolecules. 3: 1336-1342.

[147] Berth G, Dautzenberg H (2002) The degree of acetylation of chitosans and its effect on the chain conformation in aqueous solution. Carbohydrate Polymers. 47: 39-51.

[148] Brugnerotto J, Lizardi J, Goycoolea F M, Arguelles-Monal W, Desbrieres J, Rinaudo M (2001) An infrared investigation in relation with chitin and chitosan characterization. Polymer. 42: 3569-3580.

[149] Baxter A, Dillon M, Taylor K D A, Roberts G A F (1992) Improved Method for Ir Determination of the Degree of N-Acetylation of Chitosan. International Journal of Biological Macromolecules. 14: 166-169.

[150] Pelletier A, Lemire I, Sygusch J, Chornet E, Overend R P (1990) Chitin Chitosan Transformation by Thermomechanochemical Treatment Including Characterization by Enzymatic Depolymerization. Biotechnology and Bioengineering. 36: 310-315.

[151] Muzzarelli R A A, Rocchetti R (1985) Determination of the Degree of Acetylation of Chitosans by 1st Derivative Ultraviolet Spectrophotometry. Carbohydrate Polymers. 5: 461-472.

[152] Rinaudo M, Ledung P, Gey C, Milas M (1992) Substituent Distribution on O, N-Carboxymethylchitosans by H-1-Nmr and C-13 Nmr. International Journal of Biological Macromolecules. 14: 122-128.

[153] Raymond L, Morin F G, Marchessault R H (1993) Degree of Deacetylation of Chitosan Using Conductometric Titration and Solid-State Nmr. Carbohydrate Research. 246: 331-336.

[154] Lamarque G, Lucas J M, Viton C, Domard A (2005) Physicochemical behavior of homogeneous series of acetylated chitosans in aqueous solution: role of various structural parameters. Biomacromolecules. 6: 131-42.

[155] Wang Q Z, Chen X G, Liu N, Wang S X, Liu C S, Meng X H, Liu C H (2006) Protonation constants of chitosan with different molecular weight and degree of deacetylation. Carbohydrate Polymers. 65: 194-201.

[156] Delben F, Muzzarelli R A A, Terbojevich M (1989) Thermodynamic Study of the Protonation and Interaction with Metal-Cations of 3 Chitin Derivatives. Carbohydrate Polymers. 11: 205-219.

[157] Anthonsen M W, Smidsrod O (1995) Hydrogen-Ion Titration of Chitosans with Varying Degrees of N-Acetylation by Monitoring Induced H-1-NMR Chemical-Shifts. Carbohydrate Polymers. 26: 303-305.

[158] Domard A (1987) pH and CD measurements on a fully deacetylated chitosan - application to Cu-Li-polymer interactions. International Journal of Biological Macromolecules. 9: 98-104.

[159] Schatz C, Viton C, Delair T, Pichot C, Domard A (2003) Typical physicochemical behaviors of chitosan in aqueous solution. Biomacromolecules. 4: 641-648.

Chapter 10

Concept of Template Synthesis of Proteoglycans

Aleksandr N. Zimnitskii

Additional information is available at the end of the chapter

http://dx.doi.org/10.5772/48085

1. Introduction

It is accepted in modern genomics that only a minor portion (1-5%) of the genome contains information about the primary structure of protein molecules. The higher is the organism on the evolutionary ladder, the lower the density of genetic information, in the modern sense of the term, per kilobase (kb). For instance (Table 1), this parameter is 1-1.7 kb per gene in bacteria and more than 30 kb per gene in human.

Organism	Genome size, Mb	Total genes, 10^3	Density, kb/gene
Bacteria	0.5-5	0.47-4.29	1-1.7
Yeasts	12	6	2
Nematode	97	19	5
Human	3000	80-100	>30

Table 1. Comparison of the genome size and gene number (Kiselev, L.L., 2000)

The genome can be conventionally divided into three groups of nucleotide sequences: unique sequences, moderate repeats (DNA segments repeated frequently), and fast repeats (satellite DNA). This division is exclusively conventional; the assignment to a particular group is determined by the capability of fragmented and denatured DNA to find complementary regions upon reassociation in solution. Yet such a division makes it possible to study the functional role of individual genome regions. It was shown, for instance, that unique DNA sequences are responsible mostly for encoding the primary structure of protein molecules. Such sequences occur in a few copies in the genome. Part of information of repetitive DNA systems (histone, rRNA, and tRNA genes) is encoded by so-called slow DNA repeats

The role of fast repeats and satellite DNA is still enigmatic to a great extent. Such DNA is often called selfish or waste, suggesting a lack of information valuable for the organism.

INTECH
open science | open minds

© 2012 The Author(s). Licensee InTech. This chapter is distributed under the terms of the Creative Commons Attribution License http://creativecommons.org/licenses/by/3.0), which permits unrestricted use, distribution, and reproduction in any medium, provided the original work is properly cited.

Surprisingly, it is this DNA that increases in amount in organisms with the highest level of organization (for instance, repetitive sequences account for more than 95% of the human genome). Data accumulating in recent decades for inverted repeats, regulatory gene regions, centromeric and telomeric satellite DNAs, introns, and mobile genetic systems still fail to improve the understanding of the functional role of tandem repeats, which quantitatively constitute the basis of the genome. The hypothesis that repeats are just dilutors of genetic information is below the level of current knowledge, testifying to a poor understanding of the phenomenon rather than helping to elucidate their role (Singer M., Berg P. 1998).

Some authors believe that tandem repeats play an important, though still unknown, part in evolutionary improvement of organisms. Thus repeats show species specificity (Mednikov B.M., et al., 2001) and occur in hundred of thousands or even millions of copies in the genome (Table 2). Simple tandem repeats, such as 5-CA, 5-GA, 5-AT, and 5-GC, are present in virtually all eukaryotes and have huge copy numbers.

Repeat	Organisms	Purine:pyrimidine ratio
5-GGAAG	birds	2:0
5-CA	many eukaryotes	1:1
5-GA	many eukaryotes	2:0
5-GT	many eukaryotes	1:1
5-AT	many eukaryotes	1:1
5-CAAA GTTTT GTTTGA	*Xenopus laevis*	1:0.9
5-TCTCC	birds	0:2

Table 2. Some tandem repeats dispersed through the genomes of various eukaryotes (Singer, M., Berg, P., 1998)

In terms of base groups, the sequences shown in Table 2 can be presented as multiple repeats of (Pyr-Pyr), (Pur-Pur), and (Pur-Pyr), where Pyr is a pyrimidine (T or C) and Pur is a purine (A or G). Thus, the genome contains extended monopurine and/or monopyrimidine arrays as well as tandem repeats whose averaged unit contains Pur and Pyr bases in nearly equal proportions. It is also beyond doubt that excess information is read from DNA during transcription yielding gnRNA, part of which is excised during RNA maturation. Information is transferred from DNA to mRNA almost directly, without processing, in prokaryotes, whereas the situation observed in higher organisms is similar to that with DNA repeats. The higher the position of an organism on the evolutionary ladder, the greater the amount of information that is transmitted to RNA and is senseless in terms of the polypeptide sequence (Singer M., Berg P., 1998).

While DNA determines the ontogenetic and phylogenetic development of organisms and is structurally described as a polymer of dimeric units, a natural question is whether the cell possesses another structurally similar biopolymer (apart from RNA). As is well known, proteins are unfit for this role, because the triplet code of polypeptides requires at least three different monomers for their implementation. There is only one group of tandem biopolymers playing a key role in the functioning of higher organisms as multicellular entities. This group is polysaccharides.

Polysaccharides are biopolymers that consist of monosaccharides linked to each other. It was believed until the 1960s (Colman Y., Rem C.-G. 2000), that carbohydrates serve only as an energy source (monosaccharides and storage polysaccharides such as starch and glycogen) and structural material (e.g., cellulose in plants and chitin in insects). Interest in carbohydrates was moderated by their extreme structural complexity. While there is only one way of linking together monomers of nucleic acids (nucleotides) or proteins (amino acids), monosaccharide units of oligosaccharides and polysaccharides can be linked in several ways at several different positions. Thus only 256 tetranucleotides can be obtained with four different nucleotides, whereas four different monosaccharides can form 35,560 unique tetrasaccharides (Colman Y., Rem C.-G. 2000). Consequently, biological polymorphism of polysaccharides is immeasurably higher than that of proteins or nucleic acids (NA).

By composition, polysaccharides are usually divided into homopolysaccharides (homoglycans) and heteropolysaccharides (heteroglycans). In almost all cases, the chemical structure can be reduced to a dimeric unit repeated many times (Fig. 1, Table 3). Glycosaminoglycans (GAG) are the most common natural heteroglycans.

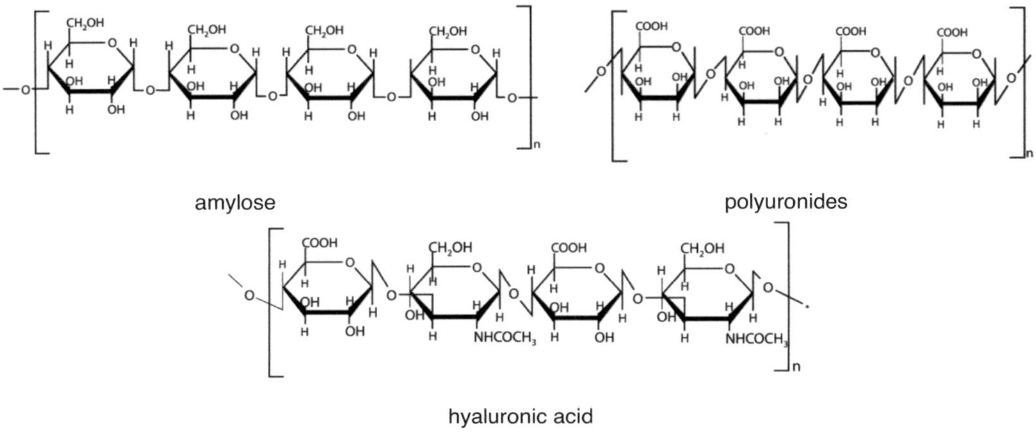

Figure 1. Structures of common polysaccharides.

GAG is a polymer that consists of hexuronic acid and hexosamine residues linked by O-glycoside bonds. The three major GAG classes are nonsulfated hyaluronic acid (HA), moderately sulfated chondroitin sulfates (CS), and highly sulfated heparan sulfates (HS). GAG of the last two classes belong to proteoglycans. GAG contained in proteoglycans play various vital parts in the intercellular matrix and within the cell (Zimnitskii A.N., Bashkatov S.A. 2004).

PS, source	PS formula	Uronic acid:hexose ratio
Polysaccharides of microorganisms	GlcUA - Glc	1:1
Polyuronides of plants	GlcUA - GlcUA	2:0
Amylose and cellulose of plants, glycogen of animals	Glc-Glc	0:2
Chitin of insects and crustaceans	GlcNAc – GlcNAc	0:2
Heparan of animals	GlcUA - GlcN	1:1
Hyaluronic acid of animals	GlcUA - GlcNAc	1:1
Dermatan sulfate of animals	IdoUA - GalNAc	1:1
Chondroitin sulfate (CS) of animals	GlcUA - GalNAc	1:1
CS A of animals	GlcUA – GalNAc4SO_4	1:1
CS C of animals	GlcUA – GalNAc6SO_4	1:1
CS of animals	GlcUA3S – GalNAc4SO_4	1:1
CS of animals	GlcUA2S – GalNAc6S	1:1
CS of animals	GlcUA - GalNAc(4,6-diSO_4)	1:1
CS of animals	IdoUA – GalNAc4SO_4	1:1
CS of animals	IdoUA2(3)SO_4 - GalNAc	1:1
CS of animals	IdoUA2(3)SO_4 - GalNAc4SO_4	1:1
CS of animals	IdoUA2SO_4 – GalNAc6SO_4	1:1
CS of animals	IdoUA – GalNAc(4,6-diSO_4)	1:1
CS of animals	IdoUA2(3)SO_4 – GalNAc(4,6-diSO_4)	1:1
Heparan sulfate (HS) of animals	IdoUA - GlcNSO_4	1:1
HS of animals	IdoUA – GlcN$SO_4$6SO_4	1:1
HS of animals	IdoUA2(3)SO_4 - GlcNSO_4	1:1
HS of animals	IdoUA2(3) – bGlcN$SO_4$6SO_4	1:1
HS of animals	GlcUA - GlcNSO_4	1:1
HS of animals	GlcUA – GlcNAc6SO_4	1:1
HS of animals	GlcUA – GlcNS6SO_4	1:1
HS of animals	GlcUA2SO_4 – GlcN$SO_4$6SO_4	1:1
HS of animals	GlcUA2SO_4 - GlcNSO_4	1:1
HS of animals	GlcUA2SO_4 -GlcN$SO_4$$SO_4$	1:1

Table 3. Uronic acid/hexose ratio in storage and structural polysaccharides (PS)

HA biosynthesis was studied in group A hemolytic streptococci (Dorfman A 1958, 1964,1965). Similar synthetic processes occur in higher animals (Glaser L., Brown D.H. 1955). Roseman (Roseman S., Ludowieg J., et. al. 1953, Roseman S., Moses F.E., et. al. 1953, Roseman S., Moses F.E, et. al. 1954) have shown that D-glucose is a precursor of glycosamine and glucuronic acid, which are HA monomers. The immediate precursors of HA are uridine diphosphate (UDP) derivatives of glucuronic acid and N-acetylglucosamine, which act as donors of carbohydrate residues during HA synthesis (Cifonelli J.A., et al. 1957, Markovic O., et al. 1964, Markovitz A., et al. 1958, Lorenzel I. 1959). A cell-free system allowing HA synthesis in the presence of UDP-glucuronic acid, UDP-glucosamine, and Mg^{2+} was obtained by sonicating a microbial culture producing HA (Markovic O., Huttl S. 1964). The enzyme responsible for HA synthesis sedimented at 100.000 g. A similar enzyme was found in rat embryo skin by Schiler (Schiller S. 1964, Schiler S. 1965) and in various chicken, rodent, and human tissues by Altschuler (Altschuler C.H., et al. 1963). As in microorganisms, nucleotide derivatives result from the reactions catalyzed by pyrophosphorylases in higher animal tissues producing HA (Strominger J.L. 1964). Acetylation of UDP-glucose involves acetyl coenzyme A. Glucose is transformed into glucosamine before being linked to UDP.

As Slutskii (Slutskii L.I. 1969). showed in the late 1970s, when biosynthesis of heteroglycans of the intracellular matrix were studied most intensely in higher animals, and modern data demonstrate, the enzyme involved in synthesis of polysaccharide fragments of GAG is associated with membrane structures in the intact cell. One is forced to accept the fact that little is still understood about the mechanism of final steps of GAG biosynthesis: how it is possible that a strict alternation of acetylglucosamine and glucuronic acid residues in the macromolecular glycan chain is achieved in the absence of a template.

Studies showed that synthesis of sulfated GAG is essentially similar to synthesis of nonsulfated HA and that D-glucose is a precursor of monosaccharide residues as in the case of HA. However, a necessary step is epimerization of glucosamine into galactosamine in the case of CS and glucuronic acid into iduronic acid in the case of dermatan sulfate. Several works demonstrated epimerization of UDP-monosaccharides in the presence of $NADPH^{+}$ and specific epimerases (Maley F., et al. 1959, Roden L., et al. 1958, Rondle C.J.M., et al. 1955). White (White B.N., et al. 1965) observed the reversible character of epimerization with the example of transformation of galactosamine into glucosamine during GAG synthesis. During synthesis of HS and, in particular, heparin, nucleotide derivatives of hexuronic acids and hexosamines form not only β-glycosidic bonds, characteristic of HA and CS, but also α-glycosidic bonds, characteristic of HS (Silbert J.J. 1963).

According to current views, the first step of proteoglycan synthesis is translation of the protein core mRNA on ribosomes of the rough endoplasmic reticulum. Then glycosylation of the protein core is initiated: a site is generated at which the prospective GAG is to be linked to the polypeptide chain. This process has not been localized so far; it is known only that the protein core already has a polysaccharide fragment when delivered into the endoplasmic reticulum (Zimina N.P., et al. 1992, Zimina N.P., et al. 1987, White J, et al. 1978, Silbert J.E. et al, 1995). It is thought that synthesis of GAG chains is template-independent:

glycosyltransferases just consecutively add monosaccharides of donor UDP-saccharides to the growing carbohydrate chain. The substrate specificity of glycosyltransferases is determined by the monosaccharide sequence. The growth of GAG chains is initiated by xylosyltransferase. However, only some serine residues of the protein core are subject to xylosylation. The site of attachment of a GAG chain is presumably selected depending on the amino acid sequence of the region to be glycosylated (Zimina N.P., et al. 1992, Zimina N.P., et al. 1987, Jeffrey D Esko, et al. 1996). After a xylose residue is linked to the protein, a carbohydrate chain is formed via adding consecutively two galactose residues and one glucuronic acid residue by galactosyltransferases I and II and glucuronyltransferase I. The newly synthesized tetrasaccharide serves as an acceptor in the first reaction of GAG synthesis. Polymeric GAG chains result from multiple reactions performed alternately by two enzymes associated with the membrane Golgi complex. These enzymes are N-acetylgalactosaminyltransferase and glucuronyltransferase II in the case of CS and dermatan sulfates. In the case of HS and heparin, polymerization is catalyzed by N-acetylglucosaminyltransferase and glucuronyltransferase (Zimina N.P., et al. 1992, Zimina N.P., et al. 1987, Sugahara K., et al. 2000, Silbert J.E. 1982). The structure of the protein core is critical for the structure of GAG chains (Jeffrey D Esko, et al. 1996).

The immediate substrates of GAG synthesis are nucleotide (UDP) derivatives of monosaccharides; these derivatives are generated in the reactions catalyzed by pyrophosphorylases. Glucuronic acid is produced from UDP-glucose via two-step oxidation of C6, with transformation of the hydroxyl group into a carboxyl group. In dermatan sulfate synthesis, UDP-glucuronic acid is epimerized into UDP-iduronic acid by specific epimerase.

Thus, during GAG biosynthesis, one enzyme complex elongates the heteroglycan chain by adding consecutively monosaccharide units. Treatment with ribonucleases or deoxyribonucleases did not prevent chain elongation in a model HA-producing system. This finding was interpreted as demonstrating the template-independent character of GAG synthesis. We think, however, that template independence of GAG synthesis is not evident from this finding, because the glycans produced in the model system were not compared with natural glycans. It is beyond doubt that, in the absence of a template, the enzyme complex is capable of producing a heteroglycan chain at random, as is the case with NA synthesis in *in vitro* systems.

To eliminate the contradiction between the clearly ordered GAG structure (Zimina N.P., et al. 1992, Zimina N.P., et al. 1987) and the concept of non-template GAG synthesis, attempts were made over the past decade to associate the ordered character of the GAG structure with template synthesis of the protein core of proteoglycans; i.e., the amino acid sequence of the protein core was considered as a kind of a template for GAG synthesis. Although some achievements were made in the field, they mostly revealed statistical, rather than cause-and-effect, relationships. It seems that this situation has been accepted as satisfactory, and the concept of non-template GAG synthesis associated with the protein core is now described in textbooks. Yet the concept is shattered completely by the fact that that HA, reaching a molecular weight of 10^7 Da, is an individual GAG lacking any protein core, in contrast to CS and HS.

Current views of proteoglycan biosynthesis are related to studies of the role of intracellular membranes. It is commonly known that cell proteoglycans are almost always associated with membrane structures. The processes of proteoglycan biosynthesis, intracellular transport to organelles, and transfer onto the cell surface in the intercellular matrix are coupled with the functioning of membranes. GAG accumulation in lysosomes and subsequent cleavage to monosaccharides are also connected with the functioning of membrane structures. Free proteoglycans and GAG that are not associated with membranes are detectable only at the last degradation stages in lysosomes and some other cell structures such as the nucleus and mast cell granules.

Scarce, if any, information is available concerning the early steps of GAG synthesis and GAG transport to the sites where polysaccharide chains are generated (Silbert J.E. et al. 1995). As the views of the structure of the protein core are generally discrepant, it is still unclear whether the protein core determines the formation of proteoglycans. The primary structure of a core protein with a potential serine xylosylation site suggests that the protein molecule plays no crucial role in recognition of the xylosylation site. Hence it is probable that the conformation of membrane structures is a factor determining the recognition of xylosylation sites in protein molecules of proteoglycans. This assumption can be advanced because the main determinant of xylosylation is probably associated with membrane structures of the endoplasmic reticulum and Golgi complex, where glycosyltransferase activity is high (Silbert J.E. et al. 1995).

Synthesis of the linker tetrasaccharide (an uronic acid residue is added to the trisaccharide) initiates addition of a particular N-acetylhexosamine and thereby determines which GAG chain is to be generated. As a result, the nascent proteoglycan is transferred into the corresponding site, containing either galactose N-acetyltransferases or glucose N-acetyltransferases; this determines the chain to be synthesized. It is the amino acid sequence of the peptide moiety that determines the position of the nascent proteoglycan on membrane structures and, consequently, the direction of synthesis and modification of the polysaccharide chain by the corresponding enzyme systems.

The primary structure of GAG in proteoglycan molecules is an important factor determining their function. For instance, the GAG chain size, the number and location of sulfo groups, and epimerization affect the properties and biological role of proteoglycans. The extent of modification of the polysaccharide chain and its size probably depend on the structure of membrane-associated complexes responsible for proteoglycan synthesis. Membrane enzyme complexes play a key role in epimerization and/or sulfation of GAG polysaccharide chains (Silbert J.E. et al. 1995).

In 1987, Lindahl (Lindahl U., et al. 1987) showed convincingly that the results of GAG biosynthesis *in vitro* cannot be extrapolated to the *in vivo* situation. Studying heparin biosynthesis, these researchers observed that polysaccharides obtained with intact microsomes contain extended sequences of both N-sulfated and N-acetylated disaccharides, suggesting a nonrandom character of their synthesis. In contrast, products obtained with a solubilized microsomal system displayed a random distribution of such groups with a greater portion of N-acetylated and N-sulfated disaccharides (Lindahl U., et al. 1987). In fact, this finding demonstrates that synthesis of the glycoside moiety of proteoglycans is genetically determined *in vivo*, because their primary structure is tissue- and species-specific.

A basic unit monomers of polysaccharides is glucose, which occurs in solution both in the α and in the β form owing to the mutarotation reaction. In addition, glucose can be epimerized to other hexoses and be oxidized to yield glucuronic and other hexuronic acids. Owing to such lability of glucose, biosynthetic systems always contain sufficient amounts of the α and β forms of hexoses and hexuronic acids. Modified with UDP at C_1, these monosaccharides provide the main components for synthesis of polysaccharide chains.

As Fig. 1 and Table 3 demonstrate, the majority of known structural and storage polysaccharides each consist of two monomers, a hexose and hexuronic acid, which occur in the α or β form, are linked through (1-3), (1-4) O-glycoside bonds, and are modified to a various extent at various carbon atoms as a result of acetylation, amination, sulfation, etc. All these biopolymers can be combined in one group with a universal structure of multiply repeated (A-B), (A-A), or (B-B) units, where A is a hexose and B is a hexuronic acid.

We think that the periodicity of the primary structure is similar between polysaccharides and DNA tandem repeats. In view of this structural similarity, it was justified and important to study a possible complementarity between monosaccharides of glycans and bases of NA.

To check the hypothesis of complementarity of NA bases to hexoses and hexuronic acids, quantum chemical methods were used in our lab for a particular case of glucose and glucuronic acid contained in the heteropolysaccharide HA. It is clear that such an approach is computational and that the relevant conclusions need experimental verification. In view of this, it was necessary to obtain empirical information supporting or contradicting the results of quantum chemical computations. For this purpose, we employed UV spectrophotometry and dot hybridization, which allow detection of specific complexes of biopolymers.

Since the mechanism initiating GAG biosynthesis is unclear, it was expedient to study glycan synthesis in the rat liver upon administration of elevated doses of glucose. The use of $^{35}SO_4^{2-}$ as a radioactive label is inadequate for studying GAG synthesis, and we decided to label a precursor of the glycan polysaccharide chain. We used glucose as such a precursor: glucose is transformed into UDP-glucose and then into UDP-glucuronic acid, which is utilized in GAG synthesis. It should be noted that ribose 5-phosphate, which is formed from glucose 6-phosphate, is incorporated in NA. Glucose is converted into ribose in the pentose phosphate cycle by eliminating C1, which is released as carbon dioxide. Hence, we used [^{14}C]glucose labeled at C_1 to prevent generation of radiolabeled NA in our experiments. We studied the composition of rat liver polysaccharides in cell nuclei, microsomes, and in a total liver homogenate. In addition, the nuclear and microsomal fractions were used to monitor the accumulation of radiolabeled saccharides, which are polysaccharide precursors.

2. Experimental

Quantum chemical modeling of biological structures, in particular, the geometric and electron structures of NA and polysaccharides, requires that the methods used report adequately the effects of weak intermolecular interactions, such as hydrogen bonds. The MP2 *(ab initio)* procedure, which utilizes bases with diffuse and polarization functions, meets this requirement quite well (Cybulski S.M., et al. 1989, Latajka Z. et al. 1990) but is hardly suitable for our objects because of their size: their analysis would be extremely time-consuming and requiring excessive computational resources. A possible alternative in this situation is provided by the corresponding semiempirical methods. Early semiempirical methods (MINDO/3 and MNDO) considerably underestimated the energy of hydrogen bonds and, consequently, were unsuitable for studying biopolymers (Williams I.H. 1987). To eliminate such drawbacks, the AM1 (Dewar M.J.S., et al. 1985) and PM3 (Stewart J.J.P. 1989) methods were developed on the basis of neglect of diatomic differential overlap (NDDO). These methods were expected to adequately describe systems with hydrogen bonds. This was not the case with AM1: while computed energies of hydrogen bonds agreed well with experimental estimates, geometric parameters failed to represent the facts (Jurema J.M.W., et al. 1993). Parametrization was performed with a far greater body of experimental data in PM3 than in other semiempirical methods, which allowed PM3 to describe well the geometric structure of molecules and the heat of their generation. PM3 is indeed the first method yielding semiempirical estimates that agree with the results of experiments and *ab initio* calculations for hydrogen-bonded systems (Kallies B., et al.1995).

2.1. Method of calculation *Ab initio*

Ab initio calculations are the main computational procedure in quantum chemistry and consists in solving Hartree-Fock one-electron equations. As initial data, the method utilizes the charges of nuclei, their positions in the molecule, and Slater- or Gaussian-type basis function sets. The method involves none of the observed physico-chemical properties of substances and, accordingly, is known as unempirical calculations. *Ab initio* calculations most commonly employ the MO LCAO (molecular orbitals as a linear combination of atomic orbitals) approximation, which takes account of all electrons of the system in question. This method is most accurate in quantum chemical computations, especially with intermediate (6-31G*) and large (6-31**) bases, and allows correct estimation of the electron and geometric structures of hydrogen bonds, which play an important role in biological objects.

Both unempirical *ab initio* calculations with basis 6-31G* and the PM3 semiempirical method were employed in theoretical computations in this work. Geometric parameters and energy characteristics were computed by the unempirical and semiempirical methods for interacting nucleotides and saccharides and by the semiempirical methods for structures containing several pairs of nucleotides and saccharides. All computations were performed with complete optimization of geometrical parameters, using the GAMESS program (Schmidt M.W., et al. 1993). A global minimum of the total electron energy was sought by the Newton-Raphson method with an energy gradient of 0.010 kcal/mol·Å, starting from various initial approximations of the complex structure.

2.2. Isolation of oat polyuronides

Fresh oat seedlings grown for 10–12 days were thoroughly ground and hydrolyzed with 2 M NaOH (1:10 w/v) at room temperature with occasional stirring for 1 day. The hydrolysate was neutralized with concentrated HCl to pH 7.0 and filtered through filter paper. Gel filtration on Sephadex G-25 was used to remove salts and low-molecular-weight components and to isolate a high-molecular-weight fraction containing polysaccharides. The preparation collected to obtain polyuronides was fractionated by anion-exchange chromatography on DEAE cellulose. The preparation was applied onto a column with fibrous DEAE cellulose (a chloride form) equilibrated with distilled water. The column was washed consecutively with distilled water and with 0.15 and 0.5 M NaCl to the minimal A_{220} of the eluate . Then, polyuronides were eluted with 1.5 M NaCl. To remove NaCl, the resulting fraction was subjected to gel filtration on Sephadex G-25; the presence of chlorides was checked using silver nitrate. The preparation was dried under vacuum at 45^{0}C and proved to contain glucuronoxylans with a purity of no less than 95% and a proportion of hexuronic acids of no less than 95%. The quality of the product was checked by the Dishe test for uronic acids. The product was dissolved to 0.1 mg/ml. A 0.5-ml aliquot of the solution was combined with 3 ml of concentrated sulfuric acid containing 0.025 M sodium tetraborate and thoroughly mixed. The mixture was heated in boiling water for 10 min, chilled to room temperature, combined with 0.1 ml of 0.1% carbazole (ethanol solution), heated in boiling water for 15 min, and chilled. A_{530} was measured against a control sample. The polyuronide concentration (mg/ml) was estimated using a calibration plot, which was constructed using aqueous solutions of glucuronic acid.

2.3. Isolation of garlic glucuronoxylans

Unpeeled garlic bulbs were thoroughly ground and hydrolyzed with 2 M NaOH (1:10 w/v) at room temperature with occasional stirring for 1 day. The hydrolysate was neutralized with concentrated HCl to pH 7.0 and filtered through filter paper. Gel filtration on Sephadex G-25 was used to remove salts and low-molecular-weight components and to isolate a high-molecular-weight fraction containing polysaccharides. The preparation collected to obtain glucuronoxylans was fractionated by anion-exchange chromatography on DEAE cellulose. The preparation was applied onto a column with fibrous DEAE cellulose (a chloride form) equilibrated with distilled water. The column was washed consecutively with distilled water and with 0.15 and 0.3 M NaCl to the minimal A_{220} of the eluate. Then, glucuronoxylans were eluted with 0.5 M NaCl. To remove NaCl, the resulting fraction was subjected to gel filtration on Sephadex G-25; the presence of chlorides was checked using silver nitrate. The preparation was dried under vacuum at 45^{0}C, yielding glucuronoxylans in a dry form. The quality of the product was checked by the Dishe test for uronic acids as above. The purity of the product was no less than 95%; the proportion of hexuronic acids was 20%.

2.4. Isolation of GAG (HA, CS, and HS) from human placenta

Placenta was thoroughly ground and hydrolyzed with 2 M NaOH (1:10 w/v) at room temperature with occasional stirring for 3 days. The hydrolysate was neutralized with concentrated HCl to pH 7.0. To precipitate nonhydrolyzed proteins, concentrated trichloroacetic acid (TCA) was added slowly to a final concentration of 5%. The preparation was incubated at 4^0C for 1 h, and the precipitate was removed by centrifugation at 1500 g for 10 min. Concentrated NaOH was added to the supernatant to pH 10-11, because some proteins withstanding TCA treatment precipitated in an alkaline milieu. The precipitate was allowed to form and was removed by centrifugation (1500 g, 10 min). The supernatant was filtered through filter paper. Gel filtration on Sephadex G-25 was used to remove salts and low-molecular-weight components and to isolate a high-molecular-weight fraction containing GAG.

The preparations used to isolate HA and CS were fractionated by anion-exchange chromatography on DEAE cellulose. Each preparation was applied individually onto a column with fibrous DEAE cellulose (a chloride form) equilibrated with distilled water. The column was washed consecutively with distilled water and 0.15 M NaCl to the minimal A_{220} of the eluate. Then HA was eluted with 0.25 M NaCl; CS, with 0.7 M NaCl; and HS, with 1.5 M NaCl. To remove NaCl, each fraction was subjected to gel filtration on Sephadex G-25; the presence of chlorides was checked with silver nitrate. The resulting fractions were dried under vacuum at 45^0C to obtain HA, CS, and HS in a dry form. The purity was no less than 95%. The quality of the products was checked by the Dishe test for uronic acids as above.

2.5. NA and amylose

The single-stranded probe $d(T)_{16}$ was obtained from Sintol (Russia). Garlic DNA was isolated from root meristem tissue of seedlings. Human DNA was isolated from the placenta by phenol-detergent extraction (Belozersky A.N. 1970). The purity and concentration of DNA were checked by optical density and by Hoechst 33258 fluorescence, using a DNA minifluorimeter (Hoefer Scientific Instruments, United States). Fragmented calf thymus double-stranded DNA (MW ~1 x 10^6 Da) and single-stranded polynucleotides $d(A)_{350}$, $d(C)_{350}$, and $d(GC)_{350}$ were purchased from Amersham Biosciences. Amylose was purchased from Serva Feinbiochemica GMBH.

2.6. Spectrophotometric assays

Working solutions of polynucleotides were prepared in a quartz cuvet with a light path of 2 mm. Small (5-25 μl) aliquots of a stock polynucleotide solution were added to 0.3M NaCl, 0.002M Na-phosphate (pH 6.8) to the final volume 400 μl. After recording the absorption or circular dichroism (CD) spectra of a polynucleotide solution, polynucleotide-polysaccharide complexes were obtained. For this purpose, 400 μl of the polynucleotide solution were combined in the cuvet with an aliquot of a stock polysaccharide solution with continuous stirring. Then the absorbance or CD spectra were recorded again.

The annealing of double-stranded DNA with the corresponding polysaccharide was carried out in a closed quartz cuvet with a light path of 2 mm. The cuvet containing DNA and the polysaccharide in certain proportions was heated in boiling water for 2 min and immediately chilled in ice-cold water (0°C) for 30 s. Then the absorbance or CD spectra were recorded.

The absorbance spectra were obtained using a Specord M-40 spectrophotometer. The CD spectra of complexes were recorded with an SKD-2 portable dichrometer (Institute of Spectroscopy, Russian Academy of Sciences, Troitsk). The CD spectra were presented as a wavelength dependence of the CD value $\Delta A = A_L - A_R$ (Dunin V.V., et al. 1979).

2.7. NA radiolabeling

NA were 5'-endlabeled *in vitro* with γ-[^{32}P]ATP and T4 polynucleotide kinase (Maxam A., et al. 1977). Dephosphorylation was carried out in 50 mM Tris-HCL (pH 9.0), 10 mM $MgCl_2$, 10 mM spermidine, 1 mM $ZnCl_2$, 0.1 units of alkaline phosphatase at 37°C. The enzyme was removed by treating the samples twice with phenol. NA was precipitated with ethanol and collected by centrifugation in a minicentrifuge at 12,000 rpm for 5 min. The pellet was washed with 70% ethanol to remove phenol, dried, and dissolved in a necessary volume of bidistilled water. The resulting solution was combined with a kinase buffer (50 mM Tris-HCl, pH 7.6), 10 mM $MgCl_2$, 50 mM dithiothreitol, 1 mM spermidine), 1-2 MBq of γ-[^{32}P]ATP, and 2 units of T4 polynucleotide kinase. The mixture was incubated at 37°C for 30 min. The reaction was terminated by adding EDTA to 20 mM, and the enzyme was removed by phenol treatment. NA was precipitated with ethanol, and the nonbound label was removed by repeated precipitation with ethanol or by gel filtration on Sephadex G-25.

2.8. Dot hybridization

BA83 and BA85 filters (Schleicher & Shuell, Germany) with immobilized DNA and polysaccharides (20 μg per dot) were hybridized with radiolabeled NA probes at 37°C for 18-24 h. The hybridization buffer contained 5 x Denhardt's solution (0.02% bovine serum albumin, 0.02% polyvinylpyrrolidone, 0.02% Ficoll), 0.1% SDS (Denhardt D.T. 1966). Before being applied onto a filter, DNA was denatured in boiling water for 5 min. Filters were dried and exposed with a PM-B X-ray film for 1-7 days.

2.9. Neutral sugar assays

Neutral sugars (monosaccharides contained in protein-polysaccharide complexes) were assayed by the anthrone method according to (Jermyn M.A. 1975, Chaplin M.F., et al. 1986). A sample (0.5 ml) was combined with 4 ml of an anthrone reagent (20 mg of anthrone, 100 ml of 80% sulfuric acid). The mixture was heated in boiling water for 5 min, chilled in water to room temperature, and tested for A_{630}. As a control, we used an aqueous solution water treated with the anthrone reagent as above. Neutral sugars were quantitated using a calibration plot, which was constructed with galactose solutions.

To study the GAG composition, a homogenate and the nuclear fraction of the liver were obtained from intact adult rats after fasting over 24 h. To estimate the contents of glycans (HA, CS, and HS), alkaline lysis, deoxyribonuclease treatment, gel filtration on Sephadex G-25, and ion-exchange chromatography on DEAE cellulose were performed as above. Polysaccharide elution profiles were obtained with a gradient of NaCl concentration (0-1.5 M). DNA was assayed in nuclei and liver homogenates as in (Slutskii L.I. 1969).

In the next series of experiments, non-inbred white rats weighing 190-210 g were injected intraperitoneally with [^{14}C]glucose, labeled at C1, at 1 g per kg body weight (4 x 10^8 Bq per rat). Rats were sacrificed and the liver frozen in liquid nitrogen 15, 30, 60, 120, and 360 min after the injection. Differential centrifugation was used to obtain the nuclear and microsomal fractions (Orekhovich V.N. 1968), which were tested for radioactivity. Tissue homogenates were obtained similarly; their GAG fractions were isolated and tested for radioactivity.

The results were statistically processed using the software package STATISTICA 6.0 for WINDOWS. The processing included computation of the mean and standard deviation, tests for significance of differences; and correlation, regression, and factor analyses as well as analysis of variance.

3. Results

3.1. Quantum chemical analysis

3.1.1. Analysis of the electron and geometric structures of hydrogen bonds in complementary pairs of NA by the PM3 method

As a control, computations were performed for the interactions of bases in classical complementary pairs AT and CG, which are DNA components. Selection of computation models showed that substitution of methyl groups for the sugar-phosphate backbone has no effect on the parameters of hydrogen bonds, but considerably reduces the computation time. It was found that the PM3 method adequately describes the geometric and electron structures of hydrogen bonds in AT and CG pairs. The two interacting molecules are lying approximately in the same plane.

In the AT pair, the bases form two hydrogen bonds, one between the amino and keto groups and the other between nitrogen atoms of the purine and the pyrimidine. The mean length of the bonds is about 1.8 Å, and their total energy is –5.55 kcal/mol (Fig. 2). In the CG pair, three hydrogen bonds are formed, including two between the amino and keto groups of the bases and one between nitrogen atoms. The mean length of the hydrogen bonds is also about 1.8 Å (Fig. 2), similar to that reported in (Singer M., Berg P., 1998). The total energy of the three hydrogen bonds is –11.73 kcal/mol (Fig. 2). Thus, the PM3 method can be used to study other types of hydrogen bonds formed by nucleotide bases.

The results of computations performed by the PM3 methods for pairs of noncomplementary nucleotides are shown in Table 4.

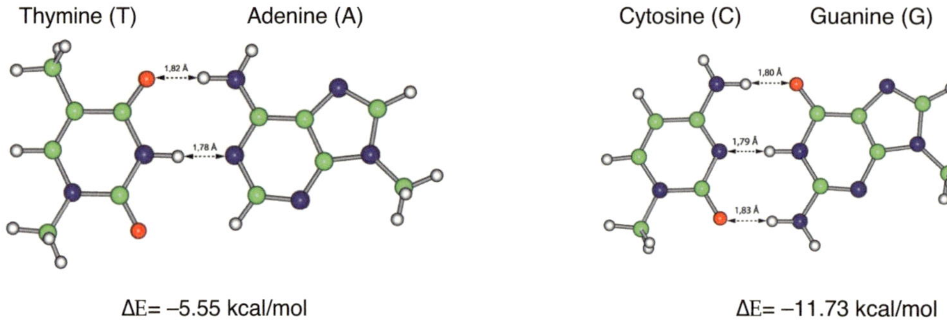

Figure 2. Hydrogen bonding in the AT and CG pairs of complementary nucleotides. Quantum chemical analysis by the PM3 method. Here and below, carbon atoms are shown green or black; oxygen atoms, red; nitrogen atoms, blue; and hydrogen atoms, gray.

	T	A	C	G
T	-2.79			
A	-5.55	-2.71		
C	-4.46	-6.07	-9.24	
G	-6.12	-7.08	-11.73	-10.57

Note: Energy (kcal/mol) was computed by the PM3 method.

Table 4. Energy of hydrogen bonds in all possible pairs of thymine (T), adenine (A), cytosine (C), and guanine

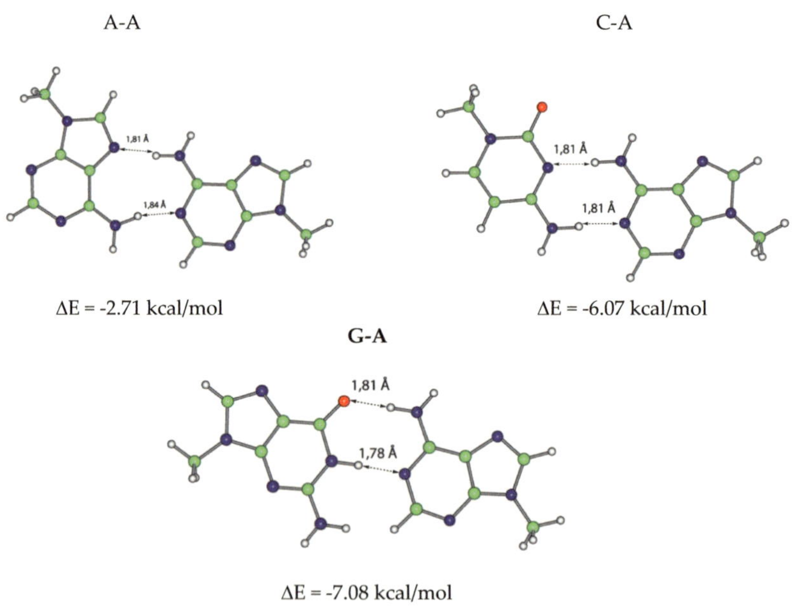

Figure 3. Hydrogen bonding in the noncomplementary pairs AA, GA, and CA. Quantum chemical computations by the PM3 method.

As follows from Table 4, A is theoretically capable of forming pairs with a total energy of -2.71 (A), -5.55 (T), -6.07 (C), and -7.08 (G) kcal/mol. Based on the energy, the preferential partner of A is G or at least C, but it is actually T. Let us consider in more detail the results of computations performed for the interactions of A with A and G with C (Fig. 3).

As Fig. 3 demonstrates, hydrogen bonding between noncomplementary nucleotides is possible, but the pairs are formed so that the angle between the interacting nucleotides is distorted in the case of AA and GA (Fig. 4) or the necessary distance between two strands in the NA double helix is not maintained in the case of CA (Fig. 4). The distance is higher in CA (Singer M., Berg P., 1998).

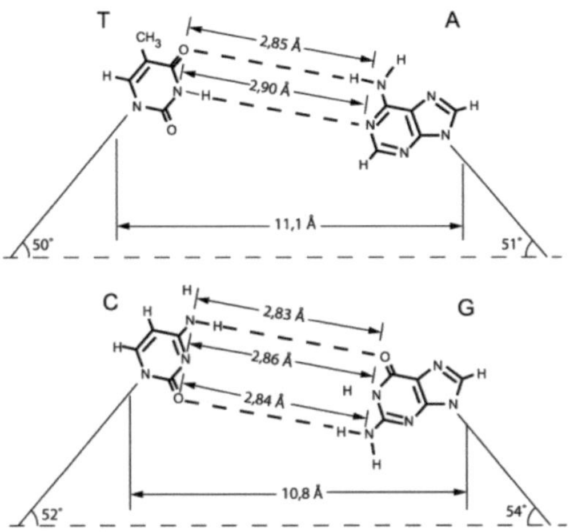

Figure 4. Geometric structure of hydrogen bonds in the complementary pairs AT and CG.

The above data allow the following conclusions.

1. As found by the PM3 method, complementarity in the AT and GC pairs is determined not only by the energy of hydrogen bonding.
2. Complementarity is mostly due to the orientation of hydrogen bonds formed between the interacting nucleotides and the distance between deoxyribose residues in the NA double helix.

3.1.2. Analysis of the electron and geometric structures of hydrogen bonds in complementary NA pairs by non-empirical ab initio calculations in the 6-31G basis set*

Calculations were performed again by the ab initio method with the 6-31G* basis set with complete optimization of the geometric structure (Fig. 5).

Ab initio calculations in the 6-31G* basis also demonstrated the formation of proper hydrogen bonds in complementary pairs. The energy estimates of the bonds were about twice higher than with the PM3 method, the lengths of the bonds were about 0.2 Å greater, but their bond orders remained the same. Electron density analysis according to Mulliken

(Mulliken R.S.,1955) showed that the charge of the hydrogen atoms is positive, about +0.5e, while the interacting O and N atoms have a negative charge ranging from –0.68 to –0.83e. As above, these results suggest quite stable hydrogen bonds.

Then, we performed *ab initio* calculations for noncomplementary pairs of adenine with other nucleotides (Table 5).

T-A C-G

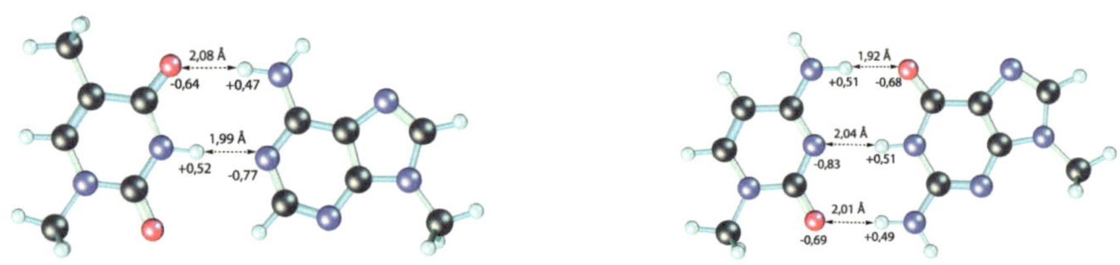

ΔE = -11.78 kcal/mol ΔE = -25.39 kcal/mol

Figure 5. Hydrogen bonding in the complementary pairs AT and CG. Quantum chemical computation by the *ab initio* method in the 6-31G* basis. The charges computed according to Mulliken are shown for the atoms involved in hydrogen bonding.

Nucleotides	T	A	C	G
A	**-11.78**	**-8.93**	**-12.51**	**-13.84**

Note: Energy (kcal/mol) was estimated by *ab initio* calculations in the 6-31G* basis.

Table 5. Energy of hydrogen bonds in pairs of A with T, A, C, and G

As Table 5 demonstrates, selection of complementary T for A is not most advantageous in terms of energy. The interactions AC and AG are more advantageous. The selectivity of nucleotide matching is due to the specific geometric structure of the NA double helix (Fig. 4) (Singer M., Berg P. 1998). Thus, *ab initio* calculations in the 6-31G* basis confirmed that complementarity in the AT and CG pairs is determined mostly by the orientation of hydrogen bonds formed between the interacting nucleotides and the distance between deoxyribose residues in the NA double helix, rather than by the energy of hydrogen bonding.

3.1.3. Analysis of the electron and geometric structures of hydrogen bonds in complementary pairs within the NA double helix by the PM3 method

Let us construct a model NA double helix of six nucleotide pairs (C-G, A-T, T-A, G-C, C-G, and A-T) and fully optimize their electron structure by the PM3 method.

The results showed that a right-turn helix is produced (Fig. 6), which agrees with experimental data. The lengths of the hydrogen bonds in the DNA double helix coincide within 0.02 Å with those computed for isolated nucleotides. The energy of the hydrogen bonds is ΔE = -8.82 kcal/mol per nucleotide pair, which is approximately the average of the energies in two different nucleotide pairs (-5.55 and -11.73 kcal/mol). Thus, the results of computations make it possible to assume that only the above interactions in nucleotide pairs occur in DNA and that there are no other significant interactions.

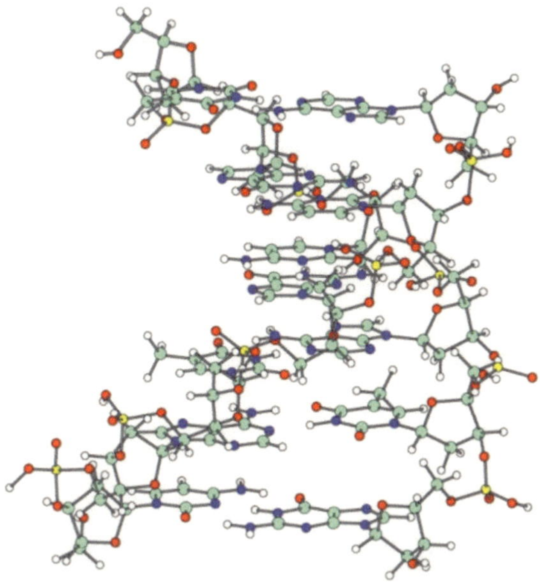

ΔE = -8.82 kcal/mol per nucleotide pair

Figure 6. Hydrogen bonding in complementary pairs AT and CG contained in the DNA double helix (exemplified by a 6-bp duplex). Quantum chemical computation by the PM3 method.

Computations performed for mismatching nucleotide pairs contained in the NA double helix yielded the following picture. In the case of two pyrimidines (T and C), the energy of interaction is equal to zero, because the internucleotide distance is too high. In the case of two purines (A and G), there is not sufficient room for the nucleotides: their rings are bent and the sugar-phosphate backbones forced apart. This is disadvantageous in terms of energy; the loss in electron energy is about 120 kcal/mol. The interactions of T with G and A with C are also disadvantageous, because the parallel arrangement of nucleotides is distorted.

Thus, the quantum chemical computations allow the following conclusions.

1. Hydrogen bonds are formed in the complementary pairs AT and GC, but nucleotide selectivity during NA synthesis is not due to the difference in energy of hydrogen bonds.
2. Selectivity is due to the geometric structure of the NA double helix, which allows the interaction for only those nucleotides that match each other by geometric parameters (AT and GC pairs). All other interactions are disadvantageous in terms of energy.

At the next step, we studied the possibility of complexation of NA bases with carboxyl or hydroxymethyl groups of polysaccharides. The questions were whether hydrogen bonds can be formed between NA bases and sugars and how the four NA bases interact with polysaccharide structural units (UDP-glucose and UDP-glucuronic acid). In addition, it was important to analyze how selectivity may be achieved in template synthesis of a polysaccharide (e.g., HA) on NA.

3.1.4. Quantum chemical analysis of the electron and geometric structures of hydrogen bonds in complementary pairs of NA and polysaccharides by the PM3 method

First, a model was selected for calculations. We found that substitution of glucuronic acid for UDP-glucuronic acid does not affect the hydrogen bonding parameters, but substantially reduces the computation time. Hence the UDP moiety attached to a sugar was omitted in further computations.

Let us assume that NA interacts with the carboxyl or hydroxymethyl group linked to C5 of a monosaccharide unit. Complete optimization of the geometric structure was performed for complexes formed with various initial arrangements of NA bases and glucuronic acid, and several local minima of potential energy were revealed. Among the resulting structures, those with the highest energy of interactions between the components were selected for each NA base–glucuronic acid pair (Fig. 7).

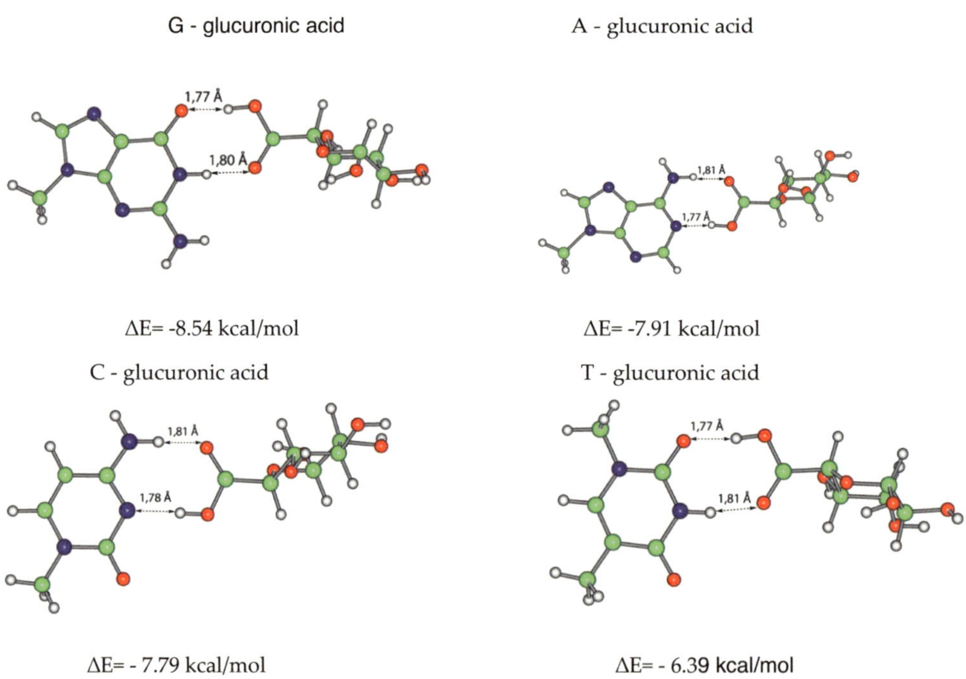

Figure 7. Formation of hydrogen bonds between the carboxyl group of glucuronic acid and DNA bases. Quantum chemical computations by the PM3 method.

The complex structure shows clearly that two hydrogen bonds are formed and are nearly in the same plane, as in the case of the T–A interaction. The most important in this situation is that the bonds have approximately the same lengths and binding energies as in the complementary pairs AT and GC. In addition, Fig. 7 shows that the interactions of the carboxyl group with the nucleotides are nonequivalent. The interactions with G and A are more efficient than with T and C, which is determined by the geometry of the hydrogen bonds. A hydrogen bond should be of a certain length and orientation relative to other atoms, and the conditions for hydrogen bonding are better in some cases and poorer in some others.

Then, the PM3 method was used to study the possibility of interactions between the bases and the hydroxymethyl group of N-acetylglucosamine. Several initial arrangements of a base and N-acetylglucosamine were analyzed for each of the four bases. Optimization of the initial arrangement allowed several modes of binding. Among the resulting structures, those with the highest energy of interactions were selected. The results are shown in Fig. 8.

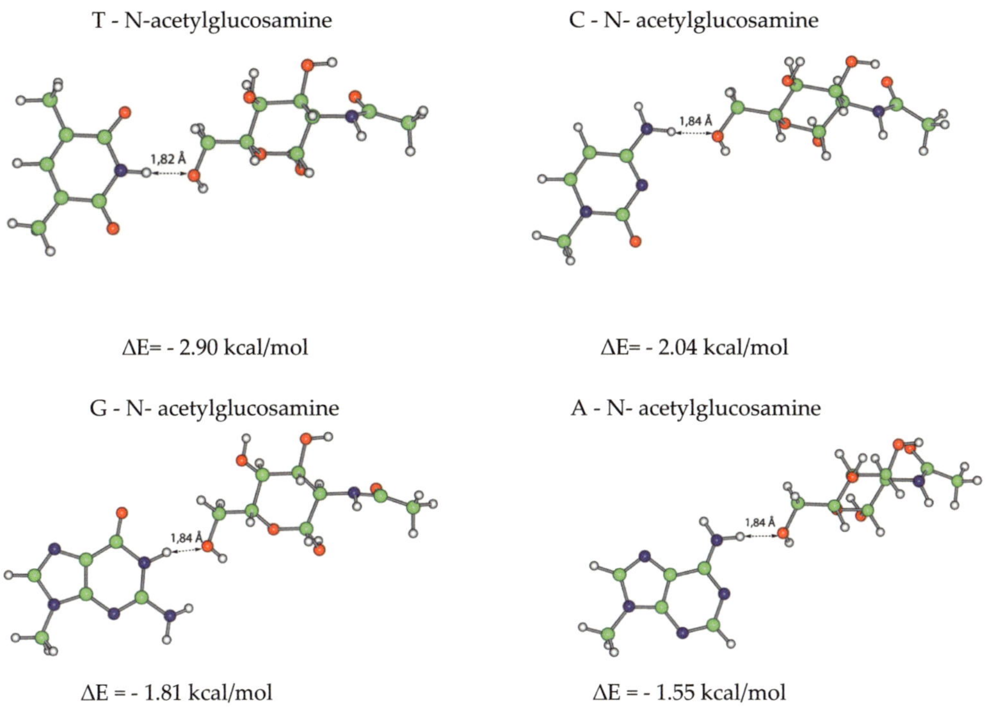

Figure 8. Formation of hydrogen bonds between the hydroxymethyl group of N- acetylglucosamine and the DNA bases. Quantum chemical computations by the PM3 method.

Let us consider the specifics of hydrogen bonding between the nucleotides and the hydoxymethyl group of N-acetylglucosamine. It is known that hydrogen bonds formed by the hydroxymethyl group are less stable than those formed by the carboxyl group (Roberts J., et al. 1978). As complete optimization of the geometric structures showed, the energy of hydrogen bonds was indeed higher in the complexes of the NA bases with the hydroxymethyl group of N-acetylglucosamine (Fig. 8) than in the complexes with the carboxymethyl group (Fig. 7). Optimization of the geometric structure for complexes of the bases with N-acetylglucosamine with various initial arrangements of the interacting groups yielded complexes with one or two hydrogen bonds. Complexes containing one hydrogen bond were most advantageous in terms of energy according to PM3 computations. The nucleotides proved to vary in energy of interaction with the hydroxymethyl group: the interaction was more efficient with T and C than with G and A. The conclusions are as follows.

1. Quantum chemical computations showed that NA bases and the carboxyl or hydroxymethyl group of sugars are capable of forming hydrogen bonds, which are comparable in energy with those occurring in the complementary AT and GC pairs.
2. The bonds are nonequivalent. The carboxyl group forms tighter bonds with purines and less tight bonds with pyrimidines. In contrast, the hydroxymethyl group forms more stable hydrogen bonds with pyrimidines and less stable bonds with purines.
3. The differences in energy computed for hydrogen bonds are insufficient for selection of monosaccharides with necessary structures.

3.1.5. Quantum chemical analysis of the electron and geometric structures of hydrogen bonds in complementary pairs of NA bases and polysaccharides by Ab initio calculations in the 6-31G basis*

Let us optimize the geometry of the above complexes by *ab initio* calculations in the 6-31G* basis (Table 6).

	T	A	C	G
Carboxyl group of glucuronic acid	-12.95	-13.77	-15.98	-17.89

Note: Energy (kcal/mol) was obtained by *ab initio* calculations in the 6-31G* basis.

Table 6. Energy of hydrogen bonds between the nucleotides T, A, C, and G and the carboxyl group of glucuronic acid

In general, *ab initio* calculations confirmed the results obtained by the PM3 method and described in the previous section. The interaction of the carboxyl group of glucuronic acid and the nucleotides is possible, and its energy is sufficient for hydrogen bonding similar to that in the classical complementary DNA pairs. However, the energy of interaction changes with nucleotide in another order: the interaction with the carboxyl group of glucuronic acid is more advantageous with C and G and less advantageous with T and A (Table 6). As for the interaction of the hydroxymethyl group of N-acetylglucosamine with NA bases, the results differed from those obtained by the PM3 method. If the two fragments were initially bonded by one hydrogen bond, like in the case of the interaction between the carboxymethyl group and hydrogen of a base (Fig. 8), then, during optimization, the groups were rotated relative to each other so that the interaction involved the atoms of the monosaccharide ring that by no means can bind to the nucleotide. This is probably explained by the fact that *ab initio* calculations yield higher estimates of hydrogen bond energy as compared with the PM3 method. As a result, greater energy estimates are obtained for the interactions of H with distant O and N and the atoms are brought close together during optimization to form new hydrogen bonds, which are impossible in the double helix.

Thus, *ab initio* calculations in the 6-31G* basis confirmed that the carboxyl group of glucuronic acid is capable of hydrogen bonding to the bases T, A, G, and C; the two bonds formed in each case are similar to those occurring in the complementary pairs AT and GC.

3.1.6. PM3 Analysis of the electron and geometric structures of hydrogen bonds in complementary pairs of a NA–glucuronic acid double helix. Selectivity in template synthesis of polysaccharides

Consider the possibility of polysaccharide synthesis on single-stranded NA. It is clear that, even if some hydrogen bonds are formed between NA bases and monosaccharides, such bonds are not necessarily formed in a polysaccharide–NA double helix because of geometric limitations.

Let us construct two oligomers each consisting of five monomeric units: NA (GCGCA) and oligosaccharide (a HA fragment). As NA, we use the chain analyzed above (Fig. 6). Constructing a chain of five HA monosaccharides, let us rotate the units around the bonds between them so that the carboxyl and hydroxymethyl groups of the resulting chain can be hydrogen bonded to nucleotides. Let us bring the two chains close together so that the distance between groups was 1.5-2.5 Å as in Figs. 9 and 10, with G and A interacting with the carboxyl group (Fig. 9) and C, with the hydroxymethyl group (Fig. 10). Let us fully optimize the resulting double helix by the PM3 method. The energy of hydrogen bonding in the resulting construct (Figs. 9, 10) is -7.04 kcal/mol, rather high and almost reaching the energy of interaction in the classical DNA double helix (-8.82 kcal/mol, Fig. 6).

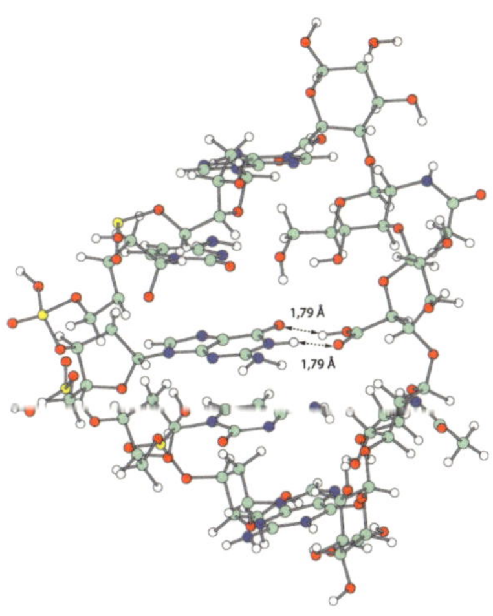

ΔE = -7.04 kcal/mol per complementary pair

Figure 9. Hydrogen bonding in the complementary pairs G–glucuronic acid, A–glucuronic acid, and C–N-acetylglucosamine in a DNA–polysaccharide double helix exemplified by a five-unit chain. Quantum chemical computation by the PM3 method.

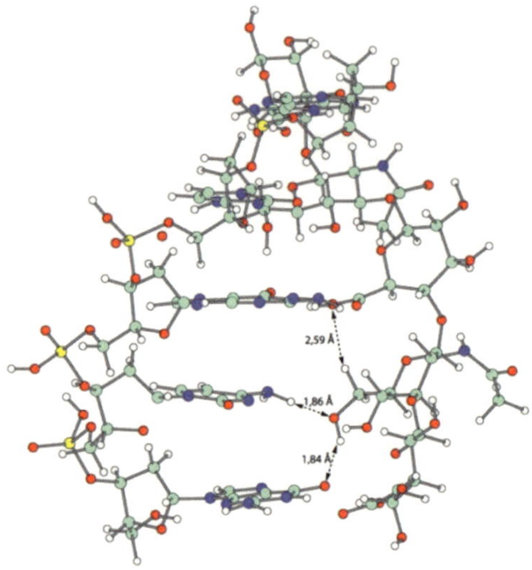

ΔE = -7.04 kcal/mol per complementary pair

Figure 10. Hydrogen bonding in the complementary pairs G–glucuronic acid, A–glucuronic acid, and C–N-acetylglucosamine in a DNA–polysaccharide double helix exemplified by a five-unit chain. Quantum chemical computation by the PM3 method. (Another view point).

The interaction purine–glucuronic acid in the NA–polysaccharide double helix was much the same as in isolation: the lengths of the two hydrogen bonds were 1.79 Å. However, interesting differences were observed for the pyrimidine–N-acetylglucosamine interaction in the double helix and in isolation. In addition to the only hydrogen bond (1.86 Å) formed in isolation (Fig. 10), another hydrogen bond (1.84 Å) was formed with O of the neighboring nucleotide in the double helix. Moreover, there was one more, weaker hydrogen bond (2.59 Å) directed oppositely. Thus, N-acetylglucosamine also forms two hydrogen bonds with nucleotides of an NA strand, which ensures its sufficiently tight and highly selective interaction with nucleotides of a native NA molecule. In addition, computations were performed for incorrect sequences of monosaccharides. As in the case of incorrect sequences of bases in the DNA double helix, low-energy hydrogen bonds were obtained in this variant or the NA–polysaccharide double helix was not formed at all because the bonds were disadvantageous in terms of energy. The conclusions are as follows.

1. Quantum chemical calculations showed that hydrogen bonds are formed in the complementary pairs purine–glucuronic acid and pyrimidine–N-acetylglucosamine, allowing the formation of a NA–polysaccharide double helix.
2. Geometric parameters of the interacting NA and polysaccharide strands play the major role in selection of particular monosaccharides during putative template synthesis of polysaccharides. The difference observed for the energy of hydrogen bonds is insufficient for selection of monosaccharides with necessary structures.
3. Quantum chemical modeling confirmed that purines are complementary to the carboxyl group and pyrimidines, to the hydroxymethyl group of UDP-monosaccharides.

3.2. Spectral analysis and dot hybridization

Table 7 shows the polysaccharide composition determined biochemically. It should be noted that all polysaccharides were end products of modification, that is, mature glycans. Polyuronides, which consist mostly of uronic acid residues, and amylose, which consists only of glucose residues, can be considered as virtually nonmodified glycans. The hydroxymethyl group at C5 of monosaccharide is substituted with hydrogen in glucuronoxylans. HA and CS have the acetoamide group at C2 of hexose residues. In CS, hydrogen of hydroxyl at C6 is substituted with the sulfo group, which provides an additional partial negative charge to the molecule (Table 7.)

Polysaccharide	mkg on mg				in %			
	UA	NS	AAH	OM	UA	NS	AAH	OM
Glucuronoxylan	220	771	0	9	22	77	0	1
Polyuronide	747	33	0	220	75	3	0	22
Hyaluronic acid	535	0	465	0	54	0	46	0
Chondroitin sulfate	413	0	587	0	41	0	59	0
Amylose	0	1021	0	0	0	100	0	0

Note: UA, uronic acid (on Dishe); NS, neutral sugar (on anthrone method); AAH, aminoacetylhexoses(AAH=1000-UA); OM, other monosaccharides (OM=1000-UA-NS).

Table 7. Characterization of the composition of polysaccharides used in experiments

Electrophoretic analysis of polysaccharides used in experiments is illustrated in Fig. 11. It is seen that polysaccharides were sufficiently pure, which was important for our study.

The results shown in Figs. 12A, 12B, and 13A demonstrate that the absorption and CD spectra remained much the same when poly(dC) was combined with oat polyuronides or poly(dA), with amylose. This finding suggests that these polynucleotides and polysaccharides do not form sufficiently tight complexes.

When poly(dA) was combined with oat polyuronides or poly(dC), with amylose, a decrease in amplitude was observed on both the absorption and CD spectra (Figs. 12C, 13B, 13C). The decrease in the peak intensity of the CD spectrum and the simultaneous decrease in the amplitude of peak absorption (a hypochromic effect) can result from an increase in the degree of spiralization and the twist of the polynucleotide (Blagoi Yu.P., et al. 1999), suggesting the formation of a single-stranded polynucleotide–polysaccharide complex. It should be noted that the amplitude of the absorption spectra changed by 9-11% in this case. Since the hyperchromic effect of double-stranded DNA (poly(dA)–poly(dT)) in 0.3 M NaCl was 27-30% (Figs. 14A, 14B), which agreed with published data (Yang J.T., et al. 1969), the change observed in our experiment accounted for 30-35% of the hyperchromic effect characterizing complete base pairing.

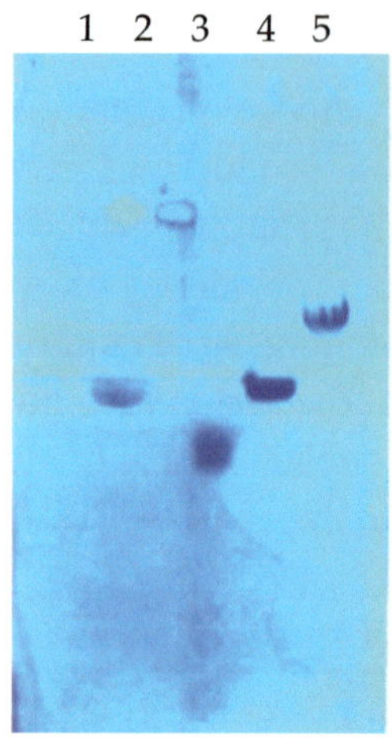

1, oat polyuronides
2, garlic glucuronoxylans
3, HS
4, CS
5, HA

Figure 11. Electrophoresis of polysaccharides in cellulose acetate bands in 0.1 N HCl.

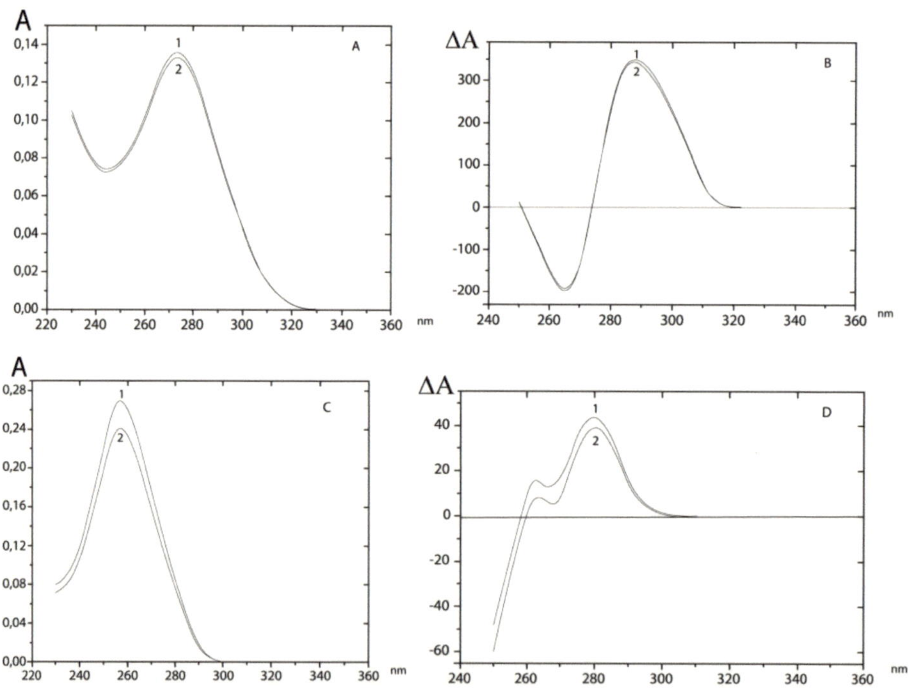

Figure 12. A. Absorption spectrum of poly(dC) in the absence (curve 1) or presence (curve 2) of polyuronide (PU), r = 2. (r - is hereafter the ratio of polysaccharide units to the molar concentration of nucleotides.) B. CD spectrum of poly(dC) in the absence (curve 1) or presence (curve 2) of PU, r = 2. C. Absorption spectrum of poly(dA) in the absence (curve 1) or presence (curve 2) of PU, r = 1. D. CD spectrum of poly(dA) in the presence (curve 1) or absence (curve 2) of PU, r = 1.

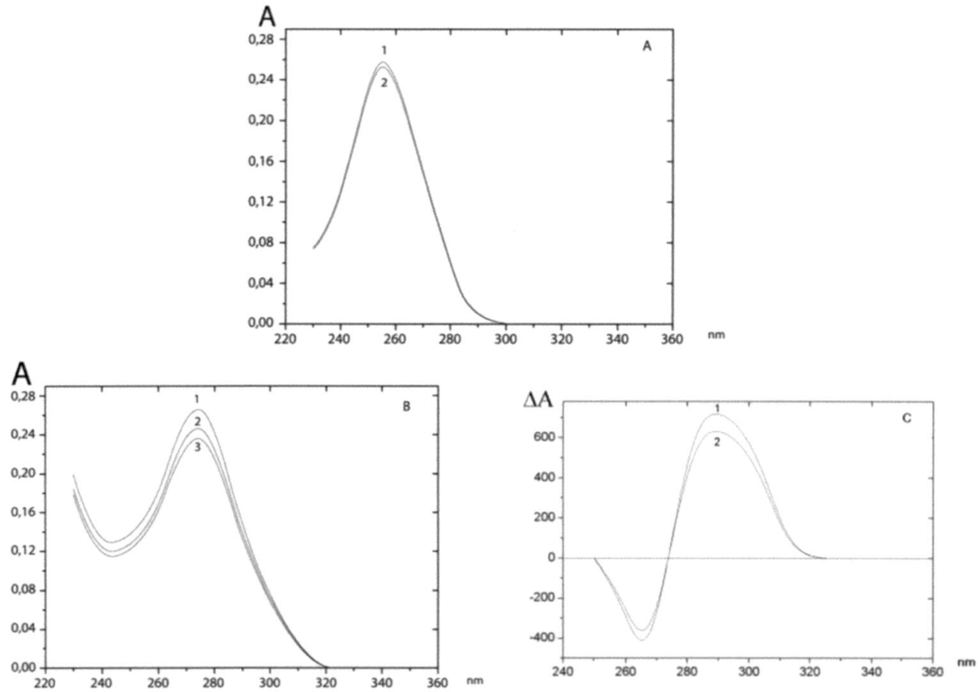

Figure 13. A. Absorption spectrum of poly(dA) in the absence (curve 1) or presence (curve 2) of amylose (AM), r = 2.B. Absorption spectrum of poly(dC) in the absence (curve 1) or presence (curves 2, 3) of AM; r = 0.5 (1), r = 1 (2), r = 2 (3). C. CD spectrum of poly(dC) in the absence (curve 1) or presence (curve 2) of AM, r = 2.

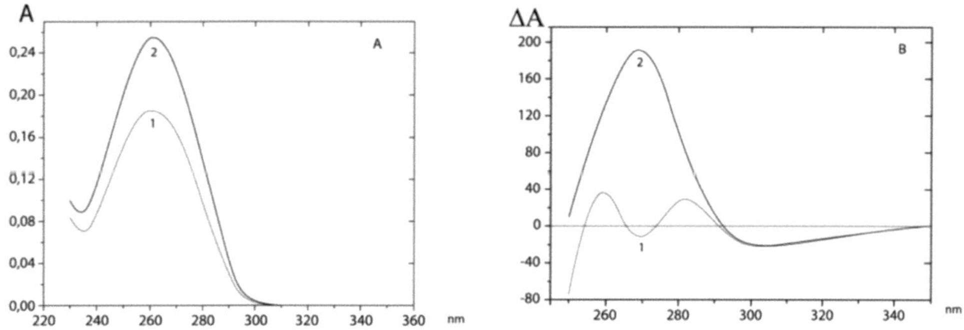

Figure 14. A. Absorption spectra of poly(dA)–$(dT)_{16}$ (curve 1) and an equimolar mixture of poly(dA) and $(dT)_{16}$ (curve 2), r =1. B. CD spectra of of poly(dA)–$(dT)_{16}$ (curve 1) and an equimolar mixture of poly(dA) and $(dT)_{16}$ (curve 2), r = 1.

The results shown in Figs. 15A and 15B demonstrate that CS does not form a tight complex with DNA, because the DNA absorption spectrum was almost completely restored after annealing in the presence of CS. Another situation was observed with HA (Fig. 15A): a hyperchromic effect of about 10-11% was observed in this case. This finding suggests complexation of HA with DNA regions, because such complexation should prevent the DNA duplex from being restored after annealing in the presence of HA.

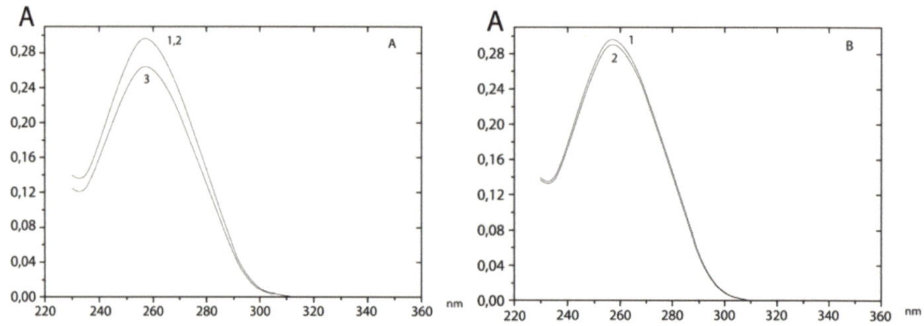

Figure 15. A. Absorption spectra of calf thymus DNA in the absence (curve 1, before annealing; curve 2, after annealing) or presence (curve 3, after annealing) HA; r = 1. B. Absorption spectra of calf thymus DNA in the presence of CS before (curve 1) and after (curve 2) annealing, r = 1.

Dot hybridization showed that DNAs of all organisms specifically binds with oligodeoxyribonucleotide probes (Fig. 16). Although the method is semiquantitative, it is possible to state that the GC tandem repeat has more complementary sequences in DNA as compared with monopurine (A) or monopyrimidine (C) homopolymers. For instance, poly(dT) and poly(dGC) arrays are rather abundant in human DNA, while poly(dG) occur in a smaller amount. Garlic DNA showed a greater amount of poly(dG) as compared with human DNA. This result agrees with the modern views of the plant genome (Zelenin A.V. 2003).

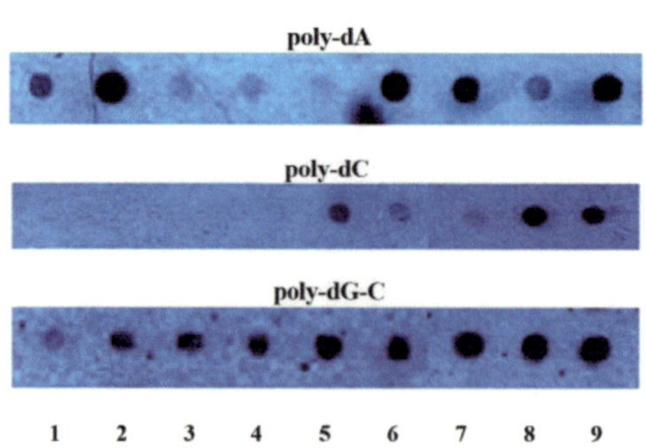

1. glucuronoxylans
2. polyuronides
3. CS
4. HA
5. amylase
6. garlic DNA
7. human DNA
8. fish DNA
9. calf thymus DNA

Figure 16. Dot hybridization of polysaccharides and DNA with oligodeoxyribonucleotide probes. The membrane was autoradiographed.

Since we aimed at studying the possibility of specific bonding between NA and polysaccharides, the results obtained with DNA were used exclusively as a control.

Biopolymer in a dot	Dot radioactivity, cpm			% of maximal count		
	Poly(dA) probe	Poly(dC) probe	Poly(dGC) probe	Poly(dA) probe	Poly(dC) probe	Poly(dGC) probe
Glururonoxylans	165	47	40	33	5	26
Polyuronides	496	18	107	**100**	2	69
HA	102	8	155	20	0	**100**
CS	2	65	97	0	7	63
Amylose	61	868	40	12	**100**	69
Garlic DNA	860	466	390	34	15	38
Human DNA	925	182	394	36	6	38
Fish DNA	2556	2990	1034	100	100	100
Calf thymus DNA	2161	1570	979	85	52	95

Table 8. Radioactivity of dots of polysaccharides and DNA after hybridization with purine-pyrimidine DNA probes

The results of autoradiography (Fig. 16, Table 8) demonstrated that the poly(dA) probe specifically hybridized with oat polyuronides to an extent comparable with that of DNA (garlic DNA). The poly(dC) probe specifically hybridized to a high extent with potato amylose. Since oat polyuronides consist almost exclusively of hexuronic acid residues and potato amylose, of hexose residues, we observed significant specific hybridization of pyrimidines (exemplified by C) with polyhexoses and purines (exemplified by A) with polyuronic acids.

The data obtained with glucuronoxylans were interpreted according to the same principle. It is known that glucuronoxylans contains about one carboxyl group per four neutral saccharide residues; i.e., the content of carboxyl groups is three or four times lower than in polyuronides. Hybridization of glucuronoxylans with poly(dA) was indeed about threefold lower than that of polyuronides.

The regular tandem poly(dGC) probe efficiently hybridized with all polysaccharides and especially with GAG. The highest hybridization efficiency was observed with HA. This finding agrees with the tandem structure of HA consists of glucose–glucuronic acid dimers, which interacted with the purine–pyrimidine dimers of the synthetic probe NA.

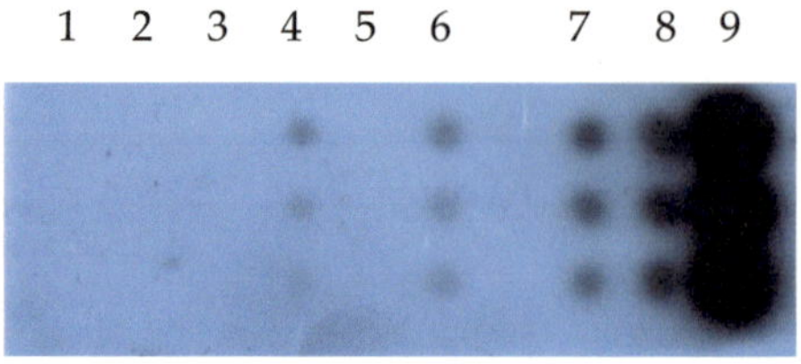

1, garlic glucuronoxylans
2, oat polyuronides
3, animal CS
4, animal HA
5, potato amylose
6, garlic DNA
7, fish DNA
8, phage DNA
9, calf DNA

Figure 17. Dot hybridization of polysaccharides and DNAs with calf thymus DNA. The membrane was autoradiographed.

Figure 17 shows the results of dot hybridization of polysaccharides and DNAs of various organisms with fragmented calf thymus DNA. The highest radioactivity was observed in dot 9, i.e., for self-hybridization of calf genomic DNA. In addition, repetitive elements of calf thymus DNA detected homologous sequences in virtually all genomic DNAs examined. Of all polysaccharides, only HA (dot 4) had matching structures in calf thymus DNA and formed specific complexes similar in radioactivity to DNA–DNA complexes.

3.3. Glycan biosynthesis in rat liver

The profile of elution of rat liver and marker polysaccharides from DEAE cellulose with a NaCl gradient is shown in Fig. 18. As markers, we used homopolymers of uronic acid (oat polyuronides) and glucose (potato amylose).

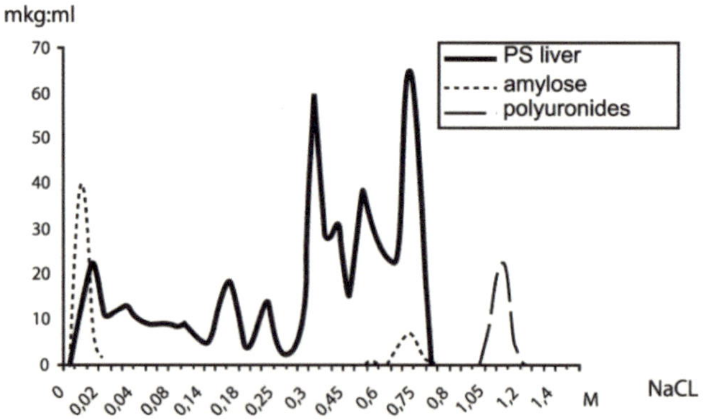

Figure 18. Profiles of elution of rat liver and marker polysaccharides from DEAE cellulose with a NaCl gradient.

As evident from the elution profile, the rat liver contains a complex spectrum of polysaccharides. Analysis with the marker homoglycan amylose, consisting of glucose residues, showed that polymers with a similar structure of neutral sugars are absent from rat liver glycans isolated by our method (such polymers were eluted from the column during washing). Glucose homoglycans were eluted with 0.01 M NaCl. The first peak of liver glycans was detected at 0.02-0.03 M NaCl. With NaCl increasing to 0.15 M, we eluted polysaccharides with a negative charge greater than in polymers of neutral sugars and lower than in the 0.25 M NaCl fraction, which mostly consisted of HA, as commonly known (Zimnitskii A.N., et al. 2004). The fractions 0.7 and 1.5 M NaCl exemplified a classical separation of CS and HS. As expected, highly charged oat polyuronides were eluted at the highest ionic strength, at about 1.2 M NaCl.

The results of quantitating uronic acids, hexosamines, and NS in the chromatographic fractions are summarized in Table 9.

Fraction	Uronic acids	Hexosamines	Neutral sugars
0.15 M	34.1±0.3	2.0±0.1	48.5±0.3
0.25 M	8.3±0.1	7.1±0.9	-
0.7 M	61.5±0.7	63.7±0.6	2.1±0.2
1.5 M	64.2±0.5	57.9±6.4	2.2±0.2

Table 9. Concentrations (μg/ml) of uronic acids, hexosamines, and NS in fractions of liver polysaccharides of intact rats (rats were not subjected to external influence)

In the 0.25, 0.7, and 1.5 M fractions, the ratio uronic acids : hexosamines was about 1:1, which is characteristic of HA, CS, and HS and agrees with published data. A minor amount of NS detected in the 0.7 and 1.5 M fractions could be explained by the presence of a trisaccharide fragment linking GAG with the core protein in proteoglycans.

Subfraction	Nuclei	Microsomes	Homogenate
0.02 M	265±18 / 67±7	108±12 / 52±7	128±15 / 67±8
0.06 M	36±5 / 7±2	24±3 / 40±6	34±6 / 22±3
0.15 M	20±3 / 6±2	18±2 / 50±6	33±5 / 81±11

Table 10. Contents (μg per g tissue) of NS/uronic acids in the 0.15 M NaCl saccharide fractions of cell nuclei, microsomes, and a homogenate of the liver of intact rats

Subfraction	Nuclei	Microsomes	Homogenate
0.02 M	0.76	1.40	1.60
0.06 M	1.44	5.0	1.94
0.15 M	1.57	8.30	7.4

Table 11. Uronic acids per neutral trisaccharide in the 0.15 M NaCl saccharide fractions of cell nuclei, microsomes, and a homogenate of the liver of intact rats

The 0.15 M fraction contained glycans enriched in neutral sugars (Table 9). Three subfractions could be isolated in this fraction: the first peak eluted up to 0.02 M NaCl, the second peak eluted up to 0.06 M NaCl, and a plateau eluted up to 0.15 M NaCl. The primary structure of polysaccharide fragments of this fraction is characterized in Tables 10 and 11. As evident from Table 11, nuclear oligosaccharides had the NS : uronic acid ratio similar to 3:1, suggesting the presence of neutral trisaccharides containing no more than one uronic acid residue. Both microsomal and homogenate fractions harbored neutral oligosaccharides containing more than one uronic acid residue per trisaccharide, the proportion of uronic acids increasing sixfold from the 0.02 to the 0.15 M subfraction. This finding testifies that uronic acid is linked to trisaccharides mostly in cell microsomes. Only one uronic acid residue per trisaccharide is linked in cell nuclei.

Substance	Content
DNA_n	9.26±0.30
0.15 M_n fraction of PS (Fig. 18)	1.42±0.12
HA_n	1.46±0.14
CS_n	2.56±0.16
HS_n	0.68±0.07
DNA_h	167.61±6.94
0.15 M_h fraction of PS (Fig. 18)	42.02±4.13
HA_h	23.74±3.35
CS_h	23.60±3.31
HS_h	13.45±1.22

Note: Subscripts n and h correspond to the nuclear fraction and the homogenate, respectively.

Table 12. Contents (μg per g tissue) of DNA and GAG in the nuclear fraction and the homogenate of the liver of intact rats

The DNA content and the GAG composition in the nuclear fraction and the homogenate of the rat liver are shown in Table 12. The total GAG content was 6.12 μg per g tissue in the nuclear fraction and 102.81 μg per g tissue in the homogenate. Nuclear GAG contained 0.15 M fraction PS to 23.20%, HA to 23.86%, CS to 41.83%, and HS to 11.11%. GAG of the homogenate contained 0.15 M fraction PS to 40.87%, HA to 23.09%, CS to 22.95%, and HS to 13.08%.

Our isolation procedure yielded about 5% of nuclei in the intact form. The remaining nuclei, along with their contents, were in the liver homogenate.

Correlation analysis was performed to study the association between the biochemical parameters and the time after a glucose load. Since the empirical sample was small ($n<30$), we computed Pearson parametric empirical correlation coefficients and Spearman nonparametric rank correlation coefficients. The results are summarized in Tables 13-15.

As Table 13 shows, a strong positive correlation was observed between time and HS content in the homogenate ($r = 0.81$) and between the contents of HS in nuclei and the 0.15 M fraction in the homogenate ($r = 0.95$) and a strong negative correlation between the contents of the 0.15 M fraction and DNA in nuclei ($r = -0.71$).

	t (min)	DNAn	0.15 Mn	HAn	CSn	HSn	DNAh	0.15 Mh	HAh	CSh	HSh
t (min)	1.00	-0.01	-0.28	-0.31	-0.35	-0.54	-0.53	-0.34	-0.24	-0.58	**0.81**
DNAn	-0.01	1.00	**-0.71**	-0.41	0.62	0.54	-0.09	0.53	-0.19	0.09	0.35
0.15 Mn	-0.28	**-0.71**	1.00	0.45	-0.08	-0.12	0.29	-0.05	-0.13	0.37	-0.47
HAn	-0.31	-0.41	0.45	1.00	0.24	0.07	-0.18	-0.09	0.30	0.37	-0.55
CSn	-0.35	0.62	-0.08	0.24	1.00	0.66	0.06	0.54	0.03	0.50	-0.30
HSn	-0.54	0.54	-0.12	0.07	0.66	1.00	0.61	**0.95**	0.24	0.69	-0.24
DNAh	-0.53	-0.09	0.29	-0.18	0.06	0.61	1.00	0.64	0.37	0.43	-0.42
0.15 Mh	-0.34	0.53	-0.05	-0.09	0.54	**0.95**	0.64	1.00	0.12	0.62	-0.03
HAh	-0.24	-0.19	-0.13	0.30	0.03	0.24	0.37	0.12	1.00	-0.18	-0.50
CSh	-0.58	0.09	0.37	0.37	0.50	0.69	0.43	0.62	-0.18	1.00	-0.34
HSh	**0.81**	0.35	-0.47	-0.55	-0.30	-0.24	-0.42	-0.03	-0.50	-0.34	1.00

Note: Significant correlation coefficients are in bold.

Table 13. Pearson empirical correlation coefficients (r) between the time after a glucose load and the DNA and GAG contents in the rat liver

	t (min)	DNAn	0.15 Mn	HAn	CSn	HSn	DNAh	0.15 Mh	HAh	CSh	HSh
t (min)	1.00	-0.15	-0.15	-0.15	-0.34	-0.47	-0.68	-0.15	-0.32	-0.49	0.63
DNAn	-0.15	1.00	**-0.74**	-0.36	0.54	0.48	-0.20	0.35	0.01	-0.01	0.34
0.15 Mn	-0.15	**-0.74**	1.00	0.53	0.07	-0.10	0.31	0.10	-0.04	0.41	-0.40
HAn	-0.15	-0.36	0.53	1.00	0.45	0.23	0.19	0.12	0.40	0.40	-0.50
CSn	-0.34	0.54	0.07	0.45	1.00	0.67	0.14	0.62	0.26	0.50	-0.07
HSn	-0.47	0.48	-0.10	0.23	0.67	1.00	0.30	**0.84**	-0.04	**0.84**	0.16
DNAh	-0.68	-0.20	0.31	0.19	0.14	0.30	1.00	0.19	0.61	0.38	-0.45
0.15 Mh	-0.15	0.35	0.10	0.12	0.62	**0.84**	0.19	1.00	-0.24	**0.74**	0.36
HAh	-0.32	0.01	-0.04	0.40	0.26	-0.04	0.61	-0.24	1.00	-0.17	-0.49
CSh	-0.49	-0.01	0.41	0.40	0.50	**0.84**	0.38	**0.74**	-0.17	1.00	0.00
HSh	0.63	0.34	-0.40	-0.50	-0.07	0.16	-0.45	0.36	-0.49	0.00	1.00

Note: Significant correlation coefficients are in bold.

Table 14. Spearman rank correlation coefficients (R) between the time after a glucose load and the DNA and GAG contents in the rat liver

Table 14 shows three strong correlations: a negative one between the contents of the 0.15 M fraction and DNA in nuclei and positive correlations between the HS content in nuclei and the CS content in the homogenate and between the contents of the 0.15 M fraction and CS in the homogenate (R = 0.74).

To identify latent variables uniting the parameters under study, factor analysis was performed using the Varimax procedure (Table 15).

Parameter	Factor 1	Factor 2	Factor 3
t (min)	-0.490	0.494	-0.473
DNAn	0.656	0.656	-0.189
0.15 Mn	-0.064	**-0.851**	-0.009
HAn	0.054	**-0.768**	0.019
CSn	**0.804**	-0.064	-0.063
HSn	**0.913**	0.061	0.361
DNAh	0.386	-0.106	**0.728**
0.15 Mh	**0.863**	0.159	0.279
HAh	-0.112	0.001	**0.866**
CSh	**0.776**	-0.500	-0.041
HSh	-0.146	0.692	-0.573
Proper factor value	3.687	2.762	2.083
Variance portion accounted for by the factor	0.335	0.251	0.189

Note: Significant correlations between the parameter and the factor are in bold.

Table 15. Factor analysis (Varimax) of the time after a glucose load and the DNA and GAG contents in the rat liver

As evident from Table 15, Factor 1 included CSn, HSn, the 0.15 Mh fraction, and CSh. Factor 2 included the 0.15 Mn fraction and HAn. Factor 3 included DNAh and HAh.

The most important result of correlation analysis is that strong linear correlations were observed between the time after a glucose load and the content of HS in the liver homogenate and between the contents of the 0.15 M fraction and DNA in the nucleus. In the first case, we record an increase in the time of synthesis of GAG modified to the greatest extent (at least two sulfo groups per disaccharide fragment). This result was conformed by factor analysis, since the time and the HS content in the homogenate are components of one factor. In the second case, we possibly establish a functional relationship between synthesis of nuclear oligosaccharides contained in the 0.15 M fraction and the genetic apparatus of the cell. The two biopolymers involved belong to different classes, but their biosynthesis utilizes the same initial substrate, glucose. The strong negative correlation suggests competition for the initial substrate between the two biosynthetic pathways. Since the nuclear DNA content was constant under the conditions of our experiments, the negative correlation suggests that synthesis of saccharides belonging to the 0.15 M fraction increases when free glucose enters cells in considerable amounts. This assumption is supported by the results shown in Fig. 19.

The composition of the 0.15 M fraction and GAG in nuclei, microsomes, and the homogenate is shown in Fig. 20. It is seen that nuclear GAG are enriched in CS (about 42%), which agrees with the accepted views (Silbert J.E. Sugumaran G ., 1995). Homogenate GAG consist mostly of HA and CS (about 23%). In microsomes, all GAG types occur in nearly equal proportions (about 12%). The results obtained for the 0.15 M fraction, which consists of three main subfractions (Table 11), are shown in Fig. 21. The fraction occurred at a high content (about 60%) in microsomes, while its content in the nucleus was no more than 23%. The homogenate was intermediate between nuclei and microsomes in the content of this fraction.

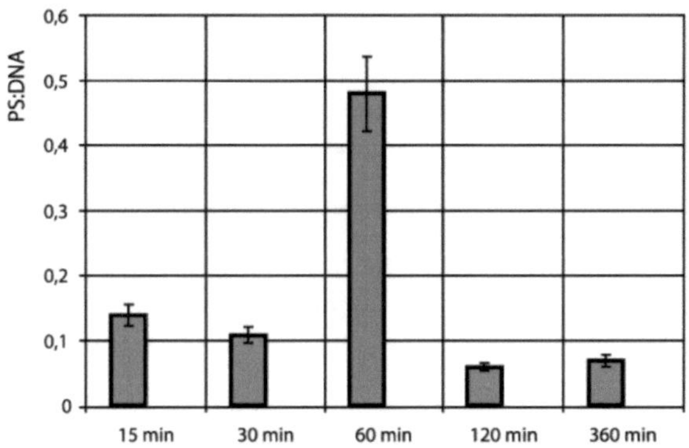

Figure 19. Accumulation of polysaccharides (PS) of the 0.15 M fraction in the nucleus. The content of the fraction was normalized with respect to the DNA content.

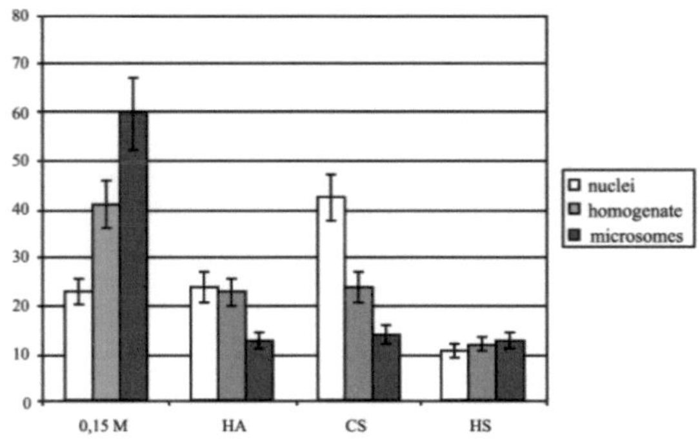

Figure 20. Relative contents (%) of polysaccharides in nuclei, microsomes, and the homogenate of the rat liver.

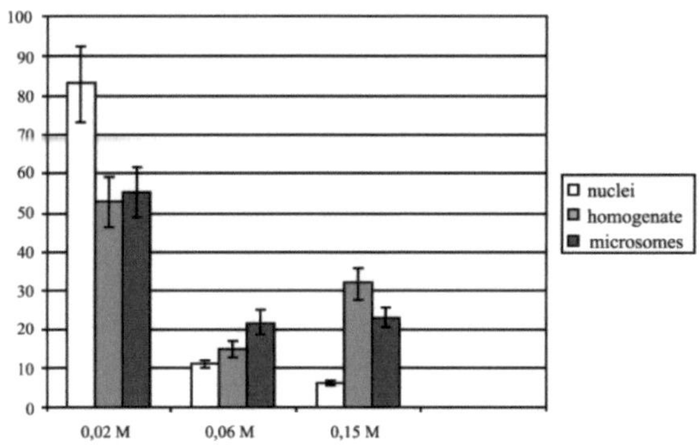

Figure 21. Relative contents (%) of polysaccharides differing in the portion of neutral sugars in the 0.15 M fractions of nuclei, microsomes, and the homogenate of the rat liver.

To study biosynthesis of polysaccharide fragments, we analyzed the time course of incorporating radiolabeled glucose in polysaccharide chains (Fig. 22).

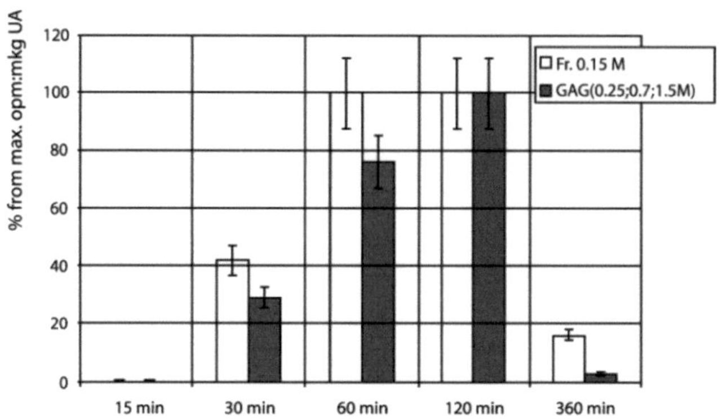

Figure 22. Growth to the maximal specific radioactivity of polysaccharides as dependent on the time of label presence in rats. The maximal specific radioactivity was taken as 100% for each fraction.

The highest radioactivity of the 0.15 M fraction was detected 60-75 min after a glucose load, while mature GAG (fraction 0.25-1.5 M) showed the highest radioactivity after 120 min. Thus, initiation of glycan synthesis is probably associated with the 0.15 M fraction, enriched in neutral oligosaccharides, which is consistent with common views of proteoglycan biosynthesis.

To identify the subcellular structures where generation of a polysaccharide chain is initiated, we studied the dynamics of label incorporation in the nuclear and microsomal fractions of the rat liver. The results of this experiment are shown in Fig. 23.

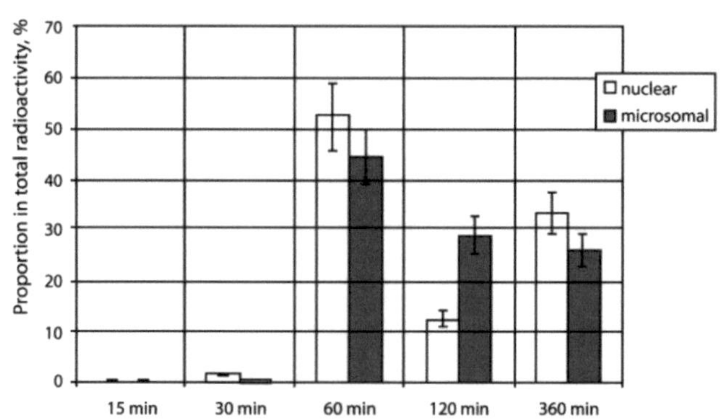

Figure 23. Dynamics of accumulation of radiolableled glucose in the nuclear and microsomal fractions of the rat liver. As 100%, we used the radioactivity averaged over all fractions and all time points.

As Fig. 23 demonstrates, the label became detectable in cell structures 30 min after a glucose load. The label accumulated to the highest content in both fractions within 1 h of the experiment. It should be noted that, in the first hour, the accumulation rate was higher in the nuclear fraction (radioactivity 53% compared with 44% in the microsomal fraction).

After 120 min of the experiment, the microsomal fraction accumulated a greater amount of the label as compared with nuclei (29 and 12%, respectively). After 3 h, the label was again accumulated more intensely in nuclear structures (33% compared with 26% in microsomes).

4. Discussion

Our results demonstrate that polysaccharides are similar to NA in many aspects. To a great extent, this is a consequence of the fact that NA is a polymer of ribose or deoxyribose having a base as a side-chain substituent. Some polysaccharides (e.g., HA, amylose, cellulose) are themselves capable of forming helical structures similar to NA.

Quantum chemical computations revealed the possibility of selective bonding between UDP-uronic acids and purines and between UDP-hexoses and pyrimidines contained in NA. The bonding strength per monomer unit in the polysaccharide–NA complex is similar to that computed for the NA double helix. This selectivity allows us to assume a binary genetic code for polysaccharides on the basis of the above principles in addition to the commonly accepted genetic code for amino acids. Quantum chemical computations were supported by the results of dot hybridization and spectral analysis of NA–polysaccharide complexes.

The spectrophotometric studies showed that amylose selectively binds with polypyrimidines (poly(dC)) while polyuronides bind with polypurines (poly(dA)). As well known, amylose is a homopolysaccharide of hexose and polyuronide, of hexuronic acid. The difference between these monosaccharides is that glucose contains the hydroxymethyl group at C5, while hexuronic acid has the carboxyl group at C5. Apart from this, the two monosaccharides are identical. Hence the observed difference in physicochemical properties of the above polysaccharides can be interpreted in terms of the structural difference at C5 of their units. Consequently, our results suggest that purines (exemplified by A) selectively bind to the carboxyl group of uronic acid residues, while pyrimidines interact with the hydroxymethyl group of hexose residues contained in polysaccharides. These experimental findings agree with the results of quantum chemical computations performed for the NA–polysaccharide complexes, testifying again to their adequacy.

The bonding of nucleotides and glycans was detected in experiments aimed at studying the interaction between mature GAG (HA and CS) with fragmented calf thymus DNA. In primary structure, HA is a multiple tandem repeat of a unit consisting of a hexose and an uronic acid residue. In turn, tandem repeats of purine–pyrimidine account for a considerable proportion of DNA in higher organisms, including calf thymus DNA (Kiselev L.L. 2000, Singer M., et al. 1998). As our results demonstrate, HA did find complementary regions in calf thymus DNA. CS showed no complementary interactions with calf thymus DNA. It is known that C6 of hexose residues of CS is modified with the sulfo group, which provides an additional partial negative charge. This modification of hexoses dramatically affects the physicochemical properties of the polysaccharide and probably prevents hydrogen bonding.

Dot hybridization confirmed possible complementarity of hexoses to pyrimidines and uronic acid to purines. The specific bonds between NA and polysaccharides are comparable in strength with the bonds between complementary nucleotides in DNA.

Thus, analysis of the interactions between polysaccharides and NA showed that purines of NA are complementary to uronic acids of polysaccharides and that pyrimidines are complementary to hexoses. The relationships between the genetic apparatus of the cell and polysaccharides deserve further studying in terms of the above complementarity principle. Such studies will probably yield a fundamentally novel view as to whether microheterogeneity of polysaccharide moieties of proteoglycans is genetically determined and related to purine–pyrimidine DNA repeats. Yet it is clear that the nature of these relationships need additional comprehensive studies.

The idea that polysaccharide moieties of proteoglycans contain information is not new. For instance, Zimina (Zimina N.P., et al. 1992, Zimina N.P., et al. 1987, Zimina N.P., et al. 1986, Zimina N.P., Rykova V.I., Dmitriev I.P. 1987) demonstrated that carbohydrate chains of proteoglycans are chemically heterogeneous and structurally irregular and concluded, quite justifiably, that the information content of proteoglycans provides a chemical basis for their intricate and highly specific functions in the cell. There is still no method for GAG sequencing, and data on the structure of their chains are circumstantial. Since chemical heterogeneity is widespread, these data make it possible to assume that proteoglycan chains are of irregular structure with a cluster arrangement of disaccharides. The clustering of bonds sensitive to testicular hyaluronidase and, consequently, of glucuronic acid residues was demonstrated for pig skin dermatan sulfate. The clusters are arranged along the chain without any distinct regularity (Fransson L.A., et al. 1982). Likewise, glucuronic acid-containing disaccharides are clustered at random along the chain of dermatan sulfates from the pig intestinal mucosa and the umbilical cord (Fransson L.A., et al. 1982). The clustering of various uronic acid residues is characteristic of dermatan suflate from the human uterine neck. The irregularity of dermatan sulfate chains was suggested from chromatographic elution profiles of dermatan sulfate isolated from the bovine aortic intima and digested with chondroitinase AC (Oegema T.R., et al. 1979). Structural studies revealed a cluster organization of HS and heparin chains. Their molecules include extended regions consisting of low-sulfated disaccharides that contain glucuronic acid residues as a main component and only a minor amount of glucose residues (Bjork I., et al. 1982). Clusters sulfated to a high extent were also found: they consist of highly charged disaccharides harboring glucose residues as a main component. These clusters are possibly separated by less ordered regions where alternating sugars occur in similar proportions. Molecules vary in size of highly and low-charged regions.

It should be noted that the structure of oligosaccharide sites is an informative element of GAG as opposed to NA, which contain information in the form of a strict sequence of monomeric units (nucleotides). For instance, the capability of self-association of proteoHS and proteodermatan sulfates is due to so-called contact zones, specific regions with alternating disaccharides containing glucuronic acid and glucose residues with a certain arrangement of sulfo groups (Franson L.-A. 1982). It is noteworthy that polysaccharides are characterized by a determined structural–functional interdependence similar to that of NA (Zimina N.P., et al. 1992).

The concept of nontemplate synthesis of polysaccharide components of proteoglycans does not allow a genetically grounded explanation of microheterogeneity (polymorphism) of proteoglycans. Yet data are continuously accumulating that polysaccharide components of proteoglycans are highly polymorphic and that their polymorphism shows distinct tissue, organ, and species specificities. It should be noted that glycans are still poorly understood and, consequently, the current views of glycans are similar to the views of NA in early molecular biology. Compounds belonging to one class strikingly differ in primary structure. We know today that, for instance, every mRNA is a unique element in realization of genetic information, but such functions are still to be elucidated in the case of glycans.

We characterized the fraction composition of polysaccharides contained in the rat liver homogenate. As Fig. 18 shows, rat liver polysaccharides include polymers of neutral sugars, hexoses, and hexuronic acids and vary in proportions of monomeric units and the degree of their modification (amination, sulfation, epimerization, etc.).

Correlation analysis revealed an association between time and synthesis of HS, which are GAG modified to the greatest extent. This result is beyond doubt because their modification is known to result in a high degree of sulfation and, consequently, their maturation takes more time as compared with maturation of other GAG groups.

In addition, the analysis results (Pearson empirical correlation coefficients and Spearman rank correlation coefficients) showed a strong linear correlation between cell nuclear structures (DNA) and the fraction of nuclear saccharides eluted from DEAE cellulose until 0.15 M NaCl. Such a correlation was not observed for total DNA and the 0.15 M fraction of the homogenate (oligosaccharides of microsomes, lysosomes, nuclei, other cell structures, and the intercellular matrix). The correlation testifies that nuclear synthesis of saccharides of this group increases after a glucose load, especially 1 h after loading. This finding was confirmed by quantitating polysaccharides with normalization with respect to the DNA content (Fig. 19).

The 0.15 M fraction of the homogenate distinctly correlated with the contents of mature proteoglycans in the homogenate and the nucleus, implicating saccharides of this fraction in GAG maturation. In other words, an increase in synthesis of this saccharide fraction of the homogenate increased synthesis of proteoglycans (CS and HS). The fact that proteoglycan synthesis starts with this fraction was evident from the data on incorporation of radiolabeled glucose in GAG. It should be noted that the 0.15 M fraction varies in composition among cell structures. The nuclear fraction harbors neutral sugar components and uronic acids at 3:0.75, suggesting incomplete addition of uronic acids to neutral trisaccharide units, which takes place in the nucleus. The 0.15 M fraction of nuclear saccharides consists mostly (to 83%) of components eluted at 0.02 M NaCl. The 0.06 and 0.15 M subfractions occur at 11 and 6%, respectively. In microsomes, the proportion of the 0.02 M subfraction is decreased to 55%, while the proportions of the 0.06 and 0.15 M subfractions are increased to 22 and 23%, respectively. The changes observed in the homogenate are similar to those in microsomes, but the proportion of the 0.15 M subfraction is greater (up to 32%).

These findings associate the genetic apparatus of the cell with synthesis of oligosaccharides similar in composition to the universal tetrasaccharide of proteoglycans. The association is also evident from the results obtained with radiolabeled glucose: the label accumulation rate in nuclear structures was higher than in other structures 1 h after a glucose load.

Oligosaccharides with a neutral sugar : uronic acid ratio of 3:1 are most probably synthesized in the nucleus. Heteroglycan chains with a tandem arrangement of uronic acid–hexose units (1:1) are formed in structures of the microsomal fraction.

Analysis of incorporation of radiolabeled glucose in the nuclear and microsomal fractions of the cell (Fig. 23) showed that the incorporation rate is more intense in the nucleus early (at the end of the first hour) after a glucose load. After 2 h, microsomes accumulate radiolabeled glucose to a high extent, which is probably due to a transfer of radiolabeled oligosaccharides from the nucleus and the formation of GAG chains. After 3 h, the content of radiolabeled saccharides in the cell nucleus increased again owing most probably to a transfer of mature GAG from the EPR. This scenario agrees with modern views of proteoglycan synthesis.

It is of interest that the label appeared first in the 0.15 M fraction and then in mature GAG (the 0.25-1.5 M fraction) of the homogenate (Fig. 22). The initially high radioactivity of this fraction can be explained by the presence of glycans involved in initiating synthesis of the polysaccharide chain of GAG. GAG synthesis starts with xylosylation of core proteins and generation of a linker tetrasaccharide; then, the polysaccharide chain is formed by consecutive addition of hexoses and uronic acids in almost equal proportions.

Comparison of the data on radiolabeled glucose incorporation in GAG and fractions of cell structures showed that the high rate of label accumulation in the nucleus early (within the first hour) after a load was associated with the dynamics of label incorporation in polysaccharides eluted until 0.15 M NaCl. This was not the case with mature GAG, which showed the highest radioactivity only 2 h after a load. At this time, the label content is minimal in the nucleus and higher in microsomes. Hence, we can state that generation of the GAG chain in the cell is associated with microsomes, while synthesis of the linker tetrasaccharide, which belongs to the 0.15 M fraction, is associated with nuclear structures.

Saccharides of the 0.15 M fraction differ from classical GAG in having another hexosamine content and probably harbor fragments with linker tetrasaccharides. Note that, until uronic acid–hexosamine ratio reaches 1:1 (as characteristic of HA), such fragments, having a considerable portion of neutral sugars, are eluted from the column at an ionic strength lower than necessary for HA elution (0.15 M NaCl). Since proteoglycan synthesis starts with generation of the linker tetrasaccharide, the radioactivity of this fraction should be higher early after a radiolabeled glucose load. Our results fully agreed with this expectation.

Thus, the glycoside moiety of proteoglycans is probably synthesized in a stepwise manner in the cell. The linker tetrasaccharides are synthesized in structures associated with the cell nucleus, which is also evident from the results of correlation analysis. Then the fragments are transferred into the EPR, where the main glycan chain is synthesized, biochemically modified, and used to form proteoglycans. The GAG chain is synthesized in the EPR. Mature GAG are delivered into the nucleus in a small amount and are mostly exported into the intercellular space to produce the intercellular matrix. This scenario agrees with data of many studies that core proteins entering the EPR and Golgi system already have the linker tetrasaccharide (xylose-galactose-galactose-uronic acid) (Colman Y., et al. 2000, Zimina N.P., et al. 1992,

Zimina N.P., et al. 1987, Silbert J.E. et al. 1995). These data indirectly support our assumption that DNA plays a role in synthesis of the universal tetrasaccharide of proteoglycans.

Based on our data, we suggest a conceptual mechanism underlying the functional association between the nuclear apparatus and the formation of the tetrasaccharide linking the protein core with the GAG moiety in proteoglycans.

Taken together, the results obtained with different methods allow us to propose a concept of template synthesis of proteoglycans with the involvement of tandem DNA repeats.

It is clear that proteoglycan synthesis starts with DNA transcription to yield pre-mRNA. The informative region of the mRNA for the core protein is formed according to the commonly accepted mechanism up to the serine codon. When the serine codon is followed by a triplet complementary to the trisaccharide uronic acid-hexose-hexose, the mode of transcription changes. It should be noted that the complementarity requirements are met only by eight DNA triplets (ACC, ACT, ATC, ATT, GCC, GCT, GTC, and GTT). The energy of bonding during NA synthesis is distributed as follows among these triplets: it is about 11.7 kcal/mol in one triplet (GCC), 9.6 kcal/mol in three triplets (GCT, GTC, and ACC), 7.6 kcal/mol in three triplets (GTT, ACT, and ATC), and 5.5 kcal/mol in one triplet (ATT). Thus, the ATT triplet has the bonding energy of only 5.5 kcal/mol per monomer, lower than in the triplet duplex DNA-polysaccharide (more than 7 kcal/mol).

Under certain conditions, this situation with the ATT codon may lead to appreciable competition with UDP-sugars in the formation of complementary pairs during RNA synthesis on DNA. As a result, synthesis of a trisaccharide may be more advantageous in terms of energy than addition of three structural units to RNA. Hence the following process is possible. Glycosyltransferase utilizes the first uronic acid residue bound to DNA adenine via complementary interactions and generates a bond between the saccaride and the last ribonucleotide of the serine codon of RNA, simultaneously converting uronic acid to xylose via decarboxylation at C_5. Decarboxylation initiates separation of the saccharide unit from DNA. Then, Gal transferases generate glycoside bonds between xylose and two hexose (galactose) residues, which are hydrogen-bonded to the thymine tandem of DNA according to the above scheme.

It is clear that some pre-mRNAs synthesized *in vivo* under such conditions have the trisaccharide xylose-galactose-galactose in place of a ribonucleotide triplet at the site of RNA branching on the ATT codon of DNA. It is well known that the ATT triplet determines a stop codon (UAA) terminating synthesis of polypeptide chains, which indirectly supports our assumption. If this sequence is followed by tandem repeats (purine-pyrimidine, such as CA) with a strength of bonding in NA synthesis about 8.8 kcal/mol, synthesis of oligosaccharide is disadvantageous and the RNA is extended according to the DNA-RNA complementarity rule, which is more advantageous in terms of energy. Although the energy of bonds plays an important role in this situation, we assume that the process is still far more intricate.

Thus, it is possible that hnRNA contains pre-mRNAs that have a trisaccharide followed by a tandem ribonucleotide repeat consisting of more than 300 monomers (which is sufficient for generating the glycoside moiety of proteoglycans) and starting from 5′-P-terminal guanosine at the nuclear RNA branching site. This can explain why the corresponding RNA regions block reverse transcription and are resistant to some RNases. Guanine acts as a hnRNA branching site and, as a purine, is capable of ensuring further addition of the first uronic acid residue, which always follows the trisaccharide. With the above system of complementary interactions, the trisaccharide is analogous to UAA and, when immediately followed by GU, allows the processing and splicing of pre-mRNA according to the commonly accepted scheme. Like adenines, the two galactose residues may interact with two uridines of the spliceosome U-RNA through their hydroxymethyl groups. Thus, the sequence Gal-Gal-G-U provides a binding site for spliceosome structures, as characteristic of the splicing tetranucleotide consensus sequence (AAGU).

In such an RNA strand, saccharides are linked by the 3′-5′ bond up to the end of the serine codon. C_2 of the saccharide of the last nucleotide of the serine codon interacts with C_5 of xylose, which is thereby capable of binding through C_1 with serine when xylosylation is initiated. Then xylose C_4 binds with C_1 of a galactose dimer. C_2 of the last galactose interacts with C_5 of the saccharide moiety of guanosine. Thus, C_3 of the last nucleotide of the serine codon and C_3 of the last galactose are free for bonding, which allows a 3-1 bond with the first glucuronic acid residue. Uridine following guanosine in an RNA intron determines the addition of a hexose to the glucuronic acid residue through the 4-1 (or 3-1) bond during template synthesis of a polysaccharide (Fig. 24).

Invariant GG at the 5′ end of the next exon, continuing the protein-coding sequence, represents the first two nucleotides of a glycine codon (GGA, GGG, GGC, or GGU). This GG is in the trisaccharide site (the RNA branching site) and, as the flanking intron is excised in the processingosome during RNA maturation, continues the mRNA coding region after the serine codon in the 5′-3′ direction. The obligatory presence of invariant splicing-site GG in the triplet following the serine codon can explain the conservation of the xylosylation site (serine-glycine tandem) among core proteins, because only glycine codons start with two guanines.

The resulting transcript is capable of directing synthesis of a polypeptide chain wherein serine is covalently bound to the trisaccharide and then with glycine. Tandem RNA sequences, representing the intron side chain following the tetrasaccharide (xylose-galactose-galactose-uronic acid), allow a polysaccharide fragment to be synthesized according to the NA base–monosaccharide complementarity by glycosyltransferases that form 4-1 or 3-1 glycoside bonds, starting from the last saccharide. After splicing occurs and the two guanines find themselves in the exon part of the molecule, the tandem RNA fragment is linked to C_2 of uronic acid through the terminal uridine. The fragment determines an ordered arrangement of monosaccharides for synthesis of a polysaccharide chain. This process is advantageous (bonding energy is about 7 kcal/mol) in systems having an excess of UDP-saccharides and glycosyltransferases and lacking mononucleotides and NA polymerases, as characteristic of membrane structures of EPR and the Golgi complex. We propose that the process is termed glycotranscription, because information contained in NA is directly transferred to the polysaccharide chain.

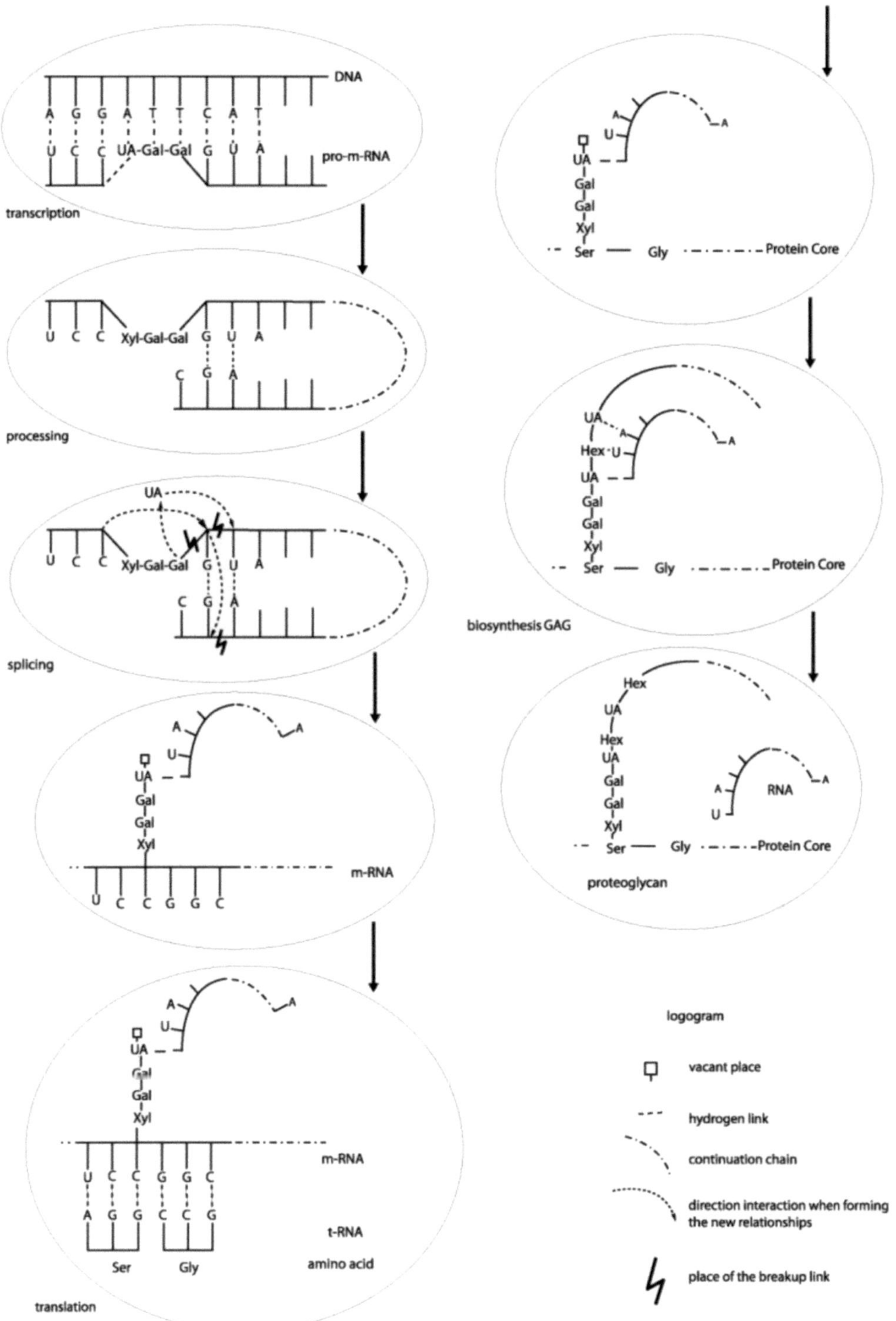

Figure 24. Hypothetical scheme of genetically determined template synthesis of proteoglycans.

It is clear that synthesis of the core protein chain on ribosomes is quite possible, because the tetrasaccharide does not occupy the 3'-5' bond of the last nucleotide of the serine codon and acts as a spacer linking the RNA intron. As a result, the newly synthesized core protein contains the linker tetrasaccharide and the RNA intron attached to the serine. The protein is delivered after synthesis into the site where the corresponding polysaccharide fragment is generated in membrane structures of the EPR and the Golgi complex. Serine is xylosylated in the EPR during biosynthesis of the core protein on ribosomes. This scenario does not contradict the modern views of GAG synthesis (Silbert J.E. et al. 1995).

It should be noted that pre-mRNA splicing is tissue-specific. Moreover, selection of the splicing site depends on the developmental stage in some cases (Singer M., et al. 1998). In other words, the primary structure of mRNA introns varies with tissue and ontogenetic stage. The same is true for the primary structure of the glycoside moiety of proteoglycans. It is known that proteoglycans are completely absent from unicellular and prokaryotic organisms, as well as pre-mRNA processing is. This fact suggests a relationship between glycan biosynthesis and mRNA maturation in terms of biological significance for the cell.

This assumption is supported by data obtained for mRNAs of CS core proteins. Translated in a cell-free wheat germ system, cartilage mRNA directed synthesis of a 340-kDa core protein. Immediately after translation, the protein already contained glycosylation signals for subsequent glycan synthesis (Hook M., et al. 1984), i.e., the linker tetrasaccharide was already present.

Based on quantum chemical analysis of the advantage of bonding, the hypothesis of proteoglycan synthesis of the NA template (Fig. 24) agrees with the views of the processes involved in realizing genetic information (genome structure, transcription, hnRNA processing and splicing, translation) (Singer M., Berg P. 1998) and eliminates the main contradictions of the existing concept of proteoglycan metabolism. The hypothesis explains why the site where the linker tetrasaccharide is attached to the protein core has not been identified in more than fifty years of studies on proteoglycans: the tetrasaccharide is already contained in RNA before translation. There is convincing evidence that the tetrasaccharide finds its way in the EPR as covalently bound to the core protein (Colman Y., et al. 2000, Silbert J.E. et al. 1995). A role of protein core structures in determining the serine xylosylation site was rejected in recent studies. A "vitalistic" hypothesis has been formulated that ascribes this role to intracellular membranes. The hypothesis is based on the fact that glycosyltransferases are mostly in membrane structures of the EPR and Golgi complex (Silbert J.E. et al. 1995). Yet this fact alone does not prove synthesis of the linker tetrasaccharide in these structures, because glycosyltransferases are also detectable in nuclei (nuclear membranes). Our biochemical studies implicate structures of the cell nucleus in initiation of synthesis of the linker tetrasaccharide.

According to our hypothesis, a serine is subject to xylosylation only when its codon is followed in DNA consequently by the ATT stop codon and CA (GU in RNA) of an intron, responsible for hnRNA branching and mRNA processing.

The above scheme of proteoglycan synthesis allows generation of a linear heteroglycan of a particular size (about 300 monomers). The question is how template synthesis following this scheme proceeds in the case of HA, consisting of thousands of monomers. It is clear that RNA templates of a corresponding size are absent from EPR structures. It is possible that GAG synthesis in this case utilizes lasso-like intron RNA, which result from cis-splicing and are potentially capable of directing cyclic synthesis of HA with any number of monomeric units. It is known that mRNA trans-splicing yields linear introns. Such RNA structures may direct template synthesis of branched homoglycans whose branching period is a multiple of the full NA helix turn (10-12 units). Monosaccharide 10 (12) overlies monosaccharide 1 of the nascent strand under these conditions, which allows an additional bond between them. Then synthesis proceeds through several such rounds to yield a branched polysaccharide structurally similar to glycogen. This hypothetical mechanism generating branched polysaccharides is supported by the fact that their branching period is a multiple of the full NA helix turn.

RNAs with the above structures may be components of small cytoplasmic RNA, the role of which is still poorly understood. It is known that scRNA account for less than 1% of total cell RNA. Some scRNAs contain tandem repeats, in particular, *Alu* sequences. These scRNAs are associated with membrane structures of the EPR. Some scRNAs (7SL RNA) are involved in transmembrane transport of polypeptides across the EPR lipid bilayer. The scRNA size varies from 90 to 330 nt, which falls well into our concept. It should be noted that synthesis of storage homoglycans probably requires no template and proceeds through a simpler mechanism, because their chains carry no information.

Many glycoproteins act as antigens on the plasma membrane. Immunohistochemical comparison of glycoproteins isolated from hepatocyte membranes of intact, embryonic, and regenerating liver and from hepatoma showed that the protein component of antigens is detectable on all hepatocytes regardless of the state of the liver. Moreover, the protein component is tissue-nonspecific and universal for most cell groups. The carbohydrate component of antigens proved to be specific. It is the structure of the polysaccharide component that determines the immunogenicity of a proteoglycan. Zimina (Zimina N.P., et al. 1987) studied the specifics of GAG synthesis in the liver of adult rats and embryos and in hepatoma and showed that GAG of tissues with active cell proliferation are sulfated to a lower extent as compared with the corresponding GAG of quiescent tissues. That is, polysaccharide fragments of the same proteoglycans differ depending on the state of hepatocytes.

According to the concept of nontemplate synthesis of GAG, the protein core is synthesized on polysomes of the rough EPR. The same protein components of proteoglycans were synthesized in all above cases. The identity of the protein core suggests that the glycoside components of these proteoglycans are synthesized in one site of the smooth EPR and, consequently, are also identical. Experimental results contradict this assumption. The variation of glycoside components is possible only when glycosyltransferase complexes of EPR membranes vary depending on the physiological state of hepatocytes. A glycosyltransferase complex of the smooth EPR should contain about 300 molecules of the enzyme generating the glycoside bond in a certain sequence. For instance, HS synthesis requires a complex of about 300 N-acetylglucosaminylatransferases and glucuronyltransferases occurring in equal proportions. Each of the 150 enzyme molecules, e.g., N-acetylglucosaminyltransferases, should be encoded by a separate gene with a nonconserved sequence coding for the site of enzyme attachment to a strongly specific membrane site of the smooth EPR. If GAG synthesis is template-independent, at least three types of such complexes are necessary for HS generation. About 450 N-acetylglucosaminyltransferase genes are required for this process. Since the enzyme is also involved in synthesis of heparin and other polysaccharides, the number of N-acetylglucosaminyltransferase genes should occur at thousands of copies per genome. Yet it is known that DNA regions complementary to mRNAs have unique sequences and occur at a few copies per genome. Glycosyltransferase genes are not exceptions to this rule. Thus, the commonly accepted hypothesis of nontemplate synthesis of GAG cannot explain the genetically determined heterogeneity of polysaccharide components of proteoglycans. This hypothesis does not stand up, since recent works have demonstrated the informative value of the glycoside moiety of proteoglycans (Zimina N.P., et al. 1992, Zimina N.P., et al. 1987).

Our scheme of template synthesis of GAG solves this problem, because only a few glycosyltransferase genes are sufficient for generating any diversity of polysaccharide chains in this case.

Synthesis of some glycans with a high information content according to the glycotranscription principle makes it possible to assume, by analogy with RNA, the existence of reverse glycotranscription, whereby information is transferred from polysaccharide to RNA and then, by reverse transcriptase, to DNA fragments, which can be inserted into the genome to preserve the acquired information about new glycans in the genome structure. Such information is contained in specific tandem DNA repeats, which are unique for each individual. The relevant genetic systems are probably capable of being transmitted to the progeny and being fixed as a hereditary or acquired character by selection. It seems that both nontemplate and genetically determined template synthesis of polysaccharides exist in nature and are closely associated with each other through information flows.

It should be noted in this connection that Ronichevskaya and Rykov (Ronichevskaya G.M., Rykova V.I. 1977) observed that GAG suppress DNA replication in proliferating cells in a keilon-like manner. This observation can be explained by homology of polysaccharides to NA. We think that polysaccharide fragments are capable of blocking DNA polymerases via specific binding to complementary DNA repeats. As reported earlier, the substances examined in (Ronichevskaya G.M., et al. 1977) were isolated from RNA preparations, which provides indirect evidence for a role of RNA in glycan synthesis. According to our results, HA fragments form stable bonds with NA.

Further studies of proteoglycan polymorphism (species and tissues specificity, age- and pathological changes) and development of sequencing technique will probably demonstrate the specificity of proteoglycans at the level of individual organisms. By analogy, polymorphism of highly repetitive genome sequences is now beyond doubt and is widely used in genome fingerprinting. We assume that it is genomic repeats that are responsible for the information structure of glycans, whose polymorphism (microheterogeneity) has come to be commonly recognized.

The results of many studies implicate proteoglycans in intricate processes regulating cell proliferation (Ronichevskaya G.M., et al. 1977) and differentiation (Kinoshita S., et al. 1979, Kinoshita S., Yoshii K. 1979), cell recognition, and the organization of intercellular interactions (Henkart P., et al. 1973). In evolution, polysaccharides played a key role in the transition from unicellular organisms to multicellular entities. It is owing to polysaccharides that the individual cell functions as a structure with a particular function in the cell ensemble of the total organism. In the ontogeny, proteoglycans are associated with the aging of cells, tissues, and organs (Zimnitskii A.N., et al. 2004). A genetic defect in synthesis of GAG polysaccharide chains leads to severe hereditary disorders (Zimnitskii A.N., Bashkatov S.A. 2004). It is beyond doubt that the relevant processes are controlled by DNA, which carries genetic information.

To conclude, life is a mode of existence of not only protein and nucleic, but also of polysaccharide bodies, because all three biopolymers determine life that we observe. Analysis of the distinct association of NA and polysaccharides will clarify the role of both polysaccharides and DNA repeats. The latter cannot be described now as waste or selfish, because some of them are potentially capable of directing polysaccharide synthesis.

We are grateful to prof. N.K. Yankovskii (Institute of General Genetics, Russian Academy of Sciences, Moscow) and Dr. V.I. Salyanov (Institute of Molecular Biology, Russian Academy of Sciences, Moscow) for help in collecting and interpreting empirical data.

Author details

Aleksandr N. Zimnitskii
«NanoDerm-profi» Ltd., Moscow, Russia

5. References

Altschuler C.H., Kinsman G., Bareta J.(1963). Connective tissue metabolism: I. Hyaluronic acid synthesis by cell-free extracts of human tissues. *Arch Pathol.* Feb;75:206-211.

Belozersky A.N. (1970). Methods for Studying Nucleic Acids. Moscow: Mir.

Blagoi Yu.P., Galkin V.L., Gladchenko G.O., Kornilova S.V. (1999). Metallocomplexes of Nucleic Acids in Solutions. Kiev: Naukova Dumka, 199p.

Bjork I., Lindahl. U. (1982). Mechanism of the anticoagulant action of heparin. *Mol Cell Biochem.*18:161-82..

Chaplin M.F., Kennedy J.F.(1986). Carbohydrate Analysis: A Practical Approach. Oxford, Washington.

Cifonelli J.A., Dorfman A.J.(1957). The biosynthesis of hyaluronic acid by group A streptococcus: V. The uridine nucleotides of group A streptococcus. *J Biol Chem.* Oct; 228(2):547-57.

Colman Y., Rem C.-G. (2000). Illustrative Biochemistry. Moscow: Mir.

Cybulski S.M., Scheiner S.J.(1989). Hydrogen bonding and proton transfers involving the carboxylate group. *J Am Chem Soc.*, 111, p.23-31.

Denhardt D.T.(1966). A membrane-filter technique for detection of complementary DNA. *Biochem Biophys Res Commun.*, 23:641-52.

Dewar M.J.S., Zoebisch E.G., Healy E.F., Stewart J.J.P. (1985). Development and use of quantum mechanical molecular models. 76. AM1: A new general purpose quantum mechanical molecular model. *J Am Chem Soc.*,107, p.3902-3909.

Dorfman A. (1958). Studies on the biochemistry of connective tissue. *Pediatrics.* Sep;22(3):576-89.

Dorfman A. (1964). Metabolism of acid mucopolysaccharides. *Biophys J.* Jan; 71:SUPPL 155-65.

Dorfman A.(1965). In: Structure and Function of Connective and Skeletal Tissue. London,178, 297.

Dunin V.V., Rukhadze E.G., Potapov V.M.(1979). Generation and Analysis of Optically Active Compounds. Moscow: Mosk. Gos. Univ.

Fransson L.A., Havsmark B. (1982). Structural requirements for heparan sulphate self-association. *Carbohydr Res.*, 16:135-44.

Franson L.-A.(1982). Structural features of the contact zones for heparan sulphate self-association. *Carbohydr Res.*,110:127-33.

Glaser L., Brown D.H.(1955). Purification and properties of d-glucose-6-phosphate dehydrogenase. *Proc. Natl. Acad. Sci. USA*, 41:253.

Henkart P., Humphreys S., Humphreys T.(1973). Characterization of sponge aggregation factor: A unique proteoglycan complex. *Biochemistry.*, 12:3045-50.

Hook M., Kjellen L., Jahansson S., Robinson J.(1984). Cell-surface glycosaminoglycans. *Annu Rev Biochem.*, 53:847-69.

Jeffrey D Esko, Zhang L.(1996). Influence of core protein sequence on glycosaminoglycan assembly *Curr Opin Struct Biol.* 5:663-70.

Jermyn M.A.(1975). Increasing the sensitivity of the anthrone method for carbohydrate. *Anal Biochem.*, 68:332-5

Jurema J.M.W., Shields G.C.(1993). Ability of the PM3 quantum-mechanical method to model intermolecular hydrogen bonding between neutral molecules. *J Comput Chem.* 14, p.89-104.

Kallies B., Mitzner R. (1995). The ability of the semiempirical PM3 method to model proton transfer reactions in symmetric hydrogen bonded systems. *J Mol. Model.*, 1, p.68-78.

Kinoshita S., Yoshii K.(1979). The role of proteoglycan synthesis in the development of sea urchins: II. The effect of administration of exogenous proteoglycan. *Exp Cell Res.*,123:361-9.

Kinoshita S., Saiga H. (1979). The role of proteoglycan in the development of sea urchins: I. Abnormal development of sea urchin embryos caused by the disturbance of proteoglycan synthesis. *Exp Cell Res.*, 123:229-36.

Kiselev L.L. (2000). Human genome and biology of the 21st century. *Vestn Ross Akad Nauk.*, 70: 414-24.

Latajka Z. Scheiner S.J. (1990). Structure, energetics, and vibrational spectrum of ammonia...water. *J Phys Chem.* 94, p.217-221.

Lindahl U., Kjellen L.(1987). Biology of Proteoglican. Wight T.N., Mecham R.P., Eds., Orlando: Acad. Press. p. 59.
Lorenzel I.(1959). Epinephrine-induced alterations in connective tissue of aortic wall in rabbits. *Proc Soc Exp Biol Med*. Nov;102:440-2.
Maley F., Maley G.F. (1959).The enzymic conversion of glucosamine to galactosamine. *Biochim Biophys Acta*. Feb;31(2):577-8.
Markovic O., Huttl S.(1964). Contribution to a methodical approach to determination of protein and glycoprotein components in synovial fluid. *Cas Lek Cesk*. May 8;103:522-5. Czech.
Markovitz A., Cifonelli J.A., Dorfman A. (1958). Biosynthesis of hyaluronic acid by cell-free extracts of group-A streptococci. *Biochim Biophys Acta*. May;28(2):453-5.
Maxam A., Gilbert W.(1977). A new method for sequencing DNA. *Proc Natl Acad Sci USA*. 74:560-4.
Mednikov B.M., Shubina E.A., Mel'nikova M.N. (2001). Molecular mechanisms of genetic isolation. *Priroda*, 5: 19-28.
Mulliken R.S.(1955). Electron population analysis on LCAO-MO molecular wave functions: II. Overlap populations, bond orders, and covalent bond energies. *J Chem Phys*. 23:1841-6.
Oegema T.R., Jr, Hascall V.C., Eisenstein R.(1979). Characterization of bovine aorta proteoglycan extracted with guanidine hydrochloride in the presence of protease inhibitors. *J Biol Chem*. 254:1312-8.
Orekhovich V.N., (1968). Ed Modern Methods in Biochemistry., Moscow: Meditsina.
Roberts J., Kaserio M.(1978). Basics of Organic Chemistry. Moscow: Mir.
Roden L., Dorfman A. (1958). The metabolism of mucopolysaccharides in mammalian tissues: V. The origin of L-iduronic acid. *J Biol Chem*. Nov;233(5):1030-3.
Rondle C.J.M., Morgan W.T. (1955). The determination of glucosamine and galactosamine. *Biochem J*. Dec;61(4):586-9.
Ronichevskaya G.M., Rykova V.I. (1977). Organic specificity of the antimitotic effect of glycopeptides isolated from RNA preparations. Dokl. Akad. Nauk SSSR., 237:481-3.
Roseman S., Ludowieg J., Moses F., Dorfman A.(1953). The biosynthesis of the glucuronic acid moiety of hyaluronic acid. *Arch Biochem Biophys*. Feb; 42(2): 472-3.
Roseman S., Moses F.E., Ludowieg J., Dorfman A. (1953).The biosynthesis of hyaluronic acid by group A streptococcus: I. Utilization of 1-C14-glucose. *J Biol Chem*. Jul; 203(1):213-25.
Roseman S., Moses F.E, Ludowieg J, Dorfman A.(1954). The biosynthesis of hyaluronic acid by group A streptococcus: II. Origin of the glucuronic acid. *J Biol Chem*.Feb; 206(2):665-9.
Schiller S.(1964). Synthesis of hyaluronic acid by a soluble enzyme system from mammalian tissue. *Biochem Biophys Res Commun*. Mar 26;15(3):250-255.
Schiler S.(1965). Structure and Function of Connective and Skeletal Tissue. London, p.302.
Strominger J.L.(1964). Nucleotide intermediates in the biosynthesis of heteropolimeric polysaccharides. *Biophys J*. Jan;71:SUPPL139-53.
Schmidt M.W., Baldridge K.K., Boatz J.A., Elbert S.T., Gordon M.S., Jensen J.H., Koseki S., Matsunaga N., Nguyen K.A., Su S.J., Windus T.L., .Dupuis M., Montgomery J.A.(1993). General atomic and molecular electronic structure system. *J Comput Chem*. 14, p. 1347-1363.

Silbert J.J. (1963). Incorporation of 14C and 3H from nucleotide sugars into a polysaccharide in the presence of a cell-free preparation from mouse mast cell tumors. *J Biol Chem.* Nov;238:3542-6.

Silbert J.E. Sugumaran G .(1995). Intracellular membranes in synthesis, transport, and metabolism of proteoglycans. *Biochim Biophys Acta.* 1241(3):371-84.

35. Silbert J.E. (1982). Structure and metabolism of proteoglycans and glycosaminoglycans. *J Invest Dermatol.* 79(1):31-7.

Singer M., Berg P. (1998). Genes & Genomes. A changing perspective. University science Books, Mill Valley, California.

Slutskii L.I.(1969). Biochemistry of Normal and Pathologically Changed Connective Tissue. Leningrad: Meditsina.

Strominger J.L.(1964). Nucleotide intermediates in the biosynthesis of heteropolimeric polysaccharides. *Biophys J.* Jan;71:SUPPL139-53.

Stewart J.J.P.(1989). Optimization of parameters for semiempirical methods: II. Applications. *J Comput Chem.* 10, p.221-261.

Sugahara K., Kitagawa H.(2000). Recent advances in the study of the biosynthesis and functions of sulfated glycosaminoglycans. *Curr Opin Struct Biol.* 10(5):518-27.

White J, Handler P., Smith E.L., Hill R.L., Lehman I.R., (1978). Principles of Biochemistry, 6th ed. , New York: McGraw-Hill, p.215.

Williams I.H.(1987). Theoretical modeling of specific solvation effects upon carbonyl addition. *J Am Chem Soc.* 109, p.6299-6307.

White B.N., Shetlar M.R., Shurley H.M., Shilling J.A. (1965). Incorporation of (1-14C)Glucosamine into mucopolysaccarides of rat connective tissue. *Biochim Biophys Acta.* Mar 1;101:97-105.

Yang J.T., Samejima T.(1969). Optical rotatory dipersion and circular dichroism of nucleic acids. In Progress in Nucleic Acids Research and Molecular Biology, Lavidson J.N., Cohn W.E., Eds., New York: Academic, pp.223-38.

Zelenin A.V.(2003). Plant genome. *Vestn Ross Akad Nauk.* 73:797-806.

Zimina N.P., Rykova V.I.(1986). Proteoglycans of animal tissues and their effect on DNA synthesis. Biokhimiya. 51:1555-61.

Zimina N.P., Dmitriev I.P., Rykova V.I.(1987). Composition and sulfation degree of glycosaminoglycans from tissues of various animals: Heterogeneity and tissue specificity of heparan sulfates. *Biokhimiya.* 52(6):984-90.

Zimina N.P., Rykova V.I., Dmitriev I.P.(1987). Comparison of composition and sulfation extent for glycosaminoglycans of quiescent and actively proliferating tissues. *Biokhimiya,* 52:856-61.

Zimina N.P., Rykova V.I., Arkhipov I.A. (1992). Animal tissue proteoglycans as irregular biopolymers: Informativeness of structure and control of biosynthesis. *Usp Sovrem Biol.* 112(4):571-88.

Zimnitskii A.N., Bashkatov S.A., Petrenko E.G., Khusnutdinova S.B.(2004). Age-related changes in skin glycosaminoglycan content in women, Biomed Khim., 50:309-13.

Zimnitskii A.N., Bashkatov S.A.(2004). Glycosaminoglycans in biochemical mechanisms of adaptation to some physiological and pathological states. Moscow, Globus: Farmatsevticheskii Byulleten.

Chapter 11

Coupled Mass Spectrometric Strategies for the Determination of Carbohydrates at Very Low Concentrations: The Case of Polysaccharides Involved in the Molecular Dialogue Between Plants and Rhizobia

V. Poinsot, M.A. Carpéné and F. Couderc

Additional information is available at the end of the chapter

http://dx.doi.org/10.5772/50069

1. Introduction

Polysaccharides play important biological roles in all living beings. They act as membrane and cell wall constituents, afford physical and chemical protection from the environment, ensure nutrition storage or compose the antigens that increase or repress defenses during infection processes, protein folding, molecular recognition, cell adhesion, etc. [1,2]. With regard to their major biological roles it appears a priority to understand their structure/activity relationship. Unfortunately, even though their relevance has been clearly demonstrated, the structural analysis of such compounds remains difficult. Actually, even though they are often highly soluble in water, their polarity due to the presence for instance of hydroxyls, carboxyls and sulfates, makes them quite difficult to handle. These polar groups are involved in intra and inter-molecular hydrogen bonding, ionic interactions and for lipopolysaccharides (LPS) hydrophobic interactions. Such molecular interactions are sources of solubility problems and can lead to the formation of aggregates, making their analysis more difficult (enormous apparent size, broadening of NMR signals, ionization difficulties, and so on).

In the past, analysis of polysaccharides was mostly achieved through Gas Chromatography coupled to Mass Spectrometry (GC-MS) of the PS hydrolysates. This technique remains widely used, due to its simplicity, but the development of high field Nuclear Magnetic

INTECH
open science | open minds

© 2012 The Author(s). Licensee InTech. This chapter is distributed under the terms of the Creative Commons Attribution License http://creativecommons.org/licenses/by/3.0), which permits unrestricted use, distribution, and reproduction in any medium, provided the original work is properly cited.

Resonance (NMR) and cryoprobes have opened new avenues. Unfortunately, NMR presents major drawbacks such as sample purity, sample complexity and concentration requirements. However, in the last decade, the development of highly sensitive, high-throughput Mass Spectrometers (MS) and software able to make global approaches ("omics") on very complex mixtures, have induced a new interest for the MS analyses (especially for the study of protein glycosylation). This time, the instruments mostly use soft ionization techniques associated to powerful $MS^{2 \text{ or } n}$ capacity analyzers experiments on electrospray ionization coupled to tandem mass spectrometers (ESI-MS/MS) or matrix assisted laser desorption and ionization coupled to a time of flight analyser (MALDI-ToF) [3]. Finally, the recent appearance of ToF and FTICR (Fourier transform ion cyclotron resonance) analyzers on the market allowed access to high resolution measurements enabling the determination of the exact molecular masses (with ToF error of less than 5 ppm and FTICR less than 0.5 ppm). Exact masses give access to the elementary composition of the molecule that is extremely precious when the analyte is present in a complex biological matrix at low concentrations precluding study by MS/MS.

Beside the evolution of structural analysis, separations techniques have also been enhanced especially by using higher resolution techniques (narrow bore columns, high pressure tolerant or acido-basic tolerant chromatographic phases). This progress has made it possible to couple chromatographic and MS systems on-line.

In this chapter, we will describe the classical MS couplings like GC-MS or HPLC-MS, but also focus on unusual but useful polysaccharide analysis systems like capillary electrophoresis coupled with mass spectrometry (CE-MS) or high performance anion exchange chromatography coupled with mass spectrometry (HPAEC-MS) that we have developed. Finally, we will present the practical uses of these techniques in a fundamental application: the analysis of the exo-, lipo- and capsular polysaccharides of *Sinorhizobium meliloti.*

2. Structural analyses using mass spectrometry

2.1. Gas chromatography coupled to mass spectrometry

The analysis of polysaccharides is relatively easy to run using ESI-MS and ESI-MS/MS to identify the concatenation of polysaccharide in hexose, deoxy-hexose, pentose, acidic sugars (polyhydroxy carboxylic acids), but it does not provide the polysaccharide composition (glucose, mannose, fructose…) mainly because many monosaccharides have isobaric masses, and often lack an ion (X and A) to identify how the monosaccharides are branched. GC-MS using electron ionization (EI) and chemical ionization (CI) is well suited for such a purpose. Some recent reviews have been published which gather and summarize the different numerous strategies developed in the literature for sugar analysis [4, 5, 6].

Because polysaccharides are mainly composed of aldoses, ketoses, deoxymonosaccharide, amino-monosaccharide and acidic monosaccharide, GC-MS analysis is based on the hydrolysis of the polysaccharides followed by derivatization and GC/MS analysis to identify the different sugar units through their migration times and/or mass spectra. Among the hydrolyzed monosaccharides some cannot be observed like uronic acids, ulosonic acids (*e.g.* Kdo), 4-amino-monosaccharide, and other charged species. Acids cannot be observed because their sodium salts are not volatile. Their treatment must be adapted for study by GC-MS. We describe below the main treatments for the study of polysaccharides: hydrolysis, derivatization, and we evoke the case of the acidic sugars.

Scheme 1 presents the general strategy which is used in GC-MS to analyze polysaccharides by determining the nature of the sugars, the relative quantities and the mode of attachment.

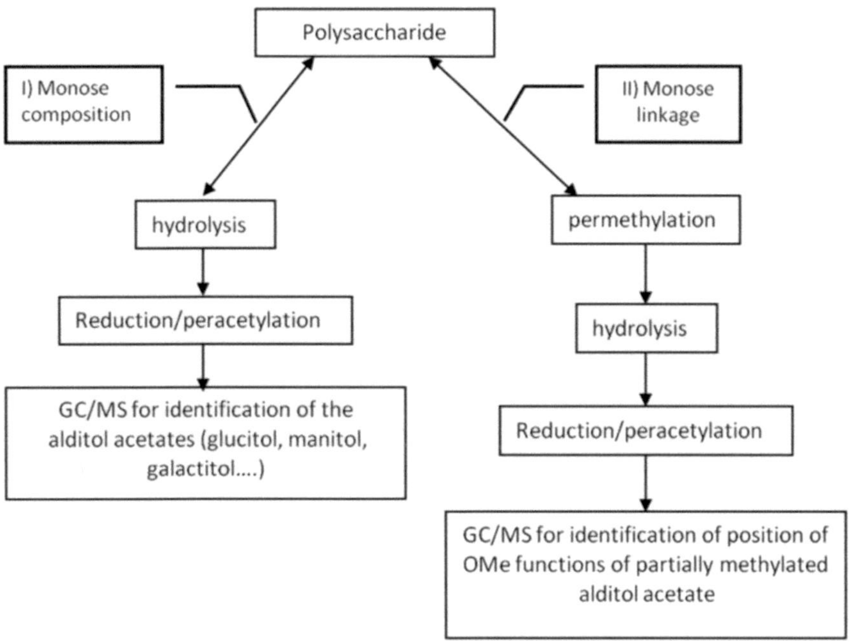

Scheme 1. General strategy used in GC-MS to analyze polysaccharides

2.1.1. Quality and relative quantities of monosaccharides: alditol acetate

To identify the nature of the monosaccharide, hydrolysis using H_2SO_4, HCl or trifluoroacetic acid (weaker than the others but easy to evaporate) is performed. After a reduction step using sodium borohydride [7], acetylation with acetic anhydride [8] results in alditol acetates which can be identified using GC retention time. GC-MS with chemical ionization using ammonia confirms the mass of the pseudo molecular ions $(M+H)^+$ and $(M+NH_4)^+$ of the alditol acetates. The chromatographic peaks help to identify the relative number of monomers composing the polysaccharide. Reduction with $NaBD_4$ allows easy identification of the anomeric carbon.

The difference in the hydrolysis rate between the sugars results in certain difficulties i.e. uronic acids being partially hydrolyzed, the monosaccharide linked to it is underrepresented. The same difficulties also exist for 2-acetamidohexoses which are partially *N*-deacetylated under excessive acid concentrations. The 2-aminohexoses obtained are not hydrolyzed and the carbohydrates linked to them are underrepresented [9].

For the acidic sugars, a modified strategy can be used, it consists first in methyl esterification of the uronic acids using DMSO and methanol or diazomethane. The methoxy group (or any other substitution) is a good leaving group for reduction of esters with sodium borohydride to yield aldehydes and, subsequently, their respective neutral sugars. This way, the reduction of the esters by $NaBD_4$ prior to hydrolysis yields a neutral polysaccharide allowing total hydrolysis and, after acetylation, identification by GC-EI-MS of the CD_2OCOCH_3 moiety bearing the monosaccharides [10,11,12] characteristic of the initial acidic sugars. Alternatively, methanolysis is also used to release monosaccharides as methylglycosides with esterified carboxyl groups. Conditions of methanolysis are presented in table 1.

Table 1 presents different hydrolysis conditions, reduction and acetylation on different kinds of polysaccharides.

Hydrolysis conditions	**Reduction and acetylation conditions**	**Polysaccharide**	**ref**
10 M HCl (80 °C, 30 min)	Excess of NaBH4 (2 h, 20 °C), acetylated with a 1:1 Ac2O–pyridine mixture (100 °C, 1 h)	LPS of *Providencia alcalifaciens* O12	[13]
100 μL sample, 50 μL of 0.7 M 4-methylmorpholine borane (MMB) and 100 μL of 6 M TFA , 80 °C, 30min.Then, 50 μL of 0.7M MMB. evaporation and addition of 200 μL 2 M TFA and 500 μL of acetonitrile. Dried again.	Dried sample +100 μL acetic anhydride and 100 μL TFA, reaction at 50 °C for 10 min. Then 1mL toluene to each dried sample . After dissolution in 1.5 mL CH_2Cl_2 and wash with 0.5M Na_2CO_3 and water. Sample are then dissolved in CH_2Cl_2 spiked with internal standards	Agar containing 3,6-anhydrogalactose	[14]
Sample was suspended in 2 N H_2SO_4 heated under vacuum at 100 °C for 3 h. 50 μL of water containing 500 ng monosaccharide internal standard and1 mL H_2O was added. Then,2 mL 50% *N,N*-dioctylmethylamine are added to neutralize the remaining acid. After centrifugation, the top layer was purified on a C18 column.	Add 50 μL of 100 mg/mL $NaBD_4$ and store at 4 °C overnight. Evaporate with methanol/acetic acid under vacuum. Then add 300 μL of acetic anhydride, react at 100 °C for 13–16 h. Quench with 750 μL H_2O for 1 h at room temperature. Extract with 1 mL of chloroform ($CHCl_3$)Remove aqueous phase, add 0.8mL of cold 80% NaOH, mix, elute with $CHCl_3$ on a Chem-Elut column.	Agar containing 3,6-anhydrogalactose	[14]

Hydrolysis conditions	**Reduction and acetylation conditions**	**Polysaccharide**	**ref**
2 mg sample treated with 0.5 mL TFA 2 M for 1 h at 121 °C	TFA removed by evaporation, then $NaBH_4$ reduction and then acetylation with Ac2O–pyridine (1:1, v/v; 2 mL) at room temperature for 12 h	Exopolysaccharide containing glucuronic acid	[15]
Methylesterification: 10mg sample dissolved in 5mL DMSO add diazomethane. Freeze- dry. For reduction, dissolve 2mL imidazol/HCl 1M pH 7 at 0°C, 2 drops of octanol add 200mg KBH_4 2h. Add acetic acic until pH 6, dialyze and freeze dry. Add 0.5mL TFA 4N at 100°C, 4h.	Evaporate sample, add NH_4OH 0.1N and 1mg KBH_4 2h 20°C. Then add acetic acid until pH 6, co-distillate with methanol. Acetylate dried residue with pyridine/acetic anhydride (Ac_2O) (50-200μL) 12h, 20°C. Dry the sample in presence of toluene.	Polysaccharide from Bacillus circulans	[12]
1 mg sample hydrolyzed with 1 M TFA at 100°C for 8 h, evaporate to dryness.	Reduce the dried sample with excess of $NaBH_4$ and acetylate with Ac_2O–pyridine (1:1, v/v; 2 mL) at room temperature for 12 h.	Exopolysaccharide from surface-liquid culture of Clonostachys rosea	[16]
1mg polysaccharide fraction in 2 M TFA at 100 °C for 8 h, followed by evaporation to dryness.	Dry residue reduced $NaBH_4$ or NaB_2H_4 and acetylated with Ac_2O-pyridine (1:1, v/v) at room temperature for 12 h.	Polysaccharide containing 3-O-methylated mannogalactan	[17]
1 mg mycelium 6 h at 100 °C with 3 mL 6 M HCl. Dry at 40°C	Add to dry sample, 20 μg inositol in 0.8 mL water, and 50 μl $NaBH_4$ (10 mg/ mL in 2 M ammonia), at 37 °C, 90 min. Add 2 mL acetic acid in methanol (1:200 v/v). After drying, add 300 μl Ac_2O, 100 °C for 15 h. In ice-bath add 750 μl of water. Extract with $CHCl_3$.	Chitin polysaccharide from mycelium.	[18]
Methanolysis	**Derivatization**		
4.5mg of freeze-dried fibers and pulps suspended in 2mL of 2M solution of HCl in anhydrous CH_3OH, 100 °C 4 h. Add 80μL pyridine. Add 1mL sorbitol 0.1mg/mL.	Dissolve dried sample in 70μL pyridine. Add 150μL hexamethyldisiloxane and 80μL trimethylchlorosilane for 12 h at room temperature.	O-acetyl-(4-O-methylglucurono)xylan from *Agave sisalana*	[19]

Hydrolysis conditions	Reduction and acetylation conditions	Polysaccharide	ref
Add MeOH 0.5M HCl (acetyl chloride 140mL + 4mL MeOH) at 80 ◦C for 16 h. dry under a stream of nitrogen gas.	300 µl of TriSilR reagent are added to dried material, at 80 ◦C for 20min and the reagents removed with a nitrogen stream.1mL hexane extraction of the sample, dry and add hexane.	Major grape polysaccharides	[20]
2mL of 0.01, 0.125, 0.25 or 0.5M HCl in MeOH added to 10mg sample, heated at85°C. Neutralize with Ag_2CO_3, dry the supernatant		Gelling carrageenans and agarose	[21]
Add 1.5mg sample to 0.5mL of MeOH/HCL (15mL MeOH and 0.4mL acetyl chloride). Heat at 80°C for 24h and dry.	Add to the dry sample 0.5mL of pyridine, hexamethyldisilazane , trimethylchlorosilane (9:3:1, v/v/v), 80°C 2h. Dry and add 0.5mL hexane.	Plant gums	[22]
Add 2mL 0.5M HCl/MeOH to 5mg sample at 80°C. Neutralize to pH 6.	30µL/30µLpyridine and *BSTFA* (N,O-bis(trimethylsilyl) trifluoroacetamide) added to dry sample, 2h 20°C.	Polysaccharides containing uronic acids.	[12]

Table 1. asd

Scheme 2 presents different ions obtained for alditol acetate in EI. Generally, no molecular ion and only very low abundance of $(M\text{-}CH_3COO^{\cdot})^+$ are observed in the spectrum. Stereoisomers give very near spectra and the base peak is the acetylium ion (CH_3CO^+) at m/z 43. The primary fragments are formed by cleavage of the carbon chain and then these odd mass number ions may lose an acetic acid (loss of 60 mass units) or a ketene (loss of 42 mass units) or an acetic anhydride (loss of 102 mass units) [23].

If deoxycarbons are present, alfa cleavage resulting in methyl radical loss is insignificant [24]. Some even mass number ions come from rearrangements resulting in the loss of acetaldehyde (loss of 44 mass units), acetic acid, acetic anhydride and ketene from the same mass number molecular ions [25].

When an acetamido group is present, the main ions are due to alfa cleavages induced by the N atom [26].

Consequently comparison of the mass spectrum (EI or CI) and the migration time to standards allows the monosaccharides to be identified.

Trimethylsilyl derivatives can also be used for alditol derivatives [27, 28]. No molecular peaks can be detected. As for most TMS derivatives in EI-MS, small M-15 ions are identified and m/z 103 ($(CH3)_3SiOCH_2^+$) and m/z (nx102)+1 (n being the number of carbon atoms in the chain of the alditol fragment) are recorded with their corresponding ion coming from a loss of 90 amu ((CH3)3SiOH).

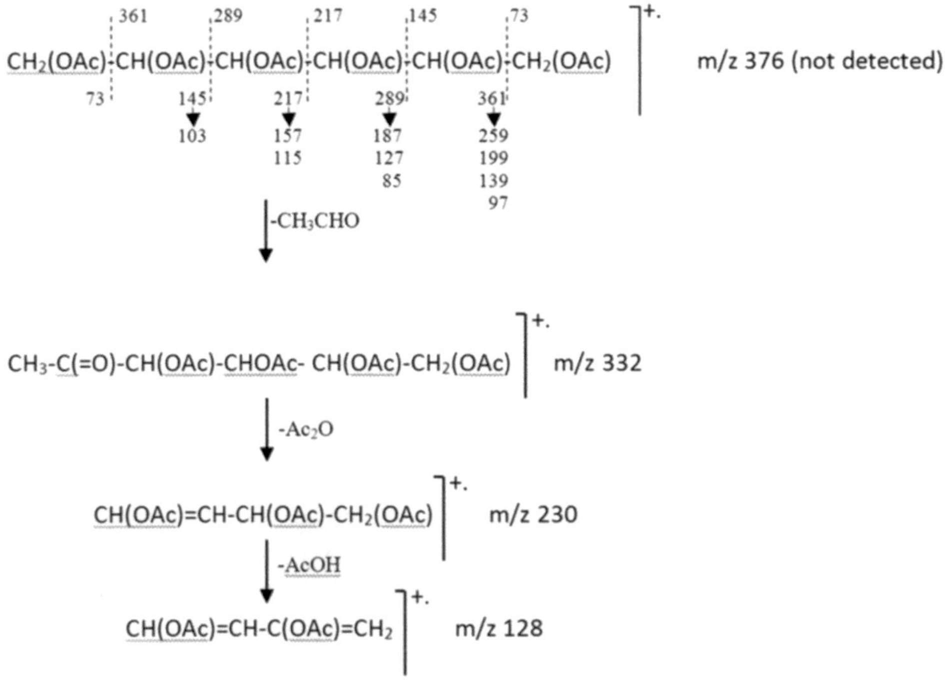

Scheme 2. Main fragmentation of hexitol peracetate.

Monosaccharides can also be analyzed prior to reduction by $NaBH_4$ or $NaBD_4$, under their cyclic forms, but the chromatograms are more complicated due to the hemiacetal group in sugars which leads to multiple structures [29].

2.1.2. *Quality and relative quantities of monosaccharides: derivatives of cyclic forms of monosaccharides*

Most of time the polysaccharide is submitted to a methanolysis and after the sample is silylated (see Table 1). With ammonia, CI can be used to identify the $(M+H)^+$ and $(M+NH_4)^+$ ions and can be identified by comparison of the migration times of standard the sugar. It is known that the relative intensity of the $(M+NH_4\text{-}CH_3OH)^+$ ion allows the assignment of pyranoside and furanoside structure [30]. Other fragmentation ions exist resulting from the loss by $(M+NH_4)^+$ or $(M+H)^+$ of CH_3OH or a single or two $(CH_3)_3SiOH$ molecules.

EI is frequently used and the resulting fragmentation has been extensively reported [30, 31]. If only very rare molecular ions are present in the spectrum, pentose and hexose are characterized by $(M\text{-}CH_3)^+$ ions and losses of CH_3OH and $(CH_3)_3SiOH$ from this last ion. Deoxyhexoses give $M^{+\cdot}$ and $(M\text{-}CH_3O\cdot)^+$. Glucuronic and galacturonic acids also exhibit $(M\text{-}CH_3)^+$ ions in the spectrum and losses of CH_3OH and $(CH_3)_3SiOH$ but also an *m/z* 234 ion is characteristic of acidic derivatives resulting from a complex McLafferty-type rearrangement of trimethylsilyl group to the carboxyl function. Finally, an *m/z* 204 ion is closely related to ring size, being favored by a six-atom cyclic structure [32].

For pyranose forms the relative intensity of this ion (compared to base peak) varies between 30 and 100% while for furanose it is lower (5%), while *m/z* 217 is very intense for furanose forms.

These fragmentations and migration times of the TMS derivatives will help to determine the species that are difficult to characterize owing to their migration times.

2.1.3. The attachment mode

2.1.3.1. Synthesis of partially methylated alditol acetate

The use of methylated alditol acetate is a convenient protocol to identify the mode of attachment, it has been well reviewed by Hellerqvist and Sweetman [33]. The first step of this study consists in a permethylation of the polysaccharide, using finely powdered sodium hydroxide and methyl-iodide. A recent article shows that presolubilization of 100-300µL of a solution of 1mg/mL polysaccharide with 5 µL of anhydrous glycerol dried prior to derivatization, offers better methylation yields [34]. Recently, solid-phase spin column permethylation and solid phase capillary permethylation were described [35, 36] and are presented in Table 2. These permethylation processes are derived from the classical method described by Costello [37, 38]. After the permethylation, the polysaccharide is hydrolyzed. For example: the polysaccharide is dried, then 1.8mL glacial acetic acid is added and the sample is briefly sonicated; 0.20mL of 2M sulfuric acid are added and heated at 100°C for 9h. Then acetic acid can be removed in a rotary evaporator several times by adding water and keeping the temperature below 40°C. An equimolar amount of $BaCO_3$ is added to remove the sulfuric acid, and the precipitate is filtered and washed in a minimum volume of water. Then washings and filtrates are combined and about 5mg of $NaBH_4$ (or $NaBD_4$) are added, the reaction is driven for at least 1h. Finally, per-acetylation is driven as reported in table 1.

GC/MS using ammonia CI and then EI are performed, to determine the molecular mass and the structure of the partially methylated acetate alditol respectively. Following this protocol, the positions of the *O*-methyl functions reflect the initial position of the free OH functions of the polysaccharides.

2.1.3.2. Fragmentation rules of partially methylated alditol acetate in EI

In EI, no molecular ions can be observed. The rules include amino sugars and predict the formation of ions formed by cleavage of the C-C chain. i) The major ion corresponds to the cleavage between acetyl-N-methylaminated carbon and *O*-methylated carbon, with charge retention predominantly on the aminylated carbon, ii) an intense ion comes from two *O*-methylated carbons with charge retention on either carbon (the intensities of both ions are inversely proportional to the size of the fragment), iii) weaker ions are due to the cleavage between two C atoms one acetylated the other methoxylated which keeps the charge. These ions may lose ketene, acetic acid or methanol molecules. Table 3 gives the different masses of the different possible ions. With the knowledge of the $(M+H)^+$ or/and $(M+NH_4)^+$ ion, obtained in GC:CI/MS and the fragment ions from the EI spectrum, the position of the methyl and the attachment of the different sugars can be identified. Combining these data with ESI-MS/MS gives the correct sequence of the polysaccharide.

Permethylation Conditions	Polysaccharides	ref
Dissolve in 200µL DMSO, powdered NaOH (5 mg) and 200µL CH_3I. Stir vigorously for 30 min and left overnight at room temperature. extract with $CHCl_3$.	matesaponin	[39]
Microspin column are placed in microcentrifuge vial, filled with acetonitrile. Pack with NaOH beads in acetonitrile. Centrifuge the spin column with acetonitrile then DMSO. Suspend dried polysaccharide in DMSO, CH_3I, H_20 (70.8; 26.4; 2.8%) spin the column at 800rpm for 30s. Recycle the 8x. Washed it with 100-µL aliquot of acetonitrile, at 10 000 rpm for 0.5 min. Extract column with $CHCl_3$.	Nglycans from serum proteins	[40,41]
Add 10 mg polysaccharide in 1mL DMSO to 1mL CH_3I and 20mg powdered NaOH for 30min. Stir vigorously and maintain at 25°C overnight. Stop with H_2O, neutralize with acetic acid, dialyse (8 kDaMr cut off) against distilled H_2O and freeze-dry. Extract with $CHCl_3$.	3-O-methylated mannogalactan from *Pleurotus pulmonarius*	[42]
Suspend 100-300µg of dry polysaccharide in 100µL DMSO. Mix 400µL CH_3OH and 100µL 50% NaOH, and 2mL DMSO and vortex. Centrifuge at 3500 rpm for 1 min to get the NaOH pellets. Wash 5x with 2 mL DMSO. Mix the polysaccharide and the 500µL NaOH slury solution with 100µL CH_3I. Evaporate, and clean with 2mL H_2O and 2mL CH_2Cl_2.	Pn6B, Pn14, Dextran polysaccharides	[43]

Table 2. Permethylation of polysaccharides.

$CH_2{=}O^+CH_3$ m/z 45	CH_2OCH_3 CHO^+CH_3 m/z89	CH_2OCOCH_3 CHO^+CH_3 m/z 117	CH_2OCOCH_3 $CH{=}NH^+(COCH_3)$ m/z 144	CH_2OCOCH_3 $CH{=}NH^+(COCH_3)CH_3$ m/z 158
CH_2OCH_3 $CHOCH_3$ $CH{=}O^+CH_3$ m/z 133	CH_2OCOCH_3 $CHOCH_3$ $CH{=}O^+CH_3$ m/z 161	CH_2OCH_3 $CHOCOCH_3$ $CH{=}O^+CH_3$ m/z 161	CH_2OCOCH_3 $CHOCOCH_3$ $CH{=}O^+CH_3$ m/z 189	CH_2CH_3 $CHOCOCH_3$ $CH{=}O^+CH_3$ m/z 131
CH_2OCH_3 $CHOCH_3$ $CHOCH_3$ $CH{=}O^+CH_3$ m/z 177	CH_2OCOCH_3 $CHOCH_3$ $CHOCH_3$ $CH{=}O^+CH_3$ m/z 205	CH_2OCH_3 $CHOCOCH_3$ $CHOCH_3$ $CH{=}O^+CH_3$ m/z 205	CH_2OCH_3 $CHOCOCH_3$ $CHOCOCH_3$ $CH{=}O^+CH_3$ m/z 233	CH_2OCOCH_3 $CHOCOCH_3$ $CHOCOCH_3$ $CH{=}O^+CH_3$ m/z 261

Table 3. Principal fragmentations in EI/MS of the C-C chain of permethylated alditol acetate
We do not present the ions which can lose CH_3OH, CH_2CO, CH_3COOH.

2.2. ElectroSpray ionisation mass spectrometry (ESI-MS) analyses

Using electrospray technology, the analyte is introduced into the mass spectrometer as a solution delivered by a syringe pump (direct input) or as a fraction eluted from a liquid chromatography (HPLC/UPLC). The analyte solution passes through a needle on which a high potential difference is applied (classically 3 kV). This produces a spray of droplets with a surface charge of the same polarity as the needle. The charged droplets shrink as the solvent evaporates. Charge promiscuity then produces continual explosions of the droplets into smaller ones until they reach the gas phase. During this process, the molecules present in the solution are concentrated and this often produces a suppression effect. This effect consists in the masking of one compound by another present in a mixture. The suppression effect is minimal when all the compounds together are chemically equivalent in terms of hydrophobicity or acido-basicity. As hydrophilic compounds stay longer in the dissolved form during the ESI process, the more hydrophobic compounds will generally suppress the more hydrophilic compounds. In the positive ion mode, basic compounds will easily suppress more acidic compounds. As carbohydrates are very hydrophilic and often acidic, only two possibilities exist to analyze them without suppression: a separation of the sugars based on their acidity (ion exchange chromatography) or derivatization of the hydroxyls to generate more hydrophobic compounds. The latter techniques include derivatization such as permethylation [44;45] or metal adduction [46;47] which has been extensively studied. Structural analysis of underivatized saccharides using negative mode ion electrospray ionization has also been investigated [48].

A systematic nomenclature for labeling fragment ions observed in MS/MS has been introduced by Domon and Costello [49] (Figure 1).

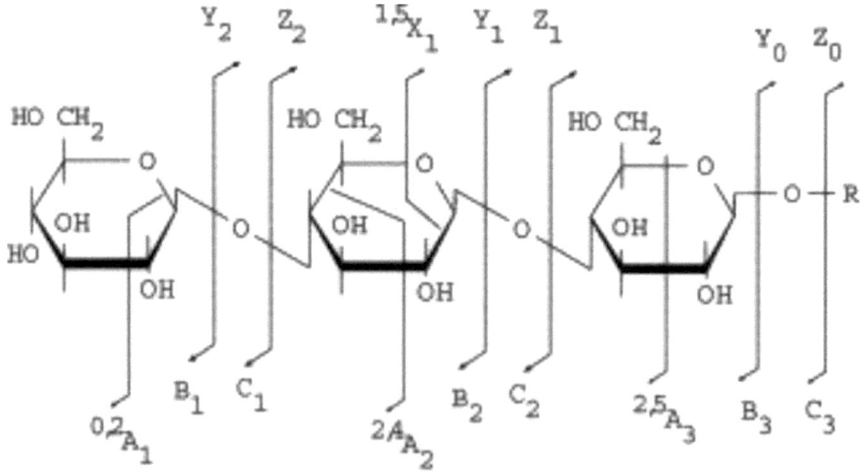

Figure 1. Systematic nomenclature for labeling fragment ions observed in MS/MS following Domon and Costello.

Acidic moieties strongly influence the fragmentation energetics and patterns of tandem mass spectrometric daughter ion formation. Non-sulfated compounds dissociate in the positive mode into abundant ions through B and Y type fragmentation resulting from the cationisation of the glycosidic oxygen [50], these fragmentations being facilitated on 4- and 6-linked reducing terminal residues [51]. In the negative mode deprotonated neutral carbohydrates produce C- and Z-type ion fragments [52], when the acidic sugars undergo B- and Y-type fragmentation [53]. Therefore, interpretation of the CID spectra implies knowledge of the fragmentation pathway, and the use of mass tables of the common monosaccharide building blocks is often essential to build the sequence of unknown structures. Of course, understanding the biosynthetic pathway of oligosaccharides helps to avoid making mistakes in the sequence [54].

2.2.1. Direct Input (DI/MS)

As explained just before, analyzing carbohydrate mixtures by direct introduction is a challenge due to the easy suppression of these compounds by numerous other natural compounds like proteins, aminoacids, lipids or other differently charged or decorated saccharides.

In comparison to MALDI, ESI produces less in-source fragmentation of acidic glucans and other fragile ions and is easily coupled on one side to tandem MS allowing structural investigations and on the other side on-line to liquid chromatography. As this ionization technique is very soft, true MS (without fragmentation) can be obtained. However, MALDI produces less complex charge state patterns and less multiple cation adducts and suppression effects. Actually, the analyte is not in solution with the salts and the deposit is quite heterogenic indicating places where suppression effects could be less important [55]. The polymeric complexity of carbohydrates often produces an overlap of the ESI charge state patterns making them extremely difficult to analyze [56] even with software such as MaxEnt (1 and 3). As a result, except for acidic glucans, ESI analysis of glycans requires chromatographic pre-separation of the sample.

2.2.2. High-resolution liquid chromatography coupled to Mass Spectrometry (HPLC-MS)

Normal phase HPLC on naked inorganic oxides was the technique first developed but it exhibited numerous drawbacks (peak tailing, retention time shifts, etc.), therefore, reversed phase HPLC was developed and proved to be suitable for most all bioanalytical solutes. However, a problem remained: how to create retention in RP-HPLC for polar compounds without dewetting the C_{18} phase [57]. Lack of retention of hydrophilic compounds is almost always due to solvophilicity. Actually, the polar functions enter more favorable dipolar interactions with the polar RP-HPLC eluent, than with the stationary phase [58]. When the lack of retention is due to charge, retention can easily be achieved on ion-exchange chromatography (see next paragraph) or by using ion pairing that does not require specific apparatus. Using this technique, glycosaminoglycans polymers up to a polymerization degree of 40 can be observed on reversed phase ion pairing HPLC [59]. Unfortunately, ion pairing has been shown to reduce ESI-MS sensitivity [60].

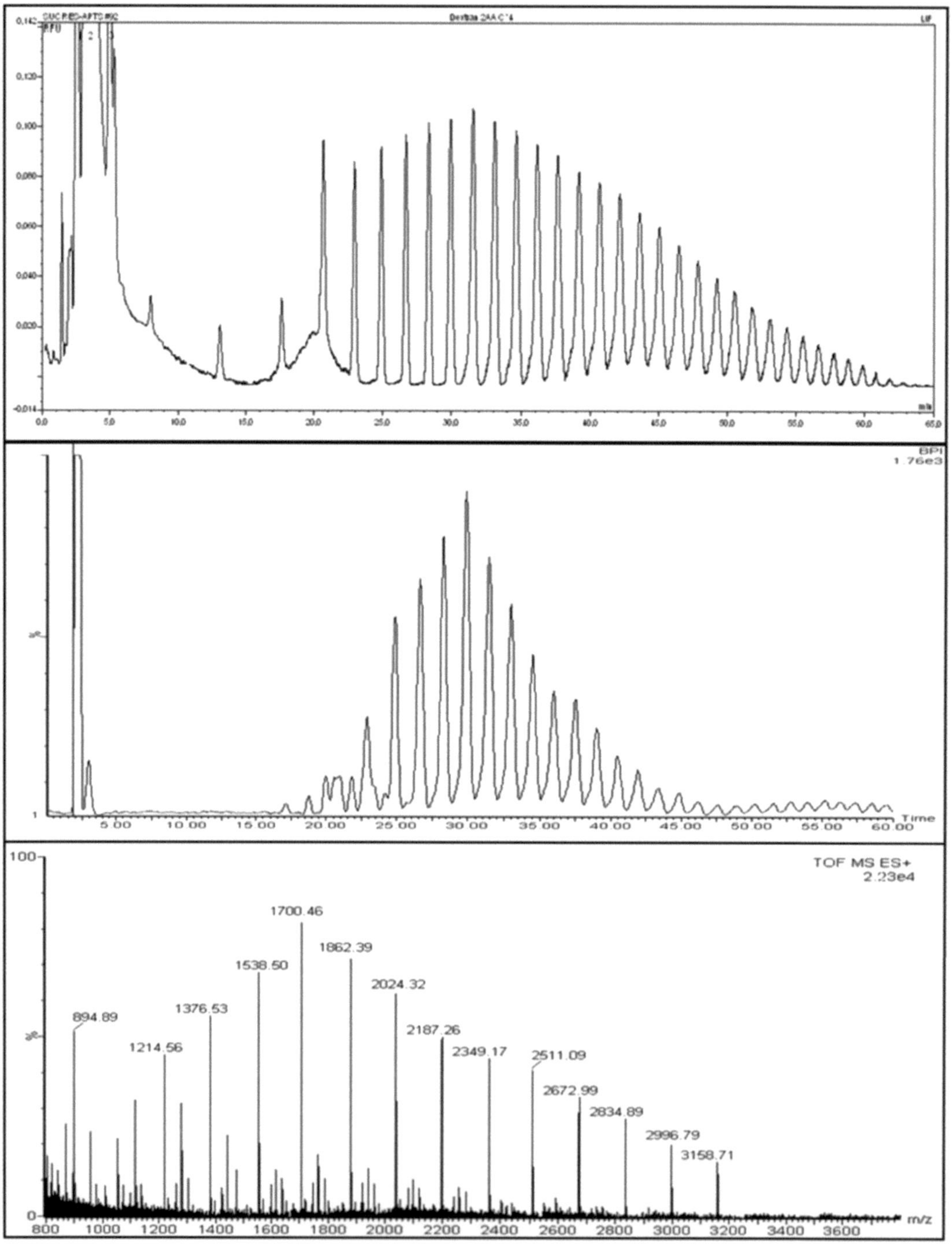

Figure 2. HILIC/LIF/ESI-MS analysis of 2-AB labeled glucose ladder. At the top: improved LIF detection of the neutral saccharides, in the middle: positive mode ESI-MS total ion current, at the bottom: spectrum obtained by summing all the spectra of the compounds eluting between 15 and 50 minutes.

The greatest problem concerns the highly hydrophilic but still neutral analytes. The classical way to analyze them is to convert them into more hydrophobic compounds by a chemical reaction [61]. In addition to the retention of the compounds, such organic derivatization improves their detection (UV absorption or fluorescence emission) (figure 2). The main drawbacks of this approach are that the separation of all the compounds is only based on the same hydrophobic tail, derivatization is time-consuming, and quantitative labeling cannot be systematically achieved. Therefore, a new strategy has been developed and is based on a new stationary phase. In practice, the stationary phase is polar and attracts the more polar part of the eluent that will act as the retentive phase. The complete eluent is relatively hydrophobic but is a sufficiently good solvent to allow distribution between the stationary and mobile phases. This technology is named hydrophilic interaction chromatography (HILIC). The quite neutral hydrophobic/hydrophilic balance of the eluent allows an easy interfacing with ESI-MS unlike the normal phase HPLC (NP-HPLC) working with totally organic solvents which are not compatible with the nebulization process [62].

2.2.3. *High Performance Anion Exchange Chromatography coupled to Mass Spectrometry (HPAEC-MS)*

High performance anion exchange chromatography (HPAEC) using sodium hydroxide or acetate based eluents is a well-established technique for determining underivatized carbohydrates [63;64] and high-performance separation of alditols, mono- and oligosaccharides ranging from 2 to 60 mers have been described using pulsed amperometric detection (PAD) [65]. The identification of individual carbohydrates is usually performed by comparison of their retention times with those of reference samples. In biological samples complex overlapping occurs: acidic sugars are eluted in the "oligosaccharide domain" making this approach very uncertain. Moreover, the nature of sugar moieties and the variability of glycosidic linkages makes the retention times of the oligosaccharides unpredictable. For these reasons, coupling of HPAEC-PAD with ESI-Q-ToF (ESI coupled to a hybrid tandem mass spectrometer constituted of a quadripole and a ToF) MS was required to collect structural information on the PS sequences. Two detections are necessary. PAD allows, from retention time, the nature of the saccharide (neutral, acidic, oligosaccharidic) to be determined while ESI-Q-ToF MS gives access to the composition. Coupling HPAEC with MS presents a technological challenge, due to the non-volatility and high conductance of the sodium hydroxide or acetate dissolved in water, used as eluent. To avoid this limitation, a commercial desalting device using a selective cation exchange membrane and a regenerant can be installed on-line between the column and the MS [71,66]. The use of this on-line desalter with oligosaccharide separations with MS has only been rarely described [67, 68, 69, 70, 71].

We have developed HPAEC and ESI-Q-ToF MS conditions to perform efficient on-line coupling of the two techniques [72]. PA1 ion exchange columns have been used for oligosaccharide analysis even though they are designed for monosaccharides, because the PA 200 column is not available in the narrowbore size. Two different gradients have been used at a flow rate of 0.3 mL/min. For neutral carbohydrates: during the first 15min NaOH concentration increased from 10 to 50mM, and during the following 5 min up to 80mM, finally, during the last 15 min the concentration of NaOAc increased from 0 to 90 mM while NaOH remains constant, return to the initial conditions was done over 10 min. A wait of 35 min between two injections is necessary to equilibrate the column. To separate acidic sugars, the gradient started with a NaOAc concentration ramp increasing from 0 to 90 mM in 15 min while the concentration of NaOH was kept constant at 80 mM, the concentration of NaOAc (90 mM) and NaOH (80 mM) were then held constant for 5 min then 5 min to return to initial conditions. A wait of 15 min between two injections was necessary to equilibrate the column.

Due to the low amounts of polysaccharide that can be dissolved, no post-column split could be used to couple ESI-Q-ToF with an analytical sized column. This methodology was never used for the analysis of heteropolysaccharides.

Performing HPAEC for carbohydrate analyses implies the use of a NaOH and NaOAc concentration gradient. At high pH (>13.5), carbohydrates are in the anionic form allowing their retention on the phase. Electrocatalytic oxidation mediated by NaOH at the surface of the gold electrode occurs by application of a positive potential. The current generated is then proportional to the carbohydrate concentration. Therefore obtaining a PAD signal, implies placing the suppressor after the PAD but before the MS (Scheme 3). Since a high ionic content is not compatible with electrospray ionization mode, the salt level was reduced using an on-line 4 mm ASRS (anion self-regenerating suppressor) desalter. The desalting efficiency and the pH of the mobile phase after desalting were checked. The peak broadening due to the additional void volume of the ASRS and the reference electrode cavity was then investigated. Desalting efficiency was followed through a conductometer as a second detector. The pH was followed each minute (using the combination pH reference electrode in the electrochemical cell of the PAD unit) to be sure that the residual conductivity was not due to acetic acid. When the regenerant used in the ASRS is pure water at a flow rate of 2 mL/min, it is possible to suppress NaOH up to 100 mM (at a flow rate 0.3 mL/min), but it does not provide the suppression of 100 mM NaOAc (the pH increased up to 13 at 25 min). In contrast, the use of 0.25% H_2SO_4 in water at a flow rate of 2 mL/min can maintain the conductivity below 300 μS for the entire analysis. Desalting of the NaOH eluent resulted in the measurement of a stable pH of 7; during the NaOAc gradient, the pH decreased regularly to pH 3. During conductimetric detection, a current is continuously applied to the membrane (50 mA). No effect of regeneration current on the membrane regeneration efficiency was observed.

The coupling of both PAD and MS detectors offers two advantages. First, the safety of the mass spectrometer, intrusion of salt in the source and in the mass analyzer can dramatically affect the performance of the mass spectrometer, because only a small quantity of salt produces signal suppression. Therefore, when signal is present on PAD but disappears on MS, it is an indication that desalting is no longer sufficient. The second advantage, as discussed in the next paragraphs, is that the MS and PAD sensitivity are complementary.

As we performed on line PAD and MS analyses, the first before and the second after desalting, we measured peak broadening, and separation efficiency. A 120 μL void volume was measured between the PAD and the ESI source with 30% peak broadening but without loss of efficiency. Since the flow rate of 0.3 mL/min is compatible with the ESI source, no split was necessary.

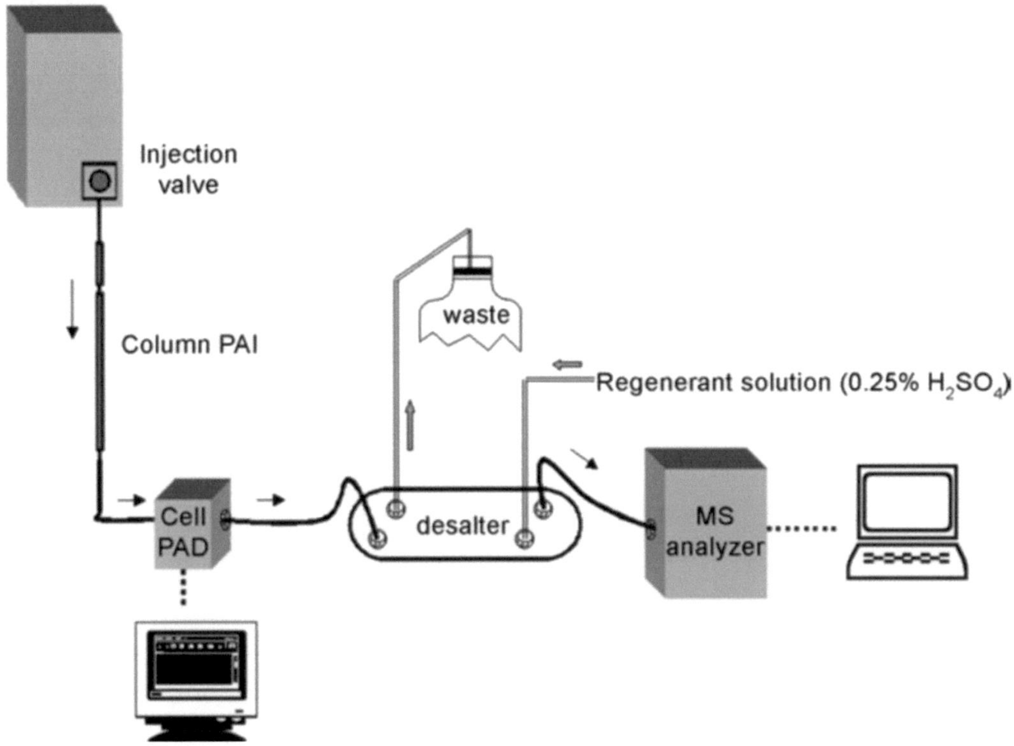

Scheme 3. Coupling scheme. The PAD is placed on-line with the MS, but before the desalting device to ensure the electrocatalytic process.

Analysis of acidic sugars opens up the possibility of using the positive or negative mode for MS detection. Theoretically, detection in the negative mode decreases the noise due to the increased specificity, and therefore increases the signal-to-noise ratio (S/N) and the sensitivity. The sugars we are looking for contain a carboxylic acid function and are easily ionisable in the negative mode. We studied the response factor of a commercial acidic oligosaccharide: 6'-sialyl-lactose; surprisingly, the TIC intensities were almost identical for ESI+ and ESI-. Only a sensitivity factor difference of two was measured at 5 µg/mL, and the positive ionization mode did not exhibit significantly more noise. In the positive mode, the presence of at most one sodium adduct was observed (as counter-ion of the acid). As both MS modes gave similar results, the behavior using MS/MS was investigated in both modes. However, collision energy is lower in the positive mode, which could be an advantage to obtain a better sensitivity, since high energy ions are difficult to refocus.

The detection sensitivity using HPAEC-PAD or HPAEC-ESI-Q-ToF MS is dependent on the nature of the sugar. The first fast oxidation step occurring on the gold electrode (PAD) involves the aldehyde of the carbohydrate, resulting in the formation of a carboxylic acid and the production of two electrons. The second fast oxidation reaction is the cleavage of the C1-C2 bond, followed by conversion of C2 and C6 to the corresponding carboxylates, resulting in the production of 6 electrons (most efficient response) [72]. Therefore the predictive response factors are: Hex>6-desoxyHex>HexA>2-desoxyHex>ulosonic acids (Figure 3). For ESI-Q-ToF MS, the ionization occurs on the glycosidic linkage and is often facilitated through the presence of acid functions close to the ionization site, implying the most sensitive response for the ulosonic acids.

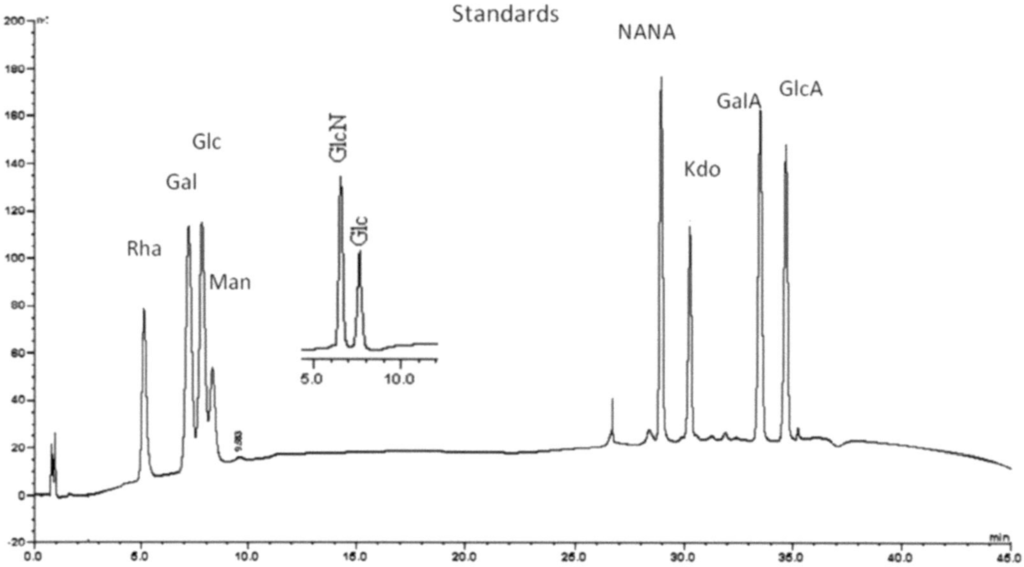

Figure 3. Chromatogram obtained by PAD detection of common sugars : Rhamnose (Rha), Galactose (Gal), Glucose (Glc), Mannose (Man), N-Acetyl Neuraminic acid (NANA), 2-Keto,3-deoxyoctuolosonic acid (Kdo), galacturonic acid (GalA) and glucuronic acid (GlcA).

The concentration response of ESI-MS is often not linear and is very variable from one sugar to the other. For this reason, a quantification curve and the limit of detection (LOD) for each type of saccharide standard were measured for the five standards in solutions ranging from 200 to 2μg/mL with an injection volume of 5 μl.

The HPAEC-ESI-Q-ToF MS response measurements surprisingly indicate that uronic acids respond weaker than expected, even less than hexoses. The LOD of GalA was not satisfying, unlike those obtained for all other saccharides.

2.3. FAB MS and FAB MS/MS of carbohydrates

The fast atom bombardment (FAB) ionization is less and less used due to the fact that the MS suppliers no more produce this type of sources. Nevertheless, this ionization process is of interest, because it is significantly more energetic than MALDI and ESI. In consequence, fragmentation can be observed in the ion source for natively charged glycans and be applied well to positive or ions. In the positive mode sodium cationized species are easily analyzed, while in negative mode it is the deprotonated ions that are observed with greatest abundances for those containing acid residues. Permethylated glycans are very easily ionized but do not result in "*in-source*" fragmentations, collision induced dissociation spectrum in MS/MS can be used to obtain such fragmentation, the nomenclature of the different fragmentations was proposed by Domon and Costello [49]. FAB ionization requires quantities of native glycans above 10nmol, and 10 times less for permethylated species [73].

Using FAB-MS and -MS/MS linkage position can be determined like with ESI, as well as the position of sulfate or other substitutions on the polysaccharide [74]. The internal fragmentation of the monosaccharide cycles (A and X) are useful for such topic.

In the eighties, periodate oxidation was reintroduced to address sequence problem. Oligosaccharides can be sequentially oxidized with periodate and reduced with sodium borodeuteride and permethylated or peracetylated [75]. The bindings of monosaccharide residues can be determined on the basis of the sequences of the primary and secondary ions.

An original attempt of determination of the anomeric configuration was done using methyl glycopyranosides and [M+Ag+]+ ions. It was observed that the unimolecular decompositions of the ions showed dramatic changes related to the relative positions of both the hydroxyl group on C2 and the anomeric methoxyl group. When these groups were in the *cis* position, methanol was exclusively expelled. When they were in the *trans* position, silver hydride was preferentially lost [76].

2.4. CE-MS

2.4.1. Capillary electrophoresis and laser induced fluorescence.

Capillary electrophoresis, due to its high resolving power and sensitivity, has been useful in the analysis of intact and derived oligosaccharides and polysaccharides affording concentration and structural characterization data essential for understanding their biological functions. As most glycans do not have any strong chromophores on their structures and have low extinction coefficient they are difficult to detect using UV absorbance. Indirect UV has been used, it is based on using a chromophore in the electrolyte resulting in negative peaks but because the chromophore must have a mobility very near that of the sugars it is often difficult to ensure an adequate detection limit [77, 78]. Labelling with a chromophore is very useful, because Laser induced fluorescence (LIF) [79,80] can be used on monosaccharides as well as on polysaccharides. CE experiments are comparable to GC experiments to identify monosaccharides. Acetylation or sylilalion of alditol is replaced by an amino reduction step using an amino fluorescent dye and $NaBH_3CN$. Most of the time 1-aminopyrene-3,6,8-trisulfonate (APTS) is used for monosaccharide, their migration time allow the identification of the different derivatized alditols, it is very convenient for small quantities of natural polysaccharides and specially polysaccharides extracted from DOC PAGE. An acidic compound is more easily prepared than with classical alditol strategies for GC analysis [81].

For polysaccharides the same labeling reaction can be used and the different compounds can be easily separated. Recently, optimization of the boric acid concentration (which increases the charge of the polysaccharide by complexation) and linear polyacrylamide concentration in the buffer yielded a separation of 12 polysaccharides of very similar structures, most with baseline separation in less than six minutes [85]. Using this technique, high throughput glycosylation can be performed [82]. Thirty-two IgG N-glycans were analyzed using a combination of weak anion exchange enrichment and exoglycosidase digestion. Aminobenzamide and Aminopyrentrisulfonate labeling of polysaccharide followed by a UPLC on 1.7 µm HILIC or CE-LIF respectively have been compared. Combination of these data demonstrated that complete structural annotation is possible within a total analysis time of 20 min due to the advantageous orthogonality of the separation mechanisms [83]. This work confirms the use of glycosidase in CE-LIF studies for the determination of the structure of polysaccharides. Significant labeling improvements were made by replacing the conventionally used acetic acid catalyst for $NaBH_3CN$ reduction with citric acid. Using a 1:10 glycan to fluorophore molar ratio resulted in a 95% derivatization yield at 55°C with only 50 min reaction time and negligible terminal sialic acid loss [84].

2.4.2. Capillary electrophoresis and mass spectrometry.

A large number of CE-MS and CE-MS/MS experiments were driven for polysaccharide structural determination and have been extensively reviewed recently by Y. Mechref [85]. Different interfaces between the CE-LIF instrument and MS are commercially available. Most works use APTS labeling and ESI ionization, although MALDI is also used but with lab-made instruments [86]. A very interesting study concerning polysaccharides from IgG was realized. Figure 4 presents a schema of the CE-MS system with on-line LIF detection adjusted 20 cm from the ESI tip. A sheath liquid consisting of a 50:50 isopropanol:water (0.2% ammonia) was added coaxially to the separation capillary at flow-rate 4 µL/min. CE was performed with PVA coated capillaries (60 cm total length x 50 cm effective length × 50 µm ID) and a running buffer of 40 mM *alpha*-aminocaproic acid, pH 4.5 (adjusted with acetic acid) + 0.02% hydroxypropylmethylcellulose (HPMC) with an applied voltage of -30 kV were used [87]. The inherent mass accuracy of the MS permitted the identification of major and minor glycans using their $(M-H)^-$ ions. This study demonstrated for the first time the ability to attain CE-MS separation efficiency somewhat comparable to that commonly observed in CE-LIF analysis. As seen in figure 4 the four early migrating species are clearly visible in both the online LIF and MS electropherograms. The MS electropherograms suffer from a shift in migration time and some loss in separation efficiency, which is due to dead volumes originating from the addition of the MS detector.

Neutral markers to label acidic polysaccharide can be used. Their main advantage will be to allow the separation of the acidic polysaccharide following the number of acid residues [88], 2-aminoacridone was proposed in place of APTS. Nakano *et al* [89] used 9-fluorenylmethyl chloroformate (Fmoc-Cl), during the digestion of a glycoprotein by PNGase which can label the released 1-amino function of the first GlcNac residue. It prevents reductive amination.

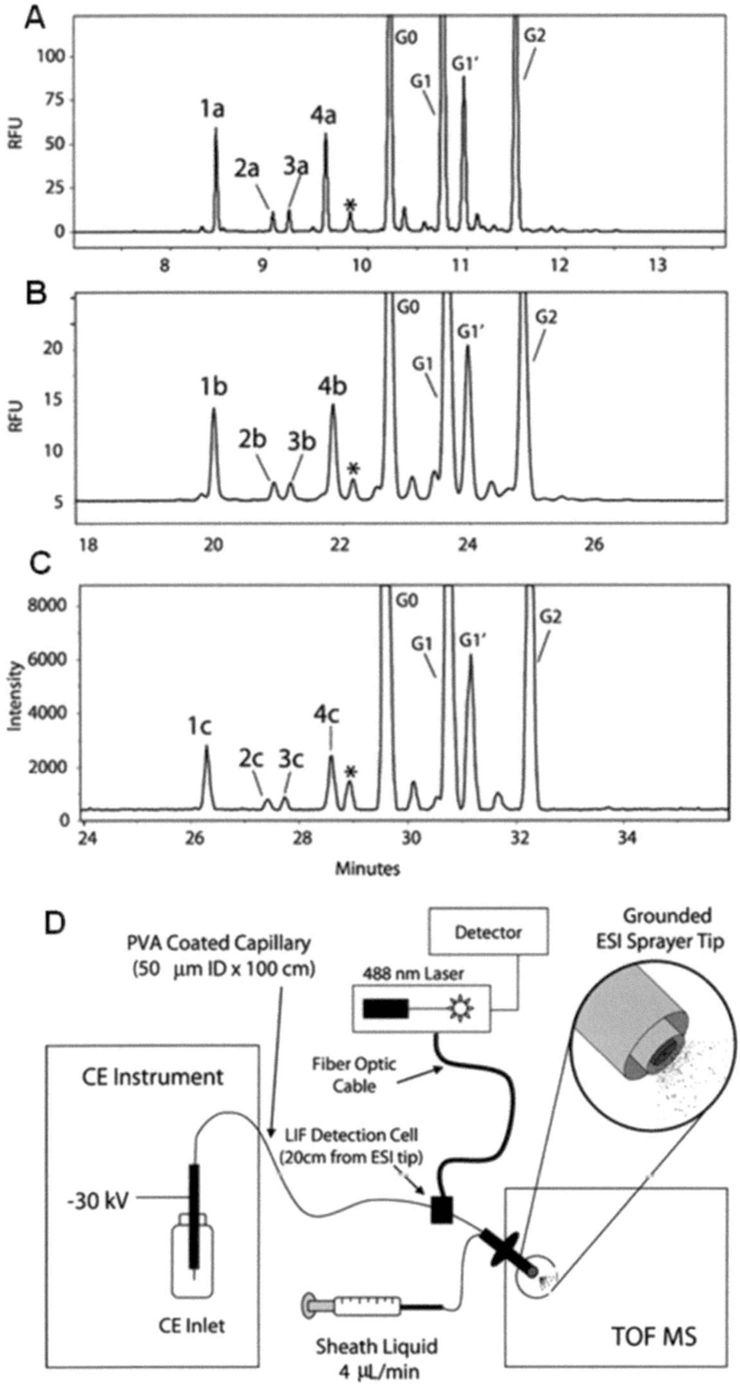

Figure 4. Expanded-scale electropherograms of rMAb 1: A) standard CE–LIF electropherogram using a 60 cm capillary, B) CE–LIF trace obtained on-line with MS detection and C) CE–MS base peak electropherogram. D) Scheme of the CE–MS system with on-line LIF detection.

Using a very simple 20 mM phosphate buffer (pH 8.5), the authors separated the different polysaccharides following the number of sialic acids, then MS and MS/MS spectra identified the composition of each polysaccharide thanks to $(M+H)^{2+}$ or $(M+H)^{3+}$ ions. An example of the mass electropherogram and the different mass spectra are presented in figure 4 which concerns the glycans from bovine fetuin. In this study the authors showed that this method can be used to identify polysaccharides from glycoprotein extracted from an SDS PAGE separation.

3. Polysaccharides of *Sinorhizobium meliloti*

Polysaccharides are commonly found at the surface of Gram negative bacteria (Figure 5). The aim of our work is to elucidate the structure of the polysaccharides from the surface of bacteria. Rhizobia are Gram negative bacteria living in soil and able to establish a symbiotic interaction with leguminous plants, known as nitrogen fixing symbiosis [90]. During this mutual interaction, bacteria bring combined nitrogen, in the form of ammonia directly transformed from atmospheric N_2, to the plant. In turn the plant provides hydrocarbons and develops new organs on its roots which host the bacteria: the nodules [91,92]. During the early stages of establishing symbiosis, a molecular dialogue takes place. First, the partners in the soil are recognized. The plant exudes flavonoid compounds and the neighboring rhizobia respond by secreting Nod Factors (lipochitooligomers) [93]. Nod factors play a major role during the physical contact between the bacteria and the root hairs, and also trigger the organogenesis of the nodule [94]. Then the bacterial threads can invade the root to colonize the nodule. At this stage, the rhizobial surface polysaccharides are essential [95,96,97,98]. To enlarge our knowledge about their role, it is necessary to determine their

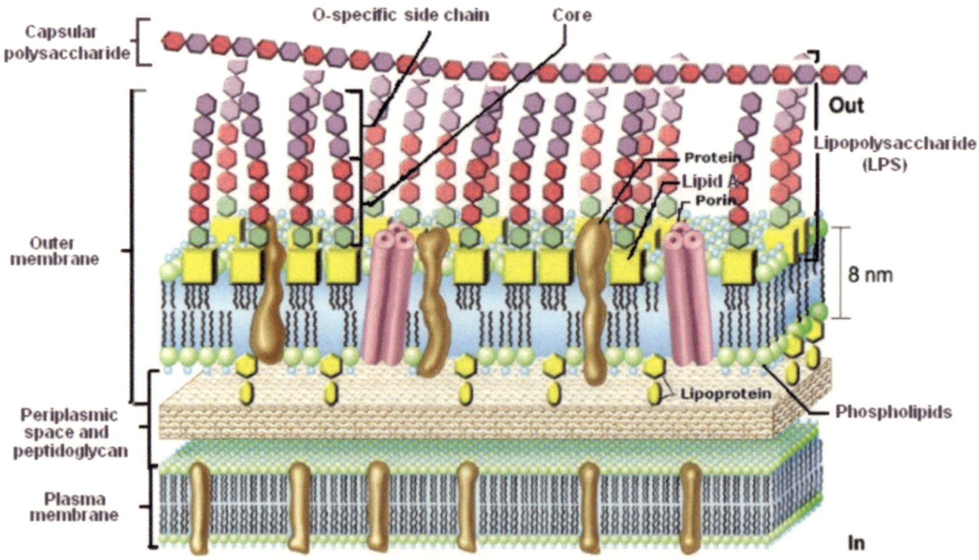

Figure 5. Scheme of the surface of Gram negative bacteria. The EPSs are not represented here; they are over the capsular polysaccharides.

structure. Their structural characterizations will be described below. They generally consist in analysis of composition by Gas Chromatography coupled with Mass Spectrometry (GC-MS), and sequence analysis made principally by Mass Spectrometry (MS). The choice of the MS coupling, mode or instrument depends on the nature of the polysaccharide (size and composition) as well as on the type of information we want to accede.

Sinorhizobium meliloti –the European model of rhizobia - has an external surface where 3 types of polysaccharides can be observed: exopolysaccharide (EPS), capsular polysaccharide (CPS) and lipopolysaccharide (LPS). Each class of PS have to be investigated alone, because their physico-chemical properties do not allow their simultaneous detection.

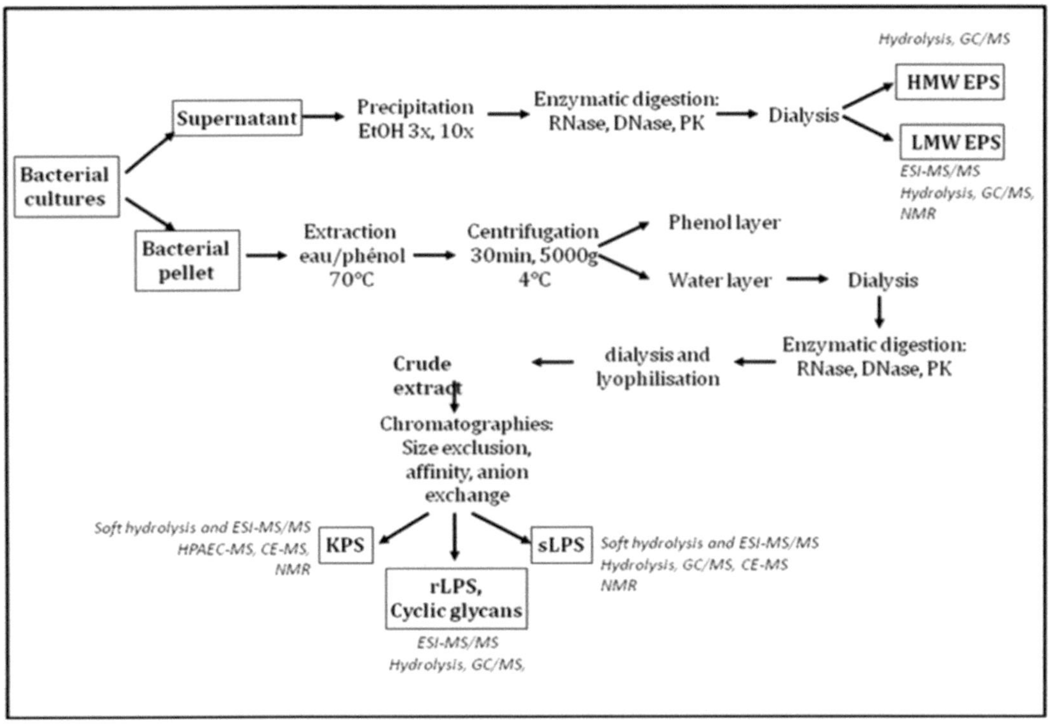

Figure 6. Isolation protocol for the different polysaccharides. In italic : caracterization methods.

3.1. Exopolysaccharides of *Sinorhizobium meliloti*

3.1.1. Isolation of the EPS

The EPS are generally composed of many repetitions of 8 to 12-mers of hexose-subunits in a linear and/or branched form. This can be more or less substituted by O-acetyl, succinyl and pyruvyl groups. Their composition and structure is species specific and depends on the growth conditions [99]. The EPS are produced by the bacteria during the stationary phase of growth [99].

When the EPS of *S. meliloti* were studied, the cultures were stopped 5 hours after the stationary phase, by centrifugation. The supernatant, containing the EPS, was isolated and lyophilized. When concentrated enough (increased viscosity but well dissolved), the EPS were precipitated, first with 3 volumes of ethanol. The pellet so formed was centrifuged and kept, the supernatant was concentrated. The pellet constitutes the high molecular weight (HMW) EPS fraction. The EPS left in the treated supernatant were then precipitated with 10 volumes of ethanol, the pellet was centrifuged and kept. This is the low molecular weight (LMW) EPS fraction. Each fraction was resuspended in water and subjected to proteinase K digestion (3.5 g/L at final concentration) for 4h at 36°C. The mixtures were then dialyzed against water. The pellets thus obtained were lyophilized, and once dried, used for structural analysis determination.

3.1.2. Characterization of the EPS

To learn about the monosaccharide composition, the polysaccharides must be first hydrolyzed. This chemical reaction is carried out in acidic conditions. A solution of EPS in water was acidified by 98% TFA to a final concentration of 10%, and the mixture was kept at 100°C for 2h. At the end of the treatment, the leaving acid has to be evaporated. To help this elimination in a N_2 stream, repetitive additions of isopropanol are necessary.

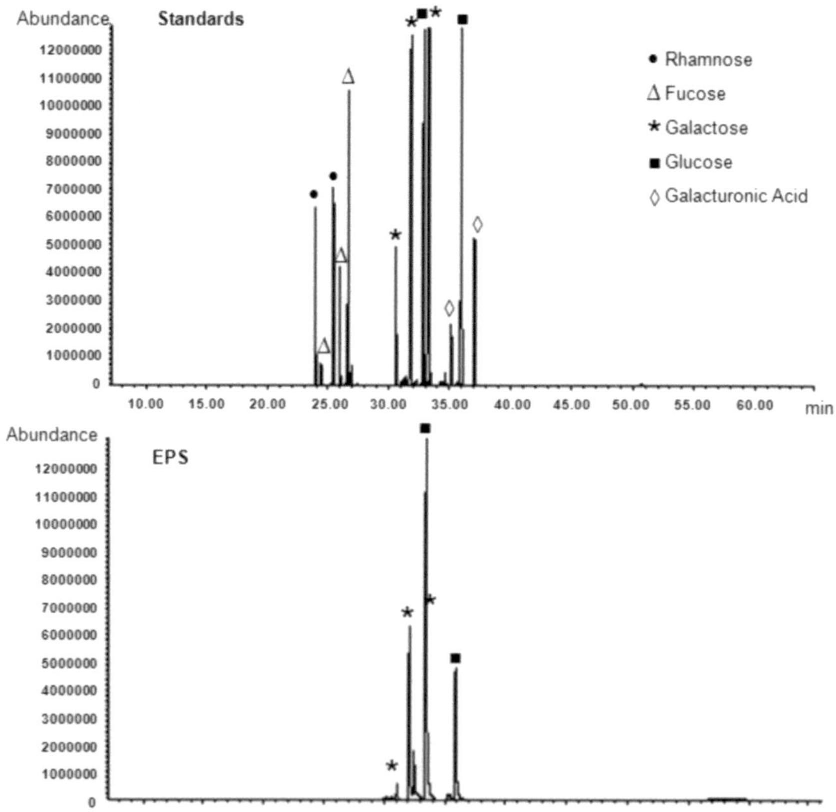

Figure 7. Chromatograms obtained by GC-MS for silylated standard saccharides and silylated hydrolyzed EPS LMW of *Sinorhizobium meliloti* 1021. EPS is made of Glucose and Galactose.

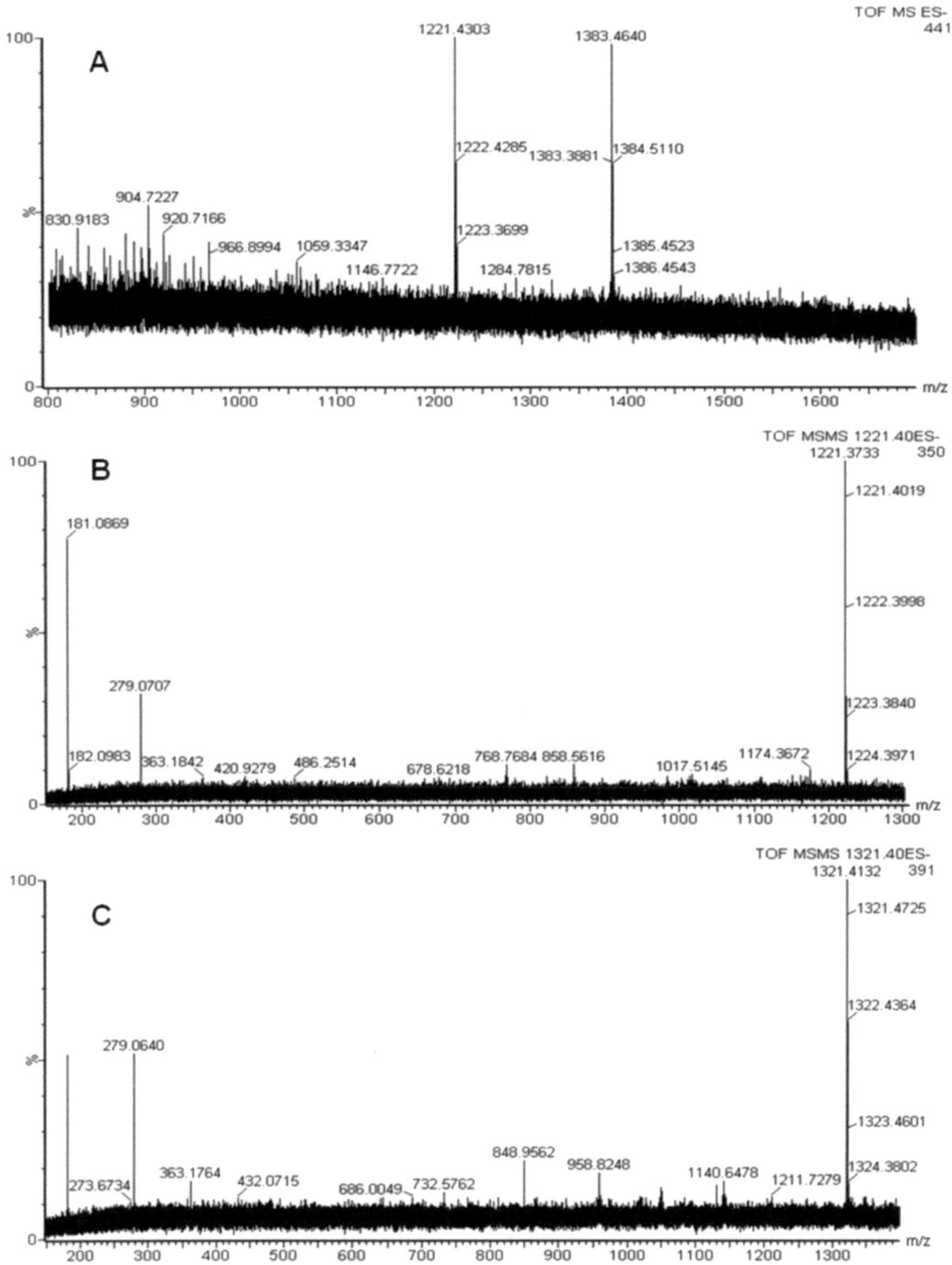

Figure 8. MS analyses of the LMW EPS of *Sinorhizobium meliloti* 1021A) ESI-MS analysis in negative mode B) ESI-MS/MS of ion *m/z* 1221.4 (the simplest form) amu. C) MS/MS of ion *m/z* 1321.4 amu.(the succinylated form), m/z 1383 corresponding to the succinylated and acetylated species.

After hydrolysis, the polysaccharides result in a mixture of monosaccharides and can be analyzed by GC-MS after derivatization. In the results presented here, we used the silylation of the sugar hydroxyl groups, which is a facile route. Actually, the protocol is simple, and a large databank is available for MS interpretations. In anhydrous pyridine, a mixture of HMDS and TMCS was added to the hydrolyzed EPS. This reaction was held for 30 min at 70°C [100]. After complete evaporation, the derivatized monosaccharides were injected directly as a hexane solution. The vector gas used was helium and the column was 95% dimethylsiloxane and 5% diphenylsiloxane. The temperature gradient was: 70°C hold for 3 min, increase at 5°C/min to 140°C and at 3°C/min to 240°C, then reach 300°C at 10°C/min, and hold at 300°C for 10 min.

This procedure applied to standard monosaccharides (glucose, galactose, mannose, glucuronic acid and galacturonic acid) allows interpretation of the chromatograms, determination of each sugar pattern (depending on alpha, beta, pyranoside, furanoside or lactone configuration) and the response factor for the different carbohydrates (Figure 7). The monosaccharide composition of the repeated units of LMW and HMW EPSs of *S. meliloti* was thus obtained.

The preparation of a simple solution of intact EPS in acidic methanol allows its ESI-MS analysis. This kind of analysis shows the purity of the EPS and the disparity of the sugars (Figure. 8A). Actually, the polysaccharide diversity is due to a variable degree of polymerization (DP) and to non-carbohydrate substituents. Here, the substituent is probably a succinyl group, adding 100 amu to ion *m/z* 1221.4 amu to yield the ion *m/z* 1321.4 amu. The ESI-MS/MS analysis of the ions found in the MS spectrum allows the sequence to be assessed by studying the fragmentation ions (Figure 8 B and C).

3.2. Lipopolysaccharides of *Sinorhizobium meliloti*

LPS have three structural domains: the lipid A, the oligosaccharide core and the O-antigen polysaccharide. The LMW LPS, named rough LPS are composed of lipid A and core polysaccharides, whereas the HMW LPS (smooth LPS) are made of the 3 associated parts [101].

3.2.1. Isolation of the LPS

As LPS are anchored in the outer membrane by lipid A, after growth of *S. meliloti* to the stationary phase, it was necessary to keep the pellet of centrifuged cultures (1L, grown until OD (600nm) 1.5). The pellet resuspended in about 40 mL of water was extracted by phenol at 60°C for 1h [102]. The extraction mixture was centrifuged and the upper phase (aqueous) contained the LPS, and all the other hydrosoluble molecules (DNAs, RNAs, proteins, carbohydrates). Enzymatic digestions were performed to eliminate DNAs, RNAs and proteins, remaining the carbohydrates. Affinity chromatography allowed the isolation of LPS. Then, gel filtration separated the LMW and HMW LPS (respectively rLPS and sLPS).

3.2.2. Characterization of the LPS

In this work, it was the HMW LPS that were studied, so both types of structure had to be determined, the saccharide parts and the lipid A. Soft hydrolysis, in 1% acetic acid at 100°C for 1h, isolated lipid A, separating out of the aqueous solution. A liquid-liquid extraction, with $CHCl_3$:CH_3OH (3:1), separated the lipidic part from the sLPS.

Structural analysis of lipid A was performed by ESI-MS. The deconvoluted spectrum of doubly-charged ions revealed two series of monocharged ions, with a mass resolution that allowed the identification of the monoisotopic molecular masses (and not only the average molecular masses), i.e.: *m/z* 2050.41, 2038.39/2036.3, 2022.41, 2010.39 for one series and 1950.41, 1938.39/1936.3, 1922.37, 1910.37 for the second series with sufficient accuracy to allow elemental composition analysis (Figure 9, note that the spectrum presented is not deconvoluted). Accurate mass analysis combined with GC-MS data led to the proposal of a first general structure for lipid A : a di-phosphorylated penta-acylated diglucosamine comprising two 3OH-C14 fatty acids, one 3OH-C18, one 3OH-C19:1 and C27OH-C28 fatty acids for a compound giving a singly charged molecular ion of *m/z* 1950.4 amu (Figure 10)

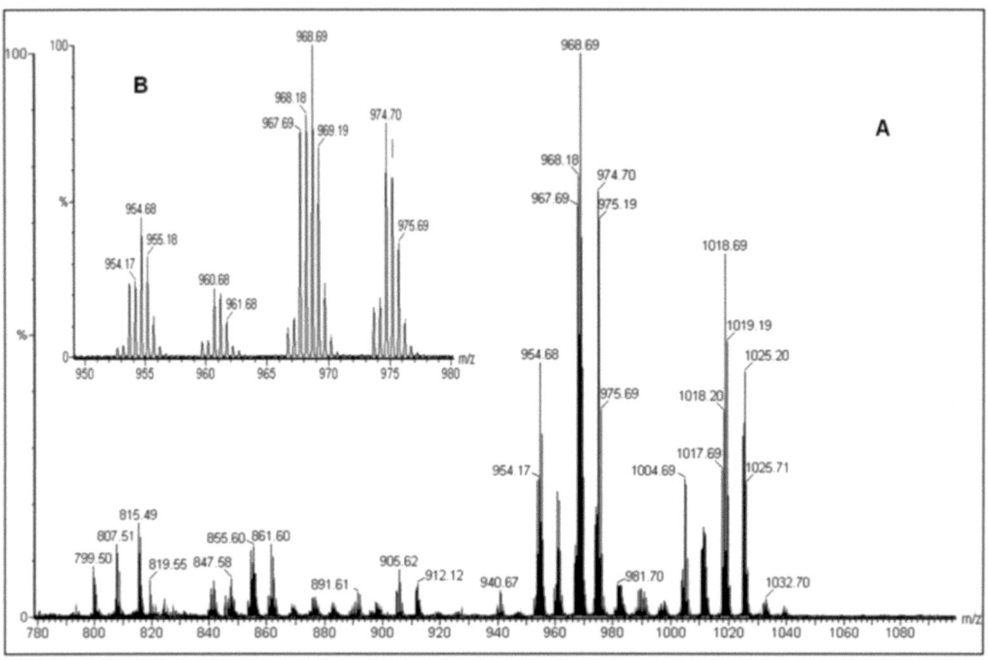

Figure 9. A) Direct ESI-MS spectrum of LPS lipid anchor. Ions ranging from 954 to 976 are assumed to be molecular species and ions observed between *m/z* 1004 and 1026 are native methoxybutyrate derivatives. Insert B) high-resolution spectrum indicating the double charged state of lipidA.

The difference of 100 amu (50 amu in spectrum shown figure 9) can be attributed to the presence of a 3-methoxybutyrate in the first series. Actually, as the second lipid A distribution was hypothesized to be 3-O-methoxy-butyrate substitution of the 27OH-C28, exact mass measurements were performed. In the mean mass difference found between the respective members of the two distributions was 100.03 amu, corresponding with high significance to a $C_5O_2H_8$. To confirm this hypothesis, MS/MS analyses of the double charged ion at *m/z* 1018.2 (Figure 11) have been performed. At low collision energy (15V) *m/z* 1018.2 generated three major fragments, respectively *m/z* 2004.4, 1992.4 and 1936.3 amu. *M/z* 2004 corresponds to the usual neutral loss of methanol (-32 amu), 1992 could be interpreted as a neutral loss of CO_2 (44 amu) after rearrangement. Finally *m/z* 1936 corresponds to the ketene loss of the 3-methoxy butyrate. Increased collision energy (up to 40V) resulted in the production of the same fragments as observed directly from the parent ion that is not substituted by a 3-O-methoxy-butyrate (*m/z* 1935.4). The MS analyses of the complete LPS provided, by comparison to lipid A, the mass of the saccharide part.

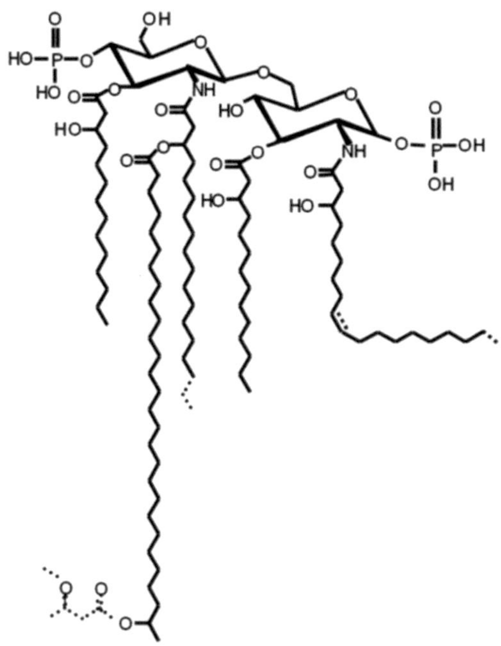

Figure 10. General Lipid A structure of *Sinorhizobium meliloti* 1021's LPS. The dotted lines correspond to the substituent variation found in the sample.

Composition analysis of the saccharide part of the sLPS was performed as for the structural characterization of EPSs. After hydrolysis of the polysaccharide, the monosaccharides obtained were derivatized into alditol acetates, because the O-antigen usually exhibits great sugar diversity.

The process consisted of a reduction by sodium borohydride in ammonia for 1h at room temperature, followed by washing with acetic acid, and an acetic acid in methanol solution [103]. A supplementary wash was made in methanol only, before acetylation. After adding acetic anhydride and pyridine, the reaction stood for 1h at 70°C [104]. Liquid-liquid extraction in water against dichloromethane allowed the isolation of the so-formed alditol acetates in the organic phase. The dried mixture of derivatized monosaccharide was directly analyzed by GC-MS. The temperature gradient started at 110°C and increased at 3°C/min to reach 300°C. Of course, the same process was performed on standard monosaccharides to establish a short database, helpful for interpreting the chromatogram (Figure 12).

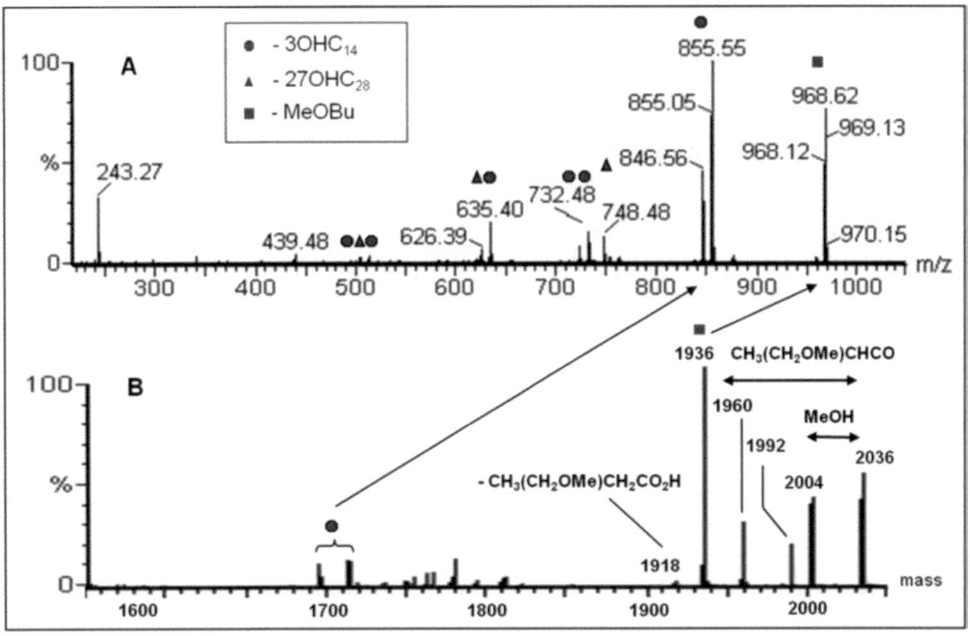

Figure 11. A) CID MS/MS high energy (Ecoll 40V) fragmentation of ion at *m/z* 1018 results principally in *m/z* 968 (loss of methoxybutyrate : -MeOBu) double charged species. B) deconvoluted spectrum indicating a characteristic methoxybutyrate fragmentation at lower collision energy (18V).

3.3. Capsular polysaccharides of *Sinorhizobium meliloti*

In rhizobia, capsular polysaccharides are generally composed of a dimer repeating unit, composed of a hexose and a 1-carboxy-2-keto-3-deoxysugar, like Kdo (2-keto-3-deoxy octulosonic acid). Such a structure looks like the K-antigen found in *E. coli*, and was therefore named Kdo-rich capsular polysaccharide (KPS) for *Rhizobia* [105,106].

3.3.1. Isolation of the K type CPS (KPS)

To isolate KPS from bacterial surface molecules, the same protocol as for the isolation of LPS was executed. Surprisingly, the *S.meliloti* 1021 KPSs were also retained by affinity chromatography. Gel filtration allowed enrichment in KPS, but their size is too close to the size of LPS, to be isolated from each other.

3.3.2. Characterization of the EPS

The analysis of carboxy-sugars such as Kdo is extremely difficult with GC-MS, due to their molecular weight and their instability. So, the composition of the KPS of *S. meliloti* was determined by GC-MS and mainly by HPAEC-PAD-MS. The latter technique requires an

HPLC system, an anion exchange column and a desalter, which is required to couple the ion chromatograph with the MS (see part II.2.c). The device is detailed above, in the methodological part. Analysis of sugars with HPAEC needs no derivatization, better for the stability of Kdo. PAD detection is not sufficient to precisely determine the composition of the mixture analyzed, so the system is coupled to a mass spectrometer. The combination of HPAEC-PAD and MS is a challenge, because of the quantity of salt used in HPAEC-PAD, hence the presence of the desalter is essential. The challenge is rewarded by the information provided by this coupling for determining the saccharidic composition, with simple and non-destructive sample preparation.

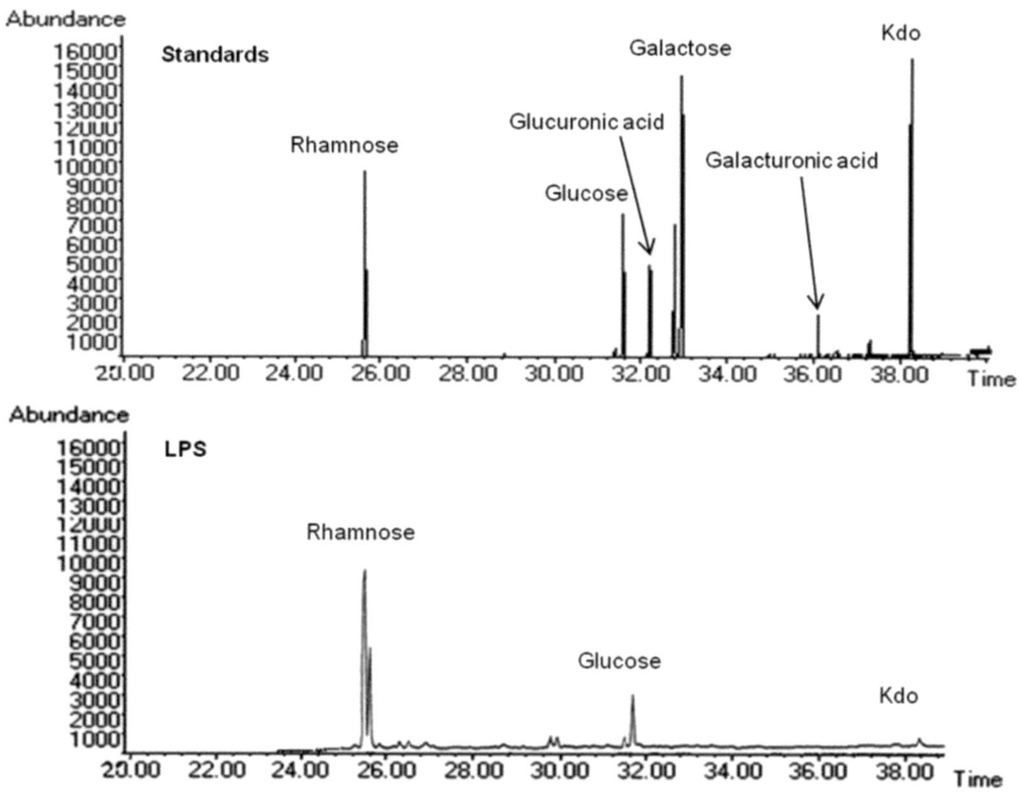

Figure 12. Chromatograms obtained by GC-MS for alditol acetates of standard saccharides and alditol acetates of hydrolyzed LPS of *Sinorhizobium meliloti* 1021. LPS is made of Rhamnose, Glucose and Kdo.

For GC-MS analysis, as the acidic sugar content was high, derivatization with heptafluorobutyric anhydride (HFBA) was necessary [107]. The methanolyzed polysaccharides were dried and resuspended in anhydrous acetonitrile, a solution of HFBA was added. The reaction was heated to 60°C for 30 min. Evaporation of the leaving HFBA involved several co-evaporations with anhydrous acetonitrile. The final solution was made up in anhydrous acetonitrile and injected directly into the GC-MS. The temperature gradient was 70°C for 3 min, 5°C/min to 100°C, 3°C/min to 240°C and 5°C/min to reach 300°C and hold for 10 min. Figure 13 indicates that the KPS enriched fraction is exclusively composed of Kdo.

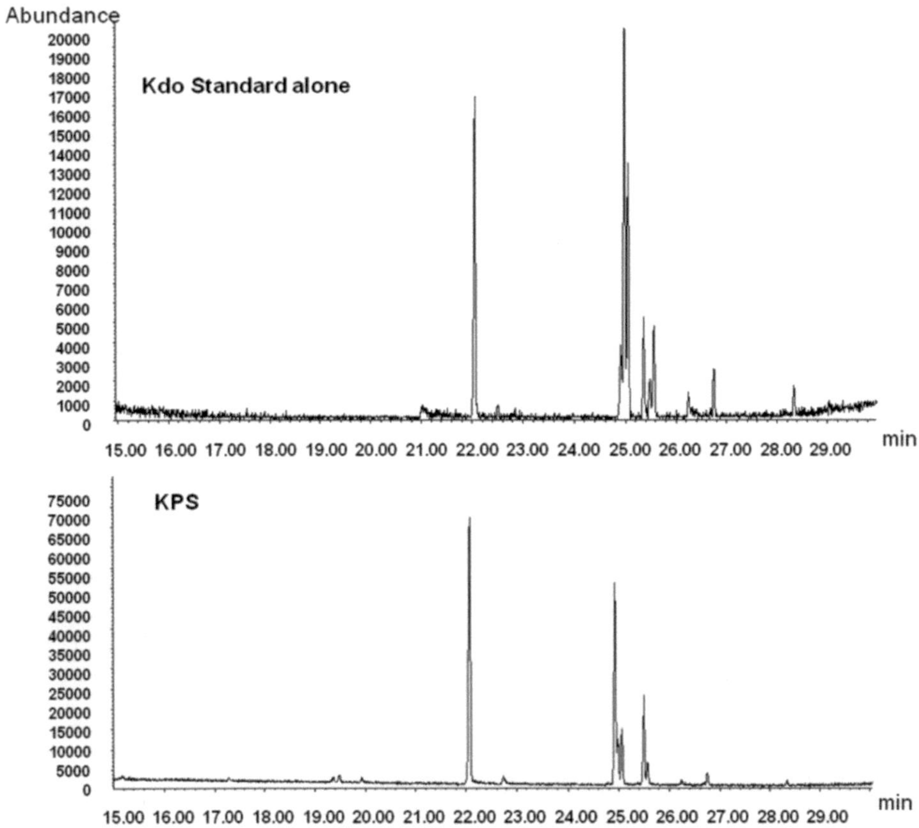

Figure 13. Chromatograms obtained by GC-MS for Kdo standard and methanolyzed KPS of *S. meliloti* 1021, both derivatized by HFBA. KPS is only composed of Kdo.

Unlike GC-MS analyses, in HPAEC-PAD-MS analyses, only soft hydrolysis of the polysaccharide mixture from the size exclusion chromatography (SEC) fractions containing KPS was implemented (1% acetic acid, 100°C, 1h). During the soft hydrolysis, a pellet appeared, indicating the presence of fatty acid on the KPS. MS analysis described bellow will detail this. The HPAEC-PAD-MS analysis revealed free Kdo and oligo-Kdo, characterized in MS by a mass difference of *m*/*z* 220 amu, when ion *m*/*z* 221 amu is extracted (Figure 14). The other compounds present in this sample are mostly composed of hexose dimers (*m*/*z* 324 amu), revealed when ion *m*/*z* 325 amu is extracted. So, the HPAEC-PAD chromatogram allows simultaneous analysis of other polysaccharides in this mixture: glycans, and substituted glycans [72].

A MS analysis confirmed that KPS is a homopolymer of Kdo. Actually, direct ESI-MS in negative mode (Figure 15 A) indicated the presence of many charged compounds. Zooming between *m*/*z* 1020 and *m*/*z* 1060, multi-charged ions (*e.g.*:1035.30 to 1036.05 are quadri-charged ions and 1028.54 to 1029.04 are discharged ions) clearly appear (Figure 15 B). A deconvolution process recovers the native mass of the compounds (Figure 16 A). This reveals a mass difference of 220 amu between the members of the series, typically corresponding to Kdo. The MS/MS analysis of one of these polymers confirms that it is exclusively composed of Kdo, but also revealed the presence of an anchor explaining the discontinuity between Kdo_{15} and Kdo_{17} (Figure 16 B).

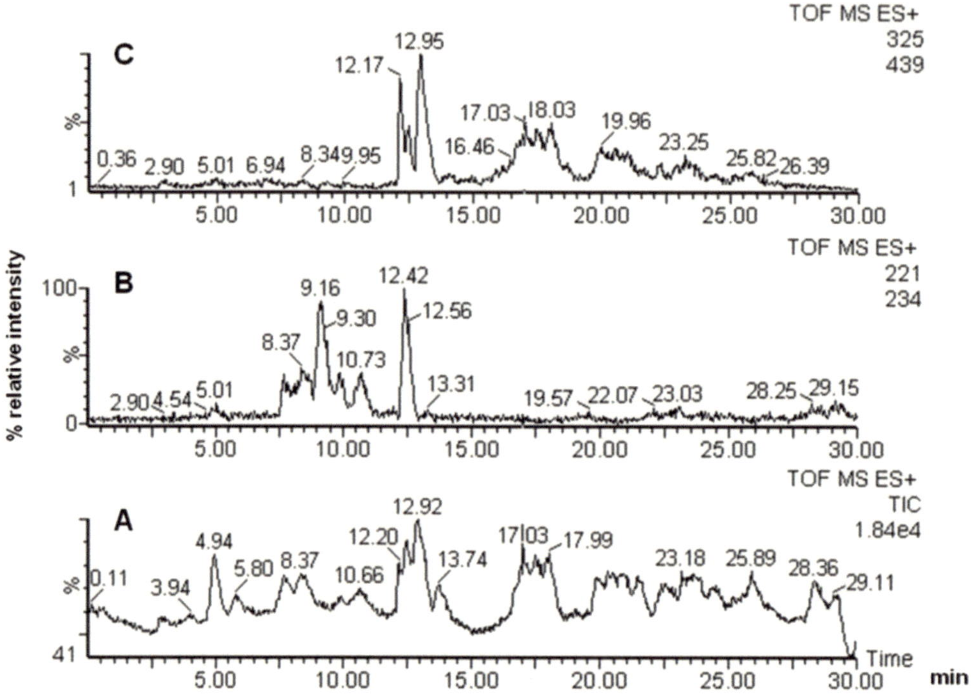

Figure 14. Chromatograms obtained by HPAEC-PAD-MS. A) Total ion current measured by MS. B) Chromatogram obtained when ion *m/z* 221 is extracted. C) Chromatogram obtained when ion *m/z* 325 is extracted. These chromatograms revealed that the KPS fraction is composed of polymers of Kdo and of glycans.

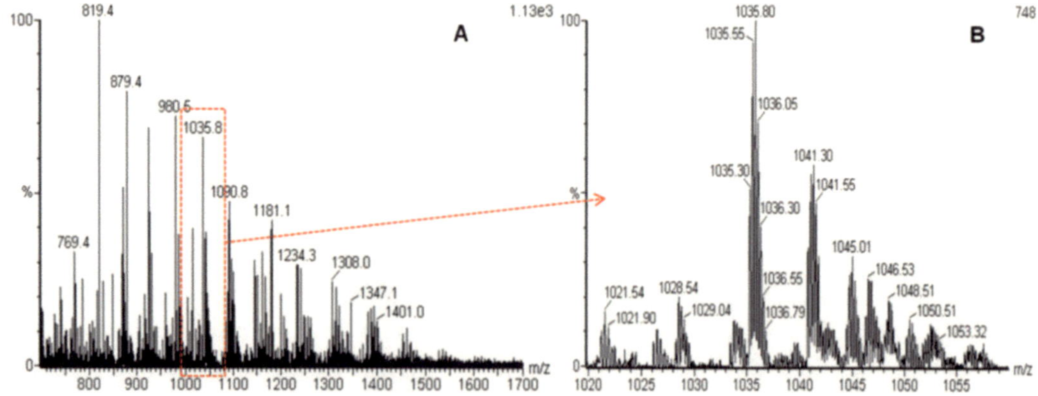

Figure 15. A) ESI-MS spectrum of KPS of *S. meliloti* 1021. B) Zoom on a group of ions revealing the multi-charged ions.

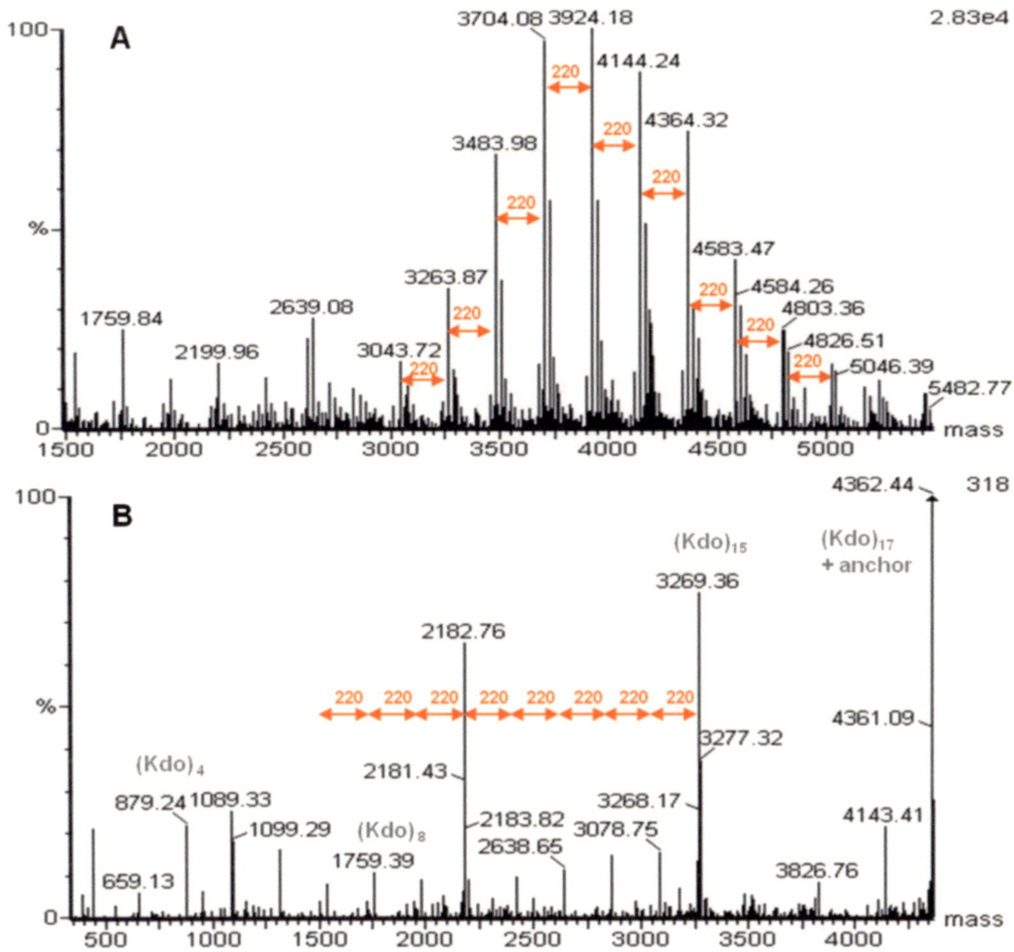

Figure 16. A) deconvoluted spectrum of KPS of S.mmeiloti 1021. B) MS/MS analysis of one Kdo-polymer, revealing an anchor on the KPS.

A MS/MS analysis of the smallest Kdo-oligomer revealed the mass of the anchor at *m/z* 622.4 amu. This ion has been studied separately by MS/MS, after mild hydrolysis on the KPS fraction (Figure 17). Ions m/z 530 corresponds to the loss of the glycerol and 548 to the loss of glycerol minus one water molecule. The interpretation of the spectra, in negative and in positive mode, determined that the anchor was lipidic and composed of a glycerol and phosphoglycerol unit, leading to the structure detailed in figure 18.

For structural characterization of polysaccharides, classical techniques are mostly used. Here, we reported results obtained with chromatography coupled to MS, MS alone and MS/MS. Depending on the characteristics, the type of polysaccharide analyzed, and which data are sought from those samples, it is necessary to adapt the coupled techniques. Advanced techniques, like NMR, are increasingly used for precise characterization. But NMR spectra of polysaccharides are very complex, therefore not easy to interpret. Actually, the heterogeneity in substitution and size is principally the origin of this complexity, especially in 2D NMR.

Moreover, because of the interpretation difficulties, detailed mapping of NMR spectra of polysaccharides in the literature are seldom found. Lastly, as detailed in part I, note that polysaccharide solubility problems can lead to poorly resolved NMR spectra. While NMR can provide useful information about the polysaccharide, MS analysis is not obsolete, because the information provided with MS is unambiguous and confirms NMR spectrum interpretation. In many examples, MS analysis is necessary for an easier interpretation of the NMR spectra. Bacterial polysaccharides are highly complex molecules and many variations occur in one family. The role of polysaccharides from the rhizobia family during nitrogen fixing symbiosis has been demonstrated, as well as the activity of other bacterial polysaccharides during pathogen infection. However, little is known about their structure/activity relationships [108], which implies a long life for polysaccharide structural characterization.

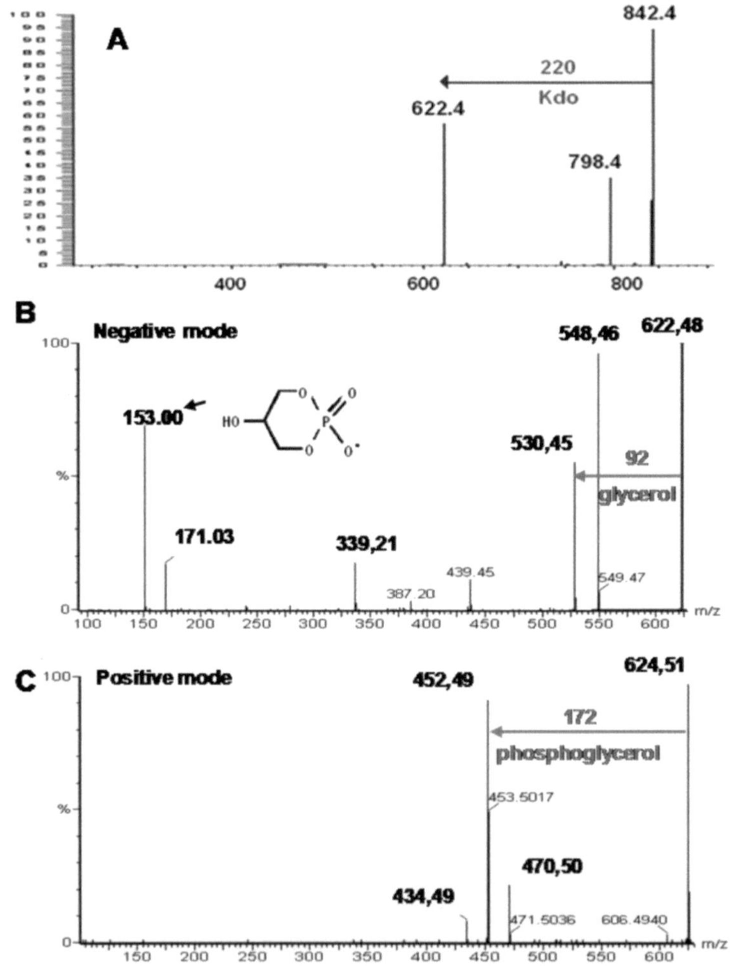

Figure 17. Top, MS/MS analysis of the smallest compound containing Kdo in the KPS, revealing the anchor at *m/z* 622 amu (A). MS/MS Fragmentation of the lipidic anchor (ion *m/z* 622 amu) obtained after soft hydrolysis in negative (B) and positive (C) mode. The presence of glycerol and phosphoglycerol in this lipid anchor was thus demonstrated.

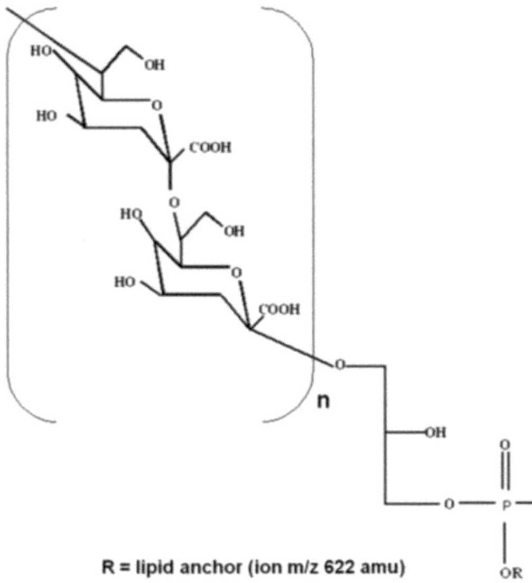

Figure 18. General structure of KPS of *S. meliloti* 1021

Author details

V. Poinsot, M.A. Carpéné and F. Couderc
Laboratoire IMRCP, UMR 5623, Université Paul Sabatier, Toulouse, France

4. References

[1] Varki A (1993) Biological roles of oligosaccharides: all of the theories are correct. Glycobiology, 1993, 3: 97-130
[2] Dweck RA (1996) Glycobiology: toward understanding the function of sugars. Chem. Rev. 96: 683-720
[3] Matamoros Fernandez LE (2007) Introduction to ion trap mass spectrometry: Application to the structural characterization of plant oligosaccharides. Carbohydr. Polym. 68: 797-807
[4] Molyneux RJ, Gardner DR, James LF, Colegate SM (2002) Polyhydroxy alkaloids: chromatographic analysis. J. Chromatogr. A. 967-957.
[5] Sanz ML, Martínez-Castro I (2007) Recent developments in sample preparation for chromatographic analysis of carbohydrates. J. Chromatogr. A. 1153: 74-89.
[6] Ruiz Matute AI, Hernández Hernández O, RodríguezSánchez S, Sanz ML, Martínez-Castro I (2011) Derivatization of carbohydrates for GC and GC–MS analyses. J Chromatogr. B, 879: 1226–1240.
[7] Wolfrom ML, Thompson A (1963) Reduction with sodium borohydride. Meth. Carbohydr. Chem. 2: 65–68.
[8] Wolfrom ML, Thompson A (1963b) Acetylation. Meth. Carbohydr. Chem. 2: 211–215.

[9] Bromund WH, Herbst RM (1945) The synthesis of oxazoline derivatives of monosaccharides and their relationship to the amino sugars. J. Org. Chem. 10: 267–276.
[10] Kim JB, Carpita NC (1992) Changes in Esterification of the Uronic Acid Groups of Cell Wall Polysaccharides during Elongation of Maize Coleoptiles, Plant Physiol. 98: 646-653.
[11] Maness NO, Ryan JD, Mort AJ (1990) Determination of the degree of methyl esterification of pectins in small samples by selective reduction of esterified galacturonic acid to galactose. Anal. Biochem. 185 : 346-352.
[12] Fontaine T, Fournet B, Karamanos Y (1994) A new procedure for the reduction of uronic acid containing polysaccharides. J. Microbiol. Meth. 20: 149-157.
[13] Parkhomchuk AA, Kocharova NA, Białczak-Kokot M, Shashkov AS, Chizhov AO, Knirel YA, Rozalski A. (2010) Structure of the O-polysaccharide from the lipopolysaccharide of Providencia alcalifaciens. O12.Carbohydr Res. 345: 1235-1239.
[14] Wunschel DS, Colburn HA, Fox A, Fox KF, Harley WM, Wahl JH, Wahl KL (2008) Detection of agar, by analysis of sugar markers, associated with Bacillus anthracis spores, after culture. J. Microbiol. Methods. 74: 57-63.
[15] Lima LFO, Habu S, Gern JC, Nascimento BM, Parada JC, Noseda MD, Gonçalves A, Nisha VR , Pandey A , Soccol VT, Soccol CR (2008) Production and Characterization of the Exopolysaccharides Produced by Agaricus brasiliensis in Submerged Fermentation. Appl. Biochem. Biotechnol. 151:283–294
[16] Viccini G, Martinelli TR , Cognialli RCR, de Faria RO, Carbonero ER , Sassaki LK (2009) Mitchell DA Exopolysaccharide from surface-liquid culture of Clonostachys rosea originates from autolysis of the biomass. Arch. Microbiol. 191:369–378
[17] Smiderle FR, Olsen LM, Carbonero ER, Marcon R, Baggio CH, Freitas CS, Santos ARS, Torri G, Gorin PAJ, Iacomini M (2008) A 3-O-methylated mannogalactan from Pleurotus pulmonarius: Structure and antinociceptive effect, Phytochemistry 69: 2731–2736
[18] Vesentini D, Steward D, Singh AP, Ball R, Daniel G, Franich R (2007) Chitosan-mediated changes in cell wall composition, morphology and ultrastructure in two wood-inhabiting fungi. Mycol. Res. 111 :875-90.
[19] Marquesa G, Gutiérreza A, del Ríoa JC, Evtuguin DV (2010) Acetylated heteroxylan from Agave sisalana and its behavior in alkaline pulping and TCF/ECF bleaching. Carb. Polym. 81: 517–523.
[20] Ayestarán B, Guadalupe Z, León D (2004) Quantification of major grape polysaccharides (Tempranillo v.) released by maceration enzymes during the fermentation process, Anal. Chim. Acta 513: 29–39.
[21] Quemener, Lahaye M, Metro F (1995) Assesment of methanolysis for the determination of composite sugars of gelling carrageenans and agarose by HPLC. Carbohydr. Res. 266: 53-64.
[22] Bleton J, Mejanelle P, Sansoulet J, Goursaud S, Tchapla (1996) Characterisation of neutral sugars and uronic acids after methanolysis and trimethylsilylation for recognition of plant gums, J Chromatogr. A 720: 27-49.

[23] Golovkina LS, Chizhov OS, Wul'fson NS Acetate of polyols. In"Mass spectrometric investigation of carbohydrate".vol 9. Bull. Acad Sci USSR, 1966.
[24] Mc Lafferty FW and Turecek F, In Interpretation of Mass Spectra, University Scince boof, Mill Valley, CA 1993
[25] Jansson PE., Lindberg B (1980) Mass spectrometry of alditol acetates: origin of the fragments having even mass numbers. Carbohydr. Res. 86: 287–292.
[26] Patoureau-Promé D, Promé JC, Carbohydrate, in GC/MS applications in Microbiolog. pp106-156, G Odham, L Larsson, PA Mardh, Plenum Press, New York 1984.
[27] Dizdaroglu M, Henneberg D, Von Sonntag C, Schuchman MN (1977) Mass spectra of trimethylsilyl-di-O-methyloximes of aldosuloses and dialdoses Org. Mass spectrum., 8: 335.
[28] Petersson G (1969) Mass spectrometry of aldonic and deoxyaldonic acids as trimethylsilyl derivatives, Tetrahedron, 25: 4437-4443.
[29] Bleton J, Mejanelle P, Sansoulet J, Goursaud S, Tchapla A (1996) Characterization of neutral sugars and uronic acids after methanolysis and trimethylsilylation for recognition of plant gums, J. Chromatogr. A, 720: 27-49.
[30] De Jongh DC, Radford T, Hribar JD, Hanessian S, Bieber M, Dawson G, Sweeley CC (1969) Analysis of trimethylsilyl derivatives of carbohydrates by gas chromatography and mass spectrometry J. Am. Chem. Soc. 91: 1728-1740.
[31] Kennedy JF, Robertson SM (1978) The structure of Cryptococcus neoformans galactoxylomannan contains β-D-glucuronic acid Carbohydr. Res. 67: 1-15.
[32] Kochetkov NK, Chizhov OS (1966) Mass Spectrometry of Carbohydrate Derivatives, Adv. Carbohydr. Chem. 21, 39-93.
[33] Hellerqvist CG and Sweetman BJ Mass spectrometry of Carbohydrates, pp 90-143. In Biomedical applications of Mass spectrometry Suelter CH, Throck Watson Ed Interscience Publication John Wiley & Sons New York 1990.
[34] Kim JS, Reuhs BL, Michon F, Kaiser RE, Arumugham RG. (2006) Addition of glycerol for improved methylation linkage analysis of polysaccharides. Carbohydr Res., 341: 1061-1064.
[35] Kang P, Mechref Y, Novotny MV. (2008)High-throughput solid-phase permethylation of glycans prior to mass spectrometry. Rapid Commun Mass Spectrom. 22: 721-734.
[36] Kang P, Mechref Y, Klouckova I, Novotny MV. (2005) Solid-phase permethylation of glycans for mass spectrometric analysis. Rapid Commun Mass Spectrom. 19: 3421-3428.
[37] Ciucanu I, Kerek F. (1984) A simple and rapid method for the permethylation of carbohydrates. Carbohydr. Res. 131: 209–217
[38] Ciucanu I, Costello CE (2003) Elimination of oxidative degradation during per-Omethylation of carbohydrates.. J. Am. Chem. Soc. 125: 16213-16219.
[39] de Souza LM, Cipriani TR, Serrato RV, da Costa DE, Iacomini M, Gorin PA, Sassaki GL. (2008) Analysis of flavonol glycoside isomers from leaves of Maytenus ilicifolia by offline and online high performance liquid chromatography-electrospray mass spectrometry. J Chromatogr A., 1207: 101-109.
[40] Kang P, Mechref Y, Novotny MV. (2008) High-throughput solid-phase permethylation of glycans prior to mass spectrometry. Rapid Commun Mass Spectrom.;22: 721-734.

[41] Mechref Y, Kang P, Novotny MV. (2009) Solid-phase permethylation for glycomic analysis. Methods Mol. Biol. 534:53-64.

[42] Smiderle FR, Olsen LM, Carbonero ER, Marcon R, Baggio CH, Freitas CS, Santos AR, Torri G, Gorin PA, Iacomini M. A (2008) 3-O-methylated mannogalactan from Pleurotus pulmonarius: structure and antinociceptive effect. Phytochemistry, 69: 2731-2736.

[43] Kim JS, Reuhs BL, Michon F, Kaiser RE, Arumugham RG. (2006) Addition of glycerol for improved methylation linkage analysis of polysaccharides. Carbohydr Res. 34:, 1061-1064

[44] Weiskopf A.S., Vouros P., Harvey P. (1997) Characterization of oligosaccharide composition and structure by quadrupole ion trap mass spectrometry. Rapid Commun Mass Spectrom, 11: 1493-1504

[45] Visieux N, De Hoffmann E., Domon B. (1998) Structural assignment of permethylated oligosaccharide subunits using sequential tandem mass spectrometry. Anal. Chem., 70: 4951-4959

[46] Hofmeister G.E., Zhou Z., Leary J.A. (1991) Linkage position determination in lithium-cationized disaccharides: tandem mass spectrometry and semiempirical calculations J. Am. Chem. Soc., 113: 5964-5970

[47] König S., Leary J.A (1998) Evidence for linkage position determination in cobalt coordinated pentasaccharides using ion trap mass spectrometry J. Am. Soc. Mass Spectrom., 9: 1125-1134

[48] Chai W., Lawson A.M., Piskarev V.J. (2002) The structural motif in chondroitin sulfate for adhesion of Plasmodium falciparum-infected erythrocytes comprises disaccharide units of 4-O-sulfated and non-sulfated N-acetylgalactosamine linked to glucuronic acid J. Am. Soc. Mass Spectrom, 13: 670-679

[49] Domon B. and Costello C.E. (1988) A systematic nomenclature of carbohydrate fragmentation in FAB-MS/MS spectra of glycoconjugates. Glycoconj. J., 5: 397-409

[50] Ngoka L.C., Gal J.F., Lebrilla C.B. (1994) Effects of cations and charge types on the metastable decay rates of oligosaccharides. Anal. Chem., 66: 692-698

[51] Caroll J., Willard D., Lebrilla C. Energetics of cross-ring cleavages and their relevance to the linkage determination of oligosaccharides. Anal. Chim. Acta, 1995, 307: 431-447.

[52] Dell A., Morris H.R., Egge H., von Nikolai H.Strecker G. (1983) Fast-atom-bombardment, negative-ion mass spectrometry of the mycobacterial O-methyl-D-glucose polysaccharide and lipopolysaccharides. Carbohydr. Res., 115: 41-52

[53] McClellan J.M., Costello C.E., O'Connor P.B., Zaia J. (2002) Influence of charge state on product ion mass spectra and the determination of 4S/6S sulfation sequence of chondroitin sulfate oligosaccharides. Anal Chem, 74: 3760-37771

[54] Tseng K, Hedrick J.L., Lebrilla C.B. (1999) Catalog-library approach for the rapid and sensitive structural elucidation of oligosaccharides, Anal Chem., 71: 3747-3754

[55] Zaia J. Mass spectrometric ionization of carbohydrates. In Encyclopedia of Mass Spectrometry, vol.6. M.G. Gross and r.M. Caprioli, eds. (Elsevier, New-York)

[56] Deery M.J., Stimson E., Chappell C.G. (2001) Mass spectrometry and glycomics. Rapid Commun mass Spectrom., 15: 2273-2283.

[57] Walter T.H., Iraneta P., Caparella P. (2005) Mechanism of retention loss when C8 and C18 HPLC columns are used with highly aqueous mobile phases J. Chromatogr. A, 1075: 177-183.

[58] Hemström P, Irgum K. (2006) Hydophilic interaction chromatography J. Sep. Sci, , 29: 1784-1821.

[59] Kuberan B. Lech M., Zhang L., Wu Z.L., Beeler D.L., Rosenberg R.D. (2002) Analysis of heparan sulfate oligosaccharides with ion pair reversed phase capillary HPLC-microelectrospray ionization time-of-flight spectrometry J. Am Chem Soc., 124:8707-8718

[60] Gustavsson, S.A., Samskog J., Markides K., Langström B.(2001) Studies of signal suppression in liquid chromatography-electrospray ionization mass spectrometry using volatile ion-pairing reagents.J. Chromatogr. A, 937: 41-47

[61] Lunn G., Hellwig L.C. Handbook of derivatization reaction for HPLC, Wiley-Interscience, New-York, 1998

[62] Wang X.D., Li W.Y., Rasmussen H.T. (2005)Orthogonal method development using hydrophilic interaction chromatography and reversed-phase high-performance liquid chromatography for the determination of pharmaceuticals and impurities. J. Chromatogr. A, ,1083: 58-62

[63] Lee Y.C. (1990) High-performance anion-exchange chromatography for carbohydrate analysis. Anal. Biochem, 189: 151-162.

[64] Rocklin R.D., T.R. Tullsen, M.G. Marucco, (1994) High-performance anion-exchange chromatography for carbohydrate analysis, J. Chromatogr. A , 671: 109-114.

[65] Thayer J.R., J.S. Rohrer, N. Avdalovic, R.P. Gearing, (1998) Improvements to in-line desalting of oligosaccharides separated by high-pH anion exchange chromatography with pulsed amperometric detection. Anal. Biochem, 256: 207-216.

[66] Torto N., Hofte A., Tjaden U.R., Gorton L., Marko-Varga G., Bruggink C., Van der Greef J. (1998) Microdialysis introduction high-performance anion-exchange chromatography/ ionspray mass spectrometry for monitoring of on-line desalted carbohydrate hydrolysates. J. Mass Spectrom., 33: 334-341.

[67] Van der Hoeven R.A.M., U.R. Tjaden, J. Van der Greef, W.H.M. Van Casteren, H.A. Schols, A.G.J. Voragen, C. Bruggink (1998), Recent progress in high-performance anion-exchange chromatography/ionspray mass spectrometry for molecular mass determination and characterization of carbohydrates using static and scanning array detection J. Mass Spectrom., 33: 377-386.

[68] Rumbold K., Okatch H., Torto N., Siika-Aho M., Gubitz G., Robra K.H., Prior B. (2002) Monitoring on-line desalted lignocellulosic hydrolysates by microdialysis sampling micro-high performance anion exchange chromatography with integrated pulsed electrochemical detection/mass spectrometry. Biotechnol. Bioeng, 78, 822-828.

[69] Bruggink C., R. Maurer, H. Herrmann, S. Cavalli, F. Hoefler (2005) Analysis of carbohydrates by anion exchange chromatography and mass spectrometry. J. Chromatogr. A, 1085: 104-109.

[70] Bruggink C., Wuhrer M., Koeleman C.A., Barreto V., Liu Y., Pohl C., Ingendoh A., Hokke C.H., Deelder A.M. (2005) Oligosaccharide analysis by capillary-scale high-pH

anion-exchange chromatography with on-line ion-trap mass spectrometry. J. Chromatogr. B., 829: 136-143.

[71] Johnson D.C., Lacourse W.R. (1990) LC with pulsed ECD at gold and platinum Anal. Chem., 62: 589A-597A.

[72] Chataigné G, Couderc F, Poinsot V. (2008) Polysaccharides analysis of sinorhizobial capside by on-line anion exchange chromatography with pulsed amperometric detection and mass spectrometry coupling. J Chromatogr A. 1185:241-250.

[73] Peter Katalinic J. (1994), Analysis of glycoconjugates by fast atom bombardment mass spectrometry and related MS techniques, Mass Spectrom. Rev., 13: 77-98.

[74] Treilhou M., Ferro M., Monteiro C., Poinsot V., Jabbouri S., Kanony C., Promé D., Promé JC. (2000) Differentiation of O-acetyl and O-carbamoyl esters of N-acetyl-glucosamine by decomposition of their oxonium ions: Application to the structure of the non-reducing end of Nod factors. *JAS mass*, 11: 301-311.

[75] Angel AS., Lindh F., Nilsson B. (2007) Determination of binding positions in oligosaccharides and glycosphingolipids by fast-atom-bombardment mass spectrometry. Carbohydr Res. 16:, 15-31.

[76] Berjeaud JM., Couderc F., Promé JC. (1993), Stereochemically controlled decomposition of silver-cationized methyl glycosides, Org. Mass Spectrom, 28, 455-458.

[77] Grimshaw J (1997) Analysis of glycosaminoglycans and their oligosaccharide fragments by capillary electrophoresis Electrophoresis, 18, 2408-2414.

[78] Lagane B, Treilhou M, Couderc F (2000) Capillary Electrophoresis: Theory, Teaching Approach and Separation of oligosaccharides Using Indirect UV Detection BMBE, 28: 251-255.

[79] El-Rassi Z, Postlewait J, Mechref Y, Ostrander GK. (1997) Capillary electrophoresis of carboxylated carbohydrates. III. Selective precolumn derivatization of glycosaminoglycan disaccharides with 7-aminonaphthalene-1,3-disulfonic acid fluorescing tag for ultrasensitive laser-induced fluorescence detection. Anal Biochem. 244: 283-290.

[80] Chen FT, Evangelista RA. (1995) Analysis of mono- and oligosaccharide isomers derivatized with 9-aminopyrene-1,4,6-trisulfonate by capillary electrophoresis with laser-induced fluorescence. Anal Biochem. 230: 273-280.

[81] Fraysse N, Verollet C, Couderc F, Poinsot V. (2003) Capillary electrophoresis as a simple and sensitive method to study polysaccharides of Sinorhizobium sp. NGR234. Electrophoresis. 24: 3364-3370.

[82] Szabo Z, Guttman A, Bones J, Karger BL. (2011) Rapid high-resolution characterization of functionally important monoclonal antibody N-glycans by capillary electrophoresis. Anal Chem. 83: 5329-5336.

[83] Mittermayr S, Bones J, Doherty M, Guttman A, Rudd PM. (2011) Multiplexed analytical glycomics: rapid and confident IgG N-glycan structural elucidation. J Proteome Res. 10: 3820-3829.

[84] Szabo Z, Guttman A, Rejtar T, Karger BL. (2010) Improved sample preparation method for glycan analysis of glycoproteins by CE-LIF and CE-MS. Electrophoresis., 31: 1389-95.

[85] Mechref Y (2011) Analysis of glycans derived from glycoconjugates by capillary electrophoresis-mass spectrometry Electrophoresis, 32: 3467–3481

[86] Tegeler TJ, Mechref Y, Boraas K, Reilly JP, Novotny MV. (2004) Microdeposition device interfacing capillary electrochromatography and microcolumn liquid chromatography with matrix-assisted laser desorption/ionization mass spectrometry. Anal. Chem., 76: 6698-6706.

[87] Gennaro, L. A., Salas-Solano, O., On-line CE-LIF-MS technology for the direct characterization of N-linked glycans from therapeutic antibodies. (2008) Anal. Chem., 80: 3838–3845.

[88] Militsopoulou M, Lecomte C, Bayle C, Couderc F, Karamanos NK. (2003) Laser-induced fluorescence as a powerful detection tool for capillary electrophoretic analysis of heparin/heparan sulfate disaccharides. Biomed Chromatogr., 17: 39-41.

[89] Nakano M, Higo D, Arai E, Nakagawa T, Kakehi K, Taniguchi N, Kondo A. (2009) Capillary electrophoresis-electrospray ionization mass spectrometry for rapid and sensitive N-glycan analysis of glycoproteins as 9-fluorenylmethyl derivatives. Glycobiology., 19: 135-143.

[90] Spaink HP (2000) Root nodulation and infection factors produced by rhizobial bacteria. Ann. Rev. Microbiol. 54:257-288.

[91] Robertson JG, Wells B, Brewin NJ, Wood E, Knight CD, Downie JA (1985) The legume-Rhizobium symbiosis: a cell surface interaction. J. Cell Sci. Suppl. 2: 317-331.

[92] Gage DJ (2004) Infection and invasion of roots by symbiotic, nitrogen-fixing rhizobia during nodulation of temperate legumes. Microbiol. Mol. Biol. Rev. 68: 280-300.

[93] Dénarié J, Debellé F, Rosenberg C (1992) Signaling and host range variation in nodulation. Annu. Rev. Microbiol. 46: 497-531.

[94] Murray JD (2011) Invasion by invitation: rhizobial infection in legumes. Mol. Plant Microbe Interact. 24:631-639.

[95] Carlson RW (1982) Surface chemistry, 199-234, In W. J.Broughton (ed.), Nitrogen fixation: Rhizobium, vol. 2. ClarendonPress, Oxford.

[96] Pellock BJ, Cheng HP, Walker GC (2000) Alfalfa root nodule invasion efficiency is dependent on Sinorhizobium meliloti polysaccharides. J. Bacteriol. 182: 4310-318.

[97] Fraysse N, Couderc F, Poinsot V (2003) Surface polysaccharide involment in establishing the Rhizobium-legume symbiosis. Eur. J. Biol. Chem. 270:1365-1380.

[98] Skorupska A, Janczarek M, Marczak M, Mazur A, Krol J (2006) Rhizobial exopolysaccharides: genetic control and symbiotic functions. Microbial Cell Factories. 5:7, 1-19.

[99] Freitas F, Alves VD, Reis MA (2011) Advances in bacterial exopolysaccharides: from production to biotechnological applications. Trends Biotechnol. 29:388-398.

[100] Rojas-Escudero E, Alarcon-Jimenez AL, Elizalde-Galvan P, Rojo-Callejas F (2004) Optimization of carbohydrate silylation for gas chromatography. J. Chromatogr. A. 1027: 117-120.

[101] Carlson RW, Forsberg LS, Kannenberg EL (2010) Lipopolysaccharides in Rhizobium-legume symbioses. Subcell. Biochem. 53:339-386.

[102] Carlson RW, Garci F, Noel D, Hollingsworth R (1989) The structures of the lipopolysaccharide core components from Rhizobium leguminosarum biovar phaseoli CE3 and two of its symbiotic mutants, CE109 and CE309. Carbohydr. Res. 195:101-110.

[103] Abdel-Akher M, Hamilton JK, Smith F (1951) The reduction of sugars with sodium borohydride. J. Am. Chem. Soc. 73: 4691-4692.

[104] York WS, Darvill AG, MCneil M, Stevenson TT, Albersheim P (1985) Isolation and characterization of plant cell walls and cell wall components. Methods in Enzymol. 118: 3-40.

[105] Reuhs BL, Carlson RW, Kim JS (1993) Rhizobium fredii and Rhizobium meliloti produce 3-deoxy-D-manno-2-octulosonic acid-containing polysaccharides that are structurally analogous to group II K antigens (capsular polysaccharides) found in Escherichia coli. J. Bacteriol. 175 :3570-3580.

[106] Fraysse N, Lindner B, Kaczynski Z, Sharypova L, Holst O, Niehaus K, Poinsot V (2005) Sinorhizobium meliloti strain 1021 produces a low-molecular-mass capsular polysaccharide that is a homopolymer of 3-deoxy-D-manno-oct-2-ulosonic acid harboring a phospholipid anchor. Glycobiology. 15 :101-108.

[107] Zanetta JP, Timmerman P, Leroy Y (1999) Determination of constituents of sulphated proteoglycans using a methanolysis procedure and gas chromatography/mass spectrometry of heptafluorobutyrate derivatives. Glycoconj. J. 16: 617-627.

[108] Becker A, Fraysse N, Sharypova L (2005) Recent advances in studies on structure and symbiosis-related function of rhizobial K-antigens and lipopolysaccharides. Mol. Plant Microbe Interact. 18: 899-905.

Chapter 12

Diatom Polysaccharides: Extracellular Production, Isolation and Molecular Characterization

Ranieri Urbani, Paola Sist, Galja Pletikapić,
Tea Mišić Radić, Vesna Svetličić and Vera Žutić

Additional information is available at the end of the chapter

http://dx.doi.org/10.5772/51251

1. Introduction

Diatoms (phylum Heterokontophyta, class Bacillariophyceae) are photoautotrophic protists derived from a secondary endosymbiosis involving a heterotrophic eukaryote host and a red alga endosymbiont. The annual primary production of diatoms is estimated to be 20 Pg (10^{15} g) of carbon, equivalent to 40% of marine and one-fifth of total global primary production and diatoms are responsible for more than half of the organic carbon flux to the deep ocean. DOM (dissolved organic matter) represents the largest pool of organic matter in the sea, with substantial part still uncharacterized [1].

The extracellular polysaccharide production by marine diatoms is a significant route by which photosynthetically produced organic carbon enters the trophic web and may influence the physical environment in the sea as observed for example when massive aggregation events on basin scale occur. Macroscopic dimensions and the massive appearance of gelatinous macroaggregates known as 'mucilage events' occurring episodically in Northern Adriatic are illustrated in Figure 1.

Despite many advances in organic carbon concentration in marine diatoms, the biophysical and biochemical mechanisms of extracellular polysaccharide production remains a significant challenge to resolve at molecular scale in order to proceed towards the understanding of its function at the cellular level and interactions and fate in the ocean. Chemical characterization of diatom extracellular polymer substance (EPS) isolated from laboratory cultures revealed that EPS are predominantly heteropolysaccharides that contain substantial amounts of uronic acid and sulphate residues [4-7]. There are of course other substances derived by disruption or decay of algal cells. Such substances are both of low

INTECH
open science | open minds

© 2012 The Author(s). Licensee InTech. This chapter is distributed under the terms of the Creative Commons Attribution License http://creativecommons.org/licenses/by/3.0), which permits unrestricted use, distribution, and reproduction in any medium, provided the original work is properly cited.

and high molecular weight, present mainly as dissolved or particulate matter. Microscopic (transmission electron and atomic force microscopy) and NMR studies reveal that fibrillar polysaccharides formed the bulk of oceanic DOM [8].

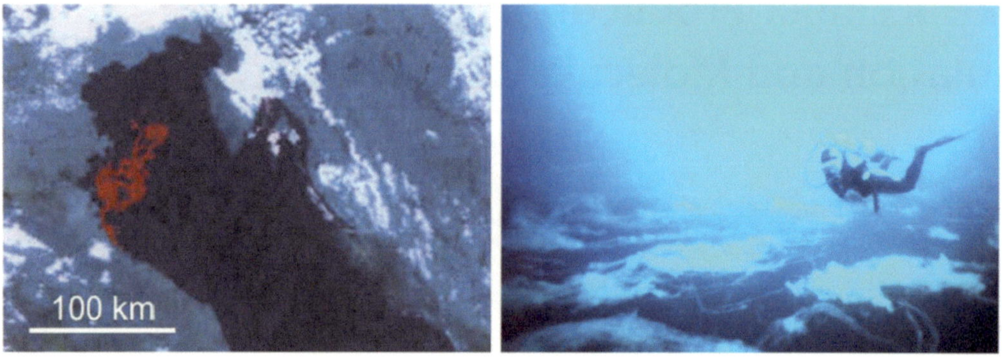

Figure 1. Macroscopic phenomenon of extracellular polysaccharide gellation in the Northern Adriatic Sea: (**a**) remote sensing by satellite showing gel phase in red color (adopted from [2]); and (**b**) at 10 m depth captured by a scuba-diver [3].

Many papers showed that the aldose signatures of marine DOM obtained from different seawater samples around the world is similar to that determined on cultured phytoplankton DOM [9,10] and that the carbohydrate production could be very different among the species selected, growth and environmental conditions [5,7,11-17]. These results are very important in order to understand the role of algal exudation in the aggregation processes observed in all of the seas and in general in carbon cycling in the euphotic zone. Many authors showed that cultured diatoms growth in P-limiting condition determines an increase of polysaccharides exudated by different diatoms species [4,6,7,11,12,15,17,18] both pelagic and benthic.

2. Techniques for primary structure characterization

Among the three major classes of biopolymers, the polysaccharides show the greatest chemical and structural variety. The nucleic acids are constructed from a handful of nucleotide bases so that the polymeric structure obtained is invariably linear. The number of amino acid building blocks used to construct the proteins is approximately twenty but, again, the proteins are always linear polymers. On the other hand, polysaccharides display a wide chemical and structural variability that is not found among the polypeptides and polynucleotides mainly due to the multiple hydroxyl functionality of the five- and six-carbon sugars. The replacement of one or more of such sugar hydroxyl functionalities by amine, ester, carboxylate, phosphate or sulfonate groups, leads to the frequent occurrence of tree-like branching and to the huge number of possible polymeric conformations of different solution behavior. For these reasons carbohydrate analysis involves, after isolation and purification, many steps, i.e. the determination of individual monosaccharides, of anomeric linkages, of branching and sequence, of anomeric configuration and, finally, of the chain conformation [19].

2.1. Isolation, purification and separation by ultra-filtration or by solvent precipitation

In the last two decades the filter (ultra-)fractionation technology has highly improved the methodologies for isolation and purification of polydisperse biopolymers, consisting of macromolecules, like algal polysaccharides, that present very often a large number of size fractions, going from oligomers of few sugar residues up to several million dalton of molecular weight. The tangential-flow filtration (known also as crossflow filtration) is one of the most useful tools for biopolymer separation and purification, both from seawater and culture medium. The principal advantage of this technique is that the residue which can obstruct the filter is substantially washed away during the filtration process by a tangential flow along the surface of the membrane. Depending on the biopolymer to be retained, membrane cut-off ranges from 1 kD to 1000 kD are used.

For large volumes, as seawater samples, a polysulfone multi-fiber system (hollow fiber tangential flow columns) is useful technique for simultaneous dia-filtration and concentration of samples given the large surface area available (on m^2 scale).

The addition of a non-solvent (or a bad-solvent for polysaccharides) to a given sample containing dissolved polysaccharides (algal cultures or seawater samples) allows the separation of the carbohydrate fraction by precipitation. This is a very common method for many advantages. It is non-destructive, inexpensive and relatively fast allowing also a fractionation in terms of polysaccharide molecular weight. Cold ethanol, isopranol or acetone are often used and added to the cell-free supernatant of cultures or filtered samples in an appropriate volume to volume ratio (about 4:1). The precipitate is usually re-dissolved in pure water and the solution dialysed exhaustively against EDTA (0.01-0.1 M) and Milli-Q water [6,7,20,21] The precipitation/re-dissolution treatment is commonly performed three-four times depending on the purity to be achieved.

2.2. Gas-chromatography of alditol acetates of neutral monosaccharides

Although the excretion of photosynthetic compounds is recognised as the major source of carbohydrates in seawater [10], so far there are not many papers reporting on the molecular composition of carbohydrates in the exudates from diatom cultures [see for examples 6-7,10,22-28]. The molecular-level characterization of dissolved polysaccharides may provide basic information on the origin, the bioreactivity and the fate of these biopolymers. For example, after monosaccharide analysis Hama and Yanagi reported that the turnover rate of dissolved glucose was the highest among dissolved neutral aldoses, while turnover rates of galactose, mannose, xylose, rhamnose and fucose were similar to each other and markedly lower than glucose [29]. This was a significant finding suggesting that the degree of degradability of autotrophic DOM depends mainly on the relative percentage of glucose with respect to other monosaccharides.

Gas-chromatographic (GC) methodologies for neutral monosaccharide analysis used to characterize the primary structure of marine polysaccharides were reported in several papers [5-7,20,30-39]. In general, the methodology is based on the acid hydrolysis of the polysaccharides and the suitable derivatisation of the saccharidic matter in order to obtain volatile compounds [40]. Thus, hydroxyl groups are subjected to chemical modifications obtaining silylated, acetylated, trifluoroacetylated, methylated or ethylated derivatives. Neutral and amino sugars are commonly analyzed after exhaustive hydrolysis with trifluoroacetic acid and the Neeser acetylation method [41]. The Neeser method provides a simple, rapid, and sensitive analytical method, which has been successfully used on glycoproteins and on plant and microbial cell-wall polysaccharide fractions in order to ensure complete release of amino sugars from glycoproteins, together with minimum losses of neutral sugars with an improved derivatization procedure by treatment with $CH_3ONH_2{\cdot}HCl$ in pyridine. In addition the occurrence of uronic acids requires a preliminar reduction process with carbodiimide and $NaBH_4$ [42]. The N-acetylated form of amino groups, that often occurs in marine polysaccharides, are removed by hydrolysis.

A very popular method of neutral carbohydrate analysis is the alditol acetate method originally described by Blakeney et al.[43], based on the four-step reaction described as:

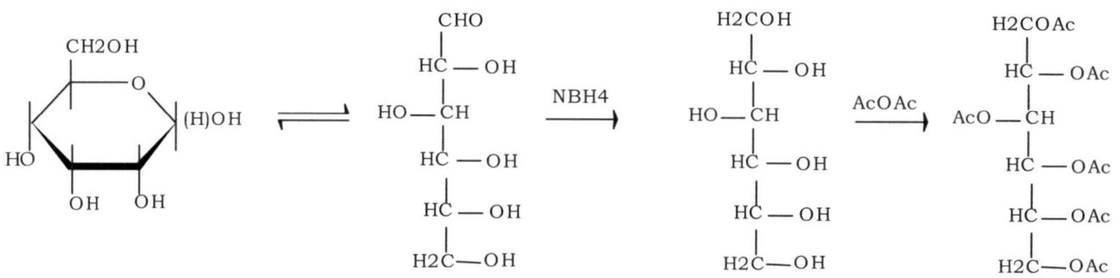

Scheme 1. Scheme 1

The complete separation of the mixture of neutral and amino sugars is usually obtained on polar capillary column (fused-silica coated with methyl silicone fluid).

As an example, the composition of extracellular polysaccharides produced by marine diatom *Chaetoceros decipiens* at the later exponential growth phase is presented in Figure 2. The exopolysaccharide fraction was isolated and purified by precipitation from bulk solution with isopropanol. The hydrolysis and the gas-chromatographic analysis yielded a suite of six neutral monosaccharides: glucose (glc), galactose (gal), mannose (man), xylose (xyl), rhamnose (rha) and fucose (fuc), present in different amount.

The molar ratio of monosaccharide presented in Table 1 shows rhamnose and fucose as the major components followed by galactose residue. The comparison with the composition of the exopolysaccharides obtained by Myklestad from *Chaetoceros decipiens* culture [5] shows a good agreement even at different growth conditions.

Exopolysaccharide molar composition	Rha	Fuc	Xyl	Man	Gal	Glc
Chaetoceros decipiens	7	7	1	4	5	1
Chaetoceros decipiens [5]	7	7	0.5	1	3	0.5

Table 1. Relative molar composition of exopolysaccharides from *Chaetoceros decipiens* cultures.

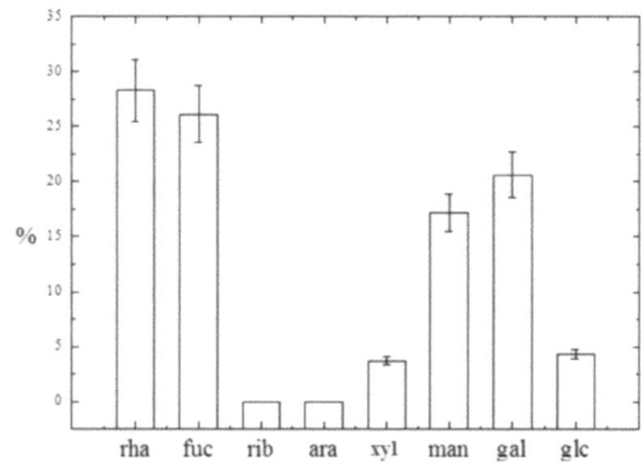

Figure 2. Monosaccharide pattern of the exopolysaccharides from cultured diatom *Chaetoceros decipiens* [44].

2.3. Anionic chromatography for charged and neutral monosaccharides

Carbohydrates in seawater include neutral sugars, aminosugars and acidic sugars, mainly uronic acids, phosphorylated and sulphated sugars [45-47]. By using high performance anion exchange chromatography (HPAEC) coupled with pulsed amperometric detector (PAD) [38] the simultaneous analysis of mixture of such substituted carbohydrates is possible after acid hydrolysis and neutralization. Chemical derivatization is not required so that time consuming gas-chromatographic methods for uronic residue analysis are avoided.

Monosaccharides are completely separated by an isocratic elution [48-50] or in addition with a gradient course of two mobile eluent phases [51]. Seawater samples and culture media require a desalting step preceding the acid hydrolysis and the use of membrane dialysis (of about 1kDa) instead of resins is strongly recommended [51,52] avoiding the losses in carbohydrate yield.

2.4. Methylation analysis and mass spectroscopy

The most widely used method for determining anomeric linkage structure of a polysaccharidic chain is the methylation analysis. The polysaccharide is partially methylated, then hydrolyzed and the resulted partially methylated monosaccharides are acetylated. These methylated alditol acetate sugars allow to establish which carbons are involved in the anomeric linkage. The advent of combined GC and mass spectrometry allows the identification of monosaccharides and provides linkage information on complex polysaccharides. The methylation reaction is commonly performed using Harris' method [53].

Unpublished data (P. Sist) on axenic culture of *Chaetoceros decipiens* are presented in Table 2. The result of the methylation analysis of the exopolysaccharide allowed to identify the linkages among monosaccharides along the polymeric chain. The results showed that fucose and rhamnose were present mainly as terminal residues (t-Rha and t-Fuc) but a lower percentage of rhamnose (8.3%) was linked in the chain backbone (2-Rha) and 5.4% of fucose

Methylation derivative	Linkage	%
2,3,4-Me_3 Rha	t-Rha	21.4
2,3,4-Me_3 Fuc	t-Fuc	10.7
2,3,4,6-Me_4 Man	t-Man*p*	3.2
3,4-Me_2 Rha	2-Rha	8.3
2,3,4,6-Me_4 Glc	t-Glc*p*	4.0
2,3,4,6-Me_4 Gal	t-Gal*p*	3.2
3,4-Me_2 Fuc	2-Fuc	2.7
3,4,6-Me_3 Man	2-Man*p*	2.3
2,3,4-Me_3 Man	6-Man*p*	3.3
3,5-Me_2 Xyl	2-Xyl*f*	1.0
Fuc	2,3,4-Fuc	5.4
3,4,6-Me_3 Gal	2-Gal*p*	13.4
2,3,6- Me_3 Gal	4-Galp	10.5
2,3,6-Me_3 Glc	4-Glc*p*	2.4
3-Me Xyl	2,4-Xyl*p*	1.8
3,6-Me_2 Man	2,4-Man*p*	1.9
4,6-Me_2 Man	2,3-Man*p*	1.0

Table 2. Methylation derivatives of monosaccharidic units of exopolysaccharide from *Chaetoceros decipiens* culture.

represented branched residues (2,3,4-Fuc). Galactose residues which were linked at carbon 2 and 4 (23.9% of 2-Galp and 4-Galp) resided predominantly in the backbone, while mannose was both a branched residue (2.9% of 2,4-Man and 2,3-Man) and mono-substituted in a linear chain (5.6% of 2-Man and 6-Man).

The relatively high percentage of galactose which could be present as α or as β anomeric configuration in the chain backbone suggested a possible extended and rigid chain conformation of the *D. decipiens* polysaccharide as also found for model polysaccharides in solution [54,55].

2.5. Gel permeation chromatography

The heterogeneity and the molecular weight polydispersity of polysaccharide fractions can be analyzed by gel permeation chromatography (GPC) [6,37,56]. This analytical technique is also called size-exclusion chromatography because the fractionation occurs according to the molecular size, V (not molecular weight!), of the polymer which easily or not permeates pores of a suitable dimension of the gel matrix packed in the column which has no specific, or weak, interactions with eluted polymers.

Unpublished data (R. Urbani) of exopolysaccharides obtained from mucilage marine samples are presented in Figure 3. The purified polysaccharides were dissolved in 0.05M NaCl and solutions were filtered on 0.45 μm filters and injected in a conditioned GPC column (1.5 m length, 310 mL internal volume), packed with Sepharose CL6B (10^4-10^6 Da). The chromatographic apparatus was pre-equilibrated using 0.05M NaCl at 25°C at a flow rate of 10 mL·h^{-1} and the column was calibrated with standard dextran solutions ($M_w = 2 \cdot 10^4$ Da and $M_w = 5 \cdot 10^5$ Da). A differential refractive index instrument was used as detector.

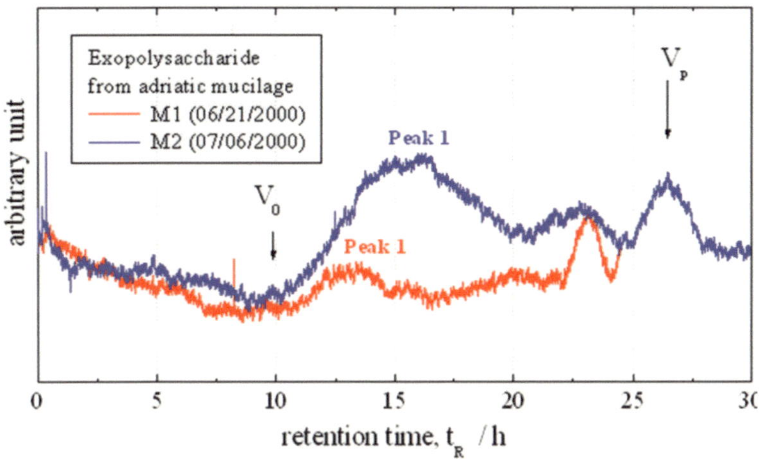

Figure 3. GPC of purified exopolysaccharides from Adriatic mucilage.

The small difference in the retention time (volumes) between peak 1 of M1 and M2 samples (t_R about 15 hours) is related to different stage (or age) of the mucilage aggregates with different degraded state [20]. The other peaks (t_R about 23 hours) very close to the (total) permeation volume, Vp, correspond to the low molecular weight fractions, very likely derived from hydrolytic activities of the bacterial pool on the native aggregates. High molecular weight fractions (peak 1 of M1 and M2) were collected, purified and derivatized for gas-chromatographic analysis. Their monosaccharidic patterns shown in Table 3 are quite similar suggesting the same origin of the high molecular weight polysaccharides even for aggregates of different degraded state [20].

Sample	Fraction	Rha	Fuc	Rib	Ara	Xyl	Man	Gal	Glc
M1	Peak 1	11.8	20.1	--	1.9	8.3	12.2	31.1	14.6
M2	Peak 1	9.3	15.8	--	2.7	13.7	13.3	29.7	15.5

Table 3. Relative monosaccharide composition (%) of unfractionated exopolysaccharide and corresponding fraction at high molecular weight.

A rapid and powerful non-preparative technique for the characterization of molecular dimensions of marine dissolved organic matter and exopolymeric material from diatoms cultures is the high performance size-exclusion chromatography (HPSEC) [57]. By using a hydrogel column system (cut-off 50.000-1.000.000 Da) HPSEC experiments were performed on purified exopolysaccharides from *Chaetoceros decipiens* cultures [44] and results are shown in Figure 4.

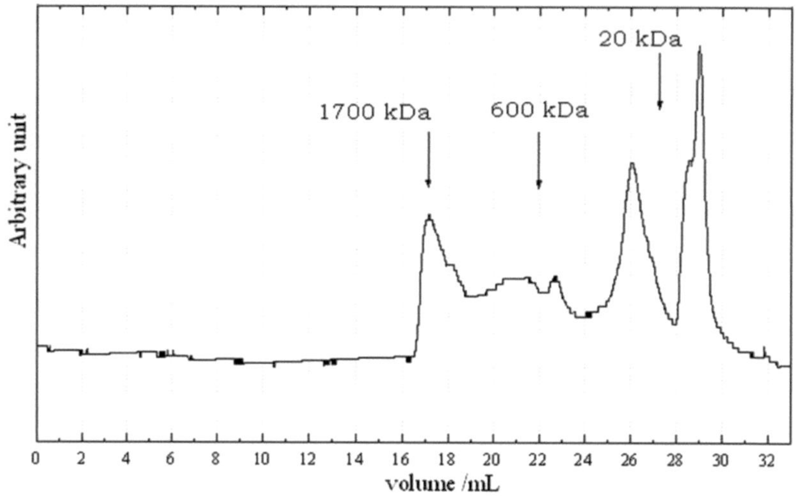

Figure 4. HPSEC chromatogram of purified exopolysaccharide from cultured *Chaetoceros decipiens.*

Two distinct peaks were observed: at lower (V=16-20 mL) and at higher retention volume (V=25-30 mL) which corresponded to about 1700 kDa and 20 kDa molecular weight substances, respectively. The low molecular size peaks were near the permeation volume (V_P) very likely due to oligosaccharides with an average number of monosaccharidic residues of about 110 units. The authors (R. Urbani, P. Sist) hypothesized that *Chaetoceros decipiens* produces a high molecular weight exopolysaccharide and some oligomers at low molecular weight probably derived from the release of storage carbohydrates by cultured diatom in given culture condition [44].

3. Solution and aggregation properties

In the macromolecular science it is well documented that going from flexible polysaccharides toward more rigid and extended chains, as well as with the presence of an ordered sequence in the primary structure, there is a general propensity of polymer chains to form chain associations/aggregations or multiple helical structures. Even if that sequence specificity were present, the absence of pronounced amphiphilic character in the sugar building blocks is not favorable for the formation of some kind of globular folding that occurs in many proteins and some nucleic acids [58]. In other words such globular structures are unknown in polysaccharides. Nonetheless, ordered conformations were proposed to represent the structure of many polysaccharides, both in solid state and in solution. In some cases the functionality is most closely associated with the occurrence of a randomly coiling polymeric character, that is, a propensity for the chain to move continuously through a vast range of nearly equally energetic conformations.

The presence of charged groups on the polysaccharidic chain confers peculiar properties to the macromolecules favoring, for example, the solubility or the association/dissociation processes in solution. On the polymer side the presence of the charged groups influences strongly all the conformational properties by enhancing the chain dimensions and increasing the hydrodynamic volume. Knowledge of polysaccharidic chain structures, from single chain up to the three-dimensional molecular shape and chain association, is essential to understand their capability to form supra-molecular structures including physical gels [54].

3.1. Viscometry and light scattering characterization

Physico-chemical techniques are widely used in order to estimate the biopolymer features such as chain stiffness and chain dimensions which are strictly related to the polymer propensity to give aggregation and highly structured systems like gels. The Smidsrød-Haug parameter B [59] measured by using capillary viscosity techniques as a function of ionic strength is related to chain stiffness: the more flexible the chain, the higher the response of the intrinsic viscosity on the ionic strength variation and the higher the B value.

Intrinsic viscosities, [η], of exopolysaccharides from Adriatic mucilage (M1 and M2) were measured in Cannon-Ubbelohde suspended-level capillary viscometers at different ionic strength and obtained by linear regression of reduced specific viscosities, η_{sp}/C ($dL \cdot g^{-1}$), as a function of polymer concentrations. By plotting the intrinsic viscosity (the polymer hydrodynamic volume) as a function of the inverse of the root of ionic strength (Figure 5), both exopolysaccharides show a linear reduction of [η] at increasing of ionic strength due to screening effect of the salt on polymer charges. From the Smidsrød-Haug equations [59]:

$$\left[\eta\right] = A + S'\underline{k}^{-1} \approx A + S\left(\sqrt{I}\right)^{-1}$$

the stiffness parameter B is obtained where $[\eta]_{0.1}$ is the intrinsic viscosity at 0.1 M ionic strength:

$$B = \frac{S}{\left(\left[\eta\right]_{0.1}\right)^{1.3}}$$

It is interesting to compare the B stiffness parameter of exopolysaccharides with those reported in the literature for other polysaccharides of similar molecular weight. This is shown in Table 4 where dextran sulphate and polyphosphate represent rather flexible polymers with a high B value and, on the other hand, DNA and xanthan represent typical stiff polymers exhibiting a low B parameter. The M1/M2 exopolysaccharides exhibited a value similar to that of alginate considered as a semiflexible polymer.

Static and dynamic laser light scattering (SLLS and DLLS, respectively) are techniques widely used for polymer characterization, measuring average chain properties and thermodynamic quantities in solution. These properties are related to the propensity of the polymer system to give elongated and stiff chain: the radius of gyration (R_G), the weight-average molecular weight (M_w), the second virial coefficient (A_2) and the hydrodynamic diameter (d_h).

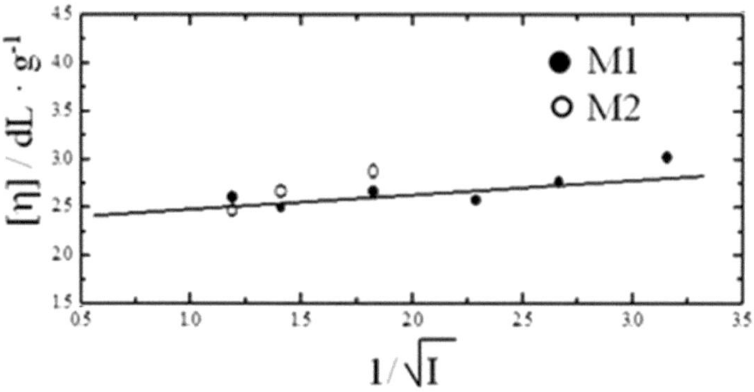

Figure 5. Plot of intrinsic viscosity *vs.* inverse of the square root of ionic strength for .M1 and M2 polysaccharides.

Polymer	**B**
Polyphosphate	0.44
Dextran sulphate	0.23
Hyaluronic acid	0.065
Alginate (mannuronic rich)	0.040
M1, M2	0.036
Alginate (guluronic rich)	0.031
DNA	0.0055
Xanthan	0.0053

Table 4. Stiffness parameters (B) of different model polymers.

Measurement of the scattering intensity at many angles allows the evaluation of the radius of gyration R_G, while by measuring the scattering intensity for many samples of various concentrations, the coefficient A_2 is obtained. Simultaneous linear least squares fits to both the angular and concentration dependence of scattering intensities are employed for the properties determination.

In Figures 6a and 6b the two Zimm plots of the M2 exopolysaccharide in 0.3M and 0.7M NaCl solutions, respectively, are presented. The results show a constancy of R_G (and M_W) at different salt concentrations, while A_2 becomes more negative for higher salt concentration as a consequence of the screening effect on polymer charges which may lead to an extensive degree of aggregation. In general, with respect to the thermodynamic stability of the polymer solution, negative A_2 values are a clear indication of the tendency of polysaccharide solution to undergo a phase separation to form a amorphous carbohydrate solid phase or a gel-like structure.

With comparison to model polysaccharides of different stiffness having the same M_w (Table 5), M2 polysaccharide possess higher R_G value than the flexible and coiled pullulan but also than the semi-rigid chain of the wellan and alginate biopolymers.

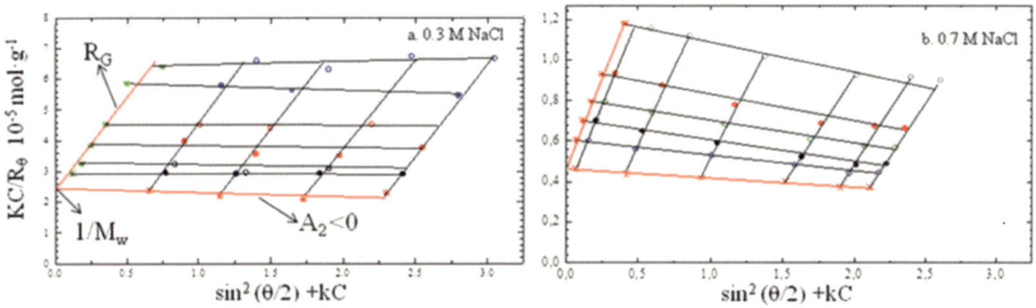

Figure 6. Zimm plot of M2 at two ionic strength (a) 0.3M and (b) 0.7M NaCl.

Sample	R_G /nm
Pullulan	24
Wellan	68
Alginate	73
M2	155[a]

[a] in 0.3M NaCl

Table 5. Radius of gyration of purified polysaccharides M1 and M2 compared to model polysaccharides with M_w=220,000 dalton in 0.1M NaCl.

Finally, taking into account all these results one may say that the M2 polysaccharide shows a marked polyelectrolytic behavior and an intrinsic chain stiffness, favoring at higher salt concentrations chain-to-chain association and/or gel formation. Thus, coupled to other favorable environmental conditions, the salinity gradient of halocline in the seawater column could play an important role in the first stage of aggregation of dissolved biopolymers.

Dynamic laser light scattering (DLLS), also called photon correlation spectroscopy, is a technique for the determination of hydrodynamic diameter of macromolecules and particulate matter in solution. In DLLS the fluctuation of the scattered light due to the Brownian motion of the molecules is detected by a photon counting detector. This technique is used nowadays to measure particle sizes at the nanometer scale and to follow the kinetics of particle formation from dissolved EPS material. As reported by Verdugo [60,61] marine aggregates may self-assemble from free DOM biopolymers and this process was easily followed by DLLS.

Following the same procedure the authors (R. Urbani, P. Sist) measured hydrodynamic diameters using Adriatic seawater samples collected during a three-years monitoring activity (1999-2001). Polymer assembly was monitored for 8 days by analyzing the scattering fluctuations detected at a 45° scattering angle. The autocorrelation function of the scattering intensity fluctuations was averaged over a 1-min sampling time and the particle size distribution calculated by the CONTIN method. Figure 7 shows an example of time dependence of hydrodynamic diameters in three seawater samples taken at different depths.

Although the aggregation kinetics was the same (asymptotic shaped curve in Figure 7) the highest value (1 μm) of hydrodynamic diameter was significantly lower than values reported by Verdugo group (several micrometers) for oceanic samples [60,61] and cultured diatoms, as well [62].

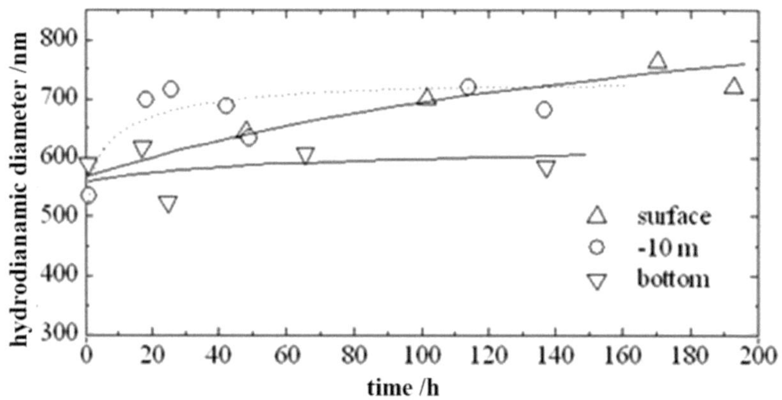

Figure 7. Hydrodynamic diameter in dependence on time for Adriatic seawater samples.

4. The supramolecular organization of polysaccharide fibrils: from cell release to giant gel networks

Diatoms excrete large quantities of extracellular polymeric substances, both as a function of their motility system and as a response to environmental conditions [63,64]. The natural occurrence of EPS is closely linked to diatom biomass, a pattern consistent over both macro (km) and micro (μm) scales. We will focus on diatom exopolysaccharides produced in cell culture and from large aggregates they form in marine environment [3]. Those polysaccharides stand out by their supramolecular organization and gelling capacity due to hydrogen bonding and electrostatic interaction resulting in formation of reversible physical gel. Unlike chemical gels that are formed by chemical reaction using a crosslinking agent, the characteristics of polysaccharide electrolytes is to form gels by physical bonds through intermolecular forces among polymer chains [65,66].

Intermolecular interactions are the basis of life, and an extremely important part of biological research, so an enormous range of techniques have been applied to their study. Among the new methodologies (experimental and theoretical) developed and applied to polysaccharide conformation and dynamics, solution properties, chain aggregation and gelation, the results obtained using atomic force microscopy (AFM) have been pointed out among those giving the most striking results [67]. AFM connects the nanometer and micrometer length scales utilizing a sharp probe tip that senses interatomic forces acting between the surface of a sample and the atoms at the apex of the tip. The physical basis behind AFM and its ability to „feel" the surface, make AFM a versatile tool in biophysics allowing high resolution imaging, nanomechanical characterization and measurements of inter and intramolecular forces in living and non-living structures [68]. Thanks to the simple principle on which it is based, the AFM is a surprisingly small and compact instrument. Its use includes electronic control unit, computer and usually two monitors for simultaneous checking of image and imaging parameters. The probe which scans the sample surface consists of a cantilever and the tip located at the free end of a cantilever. The deflection of the cantilever is measured by an optical detection system. Registered values of cantilever deflection are electronically converted into pseudo 3D image of a sample. AFM is a non-destructive method which gives real 3D images of the sample with a vertical resolution of 0.1 nm and lateral resolution of 1 nm. Measured forces range from 10^{-6} N to 10^{-11} N.

Here we will cover the recent achievements using AFM as the principal method in revealing the supramolecular organization of diatom exopolysaccharide fibrils beyond the chemical composition.

Polysaccharide samples for AFM imaging [69-72] and polysaccharide gels [70,73-77] are usually spread on freshly cleaved mica surface. The imaging of hydrated samples is preferably conducted in air to inhibit the unfavourable motion of polysaccharides in liquid medium. Such AFM studies have been validated against data obtained directly under buffers, transmission electron microscopy (TEM) studies and cryo-AFM. Balnois and Wilkinson [78] showed that when AFM is operated under ambient conditions, the thin water layer both sorbed to the biopolymers and present on the mica surface maintains molecular structure during AFM imaging.

Protocols for marine sample AFM imaging have been developed only recently: single diatom cells and released polymers [79]; isolated polysaccharides from diatom cultures [3,79]; marine gel polymers and networks [80]. The samples were prepared using the drop deposition method (5-10 μL aliquots) and mica as a substrate. Mica sheets were placed in enclosed Petri dishes for approximately 30–45 min to allow biopolymers and cells to settle and attach to the surface. Samples were then rinsed in ultrapure water (three times for 30 s) and placed in enclosed Petri dish to evaporate the excess of water on the mica. The rinsing step was necessary to remove the salt crystals that would hamper imaging. AFM imaging was performed in air at room temperature and 50–60% relative humidity. In contact mode the force was kept at the lowest possible value in order to minimize the forces of interaction between the tip and the surface. For imaging in tapping mode the ratio of the set point amplitude was maintained to the free amplitude (A/A_0) at 0.9 (light tapping).

4.1. Exopolysaccharide production at a single cell level

The ubiquitous marine diatom *Cylindrotheca closterium,* isolated from the northern Adriatic Sea, was used in AFM studies of exopolysaccharide production at a single cell level.

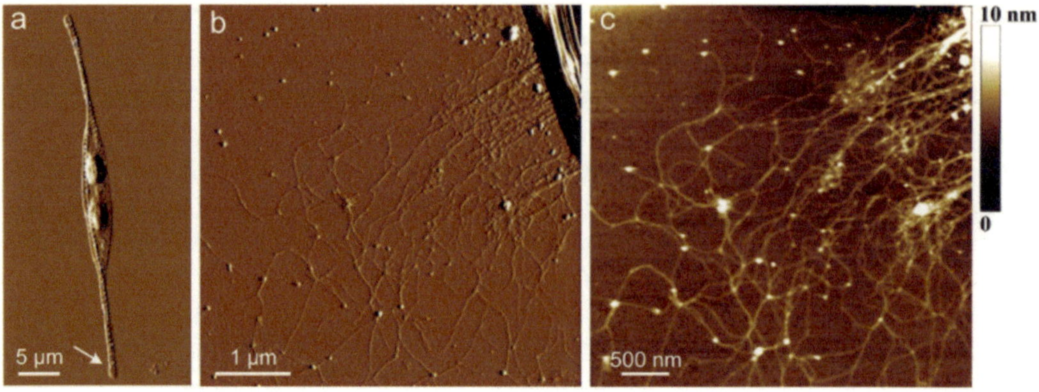

Figure 8. (**a**) AFM image of the whole *C. closterium* cell presented as deflection data. Arrow indicates the position of the polymer excretion site; (**b**) The released polymers still attached to the apex of the cell rostrum, deflection data, scan size 5 µm × 5 µm; (**c**) Released polymers presented as height data, scan size 4 µm × 4 µm and vertical scale shown as the color bar [3].

Figure 8a revealed the general features of a live *C. closterium* cell with the two chloroplasts and its drawn-out flexible rostra. Arrow indicates the position of polymer release shown in Figures 8b and 8c. The bundles of polymer fibrils extended up to10 µm from cell surface. Their heights are 5–7 nm at the position close to the site of excretion. At a distance of 1 µm the dense network is observed with fibril heights of 2–3 nm. At even larger distances the network is less dense with the fibril heights in the range of 0.4 to 1.2 nm. The lower value of fibril height corresponds to the single monomolecular polysaccharide chains [71]. At this

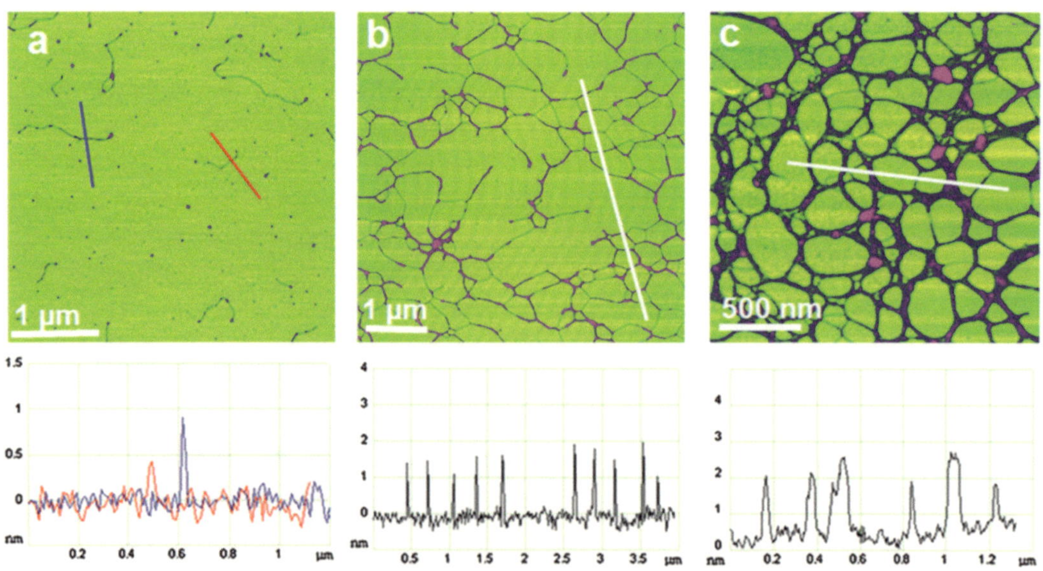

Figure 9. AFM images (tapping mode) of polysaccharides isolated from the *C. closterium* culture medium and dissolved in ultrapure water: (a) single fibrils (concentration 5 mg/L) vertical scale 2.5 nm; (b,c) fibril networks (concentration 10 mg/L). Vertical scales: 5 nm (b) and 10 nm (c) [79].

larger distance, the network appeared with incorporated spherical nanoparticles–globules. The globules are found to interconnect two or more fibrils. The globules may represent positively charged proteins whose function before the release is efficient intracellular packing of negatively charged polysaccharide fibrils, in line with molecular crowding in living cells [81].

4.2. Extracted and purified polysaccharides of C. closterium

The polysaccharide fraction isolated from the axenic *C. closterium* culture medium [7] was used to test the capacity of photosynthetically produced polymers to self-assemble into gel phase.

The isolated polysaccharides for AFM imaging were prepared in two concentrations (5 and 10 mg/L). Single fibrils prevail in samples prepared with polysaccharide concentration of 5 mg/L. Dissolved polysaccharides resumed a flexible fibrillar structure (Figure 9a) with fibril heights of 0.4 and 0.9 nm. The value of 0.4 nm corresponds to polysaccharide single molecular chain. AFM imaging of samples prepared from the solution containing 10 mg/L revealed fibrillar networks varying in the degree of fibril associations (Figures 9b,c). The fibril heights in the networks span over the same range (0.9–2.6 nm) and the mode of fibril association follows the same pattern. However, the segments forming the network shown in Figure 9c are significantly wider (50 vs. 140 nm), suggesting side-by-side associations once the maximum height of individual fibrils is reached. If we take the value of 2.6 nm observed using AFM analysis as the maximum fibril height, then up to six monomolecular fibrils can constitute a single polysaccharide fibril [79]. The absence of globules in the AFM images of isolated polysaccharides also indicates that the globules, which appeared as a constitutive component in the EPS of single cells (Figures 9b,c), are not polysaccharides.

4.3. Evolution of marine gel polymer networks

The fact that the isolated polysaccharide fraction has the capacity to self-assemble into a gel network in pure water is an important finding with implications on the mechanism of the macroscopic gel phase formation in marine systems. The marine gel is characterized as a thermoreversible physical gel and the dominant mode of gelation as crosslinking of polysaccharide fibrils by hydrogen bonding which results in helical structures and their associations [80]. This mechanism contrasts a more generally established view [82,83] that marine gel phase formation proceeds via cross-linking of negatively charged biopolymers (namely polysaccharides) by Ca^{2+} ions. Only recently, Ding *et al.* [62] reported that diatom EPS can spontaneously self-assemble in calcium-free artificial sea water, forming microscopic gels of 3–4 μm. They pointed out an overlooked issue of crosslinkers other than calcium ions in the formation of marine polymer networks.

Figure 10 represents the evolution of polymer networks of the macroscopic gel phase in the northern Adriatic Sea [84,85]. Samples were prepared from the macroaggregates with different residence time in the water column, from early stage of gel phase formation to the

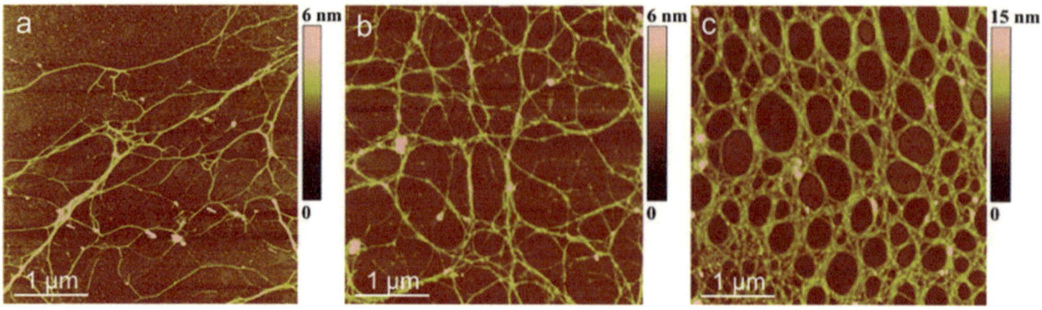

Figure 10. Evolution of polymer networks in the macroscopic gel phase from (**a**) to (**c**): early stage of gel phase formation to condensed gel network of older macroaggregate. AFM images are acquired in contact mode and presented as height data, scan size 4 µm×4 µm [3].

condensed (mature) gel network of an older macroaggregate. The long polymer strands with small patches of initial fibril associations (Figure 10a) coexisted with the continuous gel network shown in Figure 10b. With the prolonged residence time (one month) the more condensed network is formed as presented in Figure 10c. The analysis of fibril heights for early and mature gel state is given in Table 6. The fibril heights for the early stage of gel network correspond to the fibril heights produced by Adriatic *C. closterium*.

Polysaccharide fibrils	Number of fibrils analyzed	Fibril height /nm	
		Mean value	Range
Attached to the diatom cell	120	0.85 ± 0.32	0.4 - 1.8
Marine gel network: early stage	189	0.92 ± 0 .40	0.4 - 2.0
Marine gel network: mature gel	178	3.58 ± 0.76	1.6 - 5.0

Table 6. Comparison of polysaccharide fibril heights [3].

4.4. Exopolysaccharides interactions with nanoparticles

Diatom EPS production increases as a feedback response to the presence of NPs and may thus contribute to detoxification mechanisms [86,87]. Specifically, in the study of Ag NPs toxicity to the marine diatoms *C. closterium* and *C. fusiformis* [88] an increase of EPS production was documented and the incorporation of Ag NPs was clearly demonstrated. Although Ag NPs integrated in EPS–gel network are beneficial to the diatom cell (detoxification), their accumulation and persistence in microenvironments prolong their presence in the water column and make NPs available to higher organisms. Residing in a gel environment, particles are prevented from aggregation and export from the water column.

The diatom EPS component responsible for the NPs interaction was identified by bringing in direct contact polysaccharide fraction isolated from *C. closterium* EPS with Ag and SiO_2 NPs.

The polysaccharide fibril network was prepared by dissolving the polysaccharide fraction isolated from the *C. closterium* culture medium in ultrapure water at concentration of 20 µg/mL and stirred for 45 minutes before adding NPs, 10 µg/mL Ag-citrate coated NP, nominal size ~ 25 and SiO_2 NP (LUDOX® HS-40, Sigma), nominal size 15nm were used.

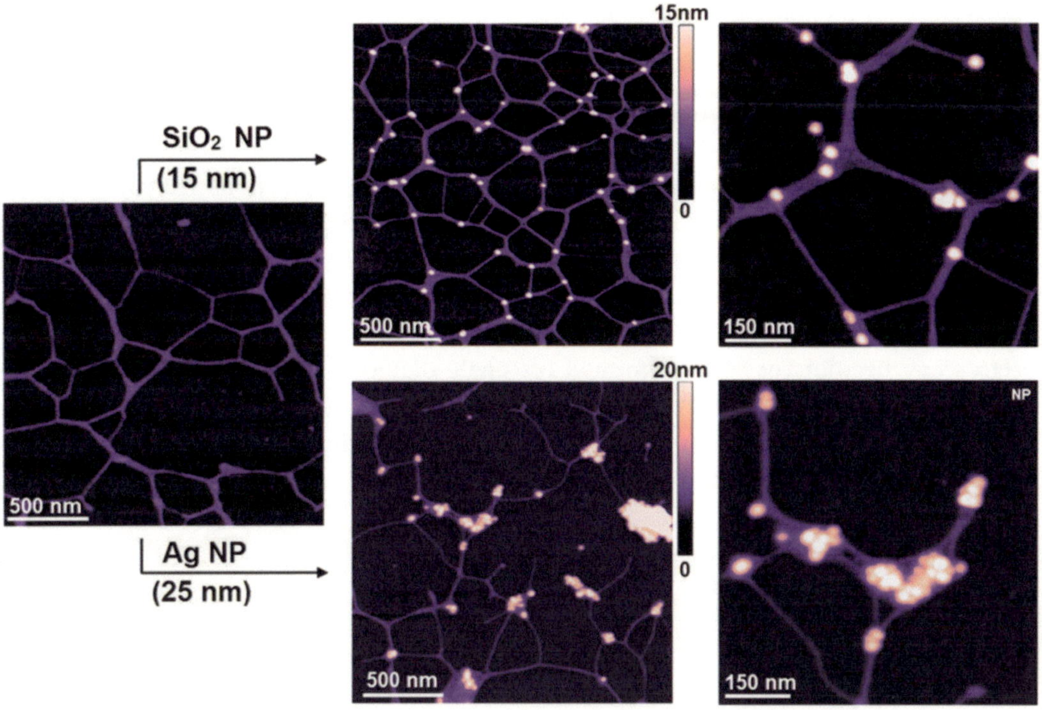

Figure 11. Interaction of polysaccharide network and nanoparticles visualized by AFM. The polysaccharide network was preformed in 20 mg/L of isolated polysaccharides in ultrapure water before addition of 10 mg/L NPs.

The Ag and SiO_2 NPs were detected exclusively on polysaccharide fibrils as single spherical particles or their agglomerates as shown by AFM images (Figure 11). The NPs did not induce cross-linking of fibrils nor change the fibril heights. Rather, particles are imbedded into the preexisting polysaccharide network. Besides the significance for the environment, such interaction of NPs and diatom polysaccharides can be applied for the design of composite materials, such as biocompatible gels with new properties [89]. Diatom extracellular polysaccharides could be also used as a capping agent giving rise to the stability of NPs in liquid environments over a broad range of ionic strength and pH.

4.5. Force spectroscopy as future prospective

In AFM force spectroscopy mode a single molecule or fiber is stretched between the AFM flexible cantilever tip and a flat substrate mounted on a highly accurate piezoelectric positioner. Polysaccharide molecule, protein or other biopolymer, is either adsorbed to the substrate or linked to it through the formation of covalent bonds. When the tip and substrate are brought together and then withdrawn, one or more molecules can attach to the tip by adsorption. As the distance between the tip and substrate increases, extension of the molecule generates a restoring force that causes the cantilever to bend. The deflection of the cantilever measures the force on the polymer with an accuracy of ~5 pN, while the piezoelectric positioner records the changes in the molecule's end-to-end length with an accuracy of 0.1nm.

AFM force spectroscopy is widely used method in polymer biophysics allowing measuring mechanical properties of single molecules, and with a possibility to directly quantify the forces involved in both intra- and inter-molecular polymer interactions [90-93]. It is also adopted in advancing diatom research into the nanotechnology era [94]. Most of the work done so far on measuring forces with AFM has distinguished non adhesive and adhesive EPS components and discovered the adhesive properties and designs that give explanation as to why diatoms have the great tendency to attach to surfaces. However, force spectroscopy has not yet been performed on single diatom polysaccharide fibrils or their networks. The data that follows are the results of exploratory experiments conducted on diatom polysaccharide molecules (single fibrils, Figure 12) and on marine gel polysaccharide network (Figure 13) in filtered seawater.

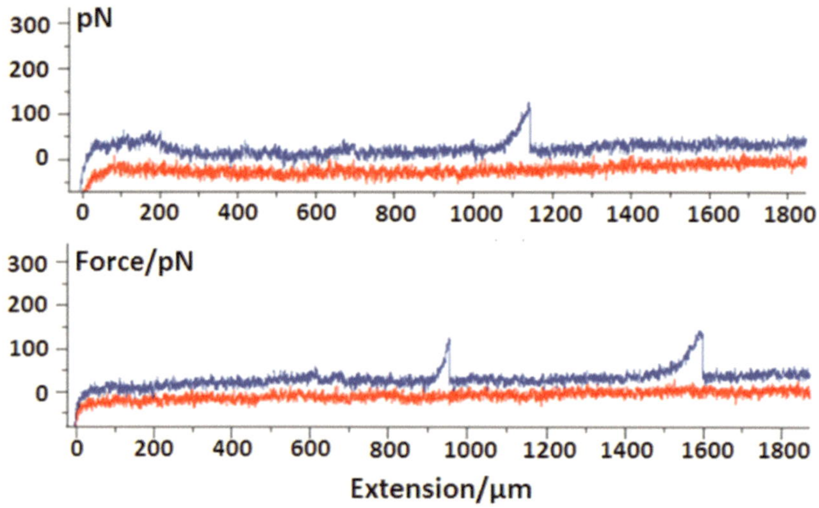

Figure 12. Force approach (in red) and extension (in blue) curves acquired for polysaccharide single fibrils in filtered seawater. Two typical curves are shown.

The extension curve with the two rupture events in Figure 12 could result from the two individual fibrils of different length (0.9 and 1.5 µm) simultaneously attached to the tip. The force spectrum signature for polysaccharide fibrils assembled in marine gel network is by far more complex (Figure 13) [95]. Assigning the underlying disentanglement events is in progress.

The force spectra can provide the critical piece of information that will allow us to characterize and quantify physical forces in polysaccharide network assemblies. Further developments will contribute to the new field of nanoecology and open the possibilities for rational design of polysaccharide gels with desired properties.

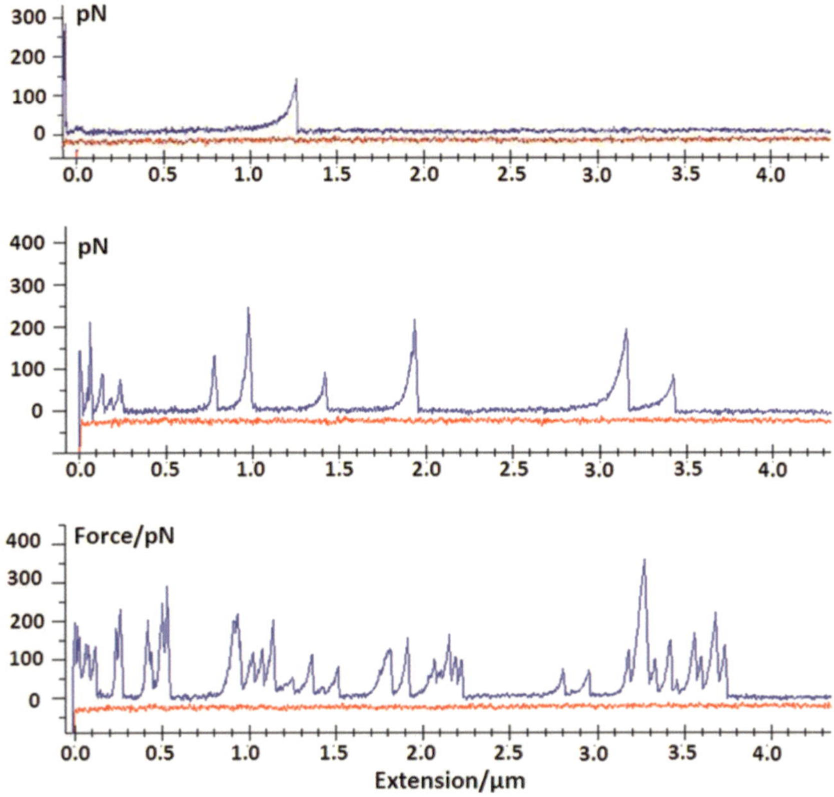

Figure 13. Force approach (in red) and extension (in blue) curves of polysaccharide fibrils assembled in marine gel network acquired in filtered seawater.

Author details

Ranieri Urbani and Paola Sist
Department of Life Sciences, University of Trieste, Trieste, Italy

Galja Pletikapić, Tea Mišić Radić, Vesna Svetličić * and Vera Žutić
2Division for Marine and Environmental Research, Ruđer Bošković Institute, Zagreb, Croatia

Acknowledgement

The work reported in the section "The supramolecular organization of polysaccharide fibrils" was performed within the scope of the project "Surface forces on atomic scale applied in marine science and nanotechnology", Croatian Ministry of Science and the project "Development of a nanotechnology-based targeted recognition system for bio-macromolecules", Croatian Science Foundation. AFM force spectroscopy of isolated polysaccharides was performed during G.P. stay in Jasna Brujić Group, Department of

---

* Corresponding Author

Physics, New York University. G.P. stay was financed through Croatian Science Foundation Doctoral Fellowship Award. Participation of Miklós Kellermayer was made possible through the networking of the COST Action "AFM for Nanomedicine and Life Science".

5. References

[1] David L Kirchman (2008) Microbial Ecology of the Oceans. New York: John Wiley & Sons. 2nd edition. 512 p.

[2] Zambianchi E, Calvitti C, Cecamore P, D'Amico F, Ferulano E, Lanciano P (1992) The mucilage phenomenon in the Northern Adriatic sea summer 1989: a study carried out with remote sensing techniques. Sci. Total Environ. (Suppl.) 581–598.

[3] Svetličić V, Žutić V, Mišić Radić T, Pletikapić G, Hozić Zimmerman A, Urbani R (2011) Polymer Networks Produced by Marine Diatoms in the Northern Adriatic Sea. Mar. Drugs 9: 666-679.

[4] Myklestad SM, Haug A (1972) Production of carbohydrate by the marine diatom *Chaetoceros affinis*: I. Effect of the conc. of nutrients in the culture medium. J. Exp. Mar. Biol. Ecol. 9: 125-136.

[5] Myklestad SM (1995) Release of extracellular products by phytoplankton with special emphasis on polysaccharides. Sci. Tot. Environ. 165: 155-164.

[6] Magaletti E, Urbani R, Sist P, Ferrari CR, Cicero AM (2004) Abundance and Chemical Characterization of Extracellular Carbohydrates Produced by the Marine Diatom *Cylindrotheca fusiformis*. Eur. J. Phycol. 39: 133-142.

[7] Urbani R, Magaletti E, Sist P, Cicero AM (2005) Extracellular Carbohydrates Released by the Marine Diatoms *Cylindrotheca closterium, Thalassiosira pseudonana* and *Skeletonema costatum* : effect of P-depletion and growth status. Sci. Tot. Environ. 353: 300-306.

[8] Santschi PH, Balnois E, Wilkinson KJ, Zhang J, Buffle J (1998) Fibrillar polysaccharides in marine macromolecular organic matter as imaged by atomic force microscopy and transmission electron microscopy. Limnol. Oceanogr. 43: 896–908.

[9] Biersmith A, Benner R (1998) Carbohydrates in phytoplankton and freshly produced dissolved organic matter. Mar. Chem. 63: 131-144.

[10] Aluwihare L, Repeta DJ (1999) A comparison of the chemical characteristics of oceanic DOM and extracellular DOM produced by marine algae. Mar. Ecol. Prog. Ser. 186: 105-117.

[11] Myklestad S (1974) Production of carbohydrates by marine planktonic diatoms I: Comparison of nine different species in culture. J. Exp. Mar. Biol. Ecol. 15: 261–74.

[12] Myklestad S (1977) Production of carbohydrates by marine planktonic diatoms II: Influence of the N/P ratio in the growth medium on the assimilation ratio, growth rate, and production of cellular and extracellular carbohydrates by *Chaetoceros affinis* var *willei* (Gran) Hustedt and *Skeletonema costatum* (Grev) Cleve. J. Exp. Mar. Biol. Ecol. 29: 161–79.

[13] Chen W, Wangersky PJ (1996) Rates of microbial degradation of dissolved organic carbon from phytoplankton cultures. J Plankton Res. 18: 1521–33.

[14] Alldredge AL (1999) The potential role of particulate diatom exudates in forming nuisance mucilaginous scums. Ann. Ist. Super Sanita` 35: 397– 400.
[15] Alcoverro T, Conte E, Mazzella L (2000) Production of mucilage by the Adriatic epipelic diatom Cylindrotheca closterium (Bacillariophyceae) under nutrient limitation. J. Phycol. 36: 1087– 1095.
[16] Staats N, Stal LJ, Mur LR (2000) Exopolysaccharide production by the epipelic diatom Cylindrotheca closterium: effects of nutrient conditions. J. Exp. Mar. Biol. Ecol. 249: 13–27.
[17] Granum E, Kirkvold S, Myklestad SM (2002) Cellular and extracellular production of carbohydrates and aminoacids by the marine diatom *Skeletonema costatum*: diel variations and effects of N depletion. Mar. Ecol. Prog. Ser. 242: 83– 94.
[18] Guerrini F, Cangini M, Boni L, Trost P, Pistocchi R (2000) Metabolic responses of the diatom *Achnanthes brevipes* (Bacillariophyceae) to nutrient limitation. J. Phycol. 36: 882–890.
[19] Gerald O Aspinall (1982) The Polysaccharides. New York: Academic Press. 248 p.
[20] Giani M, Berto D, Zangrando V, Castelli S, Sist P, Urbani R (2005) Chemical Characterisation of Different Typologies of Mucilaginous Aggregates in the Northern Adriatic Sea. Sci. Tot. Environ. 353: 232-246.
[21] Urbani R, Brant DA (1992) Kelco microbial polysaccharides S-130 (Welan) and S-657 display similar dilute aqueous solution behaviour. Carbohydr. Polym. 11: 169–191.
[22] Khodse VB, Fernandes L, Gopalkrishna VV, Bhosle NB, Fernandes V, Matondkar SGP, Bhushan R (2007) Distribution and seasonal variation of concentrations of particulate carbohydrates and uronic acids in the northern Indian Ocean. Mar. Chem. 103: 327–346.
[23] Yang GP, Zhang YP, Lu XL, Ding HB (2010) Distributions and seasonal variations of dissolved carbohydrates in the Jiaozhou Bay, China. Estuar. Coastal Shelf Sci. 88: 12-20.
[24] Khodse VB, Fernandes L, Bhosle NB, Sardessai S (2008) Carbohydrates, uronic acids and alkali extractable carbohydrates in contrasting marine and estuarine sediments: Distribution, size fractionation and partial chemical characterization. Org. Geochem. 39: 265-283.
[25] Lai J, Yu Z, Song X, Cao X, Han X (2011) Responses of the growth and biochemical composition of *Prorocentrum donghaiense* to different nitrogen and phosphorus concentrations. J. Exp. Mar. Biol. Ecol. 405: 6-17.
[26] Meon B, Kirchman DL (2001) Dynamics and molecular composition of dissolved organic material during experimental phytoplankton blooms. Mar. Chem. 75: 185-199.
[27] Hama T, Yanagi K, Hama Y (2004) Decrease in molecular weight of photosynthetic products of marine phytoplankton during early diagenesis. Limnol. Oceanogr. 49: 471-481.
[28] McCarthy M, Hedges J, Benner R (1996) Major biochemical composition of dissolved high molecular weight organic matter in seawater. Mar. Chem. 55: 281-297.
[29] Hama T, Yanagi K (2001) Production and neutral aldose composition of dissolved carbohydrates excreted by natural marine phytoplankton populations. Limnol. Oceanogr. 46: 1945 – 1955.

[30] Allan GG, Lewin J, Johnson PG (1972) Marine polymers. IV Diatoms polysaccharides. Bot. Mar. 15: 102-108.

[31] Aluwihare LI, Repeta DJ, Chen RF (1997) A major biopolymeric component to dissolved organic carbon in surface seawater. Nature 387: 166-169.

[32] Skoog A, Benner R (1997) Aldose in various size fractions of marine organic matter: Implications for carbon cycling. Limn. Oceanogr. 42: 1803-1813.

[33] Biersmith A, Benner R (1998) Carbohydrates in phytoplankton and freshly produced dissolved organic matter. Mar. Chem. 63: 131-144.

[34] Benner R, Opsahl S (2001) Molecular indicators of the sources and transformation of dissolved organic matter in the Mississippi river plume. Org. Geochem. 32: 597-611.

[35] Cotrim da Cunha L, Serve L, Blazi JL (2002) Neutral sugar as biomarkers in the particulate organic matter of a French Mediterranean river. Organic Geochem. 33: 953-964.

[36] Repeta D, Quan TM, Aluwihare L, Accardi AM (2002) Chemical characterization of high molecular weight dissolved organic matter in fresh and marine waters. Geochim. Cosmochim. Acta 66: 955-962.

[37] Sartoni G, Urbani R, Sist P, Nuccio C, Berto D, Giani M (2008) Benthic mucilaginous aggregates in the Mediterranean Sea: Origin, chemical composition and polysaccharides characterization. Mar. Chem. 111: 184-198.

[38] Kaiser K, Benner R (2009) Biochemical composition and size distribution of organic matter at the Pacific and Atlantic time-series stations. Mar. Chem. 113: 63-77.

[39] Giani M, Sist P, Berto D, Serrazanetti G, Ventrella V, Urbani R (2012) The organic matrix of pelagic mucilaginous aggregates in the Tyrrhenian Sea (Mediterranean Sea). Mar. Chem. 132-133: 83-94.

[40] Christopfer J Biermann, Gary D McGinnis (1988) Analysis of Carbohydrates by GLC and MS. Boca Raton: CRC Press. 304 p.

[41] Neeser JR (1985) G.l.c. of O-methyloxime and alditol acetate derivatives of neutral sugars, hexosamines, and sialic acids: "one-pot" quantitative determination of the carbohydrate constituents of glycoproteins and a study of the selectivity of alkaline borohydride reductions. Carbohydr. Res.138: 189-198.

[42] Taylor RL, Conrad HE (1972) Stoichiometric depolymerization of polyuronides and glycosaminoglycuronans to monosaccharides following reduction of their carbodiimide-activated carboxyl group. Biochemistry 11: 1383-1388.

[43] Blakeney AB, Harris PJ, Henry RJ, Stone BA (1983) A simple and rapid preparation of alditol acetates for monosaccharide analysis. Carbohydr. Res. 113: 291-299.

[44] Sist P (2005) Ciclo biogeochimico della sostanza organica in ambiente marino costiero: caratterizzazione e ruolo dei biopolimeri marini. PhD Thesis, University of Trieste, Italy.

[45] Mopper K (1977) Sugars and uronic acids in sediment and waters from the Black Sea and North Sea with emphasis on analytical techniques. Mar. Chem. 5: 585-603.

[46] Kaiser K, Benner R (2000) Determination of amino sugars in environmental samples with high salt content by high-performance anion-exchange chromatography and pulsed amperometric detection. Anal. Chem. 72: 2566–2572.
[47] Leppard GG (1995) The characterization of algal and microbial mucilage's and their aggregates in aquatic ecosystems. Sci. Tot. Environ. 165: 103–131.
[48] Metaxatos A, Panagiotopoulos C, Ignatiades L (2003) Monosaccharide and aminoacid composition of mucilage material produced from a mixture of four phytoplanktonic taxa, J. Exp. Mar. Biol. Ecol. 294: 203– 217.
[49] Goldberg SJ, Carlson CA, Hansell DA, Nelson NB, Siegel DA (2009) Temporal dynamics of dissolved combined neutral sugars and the quality of dissolved organic matter in the northwestern Sargasso Sea. Deep-Sea Res. 56: 672–685.
[50] Goldberg SJ, Carlson CA, Brock B, Nelson NB, Siegel DA (2010) Meridional Variability in Dissolved Organic Matter Stocks and Diagenetic State Within the Euphotic and Mesopelagic Zone of the North Atlantic Subtropical Gyre. Mar. Chem. 119: 9-21.
[51] Engel A, Händel N (2011) A novel protocol for determining the concentration and composition of sugars in particulate and in high molecular weight dissolved organic matter (HMW-DOM) in seawater. Mar. Chem. 127: 180-191.
[52] Giroldo D, Ortolano PIC, Vieira AAH (2007) Bacteria–algae association in batch cultures of phytoplankton from a tropical reservoir: the significance of algal carbohydrates. Freshwater Biol. 52: 1281–1289.
[53] Harris PJ, Henry RJ, Blakeney AB, Stone BA (1984) An improved procedure for the methylation analysis of oligosaccharides and polysaccharides. Carbohydr. Res. 127: 59-73.
[54] Urbani R, Cesàro A (2001) Chain Conformation of Polysaccharides in Different Solvents. In: Wypych G, editor. Handbook of Solvents. Toronto: Chem Tech Publishing. pp. 706-735.
[55] Ruggiero JR, Urbani R, Cesàro A (1995) Conformational Features of Galacturonans: II. Configurational Statistics of Pectic Polymers. Int. J. Biol. Macromol. 17: 213-218.
[56] Mandal SK, Singh RP, Patel V (2011) Isolation and Characterization of Exopolysaccharide Secreted by a Toxic Dinoflagellate, *Amphidinium carterae* Hulburt 1957 and Its Probable Role in Harmful Algal Blooms (HABs). Microb. Ecol. 62: 518–527.
[57] Minor EC, Simjouw JP, Boon JJ, Kerkhoff AE, Van der Horst J (2002) Estuarine/marine UDOM as characterized by size exclusion chromatography and organic mass spectrometry, Mar. Chem. 78: 75–102.
[58] Gerald O Aspinall (1982) The Polysaccharides. New York: Academic Press. 248 p.
[59] Smidsrød O, Haug A (1971) Estimation of the relative stiffness of the molecular chain in polyelectrolytes from measurements of viscosity of different ionic strengths. Biopolymers 1: 1213-1227.
[60] Chin WC, Orellana MV, Verdugo P (1998) Spontaneous assembly of marine dissolved organic matter into polymer gels. Nature 391: 568–72.

[61] Verdugo P, Santschi PH (2010) Polymer dynamics of DOC networks and gel formation in seawater. Depp-Sea Res. II 57: 1486-1493.

[62] Ding YX, Hung CC, Santschi PH, Verdugo P, Chin WC (2009) Spontaneous Assembly of Exopolymers from Phytoplankton. Terr. Atmos. Ocean. Sci. 20: 741-747.

[63] Underwood GJC, Paterson DM (2003) The importance of extracellular carbohydrate production by marine epipelic diatoms. Adv. Bot. Res. 40: 184-240.

[64] Underwood GJC, Boulcott M, Raines CA, Waldron K (2004) Environmental effects on exopolymer production by marine benthic diatoms: dynamics, changes in composition, and pathways of production. J. Phycol. 40: 293-304.

[65] Rinaudo M (2005) Advances in characterization of polysaccharides in aqueous solution and gel state. In: Dumitriu S, editor. Polysaccharides-Structural Diversity and Functional Versatility. New York: Marcel Dekker. pp. 237–252.

[66] Jacob N Israelachvili (2010) Intermolecular and surface forces. London: Academic Press.3nd edition. 700 p.

[67] Cesàro A, Bellich B, Borgogna M (2012) Biophysical functionality in polysaccharides: from Lego-blocks to nano-particles. Eur. Biophys. J. 41(4): 379-95.

[68] Peter Eaton, Paul West (2010) Atomic Force Microscopy. New York: Oxford University Press. 248 p.

[69] Kirby AR, Gunning AP, Morris VJ (1996) Imaging polysaccharides by atomic force microscopy. Biopolymers 38: 355–366.

[70] Victor J Morris, Andrew R Kirby, A Patrick Gunning (1999) Atomic Force Microscopy for Biologists. London: Imperial College Press. 332 p.

[71] Abu-Lail NI, Camesano TA (2003) Polysaccharide properties probed with atomic force microscopy. J. Microsc. 212: 217–238.

[72] Abu-Lail N.I, Camesano T.A (2004) Atomic force microscope and single-molecule force microscopy studies of biopolymers. In: Schwarz J, Contescu C, Putyera K, editors. Dekker Encyclopedia of Nanoscience and Nanotechnology. London: Taylor & Francis. pp. 119–131.

[73] McIntire TM, Brant DA (1997) Imaging of individual biopolymers and supramolecular assemblies using noncontact atomic force microscopy. Biopolymers 42: 133–146.

[74] Ikeda S, Morris VJ, Nishinari K (2001) Microstructure of aggregated and nonaggregated kappa-carrageenan helices visualized by atomic force microscopy. Biomacromolecules 2: 1331–1337.

[75] Keresztes Z, Rigó T, Telegdi J, Kálmán E (2001) Investigation of biopolymer networks by means of AFM. Appl. Phys. A. 72: 113–116.

[76] Funami T, Hiroe M, Noda S, Asai I, Ikeda S, Nishimari K (2007) Influence of molecular structure imaged with atomic force microscopy on the rheological behavior of carrageenan aqueous systems in the presence or absence of cations. Food Hydrocolloid. 21: 617–629.

[77] Noda S, Funami T, Nakauma M, Asai I, Takahashi R, Al-Assaf, S, Ikeda S (2008) Molecular structures of gellan gum imaged with atomic force microscopy in relation to

the rheological behavior in aqueous systems. 1. Gellan gum with various acyl contents in the presence and absence of potassium. Food Hydrocolloid. 22: 1148–1159.

[78] Balnois E, Wilkinson KJ (2002) Sample preparation techniques for the observation of environmental biopolymers by atomic force microscopy. Colloid. Surf. A 207: 229–242.

[79] Pletikapić G, Mišić Radić T, Hozić Zimmermann A, Svetličić V, Pfannkuchen M, Marić D, Godrijan J, Žutić V (2011) Extracellular polymer release AFM imaging of extracellular polymer release by marine diatom *Cylindrotheca closterium* (Ehrenberg) Reiman& J.C. Lewin. J. Mol. Recognit. 24: 436–445.

[80] Mišić Radić T, Svetličić V, Žutić V, Boulgaropoulos B (2011) Seawater at the nanoscale: Marine gel imaged by atomic force microscopy. J. Mol. Recognit. 24: 397-405.

[81] Falciatore A, Ribera d'Alcala M, Croot P, Bowler C (2000) Perception of environmental signals by a marine diatom. Science 288: 363–2365.

[82] Verdugo P, Alldredge AL, Azam F, Kirchmand DL, Passow U, Santschi PH (2004) The oceanic gel phase: a bridge in the DOM–POM continuum. Mar. Chem. 92: 67– 85.

[83] Verdugo P, Orellana MV, Chin WC, Petersen TW, van den Eng G, Benner R, Hedges J (2008) Marine biopolymer self-assembly: implications for carbon cycling in the ocean. Faraday Discuss. 139: 393–398.

[84] Stachowitsch M, Fanuko N, Richter M (1990) Mucous aggregates in the Adriatic Sea: an overview of stages and occurrence. Mar. Ecol. 11: 327–350.

[85] Precali R, Giani M, Marini M, Grilli F, Ferrari CR, Pećar O, Paschini E (2005) Mucilaginous aggregates in the northern Adriatic in the period 1999–2002: typology and distribution. Sci Total Environ. 353: 10–23.

[86] Navarro E, Baun A, Behra R, Hartmann NB, Filser J, Miao AJ, Quigg A, Santschi P, Sigg L (2008) Environmental behavior and ecotoxicity of engineered nanoparticles to algae, plants, and fungi. Ecotoxicology 17: 372-386.

[87] Miao AJ, Schwehr KA, Xu C, Zhang SJ, Luo ZP, Quigg A, Santschi P (2009) The algal toxicity of silver engineered nanoparticles and detoxification by exopolymeric substances. Environ. Pollut. 157: 3034–3041.

[88] Pletikapić G, Vinković Vrček I, Žutić V, Svetličić V (2012) Atomic force microscopy characterization of silver nanoparticles interactions with marine diatom cells and extracellular polymeric substance. J. Mol. Recognit. 25: 309-317.

[89] Boissière M, Tourrette A, Devoisselle JM, Di Renzo F, Quignard F (2006) Pillaring effects in macroporous carrageenan-silica composite microspheres. J Colloid Interface Sci. 294: 109-16.

[90] Bustamante C, Macosko JC, Wuite GJ (2000) Grabbing the cat by the tail: manipulating molecules one by one. Nat. Rev. Mol. Cell Biol. 1: 130-6.

[91] Bhanu P. Jena, J. K. Heinrich Hörber (2006) Force Microscopy: Applications in Biology and Medicine. New Jersey: John Wiley & Sons. 300 p.

[92] Yves Dufrene (2011) Life at the Nanoscale: Atomic Force Microscopy of Live Cells. Singapore: Pan Stanford Publishing. 460 p.

[93] Alexander Noy (2007) Handbook of Molecular Force Spectroscopy. Livermore: Springer. 300 p.

[94] Higgins M.J, Wetherbee R (2011) The Role of Atomic Force Microscopy in Advancing Diatom Research into the Nanotechnology Era. In: Dufrene Y, editor. Life at the Nanoscale: Atomic Force Microscopy of Live Cells. Singapore: Pan Stanford Publishing. pp. 405-420.

[95] Pletikapić G, Mišić Radić T, Berquand A, Murvai U, Kellermayer MSZ, Vinković Vrček I, Žutić V, Svetličić V (2012) Marine diatom cells and extracellular polymers: nanostructure, nanomechanics and interactions with nanoparticles. International Conference: From Solid State to BioPhysics VI: From Physics to Life Sciences, 9th-16th June 2012, Dubrovnik, Croatia.

Chapter 13

Exopolysaccharides of the Biofilm Matrix: A Complex Biophysical World

Pierre Lembre, Cécile Lorentz and Patrick Di Martino

Additional information is available at the end of the chapter

http://dx.doi.org/10.5772/51213

1. Introduction

Microbial biofilm development is observed on virtually all submerged surfaces in natural and industrial environments. Biofilms are also observed at interfaces as pellicles, or in the bulk of aquatic environments as flocs or granules [1, 2]. A biofilm is a complex structure made of aggregates of microbial cells within a matrix of extracellular polymeric substances (EPS) (Figure 1). The matrix structure constitutes the elastic part of the biofilm. Interstitial voids and channels separating the microcolonies contain a liquid phase, mainly constituted by water. This liquid phase is the viscous part of the biofilm. The EPS matrix provides the biofilm with mechanical stability through these viscoelastic properties [3].

All major classes of macromolecule, i.e., polysaccharides, proteins, nucleic acids, peptidoglycan, and lipids can be present in a biofilm. Although extracellular polysaccharides are considered as the major structural components of the biofilm matrix, extracellular DNA plays an important role in the establishment of biofilm structure [4]. Moreover, nucleases can be regulators of biofilm formation [5]. To get a better understanding of the role of extracellular polysaccharides in the biofilm architecture and mechanical properties, it is necessary to take a look at the properties of a limited number of components, which can be isolated. Most microbial exopolysaccharides are highly soluble in water or dilute salt solutions, and capsule-forming polysaccharides are attached to the cells surface through covalent bonds to other surface polymers. Many of the extracellular polysaccharides produced in biofilms are insoluble and not easily separated from the cells, complicating the precise determination of their chemical structures and physical properties. Jahn et al. extracted a mixture of polymers from *Pseudomonas putida* biofilm material and found it to be very heterogeneous [6]. Most bacterial exopolysaccharides can exist either in ordered or disordered forms. Elevated temperatures and extremely low ionic concentrations favour the disordered forms. Polysaccharide molecules can interact with themselves or with

INTECH
open science | open minds

© 2012 The Author(s). Licensee InTech. This chapter is distributed under the terms of the Creative Commons Attribution License http://creativecommons.org/licenses/by/3.0), which permits unrestricted use, distribution, and reproduction in any medium, provided the original work is properly cited.

heterologous ions and molecules to yield gels, often with multivalent cations playing a significant role in the process. Polysaccharides also interact with proteins molecules both as solutes and when attached to the surface of the microbial cells. The polysaccharide - protein interactions in the matrix induce both structural and functional properties. Indeed, some of these proteins are enzymes constituting an external digestion system [7].

Biofilms in differing environments can be exposed to a very wide range of hydrodynamic conditions, which greatly affect the matrix and the biofilm structure [8]. The shear rate determines the rate of erosion of cells and regions of the matrix from the biofilm. Polysaccharides of the matrix exhibit flow and elastic recovery; because of the flexibility of the matrix its shape can change in response to an applied force. The shear stress to which a biofilm is exposed also affects the physical morphology and dynamic behaviour. Biofilms grown under higher shear are more strongly adhered and have a stronger EPS matrix than those grown under lower shear [9]. Biofilm density can be influenced by the fluid shear during growth [10]. *Pseudomonas* biofilms grown under laminar flow generally consist of hemispherical mound-shaped microcolonies, which form an isotropic pattern on the surface [9]. The biofilm microcolonies grown in turbulent flow are elongated in the downstream direction to form filamentous streamers. The streamers are attached to the glass substratum by an upstream head while the downstream tails are free to oscillate in the flow. Thus, hydrodynamics conditions influence both the structure and the material properties of biofilms [9]. This may be related to the physical arrangement of individual polymer strands in the biofilm EPS matrix [11]. The constitution of the biofilm matrix of *S. enteritidis* varies with pressure forces applied to the biofilm. Indeed in the absence of pressure, the sugars in the biofilm matrix are mainly composed of glucose and very little fucose. However in the presence of power flow, the share of fucose in the biofilm matrix is increased from 11% to about 30% [12].

In this chapter, after the presentation of exopolysaccharides extraction and purification from the biofilm matrix, the structural and physical properties of bacterial alginates, cellulose and other exopoysaccharides related to biofilm formation are discussed. An illustration of the complexity of the biofilm matrix architecture and the role of exopolysaccharides in the properties of the matrix is given through biofilms formation at the surface of nanofiltration membranes used for drinking water production.

2. Exopolysaccharides extraction and purification from the biofilm matrix

This section focuses on specific extraction methods targeting exopolysaccharides. General extraction methods for exopolysaccharides are first presented, followed by a presentation of the corresponding exopolysaccharides properties and carbohydrate contents.

2.1. Methods for exopolysaccharides extraction

Exopolysaccharides constitute the main EPS in many biofilms. They form the backbone of a network where other EPS components can be included. The stability of the biofilm matrix is dominated by entanglement of EPS and weak physicochemical interactions between molecules. These interactions correspond to various binding forces such as electrostatic attractive forces, repulsive forces (preventing collapsing), hydrogen bonds, van der Waals interactions and ionic attractive forces [13].

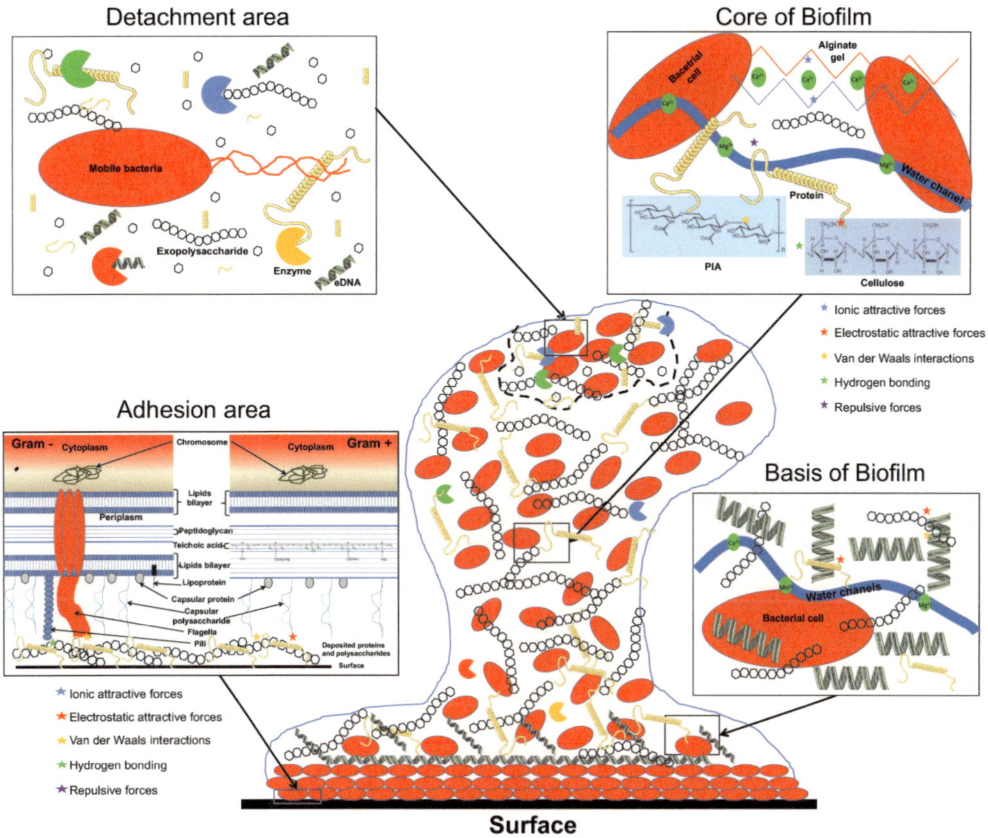

Figure 1. Schematic representation of a mature biofilm. In the centre, overall diagram of the structure of a biofilm to an interface solid / liquid: bacteria are attached to the solid surface and included in a self-induced polymer matrix. In the area of contact between bacteria and surface, the microbial cells can interact with the surface via several protein and polysaccharide appendages (pili, flagella, LPS, capsular polysaccharides) depending on the type of bacteria. On the basis of the biofilm, bacterial cells are embedded in a matrix containing high eDNA concentrations, in addition to proteins and polysaccharides. The eDNA plays a major role in early biofilm formation. In the core of the biofilm, channels of water carrying ions and nutrients cross the biofilm matrix containing high concentrations of exopolymeric substances. All these exocellular compounds form a protective gel around the microorganisms. In the biofilm detachment area, microbial enzymes destroy the exopolymeric matrix and release the cells that regain mobility, to be able to colonize new surfaces.

The exopolysaccharides recovery from the biofilm matrix in order to get a better understanding of their nature, requires to break down the interactions between EPS and selectivity separate them from other EPS and from matrix cells without cell lysis. The evaluation of cell lysis can be performed by measuring activity of the intracellular marker enzyme glucose-6-phosphate dehydrogenase (G6PDH; EC 1.1.1.49). Thus, substantial cell lysis occurring during the EPS extraction is commonly observed [14]. Regarding extraction methods, publications dealing with selective extraction of exopolysaccharides are missing, as already reviewed by Denkaus *et al.* [15]. Indeed, procedures described in the literature mainly deals with EPS extraction.

Physical and/or chemical methods are used to extract EPS from biofilms. Some EPS are tightly associated to the biofilm structure, sometimes through covalent bounds to the cells surface and are not directly extracted. Others free EPS are directly released. The easily released EPS can be separated using physical methods such as high-speed centrifugation and ultrasonication. Indeed, centrifugation is often used to separate soluble EPS from bacterial cells from pure cultures. Firmly cells-associated EPS require chemical methods of extractions. EPS cross-linked by divalent cations can be released from the biofilm matrix by complexing agents such as ethylenediamine tetraacetic acid (EDTA), by cation-exchange resins such as Dowex or by a formaldehyde treatment with or without sodium hydroxide [14, 16].

Various methods used to extract EPS can be applied to the extraction of exopolysaccharides as illustrated on Figure 2.

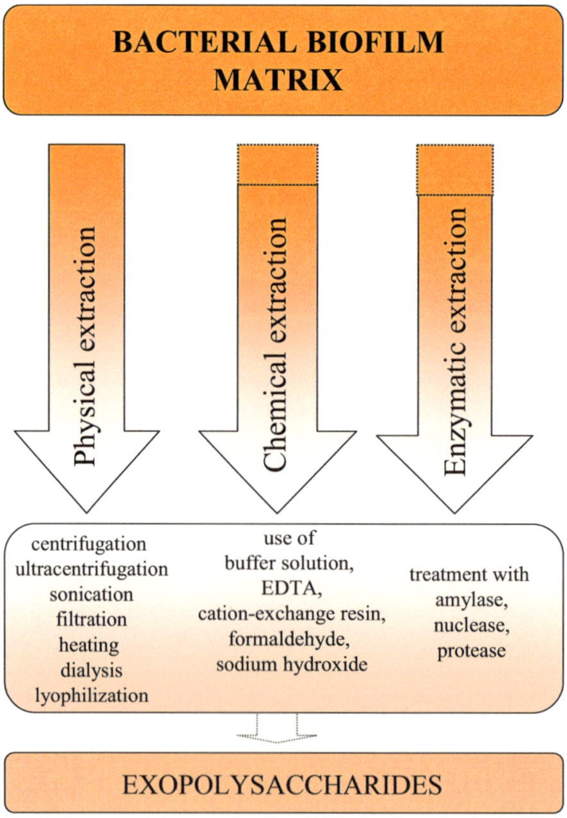

Figure 2. Pathways of exopolysaccharides extraction methods from biofilms

EPS extraction can be done from pure cultures of from complex microbial communities. For example, EPS material can be removed from *Pseudomonas aeruginosa* bacteria by centrifugation at 40 000g for 2 hours at 10°C. Purification of alginate is mostly obtained by precipitation in the presence of organic solvents from culture supernatants and treatment with enzymes such as nucleases and proteases to remove contaminating nucleic acids and proteins [14]. EPS from activated sludge samples can also be extracted by a centrifugation protocol. Then, residual bacteria can be removed from the supernatant by filtration on cellulose acetate membranes (0.2 μm). For removal of low molecular weight material, the supernatants can be dialyzed against deionised water. Depending on the studies, dialysis tubings can have various molecular weight cut-off. Then, dialysate is concentrated by lyophilisation. Other processes of exopolysaccharides extraction from various biofilm species are presented in Table 1.

Biofilm source/ support	**Sample preparation**	**Method of determination**	**Reference**
Pseudomonas fluorescens / polymethyl methacrylate plates	▪ extraction by EDTA 1.5%, 5°C, 3 h ▪ dialyse against deionized water (14 kDa) ▪ precipitation of proteins by pH adjsutement	phenol-sulfuric acid method	Oliveira *et al.*, [17]
Enterobacter cloacae /zinc selenide crystal	▪ extraction by EDTA or NaOH	phenol-sulfuric acid method	Boualam *et al.*, [18]
Leuconostoc mesenteroides strains /stainless steel	▪ scraping and washing biofilm material ▪ extraction by deionized water ▪ centrifugation at 10 000 rpm, 30 min ▪ hydrolysis (endodextranase) ▪ methylation	GC-MS	Leathers and Bischoff [19] Leathers and Cote [20]
Thermus aquaticus YT- /cellophane membrane	▪ extraction by NaCl 0.9% ▪ centrifugation, 8 000g, 15 min ▪ precipitation with alcohol 95 %, 4°C, one night ▪ centrifugation, 2500g, 0°C, 15 min ▪ dissolution in double deionized water and lyophilization	size-exclusion chromatography, GC-MS, HPAEC-PAD*, MS/MS, NMR	Lin *et al.*, [21]
aerobic activated sludge from a wastewater treatment	Comparison of five extraction processes : ▪ extraction by - EDTA 2 %, 4 C, 3 h - cation exchange resin, 4°C, 1 h - formaldehyde, 4°C,1 h - formaldehyde plus NaOH 1 N, 4 C, 3 h - formaldehyde plus ultrasonication (60 W), 2.5 min ▪ high-speed centrifugation 20 000g, 20 min, ▪ filtration (0.2 μm) ▪ dialysis (3,5 kDa), ▪ lyophilization, -50 C, 48 h	anthrone method	Liu and Fang, [16]

*HPAEC-PAD: High Performance Anion Exchange Chromatography with Pulsed Amperometric Detection

Table 1. Extraction condition and determination of exopolysaccharides from various biofilm species

The content of the EPS extracts is done by chemical analyses. The exopolysaccharide content of EPS can be determined by the phenol-sulphuric acid method described by Dubois *et al.* [22] or by using the anthrone method according to Dreywood [23], with glucose as the standard. The protein content of EPS can be determined by the Bradford method [24] or by using the bicinchoninic acid reagent [20] with bovine serum albumin as the standard.

As mentioned by several authors, yields of EPS extracted from biofilms depend on the extraction method used. Pan *et al.* [26] reported that chemical methods significantly increased the extraction yield from natural biofilm compared to physical methods. Nevertheless, the use of chemical methods can modify the composition of the EPS extracted. Indeed, added chemical extractants such as EDTA or formaldehyde can be present in the EPS extracts as contaminants and then can modify the EPS quantification efficiency. Moreover, chemicals can induce the release of intracellular components from treated cells and contaminate extracellular substances by intracellular material. These contaminants may be eliminated.

2.2. Exopolysaccharides of the biofilm matrix

The ability to synthesize exopolysaccharides is widespread among microorganisms, and microbial exopolysaccharides play important roles in biofilm formation, pathogen persistence, and have several applications in the food and medical industries. Exopolysaccharides are considered to be important components of the biofilms matrix [27]. However, some studies suggest that exopolysaccharides may not always be essential for biofilm formation [28]. Most of the matrix exopolysaccharides are very long with a molecular weight of 500-2000 kDa. They can be homo-polymers such as cellulose, curdlan or dextran, or hetero-polymers like alginate, emulsan, gellan or xanthan. Exopolysaccharide chains can be linear or branched. They are generally constituted by monosaccharides and some non-carbohydrate substituents such as acetate, pyruvate, succinate, and phosphate [29]. Various examples of exopolysaccharides encountered in bacterial biofilm are presented in Table 2.

2.3. Carbohydrate content of exopolysaccharides

Composition as well as conformation of sugar monomers may modify the properties of the exopolysaccharides and thus of the biofilm matrix. Mono-carbohydrate constituted exopolysaccharides are often D-glucose, D-galactose, D-mannose, L-fucose, L-rhamnose, L-arabinose, N-acetyl- D-glucose amine and N-acetyl-D-galactose amine as well as the uronic acids D-glucuronic acid, D-galacturonic acid, D-manuronic acid and L-guluronic acid. Other sugar monomers less frequently occurring are D-ribose, D-xylose, 3-keto-deoxy-D-mannooctulosonic acid and several hexoseamineuronic acids [29]. Some examples of carbohydrate content in biofilm are presented in Table 3.

In conclusion of this section, it is clear that the extraction of exopolysaccharides from biofilms usually require a multi-method protocol. Furthermore, there is no standard extraction procedure established, making difficult the meaning, comparison and interpretation of published results. However, recent studies tend to evaluate whether molecular diversity of EPS are potential markers for biofilm macro-scale characteristics [40].

Exopolysaccharides	Biofilm species	Reference
alginate	*Azotobacter vinelandii* *Pseudomonas aeruginosa* *Pseudomonas fluorescens*	Gorin and Spencer [30] Sabra *et al.*, [31] Donati and Paoletti [32]
curdlan 1,3-β-curdlan-type curdlan-type	*Agrobacterium sp.* ATCC 31749 *Cellulomonas flavigena* *Cellulomonas sp.*	Ruffing and Chen [33] Kenyon *et al.*, [34] Young and Reguera [35]
xanthan	*Xanthomonas citri* subsp. citr	Guo *et al.*, [36]
1,6-α-glucan 1,3-α- glucan	*Streptococcus mutans*	Aires *et al.*, [37]
glucan alternan dextran	*Leuconostoc mesenteroides* strain NRRL B-1355	Cote and Leathers [38]
pyruvated galactan	*Methylbacterium* sp. (isolated from a Finnish paper machine)	Verhoef *et al.*, [39]

Table 2. Examples of exopolysaccharides of bacterial biofilms

Biofilm source/ original source	Carbohydrate nature (concentration)	Method of carbohydrate characterization	Reference
European intertidal mudflat (Marennes-Oléron Bay, France)	galacturonic acid (20%) mannose (19,5%) glucose (19%) arabinose (15%) xylose (8%) galactose (7%)	GC-MS	Pierre *et al.*, [41]
Pseudomonas fluorescence Biovar II	mannose (14%) glucose (<5%) arabinose (28%) xylose (<5%) galactose (45%) fucose (6%) rhamnose (<5%) ribose (<5%)	GC-MS	Hung *et al.*, [42]
Pseudomonas aeruginosa PAO1, PDO300, *algD*, PA14	mannuronic acid (0-100) % glucose 0-56 % rhamnose (0- 20.7 %) galactose (0-12.4 %) mannose (0-13.9 %) xylose (0-9.7 %) ketodeoxyoctulosonate (0-9.1%) *N*-acetyl galactosamine (0-1.9 %) *N*-acetyl fucosamine (0-7.5 %) *N*-acetyl glucosamine 0-3.8 % *N*-acetyl quinovosamine 0- 18.1% unknown amino sugar 0-5 %	GC-MS	Wozniak *et al.*, [28]
Membrane bioreactor/fouling	uronic acid* glucose* mannose* fructose* xylose* ribose* arabinose* *N*-acetylglucosamine* glucuronic acid* galacturonic acid* maltose* saccharose*	HPLC-SM	Al-Halbouni *et al.*, [43]

(*) not indicated

Table 3. Carbohydrate content of various biofilms

3. Structure and function of different polysaccharides from the biofilm matrix

The most famous exopolysaccharides present inside biofilms are alginate, cellulose and poly-N-acetyl glucosamine. This section focuses on their structures and their function inside biofilms.

3.1. Bacterial alginates

Alginate, a polysaccharide which occurs in brown algae and in different bacteria like *Azotobacter vinelandii* [30] and *P. aeruginosa* [44] has been extensively studied. Alginate is an exopolysaccharide with a relatively high molecular mass (10^4-10^6 g/ml). It consists of the uronic acid residues β-D-mannuronate (M) and its C-5 epimer, α-L-guluronate (G) [45] (Figure 3).

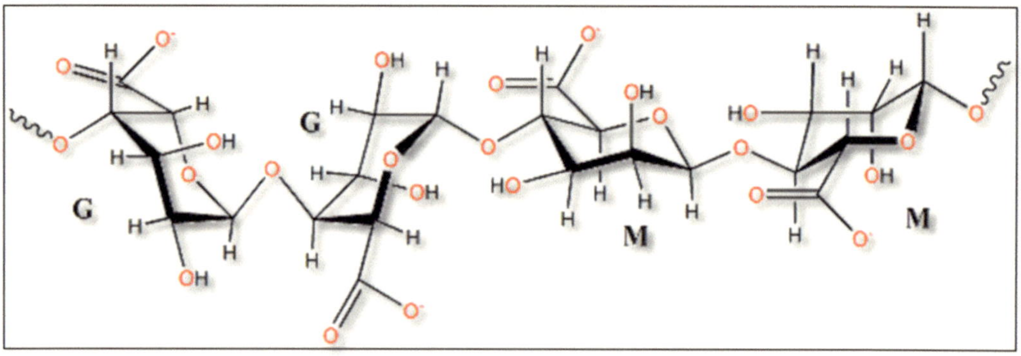

Figure 3. Structure of alginate

Generally, the monomers form a block copolymer with homopolymeric regions of poly-β-D-mannuronate (M-blocks) and poly-α-L-guluronate (G-blocks) as well as heteropolymeric regions (MG-blocks). The absence of G-blocks differentiates alginates produced by *P. aeruginosa* from alginates expressed by algae or by *A. vinelandii* [46]. The functional properties of the alginates strongly correlate with the composition (M/G ratio) and with the uronic acid sequence.

There are 24 genes located on the bacterial chromosome, involved in the production and secretion of alginate in *P. aeruginosa* [45]. Eight genes are implicated in the exportation of alginic acids (Figure 4), and twelve in the biosynthesis of the polysaccharide [47]. The four remaining genes are involved in the regulation of the synthesis.

Alginates can form a gel in the presence of chelating divalent cations. This structure formed is called a Grant "egg-box" [47]. The alginate gel is formed by ionic bonds between the G-rich blocks and divalent cations. The mechanical properties of alginate gels can vary depending on the amounts of guluronic acid present in the polymer. Moreover, alginate gels can be formed in vitro in the presence of proteins such as gelatin [48].

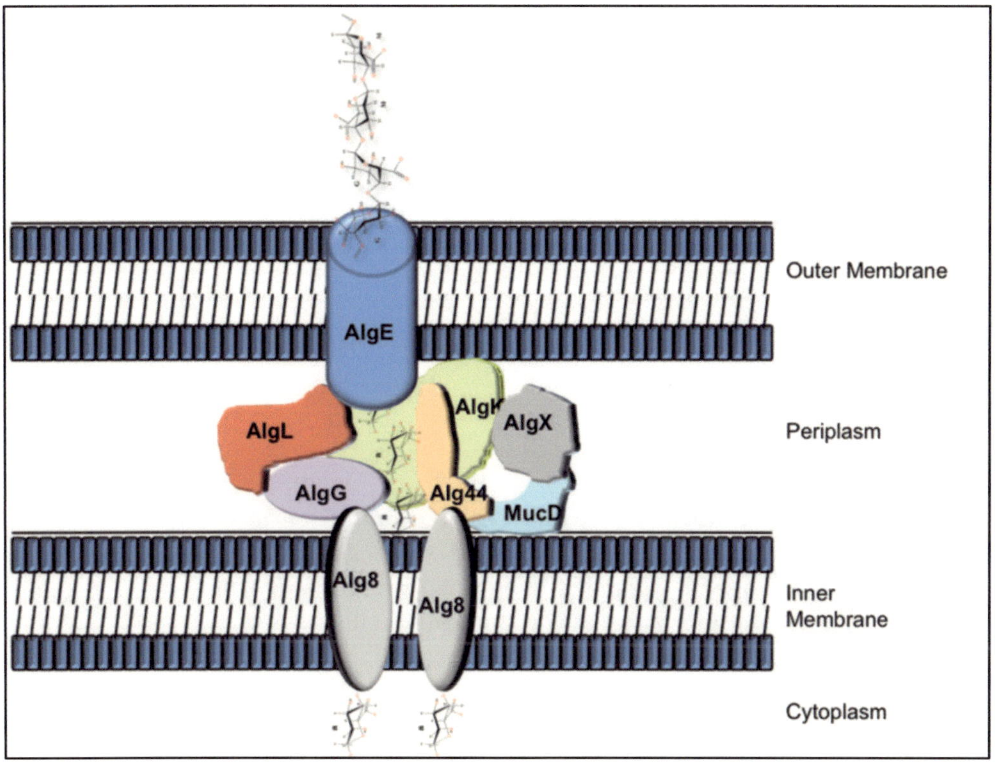

Figure 4. Biosynthesis of bacterial alginate

3.2. Cellulose

Cellulose is the most abundant sugar polymer found on the surface of the planet. It is found throughout the living world: in plants, animals, fungi and in bacteria such as *Salmonella, E. coli, Acetobacter, Agrobacterium* and *Rhizobium* [49].

Salmonella and *E. coli* produce cellulose as a crucial component of the extracellular matrix [50]. Cellulose consists of a β-1-4 linked linear glucose (Figure 5). The formation of cellulose fibers is provided by hydrogen bonds between the chains of glucose. These formed sheets are very stable and their number varies depending on the nature of the environment.

Cellulose has a crystalline structure. Each crystal of cellulose contains numerous glycan chains in parallel orientation. The reducing ends are at one terminus while the non-reducing ends are at the opposite terminus. The structure is not uniform and amorphous regions cohabit with highly crystalline regions.

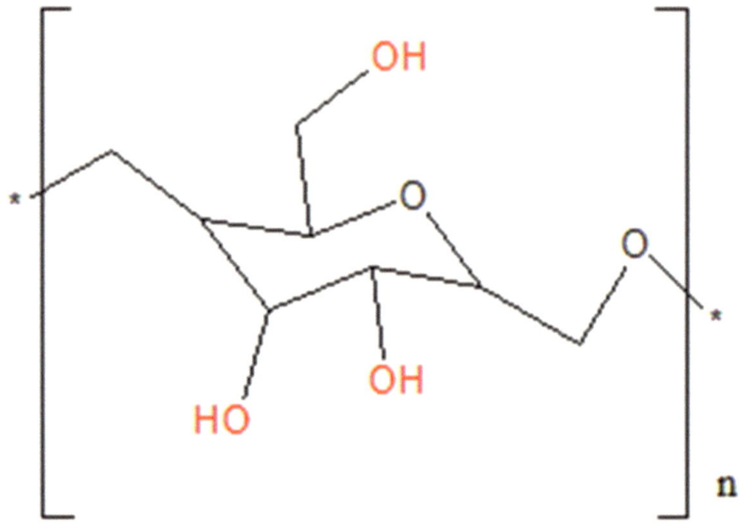

Figure 5. Structure of cellulose

Genes involved in the production of cellulose in *E. coli* and *S. typhimurium* are called *bcs* for bacterial cellulose synthesis (Figure 6). The four *bcs* genes called *bcsA*, *bcsB*, *bcsZ* and *bcsC* are organized as an operon. The *bcs* operon is partially regulated by AgfD, a thin aggregative fimbriae which increases the production of cellulose and curli [49].

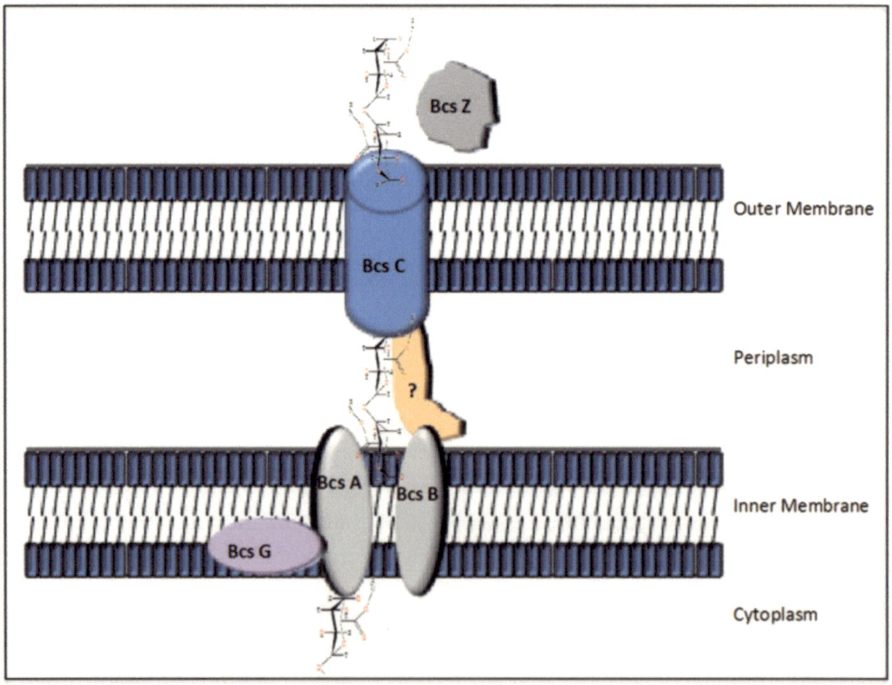

Figure 6. Cellulose biosynthesis

Cellulose can form a gel at adequate temperatures. Cellulose solutions are liquid at room temperature. Gels can form in a cellulose solution at either high temperature (above 50 °C) or low temperature (less than 10 °C). After gelification, cellulose solutions remain more or less stable in the gel state at room temperature [51, 52]. The gel structure of cellulose may explain the mechanical properties of biofilms formed by bacterial species producing this polymer.

3.3. Poly-N-acetyl glucosamine

The polysaccharide intercellular adhesin (PIA) or the related poly-N-acetyl glucosamine (PNAG) polymer is required for bacterial adherence and biofilm formation of some bacterial species. This polysaccharide family was first described in *Saphylococcus* species [53], and further in *E. coli* [54]. PNAG is a positively charged linear homoglycan composed of β-1,6-N-acetylglucosamine residues with approximately 20% deacetylated residues [55] (figure 7).

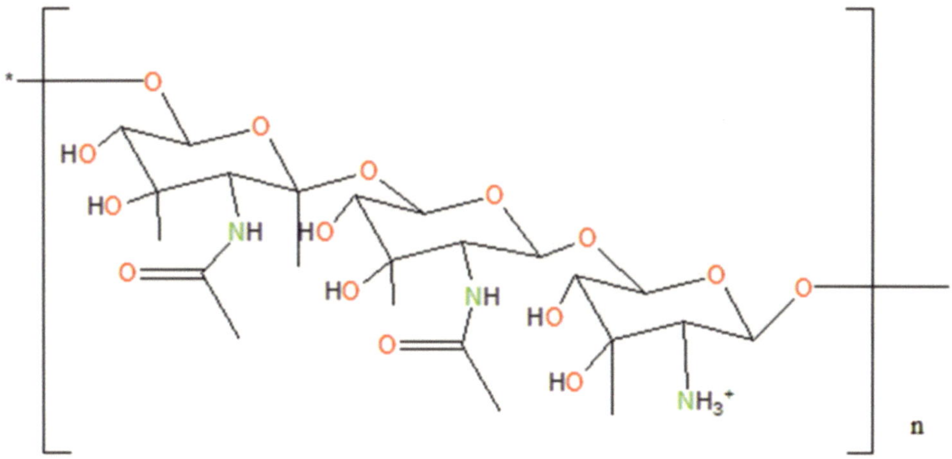

Figure 7. Structure of PNAG

The genes involved in the biosynthesis of PIA are named *ica* for intercellular adhesion. This operon is composed of a regulation gene *icaR* and four biosynthetic genes: *icaADBC* [56, 57] (Figure 8).

PNAG forms a protective matrix around bacterial cells that is also involved in cell-to-cell interactions [53, 54]. PNAG can also interact with eDNA, reinforcing the biofilm matrix structure [58].

3.4. Other polysaccharides involved in biofilm formation

Individual strains or one strain put in different environmental conditions, are able to produce several different extracellular polysaccharides. In mucoid strains of *P. aeruginosa* isolated from patients with cystic fibrosis, mucoidy is due to the overproduction of alginate that is the major constituent of the biofilm matrix. Nevertheless, biofilms formed by non mucoid strains do not contain alginate. A glucose-rich polymer named PEL (pellicle), and a mannose-rich polymer named PSL (polysaccharide synthesis locus), makes significant contribution to *P. aeruginosa* biofilm structure [59]. The *pel* genes are present in several strains but some commonly used reference strains, such as PAK and PAO1 do not express strongly these genes under common laboratory conditions. The *psl* genes are only present in some strains, but not in the well known PA14 laboratory strain. In some instance, PEL, and PSL can be present together in the biofilm matrix of *P. aeruginosa.*

Other polymers are present in the matrix of the biofilm of *S. epidermidis* for example: teichoic acid [60]. There are two types of teichoic acid in *S. epidermidis*: teichoic acid associated with the bacterial membrane (CW TA) and extracellular teichoic acid (EC TA). The EC TA is responsible for the increased viscosity of the colony. The EC TA is a (1-3)-linked poly(glycerol phosphate), substituted at the 2-position with α-glucose, α-N-acetylglucose, D-alanine and α-glucose-6-alanine (Figure 9).

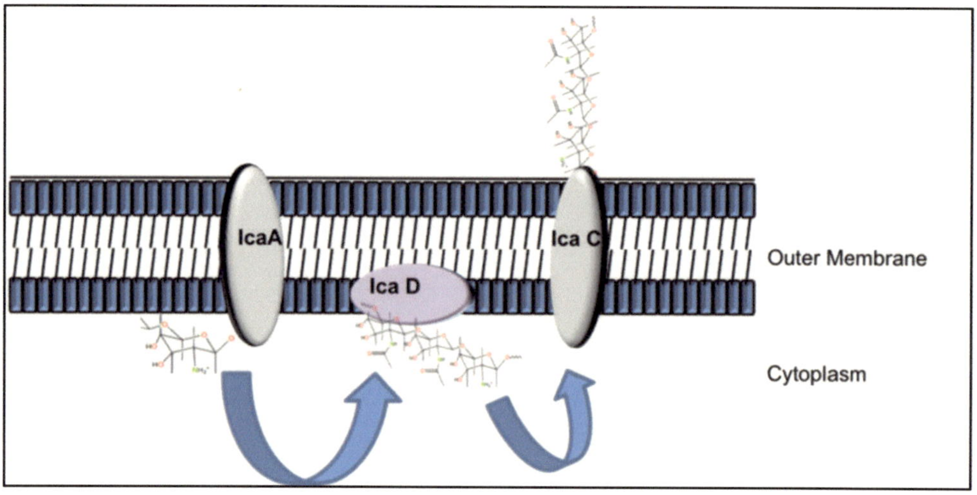

Figure 8. PNAG biosynthesis

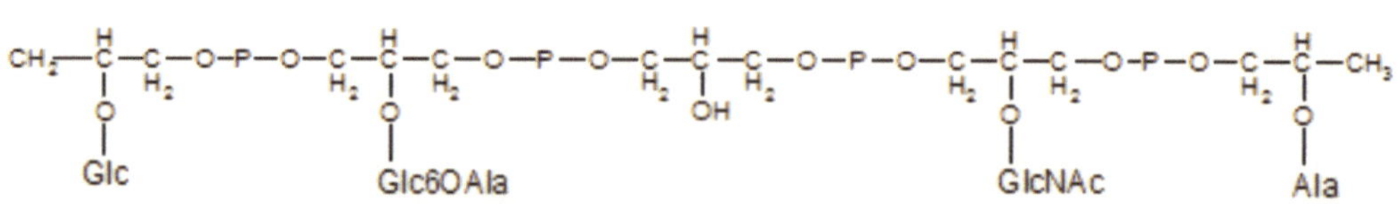

Figure 9. Structure of teichoic acid

In *E. coli,* colanic acid, a sugar polymer composed of galactose, fructose and glucose, is regularly found in the biofilm matrix [61] (Figure 10).

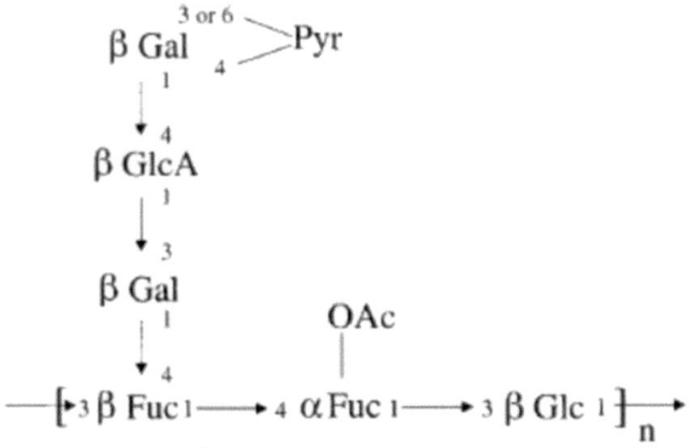

Figure 10. Structure of colanic acid

It must be remembered that although different strains can apparently synthesize the same EPS, there can be differences in physical properties especially with respect to viscosity and gel formation. Several biofilm studies have used colanic acid-producing *E. coli* [61]. Prigent-Combaret, C. et al. [62], yet this polymer can vary greatly in mass and viscosity, as can bacterial alginates.

4. An example of complex biofilm: biofilm formation at the surface of nanofiltration membranes used for drinking water production.

We and others have previously studied very complex biofilms formed on nanofiltration (NF) membranes during surface water filtration in drinking water production processes [63, 64]. After several years of filtration, the foulant consists in a brown viscous layer covering the entire surface of the membrane [65] (Figure 11).

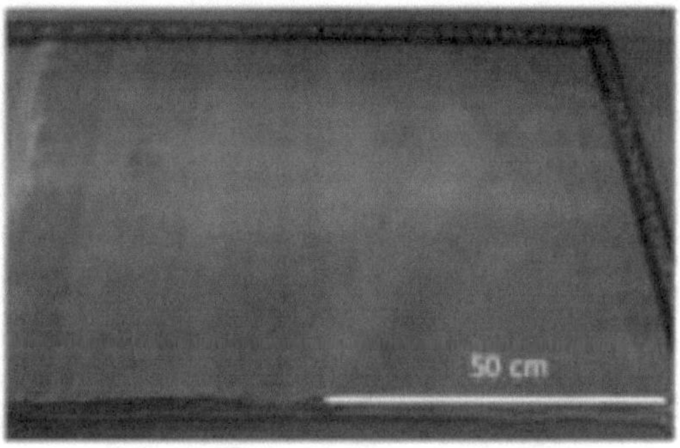

Figure 11. Visual examination of a fouled NF membrane

Dry weight of the foulant is about 2 g/m^2. The NF biofilms harbours mainly exopolysaccharides and proteins, as shown by characteristic ATR-FTIR signals near 1650 cm^{-1} (amide I), 1550 cm^{-1} (amide II), 1450 cm^{-1} (due in part to C-H deformation), 1400 cm^{-1} (due in part to symetric stretch for the carboxylate ion), 1250 cm^{-1} (P=O and C-O-C stretching and/or amide III), and in a broad complex region from 1250 to 900 cm^{-1} (due in part to C-O-

C, C-O, ring-stretching vibrations of polysaccharides and the P=O stretch of phosphodiesters) (Figure 12).

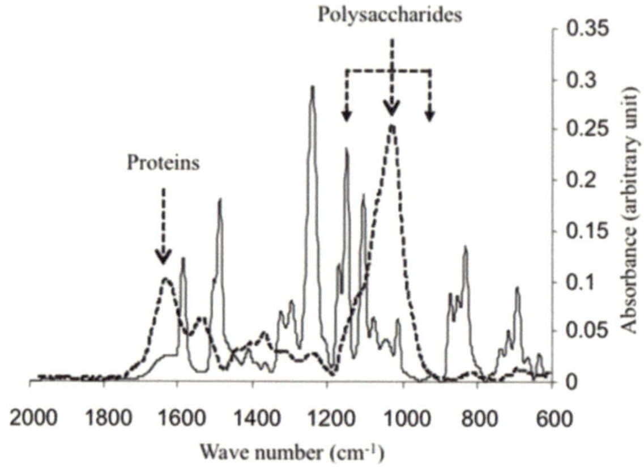

Figure 12. ATR-FTIR spectra of a virgin membrane (plain line) and of a fouled membrane (dotted line)

Fluorescence microscopy observations after nucleic acid staining with DAPI and polysaccharides staining with lectins labelled with fluorescein isothiocyanate or tetramethylrhodamine isothiocyanate indicate a high spatial heterogeneity inside the foulant matter with a mean thickness of 32.5 ± 17.7 μm [66] (Figure 13). Examples of lectins that can be used for such polysaccharides staining experiments are peanut agglutinin (PNA) targeting β-gal(1->3)galNAc residues, wheat germ agglutinin (WGA) targeting $(glcNAc)_2$ and NeuNAc residues, *Bandeiraea simplicifolia* (BS-1) agglutinin targeting α-gal and α -galNAc residues and Concanavalin A (ConA) targeting α-man and α-glc residues.

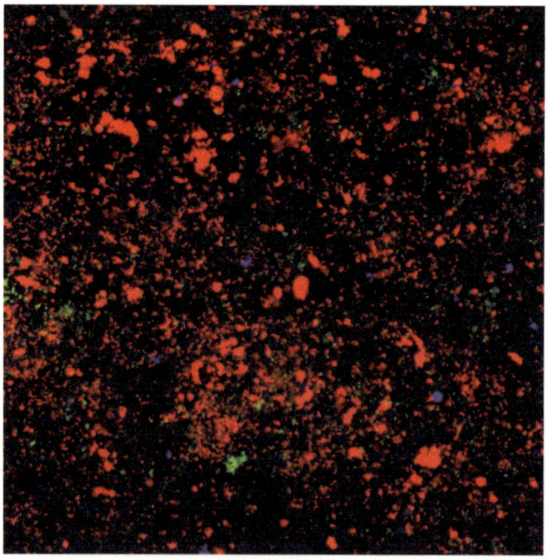

Figure 13. CLSM visualization of the heterogeneity of a NF biofilm after staining with DAPI, TRITC and TITC-labelled lectins. Magnification x630

The microbial cells, mainly composed of bacteria, are localized in the superficial layer of the fouling material and are organized as microcolonies interspersed at the membrane surface. Some algae are also present, as shown by autofluorescence properties. The presence of a dense and wide polysaccharide matrix harbouring few microbial cells at the NF membrane surface has been associated with differences in the efficiency of cleaning procedures against different foulants categories [65, 67]. Polysaccharide residues are found in areas where microcolonies are present and in areas devoid of microbial cells. This polysaccharide organization has been previously observed with environmental biofilms grown in vitro with river water as the sole source of carbon and nutrients [68]. High staining with PNA and BS-1, respectively reveals high occurrence of galactosides residues in the polysaccharide components of the foulants. The BS-1 lectin staining pattern indicates a high degree of spatial organisation with the observation of long and entangled fibers. WGA staining shows short fibers and cloud stained areas. PNA and ConA lectin staining are more interspersed. The polysaccharide composition of the fouling layer changes quantitatively and qualitatively during spring and summer [64]. Lectin staining increases from March to September for all the lectins used. Staining with BS-1 increases constantly in March, June and September. A high increase of binding with PNA, and ConA is observed between March and June, but the binding of these two lectins does not change between June and September. Staining with the WGA is weak in March and June and is higher in September. The lectin-binding changes with time may be linked to an increase of the biomass attached at the membrane surface and to changes among the populations of attached cells. Nutrients, oxygen level and the concentration of metals can influence the exopolymer abundance of environmental model biofilms grown in vitro with river water as the sole source of carbon and nutrients [69]. The modification of these parameters leads to a shift in the glycoconjugate makeup of the biofilms.

Biofilms may be considered to be highly porous polymer gels [70] and diffusion studies demonstrate gel-like characteristics [71]. Previous work has suggested that laboratory-grown and some natural biofilms are viscoelastic in nature [3, 8, 72]. During rotation analysis, a rheofluidification behaviour is observed for NF biofilms [66]. Different mechanisms can explain shear thinning of a biofilm. Break down of links between polymers in the biofilm matrix or deflocculation of particles corresponding to an irreversible modification of the biofilm structure can occur. Such irreversible modifications are unlikely in the experimental conditions published because of the reversibility of viscosity changes with shear rate [66]. Shear thinning of NF biofilms may be related to the polymeric composition of the biofilm matrix. With shear acceleration, polymers may follow the direction of the flow leading to viscosity decrease. This has been previously observed with purified polysaccharides like cellulose [73]. Moreover, bending of biofilm structures in the shear direction during the application of shear stress has been mentioned to explain the viscoelastic response of a mixed culture biofilm [72]. NF biofilms have been submitted to oscillation analysis with a cone-plate rheometer [66]. In such experiments, a sinusoidal oscillation of defined maximum strain and oscillatory frequency is applied to a sample and the storage (G') and lost (G'') modulus are measured. The storage modulus characterizes the ability of the material to store energy, whereas the loss modulus characterizes energy dissipation in the material under dynamic excitations. If the material is perfectly elastic then the resultant stress wave is exactly in phase with the strain wave. By contrast, when the rate of change of the sinusoidal oscillation is a maximum and the strain is zero, for purely viscous systems, the resultant stress wave will be exactly 90° out-of phase with the imposed deformation. For NF biofilms, during oscillation analysis, values of storage modulus (G') stay higher than values of loss modulus (G'') over the entire range of frequencies covered, indicating that the NF biofilm behave like a highly elastic physical gel [74]. Polysaccharides alone, like alginates, are known to realize a sol-gel transition under adequate physicochemical conditions [75]. The physicochemical microenvironment inside NF biofilms may be permissive to exopolysaccharides sol-gel transition. The gel state is resistant enough and presents a micro porosity favourable for resistance to flow forces, microcolonies development and cell nutrition inside the biofilm structure. This model of sol-gel transition of polysaccharides inside biofilms is consistent with rheological properties previously demonstrated for other biofilms: *Streptococcus mutans* biofilms have elasticity and viscous behaviour analogous to NF biofilms for a range of frequencies between 0.1 to 20 Hz [76]. The rheofluidification behaviour and gel-type rheological properties shared by different type of biofilms and purified polysaccharides suggest that the critical components of the biofilm matrix determining the biofilm texture are polysaccharides.

The time-dependent strain response observed in the creep curves clearly indicated that NF biofilms exhibited viscoelastic behaviour. Viscoelasticity is thought to be a general mechanical property of biofilms. A very wide range of elasticity and viscosity values has been previously observed for a wide sample of biofilms formed artificially in laboratory experiments or coming from natural aquatic environments [4, 72, 76]. Thus, it wasn't surprising to observe that the rheological properties of NF biofilms are different from the ones of natural biofilms from different aquatic environments like Nymph Creek (Yellowstone National Park) and Chico Hot springs (Montana) algal biofilms [4]. These differences in viscosity and elasticity between biofilms can be related to different exopolysaccharide contents and to different shearing strains. Bacterial and algal alginates are known to have different monomeric composition leading to a stronger binding of cations for bacteria, a property involved in the formation of a stable gel in the presence of ambient Ca^{2+} cations [77].

The specificity of NF biofilms may be the necessity to resist shear forces applied to the membrane during the filtration process. In the Méry-sur-Oise plant, NF membranes are operated at feed pressure of approximately 10 bars [78]. The high membrane feed pressures may influence the rheological properties of NF biofilms by increasing cohesive forces in the biofilm bulk, increasing forces, which keep the exopolymers to the membrane surface, and thus strengthening the mechanical stability of the biofilm. This may explain at least in part the NF biofilms resistance to industrial cleaning [65].

Shaw et al. have previously shown that the elastic relaxation time varied much less between biofilms of different origins. λ was estimated to be the time required for viscous creep length to equal elastic deformation length (so that memory of initial conditions is lost), i.e., $\lambda \approx \eta / G$.

The elastic relaxation time of about 30 minutes lies within the range previously determined for various biofilms [4]. The universality of the viscoelastic transition of biofilms has been suspected to have critical survival impact [4]. The ability of biofilms to deform in response to mechanical stress may be a conserved strategy of defence to enable persistence on surfaces in different flow conditions.

5. Conclusion

Extracellular polysaccharides are considered as the major structural components of the biofilm matrix. A large variety of polysaccharides required for bacterial adherence and biofilm formation have been described. Polysaccharide molecules can interact with themselves or with ions and proteins. These interactions result from electrostatic attractive forces, repulsive forces, hydrogen bonds, van der Waals interactions and ionic attractive forces. All these forces influence the structure and the stability of the biofilm matrix and the way EPS and polysaccharides can be extracted from the biofilm bulk. A universal protocol for extracellular polysaccharide extraction from the biofilm matrix does not exist. Each study may adapt usual extraction procedures to biofilm specificities and to the nature of the polysaccharide studied. The viscoelasticity nature of biofilms is universal but biofilms in differing environments exposed to different hydrodynamic conditions will encounter changes in the structure, composition and then physical properties of their matrix. Biofilm science is highly exciting since it is a mixture of biology, biophysic, chemistry and much more.

Author details

Pierre Lembre, Cécile Lorentz and Patrick Di Martino*
Groupe Biofilm, Laboratoire ERRMECe-EA1391, Institut des Matériaux-FD4122,
Institut Universitaire de Technologie, Université de Cergy-Pontoise, Cergy-Pontoise, France

6. References

[1] Marti S, Nait Chabane Y, Alexandre S, Coquet L, Vila J, Jouenne T, Dé E (2011) Growth of *Acinetobacter baumannii* in pellicle enhanced the expression of potential virulence factors. PLoS One 6(10), e26030.
[2] Wan J, Bessière Y, Spérandio M (2009) Alternating anoxic feast/aerobic famine condition for improving granular sludge formation in sequencing batch airlift reactor at reduced aeration rate. Water Res. 43(20), 5097-108.
[3] Shaw T, Winston M, Rupp C J, Klapper I & Stoodley P (2004) Commonality of elastic relaxion times in biofilms. Phys rev lett 93: 1-4.
[4] Whitchurch C B, Tolker-Nielsen T, Ragas P C, Mattick J S (2002) Extracellular DNA required for bacterial biofilm formation. Science 295, 1487.

* Corresponding Author

[5] Kiedrowski M R, Kavanaugh J S, Malone C L, Mootz J M, Voyich J M, Smeltzer M S, Bayles K W, Horswill A R (2011) Nuclease modulates biofilm formation in community-associated methicillin-resistant *Staphylococcus aureus*. PLoS One. 6(11), e26714.
[6] Jahn A, Griebe T, Nielson P H (2000) Composition of *Pseudomonas putida* biofilms: accumulation of protein in the biofilm matrix. Biofouling 14, 49–57.
[7] Wicker-Böckelmann U, Wingender J, Winkler U K (1987) Alginate lyase releases cell-bound lipase from mucoid strains of *Pseudomonas aeruginosa*. Zentralbl Bakteriol Mikrobiol Hyg A. 266(3-4), 379-89.
[8] Stoodley P, Dodds I, Boyle J D, Lappin-Scott H M (1999) Influence of hydrodynamics and nutrients on biofilm structure. J. Appl. Microbiol. 85, S19–S28.
[9] Stoodley P, Cargo R, Rupp C J, Wilson S, Klapper L (2002) Biofilm material properties as related to shear-induced deformation and detachment phenomena. J Industrial Microbiol. Biotechnol. 29, 361-367.
[10] Vieira M J, Melo J F, Pinjeiro M M (1993) Biofilm formation: hydrodynamic effects on internal diffusion and structure. Biofouling 7, 67-80.
[11] Flemming H C, Wingender J, Mayer C, Koerstgens V, Borchard W (2000) Cohesiveness in biofilm matrix polymers. In: Allison D., Gilbert P., Lappin-Scott H.M., Wilson M. (Eds), Community structure and cooperation in biofilms. SGM Symposium Series 59. Cambridge Univ. Press, Cambridge, 87-105.
[12] Mangalappalli-Illathu A K, Lawrence J R, Swerhone G D W, Korber D R (2007) Architectural adaptation and protein expression patterns of *Salmonella enterica* serovar enteritidis biofilms under laminar flow conditions. International journal of food Microbiology, 123: 109-120.
[13] Flemming H C and Wingender J (2010) The biofilm matrix. Nature Reviews Microbiology 8(9): 623-633.
[14] Wingender J, Strathmann M, Rode A, Leis A and Flemming H C (2001) Isolation and biochemical characterization of extracellular polymeric substances from *Pseudomonas aeruginosa* Methods in Enzymology 336:302-14.
[15] Denkhaus E, Meisen S, Telgheder U and Wingender J (2007).Chemical and physical methods for characterisation of biofilms. Microchimica Acta. 158: 1-27.
[16] Liu H, Fang H H P (2002) Extraction of extracellular polymeric substances (EPS) of sludges. J. Biotechnol. 95, 249-256.
[17] Oliveira R, Marques F and Azeredo J (1999) Purification of polysaccharides from a biofilm matrix by selective precipitation of proteins. Biotechnology Techniques. 13: 391-393.
[18] Boualam M, Quiles F, Mathieu L, Block J C (2002) Monitoring the effect of organic matter on biofilm growth in low nutritive waters by ATR-FT-IR spectroscopy. Biofouling 18(1): 73-81.
[19] Leathers T D, Bischoff K M (2011) Biofilm formation by strains of *Leuconostoc citreum* and L. mesenteroides. Biotechnology letters 33(3): 517-523.
[20] Leathers T D and Cote G L (2008) Biofilm formation by exopolysaccharide mutants of *Leuconostoc mesenteroides* strain NRRL B-1355. Applied Microbiology and Biotechnology 78(6): 1025-1031.

[21] Lin M H, Yang Y L, Chen Y P, Hua K F, Lu C P, Sheu F, Lin G H, Tsay S S, Liang S M and S H Wu (2011) A novel exopolysaccharide from the biofilm of *Thermus aquaticus* YT-1 induces the Immune response through toll-like receptor 2 Journal of Biological Chemistry J Biol Chem. 286(20): 17736-45.
[22] Dubois M, Gilles K A, Hamilton J K, Rebers PA, Smith F (1956) Colorimetric method for determination of sugars and related substances. Analytical Chemistry 38: 350–356.
[23] Dreywood R (1946) Qualitative test for carbohydrate material. Ind. Eng. Che. (Anal. ed.), 18, 8, 499.
[24] Bradford M M (1976) A rapid and sensitive method for the quantitation of microgram quantities of protein utilizing the principle of protein-dye binding. Anal. Biochem. 72, 248 - 256.
[25] Smith P K, Krohn R I, Hermanson G T, Mallia A K, Gartner F H, Provenzano M D, Fujimoto, E K, Goeke N M, Olson B J and Klenk D C (1985) Measurement of protein using bicinchoninic acid. Anal. Biochem. 150 (1): 76–85.
[26] Pan X, Liu J, Zhang D, Chen X, Li L, Song W and Yang J (2010) A comparison of five extraction methods for extracellular polymeric substances (EPS) from biofilm by using three-dimensional excitation-emission matrix (3DEEM) fluorescence spectroscopy. Water SA 36 111-116.
[27] Sutherland I W (2001) Biofilm exopolysaccharides: a strong and sticky framework. Microbiology 147(1): 3-9.
[28] Wozniak D J, Wyckoff T J O, Starkey M, Keyser R, Azadi P, O'Toole G A and Parsek M R (2003) Alginate is not a significant component of the extracellular polysaccharide matrix of PA14 and PAO1 *Pseudomonas aeruginosa* biofilms Proceedings of the National Academy of Sciences 100 (13): 7907-7912.
[29] Andersson S (2009) Characterization of bacterial biofilms for. wastewater treatment. Royal Institute of Technology. Thesis of Stockholm University, School of Biotechnology.
[30] Gorin J P A, Spencer T J F (1966) Exocellular alginic acid from *Azotobacter vinelandii.* Canadian journal of chemistry, 44: 993-998.
[31] Sabra W, Zeng A P, Deckwer W D (2001) Bacterial alginate: physiology, product quality and process aspects, Applied Microbiology and Biotechnology, 56: 315-325.
[32] Donati I, and Paoletti S (2009) Material Properties of Alginates. Alginates: Biology and Applications, Microbiology Monographs 13, 1-54, Springer.
[33] Ruffing, A M and Chen R R (2012) Transcriptome profiling of a curdlan-producing *Agrobacterium* reveals conserved regulatory mechanisms of exopolysaccharide biosynthesis. Microbial cell factories 11.
[34] Kenyon W J, Esch S W.and.Buller C S (2005) The curdlan-type exopolysaccharide produced by *Cellulomonas flavigena* KU forms part of an extracellular glycocalyx involved in cellulose degradation. International Journal of General and Molecular Microbiology 87(2): 143-148.
[35] Young, J. M., S. B. Leschine and G. Reguera. (2012). Reversible control of biofilm formation by *Cellulomonas* spp. in response to nitrogen availability. Environmental microbiology 14(3).

[36] Guo Y P, Sagaram U S, Kim J S and Wang N (2010) Requirement of the galU Gene for Polysaccharide Production by and Pathogenicity and Growth In Planta of *Xanthomonas citri* subsp *citri*. Applied and Environmental Microbiology 76(7): 2234-2242.

[37] Aires C P, Tenuta L M, Carbonero E R, Sassaki G L, Iacomini M and Cury J A (2011) Structural characterization of exopolysaccharides from biofilm of a cariogenic *streptococci*. Carbohydrate Polymers 84(4): 1215-1220.

[38] Cote G L and Leathers T D (2009) Insoluble glucans from planktonic and biofilm cultures of mutants of *Leuconostoc mesenteroides* NRRL B-1355. Applied Microbiology and Biotechnology 82(1): 149-154.

[39] Verhoef R, Waard P D, Schols H A, Siika-aho M and Voragen A G J (2003) *Methylobacterium* sp. isolated from a Finnish paper machine produces highly pyruvated galactan exopolysaccharide. Carbohydrate Research 338(18): 1851-1859.

[40] Ras M, Lefebvre D, Derlon N, Paul E, Girbal-Neuhauser E (2011) Extracellular Polymeric Substances diversity of biofilms grown under contrasted environmental conditions. Water Res. 45 (4), 1529-38.

[41] Pierre G, Graber M, Rafiliposon BA, Dupuy C, Orvain F, De Crignis M, Maugard T (2012) Biochemical composition and changes of extracellular polysaccharides (ECPS) produced during microphytobenthic biofilm development (Marennes-Oléron, France). Microb Ecol. 63(1): 157-69.

[42] Hung C C, Santschi P H, Gillow J B (2005) Isolation and characterization of extracellular polysaccharides produced by *Pseudomonas fluorescens* Biovar II, Carbohydrate Polymers, Volume 61(2): 141-147.

[43] Al-Halbouni D, Dott W and Hollender J (2009) Occurrence and composition of extracellular lipids and polysaccharides in a full-scale membrane bioreactor. Water Research 43(1): 97-106.

[44] Linker A, Jones R S (1964) A Polysaccharide resembling Alginic Acid from a *Pseudomonas* Micro-organism, Nature, 204: 187-188.

[45] Gacesa P (1998) Bacterial alginate biosynthesis-recent progress and future prospects, Microbiology, 144: 1133-1143.

[46] Remminghorst U, Rehm B H A (2006) Bacterial alginates: from biosynthesis to applications, Biotechnology letters, 28:1701-1712.

[47] Stokke B T, Draget K I, Smidsrod O, Yuguchi Y, Urakawa H, Kajiwara K (2000) Small-Angle X-ray scattering and rheological characterization of alginate gels. Macromolecules, 33: 1853-1863.

[48] Doumèche B, Picard J, Larreta-Garde V (2007) Enzyme-catalyzed phase transition of alginate gels and gelatin-alginate interpenetrated networks. Biomacromolecules, 8: 3613-3618.

[49] Zogaj X, Nimtz M, Rhode M, Bokranz W, Romling U (2001) The multicellular morphotypes of *Salmonella typhimurium* and *Escherichia coli* produce cellulose as the second component of the extracellular matrix. Molecular Microbiology, 39: 1452-1463.

[50] Solano C, Garcia B, Valle J, Berasain C, Ghigo J M, Gamazo C, Lasa I (2002) Genetic analysis of *Salmonella enteritidis* biofilm formation : critical role of cellulose, Molecular Microbiology, 43 : 793-808.

[51] Weng L, Zhang L, Ruan D, Shi L, Xu J (2004) Thermal gelation of cellulose in a NaOH/thiourea aqueous solution, Langmuir, 20: 2086-2093.
[52] Cai J, Zhang L (2006) Unique gelation behavior of cellulose in NaOH/Urea aqueous solution, Biomacromolecules, 7: 183-189.
[53] Mack D, Fischer W, Krokotsch A, Leopold K, Hartmann R, Egge H, Laufs R (1996) The intercellular adhesion involved in biofilm accumulation of *Staphylococcus epidermidis* is a linear β-1,6-linked glucosaminoglycan: purification and structural analysis. Journal of Bacteriology, 178: 175-183.
[54] Cerca N, Maira-Litran T, Jefferson K K, Grout M, Goldmann D A, Pier G B (2007) Protection against *Escherichia coli* infection by antibody to the *Staphylococcus aureus* poly-N-acetylglucosamine surface polysaccharide. PNAS, 104:7528-7533.
[55] Gotz F (2002) *Staphylococcus* and biofilm, Molecular Microbiology, 43: 1367-1378.
[56] Heilmann C, Gerke C, Perdreau-Remington F, Gotz F (1996) Characterization of Tn*917* insertion mutants of *Staphylococcus epidermidis* affected in biofilm formation. Infection and immunology, 64: 277-282.
[57] Gerke C, Kraft A, Sussmuth R, Schweitzer O, Götz F (1998) Characterization of the N-acetyglucosaminyl-transferase activity involved in the biosynthesis of the *Staphylococcus epidermidis* polysaccharide intercellular adhesion. The journal of biological chemistry, 273: 18586-18593.
[58] Izano E A, Amarante M A, Kher W B, Kaplan J B (2008) differential roles of poly-N-acetylglucosamine surface polysaccharide and extracellular DNA in *Staphylococcus aureus* and *Staphylococcus epidermidis* biofilms. Applied and environmental microbiology, 74: 470-476.
[59] Ryder C, Byrd M, Woznik D J (2007) Role of polysaccharides in *Pseudomonas aeruginosa* biofilm development. Current opinion in Microbiology, 10: 644-648.
[60] Sadovskaya I, Vinogradov E, Li J, Jabbouri S (2004) Structural elucidation of the extracellular and cell-wall teichoic acids of *Staphylococcus epidermidis* RP62A, a reference biofilm-positive strain. Carbohydrate research, 339: 1467-1473.
[61] Danese P N, Pratt L A, Kolter R (2000) Exopolysaccharide production is required for development of *Escherichia coli* K-12 biofilm architecture. Journal of bacteriology, 182: 3593-3596.
[62] Prigent-Combaret C, Prensier G, Le Thi T T, Vidal O, Lejeune P, Dorel C (2000) Developmental pathway for biofilm formation in curli-producing *Escherichia coli* strains: role of flagella, curli and colanic acid. Environ. Microbiol. 2, 450–464.
[63] Liikanen R, Yli-Kuivila J & Laukkanen R (2002) Efficiency of various chemical cleanings for nanofiltration membrane fouled by conventionally-treated surface water. J Membr Sci, 195: 265-276.
[64] Doumèche B, Galas L, Vaudry H, Di Martino P (2007) Membrane Foulants Characterization in a Drinking Water Production Unit. Food and Bioproducts Processing, 85, 42-48.
[65] Di Martino P, Doumeche B, Galas L Vaudry H, Heim V, Habarou H (2007). Assessing chemical cleaning of nanofiltration membranes in a drinking water production plant: a

combination of chemical composition analysis and fluorescence microscopy. Water Science & Technology. 55, 219–225.

[66] Houari A, Picard J, Habarou H, Galas L, Vaudry H, Heim V, Di Martino P (2008) Rheology of biofilms formed at the surface of NF membranes in a drinking water production unit. Biofouling. 24(4): 235-240.

[67] Houari A, Seyer D, Couquard F, Kecili K, Démocrate C, Heim V, Di Martino P (2010) Characterization of biofouling and cleaning efficiency of nanofiltration membranes. Biofouling. 26 (1): 15 - 21.

[68] Neu T R, Swerhone G D W, Lawrence J R, (2001), Assessment of lectin-binding analysis for in situ detection of glycoconjugates in biofilm systems. Microbiology, 147, 299-313.

[69] Lawrence J R, Chenier M R, Roy R, Beaumier D, Fortin N, Swerhone G D W, Neu T R, Greer C W (2004) Microscale and molecular assessment of impacts of nickel, nutrients, and oxygen level on structure and function of river biofilm communities. Appl. Environ. Microbiol., 70, 4326-4339.

[70] Christensen B E, Characklis W W (1990) Physical and chemical properties of biofilm. In: Characklis W.G., Marshall, K.C., editors. Biofilms. New York: Wiley, 93-130.

[71] de Beer D, and Stoodley P (1995) Relation between the structure of an aerobic biofilm and transport phenomena. Water Sci Technol 32(8): 11-18.

[72] Towler B W, Rupp C J, Cunningham Al B & Stoodley P (2003) Viscoelastic properties of a mixed culture biofilm from rheometer creep analysis. Biofouling 19: 279-285.

[73] Cancela M A, Alvarez E, and Maceiras R (2005) Effects of temperature and concentration on carboxymethylcellulose with sucrose rheology. J Food Engineer 71, 419-424.

[74] Kavanagh G N, Ross-Murphy S B (1998) Rheologicals characterisation of polymer gels. Prog Polymer Sci. 23: 533-562.

[75] Draget KI, Steinsvå K, Onsøyen E, Smidsrød O (1998) Na- and K-alginate; effect on Ca^{2+}-gelation *Carbohyd Polymers* 35: 1-6.

[76] Vinogradov A M, Winston M, Rupp C J & Stoodley P (2004) Rheology of biofilms formed from dental plaque pathogen *Streptococcus mutans*. Biofilms 1: 49-56.

[77] Schürks N, Wingender J, Flemming H C & Mayer C (2002) Monomer composition and sequence of alginates from *Pseudomonas aeruginosa*. Int J Biol Macromol 30: 105-111.

[78] Cyna B, Chagneau G, Bablon G and Tanghe N (2002) Two years of nanofiltration at the Méry-sur-Oise plant, France. Desalination 147: 69-75.

Section 3

Applications in the Food Industry

Chapter 14

Polysaccharide-Protein Interactions and Their Relevance in Food Colloids

Amit K. Ghosh and Prasun Bandyopadhyay

Additional information is available at the end of the chapter

http://dx.doi.org/10.5772/50561

1. Introduction

Polysaccharides and proteins are natural polymers that are widely used as functional ingredients for various food colloids or emulsion formulations. Majority of food emulsions are constituted with polysaccharide and protein combinations. They are the essential ingredients of any food colloid formulation mainly due to their ability to change product shelf life by varying food texture (Schmidt & Smith, 1992; Schorsch, Jones & Norton 1999). Their interaction in the formulation thus finds many applications particularly in new food formulation development. Due to complex formation and creation of nano or micro structures (aggregation and gelation behavior) they generally change the rheological properties of food colloids which may affect the food product texture and colloidal stability (Benichou, 2002; McClements, 2005, 2006, 2007; Dickinson, 2003). Polysaccharide and protein interactions in solution and interfaces have been studied by several groups (Dickinson, 2003, 2008; Bos &Van Vliet 2001; Carrera & Rodríguez Patino 2005; Krägel, Derkatch, & Miller, 2008; Koupantis, & Kiosseoglou, 2009; Mackie, 2009). However, despite the vast advancement made in the recent past, polysaccharide and protein interactions in food hydrocolloids continue to be one of the most challenging topics to understand.

Proteins, being surface active can play major role in the formation and stabilization of emulsions in the presence of polysaccharide, while interacting through electrostatic or hydrophobic-hydrophobic interactions. On the other hand, polysaccharides being hydrophilic in nature generally remain in aqueous phase thus help in controlling the aqueous phase rheology like thickening, gelling and acting as stabilizing agents. The formation and deformation of polysaccharide-protein complexes and their solubility depend on various factors like charge and nature of biopolymers, pH, ionic strength and temperature of the medium and even the presence of surfactant of the medium (Ghosh & Bandyopadhyay, 2011). If pH of the medium is reduced below isoelctric point (p*I*) of the

INTECH
open science | open minds

© 2012 The Author(s). Licensee InTech. This chapter is distributed under the terms of the Creative Commons Attribution License http://creativecommons.org/licenses/by/3.0), which permits unrestricted use, distribution, and reproduction in any medium, provided the original work is properly cited.

protein present then net positive charge of the protein will become prominent which will interact with negatively charged polysaccharide to form stable electrostatic complex. Similarly, if solution pH increased more than protein p*I*, the net negative charge of protein will tend to form complex with positively charged polysaccharides (Xia & Dubin, 1994; Dickinson, 2008; Turgeon, Schmitt & Sanchez, 2007). Generally, chances of weaker complex formation is more when solution pH is almost equal to protein p*I* , because at that pH range surface charge of protein becomes nearly zero. However, at very high concentration, similarly charged biopolymers repel each other and the net repulsion make the system unstable (separate as two distinct phases) which is known as *thermodynamic incompatibility*. Incompatibility in the system occurs at pH higher than the protein p*I* and at higher ionic strength (Grinberg & Tolstoguzov, 1997). Thus by varying pH and ionic strength of the medium one can achieve a control on the polysaccharide-protein interactions.

Polysaccharides and proteins both contribute to the structural and textural properties of food by changing rheology of food emulsions through their gelling networking system (Dickinson, 1992). Non-covalent interactions between polysaccharide and protein in any emulsion formulation play a major role to change the interfacial behavior and stability of the food colloids. The driving force for these non-covalent interactions is electrostatic interactions, hydrophobic interactions, H-bonding and Van der Waals interactions. Recent literatures also focus on how protein and polysaccharide molecules can be linked together by covalent bond. At pH close to protein p*I* this Maillard-type conjugates were used to improve the colloidal stability and interfacial structure of proteins in certain conditions (Jiménez-Castaño, Villamiel, & López-Fandiño, 2007; Benichou, Aserin, Lutz & Garti, 2007)). Recent developments in the field describe interfacial physico-chemical properties of polysaccharide-protein mixed systems (Rodríguez Patino & Pilosof 2011). In this chapter, we would like to focus more on polysaccharide and protein non-covalent interaction studies and their effect towards food colloids stability.

2. Nature of polysaccharide-protein complex

Polysaccharide and protein complex formation is mainly driven by various non-covalent interactions, like electrostatic, H-bonding, hydrophobic, and steric interactions (Kruif et al 2001). Protein carries +ve or –ve zeta potential based on the pH of the medium (+ve at pH lower than p*I* and vice-versa). This +ve or –ve electrical charge on the protein chain point towards the presence of different amino acids in the protein molecules and their mode of ionization at different pH ranges (Fig. 1). Carboxylate polysaccharides get deprotonated (become anionic) at a pH range higher than its pKa (Fig. 1). This electrical charge on the back bone of protein or polysaccharide chain is responsible for electrostatic attraction or repulsion between them. Again, presence of -COOH group on the polysaccharide and $-NH_3$, -COOH groups on the protein chain are the sources of hydrogen bonding between these two bio-polymers. Extent of both of this hydrogen bonding and electrostatic interaction depends on the solution parameters such as pH, ionic strength, temperature etc. Except these ionic patches on the bio-polymers, few non-polar segments are also present on the bio-polymers, which are responsible for the hydrophobic staking with each other. Even though solution parameters are important factors to control the different mode of interactions between protein and polysaccharide, type of proteins/polysaccharides, molecular weight, charge density, and hydrophobicity of the bio-polymers are also play significant role towards the extent of complexation between two bio-polymers at a fixed condition.

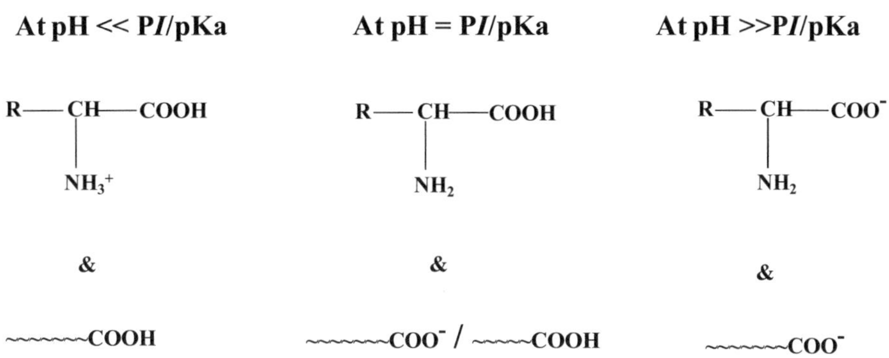

Figure 1. Variation of charge density on the polysaccharide and protein chain at various pH ranges.

In general, interactions between proteins and polysaccharides are quite explored where large numbers of report have been published based on the interactions between oppositely charged "protein-polysaccharide" systems (Dmitrochenko et al 1989; Bengoechea et al 2011, Stone & Nickerson 2012). Although electrostatic attraction is the main driving force for the complexation between protein and polysaccharide, but it is also reported that hydrogen bonding and hydrophobic interaction plays a secondary role for stability of the "protein-polysaccharide" aggregates (McClements, 2006). The extent of hydrogen bonding and hydrophobic interaction also depends on temperature (Weinbreck et al, 2004). In 2009 Nickerson and co-workers(Liu, Low, & Nickerson, 2009) have reported that pea protein and gum acacia complex stabilize at low temperature due to increase in hydrogen bonding interactions and destabilize at high temperature due to decline in hydrogen bonding interactions. Temperature also plays an important role to decide the protein conformations (folded or unfolded). In 2007, Pal (Mitra, Sinha & Pal, 2007) and coworkers have reported that human serum albumin unfolds at higher temperature and undergoes in reversible refolding conformations upon cooling (below 60^0 c). Unfolded conformations of protein expose more reactive sites (amino acids) to the solvent phase, thus more chances of interactions (or binding) with polysaccharide. Binding of anionic polysaccharides ($pH \sim pK_a$) to the cationic proteins (at $pH<pI$) result both soluble and insoluble complexes (Magnusson & Nilsson, 2011). Initial binding of polysaccharides (anionic) to the proteins (cationic) cause charge neutralizations, which lead to the formation of insoluble "protein-polysaccharide" aggregates (Schmitt et al, 1998). Further binding of anionic polysaccharides to those neutral aggregates make it effectively anionic, which leads to formation of soluble complexes. But binding of anionic polysaccharides with anionic proteins ($pH>pI$) are also known and governed by the interactions between anionic reactive sites of polysaccharide and small cationic reactive sites of protein (Fig. 2). Binding of anionic polysaccharides to the cationic side of proteins (at pH>p*I*) result in formation of anionic "protein-polysaccharide" aggregates, thus soluble complexes. Therefore, concentration of polysaccharides and pH play an important role towards the solubility of "protein-polysaccharide" aggregates.

Two bio-polymers can exist either in a single phase systems or in a phase separated systems depending on the nature of bio-polymers, their concentration, and solution conditions. When two bio-polymers carry opposite charge, then either they agglomerates to form soluble complexes (single phase) or insoluble precipitates (2-phase system). On the other hand, when two non-interacting bio-polymers mixed together, either they exist in a single phase system (where two separate entities distributes uniformly throughout the medium) or exist as two distinct phases (each phases comprise different bio-polymer). Therefore, in the protein-polysaccharide system, phase separation occurs through two different mechanisms which are *associative phase separation* and *segregative phase separation* (Tolstoguzov, 2006). *Associative phase separation* is the aggregation between two oppositely charged bio-polymers (electrostatic attraction driven), leads to the phase separation, where one phase is enriched with two different bio-polymers (coacervation or precipitation) (Fig. 3). *Segregative phase separation* occurs either due to strong electrostatic repulsion (between two similarly charged bio-polymers) or because of very high steric exclusion (between two neutral bio-polymers). In this case, at low concentration, two biopolymers can co-exist in a single phase whereas at higher concentration, it starts phase separation. (Fig. 3).

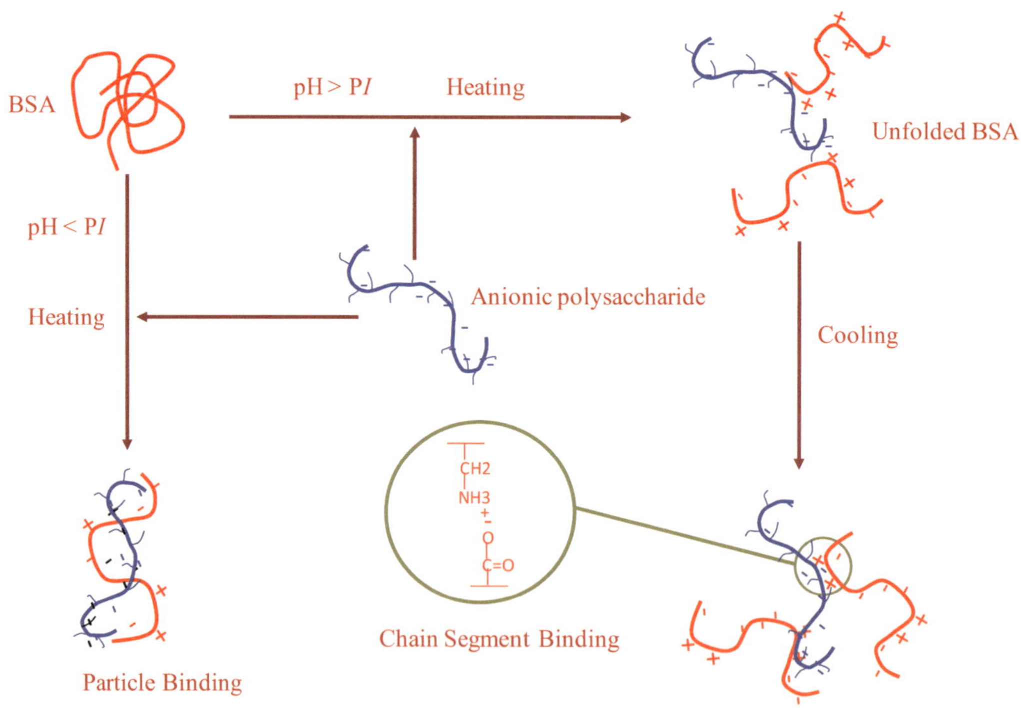

Figure 2. Interaction between polysaccharide and protein at various pH.

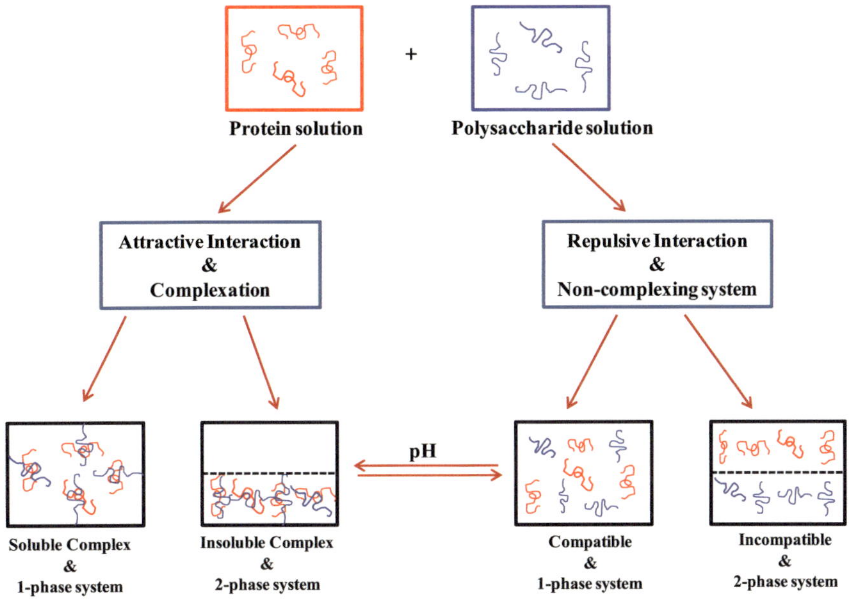

Figure 3. Schematic representation of the possible mode of interaction between polysaccharides and proteins.

3. Functional properties of polysaccharide-protein complexes related to food application

Polysaccharide-protein complexes exhibit a wide range of interesting properties, such as surface activity to stabilize air-water or oil-water interfaces, viscosifying, and gelling properties etc. Viscosifying and gelling ability of polysaccharide-protein complexes help to obtain gel-like processed food products without any thermal treatment in the process. The interfacial properties of these complexes help to impart stability into the emulsion food products. Also, protein–polysaccharide complexes are able to encapsulate several active ingredients; hence they act as delivery systems for many bioactives or sensitive molecules in food formulations. These complexes are also known to vary bulk/interfacial structures, textures and shelf-life stability of the food colloids. In the following section we will discuss this polysaccharide-protein interaction in light of their functional properties.

3.1. Viscosity of polysaccharide-protein complex and air-water foam stability

Viscosity and gelling are the rheological property which depends on the molecular characteristics of the biopolymers, such as their molecular weight, shape, chain flexibility. Other factors are the concentration, interaction between the biopolymers and water, as well as solution parameters like: pH, ionic strength and presence of other components/ligands etc. Interactions between polysaccharide and protein have been proven to widen the functional properties of each individual biopolymer. Rheological properties of polysaccharide-protein complex lead to new rheological behaviors different from each individual biopolymer.

Association of two bio-polymers is expected to increase the bulk viscosity of the system as entities of larger sizes are formed. The rheological behavior of several protein–polysaccharide mixed systems have been studied and ranged from viscous to viscoelastic properties showing elastic behavior is reported. The hydrated polysaccharide-protein complexes increase viscosity and rheology of the system found to be depends on the nature and structure of polysaccharides. Viscous property of gum acacia-protein coacervates attributes to the globular conformation of the polysaccharide, whereas the same protein with linear pectin results in gel-like system. Beside the nature of the individual bio-polymer solution property and concentration of bio-polymers are also known to affect the rheology of the system (Dickinson, 2011). For example, it was found that pH played a major role in the viscosity of the coacervate phase. A maximum viscosity was obtained at pH 4.0, where concentration of whey protein and gum arabic in the coacervate phase was maximum and extent of electrostatic attraction was highest. This suggests that the electrostatic interactions between whey protein and gum arabic were responsible for the highly viscous behavior of the coacervates. Whereas, the same composition of whey protein and gum arabic at pH above protein p*I* (i.e. comparatively lower electrostatic interactions) showed more elastic nature than viscous. Ionic strength and protein/polysaccharide ratio is also known to play an important role towards the rheology of polysaccharide-protein systems. For example, optimal salt concentrations (0.21 M NaCl) favor the coacervation of β-lactoglobulin with pectin at higher concentration and produce much stronger gel strength. For better gelling property, it is necessary to control the parameters which required to form coacervate, because strong associative interaction decrease the solubility of complexes and hence lower the hydration capacity of the complex, which leads to decrease in the viscosity (Schmitt & Turgeon, 2011; Kruif, et al 2004).

Viscoelastic properties of polysaccharide-protein complexes also play an important role towards the foam stability in variety of food products. In case of air-water system foam can be define as the air entrapment by a thin liquid film (water), where this liquid film is stabilized by some surface active molecules. Stability of the foam increases with the increase in the stability of the interfacial liquid film, because lower stability of this interfacial liquid film can lead to the diffusion of air entrapped inside the foam. Viscosity of this liquid film is another parameter by which one can control the diffusion rate of air entrapped inside the foam. Therefore, higher stability and viscosity of the interfacial liquid film leads to lower diffusion of air entrapped inside the foam and increase the foam stability. Schmitt and co-workers have studied the air-water interfacial property of β-lactoglobulin-acacia gum complexes at pH 4.2 [Schmitt et al 2005]. The group has reported that although surface activity of the complex is similar with the pure protein, but complex forms much stronger viscoelastic interfacial film with thickness of about 250 Å. As a result, gas permeability of thin-film stabilized by the complexes was significantly reduced (0.021 cm s^{-1}) compared to pure β-lactoglobulin (0.521 cm s^{-1}). This phenomenon suggests that stability of foam (stabilized by protein-polysaccharide complex) is higher compared to the foam stabilized by protein alone.

The likely explanation of the higher foam stability and different interfacial properties of coacervate is that protein-polysccharide complexes are able to re-organize at the interface by coalescence, forming interfacial microgel. These findings were applied for the ice cream formulation for improved air bubble stability (Schmitt C, Kolodziejczyk E. 2010). Similarly, gelatin has been replaced by whey protein isolate-gum acacia complexes to improve the bubble stability in chilled dairy products (Schmitt C, Kolodziejczyk E. 2010). In case of stabilization by complexes, variation in ratio of biopolymers could be used to control the size of the complexes, hence their surface activity. In addition to that, viscoelastic properties of the air-water interfacial film is possible to tune by either adsorbing two biopolymers simultaneously or by the sequential adsorption of protein followed by polysaccharide. As for example, β-lactoglobulin-pectin complexes are known to stabilize the air-water interface. In this case, thickness of the film obtained from the sequential adsorption of protein and polysaccharide was higher (450 Å) than the adsorption of complexes (250 Å) (Ganzelves et al 2008).

In contrary to air-water foam stability, use of polysaccharide-protein complexes for the stabilization of oil-water emulsion (Martínez et al 2007) has received much more attention. Use of these polysaccharide-protein hydrocolloids as an emulsion stabilizer will be discussed in the next section.

3.2. Oil-water emulsion stability

Emulsion is a uniform dispersion of liquid droplets within a continuous matrix of a second immiscible liquid, stabilized by surface active molecules. These stabilizers are termed as emulsifier. In the context of the present topic, we will limit our discussion to the role of bio-polymers as emulsifier. Generally, emulsifier has the amphiphilic character to adsorb onto the interface of liquid droplets, which can prevent the phase separation of two immiscible liquids. For a fixed emulsifier, stability of the emulsion depends on few factors, such as rate of adsorption of the emulsifier, concentration of emulsifier, etc. At low concentration of emulsifiers, emulsion system fails to retain its initial droplet size. This destabilization can take place through different mechanisms. In case of poor coverage of the interface by liquid droplets, they can coalesce with each other to form a bigger droplet (Fig. 4). Few examples are also reported, where polymer adsorbed onto the interface of liquid droplets thus bridge between two such liquid droplets and initiates bridging flocculation. Interestingly, emulsions at high emulsifier concentration produces stable oil droplets due to better coverage of the interfaces of the liquid droplets (Liu & Zhao, 2011).

Emulsion is possible to achieve by using many surface active agents, such as small surfactant molecules, bio-polymers (proteins or polysaccharides, hydrocolloids (protein-polysaccharide complexes), and inorganic particles. Stability of those emulsion systems mainly governs by the two important factors. First, repulsive force between two closely approaching liquid droplets; second, Ostwald ripening, which involves disappearing of smaller droplets in expense of the formation of larger one. Higher degree of repulsion between the two neighboring droplets results in maximum stability due to least chances of coalescence. Repulsive force between the two liquid droplets govern by the inter droplet distance, i.e. thickness of the thin liquid film between two closely approaching droplets. Thickness of this liquid film depends on the space occupied by the adsorbed molecules (emulsifier) at interface of the droplets. Emulsion generally get stabilized by different emulsifiers present in the formulations such as surfactants, proteins, or hydrocolloids (protein-polysaccharide complexes) and the relative thickness of the liquid film between two closely spaced droplets lies in the order of hydrocolloids (5-10 nm) > proteins (1-5 nm) > surfactants (0.5-1 nm) (Fig. 5). Therefore, stability of the emulsion droplets expected to be higher when they are stabilized by protein-polysaccharide complex compared to the same stabilized by protein or surfactant molecules. In addition to the thickness of the liquid film between two closely spaced droplets, rate of desorption of the emulsifiers from the interface is another important factor. Adsorption of emulsifier molecules (like surfactants, proteins etc.) at the liquid interface is highly reversible. Desorption of the emulsifiers from the liquid interface governs the instability of the system. According to this fact, emulsion stabilized by particles (size ranges from 10 nm to several μm) is likely to have indefinite stability, because of the maximum thickness of thin liquid film in between two closely spaced droplets and maximum desorption energy of the particles from the liquid interface. Despite of this theoretical consideration, experimental evidence by Tcholakova et al. does not support this hypothesis that particle stabilized emulsion are more stable compared to surfactant or protein or hydrocolloids stabilized emulsion (Tcholakova et al. 2008).

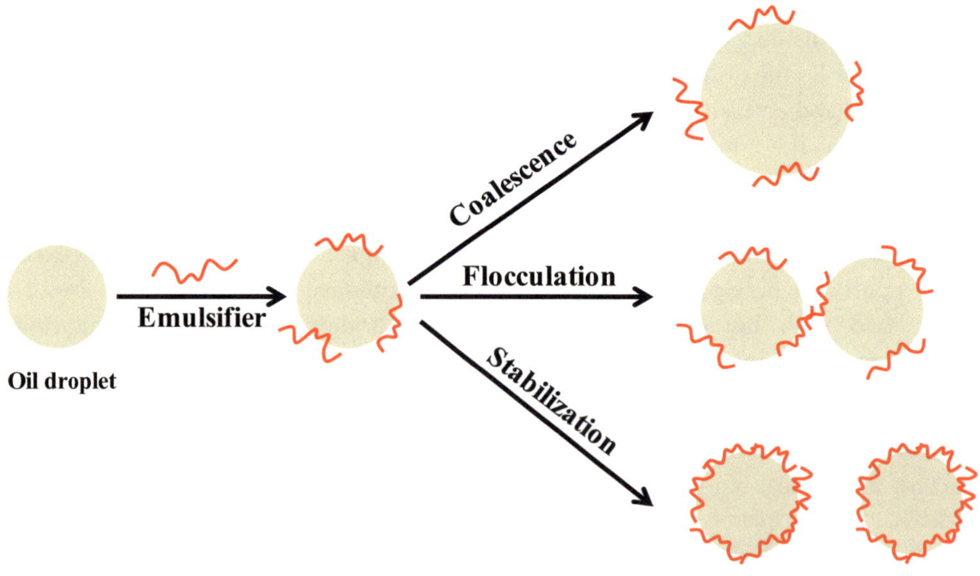

Figure 4. Schematic representation of mode of stabilization and destabilization of oil droplets in an oil-water emulsion.

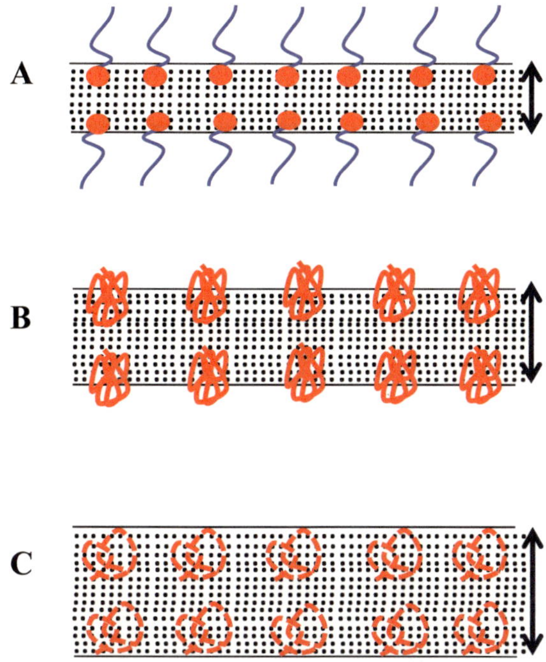

Figure 5. Schematic representation of the relative thickness of the thin liquid film between two closely spaced droplets, stabilized by A) surfactant, B) protein, and C) protein-polysaccharide hydrocolloids.

Another factor guides the emulsion instability is Ostwald ripening, which is disappearance of small size droplets at the expense of the larger droplets formation. Driving force for the Ostwald ripening is the difference in the chemical potential of the smaller and larger droplets. Mass transfer takes place between the droplets by diffusion process. Therefore, Ostwald ripening process requires the solubility of the dispersed phase into the continuous phase to initiate the diffusion process. Type of the emulsifier also plays an important role towards the Ostwald ripening process. Emulsion stabilized by the water soluble surfactant molecules has lower interfacial tension, which reduces the thermodynamic driving force of Ostwald ripening. Adsorption of the surfactant molecules at the droplet interface is a reversible phenomenon. Reversible desorption and adsorption of the surfactant molecules from the interface of the liquid droplet increases the rate of mass transfer between the dispersed droplets, hence increase the Ostwald ripening. The chances of desorption of emulsifier is less in case of the emulsions stabilized by protein molecule because it provides a thicker layer (elastic layer) around the droplets and greater surface coverage of the interfacial area. These factors reduce the ripening process in the protein stabilized emulsion. Ostwald ripening process is possible to avoid completely, only if emulsion is stabilized by insoluble particles (due to very high desorption energy) or thickness of the elastic layer around the dispersed droplets is equal to the droplet radius (Kabalnov, 2001). For this reason particulate emulsions are able to prevent the ripening process completely. Whereas hydrocolloid (protein-polysaccharide complexes) mostly behaves like a soft polymer, more resembles with the protein structures compared to solid particles, which cannot completely avoid the ripening process.

Protein–polysaccharide complex stabilized emulsions are possible to obtain by using two alternative ways. One of them involves addition of charged polysaccharide solution to a primary emulsion which is already stabilized by the protein as single emulsifier, to produce emulsion droplets having a protein–polysaccharide 'bilayer' surface coating (Fig. 6B). Another method involves addition of an aqueous solution containing the protein–polysaccharide complexes as an emulsifying agent following homogenization (Fig. 6A). For convenience, the first method is termed as 'bilayer emulsions' and second one is termed as 'mixed emulsions'. The bilayer approach is also commonly known as 'layer-by-layer' approach. Recently, it has attracted significant importance because of its use in nano-encapsulation and protection of emulsions against severe environmental stresses. The major problem lies in 'layer-by-layer' approach, where emulsion droplet tends to flocculate. Flocculation during the 'layer-by-layer' adsorption takes place because of two different mechanisms: a) bridging flocculation, b) depletion flocculation. *Bridging flocculation* takes place at low polysaccharide concentration when droplet collisions occur faster than the rate of polysaccharide saturation of the protein-coated droplet surfaces. *Depletion flocculation* occurs at higher polysaccharide concentration when unadsorbed polysaccharide exceeds a critical value. For this reason it is convenient to make emulsions with protein and polysaccharide present together before homogenization compared to 'layer-by-layer' approach. Recently direct comparison between the two techniques has been demonstrated experimentally, which shows that the more convenient mixed emulsion approach leads to better stability behavior than the bi-layer approach(Camino et al 2011).

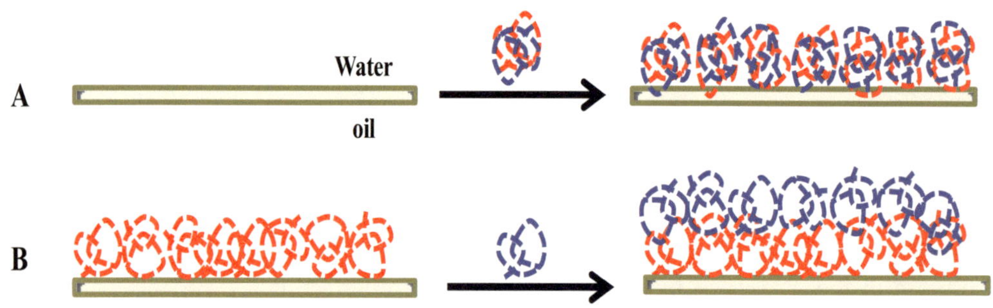

Figure 6. Schematic illustration of two alternative approach of preparing oil-water emulsion using protein-polysaccharide complexes as an emulsifier. A) Mixed emulsions, both protein and polysaccharide present together during emulsification. B) Bilayer emulsion, polysaccharide added to emulsion prior stabilized by protein.

3.3. Encapulation and release of active molecules

Generally, encapsulation includes all aspects of protection or stabilization of active molecules (flavors and bio-actives) against several external drastic conditions (such as heat, redox potential, shear, temperature, light, oxygen, moisture, etc.). Controlled release facilitates the delivery of the encapsulated material to the targeted place with the optimal kinetics. Conditions for the encapsulation of active molecules depends on the sensitivity (thermal and redox stability) and nature (solubility in oil and water) of the active components but release can be controlled by mechanical process, pH variations (acidic conditions in the stomach, neutral in the intestine) or enzymatic actions etc. As for example, Peniche et al. has shown the encapsulation of shark liver oil (rich in poly unsaturated fatty acids) using chitosan-alginate system to mask the unpleasant taste of oil. These capsules disrupts by enzymes, like lipase or pancreatine. Initially it was resistant to the acid environment of the stomach, but after 4 hour in the intestinal pH (pH 7.4), the capsule walls weakened and delivers the active molecules.

Another important application of this aspect is the encapsulation of flavor molecules. Recently, Yeo et al. has shown that gelatin–acacia gum coacervate can encapsulate flavors which can be released during cooking in baked goods (Yeo et al. 2005). Weinbreck et al. (Weinbreck et al. 2004) has shown that Whey protein–acacia gum coacervates can encapsulate lemon and orange flavors and their release under mechanical action like chewing. Encapsulation was one of the first applications of gum arabic-gelatin coacervates (Bungenberg de Jong and Kruyt, 1929). Viscous coacervate was made at a temperature higher than the gel point of gelatin and during cooling, interfacial rigidity increases which lead to a stable gelled shell around the microcapsule. This rigid shell disrupts after consumption, gelatin melts easily in the mouth and therefore releases the encapsulated actives. In addition to gelatin-acacia gum, several other protein-polysaccharide systems have been evolved (whey proteins, plant proteins, pectin, and xanthan gum) to broaden the encapsulation techniques. Beside these polysaccharide-protein combinations, process parameter for encapsulation (pH, ionic strength, macromolecular ratio, and drying/homogenization procedure) also plays an important role to modulate the physical properties (thickness, swelling rate, etc.) of the coacervate layer in the microcapsules (Savary et al 2010). Use of cross-linking agents can further harden the coacervate layer after formation of the microcapsules. As for example, use of trans-glutaminase can introduce covalent linkages between carboxyl group of a glutamine and amino group of lysine in the protein molecules. Alternatively, formaldehydes and glutaraldehydes have also been studied although they are non-food grade reagents. Recently, tannic acid, plant phenolics, citral molecules and glycerin have been studied as food grade alternatives (McClements, 2010).

Contrary to the encapsulation through coacervation, the bi-layer emulsion technique (formed by successive adsorption of biopolymers at the interface) is another way to study the microcapsule properties. Recently, McClements group has described that bi-layer approach of encapsulation has a better control of the interface structure, charge, thickness and permeability with improved stability and controlled release of actives. Group has reviewed this research area and discussed multilayer emulsions in light of bioavailability control and release of actives to the specific site of action depending on layer composition and properties (McClements, 2011). Sagis et al. has used high molecular weight pectins and pre-heated whey proteins (denaturation of protein) for the encapsulation through multilayer approach. (Sagis et al 2011).

4. Summary

Basic understanding of supramolecular chemistry which span among origin and nature of the various non-covalent molecular interactions between polysaccharides and proteins can be widely used to create various desirable nano and macro structures which are quite significant in food colloids/formulations. Food product texture modulation and colloidal stability can be achieved by controlling protein-polysaccharide interactions. Modulation of this interaction by varying medium conditions like pH, ionic strength etc. one can create many possibilities towards rheological properties of food colloids which may affect the emulsion stability. Protein-polysaccharide interactions are well characterized in various pH conditions. Although, the interaction depends on type and nature of biopolymers, no structure-activity correlation has been established until now. Moving forward, there is a huge demand to establish a correlation between biopolymer structures and interaction efficiency. The creative manipulation of polysaccharide-protein interactions can open up a completely new dimension in health and nutrition platform. Food particle travels from mouth to gut in various pH environments (for example pH decreases when it moves from mouth saliva to stomach and increases when partially digested food particle passes from stomach to small intestine), thus one can design a smart polysaccharide-protein complex system which can encapsulate or slow down and trigger or release of nutrients in various stages of digestion process depending on the pH of the system and any particular health demand. Such kind of pH sensitive system design and product development which works in vivo is a real challenge for the food scientists. Unfortunately, general understanding of protein-polysaccharide interactions, their bulk and interfacial properties toward emulsion stability is not completely understood and requires more systematic investigation in future to unveil its full potential.

Author details

Amit K. Ghosh and Prasun Bandyopadhyay*
Unilever R & D Bangalore, Whitefield, Bangalore, India

5. References

Bengoechea, C., Jones, O. G., Guerreo, A., & McClements, D. J. *Food Hydrocolloid*, Vol. 25, 2011, pp. 1227-1232

Benichou, A., Aserin, A. & Garti, N. *Journal of Dispersion Science and Technology*. Vol. 23, 2002, pp. 93-123

Benichou, A., Aserin, A., Lutz, R. & Garti, N. *Food Hydrocolloid*, Vol. 21, 2007, pp. 379-391

Bos, M. A, & Van Vilet, T. *Advances in Colloid & Interface Science*. Vol. 91, 2001, 437-471

Bungenberg, H. G. & Kruyt, H. R. Procedings K. Ned. Akad. Wet. Vol. 32, 1929, pp. 849-856

* Corresponding Author

Camino, N. A., Sánchez, C. C., Rodríguez Patino, J. M. & Pilosof, A. M. R. *Food Hydrocolloid,* Vol. 25, 2011, pp. 1-11

Carrera, C. & Rodríguez Patino, J. M. *Food Hydrocolloid.* Vol. 19, 2005, pp. 407-416

Choi, S. J., Decker, E. A., Henson, L., Popplewell, L. M. & McClements, D. J. *Food Chemistry* Vol. 122, 2010, pp. 111-116

Dickinson, E. *An introduction to food colloids.* 1992, Oxford University Press

Dickinson, E. *Colloid & Surfaces B: Biointerfaces.* Vol. 20, 2001, pp. 197-210

Dickinson, E. *Food Hydrocolloid,* Vol. 25, 2011, pp. 1966-1983

Dickinson, E. *Food Hydrocolloid.* Vol. 17, 2003, pp. 25-40

Dickinson, E. *Soft Matter.* Vol. 4, 2008, pp. 932-942

Dmitrochenko, A.P., Antoonov, Yu. A., & Tolseoguzov, V. B. *Applied Biochemistry and Microbiology,* Vol 25, 1989, 353-360

Ganzelves, R. A., Fokkink, R., van Vilet, T.,Cohen Stuart, M. A. de Jongh, H.H.J. *Journal of Colloid and Interface Science* Vol. 317, 2008, 137-140

Ghosh, A, & Bandyopadhyay, P. *Chemical Communications.* Vol. 47, 2011, pp. 8937-8939

Grinberg, V. Y. & Toolstoguzov, V. B. *Food Hydrocolloid.* Vol. 11, 1997, pp. 145-158

Jiménez-Castaño, L., Villamiel, M & López-Fandiño, R. *Food Hydrocolloid,* Vol. 21, 2007, pp. 433-443

Kabalnov, A. *Journal Dispersion Science Technology* Vol. 22, 2002, pp. 1-12

Koupantis, T. & Kiosseoglou, V. *Food Hydrocolloid.* Vol. 23, 2009, pp. 1156-1163

Krägel, J. Derkatch, S. R. & Miller, R. *Advances in Colloid & Interface Science.* Vol. 144, 2008, 38-53

Kruif, de C. G. & Tuinier, R. *Food Hydrocolloid,* Vol. 15, 2001, pp. 555-563

Kruif, de C. G., Weinbreck, F. & Vries, de R. *Current Opinion in Colloid & Interface Science.* Vol. 9, 2004, 340-349

Liu, L., Zhao, Q., Liu, T. & Zhao, M. *Food Hydrocolloid,* Vol. 25, 2011, pp. 921-927

Liu, S., Low, N. H., & Nickerson, M. T. *Journal of Agricultural and Food Chemistry,* Vol. 57, 2009, pp. 1521-1526

Mackie, A. *Current Opinion in Colloid & Interface Science.* Vol. 9, 2009, 357-361

Magnusson, E., & Nilsson, L. *Food Hydrocolloid,* Vol. 25, 2011, pp. 764-772

Martínez, K. D., Sánchez, C. C., Ruíz-Henestrosa, V. P. Rodríguez Patino, J. M. & Pilosof, A. M. R. *Food Hydrocolloid,* Vol. 21, 2007, pp. 813-822

McClements, D. J. *Biotechnology Advances,* Vol. 24, 2006, pp. 621-625

McClements, D. J. *Biotechnology Advances,* Vol. 24, 2006, pp. 621-625

McClements, D. J. *Food Emulsions.* (2nd Ed), 2005, Boca Raton, FL: CRC Press

McClements, D. J. *Soft Matter.* Vol. 7, 2011, pp. 2297-2316

McClements, D. J. *Understanding and controlling the microstructure of complex foods.* Abington: woodhead Publishing 2007

Mei, L. Y., Choi, S. J., Alamed, J., Henson, L., Popplewell, L. M. & McClements, D. J. *Journal of Agricultural and Food Chemistry* Vol. 57, 2009, pp. 11349-11353

Mitra, R. K., Sinha, S. S. & Pal, S. K. *Langmuir,* Vol. 23, 2007, pp. 10224-10229

Peniche, C., Howland, I., Carrillo, O., Zaldı´var, C., Argqelles-Monal, W., *Food Hydrocoll.* 18 2004, 865– 871.

Rodríguez Patino J. M., & Pilosof A. M. R. *Food Hydrocolloid,* Vol. 25, 2011, pp. 1925-1937

Salgam, D., Venema, P., Vries, de R. J., Sagis, L. M. C., Linden, E. van der. *Food Hydrocolloid,* Vol. 25, 2011, pp. 1139-1148

Savary, G., Hucher, N., Bernadi, E., Grisel, M. & Malhiac, C. *Food Hydrocolloid,* Vol. 24, 2010, pp. 178-183

Schmidt, K. A & Smith, D. E. *Journal of Dairy Science.* Vol.5, 1992, pp. 36-42

Schmitt C, Kolodziejczyk E. In: Williams PA, Phillips GO, editors. Gums and stabilizers for the food industry, 15. Royal Society of Chemistry; 2010. p. 211–21

Schmitt, C. & Turgeon, S. L. *Advances in Colloid & Interface Science.* Vol. 167, 2011, 63-70

Schmitt, C., Kolodziejczyk E., Lesser, M. E. *Food Colloids: interactions, microstructure and processing,* RSC, 2005, pp. 284-300

Schmitt, C., Sanchez, C., Desobry-Banon, S., & Hardy, J. *Critical Reviews in Food Science and Nutrition,* Vol. 38, 1998, pp. 689-753

Schorsch, C., Jones, M. G. & Norton, I. T. *Food Hydrocolloid.* Vol. 13, 1999, pp. 89-99

Stone, A. K., Nickerson, M. T. *Food Hydrocolloid,* Vol. 27, 2012, pp. 271-277

Tcholakova, S., Denkov, N. D., & Lips, A. (2008). *Physical Chemistry Chemical Physics,* 10, 1608–1627

Tolstoguzov, V. *Food Polysaccharides and their Application,* Taylor & Francis, 2006, pp. 589-627

Turgeon, S. L., Schmitt, C., & Sanchez, C. *Current Opinion in Colloid & Interface Science.* Vol. 12, 2007, 166-178

Weinbreck, F., Minor, M., de Kruif, C., G., *J. Microencapsul.,* 21, 2004, 667

Weinbreck, F., Nieuwenhuijse, H., Robijn, G. W, de Kruif, C. G. *Agricultural and Food Chemistry,* Vol. 52, 2004, pp. 3550-3555

Xia, J, & Dubin, P. L. *Macromolecular Complexes in Chemistry and Biology,* Springer-Verlag, Berlin, 1994, pp. 247

Yeo, Y., Bellas, E., Firestone, W., Langer, R., Kohane, D. S., *J Agric Food Chem.,* 53, 2005, 7518

Chapter 15

Chitosan: A Bioactive Polysaccharide in Marine-Based Foods

Alireza Alishahi

Additional information is available at the end of the chapter

http://dx.doi.org/10.5772/48199

1. Introduction

Since people are increasingly conscious of the relationship between diet and health, the consumption of marine-based foods has been growing continuously. Consumers identified seafoods as nutritious and complete foods. Hence, they are perceived them as an excellent source of high quality proteins, valuable lipids with high amounts of PUFA. These compounds are well known to contribute to the enhancement of human health by different alternatives such as reducing the risk of cardiovascular disease, coronary disease and hypertension. Additionally, marine-based food products are easily digested and constitute excellent source of essential minerals. Recently, seafoods have been recognized as nutraceuticals or functional foods. Functional foods, first evolved in Japan in 1980, are defined as foods demonstrating beneficial effect on one or more targeted functions on human organism (Ross, 2000). Marine-based functional foods or nutraceuticals, include omega-3 fatty acids, chitin and chitosan, fish protein hydrolysates, algal constituents, carotenoids, antioxidants, fish bone, shark cartilage, taurine and bioactive compounds (Kadam & Prabhasankar, 2010).

Despite the aforementioned desirable properties, seafood products are highly susceptible to quality deterioration, mainly due to the lipid oxidative reactions, particularly involving polyunsaturated fatty acids (PUFAs). These reactions are enhanced (catalyzed) by the presence of high concentrations of heme and non-heme proteins. These proteins are known to contain iron and other metal ions in their structures (Decker & Hultin, 1992). Moreover, marine-based food quality is highly influenced by autolysis, bacterial contamination and loss of protein functionality (Jeon, Kamil & Shahidi, 2002). More recently, pollution of seafood with different hazardous materials such as refinery, industrial wastes and heavy metals has resulted in elevated concern about the consumption of seafood (Kadam & Prabhasankar, 2010). Additionally, aquaculture industry has increasingly attracted much

INTECH
open science | open minds

© 2012 The Author(s). Licensee InTech. This chapter is distributed under the terms of the Creative Commons Attribution License http://creativecommons.org/licenses/by/3.0), which permits unrestricted use, distribution, and reproduction in any medium, provided the original work is properly cited.

attention for the intensive farming of fish and shellfish, mainly due to the depleting of wild fish and shellfish stocks worldwide. However, this intensive farming entails several difficulties such as stress, which is the most important factor affecting the immunity system of fish (Ledger, Tucker & Walker, 2002). To address the aforementioned problems, the use of chitosan as protective material appears to be a potential alternative.

In nature, chitosan is found in the cell walls of fungi of the class *Zygomycetes*, in the green algae *Chlorella sp.*, yeast and protozoa as well as insect cuticles and especially in the exoskeleton of crustaceans. Chitosan is a deacetylated derivative of chitin, the second abundant polysaccharide in nature after cellulose. In 1811, the French scientist, Henri Braconnot first discovered chitin in mushroom. In 1820, chitin was isolated from insect cuticles (Bhatnagar & Sillanpa, 2009). In 1859, Rouget reported finding chitosan after boiling chitin in potassium hydroxide (KOH). This treatment rendered the material soluble in organic acids. Hoppe-Seyler named it chitosan in 1894 (Khor, 2001). Chemically, chitosan is a high molecular weight, linear, polycationic heteropolysaccharide consisting of two monosaccharides: N-acetyl-glucosamine and D-glucosamine. They are linked by β-(1→4) glycosidic bonds. The relative amount of these two monosaccharides in chitosan vary considerably, yielding chitosans of different degrees of deacetylation varying from 75% to 95%, molecular weight in the range of 50-2000 KDa, different viscosities and pKa values (Tharanathan & Kittur, 2003). In addition, chitosan has three functional moieties on its backbone; the amino group on the C2, the primary and secondary hydroxyl groups on the C3 and C6 positions, respectively. These functional groups play important roles in different functionalities of chitosan. The amino group is the most important among the other moieties, especially in acidic conditions, due to the protonation phenomenon, rendering it able to interact with negatively charged molecules (or sites). Additionally, chitosan polymer interacts with the metal cations through the amino groups, hydroxyl ions and coordination bonds.

Commercially, chitosan is produced from chitin by exhaustive alkaline deacetylation, involving boiling chitin in concentrated alkali for several hours. Because this N-deacetylation is almost never complete, chitosan is classified as a partially N-deacetylated derivative of chitin (Kumar, 2000). From a practical point of view, many of commercial interests and applications of chitosan and its derivatives originate from the fact that this polymer combines several features, such as biocompatibility, biodegradability, nontoxicity and bioadhesion, making it as valuable compound for pharmaceutical (Dias, Queiroz, Nascimento & Lima, 2008), cosmetics (Pittermann, Horner & Wachter, 1997), medical (Carlson, Taffs, Davison & Steward, 2008), food (Shahidi, Kamil & Jeon, 1999; No, Meyers, Prinyawiwatkul & Xu, 2007; Kumar, 2000), textile (El Tahlawy, Bendary, El Henhawy & Hudson, 2005), waste water treatment (Che & Cheng, 2006), paper finishing, photographic paper (Kumar, 2000), and agricultural applications (Hirano, 1996).

Although there have been several prior reviews on the use of chitosan in food applications (No et al., 2007; Shahidi et al., 1999), the use of chitosan in seafood applications, especially its novel application in the form of nanocarriers for bioactive compounds for shelf life extention, has not yet been reported. Recently, a study was published on the use of chitosan nanoparticle for stability enhancement of vitamin C in rainbow trout diet (Alishahi, Mirvaghefi, Rafie-Tehrani, Farahmand, Shojaosadati, Dorkoosh & Elsabee, 2011). Hence, this chapter attempts to survey the applications of chitosan in various fields of marine-based products.

2. Antibacterial activity

The modern era of chitosan research was heralded by publications in the 1990s, that described the antimicrobial potentials of chitosan and its derivatives, exhibiting a wide spectrum of activities against human pathogens and food-borne microorganisms (Chen, Xing & Park & Kong, 2010; No, Park, Lee & Meyers, 2002; Rabea, Badway, Stevens, Smagghe & Steurbaut, 2003: Raafat, Bargen, Haas & Sahl, 2008; Raafat & Sahl, 2009). , The first study reporting antibacterial properties was reported by Allan & Hardwiger (1979). They reported that chitosan showed a broad range of activities and a high inactivation rate against both Gram-positive and Gram-negative bacteria, (Allan & Hardwiger, 1979). However, although several studies have been published in this area, the exact mechanism of the antimicrobial activity of chitosan remains ambiguous.

Six major mechanisms have been proposed in the literature, as follows (Kong et al., 2010; Raafat & Sahl, 2009; Rafaat et al., 2008): (1) the interaction between the positively charged chitosan amine groups and the negatively charged microbial cell membranes, leadingto the leakage of proteinaceous and other intracellular constituents; (2) the activation of several defense processes in the host tissue by the chitosan molecule acting as a water-binding agent and inhibiting various enzymes by blocking their active centers; (3) the action of chitosan as a chelating agent, selectively binding metals and then inhibiting the production of toxins and microbial growth; (4) the formation, generally by high molecular weight chitosan, of an impervious polymeric layer on the surface of the cell, thereby altering cell permeability and blocking the entry of nutrients into the cell; (5) the penetration of mainly low-molecular weight chitosan into the cystosol of the microorganism to bind DNA, resulting in interference with the synthesis of mRNA and proteins; and (6) the adsorption and flocculation of electronegative substances in the cell by chitosan, distributing the physiological activities of the microorganisms, causing their death.. However, it is very important to mention that chitosan is soluble only in acidic media and therefore, the effect of pH on microorganisms must be considered together with the effect of chitosan. Thus, the synergetic effect of chitosan/pH together is probably the most evident explanation of the antimicrobial effect of chitosan.

Complicating the issue, a number of studies aimed at determining the antibacterial activities of chitosan on Gram-positive and Gram-negative bacteria have been reported antithetical outcomes, making their interpretation difficult. More recently, Kong et al. (2010) showed that chitosan and its derivatives are more powerful antibacterial agent against Gram-negative bacteria than against Gram-positive microorganism. Conversely, Raafat and Sahl (2009) reported a study in which they demonstrated that Gram-positive bacteria are more sensitive to the antibacterial effect of chitosan than Gram-negative bacteria. Therefore, the interpretation of the sensitivity of bacteria to chitosan is quite difficult.

To address this problem, Kong et al. (2010) proposed that the variation in the bactericidal efficacy of chitosan arises from different parameters that must be considered when evaluating the antibacterial activity of chitosan. These factors can be categorized into four classes as follows: (1) intrinsic microbial factors, including microbial species and cell age, (2) intrinsic factors of chitosan molecules, namely, the positive charge density, protonation level of the amine group, the chitosan molecular weight and concentration, hydrophilic/hydrophobic characteristics and chelating capacity of the chitosan molecule; (3) its physical state, i.e., either water soluble or solid chitosan; and (4) environmental factors including the ionic strength of the testing medium, pH, temperature and contact time between chitosan and bacterial cells. In addition, it is worth noting that, despite the widely reported antimicrobial properties of chitosan in the literature, the results are mainly based on in vitro experiments. In real-world appliactions, it is important to consider that most foods are complex matrices composed of different compounds (proteins, carbohydrates, lipids, minerals, vitamins, salts and others) and many of them may interact with chitosan to varying levels, possibly leading to a loss or enhancement of its antibacterial activity (Devlieghere, Vermeulen, & Debevere, 2004).

Taking into account the above insights about the antibacterial characteristics of chitosan, the following applications of chitosan in seafood products were considered. Due to the high perishability characteristics of marine-based products, there has been an increased interest in the application of chitosan to extend the shelf life of the products. In this context, chitosan has increasingly gained attention as an antibacterial additive in seafood from both seafood processors and consumers, largely due to a desire to reduce the use of synthetic chemicals in seafood preservation. Cao, Xue and Liu (2009) reported that chitosan at 5 g/L extended the shelf life of oyster (*Crossostrea gigas*) from 8-9 days to 14-15 days. They explained that Pseudomonas and Shewanella are the most prolific microorganisms during cold storage of fish and shellfish and these bacteria can easily be reduced or eliminated with the addition of chitosan at this concentration.

Fish balls have a high water activity and are prone to the growth of microorganisms, with a relatively short shelf life of 4-5 days at a storage temperature of approximately 5 °C. Kok and Park (2007) reported that fish ball shelf life has been increased by adding chitosan which maintained both aerobic and yeast counts at < 1 log CFU/g over 21 days of storage. Kok and Park (2007) also reported that physical state of chitosan is an important parameter for its antibacterial activity. In the dissolved state, chitosan showed excellent antibacterial activity and contributed to the extension of the product shelf life. However, in the study reported by Lopez-Caballero et al. (2005a and b), it was demonstrated that the addition of powdered chitosan to fish patties had no effect on bacterial growth. Roller and Corvill (2000) reported that the spoilage flora was inhibited from log 8 CFU/g in the control sample (without chitosan) to log 4 CFU/g over the 4-week study at 5 °C with the use of chitosan combined with acetic acid in shrimp salad. However, it is important to consider the effect of acetic acid on the ability of chitosan to extend shelf life. Fernandez-Saiz, Soler, Lagaron and Ocio (2010) studied bacterial growth in two conditions. The first one was a fish soup (ANETO ® brand, packaged in TetraBrik ® and fabricated by Jamon Aneto, S. L., Barcelona, Spain). The other medium was a model laboratory growth medium, tryptone soy broth. They reported a significant reduction of the growth of *Listeria monocytogenes, Staphylococcus aureus* and *Salmonella spp* when the products were stored in the presence of chitosan. In fish soup and under laboratory conditions, the effect of chitosan in the tested medium at 4, 12, and 37 °C depended significantly on the bacteria type, incubation temperature and food matrix (substrate). The antibacterial effectiveness of chitosan was decreased in the fish soup, suggesting that the constituents of the fish soup had high influence on the antimicrobial efficacy of chitosan. The authors reported that chitosan was probably irreversibly bound by microbial cells or negatively charged compound in the soup and therefore rendered it inactive against the remaining unbound microorganisms. In conclusion, the effect of chitosan-to-cell ratio must be considered. Lopez-Caballero et al. (2005a-b) showed that the addition of chitosan to sausage treated at high pressure yielded a 2-log cycle decrease of total bacterial counts of *Pseudomonas* and *Enterobacteria* at 8-11 days storage. Ye, Neetoo and Chen (2008) stated that chitosan has antibacterial activity that is effectively expressed in aqueous system; however, its antibacterial properties against *L. monocytogenes* in cold-smoked salmon were negligible when chitosan was in the form of an insoluble film. The growth of *L. monocytogenes* in salmon samples wrapped in the chitosan-coated film and plain films was similar. The authors demonstrated that it is possible that chitosan is ineffective in films because it is unable to diffuse through a rigid food matrix such as salmon. Regarding microbial counts, Lopez-Caballero et al. (2005b) showed that chitosan in the soluble state had no significant effect in high-pressure treated cod sausages. However, when chitosan was added in a dry form, higher counts of microorganisms were recorded. This is an indication that chitosan in soluble form contributed to maintaining product safety. The microbial counts in their study were for lactic acid bacteria, *Enterobacteria, Pseudomonas* and *Staphylococcus*. Duan, Jiang, Cherian and Zhao (2010) reported that the initial total plate count (TPC) of fresh lingcod was 3.67 log CFU/g, which then rapidly increased to 6.16 and 8.36 log CFU/g on day 6 and day 14, respectively. When chitosan coatings were used, the results showed a 0.15-0.64 reduction in TPC. Moreover, the TPCs of chitosan-coated samples stored under vacuum or modified-atmosphere packaging were significantly lower than those of the control sample during the subsequent cold storage. The combination of chitosan coating and vacuum or modified atmosphere packaging (MAP) resulted in 2.22-4.25 reductions in TPC for the first 14 days of cold storage. The TPCs of chitosan-coated and MPA samples were lower than 10^5 CFU/g even after 21 days of cold storage. This result indicated a significant delay of microbial spoilage. Qi, Zhang and Lan-Lan (2010) reported that because non-fermenting Gram-negative bacteria are dominant in the initial microbial flora of fish and shellfish sourced from cold seawater, controlling the growth of these Gram-negative bacteria may be important for the preservation of oysters. They demonstrated that combined treatment with chitosan and ozonated water had better antibacterial effect than either treatment alone. When only aerobic plate count was measured, the authors showed that the product shelf life with the combination of chitosan with ozonated water was at least

20 days, whereas it was only 8 days for the control sample, 10 days for the ozonated samples, and 14 days for the chitosan-treated samples. Duan, Cherian and Zhao (2009) indicated that fish oil incorporated in chitosan coatings lowered significantly the total and psychrotrophic counts in frozen lingcod fillets over three months cold storage. Ojagh, Rezaei, Razavi and Hosseini (2010) also reported that chitosan coatings enriched with cinnamon oil decreased effectively total viable counts and psychrotropic bacteria in rainbow trout (*Oncorhynchus mykiss*) during 16-day cold storage.

3. Antioxidant activity

Seafood is considered as excellent sources of functional foods for balanced nutrition favorable for promoting good health. The beneficial health effects of marine foods are ascribed to their lipids, mainly the long-chain omega-3 PUFA such as eicosapentaenoic acid (EPA) and docosahexaenoic acid (DHA) (Newton, 2001). However, these valuable compounds in seafood are highly sensitive to oxidative reactions and development of off-flavor even during cold storage (Cadwalladar & Shahidi, 2001). It has been proposed that lipid oxidation in fish and shellfish may be initiated and propagated by a number of mechanisms, namely autooxidation, photosentized oxidation, lipoxygenase, peroxidase and microsomal enzymes (Hsieh & Kinsella, 1989). Additionally, fish and shellfish muscles contain protein-bound iron compounds, for example, myoglobin, hemoglobin, ferritin, transferrin and haemosiderin, as well as other metals. This is a factor that plays an important role in initiating lipid oxidation in marine-based products (Decker & Hultin, 1992). Castell, Maclean and Moore (1965) showed that the relative pro-oxidant activity of metal ions in fish and shellfish muscles decreased in the order of $Cu^{+2} > Fe^{+2} > Cd^{+2} > Li^{+2} > Mg^{+2} > Zn^{+2} > Ca^{+2} > Ba^{+2}$. The iron-bound proteins and other metal ions may be released during the storage period and can thus activate and/or initiate lipid oxidative reactions (Decker & Hultin, 1992).

Along with the growing consumer demand for seafood devoid of synthetic antioxidants, chitosan has been a booming antioxidant agent in fish and shellfish. The antioxidant activity of chitosans of different viscosities (360, 50 and 14 cP) in cooked, comminuted flesh of herring (*Clupea harengus*) was investigated (Kamil, Jeon and Shahidi, 2002). The oxidative stability of fish flesh during cold storage at 4 °C with the addition of chitosan at concentrations of 50, 100 and 200 ppm was compared with that of fish treated with conventional antioxidants, such as butylated hydroxyanisole (BHA) and butylated hydroxytoluene (BHT) and *tert*-butylhydoquinone (THQ) (all at 200 ppm). Among the three chitosan samples tested, the 14 cP chitosan was the most effective in preventing lipid oxidation. The formation of thiobarbituric acid test reactive substances (TBARS) in herring samples containing 200 ppm of 14 cP chitosan was reduced by 52% as after 8 days of storage compared with that the control sample without chitosan. At a chitosan concentration of 200 ppm, the 14 cP chitosan exerted an antioxidant effect similar to that of commercial antioxidants in reducing TBARS values in comminuted herring flesh. A study by Kamil, Jeon and Shahidi (2002) indicated that the antioxidant capacity of chitosan added to fish muscle depends on its molecular weight (MW) and concentration in the product. Similarly, Kim and Thomas (2007) observed that the antioxidative effects of chitosan in salmon depended on its molecular weight (tested at MW = 30, 90 and 120 kDa) and concentration (evaluated at 0.2%, 0.5% and 1% w/w). The authors reporteda that the 30 kDa chitosan showed the highest radical-scavenging activity. The scavenging activities of chitosan werer increased by increasing its concentration. However, varying the concentration showed no significant effects when 120 KDa chitosan was used. Ahn and Lee (1992) studied the preservative effect of chitosan film on the quality of highly salted and dried horse mackerel. The product was prepared by soaking the fresh horse mackerel in 15% salt solution for 30 min, coating with or without (control) chitosan, and drying for 3h at 40 °C in a hot air dryer. During cold storage at a temperature of 5 °C for 20 days, the chitosan-coated samples had lower TBARS and peroxide values (PV) than the control samples. Similarly, Lopez-Caballero et al. (2005) also found that coating codfish patties with a chitosan-gelatin blend considerably lowered lipid oxidation. However, being non soluble in powder form at neutral pH, chitosan had no effect on the prevention of lipid oxidation. Shahidi, Kamil and Jeon (2002) reported that chitosans with different molecular weights and viscosities (14, 57 and 360 cP) were effective in controlling the oxidation of lipids in comminuted cod (*Gadus morhua*) following cooking. Both PV and TBARS values were reduced as a result of treatment of the fish prior to cooking with 50, 100 and 200 ppm of 14, 57, and 360 cP chitosans. Inhibition of the oxidation was concentration-dependent and was the highest for the 14 cP chitosan. The authors stated that the antioxidant activity of chitosans of different viscosity in cooked comminuted cod may be attributed to their metal-chelating capacities. Chitosans with different viscosities were found to protect cooked cod from oxidation at various levels. The observed differences were presumably due to differing degrees of deacetylation and molecular weights of the chitosan molecules. In the study reported by Qin (1993), it has been indicated that the ion-chelating ability of chitosan is strongly affected by the degree of deacetylation. The highly acetylated chitosan has very little chelating activity. In addition, high molecular weight chitosan has a compact structure and the effect of intramolecular hydrogen bonding is stronger, which weakens the activities of the hydroxyl and amino groups. As result, the probability of the exposure of these active moieties may be restricted, resulting in less radical scavenging activity. Obviously, low molecular weight chitosan exhibits higher hydroxyl radical scavenging activity, which is partially attributable to its metal chelating ability. The Fe^{+2} chelating ability of chitosan is mainly attributed to the presence of amino groups, which contain free electron pairs that contribute to form chitosan/Fe^{+2} complex. The Fe^{+2} chelating ability of low molecular weight chitosan is more pronounced than that of high molecular weight chitosan. Consequently, the amino groups in chitosan can react with free radicals to form additional stable macroradicals. Therefore, the active hydroxyl and amino groups in the polymer chains are the origin of the scavenging ability of chitosan (Jeon, Shahidi & Kim, 2000; Feng, Du, Li, Hu & Kennedy, 2008; No et al., 2007). Kamil et al. (2002) explained that in the charged state (protonated amino groups), the cationic groups of chitosans impart intramolecular electric repulsive forces. This phenomenon may be responsible for lesser chelating ability of high viscosity (high Mw) chitosans. Jeon et al. (2002) demonstrated that the antioxidant activity of chitosan is also effective when it is applied as a protective film. In this kind of application, it retards lipid oxidation by acting as a barrier against oxygen penetration. Sathivel, Liu, Huang and Prinyawiwatkul (2007) showed that the TBARS value of coated pink salmon (*O. gorbuscha*) fillets glazed with chitosan (1.3 mg MDA(malondialdehyde)/kg sample) was significantly lower than that of fillets glazed with lactic acid (3 mg MDA/kg sample) or distilled water (1.8 mg MDA/kg sample). The results indicated that chitosan glazing was more effective at reducing lipid oxidation among the studied alternatives. Sathivel (2005) also reported that the TBARS value of pink salmon fillets coated with 1% and 2% chitosan was significantly lower than in both the control sample and protein-coated product after 3 months of frozen storage. The author stated that a higher the concentration of chitosan, resulted in a lower TBARS value. The latter implies that the antioxidant effect of chitosan in the coating state is highly correlated to the coating material thickness, thereby hindering the entrance of oxygen into pink salmon fillet and initiating the oxidative process. Moreover, the primary amino groups of chitosan would form a stable fluorosphere with volatile aldehydes such as malondialdehyde which are derived from breakdown of fats during lipid oxidation (Weist & Karel, 1992).

Duan et al. (2009; 2010) showed that the combination of chitosan with modified atmosphere packaging enhanced the lipid stability of lingcod (*O. elongates*) within 21 days of cold storage. When applied on the surface of lingcod fillets, chitosan coatings may act as a barrier between the fillet and the surrounding atmosphere. This is mainly due to the good oxygen barrier properties of chitosan films, which slow down the diffusion of oxygen from the surrounding air to the surface of fillet and retard lipid oxidation (Aider, 2010). Additionally, Ojagh et al. (2010) reported that chitosan coatings enriched with cinnamon oil could suitably delay lipid oxidation in the refrigerated rainbow trout during 16 days of storage and markedly reduced the TBARS and PV values as compared with the control product. Mao and Wu (2007) showed that lipid oxidation of kamabako gel from grass carp (*Ctenopharyngodon idellus*) significantly decreased when a 1% chitosan solution was added.

4. Bioactive coatings

The modern marine-related food industries are encountering challenges and require for specific alternatives to surmount them. Among these, issues related to seafood packaging for products with a short shelf life are of pivotal importance. Although the utilization of conventional packaging materials such as plastics and their derivatives are effective for seafood preservation, they create serious and hazardous environmental problems, a situation which presents the seafood industry as a source of pollution and social concerns. This problem requires that all stakeholders in this industry and especially scientists specializing in the food engineering and packaging field to seek alternatives to address this serious problem related to the packaging material. A non-negligible aspect, which is the total cost of the final product, is also related to the packaging materials because it is well known that the contribution of the packaging to the product total cost is highly significant. So, the search for more economical packaging materials is a very important subject in the seafood industry (Aider, 2010).

Edible bio-based films have been investigated for their abilities to avoid moisture or water absorption by the seafood matrix, oxygen penetration to the food matrix, aroma loss and solute transport out of the product (Dutta, Tripathi, Mehrotra & Dutta, 2009). Based on this consideration, one of the most perspective active bio-film is the one based on chitosan. More recently, two review studies have reported the application of chitosan as bioactive film in the food industry (Aider, 2010; Dutta et al., 2009). Chitosan film, like many other polysaccharide based films, tend to exhibit resistance to fat diffusion and selective permeability to gases. However, they have a serious lack in terms of resistance to water and water vapor transmission. This behavior is mainly due to the strong hydrophilic character of these biopolymers, a property that leads to high interaction with water molecules (Bordenave, Grelier, & Coma, 2007). Owing to this, polymer blending or biocomposites and multilayer systems are potential approachs to prepare chitosan based bioactive coatings or films with desirable characteristics. In this context, Ye et al. (2008) stated that since edible film formed by chitosan is brittle and does not have good mechanical properties, coating chitosan onto a plastic film would overcome this problem. These authors have used chitosan-coated plastic films in which they have incorporated five antibacterial agents, namely nisin, sodium lactate (SL), sodium diacetate (SD), potassium sorbate (PS) and sodium benzoate (SB) as a novel antibacterial edible film against *Listeria monocytogenes* on cold-smoked salmon. This approach solved problems related to food safety since it is well known that *L. monocytogenes* could grow to high levels on cold-smoked salmon, even at normal refrigeration temperature. The risk related to *L. monocytogenes* is particularly high at abusive storage temperatures. Chtiosan-coated films containing 4.5 mg/cm^2 SL, 4.5 mg/cm^2 SL-0.6mg/cm^2 PS and 2.3 mg/cm^2 SL-500 IU/cm^2 nisin were the most effective treatments against *L. monocytogenes* at ambient temperature. These treatments showed long term antilisterial efficacy during refrigerated storage on vacuum-packed cold-smoked salmon. However, it is important to consider the fact that since antibacterial activity of chitosan may be negligible when it is in the form of insoluble films. Under this state, chitosan is ineffective because it is unable to diffuse through a rigid food matrix such as salmon. Sathivel et al. (2007) showed that skinless pink salmon (*Oncorhynchus gorbuscha*) fillets glazed with chitosan at a solution concentration of 1% (w/w) had significantly ($p < 0.05$) higher yield and thaw yield than the lactic acid–glazed and distilled water-glazed fillets. This behavior was valid although those fillets all had similar moisture content after thawing. In addition, the rheological study showed that chitosan has pseudoplastic and viscoelastic characteristics. The glass transition temperature for the chitosan film was observed at 80.23 °C. The oxygen, carbon dioxide, nitrogen and water vapor permabilities of the chitosan film were 5.34 10^{-2} (cm^3/ m day atm), 0.17 (cm^3/m day atm), 0.03 (cm^3/m day atm) and 2.92 10^{-10} (g water m/m^2 Pas), respectively. The authors demonstrated that despite the good barrier properties of chitosan against oxygen, it maintained low water vapor transmission because of their hydrophilic nature. Likewise, they stated that chitosan film showed shear thinning and viscoelastic characteristics and temperature dependent viscosity, which allowed uniform glazing on the salmon fillets and prevented rupturing of chitosan glazing during solidification when the glazed fillets were frozen. Therefore, chitosan glazing applied on the surface of the pink salmon fillets might have acted as a barrier between the fillets and the air surrounding, thus slowing down the diffusion of oxygen from the surrounding air into the fillets. Kester and Fennema (1986) reported that chitosan coatings might act as moisture-sacrificing agents of moisture barriers. Thus, moisture loss from the product could be delayed till the moisture contained within the chitosan coating had been evaporated. Sathivel (2005) highlighted that pink salmon fillets coated with chitosan resulted in significantly higher yield, thaw yield, similar drip loss and cook yield, higher moisture content after thawing, less moisture loss than the control samples and somewhat less than protein-coated products. Besides, there were no significant ($p < 0.05$) effects of coating on color parameters (a^*, b^* and L^* values) for cooked fillets after three months frozen storage. Lopez-Caballero et al. (2005) used chitosan as a material to form a chitosan-gelatin coating for cod patties. They showed that the use of chitosan either as a coating or a powdered ingredient did not affect the product lightness at the end of the storage period. However it resulted in an increase of the product yellowness (b-color parameter). The chitosan coating increased the patty elasticity, whereas the addition of powdered chitosan to the patty mixture increased the other rheological parameters such as gumminess, chewiness, cohesiveness and adhesiveness. Moreover, the coating did prevent spoilage of cod patties as reflected by a decrease in total volatile basic nitrogen (TVBN). Conversely, none of these effects on the bacterial spoilage were observed when the chitosan was added to the patty mixture in a powdered form. Ultimately, the authors reported that the coating had good sensory properties, melted away on cooking and hence did not impart any taste to the product. They provided protection by delaying spoilage. Duan et al. (2010) produced chitosan-krill oil coating and used it in modified atmosphere packaging to extend the shelf life of Lingcod fillets. They reported that chitosan-krill oil coating increased total lipid and omega-3 fatty acid contents of the lingcod by about 2-fold. The reduced chemical changes were reflected by the TVBN values and did not change the color of the fresh fillets, did not affect consumer's acceptance of both raw and cooked lingcod fillets. Consumers preferred the overall quality of chitosan-coated, cooked lingcod fillets over the control. The preference was based on the product firm texture and less fishy aroma and flavor. Considering the lower cost of vacuum packaging, it could be applied in combination with chitosan coatings to maintain the omega-3 fatty acid content and extend shelf life of fresh lean fish such as lingcod. Duan et al. (2009) also showed that fish oil incorporated to chitosan coating decreased the drip loss of frozen samples by 14.1-27.6%. This coating also well fortified the omega-3 fatty acids in lean fish. Cao et al. (2009) and Qi et al. (2010) showed that the chitosan coating could surprisingly increase the shelf life of highly perishable pacific oyster (*C. gigas*) during 21 days storage. This affirmation was based on TVBN, pH values and sensory evaluation of pacific oyster. They stated that the discrepancies between their results and others were derived from the differences in chemical composition of fish and shellfish in which oyster contains significant levels of carbohydrate (glycogen) and a lower total quantity of nitrogen. Ojagh et al. (2010) synthesized chitosan coatings enriched with cinnamon oil to extend the shelf life of refrigerated rainbow trout and showed that sensory characteristics and TVBN of the end product were drastically improved as the coating was employed on rainbow trout fillets within 16 days cold storage. Similarly, Lopez-Caballero et al. (2005) stated that the addition of dry chitosan led to a noticeable increase in elasticity and product yellowness when cod sausages were enriched with chitosan solution. The TVBN remained stable during 25 days storage and the product elasticity was reinforced.

5. Effluent treatment

The use of chitosan as a coagulating agent for removing suspended solids from various processing streams has been widely investigated including cheese whey and dairy wash water, in the processing of poultry and seafood products (Kumar, 2000; Savant, 2001; Savant & Torres, 2000; Savant & Torres, 2003; Shahidi et al., 1999). Chitosan at a concentration of 10 mg/L reduced up to 98% the total suspended solids in shrimp processing wastewater (Bough, 1976). Protein recoveries from surimi wash water (SWW) using 150 mg/L chitosan-alginate complex per liter SWW at mixing ratio of 0.2 resulted in 78-94% adsorption after 24 h (Wibowo, 2003). This result was higher than the one obtained by using 50 mg/L, which yielded 81-90 % protein adsorption in the same treatment time (Savant, 2001). These reported findings suggest that reaction time and chitosan concentration play an important role in reducing total suspended solids and lowering solution turbidity. Moreover, the differences in molecular weights (MW) and degree of deacetylation (DD) between chitosan samples could explain the significant differences in protein recovery capacity. At the lowest concentration (20 mg/L SWW) tested in the study reported by Wibowo (2003), the experimental chitosan gave higher protein recovery than a commercial sample, which required a 5-fold higher concentration for the same effectiveness. This finding has commercial implications as it would reduce processing costs and the chitosan content in the solids recovered by the treatment (Wibowo, Velazquez, Savant & Torres, 2007a). If implemented commercially, the chitosan-alginate complex may be an effective alternative not only for the recovering of soluble proteins that would otherwise be discarded into the environment, but also as an economically viable downstream process over expensive, commercial membrane treatments and their limited use due to fouling (Savant, 2001). Surimi wash water protein (SWWP) was precipitated by using a chitosan-alginate complex. The precipitate had a crude protein content of 73.1 % and a high concentration of essential amino acids (3% histidine, 9.4% lysine, 3.7% methionine, and 5.1% phenylalanine). In a rat-feeding trial, SWWP as a single protein source showed higher modified protein efficiency and net protein rations than the casein control. Blood chemistry analysis did not reveal any deleterious effect from the full protein substitution or the chitosan in SWWP (Wibowo, Savant, Gherian, Velazquez & Torres, 2007b; Wibowo, Velazquez, Savant & Torres, 2005). Moreover, Guerrero, Omil, Mendez & Lema (1998) showed that the utilization of chitosan at a concentration of 10 mg/L and pH 7 in the process of coagulation-flocculation followed by centrifugation in fish-meal factory effluents decreased the total suspended solids up to 85%. The most important mechanisms explaining the chitosan effectiveness in seafood plant effluents treatment was mainly attributed to its positive charge and interaction with negatively charged compounds in the effluents such as protein. Furthermore, the hydroxyl groups on the chitosan molecule contribute to increase the precipitation of proteins and other suspended solids in the seafood plant effluents (Savant, 2001; Wibowo et al., 2007a,b).

6. Gelling enhancer

Surimi is a refined fish protein product prepared by washing mechanically deboned fish to remove blood, lipids, enzymes and sarcoplasmic protein. The myofibrillar proteins are concentrated in the resulting product and form an elastic gel when solubilized with NaCl and heated (Mao & Wu, 2007). Gel forming properties of myofibrillar proteins are quickly lost by degradation by the action of endogenous proteolytic enzymes if fish is not processed into surimi immediately. The utilization of frozen fish flesh for surimi production is unsuitable due to the rapid loss of protein functionality by freeze denaturation. High quality surimi is produced from fresh, unfrozen fish. Thus, processing at sea has been required in order to obtain high quality surimi. However, the cost of the processing at sea is much higher compared to the land-base processing (Lanier, Manning, Zetterling & Macdonald, 1992). In order to prepare a strong and elastic gel from fish species with low commercial value, low quality surimi is produced onshore with the aid of gel-forming biopolymers such as starch. In this way, chitosan is a good option to be incorporated into the products to improve their techno-functional quality (Kataoka, Ishizaki, & Tanaka, 1998; Mao & Wu, 2007; Li & Xia, 2010). Overall, gel-forming ability of surimi depends on both intrinsic and extrinsic factors, namely fish species, physio-chemical properties of muscle proteins, the presence of endogenous enzymes such as proteinase amd transglutaminase, and the conditions used in the product processing (Benjakul, Visessanguan, Phatchrat & Tanaka, 2003). The strength of gels prepared from low quality walleye Pollock (*Theragra chalcogramma*) was almost doubled by the addition of 1.5% chitosan when salted surimi pastes were set below 25 °C. The polymerization of myosin heavy chain accelerated in the presence of 1.5% chitosan (Kataoka et al., 1198). Along with chitosan, endogenous transglutaminase (TGase) played an important role in the formation of gel. The addition of TGase inhibitor to the salted walleye Pollock surimi inhibited the gel enhancement by chitosan. The mechanisms of chitosan effect on enhancing the gel formation in not clear. However, the participation of hydrophobic interactions, hydrogen bondings, and electrostatic interactions during the setting process has been proposed as a possible mechanism by which chitosan can enhance the formation of cross-linked myosin heavy chain components during their polymerization by endogenous enzymes (Benjakul et al., 2003: Kataoka et al., 1998; Li & Xia, 2010; Mao & Wu, 2007). Benjakul et al. (2003) reported that barred garfish (*Hemiramphus far*) surimi gel showed an increase in the breaking force when 1% chitosan was added. However, gel-forming ability of surimi containing chitosan was inhibited in the presence of EDTA due to the chelating of calcium ions that are necessary for TGase activity. Owing to this, the enhancing effect of chitosan was possibly mediated through the action of endogenous TGase during product processing, resulting in the formation of protein-protein and protein-chitosan conjugates. In conjunction with processing and the addition of calcium ions, TGase may play an important role in the cross-linking of protein-protein and protein-chitosan conjugates by means of the amino groups of chitosan as the acyl acceptor. Conversely, chitosan did not substantially modify the rheological and microstructural properties of horse mackerel gels (*Trachurus spp.*). Also, it had a slight reduction in gel elasticity obtained under high-pressure conditions (Gomez-Guillel, Montero, Sole & Perez-Mateos, 2005). Kok and Park (2007) stated that in the threadfin bream (*Nemipterus spp.*) surimi, the balance of protein-chitosan and protein-protein conjugates determined the surimi gel strength. Similarly, Mao and Wu (2007) showed that in the presence of chitosan in kamaboko gel of grass carp (*Ctenopharyngodon idellus*), protein-chitosan conjugates would be formed between the reactive amino groups of glucosamine and the glutaminyl residue of the myofibrillar proteins. The bonds between chitosan and myofibrillar proteins could be associated with the improvement of texture properties in the gels with final structure formed by both covalent and non-covalent interactions. The effect would be also due to some modifications of the endogenous TGase activity. More recently, Lia and Xia (2010) showed that molecular weight and degree of deacetylation (DD) of chitosan have different impacts on gel properties of salt-soluble meat proteins from silver carp. The gel containing chitosan with DD of 77.3% showed the highest penetration force and storage modulus. The penetration forces of gels increased with increasing the amount of molecular weight of chitosan incorporated in the gel. The interaction between chitosan and salt-soluble meat proteins was mainly stabilized by the electrostatic interactions and hydrogen bonds.

7. Encapsulation

Nowadays, the value of functional foods and bioactive compounds are increasing due to the awareness and consciousness of people about it. Despite this fact, many of these compounds are so much sensitive to environmental factors such as oxygen, light, and temperature. In addition, being incorporated into foods and drugs in delivery systems, these bioactive components are hydrolyzed by harsh conditions in the gastrointestinal tracts (Alishahi et al., 2011). Schep, Tucker, Young, Ledger and Butt (1999) stated that many of oral delivery systems of bioactive compounds in aquaculture met the three major barriers through the gastrointestinal tract, involving the enzymatic barriers from the host luminal and membrane bound enzymes, immunological cells present within both the enterocytes and underlying connective tissue and the physical barrier of the epithelial cells. Based on this consideration, the encapsulation of bioactive compounds and functional foods could be a promising way to overcome these problems. Encapsulation is a process in which thin films, generally of polymeric materials, are applied to little solid particles, liquids or gas droplets. This method is used to entrap active components and release them under controlled conditions (Deladino, Anbinder, Navarro & Martino, 2008). Several materials have been encapsulated for the use in the food industry such as vitamins, minerals, antioxidants, colorants, enzymes and sweeteners (Shahidi & Han, 1993). Chitosan can act as an encapsulating agent because of its non-toxicity, biocompatibility, mucus adhesiveness and biodegradability (Alishahi et al., 2011: Kumar, 2000). Recently, Alishahi et al. (2011) showed that chitosan/vitamin C nanoparticle system successfully increased the shelf life and delivery of vitamin C during 20 days storage of rainbow trout. They showed that shelf life of vitamin C significantly ($p <$

0.05) increased in rainbow trout feed till 20 days at ambient temperature, while the control which was feed by vitamin C alone, drastically lost its vitamin C content during few days at ambient temperature. Moreover, the controlled release behavior of vitamin C, in vitro and in vivo, showed that vitamin C was released in the gastrointestinal tract of rainbow trout in the controlled manner (up to 48 h) and chitosan nanoparticles could well maintain vitamin C against harsh conditions, acidic and enzymatic hydrolysis, in the gastrointestinal tract of rainbow trout. Also, Alishahi et al. (2011) showed that the chitosan nanoparticles containing vitamin C could significantly ($p < 0.05$) induce the non-specific immunity system of rainbow trout, as compared with the control. RajeshKumar, VenKatesan, Sarathi, Sarathbabu, Thomas and Anver Basha (2009) demonstrated that chitosan nanoparticles are able to encapsulate DNA and then favorably incorporated into shrimp feed to protect them from white spot syndrome virus. Their results showed that these nanoparticles increased the survival rates of shrimp against white spot syndrome during 30 days post-treatment. Likewise, RajeshKumar, Ishaq Ahmed, Parameswaran, Sudhakaran, Sarath Babu and Sahl Hameed (2008) incorporated chitosan nanoparticles containing DNA vaccine into Asian sea bass (Lates calcarifer) feed. Their results indicated that the sea bass orally vaccinated with chitosan-DNA (pVAOMP38) complex showed moderate protection against experimental *Vibrio anguillarum* infection. Similarly, Tian, Yu and Sum (2008) reported that chitosan microspheres loaded with plasmid vaccine was interestingly used to orally immunize Japanese flounder (*Paralichthys olivaceus*). They explained that the release profile of DNA from chitosan microspheres in PBS buffer (pH 7.4) was up to 42 days after intestinal imbibitions. Aydin and Akbuga (1996) showed that salmon calcitonin, available for clinical use, was suitably encapsulated in chitosan beads and the results confirmed that salmon calcitonin-loaded chitosan beads could be prepared by gelling the cationic chitosan with the anionic counterpart providing a controlled release property. Also, shark liver oil could be efficiently encapsulated in calcium alginate beads coated with chitosan in order to mask its unpleasant taste (Peniche, Howland, Carrillo, Zaldivar & Arguelles, 2004). The chitosan coating allowed controlling the permeability of capsules and avoiding leakage. The shark liver oil loaded chitosan/calcium alginate capsules were initially resistant to the acid environment of the stomach. But after 4 h at the intestinal pH (7.4), the capsule wall weakened and thereby was able to be easily deteriorated and disintegrated by the mechanical and peristaltic movements of the gastrointestinal tract. Likewise, Klinkesorn and Mcclements (2009) stated that the encapsulation of tuna oil droplets with chitosan affected their physical stability and digestibility when they were passed through an in vitro digestion model containing pancreatic lipase. The amount of free fatty acids released from the emulsions decreased as the concentration of chitosan increased. However the relesae was independent of chitosan Mw. These results showed that chitosan was able to reduce the amount of free fatty acids released from the emulsion, which may be attributed to a number of different physiological mechanisms, including formation of a protective chitosan coating around the lipid droplets, direct interaction of chitosan with lipase, or fatty acid binding by the chitosan. Also, they showed that pancreatic lipase was able to digest chitosan and release glucosamine, having important implications for the utilization of chitosan coatings for the encapsulation, protection and delivery of Omega-3 fatty acids. They suggested that encapsulation with chitosan could be used to protect emulsified polyunsaturated lipids from oxidation during storage. However, they will release the functional lipids after they are consumed. Industrially, tuna oil encapsulation with chitosan using ultrasonic atomizer was shown to be the promising technique in the near future (Klaypradit & Huang, 2008).

8. Conclusions

Chitosan, a deacetylated derivative of chitin, has attracted a great attention in the seafood industry due to its non-toxicity, biodegradability, biocompatibility and mucus adhesiveness properties. Chitosan has different characteristics such as antibacterial, antioxidant, film-forming ability, gel enhancer, encapsulating capacity, tissue engineering scaffold, wound dressing, and coagulating agent. Upon knowing these, chitosan could successfully be incorporated into seafood products for both seafood quality and human health enhancement. Regarding its outstanding characteristics, chitosan would be used as functional ingredients in marine-based products and it merits further researches in the future.

Author details

Alireza Alishahi
The University of Agriculture Sciences and Natural Resources of Gorgan, Gorgan, Iran

9. References

Aider, M. (2010). Chitosan application for active bio-based films production and potential in the food industry: A review. *Food Science and Technology*, 43, 837-842.

Ahn, C.B., & Lee, E.H. (1992). Utilization of chitin prepared from the shellfish crust.2. Effect of chitosan film packaging on quality of lightly-salted and dried horse mackerel. *Bulletin of Korean Fish Society*, 25, 51-57.

Alishahi, A., Mirvaghefi, A., Rafie-Tehrani, M., Farahmand, H., Shojaosadati, S.A., Dorkoosh, F.A., & Elsabee, M.Z. (2011). Shelf life and delivery enhancement of vitamin C using chitosan nanoparticles. *Food Chemistry*, In press.

Allan, C. R., & Hardwiger, L. A. (1979). The fungicidal effect of chitosan on fungi of varying cell wall composition. *Experimental Mycology*, 3, 285-287.

Aydin, Z., & Akbuga, J. (1996). Chitosan beads for delivery of salmon calcitonin: preparation and release characteristics. *International Journal of Pharmaceutics*, 131, 101-103.

Benjakul, S., Visessanguan, W., Phatchrat, S., & Tanaka, M. (2003). Chitosan affects transglutaminase – induced surimi gelation. *Journal of Food Microbiology*, 27, 53-66.

Benjakul, S., Viscessanguan, W., Tanaka, M., Ishizaki, S., Suthidham, R., & Sungpech, O. (2001). Effect of chitin and chitosan of gelling properties of surimi from barred garfish (*Hemiramphus far*). *Journal of the Science of Food and Agriculture*, 81, 102-108.

Bhatnagar, A., & Sillanpa, M. (2009). Application of chitin and chitosan-derivatives for water and wastewater- a short review. Advances in Colloid and Interface Science, 55, 9479-9488.

Bordenave, N., Grelier, S., & Cama, V. (2007). Water and moisture susceptibility of chitosan and paper-based materials: structure-property relationships. *Journal of Agriculture and Food Chemistry*, 40, 1158-1162.

Bough, W.A. (1976). Chitosan a polymer from seafood waste, for use in treatment of food processing wastes and activated sludge. *Process Biochemistry*, 11, 13-16.

Cadwallader, K.R. & Shahidi, F. (2001). Identification of potent odorants in seal blubber oil by direct thermal desorption-gas chromatography-olfactometry. In omega-3 fatty acids: chemistry, nutrition and health effects. F. Shahidi & J.W. Finley (eds) pp.221-234, symposium series 788, American chemical society, Washington, D.C.

Cao, R., Xue, C.H., & Liu, Q. (2009). Change in microbial flora of pacific oysters (*Crassostera gigas*) during refrigerated storage and its shelf life extension by chitosan. *International Journal of Food Microbiology*, 131, 272-276.

Carlson, R.P., Taffs, R., Davison, W.M., & Steward, P.S. (2008). Anti-biofilm properties of chitosan coated surfaces. *Journal of Biomaterial Science Polymer*, 19, 1035-1046.

Castell, C.H., Maclean, J., & Moore, B. (1969). Rancidity in lean fish muscle. IV. Effect of sodium chloride and other salts. *Journal of Fish Resources and Biodiversity of Canada*, 22, 929-944.

Chi, F.H., & Cheng, W.P. (2006). Use of chitosan as coagulant to treat wastewater from milk processing plant. *Journal of Polymer and the Environment*, 14, 411-417.

Decker, E.A., & Haultin, H.O. (1992). Lipid oxidation in muscle foods via redox ion. In: Angelo, A.J. (ed). Lipid oxidation in food. Washington, D.C.: American Chemical Society, pp. 33-54.

Deladino, L., Anbinder, P.S., Navarro, A.S. & Martino, M.N. (2008). Encapsulation of natural antioxidants extracted from *Ilex paraguariensis*. *Carbohydrate Polymer*, 71, 126-134.

Devlieghere, F., Vermeulen, A., & Debevere, J. (2004). Chitosan: antimicrobial activity, interactions with food components and applicability as coating on fruit and vegetables. *Food Chemistry*, 21, 703-714.

Dias, F.S., Querroz, D.C., Nascimento, R.F., & Lima, M.B. (2008). Simple system for preparation of chitosan microspheres. *Quimica Nova*, 31, 160-163.

Duan, J., Cherian, G., & Zhao, Y. (2009). Quality enhancement in fresh and frozen lngcod (*Ophiodon elongates*) fillets by employment of fish oil incorporated chitosan coatings. *Food Chemistry*, 119, 524-532.

Duan, J., Jiang, Y., Cherian, G., & Zhao, Y. (2010). Effect of combined chitosan-krill oil coating and modified atmosphere packaging on the storability of cold-stored lingcod (*Ophiodon elongates*) fillets. *Food Chemistry*,122, 1035-1042.

Dutta, P.K., Tripathi, S., Mehratra, G.M., & Dutta, J. (2009). Perspectives for chitosan based antimicrobial films in food applications. *Food Chemistry*, 114, 1173-1182.

El Tahlawy, K.F., El Benday, M.A., Elhendawy, A.G., & Hudson, S.M. (2005). The antimicrobial activity of cotton fabrics treated with different crosslinking agents and chitosan. *Carbohydrate Polymer*, 60, 421-430.

Feng, T., Du, Y. Li. J., Hu, Y., & Kennedy, J.F. (2008). Enhancement of antioxidant activity of chitosan by irradiation. *Carbohydrate Polymer*, 73, 126-132.

Hirano, S. (1996). Chitin biotechnology applications. *Biotechnology Annual Review*, 2, 237-258.

Gomez-Guillen, M., Montero, P., Solas, M.T., & Perez-Mateos, A.G. (2005). Effect of chitosan and microbial transglutaminase on the gel forming ability of horse mackerel (*Trachurus spp.*) muscle under high pressure. Food Research International, 38, 103-110.

Guerrero, L., Omil, F., Mendez, R., & Lema, J.M. (1998). Protein recovery during the overall treatment of wastewaters from fish-meal factories. *Bioresources Technology*, 63, 221-229.

Hsieh, R.J., & Kinsella, J.E. (1989). Oxidation of polyunsaturated fatty acids: mechanism, products and inhibition with emphasis on fish. *Advances in Food and Nutrition Research*, 33, 233-241.

Jeon, Y.J., Kamil, J.Y.V.A., &Shahidi, F. (2002). Chitosan as an edible invisible film for quality preservation of herring and Atlantic cod. *Journal of Agriculture and Food Chemistry*, 50, 67-78.

Jeon, J.J., Shahidi, F., & Kim, S.K. (2000). Preparation of chitin and chitosan oligomers and their applications in physiological functional foods. *Food Review International*, 16, 159-176.

Kadam, S.U. & Prabhasankar, P. (2010). Marine foods as functional ingredients in bakery and pasta products. *Food Research International*, 43, 1975-1980.

Kamil, J.Y.V.A., Jeon, J.J., & Shahidi, F. (2002). Antioxidative activity of chitosans of different viscosity in cooked comminuted flesh of herring (*Clupea harengus*). *Food Chemistry*, 79, 69-77.

Kataoka, J., Ishizaki, S., & Tanaka, M. (1998). Effects of chitosan on gelling properties of low quality surimi. *Journal of Muscle Foods*, 9, 209-220.

Kester, J.J., & Fenneema, O. (1986). Edible films and coatings: A review. Food Technology, 40, 47-59.

Kim, K.W., & Thomas, R.L. (2007). Antioxidant role of chitosan in a cooked cod (Godus morhua) model system. *Journal of Food Lipids*, 9, 57-64.

Khor, E. (2001). The relevance of chitin. In: chitin: fulfilling a biomaterial promise. Elsevier, pp. 1-8.

Klaypradit, W., & Huang, Y.W. (2008). Fish oil encapsulation with chitosan using ultrasonic atomizer. *Journal of food science and technology*, 41, 1133-1139.

Klinkesorn, U., & McClement, D.J. (2009). Influence of chitosan on stability and lipase digestibility of lecithin-stabilized tuna oil-in-water emulsions. *Food Chemistry*, 114, 1308-1315.

Kok, T.N., & Park, J.W. (2007). Extending the shelf life of set fish ball. *Journal of Food Quality*, 30, 1-27.

Kong, M., Chen, X.G., Xing, K., & Park, H.J. (2010). Antimicrobial activity of chitosan and mode of action: A state of art review. *International Journal of Food Microbiology*, 144, 51-63.

Kumar, M.N.V. (2000). A review of chitin and chitosan applications. *Reactive Functional Polymer*, 46, 1-27.

Lanier, T.C., Manning, P.K., Zetterling, T., & Macdonald, G.A. (1992). Process innovations in surimi manufacture. In T.C. Lanier & C.M. Lee (eds), Surimi technology, pp. 167-179, Marcel Decker, New York.

Ledger, R., Tucker, I.G., & Walker, G.F. (2002). The metabolic barrier of the lower intestinal tract of salmon to the oral delivery of protein and peptide drugs. *Journal of Controlled Release*, 85, 91-103.

Li, X., & Xia, W. (2010). Effect of chitosan on the gel properties of salt-soluble meat proteins from silver carp. *Carbohydrate Polymer*, 82, 958-964.

Lopez-Caballero, M.E., Gomez-Guillen, M.C., Perez-Mateos, M., & Montero, P. (2005a). A chitosan-gelating blend as a coating for fish patties. *Food Hydrocolloids*, 19, 303-311.

Lopez-Caballero, M.E., Gomez-Guillen, M.C., Perez-Mateos, M., & Montero, P. (2005b). A functional chitosan-enriched fish sausage treated by high pressure. *Journal of Food Science*, 70, 166-171.

Mao, L., & Wu, T. (2007). Gelling properties and lipid oxidation of kamabako gels from grass carp (*Ctenopharyngodon idellus*) influenced by chitosan. *Journal of Food Engineering*, 82, 128-134.

Newton, I.S. (2001). Long chain fatty acids in health and nutrition. In omega-3 fatty acids: chemistry, nutrition and health effects, F. Shahidi & J.W. Finley (eds), ACS symposium series 788, American chemical society, Washington, D.C., pp. 14-27.

No, H.K., Meyers, S.P., Prinyawiwatkul, W., & Xu, Z. (2007). Applications of chitosan for improvement of quality and shelf life of foods: A review. *Journal of Food Science*, 72, 87-100.

No, H.K., Park, N.Y., Lee, S.H., & Meyers, S.P. (2002). Antibacterial activity of chitosan and chitosan oligomers with different molecular weights. *International Journal of Food Microbiology*, 74, 65-72.

Ojagh, S.M., Rezaei, M., Razavi, S.H., & Hossieni, S.M.H. (2010). Effect of chitosan coatings enriched with cinnamon oil on the quality of refrigerated rainbow trout. *Food Chemistry*, 120, 193-198.

Pittermann, W., Horner, V., & Wachter, R. (1997). Food applications of high molecular weight chitosan in skin care applications. In: Muzzarelli, R.A.A. & Peter, M.G. (eds). chitin handbook, Grottammare, Italy: European chitin society, pp. 361.

Peniche, C., Howland, I., Carrillo, O., Zaldivar, C., & Arguelles-Monal, W. (2004). Formation and stability of shark liver oil loaded chitosan/calcium alginate capsules. *Food Hydrocolloids*, 18, 865-871.

Qi, C.R., Bang-Zhang, Y., & Lan-Ian, Z. (2010). Combined effect of ozonated water and chitosan on the shelf life of pacific oyster (*Crassostrea gigas*). *Innovativ Food Science and Emerging Technologies*, 11, 108-112.

Qin, Y. (1993). Chelating property of chitosan fibers. *Journal of Applied Polymer*, 49, 727-731.

Raafat, D., Bargen, K., Haas, A. & Sahl, HG. (2008). Insights into the mode of action of chitosan as an antimicrobial compound. *Applied and Environmental Microbiology*, 74, 3764-3773.

Raafat, D., & Sahl, H.G. (2009). Chitosan and its antimicrobial potential- a critical literature survey. *Microbial Biotechnology*, 2, 186-201.

Rajeshkumar, S., Ishaq Ahmed, V.D., Parameswaran, V., Sudhakaran, R., Sarath Babu, V., & Sahl Hameed, A.S. (2008). Potential use of chitosan nanoparticles for oral delivery of DNA vaccine in Asian sea bass (Lates calcarifer) to protect from *Vibrio anguillarum. Fish & Shellfish Immunology*, 25, 47-56.

Rajeshkumar, S., Venkatesan, C., Sarathi, M., Sarathbabu, V., Thomas, J., & Anver Basha, K. (2009). Oral delivery of DNA construct using chitosan nanoparticles to protect the shrimp from white spot syndrome virus (WSSV). *Fish & Shellfish Immunology*, 26, 429-437.

Roller, S., & Corvill, N. (2000). The antimicrobial properties of chitosan in mayonnaise and mayonnaise-based shrimp salads. *Journal of Food Protection*, 63, 202-209.

Ross, S. (2000). Functional foods: the food and drug administration perspective. *The American Journal of Clinical Nutrition*, 71, 17355S-1738S.

Sathivel, S. (2005). Chitosan and protein coatings affect yield, moisture loss, lipid oxidation of pink salmon (*Oncorhynchus gorbushcha*) fillets during frozen storage. Journal of Food Science, 70, 755-459.

Sathivel, S., Liu, Q., Huang, J., & Prinyawiwatkul, W. (2007). The influence of chitosan glazing on the quality of skinless pink salmon (*Oncorhynchus gorbuscha*) fillets during frozen storage. *Journal of Food Engineering*, 83, 366-375.

Savant, V.D. (2001). Protein absorption on chitosan-polyanion complexes: application to aqueous food processing wastes. PhD. Thesis, Food Science and Technology, Oregon State University.

Savant, V.D., & Torres, J.A. (2000). Chitosan based coagulating agents for treatment of cheddar cheese whey. *Biotechnological Progress*, 16, 1091-1097.

Savant, V.D., & Torres, J.A. (2003). Fourier transform infrared analysis of chitosan based coagulating agents for treatment of surimi waste water. *Journal of Food Technology*, 1, 23-28.

Schep, L.J., Tucker, I.G., Young, G., Ledger, R., & Butt, A.G. (1999). Controlled release opportunities for oral peptide delivery in aquaculture. *Journal of Controlled Release*, 59, 1-14.

Shahidi, F., & Han, X. (1993). Encapsulation of food ingredients. *Critical Reviews in Food Science and Nutrition*, 33, 501-547.

Shahidi, F., Kamil, J.Y.V.A., & Jeon, Y.J. (1999). Food applications of chitin and chitosan. *Trends in Food Science & Technology*, 10, 37-51.

Shahidi, F., Kamil, J.Y.V.A., & Jeon, J.J. (2002). Antioxidant role of chitosan in a cooked cod (*Gadus morhua*) model system. *Journal of Food Lipids*, 9, 57-64.

Tian, J., Yu, J., & Sun, X. (2008). Chitosan microspheres as candidate plasmid vaccine carrier for oral immunization of Japanese flounder (*Paralichthys olivaceus*). *Veterinary immunology and Immunopathology*, 126, 220-229.

Tharanathan, R.N., & Kittur, F.S. (2003). Chitin- the undisputed biomolecule of great potential. *Critical Review in Food Science and Nutrition*, 43, 61-87.

Weist, J.L., & Karel, M. (1992). Development of a fluorescence sensor to monitor lipid oxidation. 1. Florescence spectra of chitosan powder and polyamide powder affect exposure to volatile lipid oxidation products. Journal of Agriculture and Food Chemistry, 40, 1158-1162.

Wibowo, S. (2003). Effect of the molecular weight and degree of deacetylation of chitosan and nutritional evaluation of solid recovered from surimi processing plant. PhD. Thesis, Food Science and Technology, Oregon State University.

Wibowo, S., Savant, V., Cherian, G., Savange, T.F., Velaquez, G., & Torres, J.A. (2007b). A feeding study to assess nutritional quality and safety of surimi wash water proteins recovered by a chitosan-alginate complex. *Journal of Food Scinece*, 72, 179-184.

Wibowo, S., Velazquez, G., Savant, V., & Torres, J.A. (2005). Surimi wash water treatment for protein recovery: effect of chitosan-alginate complex concentration and treatment time on protein adsorption. *Bioresoures Technology*, 96, 665-671.

Wibowo, S., Velazquez, G., Savant, V., & Torres, A. (2007a). Effect of chitosan type on protein and water recovery efficiency from surimi wash water treated with chitosan-alginate complexes. *Bioresources Technology*, 98, 539-545.

Ye, M., Neetao, H., & Chen, H. (2008). Effectiveness of chitosan-coated plastic films incorporating antimicrobials in inhibition of *Listeria monocytogens* on cold-smoked salmon. *International Journal of Food Microbiology*, 127, 235-240.

Chapter 16

Polysaccharides as Carriers and Protectors of Additives and Bioactive Compounds in Foods

Rosa M. Raybaudi-Massilia and Jonathan Mosqueda-Melgar

Additional information is available at the end of the chapter

http://dx.doi.org/10.5772/50206

1. Introduction

Overall quality and shelf-life of fresh foods post-harvest or -slaughter is reduced by several factors including microbial growth, water loss, enzymatic browning, lipid oxidation, off-flavor, texture deterioration, rise in respiration rate and senescence processes, among others. In fresh-cut fruits and vegetables, these events are accelerated due to lesions of tissues during peeling, slicing and cutting [1]; whereas, in meat products these events are accelerated due to lesions of tissues during cutting and grounding.

Fresh-cut fruits and vegetables during mechanical operations are exposed to spoil because the natural protection of fruit (the peel or skin) is generally removed and hence, they become highly susceptible to microbial growth due to the leakage of juices and sugar from damaged tissues which allow the growth and fermentation of some microorganism. Likewise, during processing, enzymes such as polyphenol oxidase, polygalacturonase and pectin methylesterase are released, thus causing, browning and softening from the tissues, respectively. These enzymes come in contact with phenolic compounds for forming brown pigments, and hydrolyze the α-1,4-glucosidic bonds to degrade the tissues [2].

Appearance/color, texture, and flavor are the main quality attributes that affect consumer acceptance of meat, and lipid oxidation is one of the primary causes by quality deterioration in meat and meat products [3]. The meat once cut or sliced is exposed to the surrounding environment, and cell compounds released during mechanical operations react with the environment and cause quality deterioration of tissues. Lipid oxidation occur when oxygen come in contact with lipid present in pieces of meats, being the iron the major catalyst in lipid oxidation processes. This process is associated with the presence of free radicals that lead to the production of aldehydes, which are responsible for the development of rancid flavors and changes in the color of meat [4].

INTECH
open science | open minds

© 2012 The Author(s). Licensee InTech. This chapter is distributed under the terms of the Creative Commons Attribution License http://creativecommons.org/licenses/by/3.0), which permits unrestricted use, distribution, and reproduction in any medium, provided the original work is properly cited.

On the other hand, dairy products such as fresh and semi-hard cheeses are complex food products consisting mainly of casein, fat, and water. Such products are highly perishable due to the high content of moisture (only in fresh cheeses) and microorganisms, and some cases high fat-content [5]; therefore, off-flavor, lipid oxidation and microbial spoilage are the major quality deteriorations.

Because of happened issues in the past with fresh or fresh-cut products, new technologies have been applied to counteract these negatives effects. Among them, polysaccharide-based edible films and coatings have emerged like good alternative for enhancing the quality and safety of such foods. Edible films and coatings have been used to reduce the deleterious effect caused by minimal processing. The semipermeable barrier provided by edible coatings is focused to extend shelf-life by reducing moisture and solutes migration, gas exchange, respiration and oxidative reaction rates, as well as suppress physiological disorders on fresh and/or fresh-cut foods [6]. However, the use of edible films and coatings for a wide range of food products, including fresh and minimally processed vegetables and fruits, has received an increasing interest because films and coatings can serve as carriers for a wide range of food additives including: antimicrobials, antioxidants and antisoftening compounds into edible films or coatings to provide a novel way for enhancing the safety and shelf-life of fresh, fresh-cut or ready-to-eat foods [7-10].

The new generation of edible films and coatings is being especially designed to increase their functionalities by incorporating natural or nutraceutical/functional ingredients such as probiotics, minerals and vitamins [10-11]. In addition, the sensory quality of coated products with edible materials can be also improved [2,7-9]. On the other hand, encapsulation (microencapsulation or nanoencapsulation) are being currently applied to foods to preserve and protect the additive or bioactive compounds from the surrounding environment [12-14].

In this chapter will discuss the use of polyssacharide-based edible films and coatings as polymeric matrix to carrier additive or bioactive compounds such as antimicrobial, antioxidant, antisoftening and nutraceutical for enhancing the shelf-life, safety and sensory attributes of fresh food products, as well as methodologies of forming and application of edible films and coatings and futures trends using microencapsulation or nanotechnology.

2. Use of polysaccharide-based edible films and coatings as carriers of additives and bioactive compounds on foods

The overall quality of food products decreases from harvest or slaughter until they are consumed. Quality loss may be due to microbiological, enzymatic, chemical, or physical changes. Therefore, food additives should be added to prevent the quality loss and extend the shelf-life of foods. The use of films and coatings have been a good alternative for carrier different additives and bioactive compounds to the food, as well as, to protect them of the water loss, volatile compounds loss, discoloration, gas permeability and microbial spoilage; since, these can to guarantee the controlled supply of antimicrobial, antioxidant, antisoftening and nutraceutical compounds. Tables 1, 2, 3 and 4 show the additives (antimicrobial, antioxidant, antisoftening and nutraceutical compounds) added on foods through polysaccharide-based edible films and coatings for improving the food quality and safety. Each of these additives is studied in the following sections of this chapter.

2.1. Carriers of antimicrobial compounds

Foods may be contaminated with pathogenic and spoilage microorganisms if bad manufacture practices are carried out during handling, processing, distribution and commercialization [15]. Therefore, antimicrobial compounds should be used during processing and packaging for controlling the microbiological safety and quality, and prolonging the shelf-life of foods. Food antimicrobials are chemical compounds added or naturally occurring in foods to inhibit or inactivate populations of pathogenic and spoilage microorganisms.

Several studies have demonstrated that antimicrobials such as organic acids, enzymes, essential oils, spices and bacteriocin incorporated into polysaccharide-based edible films and coatings have been effective for controlling pathogenic and spoilage microorganisms in different foods (Table 1). In this context, different researchers have demonstrated that the incorporation of bacteriocins into alginate-based film have been effective to inactivate or delay the growth of some pathogenic microorganisms. The alginates are anionic polysaccharides from the cell walls of brown algae that can serve to prepare carriers solution of antimicrobial substances. Hence, Cutter and Siragusa [16], Natrajan and Sheldon [17] and Milette et al. [18] achieved to reduce populations of *Brochothrix thermosphacta* (> 3.0 log CFU/g), *Salmonella enterica* ser. Typhimurium (> 4.0 log CFU/g) and *Staphylococcus aureus* (> 2.5 log CFU/g) on ground beef, poultry skin and beef fillets, respectively, using calcium alginate-based film and coating, and palmitoylated alginate incorporated with nisin (from 5 to 100,000 mg/mL) during storage refrigerated. Likewise, Datta et al. [19] indicated that the growth of *Listeria monocytogenes* and *Salmonella enterica* ser. Anatum was suppressed in the range of 2.2 to 2.8 log CFU/g in smoked salmon coated with an alginate coating containing oyster lysozyme at 160,000 mg/g plus nisin at 10 mg/g during storage at 4ºC by 35 days. Neetoo et al. [20] delayed the growth of *L. monocytogenes* on cold-smoked salmon slices and fillets during the 30 days storage at 4ºC using alginate-based edible coating with sodium lactate (2.4%) and diacetate (0.25%). Marcos et al. [21] reported a bacteriostatic effect against *L. monocytogenes* inoculated in sliced cooked ham during 60 days of storage at 1ºC, when enterocins A and B (2,000 AU/cm^2) were incorporated into an alginate film.

On the other hand, essential oils and their active compounds have been also incorporated into the alginate-based films and coatings to control the growth of pathogenic and spoilage microorganisms in several foods [22-26]. Hence, Oussalah et al. [22] evaluated the effect of an alginate-based film containing essential oils of Spanish oregano, Chinese cinnamon or winter savory at 1% w/v against populations of *S. enterica* ser. Typhimurium or *E. coli* O157:H7 inoculated in beef muscle slices stored at 4 ºC by 5 days. These authors reported that films including essential oils of oregano or cinnamon were more effectives against *S enterica* ser. Typhimurium (>1 log cycle); whereas, films including essential oils of oregano

Type of polysaccharide matrix		Food	Antimicrobial compounds	Reference
Film	Coating			
Alginate	-	Ground beef	Nisin	[16]
Alginate	-	Poultry skin	Nisin	[17]
Alginate	-	Beef fillets	Nisin	[18]
Alginate	-	Smoked salmon	lysozyme / nisin	[19]
Alginate	-	Cooked ham sliced	Enterocin A and B	[21]
Alginate	-	Beef fillets	EOs of oregano, cinnamon, savory	[22]
Alginate	-	Cooked ham and bologna sliced	EOs of oregano, cinnamon, savory	[23]
-	Alginate	Cold-smoked salmon slices and fillets	Sodium lactate and diacetate	[20]
-	Alginate/ apple puree	Fresh-cut apple	Vanillin and EOs of lemongrass, oregano	[24]
-	Alginate	Fresh-cut melon	Malic acid and EOs of lemongrass, cinnamon, plamarose, eugenol, citral, geraniol	[25]
-	Alginate	Fresh-cut apple	Malic acid and EOs of lemongrass, cinnamon, clove, cinnamaldehyde, eugenol, citral	[26]
-	Alginate	Fresh-cut apple	Potassium sorbate	[27]
	Alginate	Roasted turkey	Sodium lactate and diacetate	[29]
Carrageenan	-	Fresh chicken breast	Ovotransferrin	[28]
Chitosan	-	Cooked ham, bologna and pastrami	Cinnamaldehyde, acetic, propionic and lauric acids	[32]
Chitosa	Chitosan	Mozzarella cheese	Lysozyme	[74]
Chitosan	-	Cheese	Natamycin	[38]
-	Chitosan	Whole strawberry	Potassium sorbate	[33]
-	Chitosan	Rainbow trout	Cinnamon oil	[37]
-	Chitosan	Roasted turkey	Sodium lactate and diacetate	[29]
Chitosa	Chitosan	Cold-smoked salmon	Potassium sorbate, sodium lactate and diacetate	[30]
-	Chitosan / Plastic	Ham steaks	Sodium lactate, diacetate and benzoate, potassium sorbate, or nisin	[36]
Chitosan	-	Pork sausages	Green tea extract	[39]
-	Chitosan / MC	Fresh-cut Pineapple and melon	Vanillin	[35]
CMC	-	Fresh pistachios	Potassium sorbate	[43]
Cellulose	-	Cooked ham sliced	Pediocin	[42]
Cellulose	-	Frankfurter sausages	Nisin	[41]
MC / HPMC	-	Hot Dog Sausage	Nisin	[40]
-	HPMC	Whole oranges	Potassium sorbate, sodium benzoate, sodium propionate	[44]
-	HPMC	Whole strawberry	Potassium sorbate	[33]
-	Pectin	Pork patties	Powder of green tea	[47]
-	Pectin	Roasted turkey	Sodium lactate and diacetate	[29]
-	Starch	Minimally processed carrots	Chitosan	[45]
-	Starch	Roasted turkey	Sodium lactate and diacetate	[29]
-	Starch / Gum	Fruit-based salad, romaine hearts and pork slices	Green tea extract	[46]

MC: methyl cellulose; CMC: carboxy methyl cellulose; HPMC: hydroxyl propyl methyl cellulose; EOs: essential oils

Table 1. Major antimicrobial compounds applied on foods through polysaccharide-based edible films and coatings

Type of polysaccharide matrix		Food	Antioxidant compounds	Reference
Film	Coating			
-	Alginate	Fresh-cut apple	Glutathione and N-Acetyl-cysteine	[61-62]
-	Alginate	Fresh-cut pears	Glutathione and N-Acetyl-cysteine	[63]
-	Alginate	Fresh-cut apples	Calcium chloride	[27]
-	Alginate	Bream (freshwater fish)	Vitamin C and tea polyphenols	[53]
-	Carrageenan	Fresh-cut apples	Ascorbic, citric and oxalic acids	[59]
-	Carrageenan	Fresh-cut banana	Ascorbic acid and cysteine	[60]
-	CMC	Fresh-cut apples and potatoes	Ascorbic acid and TBHQ	[57]
-	Gellan	Fresh-cut apple	Glutathione and N-Acetyl-cysteine	[61-62]
-	Gellan	Fresh-cut pears	Glutathione and N-Acetyl-cysteine	[63]
-	HPMC	Toasted almonds	Ascorbic, citric and EO ginger	[52]
-	MC	Fresh-cut apples	Ascorbic acid	[58]
-	Maltodextrin	Fresh-cut apples	Ascorbic acid	[58]
-	Pectin	Fresh-cut pears	Glutathione and N-Acetyl-cysteine	[63]
-	Pectin	Pork patties	Powder of green tea	[47]

MC: methyl cellulose; CMC: carboxy methyl cellulose; HPMC: hydroxyl propyl methyl cellulose; EOs: essential oils

Table 2. Major antioxidant compounds applied on foods through polysaccharide-based edible films and coatings

Type of polysaccharide matrix		Food	Antisoftening compounds	Reference
Film	Coating			
-	Alginate	Fresh-cut apples	Calcium lactate	[26]
-	Alginate	Fresh-cut melons	Calcium lactate	[27]
-	Alginate	Fresh-cut apples	Calcium chloride	[61-62]
-	Alginate	Fresh-cut melons	Calcium chloride	[67]
-	Alginate	Fresh-cut pears	Calcium chloride	[63]
-	Alginate	Fresh-cut apples	Calcium chloride	[27]
-	Alginate	Fresh-cut papayas	Calcium chloride	[68]
-	Alginate / Apple puree	Fresh-cut apples	Calcium chloride	[24]
-	Carrageenan	Fresh-cut apples	Calcium chloride	[59]
-	Carrageenan	Fresh-cut banana	Calcium chloride	[60]
-	Gellan	Fresh-cut papayas	Calcium chloride	[68]
-	Gellan	Fresh-cut apples	Calcium chloride	[61-62]
-	Gellan	Fresh-cut melons	Calcium chloride	[67]
-	Gellan	Fresh-cut pears	Calcium chloride	[63]
-	Pectin	Fresh-cut melons	Calcium chloride	[67]
-	Pectin	Fresh-cut pears	Calcium chloride	[63]

Table 3. Major antisoftening compounds applied on foods through polysaccharide-based edible films and coatings

Type of polysaccharide matrix		Food	Nutraceutical compounds	Reference
Film	**Coating**			
-	Alginate	Fresh-cut papayas	Probiotics	[11]
-	Gellan	Fresh-cut papayas	Probiotics	[11]
-	Chitosan	Lingcod fillets	Omega 3 and Vitamin E	[34]
-	Chitosan	Whole strawberry	Calcium	[72]
-	Chitosan	Whole strawberry and red raspberry	Calcium and Vitamin E	[73]
-	Xanthan gum	Peeled baby carrots	Calcium and Vitamin E	[71]

Table 4. Major nutraceutical compounds applied on foods through polysaccharide-based edible films and coatings

was more effective against *E. coli* O157:H7 (> 2 log cycles). Similarly, Oussalah et al. [23] studied the effect of alginate-based edible film containing essential oils of Spanish oregano, Chinese cinnamon, or winter savory at 1% (w/v) against *S. enterica* ser. Typhimurium or *L. monocytogenes* inoculated onto bologna and ham slices. These authors concluded that alginate-based films containing essential oil of cinnamon was the most effective in reducing the populations of both pathogenic microorganisms by more than 2 logs CFU/g on bologna and ham sliced. In the same way, Rojas-Grau et al. [24] studied the antimicrobial effect of essential oils of lemongrass (1and 1.5%) and oregano (0.1 and 0.5%), and vanillin (0.3 and 0.6%) incorporated into coating forming solutions based on alginate and apple puree against the naturally occurring microorganisms and *Listeria innocua* inoculated on fresh-cut apples. These authors found that all the essential oils used significantly inhibited the native flora during 21 days of storage at 4ºC, being lemongrass and oregano oils more effective against *L. innocua* than vanillin. Likewise, Raybaudi-Massilia et al. [25] reported significant reduction (3–5 log cycles) of the inoculated *Salmonella enterica* var. Enteritidis population in pieces of melon when an alginate-based edible coating containing malic acid (2.5%), alone or in combination with essential oils of cinnamon, palmarose or lemongrass at 0.3 and 0.7% or their actives compounds eugenol, geraniol and citral at 0.5%, were applied. In addition, inhibition of the native flora by more than 21 days of storage was also observed at 5ºC. Similar results were found by Raybaudi-Massilia et al. [26], who evaluated the antimicrobial effect of an alginate-based edible coating with malic acid (2.5%) incorporated, alone or in combination with essential oils of cinnamon bark, clove or lemongrass at 0.3 and 0.7% or their actives compounds cynnamaldehyde, eugenol or citral at 0.5% on fresh-cut apples. They reached to reduce population of *Escherichia coli* O157:H7 (4 log cycles) after 30 days of refrigerated storage (5ºC), as well as to inhibit the native flora by more than 30 days.

Other antimicrobial compounds such as potassium sorbate, ovotransferrin, sodium lactate and sodium acetate have been also applied to fresh-cut apples, fresh chicken breast and ready-to-eat roasted turkey through an alginate-based edible coating to inhibit the native flora growth. In such sense, Olivas et al. [27] inhibited the microbial growth of mesophilic and psychrotropic bacteria, moulds and yeasts in apple slices coated with an edible alginate coating containing 0.05% potassium sorbate during 8 days of storage at 5ºC. Likewise, Seol et al. [28] reduced populations of total microorganisms (about 2 log cycles) and *E. coli* (about 3 log cycles) on fresh chicken breast stored at 5ºC after 7 days, when a κ-carrageenan-based edible film containing ovotransferrin (25 mg) and EDTA (5 mM) was applied on its surface. Jiang et al. [29-30] showed that potassium sorbate (0.15%), sodium lactate (1.2-2.4%) and sodium diacetate (0.25-0.50%) incorporated into chitosan- or alginate-based edible coating and film were able to inactivate *L. monocytogenes* (about 1-3 log CFU/g) on ready-to-eat cold-smoked salmon and roasted turkey stored at 4°C. All these results have demonstrated that alginate- κ-carrageenan-based film and coating are excellent carriers of antimicrobial substances on meat, poultry and fruits and vegetables products for reducing populations of pathogenic microorganisms.

In the same way, chitosan which is a linear polysaccharide consisting of (1,4)-linked 2-amino-deoxy-b-D-glucan, and a deacetylated derivative of chitin, and the second most abundant polysaccharide found in nature after cellulose [31] has been used as carrier of antimicrobial compounds in other foods. In such sense, Ouattara et al. [32] evaluated the effectiveness of chitosan films incorporated with acetic or propionic acid, with or without addition of lauric acid or cinnamaldehyde to preserve vacuum-packaged bologna, cooked ham and pastrami during refrigerated storage. The efficacies of the films to inhibit the microbial growth were tested against native lactic acid bacteria, Enterobacteriaceae, and against *Lactobacillus sakei* or *Serratia liquefaciens* inoculated on the surface of products. The authors indicated that the growth of lactic acid bacteria were not affected by the antimicrobial films, but the growth of Enterobacteriaceae and *S. liquefaciens* was delayed or completely inhibited after application. Park et al. [33] showed the antifungal effect of a chitosan-based edible coating containing potassium sorbate (0.3%) to inhibit the *Cladosporium* sp. and *Rhizopus* sp, total aerobic count and coliforms growth, and in fresh strawberries stored at 5°C and 50% RH by 23 days. Coating treatment also reduced total aerobic count, coliforms, and weight loss of strawberries during storage. Duan et al. [34] reduced about 1 log cycle the populations of *L. monocytogenes, E. coli,* or *Pseudomonas fluorescens* inoculated on the surface of Mozzarella cheese using chitosan composite films and coatings incorporated with lysozyme and storage at 10 ºC. Sangsuwan et al. [35] studied the antimicrobial effect of a chitosan/MC film incorporated with vanillin against *E. coli* and *Saccharomyces cerevisiae* inoculated on fresh-cut cantaloupe and pineapple. They found that antimicrobial film inactivated populations of *E. coli* and *S. cerevisae* on fresh-cut cantaloupe by more than 5 and 0.6 log CFU/g during 8 and 20 days of storage, respectively, at 10°C. Whereas, this antimicrobial film inactivated *S. cerevisiae* on fresh-cut pineapple by more than 4 log CFU/g during 12 days of storage at 10°C, but against *E. coli* there was not significant reductions. Ye et al. [36] used a plastic film coated with chitosan and Sodium lactate (1%), diacetate (0.25%), and benzoate (0.1%), potassium sorbate (0.3%) or nisin (5 mg/cm^2) for inhibiting the growth of *L. monocytogenes* on strawberries during 10 days of storage at room temperature (20ºC). Ojagh et al. [37] extended the shelf-life of Rainbow trout (a fish native of North America) during 16 days at 4ºC incorporating cinnamon oil (at 1.5%) into a matrix of chitosan-based edible coating. Fajardo et al. [38] evaluated the antifungal activity of chitosan-based edible coating containing 0.5mg/mL natamycin on semi-hard "Saloio" cheese; and demonstrated that populations of moulds and yeasts were reduced by about 1.1 log CFU/g compared to control samples after 27 days of refrigerated storage. Jiang et al. [29] showed that a combination of sodium lactate and sodium diacetate incorporated into chitosan edible coating was able to inactivate *L. monocytogenes* on ready-to-eat roasted turkey stored at 4°C. Siripatrawan and Noipha [39] used a chitosan film containing green tea extract as active packaging for extending shelf-life of pork sausages. These authors completely inhibited the microbial growth in pork sausages refrigerated (4 °C). Hence, chitosan can be used as a natural antimicrobial coating on fresh strawberries to control the growth of microorganisms, thus extending shelf-life of the products

Films and coatings based on cellulose or derivatives such as methyl cellulose (MC), carboxy methyl cellulose (CMC) or hydroxy propyl methyl cellulose (HPMC) containing antimicrobial compounds have been used to control microbial growth and extend the shelf-life of several foods. In such sense, Franklin et al. [40] determined the effectiveness of packaging films coated with a MC/HPMC–based solution containing 100, 75, 25 or 1.563 mg/ml nisin for controlling *L. monocytogenes* on the surfaces of vacuum-packaged hot dogs. They found that packaging films coated with a cellulose-based solution containing 100 and 75 mg/ml nisin significantly decreased ($P \leq 0.05$) *L. monocytogenes* populations on the surface of hot dogs by greater than 2 logs CFU/g throughout the 60 days of storage. Nguyen et al. [41] developed and used cellulose films produced by bacteria containing nisin to control *L. monocytogenes* and total aerobic bacteria on the surface of vacuum-packaged frankfurters. Bacterial cellulose films were produced by *Gluconacetobacter xylinus* K3 in corn steep liquor-mannitol medium and were subsequently purified before nisin was incorporated into them. Cellulose films with nisin at 25 mg/ml significantly reduced ($P<0.05$) *L. monocytogenes* (approximately 2 log CFU/g) and total aerobic bacteria (approximately 3.3 log CFU/g) counts on frankfurters after 14 days of storage as compared to the control samples. Whereas, Santiago-Silva et al. [42] developed and evaluated the antimicrobial efficiency of cellulose films with pediocin (antimicrobial peptide produced by *Pediococcus* sp.) incorporated at 25% and 50% of cellulose weight on sliced ham. They found that antimicrobial films were more effective against *L. innocua* than *Salmonella* sp., since the 50% pediocin-film showed a reduction of 2 log CFU/g in relation to control treatment after 15 days of storage; whereas, the 25% and 50% pediocin-films had similar performance on *Salmonella* sp. about 0.5 log CFU/g reductions in relation to control, after 12 days of storage at 12ºC. On the other hand, Park et al. [33] achieved to inhibit the growth of *Cladosporium* sp., *Rhizopus* sp, total aerobic count and coliforms on fresh strawberry through a HPMC-based edible coating containing potassium sorbate (0.3%) stored at 5°C and 50% RH by 23 days. Sayanjali et al. [43] evaluated the antimicrobial properties of edible films based on CMC containing potassium sorbate (at 0.25, 0.5 and 1.0%) applied on fresh pistachios, and reported that all concentrations of potassium sorbate used inhibited the growth of molds. Valencia-Chamorro et al. [44] studied the antifungal effect of HPMC based coatings with potassium sorbate (2%), sodium benzoate (2.5%), sodium propionate (0.5%) and their combinations on the postharvest conservation of "Valencia" oranges. These authors reported that the application of HPMC coatings reduce significantly the effects caused by *Penicillum digitatum* and *Penicilllum italicum* inoculated in the surface of the oranges, resulting more effectives those coatings with potassium sorbate and sodium propionate combined.

Others polysaccharides-based films and coatings such as pectins and starches have been used also as carriers of antimicrobials compounds in foods. Durango et al. [45] controlled the growth of mesophilic aerobes, yeasts and moulds and psychrotrophics populations in processed minimally carrots during the first 5 days of storage at 15ºC using yam starch-based edible coatings containing chitosan (0.5 and 1.5%). In the same way, Chiu and Lai [46] studied the antimicrobial properties of edible coatings based on a tapioca starch/decolorized hsian-tsao leaf gum matrix with incorporated green tea extracts on fruit-based salads, romaine hearts and pork slices. The authors indicated that when green tea extracts at 1% were added into edible coating formulations, the aerobic count successfully decreased and growth of yeasts/molds decreases by 1 to 2 logs CFU/g in fruit-based salads. In addition, they reported that romaine hearts and pork slices coated with these antimicrobial edible coatings reduced populations of Gram positive bacteria from 4 to 6 logs CFU/g during 48 h of refrigerated storage. On the other hand, Kang et al. [47] evaluated the microbiological quality of pork hamburger coated with a pectin-based edible coating with incorporated green tea powder (0.5%), and packed in air or vacuum during 14 days at 10ºC. These authors reported that initial population of total aerobic microorganisms (10^4 CFU/mg) decreased until undetectable levels by more than 7 days under vacuum conditions; whereas, in normal conditions of atmosphere (air) a level of 10^5 CFU/mg was reached at the same time. Jiang et al. [29] showed that a combination of sodium lactate and sodium diacetate incorporated into pectin-based edible coating was able to inactivate populations of *L. monocytogenes* on ready-to-eat roasted turkey stored at 4°C.

Previous results have showed that several polysaccharides-based films and coatings (alginate, carrageenan, chitosan, cellulose derivatives, pectin, starch and apple puree) could be used as outstanding carriers of antimicrobial substances for ensuring the quality and safety of foods in the meat, poultry, seafood, dairy, fruits and vegetables industries. In addition, the incorporation of essential oils into films and coatings formulations may contribute to prevent the water vapor permeability and decreases the solubility of films and coatings in foods with high content of humidity.

3. Carriers of antioxidant compounds

Antioxidant compounds can also be incorporated into edible films and coatings to avoid the food oxidation and browning. In such sense, rosemary oleoresin, an extract of spice with antioxidant activity, has been added into starch-alginate coatings to inhibit the lipid oxidation and warmed-over flavor (WOF) development in precooked pork chops [48] and beef patties [49]. In the same way, tocopherols have been incorporated into starch-alginate coatings to retard the formation of WOF in precooked pork chops [50]. Wu et al. [51] studied the effect of starch-alginate (SA), SA-stearic acid (SAS), SA-tocopherol (SAT), SAS-tocopherol (SAST), SAT-coated (SATC), and SAST-coated (SASTC) films on moisture loss and lipid oxidation in precooked ground-beef patties. These authors reported that tocopherol-treated films were more effective ($P < 0.05$) in inhibiting lipid oxidation than those tocopherol-untreated films on ground-beef patties. However, SAS-based films were more effective ($P < 0.05$) in controlling moisture loss than lipid oxidation. Atarés et al. [52] evaluated the antioxidant efficiency of HPMC coatings with ascorbic acid, citric acid or ginger essential oil incorporated on toasted almonds to avoid the lipid oxidation. They concluded that films with ascorbic and citric acid showed a cross-linking effect, and were the most effective protectors against oxidation of almonds, due to both their antioxidant effect and the tighter structure which leads to lower oxygen permeability. Khang et al. [47] found that lipid oxidation decreased and radical scavenging increased in the pork patties coated with a pectin-based edible coating containing green tea leaf extract (0.5%) during 14 days at 10ºC. These authors indicated that coated patties held higher moisture contents than the controls in both air- and vacuum packaging. Song et al. [53] indicated that sodium alginate-based edible coating containing vitamin C (5%) or tea polyphenols (0.3%) were able to delay the chemical spoilage and water loss of bream (*Megalobrama amblycephala*), in addition to enhancing the overall sensory attributes, in comparison with uncoated bream during 21 days storage at 4 ± 1ºC.

On the other hand, the color in fresh-cut fruits and vegetables is of great importance, since oxidation and enzymatic browning take place quickly upon contact with oxygen during processing, leading to discoloration [54]. Browning phenomena in fresh-cut products are caused when, after mechanical operations (cutting, slicing, coring, shredding, etc) during processing, enzymes, which are released from wounded tissues, come in contact with phenolic components to give dark colored pigments [55]. Such phenomenon is caused by the action of a group of enzymes called polyphenol oxidases (PPOs), which can oxidize the phenolic substrates to *o*-quinones in presence of oxygen [56]. Therefore, the application of antioxidant agents incorporated into edible coatings would be a good alternative to ensure the inhibition of browning, to prevent ascorbic acid or vitamin C loss, and extend the shelf-life of fresh-cut fruits and vegetables [9]. In such sense, Baldwin et al. [57] reported that ascorbic acid (0.5%) and *ter*-butyl-hydro-quinone (0.2%) had a better effect on the inhibition of browning in fresh-cut apples and potatoes throughout storage when these antioxidants were incorporated into an edible coating based on CMC than when these were used in an aqueous dipping solution after 14 days at 4ºC. Both methods were effective during the first day of storage, but samples coated with the edible coating prevented browning for a longer time than those samples dipped in an aqueous solution alone. Brancoli and Barbosa-Cánovas [58] achieved a decreasing browning in surface of apple slices during 21 days of storage at 4ºC using maltodextrin and MC-based coatings containing ascorbic acid (1%). Likewise, Lee et al. [59] delayed the browning of fresh-cut apples using antibrowning agents such as ascorbic (1%), citric (1%), oxalic (0.05%) acid or their combinations incorporated into edible coatings based on carrageenan. These authors observed an inhibition of the enzymatic browning in fresh-cut apples during 14 days storage at 3ºC. In addition, edible coating with antioxidants obtained higher sensory scores (positive effect) during sensory evaluation than non-coated apples. In the same way, Bico et al. [60] reached to retard the browning of fresh-cut bananas using ascorbic acid and cysteine at 0.75% incorporated into an edible coating based on carrageenan during 5 days of storage at 5ºC. Rojas-Grau et al. [61-62] inhibited the browning in fresh-cut apples using edible coatings based on alginate or gellan with the addition of glutathione (up to 2%) or N-acetyl-cysteine (up to 2%), or their combination. These authors indicated that a concentration of 1% each of the antibrowning agents was needed to maintain the color of cut apples. Similar results were also obtained by Oms-Oliu et al. [63], who achieved browning inhibition of fresh-cut "Flor de invierno" pears for 14 days at 4ºC using N-acetyl- cysteine (0.75%) and glutathione (0.75%) incorporated into edible coatings based on alginate, gellan or pectin. Olivas et al. [27] delayed the development of browning in apple slices during 8 days of storage at 5ºC after applying alginate coatings containing calcium chloride (10%). Calcium chloride is an anti-browning agent known to inhibit PPO by interaction of the chloride ion with copper at the PPO active site [64].

Based on the different works reported in the bibliography is possible indicates that several polysaccharides-based films and coatings (alginate, carrageenan, cellulose derivatives, pectin, gellan and maltodextrin) could be used as excellent carriers of antioxidant substances for avoiding the lipid oxidation and enzymatic browning of meat and fruits products.

4. Carriers of antisoftening compounds

Foods more susceptible to the texture loss are fresh-cut fruits and vegetables. This fact is due to that during mechanical operations (peeling, cutting, sliced, shredded) plant tissues are breakdown and enzymes such as pectinolytic and proteolytic are released, thus causing softening [1]. In addition, these enzymes could also affect the morphology, cell wall middle lamella structure, cell turgor, water content, and biochemical components [65]. Pectinase enzymes such as polygalacturonase and pectin methylesterase are responsible for texture losses in plant tissues. Polygalacturonase hydrolyses the α-1,4-glucosidic bond among anhydrogalacturonic acid units, whereas, pectin methylesterase hydrolyses the methyl-ester bonds of pectin to give pectic acid and methanol, thus resulting in texture degradation because of hydrolysis of the pectin polymers [1]. Nonetheless, treatments with calcium can helping to counteract this problem improving the firmness of fruit tissues by reacting with pectic acid present in the cell wall to form calcium pectate, which reinforces the molecular bonding among constituents of the cell wall, thus delaying the senescence and controlling physiological disorders in fruits and vegetables [8,66]. Different studies have demonstrated that the use of polysaccharide based films and coatings (alginate, carrageenan, pectin, gellan and apple puree) as carriers of calcium chloride or lactate have resulted be a good alternative to prevent the firmness or texture loss of the fresh-cut fruits, which could be beneficial to the fresh-cut fruits industry.

In this sense, Oms-Oliu et al. [63,67] and RojasGraü et al. [24,61-62] observed that fresh-cut melons, pears, and apples coated with alginate-, gellan-, pectin- or apple-puree edible coatings containing calcium chloride (2%) maintain in excellent conditions their initial firmness during refrigerated storage (4ºC) from 14 to 21 days. The authors indicated that polysaccharide matrices with substances increased the water vapor resistance, thus preventing dehydration, and they had an inhibitory effect on ethylene production, but O_2 and CO_2 production was not affected. Similar effects were achieved by Olivas et al. [27], who preserved the firmness of apple slices stored at 5ºC for 10 days by using an alginate edible coating containing calcium chloride (10%). Raybaudi-Massilia et al. [25,26] showed that the incorporation of calcium lactate (2%) into an alginate-edible coating maintained the firmness of fresh-cut apples and melons during 21 days at 5ºC. Similarly, Tapia et al. [68] improved the firmness of fresh-cut papaya with the addition of calcium chloride (2%) into alginate- and gellan edible coating during the period studied (8 days at 4ºC). Likewise, Lee et al. [59] and Bico et al. [60] kept the firmness of fresh-cut apple and banana slices storage at refrigerated temperature using a carrageenan-based edible coating containing calcium chloride (1%).

5. Carriers of nutraceutical compounds

Nutraceuticals are chemical compounds found as natural components of foods or other ingestible forms that have been determined to be beneficial to the human body in preventing or treating one or more diseases or improving physiological performance [69]. Calcium and vitamin E are the most important nutraceutical compounds and, they can play significant roles in the human body in preventing certain diseases [70]. Nonetheless, probiotics are being used currently as a functional compound in foods, since potential health benefits and biological functions of bifidobacteria in humans like the intestinal production of lactic and acetic acids, pathogens inhibition, reduction of colon cancer risks, cholesterol reduction in serum, improved calcium absorption, and activation of the immune system, among others [11]. Thus, nutraceutical compounds carried into edible coatings and films to strengthen and increase the nutritional value of foods have been researched.

Edible coatings can provide an excellent vehicle to further enhance the health benefit of products like berry fruits where the lack of some important nutraceuticals, such as vitamin E and calcium may be compensated by incorporating them into the coatings [10]. In this way, Mei et al. [71] used xanthan gum coating as a carrier of calcium (as calcium lactate at 5%) and vitamin E (as α-tocopheryl acetate at 0.2%) for covering peeled baby carrots. The authors indicated that calcium and vitamin E contents of the coated samples (85g per serving), increased from 2.6 to 6.6% and from 0 to about 67% of the Dietary Reference Intakes values, respectively. In addition, they found that edible coatings improved the desirable surface color of carrots without significant effects on the taste, texture and fresh aroma. Hernández-Muñoz et al. [72] coated strawberries (*Fragaria* x *ananassa* Duch.) with a chitosan-based edible coating containing 1% calcium gluconate and stored during 4 days at 20°C. These authors found that strawberries coated with chitosan-based edible coating with incorporated calcium were better retained in coating (3,079 g/kg dry matter) than in strawberries dipped in calcium solutions alone (2,340 g/kg), thus resulting in increased nutritional value. Likewise, Han et al. [73] used chitosan-based edible coatings containing 5% Gluconal® CAL or 0.2% DL-α-tocopheryl acetate to enhance the nutritional value of strawberries (*Fragaria* × *ananassa*) and red raspberries (*Rubus ideaus*) stored at 2°C and 88% relative humidity (RH) for 3 weeks or at 23°C up to 6 months. They concluded that chitosan-based coatings containing calcium or Vitamin E significantly increased the content of these nutrients in both fresh and frozen fruits. These researchers also indicated that adding high concentrations of calcium or Vitamin E into chitosan-based coatings did not alter their anti-fungal and moisture barrier functions. Moreover, the coatings significantly decreased decay incidence and weight loss, drip loss and delayed the change in color, pH and titratable acidity of strawberries and red raspberries during cold storage. Duan et al. [74] increased total lipid and omega-3 fatty acid contents of fresh and frozen lingcod by about 3-fold and reduced TBARS (Thio-barbituric acid reactive substances) values in both fresh and frozen samples, incorporating 10% fish oil (containing 91.2% EPA (eicosapentaenoic acid) and DHA (docosahexaenoic acid)) plus 0.8% vitamin E into chitosan-based edible coating.

Developing edible coatings to carry high concentrations of nutraceuticals for nutritionally fortified foods can also be considered as an important way to afford functional characteristics to coated foods. In this context, Tapia et al. [11] managed to incorporate viable *Bifidobacterium lactis* Bb12 strains into alginate and gellan film-forming solutions to coat fresh-cut apples and papayas, and evaluated the effectiveness of such edible coatings to carry and support the probiotic culture. The authors reported that populations > 10^6 CFU/g of the microorganism were kept during 10 days of refrigerated storage. A viable bifidobacteria population of 5 logs CFU/g in the final product has been pointed out as the therapeutic minimum to attain health benefits [75].

In general, fruits, vegetables and seafood industries could apply different polysaccharides-based coatings (alginate, gellan, chitosan and gum) as excellent carriers of nutraceutical compounds for adding nutritive value and functional properties to the products.

6. Methodology for film and coating formation, incorporation of additives/bioactive compounds and ways of applications

6.1. Film and coating formation, and incorporation of additives/bioactives compounds

An edible film is essentially an interacting polymer network of three-dimensional gel structure. Despite the film-forming process, whether it is wet casting or dry casting, film-forming materials should form a spatially rearranged gel structure with all incorporated film-forming agents, such as biopolymers, plasticizers, other additives, and solvents in the case of wet casting. Biopolymers film-forming materials are generally gelatinized to produce film-forming solutions. Sometimes drying of the hydrogels is necessary to eliminate excess solvents from the gel structure. This does not mean that the film-forming mechanism during the drying process is only the extension of the wet-gelation mechanism. The film-forming mechanism during the drying process may differ from the wet-gelation mechanism, though wet gelation is initial stage of the film-forming process. There could be a critical stage of a transition from a wet gel to a dry film, which relates to a phase transition from a polymer-in-water (or other solvents) system to a water-in-polymer system [76].

Two processes can be used for film-production: dry and wet. The dry process of edible film production does not use liquid solvent, such as water or alcohol. Molten casting, extrusion, and heat pressing are good examples of dry process. For the dry process, heat is applied to the film-forming materials to increase the temperature to above the melting point of the film-forming materials, to cause them to flow. The wet process uses solvents for the dispersion of film-forming materials, followed by drying to remove the solvent and form a film structure. For the wet process the selection of solvents is the one of the most important factors. Since the film-forming solution should be edible and biodegradable, only water, ethanol and their mixtures are appropriated as solvents. To produce a homogeneous film structure avoiding phase separation, various emulsifiers can be added to the film forming solution. This solvent compatibility of ingredients is very important to develop homogeneous edible film and coating systems carrying active agents. All ingredients, including active agents as well as biopolymers and plasticizers should be homogeneously dissolved in solvent to produce film-forming solutions [76].

7. Ways of application of films and coatings

Different ways for film and coating application have been reported in the literature; being dipping, spraying, brushing, casting and wrapping the more commons methods [7,76-79]:

- *Dipping:* This method lends to food products that require several applications of coating materials or require a uniform coating on an irregular surface. After dipping, excess coating material is allowed to drain from the product and it is then dried or allowed to solidify. This method has been generally used to apply coating of alginate, gellan, chitosan, MC and pectin to fresh-cut fruit.
- *Spraying:* Film applied by spraying can be formed in a more uniform manner and thinner than those applied by dipping. Spraying, unlike dipping, is more suitable for applying a film to only one side of a food to be covered. This is desirable when protection is needed on only one surface, e.g., when a pizza crust is exposed to a moist sauce. Spraying can also be used to apply a thin second coating, such as the cation solution needed to cross-link alginate or pectin coatings.
- *Brushing:* This method consists in the direct application and distribution of the coating material in a liquid form using a hand brush.
- *Casting:* This technique, useful for forming free-standing films, is borrowed from methods developed for not edible films. For formation of a film the film-forming biopolymers are first dissolved in the solvent. If heating or pH adjustment enhances film formation and/or properties, this is done nest. If a composite film or coating based on an emulsion is desired, a lipid material, and possibly a surfactant, is added. Next the mixture is heated to above the lipid melting point and then homogenized. Degassing is an important step to eliminate bubble formation in the final film or coating. Finally, the edible film or coating is formed by applying the prepared formulation to the desired coating or product surface and allowing the solvent to evaporate

- *Wrapping:* this method is obtained from cast films depending on firmness and flexibility for wrapping surface. It allows films to be cut to any size, and serves as an innovative and easy method for carrying and delivering a wide variety of ingredients such as flavoring, spices and seasoning that can later be used to cover foods. This method is especially useful when applied to highly spicy materials that need to be separated from the food products.

8. Preserving and protecting bioactive compounds through microencapsulation

Microencapsulation is a technique by which solid, liquid or gaseous active ingredients are packaged within a second material for the purpose of protecting or shielding the active compound from the surrounding environment. Thus the active compound is designated as the core material, whereas the surrounding material forms the shell. This technique can be employed in a diverse range of fields such as agricultural, chemical, pharmaceutical, cosmetics, printing and food industry [13].

Microcapsules can be classified on the basis of their size or morphology. Thus, microcapsules range in size from one micron; whereas, some microcapsules whose diameter is in the nanometer range are referred as nanocapsules to emphasize their smaller size. On the other hand, morphology microcapsules can be classified into three basic categories as mono-core (also called single-core or reservoir type), poly-core (also called multiple-core) and matrix types (Figure 1). Mono-core are microcapsules that have a single hollow chamber within the capsule; Poly-core are microcapsules that have a number of different sized chambers within the shell; and matrix type are microparticles that has the active compounds integrated within the matrix of the shell material. However, the morphology of the internal structure of a microparticle depends mostly on the selected shell materials and the microencapsulation methods that are employed [12-13].

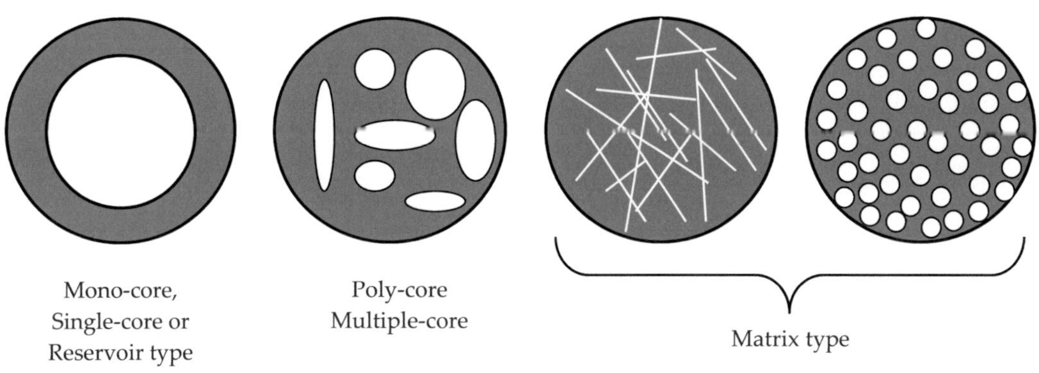

Figure 1. Morphology of microcapsules

Current trends of the consumers for eating healthy foods that preventing illness and to be low calories but rich in vitamins, minerals and other bioactive component have conduced to the researches and industrials to develop foods called "functional", where some ingredients to promote health are added. However simply adding ingredients to food products to improve nutritional value can compromise their taste, color, texture and aroma. Sometimes, they are slowly degraded and in consequence, lose their activity, or become hazardous by oxidation reactions. Active compounds can also react with components present in the food system, which may limit bioavailability. Microencapsulation is used to overcome all these challenges by providing viable texture blending, appealing aroma release, and taste, aroma and color masking. This technology enables to the food industries to incorporate minerals, vitamins, flavors and essential oils. In addition, microencapsulation can simplify the food manufacturing process by converting liquids to solid powder, decreasing production costs by allowing batch processing using low cost, powder handling equipment. Microcapsules also help at fragile and sensitive materials survive processing and packaging conditions and stabilize the shelf-life of the active compound.

On the other hand, applications of microencapsulations to foods have been increasing due to the protection of encapsulate materials of factors such as heat and humidity, allowing to maintain its stability and viability. The microcapsules help to food materials to withstand the conditions of processing and packing to improve taste, aroma, stability, nutritional value and appearance of their products. Some of the substances encapsulated have been fertilizers, oil of lemon, lipids, volatile flavors, probiotics, nutraceuticals, seeds of fruits like banana, grapes, guava, papaya, apple, blackberry, granadilla and citrus seeds. In this regard, the encapsulation offers great scope for conservation, germination and exchange of several fruit species, resulting in promising technique for the conservation, transport of transgenic plants and not seed-producing plants, lactase, colorants, enzymes, phytosterols, lutein, fatty acids, plant pigments, antioxidants, aromas and oleoresins, vitamins and mineral [14].

9. Pigments

Pigments are compounds very sensitive due to their instability in the presence of light, air, humidity and high temperatures therefore their use requires a chemical knowledge of their molecules and stability, in order to adapt them to the conditions of use during processing, packaging and distribution. One alternative for their use in the food industry is microencapsulation technology [80]. Carotenoids are used as dyes in food, beverages, cosmetics and animal feed, mainly poultry and fish. During the processing and storage, carotenoids can easily change in different isomers geometric and rust, this result in the reduction or loss of the dye and its biological properties. The main alternatives of applications to increase the stability of carotenoids and, allow its incorporation in hydrophilic environments, is the technique of microencapsulation by the method of spray called spray drying. In the same way, other pigments such as licopeno, lutein, enocianin, astaxantin, antocianins and pigments of nogal and urucú have also been encapsulated [14].

10. Vitamins

Both lipid-soluble (e.g. vitamin A, β-carotene, vitamins D, E and K) and water-soluble (e.g. ascorbic acid) vitamins can be encapsulated using various technologies. The most common reason for encapsulating these ingredients is to extend the shelf-life, either by protecting them against oxidation or by preventing reactions with components in the food system in which they are present. A good example is ascorbic acid (vitamin C), which is added extensively to a variety of food products as either an antioxidant or a vitamin supplement. Its application as a vitamin supplement is impaired by its high reactivity and, hence, poor stability in solution. It can degrade by a variety of mechanisms. For vitamin C encapsulation, both spray-cooling or spray-chilling and fluidized-bed coating can be used when the vitamins are added to solid foods, such as cereal bars, biscuits or bread. For application in liquid food systems, the best way to protect water-soluble ingredients is by encapsulation in liposomes. Liposomes are single or multilayered vesicles of phospholipids containing either aqueous-based or lipophilic compounds. Lipid-soluble vitamins such as vitamin A, β-carotene and vitamins D, E or K are much easier to encapsulate than water-soluble ingredients. A commonly-used procedure is spray-drying of emulsions [81].

11. Minerals

From a nutritional point of view, the iron is one of the most important elements, and its deficiency affects about one-third of the world's population. The best way to prevent this problem is through the iron fortification of foods. However, the bioavailability of iron is negatively influenced by interactions with food ingredients such as tannins, phytates and polyphenols. Moreover, the iron catalyses oxidative processes in fatty acids, vitamins and amino acids, and consequently alters sensory characteristics and decreases the nutritional value of the food. Microencapsulation can be used to prevent these reactions, although bioavailability should be rechecked carefully. The bioavailability of readily water-soluble iron salts such as FeSO4 or ferrous lactate is higher than that of poorly water-soluble (e.g. ferrous fumarate) or water-insoluble (e.g. FePO4) iron. Suitable encapsulation techniques depend on the water solubility of the compound. Liposome technology is the method of choice for iron fortification of fluid food products [81].

12. Polyunsaturated fatty acid

In recent years, have been developing a high consumer preference towards products that possess functional properties. Thus the addition of beneficial substances such as polyunsaturated fatty acids omega 3 and omega 6, to the daily diet of humans has increased significantly since that is associated with the prevention and treatment of heart disease, because they have antithrombotic effects. Also has been associated with inflammatory diseases, autoimmune arthritis and even cancer. However, rich in polyunsaturated fatty acids oils are susceptible to oxidative deterioration and acquire easily bad tastes and odors, also environmental factors as moisture, light and oxygen accelerate its degradation, what he has done to the food and pharmaceutical industry to seek alternatives to prevent their deterioration. The microencapsulation has been an alternative to avoid the deterioration of oils because it can increase the oxidative stability of these and avoid the formation of products of oxidation of high molecular weight, in addition to mask unwanted flavors and aromas. It also gives some properties such as ease of handling and mixing, dispersion, and improvement of the consistency of the product during and after processing [82]. Advances in technologies of microencapsulation and the strategies used in its production have resulted in an increasing number of successful products fortified with omega-3 in the market [83], such as: dietary supplements, dairy, snacks, infant formulas and foods for babies, bakery products and beverages [84].

13. Probiotic bacteria

The microencapsulation has been successfully used to improve the survival of probiotic bacteria in dairy products such as cheese. Ozer et al. [85] demonstrated that colonies of *Lactobacillus ramnosus* microencapsulated in a matrix of alginate maintained their viability over 48 h at pH 2, in comparison with free (without encapsulate) cells that were inactivated completely under the same conditions; Another example related to dairy products is the yogurt, in which Bifidobacteria are encapsulated to increase viability in this fermented beverage; Also the whey, the liquid product obtained during the preparation of the cheese can be dried by spraying for the production of whey powder and/or whey protein concentrates [14]. Bio-functional foods offer physiological health benefits and disease prevention over and above their nutritional contribution. Microencapsulation has become the recent tool used for protecting and delivering bio-actives in the development of bio-functional foods. Probiotic foods are by far the largest functional food market. They provide several health benefits including immune-stimulation. Viability, physiological and metabolic activity of these bio actives in a food product at the point of sale are important consideration for their efficacy, as they have to survive during shelf life of a food, transit through high acidic and alkaline conditions in the gastro-intestinal tract. Microencapsulation is an inclusion technique for entrapping a bio-functional nutrient or bio-active compound such as probiotic bacteria, folic acid and protease enzymes into a polymeric (gelled) matrix that may be coated by one or more semi-permeable polymers, by virtue of which the encapsulated substance become more stable than the free one [86].

14. Microencapsulation methods

Several encapsulation processes are based on making first droplets of the active compound (in gas, liquid or powder form) and these droplets are subsequently surrounded by the carrier material in a gas or liquid phase via different methods. Excellent reviews on the encapsulation processes have been published in the last years [12-14, 87-88]. For this reason, this section will only show the most commonly used methods in microencapsulation and the steps involved through a list (see Table 5).

Methods	Process steps	Morphology	Load (%)	Particle size (μm)
Spray-drying	Disperse or dissolve active in aqueous coating solution Atomize Dehydrate	Matrix	5-50	10-400
Fluid bed coating	Fluidize active powder Spray coating Dehydrate or cool	Reservoir	5-50	5-5,000
Spray-chilling/cooling	Disperse or dissolve active in heated lipid solution Atomize Cool	Matrix	10-20	20-200
Melt injection	Melt the coating Disperse or dissolve active in the coating Extrude through filter Cooling and dehydrating	Matrix	5-20	200-2,000
Emulsification	Dissolve active and emulsifiers in water or oil phase Mix oil and water phases under shear	Matrix	1-100	0.2-5,000
Preparation of emulsions with multilayers	Prepare o/w emulsions with lipophilic active in oil phase and ionic emulsifiers Mix with aqueous solution containing oppositely charged poly-electrolytes Remove excess of free poly-electrolytes (option) Repeat steps 2 and 3	Reservoir	1-90	0.2-5,000
Coacervation	Prepare o/w emulsions with lipophilic active in oil phase Mix under turbulent conditions Induce three immiscible phases Cool Crosslink (optionally)	Reservoir	40-90	10-800
Preparation of microspheres via extrusion or dropping	Dissolve or disperse active in alginate solution Drop into gelling bath	Matrix	20-50	200-5,000
Preparation of microspheres via emulsification	Emulsify water with biopolymer in oil phase	Matrix	20-50	10-1,000
Co-extrusion	Dissolve or disperse active in oil Prepare aqueous or fat coating Use an concentric nozzle, and press simultaneously the oil phase through the inner nozzle and the water phase through the outer one Drop into gelling or cooling bath	Reservoir	70-90	150-8,000
Encapsulation by rapid expansion of supercritical solution (RESS)	Create a dispersion of active and dissolved or swollen shell material in supercritical fluid Release the fluid to precipitate the shell onto the active	Matrix	20-50	10-400
Freeze- or vacuum drying	Dissolve or disperse active agent and carrier material in water Freeze the sample Drying under low pressure Grinding (option)	Matrix	Various	20-5,000
Preparation of nanoparticles	Various methods	Various	Various	0.1-1

Adapted from Zuidam and Shimoni [12]

Table 5. Overview of the most common microencapsulation processes

15. Advantages and disadvantages of the microencapsulation

According to Zuidam and Shimoni [12] the possible advantages and disadvantages of microencapsulated active compounds in the food industry could be:

Advantages

1. Superior handling of the active agent (e.g., conversion of liquid active agent into a powder, which might be dust free, free flowing, and might have a more neutral smell).
2. Immobility of active agent in food processing systems.
3. Improved stability in final product and during processing (i.e., less evaporation of volatile active agent and/or no degradation or reaction with other components in the food product such as oxygen or water).
4. Improved safety (i.e., reduced flammability of volatiles like aroma, no concentrated volatile oil handling).
5. Creation of visible and textural effects (visual cues).
6. Adjustable properties of active components (particle size, structure, oil- or water-soluble, color).
7. Off-taste masking.
8. Controlled release (differentiation, release by the right stimulus).

Disadvantages

1. Additional costs.
2. Increased complexity of production process and/or supply chain.
3. Undesirable consumer notice (visual or touch) of encapsulates in food products.
4. Stability challenges of encapsulates during processing and storage of the food product.

16. Conclusions and perspectives

Since, very good results have been obtained on enhancing overall quality, browning, oxidation and softening, shelf-life extension, control of decay, and nutraceutical benefits with a number of diverse additives and bioactive compounds incorporated into edible films and coatings designed for different foods, could be useful to consider that the use the edible films and coating as carrier of food additives and bioactive compounds represent a good alternative for food industry to improve the quality and safety of the food, as well as to offer new product to consumers.

Many companies and research institutes are looking for new ingredients with possible health benefits. Ingredients such as phytochemicals, wood-derived ingredients such as phytosterols, pro- and prebiotics, new types of carotenoids, trace minerals and polyphenols will be available in next the years. Microencapsulation will certainly play an important role in this process, although it will always make an ingredient more expensive to use and bioavailability should always be considered carefully.

Author details

Rosa M. Raybaudi-Massilia and Jonathan Mosqueda-Melgar
Institute of Food Science and Technology, Central University of Venezuela, Lomas de Bello Monte, Calle Suapure, Caracas, Venezuela

17. References

[1] Martín-Belloso O, Soliva-Fortuny R, Oms-Oliu G (2006) Fresh-cut fruits. In: Hui Y.H, editor. Handbook of fruits and fruit processing. Iowa, USA: Blackwell Publishing. pp. 129-144.

[2] Olivas G.I, Barbosa-Cánovas G.V (2005) Edible coatings for fresh-cut fruits. Crit. Rev. Food Sci. Nutr. 45(7-8): 657-670.

[3] Min B, Ahn D.U (2005) Mechanism of lipid peroxidation in meat and meat products – A review. Food Sci. Biotechnol. 14(1): 152-163.

[4] Fasseas, M.K, Mountzouris K.C, Tarantilis P.A, Polissiou M, Zervas G (2007) Antioxidant activity in meat treated with oregano and sage essential oils. Food Chem. 106: 1188-1194.

[5] Cerqueira M.A, Lima A.M, Souza B.W.S, Texeira J.A, Moreira R.A, Vicente A.A, (2009) Functional polysaccharides as edible coating for cheese. J. Agric. Food Chem. 57: 1456-1462.

[6] Gennadios A, Hanna M.A, Kurth L.B (1997) Application of edible coatings on meats, poultry and seafoods: A review. LWT. 30: 337-350.

[7] Martín-Belloso O, Rojas-Graü M.A, Soliva-Fortuny R (2009) Delivery of flavor and active ingredients using edible films and coatings. In: Embuscado M.E, Huber K.C, editors. Edible films and coating for food applications. Chapter 10. New York: Springer Science + Business Media, LLC. pp. 295-313.

[8] Raybaudi-Massilia R.M, Mosqueda-Melgar J, Tapia M.S (2010) Edible coatings as carriers of food additives on fresh-cut fruits and vegetables. Stewart Postharvest Rev. 3: 1-7.

[9] Rojas-Graü M.A, Soliva-Fortuny R.C, Martín-Belloso O (2009) Edible coatings to incorporate active ingredients to fresh-cut fruits: a review. Trends Food Sci. Technol. 20:438-447.

[10] Zhao Y (2010) Edible Coatings for enhancing quality and health benefits of berry fruits. In: Qian M.C, Rimando A.M, editors. Flavor and Health Benefits of Small Fruits. Chapter 18, Volume 1035. Washington DC, USA: ACS Symposium Series. pp 281–292.

[11] Tapia M.S, Rojas-Graü M.A, Rodríguez F.J, Ramírez J, Carmona A, Martín-Belloso O (2007) Alginate- and gellan-based edible films for probiotic coatings on fresh-cut fruits. J. Food Sci. 72: E190-E196.

[12] Zuidam N.J, Shimoni E (2010) Overview of microencapsulates for use in food products or processes and methods to make them. In: Zuidam N.J, Nedovic V.A, editors. Encapsulation technologies for active food ingredients and food processing. Chapter 2. New York: Springer Science + Business Media, LLC. pp. 3-29.

[13] Dubey R, Shami T.C, Brasker-Rao K.U (2009) Microencapsulation technology and application. Defence Sci. J. 59(1): 82-95.

[14] Parra-Huertas R.A (2011) Revisión: Microencapsulación de alimentos. Rev. Fac. Nal. Agr. Medellín. 63(2): 5669-5684.

[15] Raybaudi-Massilia R.M, Mosqueda-Melgar J (2009) Factors affecting the incidence of pathogenic microorganisms of fresh-cut produce. Stewart Postharvest Rev. 4: 1-8.

[16] Cutter C.N, Siragusa G.R (1997) Growth of *Brochothrix thermosphacta* in ground beef following treatments with nisin in calcium alginate gels. Food Microbiol. 14: 425-430.

[17] Natrajan N, Sheldon B (2000) Inhibition of Salmonella on poultry skin using protein and polysaccharide-based films containing a nisin formulation. J Food Prot. 63(9): 1268-1272.

[18] Millette M, Le Tien C, Smoragiewicz W, Lacroix M (2007) Inhibition of Staphylococcus aureus on beef by nisin-containing modified alginate films and beads. Food Contr. 18: 878-884.

[19] Datta S, Janes M.E, Xue Q.G, Losso J, La Peyre J.F (2008) Control of *Listeria monocytogenes* and *Salmonella anatum* on the Surface of Smoked Salmon Coated with Calcium Alginate Coating containing Oyster Lysozyme and Nisin. J. Food Sci. 73(2): M67-M71.

[20] Neetoo H, Ye M, Chen H (2010) Bioactive alginate coatings to control *Listeria monocytogenes* on cold-smoked salmon slices and fillets. Int. J. Food Microbiol. 136(3): 326-331.

[21] Marcos B, Aymerich T, Monfort J.M, Garriga M (2008) High-pressure processing and antimicrobial biodegradable packaging to control *Listeria monocytogenes* during storage of cooked ham. Food Microbiol. 25: 177-182.

[22] Oussalah M, Caillet S, Salmiéri S, Saucier L, Lacroix M (2006). Antimicrobial effects of alginate-based film containing essential oils for the preservation of whole beef muscle. J. Food Prot. 69: 2364-2369.

[23] Oussalah M, Caillet S, Salmieri S, Saucier L, Lacroix M (2007) Antimicrobial effects of alginate- based film containing essential oils o *Listeria monocytogenes* and *Salmonella thyphimurium* present in bologna and ham. J. Food Prot. 70(4): 901-908.

[24] Rojas-Graü M.A, Raybaudi-Massilia R.M, Soliva-Fortuny R.C, Avena-Bustillos R.J, McHugh T.H, Martín-Belloso O (2007) Apple puree-alginate edible coating as carrier of antimicrobial agents to prolong shelf life of fresh-cut apples. Postharvest Biol. Technol. 45: 254-264.

[25] Raybaudi-Massilia R.M, Mosqueda-Melgar J, Martín-Belloso O (2008) Edible alginate-based coating as carrier of antimicrobials to improve shelf life and safety of fresh-cut melon. Int. J. Food Microbiol. 121: 313-327.

[26] Raybaudi-Massilia R.M, Rojas-Graü M.A, Mosqueda-Melgar J, Martín-Belloso O (2008) Comparative study on essential oils incorporated into an alginate-based edible coating to assure the safety and quality of fresh-cut Fuji apples. J. Food Prot. 71: 1150-1161.

[27] Olivas G.I, Mattinson D.S, Barbosa-Cánovas G.V (2007) Alginate coatings for preservation of minimally processed 'Gala' apples. Postharvest Biol. Technol. 45: 89-96.

[28] Seol, K.H., Lim, D.G., Jang, A., Jo, C., y Lee, M. (2009). Antimicrobial effect of κ-carrageenan-based edible film containing ovotransferrin in fresh chicken breast stored at 5 °C. Meat Science 83(3): 479-483.

[29] Jiang Z, Neetoo H, Chen H (2011) Efficacy of freezing, frozen storage and edible antimicrobial coating used in combination for control of *Listeria monocytogenes* on roasted turkey stored at chiller temperatures. Food Microbiol. 28(7): 1394-1401.

[30] Jiang Z, Neetoo H, Chen H (2011) Control of *Listeria monocytogenes* on cold-smoked salmon using chitosan-based antimicrobial coatings and films. J. Food Sci. 76(1): M22-M26.

[31] Dutta P.K, Tripathi S, Mehrotra G.K, Dutta J (2009) Perspectives for chitosan based antimicrobial films in food applications. Food Chem. 114(4): 1173-1182.

[32] Ouattara B, Simard R.E, Piette G, Bejín A, Holler R.A (2000) Inhibition of surface spoilage bacteria in processed meats by application of antimicrobial films prepared with chitosan. Int. J. Food Microbiol. 62: 139-148.

[33] Park S.L, Stan S.D, Daeschel M.A, Zhao Y (2006) Antifungal coatings on fresh strawberries (*Fragaria* × *ananassa*) to control mold growth during cold storage. J. Food Sci. 70(4): M202-M207.

[34] Duan J, Park S, Daeschel M, Zhao Y (2007). Antimicrobial chitosan-lysozyme (CL) films and coatings for enhancing microbial safety of mozzarella cheese. J. Food Sci. 72(9): M355-M362.

[35] Sangsuwan J, Rattanapanone N, Rachtanapun P (2008) Effect of chitosan/methyl cellulose films on microbial and quality characteristics of fresh-cut cantaloupe and pineapple. Postharvest Biol. Technol. 49: 403-410.

[36] Ye M, Neetoo H, Chen H (2008) Control of *Listeria monocytogenes* on ham steaks by antimicrobials incorporated into chitosan-coated plastic films. Food Microbiol. 25(2): 260-268.

[37] Ojagh S.M, Rezaei M, Razavi S.H, Hashem-Hosseini S.M (2010) Effect of chitosan coatings enriched with cinnamon oil on the quality of refrigerated rainbow trout. Food Chem. 120(1): 193-198.

[38] Fajardo P, Martins J.T, Fuciños C, Pastrana L, Teixeira J.A, Vicente A.A (2010) Evaluation of a chitosan-based edible film as carrier of natamycin to improve the storability of Saloio cheese. J. Food Eng. 101(4): 349-356.

[39] Siripatrawan U, Noipha S (2012). Active film from chitosan incorporating green tea extract for shelf life extension of pork sausages. Food Hydrocolloids 27(1): 102-108.

[40] Franklin, N.B., Cooksey, K.D., y Getty, K.J.K. (2004). Inhibition of *Listeria monocytogenes* on the Surface of Individually Packaged Hot Dogs with a Packaging Film Coating Containing Nisin. *Journal of Food Protection* 67(3). 480-485

[41] Nguyen T, Gidley M, Dykes G (2008). Potential of a nisin-containing bacterial cellulose film to inhibit *Listeria monocytogenes* on processed meats. Food Microbiol. 25(3): 471-478.

[42] Santiago-Silva P, Soares N.F.F, Nóbrega J.E, Júnior M.A.W, Barbosa K.B.F, Volp A.C.P, Zerdas E.R.M.A, Würlitzer N.J (2009) Antimicrobial efficiency of film incorporated with pediocin (ALTA®2351) on preservation of sliced ham. Food Contr. 20: 85-89.

[43] Sayanjali S, Ghanbarzadeh B, Ghiassifar S. 2011. Evaluation of antimicrobial and physical properties of edible film based on carboxy methyl cellulose containing potassium sorbate on some mycotoxigenic *Aspergillus* species in fresh pistachios. LWT. 44(4): 1133-1138.

[44] Valencia-Chamorro S.A, Pérez-Gago M.B, del Río M.A, Palou L (2009) Effect of antifungal hydroxy propyl methyl cellulose (HPMC)–lipid edible composite coatings on postharvest decay development and quality attributes of cold-stored "Valencia" oranges. Postharvest Biol. Technol. 54(2): 72-79.

[45] Durango A.M, Soares N.F.F, Andrade N.J (2006) Microbiological evaluation of an edible antimicrobial coating on minimally processed carrots. Food Contr. 17(5): 336-341.

[46] Chiu P.E, Lai L.S (2010) Antimicrobial activities of tapioca starch/decolorized hsian-tsao leaf gum coatings containing green tea extracts in fruit-based salads, romaine hearts and pork slices. Int. J. Food Microbiol. 139(1-2): 23-30.

[47] Kang H.J, Jo C, Kwon J.H, Kim J.H, Chung H.J, Byun M.W (2007) Effect of a pectin-based edible coating containing green tea powder on the quality of irradiated pork patty. Food Contr. 18(5): 430-435.

[48] Handley D, Ma-Edmonds M, Hamouz F, Cuppett S, Mandigo R, Schnepf M (1996) Controlling oxidation and warmed-over flavor in precooked pork chops with rosemary oleoresin and edible film. In: Shahidi F, editor. Natural antioxidants chemistry, health effect, and application. Champaign, USA: AOCS Press. pp. 311-318.

[49] Ma-Edmonds M, Hamouz F, Cuppett S, Mandigo R, Schnepf M (1995) Use of rosemary oleoresin and edible film to control warmed-over flavor in precooked beef patties [abstract]. In: IFT Annual Meeting Book of Abstracts. Chicago, Illinois: Institute of Food Technologists. pp. 139. Abstract Nº 50-6.

[50] Hargens-Madsen M, Schnepf M, Hamouz F, Weller C, Roy S (1995) Use of edible films and tocopherols in the control of warmed-over flavor. J. Am. Diet. Assoc. 95: A-41.

[51] Wu Y, Weller C.L, Hamouz F, Cuppett S, Schnepf M (2001) Moisture loss and lipid oxidation for precooked ground-beef patties packaged in edible starch-alginate-based composite films. J. Food Sci. 66(3): 486-493.

[52] Atarés L, Pérez-Masiá R, Chiralt A (2011). The role of some antioxidants in the HPMC film properties and lipid protection in coated toasted almonds. J. Food Eng. 104(4): 649-656.

[53] Song Y, Liu L, Shen H, You J, Luo Y (2011) Effect of sodium alginate-based edible coating containing different anti-oxidants on quality and shelf life of refrigerated bream (*Megalobrama amblycephala*). Food Contr. 22(3-4): 608-615.

[54] Lin D, Zhao Z (2007) Innovations in the development and application of edible coatings for fresh and minimally processed fruits and vegetables. CRFSFS. 6: 60-75.

[55] Ahvenainen R (1996) Improving the shelf life of minimally processed fruit and vegetables. Trends Food Sci. Tech. 7: 179-187.

[56] Friedman, M. Food browning and its prevention: Overview. Journal of Agricultural and Food Chemistry 1996: 44(3):631–652.

[57] Baldwin E.A, Nisperos M.O, Chen X, Hagenmaier R.D (1996) Improving storage life of cut apple and potato with edible coating. Postharvest Biol. Technol. 9: 151-163.

[58] Brancoli N, Barbosa-Cánovas G.V (2000) Quality changes during refrigerated storage of packaged apple slices treated with polysaccharide films. In: Barbosa-Cánovas G.V, Gould G.W, editors. Innovations in food processing. Pennsylvania: Technomic Publishing Co. pp. 243-254.

[59] Lee J.Y, Park H.J, Lee C.Y, Choi W.Y (2003) Extending shelf-life of minimally processed apples with edible coatings and antibrowning agents. LWT. 36: 323-329.
[60] Bico S.L.S, Raposo M.F.J, Morais R.M.S.C, Morais A.M.M.B (2009) Combined effects of chemical dip and/or carrageenan coating and/or controlled atmosphere on quality of fresh-cut banana. Food Contr. 20: 508-514.
[61] Rojas-Graü M.A, Tapia M.S, Rodríguez F.J, Carmona A.J, Martín-Belloso O (2007) Alginate and gellan based edible coatings as support of antibrowning agents applied on fresh-cut Fuji apple. Food Hydrocolloids 21: 118-127.
[62] Rojas-Graü M.A, Tapia M.S, Martín-Belloso O (2008) Using polysaccharide-based edible coatings to maintain quality of fresh-cut Fuji apples. LWT. 41: 139-147.
[63] Oms-Oliu G, Soliva-Fortuny R, Martín-Belloso O (2008) Edible coatings with antibrowning agents to maintain sensory quality and antioxidant properties of fresh-cut pears. Postharvest Biol. Technol. 50: 87-94.
[64] García E, Barret D.M (2002) Preservative treatments for fresh cut fruits and vegetables. In: Lamikanra O, editor. Fresh-Cut Fruits and Vegetables. Florida: CRC Press. pp. 267-304.
[65] Harker F.R, Redgwell R.J, Hallett I.C, Murray S.H, Carter G (1997) Texture of fresh fruit. Horticultural Rev. 20: 121-224.
[66] Soliva-Fortuny R.C, Martín-Belloso O (2003) New advances in extending the shelf life of fresh-cut fruits: a review. Trends Food Sci. Technol. 14: 341–353.
[67] Oms-Oliu G, Soliva-Fortuny R, Martín-Belloso O (2008) Using polysaccharide-based edible coatings to enhance quality and antioxidant properties of fresh-cut melon. LWT. 41: 1862-1870.
[68] Tapia M.S, Rojas-Graü M.A, Carmona A, Rodriguez F.J, Soliva-Fortuny R, Martín-Belloso O (2008) Use of alginate and gellan-based coatings for improving barrier, texture and nutritional properties of fresh-cut papaya. Food Hydrocolloids. 22: 1493-1503.
[69] Mei Y, Zhao Y (2003) Barrier and Mechanical Properties of Milk Protein-Based Edible Films Containing Nutraceuticals. J. Agr. Food Chem. 51(7): 1914-1918.
[70] Elliott J.G (1998) Application of antioxidant vitamins in foods and beverages. Food Technol. 53: 46-48.
[71] Mei Y, Zhao Y, Yang J, Furr H.C (2002) Using edible coating to enhance nutritional and sensory qualities of baby carrots. J. Food Sci. 67(5): 1964-1968.
[72] Hernández-Muñoz P, Almenar E, Ocio M.J, Gavara R (2006) Effect of calcium dips and chitosan coatings on postharvest life of strawberries (*Fragaria* x *ananassa*). Postharvest Biol. Technol. 39: 247-253.
[73] Han C, Zhao Y, Leonard S.W, Traber M.G (2004) Edible coatings to improve storability and enhance nutritional value of fresh and frozen strawberries (*Fragaria* x *ananassa*) and raspberries (*Rubus ideaus*). Postharvest Biol. Technol. 33(1): 167-178.
[74] Duan J, Cherian G, Zhao Y (2010). Quality enhancement in fresh and frozen lingcod (*Ophiodon elongates*) fillets by employment of fish oil incorporated chitosan coatings. Food Chem. 119(2): 524-532.

[75] Naidu A.S, Bidlack W.R, Clemens R.A (1999) Probiotic spectraof lactic acid bacteria (LAB). Crit. Rev. Food Sci. Nutr. 38: 113-126.

[76] Giancone T (2007) Hydrocolloid-based edible films: composition-structure-properties relationship. Thesis Doctoral. Universitá degli Studi di Napoli Federico II. Department of Food Science. Available: http://www.fedoa.unina.it/1669/1/Giancone_Scienze_Tecnologie.pdf. Accessed 2012 Feb 10.

[77] Gontard N, Guilbert S (1999) Bio-packaging: technology and properties of edible and/or biodegradable material of agricultural origin. In: Mathlouthi M, editor. Food packaging and preservation. Chapter 9. Gaithersburg, USA: Aspen Publisher, Inc. pp. 159-181.

[78] Pavlath A.E, Orts W (2009). Edible films and coatings: Why, what, and how?. In: Embuscado M.E, Huber K.C, editors. Edible films and coatings for food applications. Chapter 1. New York: Springer Science + Business Media, LLC. pp. 1-24.

[79] Chien P.J, Sheu F, Yang F.H (2007) Effects of edible chitosan coating on quality and shelf life of sliced mango fruit. J. Food Eng. 78: 225-229.

[80] Parize A, Rozone T, Costa I Fávere V, Laranjeira M, Spinelli A, Longo E (2008) Microencapsulation of the natural urucum pigment with chitosan by spray-drying in different solvents. African J. Biotechnol. 7(17): 3107-3114.

[81] Schrooyen P, Meer R, Kruif C (2001) Microencapsulation: its application in nutrition. Proc. Nutr. Soc. 60(4): 475-479.

[82] Quispe-Arpasi D, Matos-Chamorro A, Quispe-Condori S (2011) Microencapsulación de aceites y su aplicación en la industria de alimentos. I Congreso Nacional de Investigación, Lima, Perú. November 2-4, 2011. Available: http://papiros.upeu.edu.pe/bitstream/handle/123456789/180/CIn33Articulo.pdf?sequence=1 Accessed 2012 Abril 10.

[83] Smith J, Charter E (2010) Functional food product development. United Kingdom: Blackwell Publishing Ltd. 505 p.

[84] Sanguansri L, Augustin M (2007) Microencapsulation and delivery of Omega-3 fatty acids. In: Shi J, editor. Functional food ingredients and nutraceuticals: Processing technologies. Florida, USA: Taylor & Francis. pp. 297-327.

[85] Ozer B, Avni H, Senel E, Atamer M, Hayaloglu A (2009) Improving the viability of Bifidobacterium bifidum BB-12 and Lactobacillus acidophilus LA-5 in white-brined cheese by microencapsulation. Int. Dairy J. 19(1): 22-29.

[86] Kailasapathy K, Madziva H, Anjani K, Seneweera S, Phillips M (2006) Recent trends in the role of micro encapsulation in the development of bio-functional foods. XIVth International Workshop on Bioencapsulation, Lausanne, USA. Available: http://impascience.eu/bioencapsulation/340_contribution_texts/2006-10-05_O7-1.pdf. Accessed 2012 April 10.

[87] Gibbs B.F, Kermasha S, Alli I, Mulligan C.N (1999) Encapsulation in the food industry: A review. Int. J. Food Sci. Nutr. 50: 213-224.

[88] Desobry S, Debeaufort F (2011) Encapsulation of flavors, nutraceuticals, and antibacterials. In: Baldwin E.A, Hagenmaier R, Bai J, editors. Edible coatings and films to improve food quality. 2nd Edition. Chapter 11. Boca Raton, USA: CRC Press. pp. 333-372.

Chapter 17

Dietary Fiber and Availability of Nutrients: A Case Study on Yoghurt as a Food Model

Marina Dello Staffolo, Alicia E. Bevilacqua,
María Susana Rodríguez and Liliana Albertengo

Additional information is available at the end of the chapter

http://dx.doi.org/10.5772/54031

1. Introduction

Dietary fibers are consumed from cereals, fruit and vegetables, but now are also added in purified form to food preparations since the roles of dietary fibers in preventing and treating some diseases have been well documented. Dietary fiber intake in Western countries is currently estimated to be 16.3-43.4 g per person per day [1]. According to current recommendations (Food and Nutrition Board, Institute of Medicine, 2001), the average daily requirement of dietary fiber is 25 g per day for women younger than 50, 21 g per day for women older than 50; 38 g per day for men younger than 50, and 30 g per day for men older than 50 [2].

The addition of dietary fibers to foods confers three different types of benefits. Their nutritional value motivates consumers to eat increased quantities of dietary fibers, which is advised by nutritionists. Their technological properties are of great interest to food manufacturers. Finally, dietary fibers may also be used to upgrade agricultural products and by-products for use as food ingredients. Consequently, both the nutritional value and technological properties of dietary fibers are important in the potential development of a wide range of fiber-enriched foods for example: bakery products, snacks, sauces, drinks, cereals, cookies, dairy products, meat products [3].

Different types of dietary fibers have different structures and chemical compositions, and correspondingly are of varying nutritional and technological interest. Although many studies have confirmed the nutritional benefits of dietary fibers (preventing diabetes mellitus, cardiovascular diseases, various types of cancer, and improving immune functions), the results depend on the types of dietary fibers studied or on the experimental conditions used [3]. The intake of dietary fiber might influence in different ways the absorption of nutrients [4]. With respect to glucose, an increase in the total fiber content of food can delay the glycemic response

INTECH
open science | open minds

© 2012 The Author(s). Licensee InTech. This chapter is distributed under the terms of the Creative Commons Attribution License http://creativecommons.org/licenses/by/3.0), which permits unrestricted use, distribution, and reproduction in any medium, provided the original work is properly cited.

[5]. However, there is fairly consistent evidence that soluble types of fiber reduce blood glucose and purified insoluble fibers have little or no effect on postprandial blood glucose [6]. In the other direction, dietary fiber has been shown to impair the absorption of minerals and trace elements in the small intestine because of their binding and/or sequestering effects [7]. This is associated with negative impacts on mineral bioavailability, particularly in high-risk population groups [8]. Glucose, calcium and iron have gained increasing interest in nutrition fields. Glucose is the key in carbohydrate and lipid metabolism influencing the management of body weight [9]. Calcium is involved in most metabolic processes and the phosphate salts of which provide mechanical rigidity to the bones and teeth. Intake of calcium is related to the prevention of osteoporosis [10,11]. Iron (Fe) deficiency is a leading nutritional concern worldwide, affecting 20–50% of the world's population [12].

There is an unequivocal need for predicting absorption of these nutrients. The aim of most of the investigations in this field is to make evident that fiber may be an important determinant of the utilisation of these nutrients in the diet. Much research has been done to better understand the physicochemical interactions between dietary fiber and these nutrients in the past decades [13-15]. Several of these investigations have applied in vitro digestive models to study their absorption in foods [16-18]. However, few works have been done to study their absorption from fermented milk products [19]. Yogurt is one of the dairy products, which should continue to increase in sales due to acceptance for the consumers and diversification in the range of yogurt-like products, including reduced fat content yogurts, yogurts with dietary fibers, probiotic yogurts, symbiotic yogurts, yogurt ice-cream, etc [20]. For a long time, yogurt has been recognised as a healthy food and as an important nutritional source [21].

The interactions between fibers added to yogurt from different sources (animal and plant fibers) and with different behaviors (soluble, insoluble and viscous fibers) and glucose, calcium and iron, have been studied using chemical experimental models of the human digestive tract to evaluate the availability of these nutrients.

2. Definition and composition of dietary fiber

The term 'dietary fiber' (DF) first appeared in 1953 and referred to hemicelluloses, cellulose and lignin [22]. Since the 1970s it has been recognised as having health benefits. Burkitt [23] recommended that individuals should increase their DF intake in order to increase their stool volume and stool softness. This was based on comparisons between Africa and the UK concerning fiber intakes and disease incidence. Trowell [24] first defined DF as 'the remnants of the plant cell wall that are not hydrolysed by the alimentary enzymes of man'. From those days the definition has undergone numerous revisions that were summarized accurately by Tungland and Meyer [4].

The Codex Alimentarius Commission adopted a new definition of fiber in July 2009, designed to harmonize the use of the term around the world. It describes fiber as elements not hydrolised by endogenous enzymes in the small intestine (indigestibility) as well as having physiological effects beneficial to health. Dietary fibers are carbohydrate polymers with ten or more monomeric units and belonging to one of three categories of carbohydrates polymers: edible carbohydrate polymers naturally occurring in food, carbohydrate polymers which have been obtained from raw food material by physical, enzymatic, or chemical means, and synthetic carbohydrate polymers [25–27].

The chemical nature of fibers is complex; dietary fibers are constituted of a mixture of chemical entities [28]. Dietary fiber is composed of nondigestible carbohydrate, lignin and other associated substances of plant origin, fibers of animal origin and modified or synthetic nondigestible carbohydrate polymers. The nondigestible carbohydrates are composed of the following polysaccharides: cellulose, β-glucan, hemicelluloses, gums, mucilage, pectin, inulin, resistant starch; oligosaccharides: fructo-oligosaccharides, oligofructose, polydextrose, galacto-oligosaccharides; and soybean oligosaccharides raffinose and stachyose [4]. Chitosan is an example of fiber of animal origin, derived from the chitin contained in the exoskeletons of crustaceans and squid pens; its molecular structure is similar to that of plant cellulose [29]. Cereals are the principal source of cellulose, lignin and hemicelluloses, whereas fruits and vegetables are the primary sources of pectin, gums and mucilage [30]. Each polysaccharide is characterised by its sugar residues and by the nature of the bond between them [28]. They are presented in Table 1.

Fibers	**Main chain**	**Branch units**
Cellulose	β-(1,4) glucose	
β-glucans	β-(1,4) glucose and β-(1,3) glucose	
Hemicelluloses		
Xylans	β-D-(1,4) xylose	
Arabinoxylans	β-D-(1,4) xylose	Arabinose
Mannans	β-D-(1,4) mannose	
Glucomanns	β-D-(1,4) mannose and β-D-(1,4) glucose	
Galactoglucomannans	β-D-(1,4) mannose, β-D-(1,4) glucose	Galactose
Galactomannans	β-(1,4) mannose	α-D-galactose
Xyloglucans	β-D-(1,4) glucose	α-D-xylose
Pectin		
Homogalacturonan	α-(1,4)-D-galacturonic acid (some of the carboxyl groups are methyl esterified)	
Rhamnogalacturonan-I	(1,4) galacturonic acid, (1,2) rhamnose and 1-, 2-, 4-rhamnose	Galactose, arabinose, xylose, rhamnose, galacturonic acid
Rhamnogalacturonan-II	α-(1-4) galacturonic acid	Unusual sugar such as: apiose, aceric acid, fucose
Arabinans	α-(1-5)-L-arabinofuranose	α-arabinose
Galactans	β-(1-4)-D-galactopyranose	
Arabinogalactanes-I	β-(1-4)-D-galactopyranose	α-arabinose
Arabinogalactanes-II	β-(1-3)-and β-(1-6)-D-galactopyranose	α-arabinose
Xylogalacturonan	α-(1-4) galacturonic acid	xylose
Inulin	β-(2-1)-D-fructosyl-fructose	
Gum		
Carrageenan	Sulfato-galactose	
Alginate	β-(1,4)-D-mannuronic acid or α-(1-4)-L-guluronic acid	

Fibers	Main chain	Branch units
e.g. 1: seed gum from Abutilon indicum	β-(1,4)-D-mannose	D-(1,6) galactose
e.g. 2: seed gum from Lesquerella fendleri	Rhamnose, arabinose, xylose, Mannose, galactose, glucose, galacturonic acid	
Oligofructose (enzymatic hydrolysis of inulin)	β-(2-1)-D-fructosyl-fructose	
Polydextrose (synthetic)	D-Glucose	
Resistant maltodextrins (heat and enzymatic treatment of starch)	α (1-4)-D-Glucose	α (1-6)-D-Glucose
Lignin	Polyphenols: Syringyl alcohol (S), Guaiacyl alcohol (G) and p-coumaryl alcohol (H)	
Chitosan	β-(1-4)-linked D-glucosamine and N-acetyl-D-glucosamine	

Table 1. Chemical composition of dietary fibers [28].

3. Classification of dietary fiber

Several different classification systems have been suggested to classify the components of dietary fiber: based on their role in the plant, on their fiber constituents (Table 2), on the type of polysaccharide, on their simulated gastrointestinal solubility, on site of digestion and on products of digestion and physiological classification. However, none is entirely satisfactory, as the limits cannot be absolutely defined. Two very accepted classifications are those which use the concept of solubility in a buffer at a defined pH, and/or the concept of fermentability in an in vitro system using an aqueous enzyme solution representative of human alimentary enzymes. Generally, well fermented fibers are soluble in water, while partially or poorly fermented fibers are insoluble [4]. However, dietary fiber is conventionally classified in two categories according to their water solubility: insoluble dietary fiber (IDF) such as cellulose, part of hemicellulose, and lignin; and soluble dietary fiber (SDF) such as pentosans, pectin, gums, and mucilage [31].

Taking into account the physiological and physicochemical behavior of fibers it could be necessary to add two subcategories among the group of soluble fibers: viscous soluble fibers (such as guar gum, glucomannans, pectins, oat β-glucan, psyllium, mucilages, etc) and nonviscous soluble fibers (such as lactulose, oligosaccharides, fructo-oligosaccharides, inulin, etc.). Jenkins et al. [6, 32] mentioned the term viscous soluble fiber in their works about fibers and low glycaemic index, blood lipids and coronary heart diseases. Dikeman and Fahey [33] also mentioned the term in their work in which they investigated the viscosity in relation with dietary fiber including definitions and instrumentation, factors affecting viscosity of solutions, and effects on health. Thus, dietary fiber could be classified in soluble (viscous and non viscous) and insoluble fiber. The latter do not form gels due to their water insolubility and fermentation is very limited [34]. However, numerous commercial fibers are available in the market for use in food technology and have both insoluble and soluble fiber components in the same product. This is due to the fact that many of these are powders which come from the extraction, concentration and drying of the fiber contained in cereals, fruits and vegetables. Therefore, it might be more appropriate to classify fibers based on their content of soluble and insoluble fractions (Table 3). It is recognised that the physiological and physicochemical effects of dietary fibers depend on the relative amount of individual fiber components, especially as regards the soluble and insoluble fractions [28].

Fractionation of dietary fibers aims to quantify those constituents, to isolate fractions of interest and to eliminate undesirable compounds. Techniques for fractionation of dietary fibers are limited in number. Several researchers have determined the cellulose, hemicelluloses and lignin contents of dietary fibers from different food sources. Claye, Idouraine, and Weber [35] isolated and fractionated insoluble fibers from five different sources. Using cold and hot water extraction, enzymatic and chemical treatment, they obtained four fractions: cellulose, hemicellulose A and B, and lignin. The fractionation methods are varied and were developed according to the material tested. Therefore, there is no global method used. The existing methods described universal techniques of fractionation. Each analyst must modify previously used approaches to develop a method optimal to the material being tested [28].

Fiber Constituents	**Principal groupings**	**Fiber components/sources**
Nonstarch polysaccharides & oligosaccharide	Cellulose	Cellulose-Plants (vegetables, sugar beet, various brans)
	Hemicellulose	Arabinogalactans, _-glucans, arabinoxylans, glucuronoxylans, xyloglucans, galactomannans, pectic substances.
	Polyfructoses	Inulin, oligofructans
	Gums & Mucilages	Seed extracts (galactomannans –guar and locust bean gum), tree exudates (gum acacia, gum karaya, gum tragacanth), algal polysaccharides (alginates, agar, carrageenan), psyllium
	Pectins	Fruits, vegetables, legumes, potato, sugar beets
Carbohydrate analogues	Resistant starches and maltodextrins	Various plants, such as maize, pea, potato
	Chemical synthesis	Polydextrose, lactulose, cellulose derviatives (MC, HPMC)
	Enzymatic synthesis	Neosugar or short chain fructooligosaccharides (FOS), transgalactooligo collagen, chondroitin saccharides (TOS), levan, xanthan gum, oligofructose, xylooligosaccharides (XOS), guar hydrolyzate, curdlan.

Fiber Constituents	Principal groupings	Fiber components/sources
Lignin	Lignin	Woody plants
Substances associated with nonstarch polysaccharides	Waxes, cutin, Suberin	Plant fibers
Animal origin fibers	Chitin, chitosan, collagen, chondroitin	Fungi, yeasts, invertebrates

Table 2. Dietary fiber constituents [4]

Category	Subcategory	Fiber fraction	Main food source
Soluble fiber	Viscous	β-glucans	Grains (oat, barley, rye)
		Pectins	Fruits, vegetables, legumes, sugar beet, potato
		Gums & Mucilages	Leguminous seed plants (guar, locust bean), seaweed extracts (carrageenan, alginates), plant extracts (gum acacia, gum karaya, gum tragacanth), microbial gums (xanthan, gellan), psylluim
	Nonviscous	Sugars	Lactulose
		Oligosaccharides	Various plants and synthetically produced (polydextrose, fructooligosaccharides, galactooligosaccharides, transgalactooligosaccharides)
		Inulin	Chicory, Jerusalem artichoke, sugar beet, onions
Insoluble fiber		Cellulose	Plants (vegetables, various brans)
		Hemicellulose	Cereal grains
		Lignin	Woody plants
		Cutin/suberin/other plant waxes	Plant fibers
		Chitin and chitosan, collagen	Fungi, yeasts, invertebrates
		Resistant starches	Plants (corn, potatoes, grains, legumes, bananas)
		Curdlan (insoluble β-glucan)	Bacterial fermentation

Table 3. Table 3. Classification of dietary fiber based on solubility

4. Analytical methods for studying dietary fibers

The complexity of fibers is given by their chemical nature and polymerisation degree that they possess. This requires various analytical methods for the measurement of dietary fiber, to precisely estimate its composition in food and food by-products. Methods for the determination of dietary fiber may be divided into three categories: non-enzymatic-gravimetric, enzymatic-gravimetric, and enzymatic-chemical methods. The latter includes enzymatic-colorimetric and enzymatic-chromatographic (GLC/ HPLC) methods [28]. Nowadays, the most commonly used methods for dietary fiber measurement are the enzymatic-gravimetric Association of Official Analytical Chemists (AOAC) method [36] and enzymatic-chemical method [37]. The method of Van Soest [38] is generally used in veterinary studies.

5. Dietary fiber and human health

5.1. Fiber, lipid metabolism and cardiovascular disease

The earliest and most widely researched topic related to dietary fiber and human health is reducing the risk factors for coronary heart disease [24]. Total serum cholesterol and low density- lipoprotein (LDL) cholesterol levels are generally accepted as biomarkers indicating of potential risk for developing the disease [5]. In consequence, research has primarily focused on their reduction as a means to diminish the risk of developing cardiovascular disease. Substantial experimental data support that blood cholesterol can be lowered using viscous soluble fibers that produce relatively high viscosity in the intestinal tract [39-41]. It is known that viscous soluble and insoluble dietary fibers can bind bile acids and micelle components, such as monoglycerides, free fatty acids, and cholesterol, which decrease the absorption and increase the fecal excretion of these entities [42,43]. For insoluble dietary fibers such as lignin or citric fiber this reducing effect is rather low compared to viscous soluble dietary fibers and is mainly based on direct binding of bile acids. In the small intestine the bile acids are bound by the insoluble dietary fibers through hydrophobic interactions and excreted from enterohepatic circulation together with the undigested insoluble dietary fibers which results in a lowering of the blood biomarkers levels [44,45]. Furthermore, free fatty acids and cholesterol bound by dietary fibers cannot be absorbed by the body and will be excreted. The biomarkers-lowering effect of viscous soluble dietary fibers such as psyllium, oat β-glucan or pectin is based on different mechanisms. The binding of water in the chyme and the resulting increase in viscosity is regarded as the main effect. This leads to a reduced diffusion rate of bile acids, which cannot be reabsorbed by the body and thus are excreted [46-48]. Besides, some studies indicate that there are also direct binding forces such as hydrophobic interactions between soluble dietary fibers and bile acids [49]. Dietary fiber also modifies lipid metabolism by influencing the expression of key genes. Acetyl-CoA carboxylase is the rate-limiting enzyme in lipogenesis and is regulated by AMP-activated protein kinase (AMPK). In a 10-week study comparing obese and lean rats, adding 5 g of P. ovata to rat chow increased the phosphorylation of AMPK, consequently inhibiting acetyl-CoA carboxylase [50]. Fructooligosaccharide (10g/100g) also has been shown to decrease the hepatic acetyl-CoA carboxylase expression in rats [51]. In view of new research [52,34] the United States Food and Drug Administration has approved health claims supporting the role of dietary fiber in the prevention of coronary heart disease [53].

5.2. Fiber, carbohydrate metabolism, and diabetes mellitus

It is known that exists a link among an elevated body mass index, waist circumference and the risk of type 2 diabetes mellitus [54,55]. The role of DF in weight reduction has been examined in animal and human studies. A reduced risk for type 2 diabetes mellitus (T2DM) appears to depend on the type and dose of dietary fiber and the study population [36]. In mice, 10% psyllium and 10% sugar cane fiber decreased the fasting blood glucose and fasting plasma insulin when added to a high fat diet for 12 weeks, compared to the insoluble fiber cellulose [56]. β-Glucan also improved the glucose tolerance and decreased the serum insulin in mice when added to a high fat diet at a 2% and 4% level [57]. In humans, muffins high in β-glucan and resistant starch lowered the postprandial blood glucose and insulin levels [58]. A prospective cohort design with 252 women was used to measure energy intake, dietary fat intake, fiber intake, body weight, body fat percentage, physical activity, season of assessment, age and time between assessments. They concluded that increasing dietary fiber intake significantly reduced the risk of gaining weight and fat in women, independent of several potential confounders such as: physical activity, dietary fat intake, and others [59]. Another study about the consumption of soluble viscous fiber, that included one hundred and seventy six men and women, reached the same conclusions [60]. Other biomarkers such as Glycemic Index (GI) and Glycemic Load (GL) when they are high were both associated with an increased risk of diabetes in a meta-analysis of observational studies [61,62]. On the other hand, numerous epidemiological studies performed to date relate to a high intake of dietary fiber with low levels of GI and GL [63-65]. Moreover, DF has also shown to be effective improving altered parameters in obesity and T2DM [66,67]. Soluble viscous fiber plays an important role in controlling satiety and postprandial glycemic and insulin responses [68] and some studies showed that insoluble dietary fiber improved the quality of life for these patients [69]

The protective effect of DF on obesity and T2DM has been historically attributed to greater satiety due to an increased mastication, calorie displacement, and decreased absorption of macronutrients [55]. This mechanism is associated with the ability of soluble fibers to form viscous solutions that prolong gastric emptying, consequently inhibiting the transport of glucose, triglycerides and cholesterol across the intestine [70-72]. Recently, it was observed that both soluble and insoluble DFs also modifies carbohydrate metabolism by influencing the expression of hormones such as glucose-dependent insulin tropic polypeptide and glucagon-like peptide-1, that stimulate postprandial insulin release, enhance glucose tolerance, and delay gastric emptying [73-76].

5.3. Fiber and gut microflora

The large intestine plays host to a large and diverse resident microflora. Over the last 10-15 years, 16S ribosomal RNA analyses has allowed a more complete characterization of the diverse bacterial species that make up this population [77]. Around 95% of human colonic microflora (as estimated from faecal sampling) appear to be within Bacteroides and Clostridium phylogenic groups, with less than 2% of the total microflora being made up of

Lactobacilli and bifidobacteria [78]. In general, the colonic microflora is partitioned from the rest of the body by the mucus layer and mucosa. Loss of this partitioning effect is associated with disease processes within the large intestine [79], but it is unsure whether this is a cause or -effect of the disease process. Within the healthy large intestine, the main way the colonic microflora interacts with the host is through its metabolites [80]. Some of these metabolites are putatively damaging to the underlying mucosa, such as indoles, ammonia and amines while others are potentially beneficial to the host, including short chain fatty acids (SCFA) [81] and lignans that the mammalian gut can absorb [82,83]. SCFA are produced by bacterial fermentation of dietary carbohydrate sources, of which dietary fiber is the main type in the large intestinal lumen.

Dietary fiber, plays a profound role on the number and diversity of bacteria that inhabit the large intestine. In the absence of dietary fiber or other luminal energy sources, resident bacteria in the colon will turn to large intestinal mucus as an energy source prior to attacking the underlying mucosa [84]. As bacteria require the necessary enzymes to break down saccharide bonds of the diverse range of dietary fibers, fiber will clearly affect microfloral population dynamics. The presence of any fermentable dietary fiber is likely to lead to an increase in microfloral bifidobacterial and Lactobacillus levels, as these bacteria ferment carbohydrates. Previous studies in humans have suggested dietary fibers like alginate [85], chitosan [86], and inulin [87] lead to a reduction in potentially harmful microfloral metabolites. A range of small human interventions with various fermentable dietary fibers have shown significant, but small, clinical benefit in a number of intestinal diseases and disorders either on their own or in combination with probiotics [88-90].

Catabolism by microbial populations may also be important for decrease the levels of cholesterol and lipids. Bacteria such as Lactobacillus and Bifidobacteria can exert a hypocholesterolemic effect by enhancing bile acid deconjugation [91,92]. Furthermore, Lactobacillus and Bifidobacteria remove cholesterol in vitro by assimilation and precipitation [93,94]. Fermentation products further affect lipid metabolism. Propionate inhibits the incorporation of acetic acid into fats and sterols, resulting in decreased fatty acid and cholesterol synthesis [95].

5.4. Fiber and Immune function

Besides its absorptive functions, the gastrointestinal tract is involved with a range of immune functions. The mucosa effectively partitions the rest of the body from digestive enzymes, large numbers of bacteria and assorted toxins that occur within the gut. The mucosa has two main roles in immunity. Firstly, the mucosa samples luminal contents to assess the threat to the body because the gut comes into contact with a wide range of external antigenic compounds. This is carried out by the gut-associated lymphoid tissue or GALT [96]. In the second place, gut epithelium must also protect itself from the luminal stress of damaging agents and shear forces [97]. To do this, protective mucus is secreted along almost the entirety of the gastrointestinal tract (excluding the oesophagus and possibly Peyer's patches). Within the mouth, mucin is secreted alongside other salivary secretions and acts as a lubricant. In the stomach and intestine, mucus is secreted as a protective barrier [98].

There is a paucity of data regarding intake of DFs and immune function associated with the gut or otherwise in humans [99]. Animal studies within this area are also sparse. Field et al. [100] carried out studies with dogs and they found that fermentable fiber intake resulted in increased intra-epithelial T-cell mitogen response [92]. In a recent study it has been observed that DF may interact directly with immunoregulatory cells. Mucosal macrophages and dendritic cells have receptors with carbohydrate-binding domains that bind β-glucans and cause a decrease in IL-12 and increase in IL-10, which is consistent with an anti-inflammatory phenotype [101]. No previous study has assessed the impact of DFs on the human mucus barrier due to the invasiveness of procedures involved with measuring the mucus barrier directly. However, the effects of different types of DFs on the intestinal mucus barrier have been studied in animal models. Fibers and fiber sources such as alginates, ispaghula husk, wheat bran, ulvan and carrageenan all appear to benefit the protective potential of the colonic mucus barrier [100,102].

5.5. Fiber and prevention of cancer

Cancer continues to be one of the number one health concerns of populations worldwide. Most cancers strike both men and women at about the same rate, with exception of cancers of the reproductive system. Of particular concern is cancer of the colon, ranking among the top 3 forms of cancer in the U.S.A., for both men and women. Colon cancer is also one of the leading causes of cancer morbidity and mortality among both men and women in the Western countries, including the U.S.A. [103]. The European Prospective Investigation of Cancer (EPIC) is a project that includes more than half a million people in 10 European countries and they results indicate that dietary fiber provides strong protective effects against colon and rectal cancer. In one of its papers, the authors clarify that methodological differences in some previous studies (e.g., study design, dietary assessment instruments, definition of fiber) may account for the lack of convincing evidence for the inverse association between fiber intake and colorectal cancer risk [104]. A careful work within the same project was conducted as a prospective case–control study nested within seven UK cohort studies which included 579 patients who developed incident colorectal cancer and 1996 matched control subjects. They used standardized dietary data obtained from 4- to 7-day food diaries that were completed by all participants to calculate the odds ratios for colorectal, colon, and rectal cancers. In this work, the researchers confirmed that the intake of dietary fiber is inversely associated with colorectal cancer risk [105]. Taking into account these studies, the United States Food and Drug Administration has approved health claims in 2010 supporting the role of DF in the prevention of cancer [106].

Human metabolic and animal model studies indicate that beneficial effects of dietary fiber in relation to colon cancer development depend on the composition and physical properties of the fiber [107,108]. The effect of soluble fiber sources is mainly based on their fermentation and on the effects of short-chain fatty acids produced, especially of butyric acid. It has been known since 1982 that the colonic mucosa uses these acids, especially butyrate, as a preferential energy source [109]. Butyric acid stimulates the proliferation of normal cell lines both in vitro and in the normal epithelium, but retards the growth of carcinoma cell lines and induces apoptosis in cultured colonic adenoma and carcinoma cells [110,111]. Insoluble fiber has been found to have a protecting effect by absorbing hydrophobic carcinogens [112-114]. A third potentially effective mechanism is that of the accompanying phenolic compounds. Several phenolic compounds, having antioxidative properties, are present especially in cereal fiber sources. They are released from their bound states by bacterial enzymatic action in the colon, and can act in the intestine locally as anticarcinogens both in preventing cancer initiation and progression [115-117].

6. Digestive and absorptive functions of the gastrointestinal tract and dietary fibers

The gastrointestinal tract (GIT) is the initial site of action from which dietary fibers produce systemic effects presented in the previous section. The physiological effects of dietary fiber depend on a myriad of variables, but generally they depend on the type (soluble or insoluble), the dose of a specific fiber consumed, the composition of the entire fiber-containing meal, and the individual physiological profile of the subject who consume the fiber-containing meal [5]. The GIT serves as an interface between the body and the external environment. The main function of the GIT is to absorb nutrients from ingested foods. The organs of the GIT are connected to the vascular, lymphatic and nervous systems to facilitate regulation of the digestive function [118]. To carry out this function digestive processes are realized by secretion of enzymes and associated co-factors, and through maintenance of the gut lumen at optimal pH for digestion [119]. Gastrointestinal secretion of enzymes and other factors, alongside control of gut motility is governed by a series of complex neurohumoral pathways (mediated by acetylcholine, gastrin, motilin, cholecystokinin, gastric inhibitory peptide (GIP), secretin, etc.) that begin to operate by luminal content. Two main features of luminal content which appear to govern gastrointestinal physiology are luminal chemical profile and luminal bulk. The nutrient/chemical profile of the gut lumen is sensed by specialised chemosensor enteroendocrine cells within the epithelium [120], while mechanoreceptors (stretch activated neural cells) occurring within the myenteric and submucosal plexusues [121] are activated as a result of mechanical pressure from luminal contents. The main absorptive area in the gut is the small intestine, which is involved in the absorption of the subunits of digestible macronutrients, as well as vitamins, minerals and other micronutrients [87]. Ingested foods must be mechanically homogenised with digestive secretions in order to allow better hydrolysis of macronutrients, and, in some cases, to allow micronutrient release. Mastication in the buccal cavity mix food with salivary secretions among them α-amylase starts digestion of starches [122]. Food boluses entering the stomach are maintained there for mixing with gastric secretions. A strongly acidic secretion allows denaturation of proteins and solubilisation of other factors. Gastric proteases (mainly pepsin) cleave bonds in proteins to form a range of shorter peptides and amino acids. Gastric lipase initiates digestion of dietary lipids [123]. By the time the majority of luminal contents leave the stomach, they have been processed into creamy, homogenous slurry, known as chyme. As luminal contents appear in the upper section of the small intestine (the duodenum), they are met with alkali (bicarbonate-rich) secretions from the liver, pancreas and intestinal crypts. Pancreatic exocrine secretions also contain a myriad of enzymes for digestion of all macronutrients [124].

Classically, dietary fiber is cited as reducing whole gut transit time, thereby increasing frequency of defecation. This effect can be explained on the one hand, due to DF increase the intestinal luminal bulk resulting in an increased peristalsis which reduce the whole gut transit time. DFs that increment the luminal bulk are those that have a high water-binding capacity [125]. Furthermore, feed-forward and feedback from other portions of the gut as a result of fiber intake could also affect motility of the different organs of the GIT. Prolongation of nutrient release into the intestinal lumen from the stomach is likely to result in a lengthened phase of hormonal feedback from the duodenum, terminal ileum and colon, leading to a delay in gastric emptying [84]. At the same time, this delay in the gastric emptying towards small intestine are likely to increased motility distally (and therefore decreased transit time). The most researched area of the effects of dietary fibers on gastric motility is linked to gastric emptying. A range of studies have demonstrated that inclusion of viscous fibers in liquid test meals results in delayed gastric emptying, and are particularly consistent in the case of pectins in human studies [126,127]. In a study comparing the physiological effects of a mixed meal containing high levels of natural fibers (fruit, vegetables and whole grains) against one without these fibers (instead containing fruit and vegetable juice and refined grains), concluded that removal of natural fiber decreased gastric emptying mean rate of approximately 45 min in a crossover feeding trial in 8 healthy adult participants [128]. The dietary fibers that raise the bulk of luminal contents of the large bowel are those that are not well fermented by the colonic microflora, and those that have a high water-binding capacity [84].

6.1. Nutrients absorption

To date, evidence has been obtained in different types of studies that dietary fiber can influence the metabolism of carbohydrates and lipids preventing the development of diabetes mellitus and cardiovascular disease. Intake of dietary fiber can influence the absorption of nutrients in different ways. It has been postulated that the presence of any dietary fiber in the upper GI tract will result in a decreased rate of intestinal uptake of a range of nutrients. However, it is necessary to consider what physicochemical factors of dietary fibers are important in these roles [84]. In previous animal studies, Kimura et al. [129], noted higher levels of cholesterol excretion in rats fed diets containing 1000 mg/kg of degraded alginates with molecular weights of 5 and 10 KDa compared to the effect of a diet with a lower molecular weight (1 kDa) alginate or a control (no fiber) diet. While such absorption-lowering effects can be beneficial in reducing energy uptake, it must also be noted that such factors are also likely to reduce the bioavailability of minerals, vitamins and phyto-chemicals. Dietary fiber fractions differ largely in their abilities to affect mineral and trace element availability and this might have negative impacts in high-risk population groups. Small human feeding studies have suggested that inclusion of food hydrocolloids like alginates [130], guar gum [131,132] and β-glucan [58,133-135] into test meals results in a blunting of postprandial glycaemic and insulinaemic responses.

7. *In vitro* chemical experimental models

To study the absorption of nutrients *in vivo*, feeding methods, using animals or humans, usually provide the most accurate results, but they are time consuming and costly, which is why much effort has been devoted to the development of in vitro procedures [136]. The *in vitro* digestive chemical experimental model enabled mimicking, in the laboratory, the in vivo reactions that take place in the stomach and duodenum. In principle, in vitro digestion models provide a useful alternative to animal and human models by rapidly screening food ingredients. The ideal in vitro digestion method would provide accurate results in a short time [137] and could thus serve as a tool for rapid food screening or delivery systems with different compositions and structures [19]. In vitro methods cannot be used alone for important decisions taken by industry or international organizations because human studies are required for such determinations, but are important for screening purposes and to project future studies.

8. Work objective

Considering the fact that dietary fibers are new ingredients widely applied in foods, it is important to know their effect on absorption of nutrients and micronutrients. For this reason, the interaction between nutrients and fibers from different sources (animal and plant fibers) and types (soluble and insoluble fibers) has been studied using chemical experimental models of the human digestive tract to evaluate the availability for absorption of glucose, calcium and iron using yoghurt as a food model.

9. Materials and methods

9.1. Dietary fibers employed

The plant fibers used in this work were: inulin (Frutafit-inulin, Imperial Sensus, The Netherlands), bamboo (Qualicel, CFF, Gehren, Germany), wheat (Vitacel WF 101, JRS, Rosenberg, Germany), apple (Vitacel AF 400-30, JRS, Rosenberg, Germany) and psyllium (Metamucil, Procter and Gamble Co., Cincinnati, OH, USA). Metamucil is a pharmaceutical formula with Plantago ovata seed husk (49.15% w/w) and sucrose (50.85%). Suppliers of wheat and apple fiber indicated that these products are free from phytic acid, and besides, the wheat fiber is gluten free. The inulin utilized in this work has a degree of polymerisation ≥ 9 as declared by suppliers.

The dietary fiber from animal source utilised in these assays was chitosan. It was obtained from crustacean chitin in the Laboratorio de Investigación Básica y Aplicada en Quitina (LIBAQ-INQUISUR-CONICET), Universidad Nacional del Sur, Bahía Blanca, Argentina. Chitin firstly was isolated from shrimp (Pleoticus mülleri) waste by the process that was described in our previous work [138]. Chitosan was prepared directly by heterogeneous deacetylation of chitin with 50% (w/w) NaOH. For the biopolymer characterisation, moisture and ash contents were determined at 100–105 °C and 500–505 °C, respectively.

Deacetylation degree was obtained using FT-IR spectroscopy (Nicolet iS10 FT-IR Spectrometer, Thermo Fisher Scientific, USA) with samples in the form of KBr at a ratio of 1:2. Viscosity of 1% chitosan in 1% acetic acid solution was measured with a Brookfield model DV-IV + viscosimeter (Brookfield, USA) with spindle 21 and a 50 rpm rotational speed at 25 °C [139].

9.2. Analysis for dietary fiber

Total, soluble and insoluble dietary fiber contents of chitosan and plant fibers were analysed according to the enzymatic–gravimetric method of the Association of Official Analytical Chemists (AOAC) Official Method 991. 43 [140]. Apple, bamboo, psyllium and wheat fibers were investigated to obtain contents of main cell wall constituents (lignin, cellulose, hemicellulose). These components were determined by modifications of the method described by Robertson and van Soest [38, 141] using ANKOM200/220 Fiber Analyzer (ANKOM Technology, Macedon, NY, USA). This method measures Acid Detergent Fiber (ADF), Neutral Detergent Fiber (NDF) and Lignin. Cellulose and hemicellulose contents were obtained by calculations. To determine ADF, duplicate samples were agitated under pressure with hot acid detergent solution for 60 min, rinsed in hot water and dried. To determine lignin content, duplicated samples were digested in 72% (v/v) sulfuric acid, following ADF analysis. Cellulose content of samples was calculated from ADF minus the lignin content. To determine NDF, duplicated samples were shaken with neutral detergent solution and heat-stable α-amylase for 60 min, rinsed and dried. Hemicellulose content of samples was calculated as NDF minus ADF [141].

9.3. Yoghurt preparation

Yoghurt was prepared using reconstituted whole milk powder (15% w/w) and 5% sucrose. This mix was homogenized and heated to 85 °C for 30 min., cooled to ambient temperature and inoculated with 0.03% starter culture. Starter was constituted by a 1:1 mixture of Streptococcus thermophilus (CIDCA collection 321) and Lactobacillus delbrueckii subsp. bulgaricus (CIDCA collection 332). Samples were incubated at 43°C to reach a pH of 4.4–4.6 and stored at 4°C, after completion of the fermentation process 1.3% (w/w) of each dietary fiber was added to samples of yoghurt [142]. The amount of fiber was selected following US regulations for fiber-fortified products [143].

To study glucose availability 0.6 g of glucose (Sigma-Aldrich Co., St. Louis, MO, USA) was added for each sample of yoghurt with each type of dietary fiber. In calcium availability studies the digestive mimicking was done without the addition of exogenous calcium because yoghurt is a source of calcium in the diet [138]. To evaluate the interactions between the fibers and iron, 0.8% (w/w) of ferrous sulfate was added to yoghurt samples with each type of fiber [139]. This addition was in accordance with local regulations governing iron supplementation in milk products. Ferrous sulfate ($FeSO_4 \cdot 7\ H_2O$) of 99.9% purity was used as purchased (Sigma-Aldrich Co., St. Louis, MO, USA).

9.4. Digestive chemical experimental model

Two types of digestive simulations were performed to study the interactions between dietary fibers used and the macro and micro nutrients tested. To evaluate the interaction of glucose and calcium with the fibers, gastric and duodenal environments were simulated. To examine the interactions between the fibers and iron was used in addition, a dialysis membrane to imitate the iron passage through the intestinal wall. Digestive enzymes were not utilized in these models because they do not hydrolyze fibers. The importance of duodenal simulation in these studies is because most dietary glucose, calcium and iron are absorbed in the duodenum.

The experiments to study the availability of glucose and calcium were performed in the following steps: a mix of 12.5 g of yogurt with 0.3 g of each fiber was stirrer in 50 mL of 0.1 M HCl (Merck) pH 1.0–2.0, 30 rpm and 37°C to reproduce the gastric environment. After 1 hour simulations were taken from the acidic medium to pH 6.8–7.2 with 15 g/L of $NaHCO_3$ (Sigma Chemical Co., St. Louis, MO, USA). The stirring speed was increased from 30 to 300 rpm and the temperature was maintained at 37°C to reproduce the duodenal environment. Then simulations were allowed to rest for 15 min until two phases separated. Samples to determine glucose and calcium concentration were taken from the supernatant. Glucose and calcium amounts, determined by this way, represent the bioavailability fraction of those nutrients. A control without fibers was made to consider glucose and calcium 100% availability [138].

Experiments to study the interaction of dietary fibers with iron were carried out in the following manner. Yoghurts with ferrous sulfate and each fiber were stirred in 50 mL of 0.1 M HCl (Merck) for 1 h at pH 1.0–2.0, 30 rpm and 37 °C to reproduce the gastric environment. During this first step of simulation pH was checked each 15 min with a pH Meter Hach model EC-30 (USA) and it remained constant (pH 1.0–2.0). To reproduce the chemical duodenal environment pH level was increased to pH 6.8–7.2 with 0.2 M NaHCO3 (Sigma-Aldrich Co., St. Louis, MO, USA), stirring speed was increased from 30 to 300 rpm to imitate the peristaltic movement and temperature was maintained at 37°C. Simulations were immediately transferred into a dialysis tubing cellulose membrane (D9527-100 FT, (Sigma-Aldrich Co., St. Louis, MO, USA). This cellulose membrane (molecular weight cut-off 12,400) was previously prepared, as indicated by suppliers, and it was cut into 28 cm length pieces. The loaded tubes were immersed in 100 mL of distilled water; at 37°C. Iron concentrations were determined from the dialysed medium at 30 and 60 minutes. Control yoghurt with ferrous sulfate without fibers was subjected to the digestive simulation and was considered as 0% iron retention to calculate iron retention percentages for each fiber [139].

9.5. Analytical techniques

To determine glucose concentration an enzymatic method was used. Glucose reacts with 10 kU/L glucose oxidase (GOD), and 1 kU/L peroxidase (POD) in presence of 0.5 mM 4-aminophenazone (4-AP) and 100 mM phosphate buffer (pH 7.0) containing 12 mM hydroxybenzoate (Wiener Lab Glicemia enzymatic AA Kit, Argentina). An amount of digestive simulation solution (10 mL) was mixed with 1.0 mL of reagent, tubes were incubated for 5 min in water bath at 37°C and developed colour were read in spectrophotometer (Spectronic 20 Genesys TM, Spectronic Instrument, USA) at 505 nm. Final reaction colour is stable for 30 min. Glucose calibration curve was carried out. The amounts of glucose used in this study correspond to available carbohydrates in the human mixed diet.

To determine calcium concentration a spectrophotometric method was used. Calcium reacts with 3.7 mmol/L cresolphtalein complexone (Cpx) at pH 11 (buffer 0.2 mol/L aminomethylpropanol (AMP) solution in 35%v/v methanol) (Wiener Lab Ca-color Kit, Argentina). Assays were carried directly in spectrophotometer test tubes: 50 lL Cpx were mixed with a plastic rod and absorbance was read in spectrophotometer (Spectronic 20 Genesys TM, Spectronic Instrument, USA) at 570 nm (internal blank), then 20 mL of each digestive mimicking sample were added, immediately mixed and read after 10 min. A standard curve was developed [138].

To determine iron concentration in the dialysates a spectrophotometric method was used, 500 μL of dialyzates was reduced with 2 mL of mercaptoacetic acid (succinic acid buffer, pH 3.7). Then, iron reacted with one drop of pyridyl bis-phenil triazine sulfonate (PBTS) producing a pink color due to the complex formed (Wiener Lab Fe-colour Kit, Rosario, Argentina). Absorbance was read on a spectrophotometer (Spectronic 20 Genesys Thermo Electron Scientific Instruments Corp., Madison, WI, USA) at 560 nm (internal blank). All glassware used in sample preparation and analysis was rinsed with 10% (v/v) concentrated HCl (37%) and deionised water before using, to avoid mineral contamination. A regression equation ($y = 2.5333x + 0.0042$, $R2 = 0.995$) derived from data generated from standards of $FeSO_4$ was used to calculate iron concentrations in the samples. Iron retention percentages for each studied fibers were calculated as a percentage of the amount of iron measured in the dialysed medium obtained with the control yoghurt without fibers [139].

9.6. Statistical analysis

Experiments were performed at least five times for each dietary fiber using freshly prepared yogurt. For total iron concentration in dialyzates, each individual sample was run in duplicate. Averages and standard deviations were calculated and expressed in each case as the mean ± SD for n replicates. Normality of the data was checked with the Lilliefors test. The influence of different dietary fibers on the retention percentages of glucose, calcium and iron were statistically analyzed by a one-way analysis of variance (ANOVA) ($p < 0.05$) to find significant differences and Tukey's test to compare means.

10. Results and discussion

10.1. Characterisation of fibers

The dietary fibers used in this study have different water solubility characteristics: inulin is a soluble fiber, bamboo and wheat are insoluble fibers, apple is partially insoluble fiber, and psyllium forms a viscous dispersion at concentrations below 1% and a clear gelatinous mass at 2%. Chitosan is a fiber of a different origin, i.e. from animal source and is soluble in an acidic medium and flocculates in an alkaline medium. We used these fibers because they present different physicochemical behaviors that have been described in literature [144,4]. The commercial fiber compositions used in this study, regarding total, soluble and insoluble fractions, are shown in Table 4. Analysis for dietary fiber using the AOAC method 991.43 showed that wheat and bamboo have high amounts of insoluble fraction.

Fiber	Insoluble fiber	% Insoluble fiber	Soluble fiber	% Soluble fiber	Total fiber
Apple	44.8 ± 0.4	77.1	13.3 ± 0.7	22.9	58.1 ± 1.0
Bamboo	91.4 ± 0.5	95.9	3.2 ± 0.8	3.4	95.3 ± 0.9
Chitosan	98.0 ± 1.0	100	nd	nd	98.0 ± 1.0
Inulin	nd	Nd	≥ 85.5	100	≥ 85.5
Psyllium	37.5 ± 0.6	82.9	7.1 ± 0.5	15.7	45.2 ± 0.8
Wheat	92.1 ± 0.6	97.6	2.3 ± 0.6	2.4	94.4 ± 1.1

nd: no detectable

Table 4. Total, soluble and insoluble fiber content (g/100g) of employed fibers

Inulin presents only soluble fraction as expected. Psyllium and apple have both soluble and insoluble fractions. Apple fiber is characterized by a well balanced proportion between soluble and insoluble fraction [145]. The total dietary fiber content is 45.2% for psyllium, which is an acceptable value, taking into account that the supplier declared a 49.15% content for Plantago ovata seed husk in Metamucil preparation. Van Craeyveld et al. [146] reported 3.4% (dm) ash and 7.1% (dm) protein contents for Plantago ovata seed husks. The total dietary fiber content is 58.1% for apple, which is about 10–14% higher than the values reported by Sudha et al. [147]; however, this value was in accordance with suppliers. The chitosan used in this study has 98% of insoluble fraction and no detectable soluble fraction. Furthermore the characteristics of this biopolymer are a deacetylation degree of 89%, a viscosity of 120 mPa.s, 6.7 g% moisture and 0.67 g% ash content.

Plant fiber characterisations were completed with the study of Acid Detergent Fiber (ADF) and Neutral Detergent Fiber (NDF), lignin, cellulose and hemicellulose contents (Table 5). Apple presents the highest lignin content. Wheat fiber mainly has cellulose. Bamboo has proportional amounts of cellulose and hemicellulose, but compared with other fibers, has the highest hemicellulose content. These results are in accordance with their plant fiber origins and previous works [145-149]. Frutafit-Inulin was not analysed because its composition was ≥85.5% (w/w) of inulin, ≤9.5% of mono and disaccharides, ≤0.1% of ash with polymerisation degree ≥9 according to suppliers. Chitosan was not analysed either, because of its animal origin.

Fibre	ADF	NDF	Lignin	Cellulose	Hemicellulose
Apple	38.6 ± 0.9	44.3 ± 0.7	8.4 ± 0.8	30.2 ± 1.7	5.7 ± 1.6
Bamboo	50.2 ± 0.7	90.4 ± 0.6	5.0 ± 0.3	45.2 ± 1.0	40.2 ± 1.7
Psyllium	7.3 ± 0.4	36.8 ± 0.9	0.8 ± 0.1	6.5 ± 0.4	29.5 ± 1.3
Wheat	74.8 ± 0.3	89.7 ± 0.6	2.6 ± 0.4	72.2 ± 0.7	14.9 ± 0.9

Table 5. Acid Detergent Fiber (ADF), Neutral Detergent Fiber (NDF), lignin, cellulose and hemicellulose (%) of fibers

Scientists who deal with animal nutrition usually use Van Soest's method to analyse feed. Scientists working on human nutrition use methods of the AOAC, because of their interest in soluble fiber. It is known that soluble fiber plays an important role in human health and the food industry. However, it could be useful in human nutrition to know the composition of insoluble fiber, as it is possible that insoluble fibers do not all have the same effect on human health. The NDF and insoluble fiber methods were applied to the same samples. Insoluble fiber includes hemicellulose, cellulose, lignin, cutin, suberin, chitin, chitosan, waxes and resistant starch. NDF includes hemicellulose, cellulose and lignin. Escarnot et al. [149] studied three wheat varieties and four spelt genotypes. They analysed three milling fractions from those grains for insoluble and soluble fiber contents, lignin, hemicellulose and cellulose. They found a very high correlation ($r^2 = 0.99$) between the two methods, showing that NDF and insoluble fiber methods cover the same types of fiber. For insoluble fiber analysis, the NDF method is faster and more thorough.

10.2. Digestive chemical model and glucose, calcium and iron retention percentages

Dietary fiber have been found to have the capacity of binding different substances like bile salts and glucose which have implications in cholesterol lipid and carbohydrate metabolism respectively, as presented in the preceding sections. However, the continuous introduction of new ingredients in the food industry requires further studies in order expand knowledge of the impact on nutrient absorption.

Figure 1 shows the behavior of samples during the digestive tract simulation to evaluate glucose and calcium retention percentages and macroscopical differences between them could be observed. Different simulated digestive contents for different fibers before dialysis in assays to determine the iron retention percentages are not shown because they are similar to those presented in Fig. 1. Simulation of gastrointestinal environment during dialysis of different yoghurts can be observed in Figure 2. Changes in pH during gastrointestinal simulation produces different behaviors depending on the type of fiber employed. The apple fiber is a fine powder with brownish color, probably due to the content of phenolics compounds [150]. When apple fiber is added to the yogurt and subjected to the gastrointestinal simulation this color persists (Figure 1). In Figure 2 it can be seen that Psyllium fiber gives a viscous dispersion [151,152]. Due to changing pH values in the

(1) Wheat, (2) psyllium, (3) apple, (4) inulin, (5) chitosan and (6) bamboo fibers.

Figure 1. Photograph of the macroscopic view of different fibers in the in vitro digestive tract simulation.

(1) Yoghurt without fiber, (2) chitosan, (3) psyllium, (4) wheat.

Figure 2. Different fiber behaviors in the dialysis step of digestive simulation.

digestive tract, Chitosan precipitates while passing through the first portion of the small intestine, forming flocculus. Chitosan, that is a positively charged polysaccharide, is insoluble in neutral and alkaline pH. It is only soluble in acidic pH because below pH 6.5 (pKa = 6.5), the amine groups of chitosan are positively charged. When it is solubilised in dilute acid, chitosan has a linear structure [153]. At pH > 6.5, the polymer loses its charges from the amine groups and therefore becomes insoluble in water and precipitate forming flocculus.

Using the model that reproduces *in vitro* gastrointestinal conditions we determined glucose availability reduction and the results are shown in Figure 3. Significant differences ($p < 0.05$) are observed in glucose availability reduction percentage for the different fiber samples. In the gastrointestinal conditions chitosan formed a flocculus that entrapped glucose so its availability reduction is the highest. Psyllium increases viscosity medium and glucose availability reduction is 15.3 ± 1.8%; wheat has 9.5 ± 2.1% of glucose retention and inulin 5.7 ± 1.8%, apple and bamboo showed no availability reduction. This *in vitro* study supports the view that certain types of dietary fiber reduce the rate of glucose absorption but chitosan has the most pronounced effect. The behavior in delaying absorption could be likely to alter the gut endocrine response both by carrying material further down the small intestine prior to absorption as well as by producing a flatter blood glucose profile.

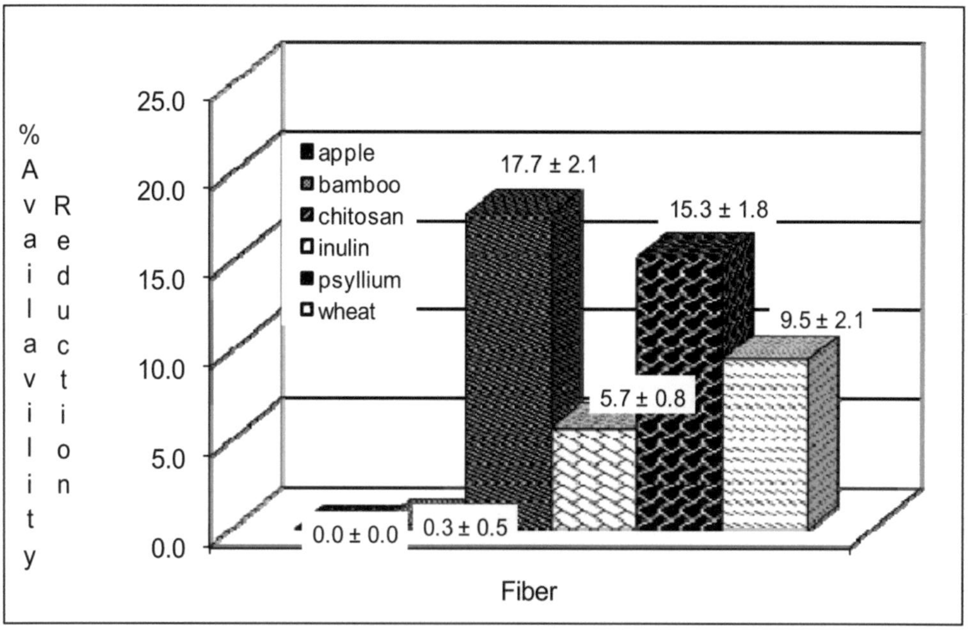

Figure 3. Glucose availability reduction

On the other hand, dietary fiber may influence the availability of minerals, such as calcium, magnesium [154] and iron [155]. Animal studies have found that dietary chitosan possibly arrests the absorption of calcium [156,157].

To study calcium availability the same model for glucose was used but without the addition of exogenous calcium because yogurt is an important source of this mineral in the human diet. Data are shown in Figure 4. Statistical analysis confirmed significant differences ($p < 0.05$) among the behavior of the different fibers with calcium. It is observed 16.5 ± 1.6% of calcium availability reduction for apple fiber that have significant differences with the others fibers. However availability reduction responses, between insoluble fibers (wheat and bamboo) and soluble ones (inulin and psyllium plantago), have no significant differences ($p < 0.05$) by Tukey's test. Again, like results obtained with glucose, this study demonstrated that the chitosan effect is more pronounced and higher than for the other studies [138].

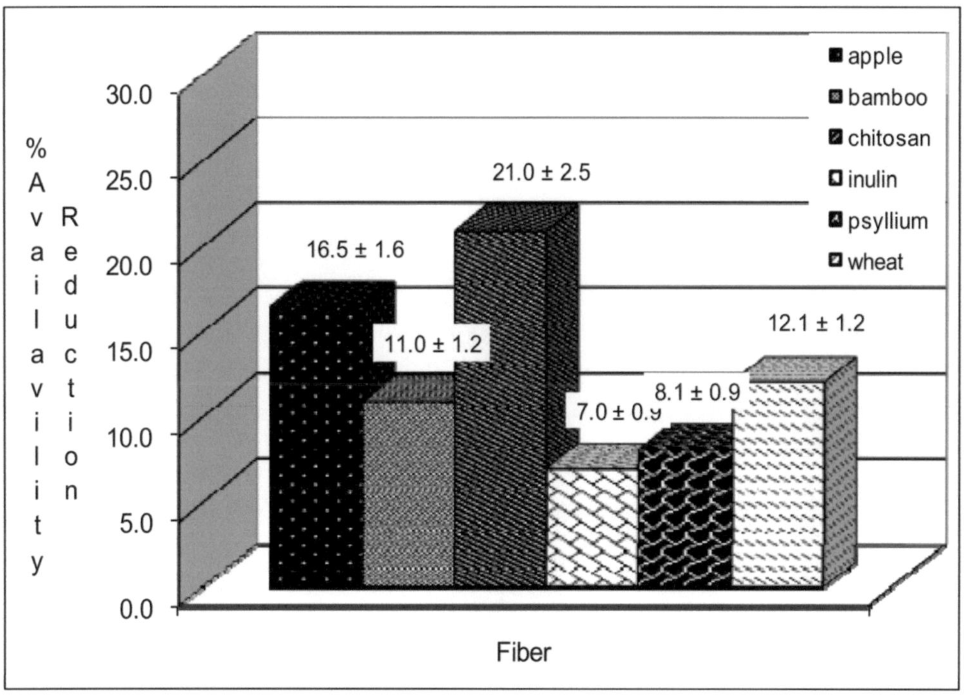

Figure 4. Calcium availability reduction.

To study iron retention percentages by the fibers tested in the present study, the introduction of cellulose dialysis tubes in the digestive chemical experimental model is utilised. The use of a membrane dialysis tube reproduces, in the laboratory, the duodenum wall and its utilisation is presumably a significant factor that determines iron absorption according to Miret et al. [158].

In yoghurt, caseins are modified as a consequence of its production process. Bioactive peptides are formed from caseins during the elaboration of milk products (cheese, yoghurt) under the action of endogenous enzymes of milk (plasmin, cathepsin, among others) or of microorganisms [159]. These peptidic fragments that are already present in yoghurt, could fix iron and calcium according to Bouhallab and Bouglé [159]. Then, these complex matrices (yoghurts with each type of fiber and iron or calcium) are subjected to the gastrointestinal simulation. Control yoghurt with ferrous sulfate without fiber was also subjected to the digestive simulation and considered to be 0% iron retention (100% iron dialyzated) to calculate iron retention percentages for each fiber. Similarly, control yoghurt without fibers was subjected to the digestive simulation to estimate calcium 100% availability. With these control yoghurts, we could consider the interaction of iron or calcium with casein peptidic fragments.

Iron retention percentages of different fibers are presented in Figure 5. Bamboo and wheat fibers, both insoluble, have low iron retention percentages between 2–5% at 30 min with a maximum of 10% at 60 min. There are no significant differences ($p < 0.05$) between them by Tukey's test. Bamboo and wheat are high in cellulose content. Cellulose could retain iron by physical adsorption according to results reported by Torre et al. [15]. They worked with high dietary fiber food materials studying the physicochemical interactions with Fe(II), Fe(III) and Ca(II) without an *in vitro* digestive model. They found that the interaction between Fe(II) and cellulose could be explained better by physical adsorption than complex formation. Inulin, a soluble fiber, has no iron retention at 30 or 60min of simulation. This result is in accordance with studies that confirm that inulin does not interfere with iron absorption [17,20,160,161].

Although psyllium and apple fiber contain both soluble and insoluble fractions, they have significantly different responses ($p < 0.05$). The apple fiber incorporated in yoghurt has no

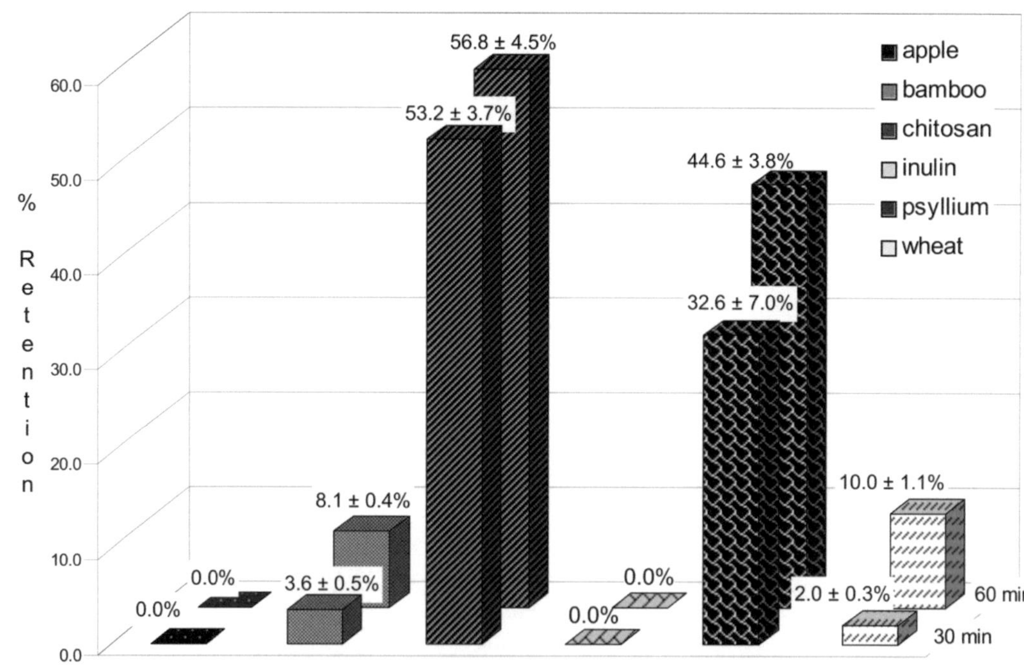

Figure 5. Iron retention percentages of yoghurts with different studied fibers

influence on iron retention. Psyllium shows, on average, 44.6 ± 3.8% iron retention at 60min, which may be mainly attributed to the formation of high viscous dispersion that could be interfering with iron absorption (Figure 2). In addition, the differing behaviors between apple fiber and psyllium could be explained by the different chemical composition of these fibers. Psyllium has high hemicellulose content and apple has the highest lignin content and cellulose. However, bamboo has a low iron retention percentage. The bamboo behavior could be explained due to the composition with equal proportions of soluble and insoluble fractions. More research is needed of this type of fiber.

Chitosan presents the highest iron retention percentages at 30 min (53.2 ± 3.7%) and 60 min (56.8 ± 4.5%), which shows significant differences ($p < 0.05$) with other fibers. This biopolymer, which has an animal origin, contains 98% insoluble fiber, and flocculates in the first portion of the small intestine. This flocculus (Figure 2), which could entrap iron, clearly decrease iron dialysis. However, certain amount of iron could go through the cellulose membrane and could be measured to calculate the iron retention percentage. Certain amount of casein-peptide-fragments interacting with iron could remain in solution. Nevertheless, their presence does not interfere with the calculation of iron retention percentages as proven by the digestive simulations performed with control yoghurts.

This study shows that the effect of chitosan on iron absorption is more pronounced and higher than those measured for the other studied plant fibers, as dietary fiber is a significant factor that influences iron absorption. The iron retention percentages of different fibers used in this work could be explained mainly as a result of physicochemical phenomena, like adsorption, formation of viscous dispersion and flocculus.

Yoghurt contains peptidic fragment s from caseins. The caseins are amphiphilic phosphoproteins and their isoelectric point (p*I*) value is 4.6. At pH above the *pI*, caseins are negatively charged and soluble in water. The caseins have an electronegative domain preferentially located in small peptidic fragments known as α_{S1}-Casein, β-casein and κ-casein. These structural features of the caseins may render these molecules adept at forming complexes with multivalent cationic macromolecules, such as chitosan [153]. In yoghurt (pH = 4.4–4.6) aggregation of the casein-peptide-fragments occur because of a reduction in the electrostatic repulsion at around their *pI* value. Anal et al. [153] studied the interactions between sodium caseinate and chitosan, under a range of conditions. This study showed that soluble or insoluble chitosan–caseinate complexes can be formed depending on the pH. The characteristics of the complexes are determined by the biopolymer types and their concentration, as well as by environmental conditions. In a certain pH range (5.0–6.0), nanocomplexes of chitosan and sodium caseinate with diameter between 250 and 350 nm were formed. The chitosan and sodium caseinate complexes associated to form larger particles, which resulted in phase separation appear when the pH was either in the range 4.0–4.5 or >6.5. At pH 3.0–3.8, where chitosan and sodium caseinate have similar charges, they may dissociate from each other and become solubilized in solution. According to these authors, yoghurts with chitosan could contain chitosan-casein-peptidic complexes apart from free chitosan molecules in solution.

Besides, calcium existing in yoghurt or the added iron could interact with free chitosan molecules and those complexes. In our work, yoghurts with chitosan are subjected to gastrointestinal simulations. In the first step, our food passes through the simulated stomach (pH = 1.0–2.0) and it could be expected that casein peptidic fragments, chitosan, iron or calcium, all remain in solution. While the food passes through the first portion of the simulated small intestine, changes in pH can lead to formation of chitosan-casein peptidic complexes and iron or calcium could be interacting with them. At pH 6.8–7.0, free chitosan molecules and chitosan-casein-peptidic complexes precipitate forming flocculus. The force of the coagulum formed is high and can be seen in Figures 1 and 2. The results reported by Ausar et al. [162] indicate that hydrophobicity of the casein-chitosan complex is the main mechanism by which the casein-chitosan flocculation is produced.

Chitosan is essentially a positively charged polysaccharide. Iron and calcium are cations. Anal et al. [153] measured zeta potential of chitosan solutions, sodium caseinate solutions and chitosan-caseinate mixtures in a range of pH (3.0–6.5). They found that the pure chitosan solutions were strongly positively charged between pH 3.0 and 6.0. The zeta potential values of chitosan solutions decreased with increasing pH and were slightly negative (approximately − 2.5 mV) at pH 6.5. In our study, in the range of pH 3.0–6.0, isolated molecules of chitosan were probably interacting with iron or calcium by adsorption rather than by electrostatic forces. Besides, Anal et al. [153] observed that the zeta potentials of the chitosan–caseinate solutions were negative at pH > 5.5. In this range of pH, in our work, electrostatic interaction could exist between chitosan-caseinate complexes and iron or calcium. However, when chitosan precipitates, it captures the iron or calcium either by electrostatic forces or by adsorption [138,139].

The behavior of chitosan with calcium and iron in the digestive simulations were similar and can be explained in the same manner. However, the behavior between the other fibers used and the same micronutrients in the digestive simulations were significantly different. The flocculus formation by chitosan is a very strong kind of behavior which is independent of the use of the dialysis membrane. Evidently other types of interactions are brought into play for the other fibers that need further studies to determine them.

11. Conclusion

Results showed that the different plant fibers decreased glucose, calcium and iron availabilities whereas the effect of chitosan (fiber from animal source) was more pronounced. These findings could be positive or negative depending on the nutrient and the nutritional stage or health of the population who would receive the food under study. However, the *in vitro* digestive chemical experimental model may be used to increase the understanding of the interactions between animal and plant fibers with nutrients and micronutrients. This knowledge is very important from the point of view of health and for food industry and technologists.

Author details

Marina Dello Staffolo
CIDCA (Centro de Investigación y Desarrollo en Criotecnología de Alimentos), CONICET–CCT La Plata, Fac. Cs. Exactas, Universidad Nacional de La Plata, La Plata, Argentina

Alicia E. Bevilacqua
CIDCA (Centro de Investigación y Desarrollo en Criotecnología de Alimentos), CONICET–CCT La Plata, Fac. Cs. Exactas, Universidad Nacional de La Plata, La Plata, Argentina
Departamento de Ingeniería Química, Facultad de Ingeniería, UNLP

María Susana Rodríguez and Liliana Albertengo
Instituto de Química del Sur (INQUISUR), CO NICET–UNS Bahía Blanca, Departamento de Química, Universidad Nacional del Sur, Argentina

Acknowledgement

Financial support from CONICET and SeCyT-Universidad Nacional del Sur is gratefully acknowledged. Marina Dello Staffolo and Alicia Bevilacqua express their gratitude to Universidad Nacional de La Plata and bamboo, inulin and wheat suppliers (Imperial Sensus, CFF and JRS).

12. References

[1] Green CJ. Fiber in enteral nutrition. South African Journal of Clinical Nutrition 2000;13(4) 150-160.

[2] Food and Nutrition Board, Institute of Medicine (2001). Dietary reference intakes. Proposed definition of dietary fiber. A report of the panel on the definition of dietary fiber and the standing committee on the scientific evaluation of dietary reference intakes. Washington, DC: National Academy Press.

[3] Thebaudin JY., Lefebvre AC.,Harrington M., Bourgeois CM. Dietary fibres: Nutritional and technological interest. Trends in Food Science & Technology 1997;81 41-48.

[4] Tungland BC., Meyer D. Nondigestible oligo and polysaccharides (dietary fibre): their physiology and role in human health and food. Comprenhensive Reviews in Food Science and Food Safety 2002;1(3) 90-109.

[5] Trout D., Behall K., Osllesl O. Prediction of glycaemic index for starchy foods. American Journal of Clinical Nutrition 1993;58 873-878.

[6] Jenkins DJA., Kendall CWC., Axelsen M., Augustin LSA., Vuksan V. Viscous and nonviscous fibres, nonabsorbable and low glycaemic index carbohydrates, blood lipids and coronary heart disease. Current Opinion in Lipidology 2000;11(1) 49-56.

[7] Luccia BHD., Kunkel ME. In vitro availability of calcium from sources of cellulose, methylcellulose, and psyllium. Food Chemistry 2002;77 139-146.

[8] Bosscher D., Van Ciaillie-Bertrand M., Deelstra H. Effect of thickening agents, based on soluble dietary fibre on the availability of calcium, iron, and zinc from infants formulas. Nutrition 2001;17 614-618.
[9] Salas-Salvadó J., Martinez-González MA., Bulló M., Ros E. The role of diet in the prevention of type 2 diabetes. Nutrition, Metabolism & Cardiovascular Diseases 2011;21, B32-B48.
[10] Nordin BEC. Calcium and Osteoporosis. Nutrition 1997;13 664-686.
[11] Fardellonea P., Cottéb F-E., Rouxc C., Lespessaillesd E., Merciere F., Gaudinf A-F. Calcium intake and the risk of osteoporosis and fractures in French women. Joint Bone Spine 2010;77 154-158.
[12] Beard J., Stoltzfus RJ. Iron-deficiency anemia: Reexamining the nature and magnitude of the public health problem. Journal of Nutrition 2001;131 563S-703S.
[13] Thompson SA., Weber CW. Influence of pH on the binding of copper, zinc and iron in six fiber sources. Journal of Food Science 1979;44 752-754.
[14] Laszlo JA. Mineral binding properties of soy hull. Modelling mineral interactions with an insoluble dietary fiber source. Journal of Agriculture and Food Chemistry 1987;35 593-600.
[15] Torre M., Rodriguez AR., Saura-Calixto F. Interactions of Fe(II), Ca(I1) and Fe(II1) with high dietary fibre materials: A physicochemical approach. Food Chemistry 1995;54 23-31.
[16] Chau C-F., Chen C-H., Lin C-Yi. Insoluble fibre-rich fractions derived from Averrhoa carambola: Hypoglycemic effects determined by in vitro methods. Lebensmittel-Wissenschaft und-Technologie 2004;37 331-335.
[17] Bosscher D., Van Caillie-Bertrand M., Van Cauwenbergh R., Deelstra H. Availabilities of calcium, iron, and zinc from dairy infant formulas is affected by soluble dietary fibres and modified starch fractions. Nutrition 2003;19 641-645.
[18] Argyri K., Birba A., Miller DD., Komaitis M., Kapsokefalou M. Predicting relative concentrations of bioavailable iron in foods using in vitro digestion: New developments. Food Chemistry 2009;113 602-607.
[19] Hur SJ., Lim BO., Decker EA., McClements DJ. In vitro human digestion models for food applications. Food Chemistry 2011;125 1-12.
[20] Laparra JM., Tako E., Glahn RP., Miller DD. Supplemental inulin does not enhance iron bioavailability to Caco-2 cells from milk- or soy-based, probiotic-containing, yoghurts but incubation at 37°C does. Food Chemistry 2008;109 122-128.
[21] Tamine A., Robinson R. Yoghurt Science and Technology, 3rd ed. Boca Raton: CRC Press; 2007.
[22] Hispley EH. Dietary 'fibre' and pregnancy toxaemia. British Medical Journal 1953;2 420-422.
[23] Burkitt DP. The role of refined carbohydrate in large bowel behaviour and disease. Plant Foods for Man 1973;1 5–9.
[24] Trowell H. Atherosclerosis, Ischemic heart disease and dietary fiber. American Journal of Clinical Nutrition 1972;25 926-932.

[25] Cummings TJ., Mann J., Nishida C., Vorster H. Dietary fibre: An agreed definition. Lancet 2009;373 365-366.
[26] Harris SS., Pijls L. Dietary fibre: Refining a definition. Lancet 2009;374 28.
[27] Phillips GO., Cui SW. An introduction: Evolution and finalisation of the regulatory definition of dietary. Food Hydrocolloid 2011;25 139-143.
[28] Elleuch M., Bedigian D., Roiseux O., Besbes S., Blecker C., Attia H. Dietary fibre and fibre-rich by-products of food processing: Characterization, technological functionality and commercial applications: A review. Food Chemistry 2011;124 411-421.
[29] Borderías AJ., Sánchez-Alonso I., Pérez-Mateos M. New application of fibre in foods: Addition to fishery products. Trends in Food Science and Technology 2005;16 458-465.
[30] Normand FL., Ory RL., Mod RR. Binding of bile acids and trace minerals by soluble hemicelluloses of rice. Food Technology 1987;41 86-99.
[31] Chawla R., Patil GR. Soluble Dietary Fiber. Comprenhensive Reviews in Food Science and Food Safety 2010;9 178-196.
[32] Jenkins DJA., Jenkins MJA., Wolever TMS., Taylor RH. Slow release carbohydrate: Mechanism of action of viscous fibers. Journal of Clinical Nutrition and Gastroenterology 1986;1 237-241.
[33] Dikeman CL., Fahey GCJr. Viscosity as Related to Dietary Fiber: A Review. Critical Reviews in Food Science and Nutrition 2006;46(8) 649-663.
[34] Lattimer J., Haub M. Effects of dietary fiber and its components on metabolic health. Nutrients 2010;2 1266-1289.
[35] Claye SS., Idouraine A., Weber CW. Dietary fibre, what it is and how it is measured. Food Chemistry 1996;57 305-310.
[36] Prosky L., Asp N-G., Scheweizer TF., DeVries JW., Furda I. Determination of insoluble and soluble, and total dietary fibre in foods and food products: Interlaboratory study. Journal of the Association of Official Analytical Chemists1988;71 1017-1023.
[37] Englyst HN., Quigley ME., Hudson GJ. Determination of dietary fiber as non-starch polysaccharides with gas–liquid chromatographic or spectrophotometric measurement of constituent sugars. Analyst 1994;119 1497-1509.
[38] Van Soest PJ. Use of detergents in the analysis of fibrous feeds II. Determination of plant cell-wall constituents. Journal of the Association of Official Analytical Chemistry 1963;48 829-835.
[39] Ripsin CM., Keenan JM., Jacobs DR., Elmer PJ., Welch RR., Van Horn L., Liu K., Turnbull WH., Thye FW., Kestin M., Hegsted M., Davidson DM., Davidson MH., Dugan LD., Demark-Wahnefried W., Beling S. Oat products and lipid lowering: a meta-analysis. Journal of the American Medical Association 1992;267 3317-3325.
[40] Olson BH., Anderson SM., Becker MP., Anderson JW., Hunninghake DB., Jenkins DJA., LaRosa JC., Rippe JM., Roberts DCK., Stoy DB., Summerbell CD., Truswell AS., Wolever TMS., Morris DH., Fulgoni VL. Psyllium-enriched cereals lower blood total cholesterol and LDL-cholesterol, but not HDL cholesterol, in hypercholesterolemic adults: results of a meta-analysis. Journal of Nutrition 1997;127 1973-1980.
[41] Anderson JW., Allgood LD., Lawrence A., Altringer LA., Jerdack GR., Hengehold DA., Morel JG. Cholesterol-lowering effects of psyllium intake adjunctive to diet therapy in

men and women with hypercholesterolemia: meta-analysis of 8 controlled trials. American Journal of Clinical Nutrition 2000;71 472-479.

[42] Chau CF., Huang YL. Effects of the insoluble fiber derived from Passiflora edulis seed on plasma and hepatic lipids and fecal output. Molecular Nutrition & Food Research 2005;49 786-790.

[43] Cho IJ., Lee C., Ha TY. Hypolipidemic effect of soluble fiber isolated from seeds of Cassiatora Linn in rats fed a high-cholesterol diet. Journal of Agriculture and Food Chemistry 2007;55 1592-1596.

[44] Chau CF., Huang YL., Lin CY. Investigation of the cholesterol-lowering action of insoluble fibre derived from the peel of *Citrus sinensis* L. cv. Liucheng. Food Chemistry 2004;87(3) 361-366.

[45] Phillips MC. Mechanisms of cholesterol-lowering effects of dietary insoluble fibres: Relationships with intestinal and hepatic cholesterol parameters. British Journal of Nutrition 2005;94 331-337.

[46] Gonzalez M., Rivas C., Caride B., Lamas A., Taboada C. Effects of orange and apple pectin on cholesterol concentration in serum, liver and faeces. Journal of Physiology and Biochemistry 1998;54(2) 99-104.

[47] Anttila H., Sontag-Strohm T., Salovaara H. Viscosity of β-glucan in oat products. Agricultural and Food Science 2004;13 80-87.

[48] Wood PJ. Relationships between solution properties of cereal β-glucans and physiological effects – A review. Trends in Food Science and Technology 2004;15 313-320.

[49] Zacherl C., Eisner P., Engel K-H. In vitro model to correlate viscosity and bile acid-binding capacity of digested water-soluble and insoluble dietary fibres. Food Chemistry 2011;126 423-428.

[50] Galisteo M., Moron R., Rivera L., Romero R., Anguera A., Zarzuelo A. Plantago ovata husks-supplemented diet ameliorates metabolic alterations in obese Zucker rats through activation of AMP-activated protein kinase. Comparative study with other dietary fibers. Clinical Nutrition 2010;29 261-267.

[51] Delzenne NM., Kok N. Effects of fructans-type prebiotics on lipid metabolism. American Journal of Clinical Nutrition 2001;73 456S-458S.

[52] Eshak ES., Iso H., Date C., Kikuchi S., Watanabe Y., Wada Y., Tamakoshi A. and the JACC Study Group. Dietary fiber intake is associated with reduced risk of mortality from cardiovascular disease among Japanese men and women. Journal of Nutrition 2010;140 1445-1453.

[53] Health claims: fruits, vegetables, and grain products that contain fiber, particularly soluble fiber, and risk of coronary heart disease. Code of Federal Regulations 2010;Secc. 101, 77.

[54] Kaczmarczyka MM., Michael J., Miller MJ., Freunda GG. The health benefits of dietary fiber: Beyond the usual suspects of type 2 diabetes mellitus, cardiovascular disease and colon cancer. Metabolism 2012, article in press. doi:10.1016/j.metabol.2012.01.017.

[55] Slavin JL. Dietary fiber and body weight. Nutrition 2005;21 411-418.

[56] Wang ZQ., Zuberi AR., Zhang XH., Macgowan J., Qin J., Ye X. Effects of dietary fibers on weight gain, carbohydrate metabolism, and gastric ghrelin gene expression in mice fed a high-fat diet. Metabolism 2007;56 1635-1642.

[57] Choi JS., Kim H., Jung MH., Hong S., Song J. Consumption of barley β-glucan ameliorates fatty liver and insulin resistance in mice fed a high-fat diet. Molecular Nutrition & Food Research 2010;54 1004-1013.

[58] Behall KM., Scholfield DJ., Hallfrisch JG., Liljeberg-Elmstahl HG. Consumption of both resistant starch and β-glucan improves postprandial plasma glucose and insulin in women. Diabetes Care 2006;29 976-81.

[59] Tucker LA., Thomas KS. Increasing total fiber intake reduces risk of weight and fat gains in women. Journal of Nutrition 2009;139 576-581.

[60] Birketvedt GS., Shimshi M., Erling T., Florholmen J. Experiences with three different fiber supplements in weight reduction. Medical Science Monitor 2005; 11 I5-I8.

[61] Barclay AW., Petocz P., McMillan-Price J., Flood VM., Prvan T., Mitchell P. Glycemic index, glycemic load, and chronic disease riskea meta-analysis of observational studies. American Journal of Clinical Nutrition 2008;87 627-637.

[62] Sluijs I., van der Schouw YT., van der A DL., Spijkerman AM., Hu FB., Grobbee DE., Beulens JW. Carbohydrate quantity and quality and risk of type 2 diabetes in the European Prospective Investigation into Cancer and Nutrition-Netherlands (EPIC-NL) study. American Journal of Clinical Nutrition 2010;92 905-911.

[63] Anderson JW., Baird P., Davis RH Jr., Ferreri S., Knudtson M., Koraym A., Waters V., Williams CL. Health benefits of dietary fiber. Nutrition Review 2009;67 188-205.

[64] Salmerón J., Ascherio A., Rimm EB., Colditz GA., Spiegelman D., Jenkins DJ. Dietary fiber, glycemic load, and risk of NIDDM in men. Diabetes Care 1997;20 545-550.

[65] Salmerón J, Manson JE, Stampfer MJ, Colditz GA, Wing AL, Willett WC. Dietary fiber, glycemic load and risk of non-insulin-dependent diabetes mellitus in women. Journal of the American Medical Association 1997;277 472-477.

[66] Sierra M., Garcia JJ., Fernandez N., Diez MJ., Calle AP. Therapeutic effects of psyllium in type 2 diabetic patients. European Journal of Clinical Nutrition 2002;56 830-842.

[67] Chandalia M., Garg A., Lutjohann D., von Bergmann K., Grundy SM., Brinkley LJ. Beneficial effects of high dietary fiber intake in patients with type 2 diabetes mellitus. The New England Journal of Medicine 2000;342 1392-1398.

[68] Salas-Salvadó J., Farrés X., Luque X., Narejos S., Borrell M., Basora J. Fiber in Obesity-Study Group. Effect of two doses of a mixture of soluble fibres on body weight and metabolic variables in overweight or obese patients: a randomised trial. British Journal of Nutrition 2008;99 1380-1387.

[69] Weickert MO., Mohlig M., Koebnick C., Holst JJ., Namsolleck P., Ristow M. Impact of cereal fibre on glucose-regulating factors. Diabetologia 2005;48 2343-2353.

[70] Raninen K., Lappi J., Mykkanen H., Poutanen K. Dietary fiber type reflects physiological functionality: comparison of grain fiber, inulin, and polydextrose. Nutrition Review 2011;69 9-21.

[71] Lairon D., Play B., Jourdheuil-Rahmani D. Digestible and indigestible carbohydrates: interactions with postprandial lipid metabolism. Journal of Nutrition Biochemistry 2007; 18 217-27.

[72] Anderson JW., Chen WJ. Plant fiber. Carbohydrate and lipid metabolism. American Journal of Clinical Nutrition 1979;32 346-363.

[73] Naslund E., Bogefors J., Skogar S., Gryback P., Jacobsson H., Holst JJ. GLP-1 slows solid gastric emptying and inhibits insulin, glucagon, and PYY release in humans. American Journal of Physiology 1999;277 R910-R916.

[74] Reimer RA., Grover GJ., Koetzner L., Gahler RJ., Lyon MR., Wood S. The soluble fiber complex PolyGlycopleX lowers serum triglycerides and reduces hepatic steatosis in high-sucrose fed rats. Nutrition Research 2011;31 296-301.

[75] Massimino SP., McBurney MI., Field CJ., Thomson AB., Keelan M., Hayek MG. Fermentable dietary fiber increases GLP-1 secretion and improves glucose homeostasis despite increased intestinal glucose transport capacity in healthy dogs. Journal of Nutrition 1998;128 1786-93.

[76] Delzenne NM, Cani P. Nutritional modulation of gut microbiota in the context of obesity and insulin resistance: potential interest of prebiotics. International Dairy Journal 2010;20 277-280.

[77] Eckburg PB., Bik EM., Bernstein CN., Purdom E., Dethlefsen L., Sargent M. Microbiology: diversity of the human intestinal microbial flora. Science 2005; 308 1635–1638.

[78] Sghir A., Gramet G., Suau A., Rochet V., Pochart P., Dore J. Quantification of bacterial groups within human fecal flora by oligonucleotide probe hybrid-ization. Applied and Environmental Microbiology 2000; 66(5) 2263–2266.

[79] Muller CA., Autenrieth IB., Peschel A. Innate defenses of the intestinal epithelial barrier. Cellular and Molecular Life Sciences 2005; 62(12) 1297–1307.

[80] Macfarlane, GT., Macfarlane S. Human colonic microbiota: ecology, physiology and metabolic potential of intestinal bacteria. Scandinavian Journal of Gastroenterology 1997; Supplement 32(222) 3-9.

[81] Cook SI., Sellin JH. Review article: short chain fatty acids in health and disease. Alimentary Pharmacology and Therapeutics 1998;12(6) 499–507.

[82] Mazur WM., Duke JA., Wahala K., Rasku S., Adlercreutz H. Isoflavonoids and lignans in legumes: nutritional and health aspects in humans. Journal of Nutritional Biochemistry 1998;9(4) 193-200.

[83] Rowland I., Faughnan M., Hoey L., Wahala K., Williamson G., Cassidy A. Bioavailability of phyto-oestrogens. British Journal of Nutrition 2003, 89(Suppl. 1) 45–58.

[84] Brownlee IA. The physiological roles of dietary fibre. Food Hydrocolloids 2011;25 238–250.

[85] Terada A., Hara H., Mitsuoka T. Effect of dietary alginate on the faecal microbiota and faecal metabolic activity in humans. Microbial Ecology in Health and Disease 1995;8(6) 259–266.

[86] Terada, A., Hara, H., Sato, D., Higashi, T., Nakayama, S., Tsuji, K. Effect of dietary chitosan on faecal microbiota and faecal metabolites of humans. Microbial Ecology in Health and Disease 1995;8(1) 15–21.

[87] Grasten S., Liukkonen KH., Chrevatidis A., El-Nezami H., Poutanen K., Mykkanen H. Effects of wheat pentosan and inulin on the metabolic activity of fecal microbiota and on bowel function in healthy humans. Nutrition Research 2003;23(11) 1503–1514.

[88] Bittner AC., Croffut RM., Stranahan MC., Yokelson TN. Prescriptassist probiotic–prebiotic treatment for irritable bowel syndrome: an open- label, partially controlled, 1-year extension of a previously published controlled clinical trial. Clinical Therapeutics 2007;29 (6) 1153–1160.

[89] Fujimori S., Gudis K., Mitsui K., Seo T., Yonezawa M., Tanaka S. A randomized controlled trial on the efficacy of synbiotic versus probiotic or prebiotic treatment to improve the quality of life in patients with ulcerative colitis. Nutrition 2009;25(5) 520–525.

[90] Molist F., de Segura AG., Gasa J., Hermes RG., Manzanilla EG., Anguita M. Effects of the insoluble and soluble dietary fibre on the physicochemical properties of digesta and the microbial activity in early weaned piglets. Animal Feed Science and Technology 2009;149(3–4) 346-353.

[91] Noriega L., Cuevas I., Margolles A., de los Reyes-Gavilán CG. Deconjugation and bile salts hydrolase activity by Bifidobacterium strains with acquired resistance to bile. International Dairy Journal 2006;16(8) 850-855.

[92] Klaver FA., van der Meer R. The assumed assimilation of cholesterol by Lactobacilli and Bifidobacteriu m bifidum is due to their bile salt-deconju gating activity. Applied and Environmental Microbiology 1993;59 1120-1124.

[93] Tahri K., Grill JP., Schneider F. Bifidobacteri a strain behaviour toward cholestero l: coprecipita tion with bile salts and assimilation. Current Microbiology 1996;33 187-93.

[94] Pereira DI., Gibson GR. Cholesterol assimilat ion by lactic acid bacteria and bifidobacteria isolated from the human gut. Applied and Environmental Microbiology 2002;68 4689-4693.

[95] Demigne C., Morand C., Levrat MA., Besson C., Moundras C., Remesy C. Effect of propionate on fatty acid and cholesterol synthesis and on acetate metabolism in isolated rat hepatocytes. British Journal of Nutrition 1995;74 209-219.

[96] Nagler-Anderson C. Man the barrier! Strategic defences in the intestinal mucosa. Nature Reviews Immunology 2001;1(1) 59-67.

[97] Strugala V., Allen A., Dettmar PW. Pearson JP. Colonic mucin: methods of measuring mucus thickness. Proceedings of the Nutrition Society 2003; 62(1) 237-243.

[98] Taylor C., Allen A., Dettmar PW., Pearson JP. Two rheologically different gastric mucus secretions with different putative functions. Biochimica et Biophysica Acta - General Subjects 2004;1674(2) 131-138.

[99] Watzl B., Girrbach S., Roller M. Inulin, oligofructose and immunomodulation. British Journal of Nutrition 2005;93(Supp) S49-S55.

[100] Field CJ., McBurney MI., Massimino S., Hayek MG., Sunvold GD. The fermentable fiber content of the diet alters the function and composition of canine gut associated lymphoid tissue. Veterinary Immunology and Immunopathology 1999;72(3-4) 325-341.

[101] Wismar R., Brix S., Frokiaer H., Laerke HN. Dietary fibers as immunoregulatory compounds in health and disease. Annals of the New York Academy of Sciences 2010;1190 70-85.

[102] Brownlee IA., Dettmar PW., Strugala V., Pearson JP. The interaction of dietary fibres with the colon. Current Nutrition and Food Science 2006;2(3) 243-264.

[103] Parker SL., Tong T., Bolden S., Wingo PA. Cancer Statistics. CA- A Cancer Journal for Clinicians 1997;47 5-27.

[104] Bingham SA., Day NE., Luben R., Ferrari P., Slimani N., Norat T., Clavel-Chapelon F., Kesse E., Nieters A., Boeing H., Tjønneland A., Overvad K., Martinez C., Dorronsoro M., Gonzalez, C.A., Key TJ., Trichopoulou A., Naska A., Vineis P., Tumino R., Krogh V., Bueno-de-Mesquita HB., Peeters PHM., Berglund G., Hallmans G., Lund E., Skeie G., Kaaks R., Riboli E. Dietary fibre in food and protection against colorectal cancer in the European Prospective Investigation into Cancer and Nutrition (EPIC): an observational study. The Lancet 2003;361 1496-1501.

[105] Dahm CC., Keogh RH., Spencer EA., Greenwood DC., Key TJ., Fentiman IS, Shipley MJ., Brunner EJ., Cade JE., Burley VJ., Mishra G., Stephen AM., Kuh D., White IR., Luben R., Lentjes MAH., Khaw KT, Bingham SA. Dietary Fiber and colorectal cancer risk: A Nested case–control Study Using Food Diaries. Journal of the National Cancer Institute 2010;102(9) 614-626.

[106] Health claims: fruits, vegetables, and grain products that contain fiber, particularly soluble fiber, and risk of coronary heart disease. Code of Federal Regulations 2010;101.77.

[107] Reddy BS., Engle A., Simi B., Goldman M. Effect of dietary fiber on colonic bacterial enzymes and bile acids in relation to colon cancer. Gastroenterology 1992;102 1475-1482.

[108] Reddy BS. Dietary fiber and colon cancer: animal model studies. Pre-Med Journal 1995;6 559-565.

[109] Roediger WEW. Utilization of nutrients by isolated epithelial cells of rat colon. Gastroenterology 1982;83 424-429.

[110] Kim YS., Tsao D., Siddiqui B., Whitehead JS., Arnstein P., Bennett J., Hicks J. Effects of sodium butyrate and dimethylsulfoxide on biochemical properties of human colon cancer cells. Cancer 1980;45 1185-1192.

[111] Hague A., Manning AM., Haneon KA., Huschstscha LL., Hart D., Paraskeva C. Sodium butyrate induces apoptosis in human colonic tumour cell lines in a p53-independent pathway implications for the possible role of dietary fibre in the prevention of large-bowel cancer. International Journal of Cancer 1993;55 498-505.

[112] Harris PJ., Roberton AM., Hollands HJ., Ferguson LR. Adsorption of a hydrophobic mutagen to dietary fibre from the skin and flesh of potato tubers. Mutation Research 1991;260 203-213.

[113] Harris PJ., Triggs CM., Roberton AM., Watson ME., Ferguson LR. The adsorption of heterocyclic aromatic amines by model dietary fibres with contrasting compositions. Chemico-Biological Interactions 1996;100 13-25.

[114] Ferguson LR., Roberton AM., Watso ME., Kestell P., Harris PJ. The adsorption of a range of dietary carcinogens by α-cellulose, a model insoluble dietary fiber. Mutation Research. 1993;319 257-266.

[115] Halliwell B., Zhao K., Whiteman M. The gastrointestinal tract: a major site of antioxidant action? Free Radical Research 2000;33 819-830.

[116] Gee JM., Johnson IT. Polyphenolic compounds: interactions with the gut and implications for human health. Current Medicinal Chemistry 2001;8 1245-1255.

[117] Mälki Y. Trends en dietary fiber research and development: a review. Acta Alimentaria 2004;33(1) 39-62.

[118] Schneeman BO. Gastrointestinal physiology and functions. British Journal of Nutrition 2002;88(2) S159-S163.

[119] Hersey SJ., Sachs G. Gastric acid secretion. Physiological Reviews 1995;75(1) 155-189.

[120] Sternini C., Anselmi L., Rozengurt E. Enteroendocrine cells: a site of 'taste' in gastrointestinal chemosensing. Current Opinion in Endocrinology, Diabetes and Obesity 2008;15(1) 73-78.

[121] Furness JB. Types of neurons in the enteric nervous system. Journal of the Autonomic Nervous System 2000;81(1–3) 87-96.

[122] Pedersen AM., Bardow A., Jensen SB., Nauntofte B. Saliva and gastrointestinal functions of taste, mastication, swallowing and digestion. Oral Diseases 2002;8(3) 117-129.

[123] Embleton JK., Pouton CW. Structure and function of gastro-intestinal lipases. Advanced Drug Delivery Reviews 1997;25(1) 15-32.

[124] Miled N., Canaan S., Dupuis L., Roussel A., Riviere M., Carriere F. Digestive lipases: from three-dimensional structure of physiology. Biochimie 2000;82(11) 973-986.

[125] Chaplin, MF. Fibre and water binding. Proceedings of the Nutrition Society 2003;62(1) 223-227.

[126] Holt S., Heading RC., Carter DC. Effect of gel fibre on gastric emptying and absorption of glucose and paracetamol. Lancet 1979;1(8117) 636-639.

[127] Sanaka M., Yamamoto T., Anjiki H., Nagasawa K., Kuyama Y. Effects of agar and pectin on gastric emptying and post-prandial glycaemic profiles in healthy human volunteers. Clinical and Experimental Pharmacology and Physiology 2007;34(11) 1151-1155.

[128] Benini L., Castellani G., Brightenti F., Heaton KW., Brentegani MT., Casiraghi MC. Gastric emptying of a solid meal is accelerated by the removal of dietary fibre naturally present in food. Gut 1995;36(6) 825-830.

[129] Kimura Y., Watanabe K., Okuda H. Effects of soluble sodium alginate on cholesterol excretion and glucose tolerance in rats. Journal of Ethnopharmacology 1996;54(1) 47-54.

[130] Torsdottir I., Alpsten M., Holm G., Sandberg AS., Tolli J. A small dose of soluble alginate-fiber affects postprandial glycemia and gastric emptying in humans with diabetes. Journal of Nutrition 1991;121(6) 795-799.

[131] Fairchild RM., Ellis PR., Byrne AJ., Luzio SD., Mir MA. A new breakfast cereal containing guar gum reduces postprandial plasma glucose and insulin concentrations in normal-weight human subjects. British Journal of Nutrition 1996;76(1) 63-73.

[132] Sierra M., Garcia JJ., Fernandez N., Diez MJ., Calle AP., Sahagún AM. Effects of ispaghula husk and guar gum on postprandial glucose and insulin concentrations in healthy subjects. European Journal of Clinical Nutrition 2001;55(4) 235-243.

[133] Behall KM., Scholfield DJ., Hallfrisch JG. Barley β-glucan reduces plasma glucose and insulin responses compared with resistant starch in men. Nutrition Research 2006;26(12) 644-650.

[134] Jenkins AL., Jenkins DJA., Zdravkovic U., Wulrsch P., Vuksan V. Depression of the glycemic index by high levels of β-glucan fiber in two functional foods tested in type 2 diabetes. European Journal of Clinical Nutrition 2002;56(7) 622-628.

[135] Tosh SM., Brummer Y., Wolever TMS., Wood PJ. Glycemic response to oat bran muffins treated to vary molecular weight of β-glucan. Cereal Chemistry 2008;85(2) 211-217.

[136] Boisen S., Eggum BO. Critical evaluation of in vitro methods for estimating digestibility in simple-stomach animals. Nutrition Research Reviews 1991;4 141-162.

[137] Coles LT., Moughan PJ., Darragh AJ. In vitro digestion and fermentation methods, including gas production techniques, as applied to nutritive evaluation of foods in the hindgut of humans and other simple stomached animals. Animal Food Science and Technology 2005;123 421-444.

[138] Rodríguez MS., Montero M., Dello Staffolo M., Martino M., Bevilacqua A., Albertengo L. Chitosan influence on glucose and calcium availability from yoghurt: *In vitro* comparative study with plants fibre. Carbohydrate Polymers 2008;74 797–801.

[139] Dello Staffolo M., Martino M., Bevilacqua A., Montero M., Rodríguez MS., Albertengo L. Chitosan Interaction with Iron from Yoghurt Using an In Vitro Digestive Model: Comparative Study with Plant Dietary Fibers. International Journal of Molecular Science 2011;12 4647-4660.

[140] Total, soluble, and insoluble dietary fibre in foods. AOAC Official Methods 991.43. In: Cunniff P. (ed.) Official Methods of Analysis (16th ed.); Arlington: AOAC International; 1995. p7–9.

[141] Robertson JB., Van Soest PJ. The detergent system of analysis and its applications to human foods. In: James WPT., Theander O., (eds.) The Analysis of Dietary Fiber and Food. New York: Marcel Dekker; 1981. p139-153.

[142] Dello Staffolo M., Bertola N., Martino M., Bevilacqua A. Influence of dietary fibre addition on sensory and rheological properties of yoghurt. International Dairy Journal 2004;14 263-268.

[143] Fernández-García E., McGregor JU. Fortification of sweetened plain yoghurt with insoluble dietary fibre. European Food Research and Technology 1997;204, 433-437.

[144] Kumar MNVR., Muzzarelli RAA., Muzzarelli C., Sashiwa H., Domb AJ. Chitosan chemistry and pharmaceutical perspectives. Chemical Reviews 2004;104 6017-6084.

[145] Gorinstein S., Zachwieja Z., Folta M., Barton H., Piotrowicz J., Zember M., Comparative content of dietary fiber, total phenolics, and minerals in persimmons and apples. Journal of Agricultural and Food Chemistry 2001;49 952-957.

[146] Van Craeyveld V., Delcour JA., Courtin CM. Extractability and chemical and enzymic degradation of psyllium (Plantago ovata Forsk) seed husk arabinoxylans. Food Chemistry 2009;112 812-819.

[147] Sudha ML., Baskaran V., Leelavathi K. Apple pomace as a source of dietary fiber and polyphenols and its effect on the rheological characteristics and cake making. Food Chemistry 2007;104 686-692.

[148] Dierenfeld ES., Hintz HF., Robertson JB., van Soest PJ., Oftedal OT. Utilization of bamboo by the giant panda. Journal of Nutrition 1982;112 636-641.

[149] Escarnot E., Agneessens R., Wathelet B., Paquot M. Quantitative and qualitative study of spelt and wheat fibres in varying milling fractions. Food Chemistry 2010;122 857-863.

[150] Sun-Waterhouse D., Farr J., Wibisono R., Saleh Z. Fruit-based functional foods I: Production of food-grade apple fibre ingredients. International Journal Food Science and Technology 2008;43 2113-2122.

[151] Chan JKC., Wypyszyk VA. Forgotten natural dietary fiber: Psyllium mucilloid. Cereal Food. World 1988;33 919-922.

[152] Fischer MH., Yu N., Gray GR., Ralph J., Andersond L., Marletta JA. The gel-forming polysaccharide of psyllium husk (*Plantago ovata* Forsk). Carbohydrate Research 2004; 339 2009-2017.

[153] Anal AK., Tobiassen A., Flanagan J., Singh H. Preparation and characterization of nanoparticles formed by chitosan–caseinate interactions. Colloids and Surfaces B: Biointerfaces 2008;64 104-110.

[154] Reinhold JG., Faradji B., Abadi P., Ismail-Beigi F. Decreased absorption of calcium, magnesium, zinc and phosphorus by humans due to increased fibre and phosphorus consumption of wheat bread. Journal Nutrition 1976;106 493-503.

[155] Vitali D., Dragojević IV., Šebečić B., Vujić L. Impact of modifying tea–biscuit composition on phytate levels and iron content and availability. Food Chemistry 2007;102 82-89.

[156] Deuchi K., Kanauchi O., Shizukuishi M., Kobayashi E. Continuous and massive intake of chitosan affects mineral and fat-soluble vitamin status in rats fed on a high-fat diet. Bioscience Biotechnology and Biochemist 1995;59 1211-1216.

[157] Yang CY., Oh TW., Nakajima D., Maeda A., Naka T., Kim CS. Effects of habitual chitosan intake on bone mass, bone-related metabolic markers and duodenum CaBP D9K mRNA in ovariectomized SHRSP rats. Journal of Nutritional Science and Vitaminology 2002;48 371-378.

[158] Miret S., Simpson RJ., McKie AT. Physiology and molecular biology of dietary iron absorption. The Annual Review of Nutrition 2003;23 283-301.

[159] Bouhallab S., Bouglé D. Biopeptides of milk: Caseinophosphopeptides and mineral bioavailability. Reproduction Nutrition Development 2004;44 493-498.

[160] Van den Heuvel EG., Schaafsma G., Muys T., van Dokkum W. Nondigestible oligosaccharides do not interfere with calcium and nonheme-iron absorption in young, healthy men. American Journal of Clinical Nutrition 1998;67 445–451.

[161] Azorín-Ortuño M., Urbán C., Cerón JJ., Tecles F., Allende A., Tomás-Barberán FA., Espín JC. Effect of low inulin doses with different polymerisation degree on lipid metabolism, mineral absorption, and intestinal microbiota in rats with fat-supplemented diet. Food Chemistry 2009; 113 1058-1065.

[162] Ausar SF., Bianco ID., Badini RG., Castagna LF., Modesti NM., Landa CA, Beltramo DM. Characterization of casein micelle precipitation by chitosans. Journal of Dairy Science 2001;84 361-369.

Chapter 18

Plant Biotechnology for the Development of Design Starches

María Victoria Busi, Mariana Martín and Diego F. Gomez-Casati

Additional information is available at the end of the chapter

http://dx.doi.org/10.5772/50195

1. Introduction

Obesity is a major public health problem due to its pronounced increase and prevalence worldwide. The World Health Organization indicated that in 2005 at least 1.6 billion of adult people were overweight and about 400 million of adults were obese. Predictions for 2015 are even more alarming because indications are that more than 700 million of people will be obese. One of the most common problems associated with obesity is the current lifestyle. Overweight is one of the main risk factors in the development of many chronic diseases, such as respiratory and heart diseases, type 2 diabetes, hypertension and some types of cancer. The increased risk of acquiring some of these diseases is associated with small changes in weight but it can be prevented if appropriate changes in lifestyle are introduced [5].

Furthermore, gastrointestinal infections remain a major health problem despite new advances in medicine. The global incidence of deaths caused by this type of disease is about of 3 million deaths per year. Although this problem is more severe in developing countries, it also occurs in industrialized countries where the incidence of intestinal infection affects about 10% of the population. In most people, the enteropathogenic bacteria cause gastroenteritis that can be treated with drugs and an adequate rehydration. However, in populations such as old people, children, people with chronic intestinal inflammation and immunodeficiencies, it could be a serious problem, leading to the production of septicemia and death. The control of intestinal infections with antibiotics has been one of the medical breakthroughs of the twentieth century. However, the misuse and abuse of these compounds, has led to increased bacterial resistance. Thus, it becomes extremely important to look for new strategies to prevent and/or treat infections. One promising approach is based on the modulation and control of the intestinal microflora through the diet [6].

INTECH
open science | open minds

© 2012 The Author(s). Licensee InTech. This chapter is distributed under the terms of the Creative Commons Attribution License http://creativecommons.org/licenses/by/3.0), which permits unrestricted use, distribution, and reproduction in any medium, provided the original work is properly cited.

Starch is a substantial component of the human diet, mainly in populations that are fed on agricultural crops, providing about 50% of daily energy uptake, mostly through unrefined cereals. In contrast, in westernized societies, the average consumption of grains is much lower, reaching only 25%. Polysaccharides such as glycogen or starch are some of the polymers which can be digested by the enzymes of the human gut. This digestion occurs in the small intestine, except a portion named resistant starch (RS) which is degraded in the large intestine. RS is defined as the set of starch and products of starch degradation (oligosaccharides and others) that are not absorbed in the small intestine but are fermented in the colon producing short-chain fatty acids (such as butyrate), and promotes the normal function of the colonocytes. Because the good function of the human gut is given by the consumption of foods rich in starch, and the change of dietary habits towards healthier eating is not a simple job, the enrichment of some foods with RS becomes the most promising option for a healthy diet [3,7,8].

RS function as dietary fibers, including pre-biotic effect on colon microflora, altering lipid metabolism, improving cholesterol metabolism, and reducing the risk of ulcerative colitis and colon cancer. Since RS is not digested in the small intestine it also reduces the glycemic index of the food [9] (Table 1).

POTENTIAL PHYSIOLOGICAL EFFECTS	POSSIBLE PROTECTIVE EFFECTS
Prebiotic and improved bowel health	Colonic health; colorectal cancer, inflammatory bowel disease, constipation, ulcerative colitis
Improve insulinaemic and glycaemic responses	Diabetes, the metabolic syndrome, impaired insulin and glucose responses
Improvement of blood lipid profile	Lipid metabolism, cardiovascular disease, the metabolic syndrome
Increased satiety and synergistic interactions with other dietary components	Obesity, improved metabolic control and enhanced bowel health
Adjunct to oral rehydration therapies and increased micronutrient absorption	Treatment of chronic diarrhea and cholera; osteoporosis
Thermogenesis	Diabetes and obesity

Table 1. Physiological effects of resistant starch (adapted from [3])

There are at least four mechanisms by which resistant starches are obtained [7,10]: RS1: physically inaccessible starch, usually encapsulated in indigestible tissues (encapsulated or embedded within a matrix of lipid and/or protein) ; RS2: starch granules resistant to degradation, with two subtypes, RS2a with low amylose (0 - 30%), which generally loses its strength when cooked, and RS2b, starches with high amylose content which retains its granular structure during processing, RS3: starch retrograde which requires cooking to be released from the granules, and the starch retrograde capacity is affected by the intrinsic biosynthetic process; finally, RS4: chemically modified starches; although this mechanism is the most used to produce resistant starch, there are no reports of changes in plant that can mimic those obtained by chemical methods. Because each of these processes is independent, it is possible that in some foods resistant starches are derived from more than one mechanism. In these classes, RS1, RS2 and RS3 can be influenced by genetic manipulation of plants. The high amylose starches have the greatest potential to generate resistant starch through two mechanisms, RS2b and RS3 [11]. To achieve this, three strategies have been proposed: reduction of branching enzyme activity; reduction of the amylopectin synthesis rate without altering the synthesis of amylose or/and the increment of the amylose synthesis without altering the synthesis of amylopectin (Table 2).

TYPE OF RESISTANT STARCH	EXAMPLES OF OCCURRENCE	RESISTANCE REDUCED BY
RS1: Physycally inaccessible	Whole or partly milled grains and seed, legumes, pasta	Chewing, milling
RS2: Resistant granules	Raw potatoes, green bananas, high-amylose starches, some legumes. Ungelatinised resistant granules, hydrolysed slowly by α-amylases	Cooking and food processing
RS3: Retrograded	Cooked and cooled potato, food products with repeated and/or prolonged heat treatment, bread, cornflakes	Processing conditions
RS4: Chemically modified	Modified starches due to cross-bonding with esters, ethers, etc. Some cakes, breads and fibre-drinks that were made with modified starches.	Less susceptible to digestibility *in vitro*

Table 2. Nutritional classification of resistant starches (adapted from [3] and [7]).

Given that each of these mechanisms is independent, it is possible that any food could contain RS derived from more than one mechanism. Moreover, RS, RS1, RS2, and RS3 content in foods can be modified by crop genetics [10]. Examples of major components of dietary RS are retrograded amylose (RS1), such as cooked and cooled starchy foods like pasta salad, and native starch granules (RS2), such as those found in high amylose maize starch and bananas [12]. On the other hand, RS3 preserves its nutritional functionality during the cooking process. Thus, it may be used as a food ingredient. RS3 is produced in two steps: gelatinization, which is a disruption of the granular structure by heating with excess of water [13] and retrogradation, a slow recrystallization of the starch molecules upon cooling or dehydration [14]. The resistant fraction may be then isolated using amylolytic enzymes such as pancreatic amylase [15], or Termamyl—heat stable α-amylase [16]. It has been shown that the later approach leads to formation of very thermally stable RS3, and to yields up to 40% [9,14].

Finally, but not least, is the role of investigation and development conducted by researchers from universities and industry. The incorporation of progress in science and the use of currently existing technology contributes to the production of healthy foods, and in this context, designing plants with biology tools to improve their current molecular nutritional qualities is a challenge [5].

The first use of transgenesis in plants in the 1980s brought the arrival of a powerful tool for the study of metabolic regulation and crop improvement. Of particular interest from a health and commercial viewpoint was the potential for increasing yield making alteration of carbon partitioning between sucrose, starch and amino acids [17]. Since that time, plant biotechnology and its commercialization are in exponential phase. Already In 1998, more than 28 million hectares of transgenic crop plants were grown worldwide. Of these 28 million hectares, the largest area was in the USA (22 million hectares) followed by Canada (1.8 million hectares), Argentina (1.8 million hectares) and China (estimated at 1.1 million hectares). It was also estimated that in the US 40% of the cotton, 24% of corn and 40% of soybean planted was transgenic [18].

Given the large amount of information available from molecular biology studies and from genomic programs about the starch biosynthetic genes from crop plants, it is now relatively simple to identify the changes at the DNA level to generate desired starch phenotypes [19-21]. Transgenic approaches to altering the composition of crop plants involve two general approaches: overexpression of an endogenous or foreign gene in the target tissue, and use of RNAi technology to specifically suppress the activity of a specific plant gene [10]. We propose in this chapter to give an overview of starch synthesis to review the potential target technologies and to summerize the successful work done by numerous research groups in different plant species using different strategies.

2. Overview of the starch biosynthesis and degradation in plants

Polyglucans are the most important and widespread carbohydrate storage compounds found in nature, with glycogen and starch being the most abundant forms. Both polysaccharides are comprised of glucose chains linked by an α-(1,4) bond, and branched at α-(1,6). Glycogen is a homogeneous water-soluble polymer with relatively uniformly distributed branches [22] and is found in organisms such as archaea, bacteria and certain eukaryotes. Starch is made up of amylose (a largely unbranched, minor component) and amylopectin (an asymmetrically branched major component) and is present in the cytoplasm of *Rhodophyceae* (red algae) and *Glaucophyta* [23], but is confined to the plastid stroma (chloroplasts in green tissues and amyloplasts in reserve organs) in green algae and higher plants. In fact, starch synthesis is restricted to the *Archaeplastida*, whose origins are thought to be via a single endosymbiotic event involving ancestors of cyanobacteria and a heterotrophic host [24], rendering the organelle known as the plastid, which is capable of oxygenic photosynthesis. Recent phylogenetic studies indicate that the plastidial starch pathway is complex, and made up of genes with both cyanobacterial and eukaryotic origins [25,26], and is in sharp contrast to the lower-complexity pathway of cytosolic starch synthesis found in the *Rhodophyceae* and *Glaucophyta* [27]. Phylogenetic analysis of the enzymes of the starch biosynthetic pathway strongly suggests that the pathway was originally cytosolic (in the common ancestor of the *Archaeplastida*), and then re-directed to plastids via three discrete steps, leaving some enzymes involved in the metabolism of malto-oligosaccharides (MOS) and amylopectin degradation in the cytoplasm. The three evolutionary steps involved are: (1) plastidial synthesis of unbranched MOS; (2) glycogen synthesis (including priming steps and branching activities); and (3) plastidial starch synthesis, resulting in the eventual loss of cytosolic starch synthesis. Interestingly, the relocation of the starch synthesis pathway to plastids coincides with the evolution of light-harvesting complexes [26,28].

There are four biochemical steps in each tissue that are required for the synthesis of starch, substrate activation, chain elongation, chain branching, and chain debranching [10] and it involves at least three enzymes such as ADP-glucose pyrophosphorylase (ADPGlc PPase, EC 2.7.7.27), starch synthase (SS, EC 2.4.1.21), and branching enzyme (BE, EC 2.4.1.18) [29,30] (Figure 1).

The first step of the starch biosynthesic pathway is the synthesis of the activated monomer ADPglucose (ADPGlc) from glucose-1-phosphate and ATP, synthesized by ADPGlc PPase. This reaction is the key step for the control of carbon flux through the starch biosynthetic pathway [29,30].

The second step of the starch biosynthesis pathway is the reaction catalyzed by starch synthase, in which the glucosyl moiety of ADPGlc is transferred to the non-reducing end of a pre-existing α-1,4 glucan polymer [10]. To date, five SS isoforms have been described based on sequence similarities: granule-bound SS (GBSS), involved mainly in amylose synthesis and the soluble isoforms: SSI (involved in the synthesis of small chains of amylopectin), SSII and SSIII (with a major role in amylopectin synthesis) and SSIV (recently found to be involved in the control of starch granule number and starch granule initiation) [31-34].

To produce an efficient clustering of the branch points and the formation of crystalline lamella, several debranching enzymes (DBE) are required [35,36]. In addition, the degradation of the crystalline granules depends on a recently discovered group of enzymes – the glucan, water dikinases (GWDs) – which phosphorylate crystalline sections of the granules. Such phosphorylation is catalyzed by two GWD types: the GWD1, involved in the tagging of the glucan chains by C-6 phosphorylation, which is a prerequisite for subsequent C-3 phosphorylation by the second isoform, the GWD3/PWD (glucan, water dikinase 3/phosphoglucan, water dikinase) [37-39]. These enzymes seem to have evolved concomitantly with the appearance of starch deposition [36,40].

A fourth obligatory step in starch biosynthesis has been identified through genetic studies but is poorly understood in terms of the biochemical mechanism that mediates the effect. This step is the cleavage of α-1,6 linkages by isoamylase-type DBE [10,41]. The DBE are crucial for the generation of longer, clustered linear segments in the amylopectin molecule that can crystallize and increase the density of the polysaccharide [42]. Plants contain four DBE genes, three of which are classified as isoamylases on the basis of their sequence homologies and substrate specificities, and one pullulanase-type debranching enzyme [10,43].

While the steps leading to the synthesis of starch are common in most cereals, there are differences in the location and engagement of enzymes, depending on whether the synthesis is in leaf or endosperm (Figure 1).

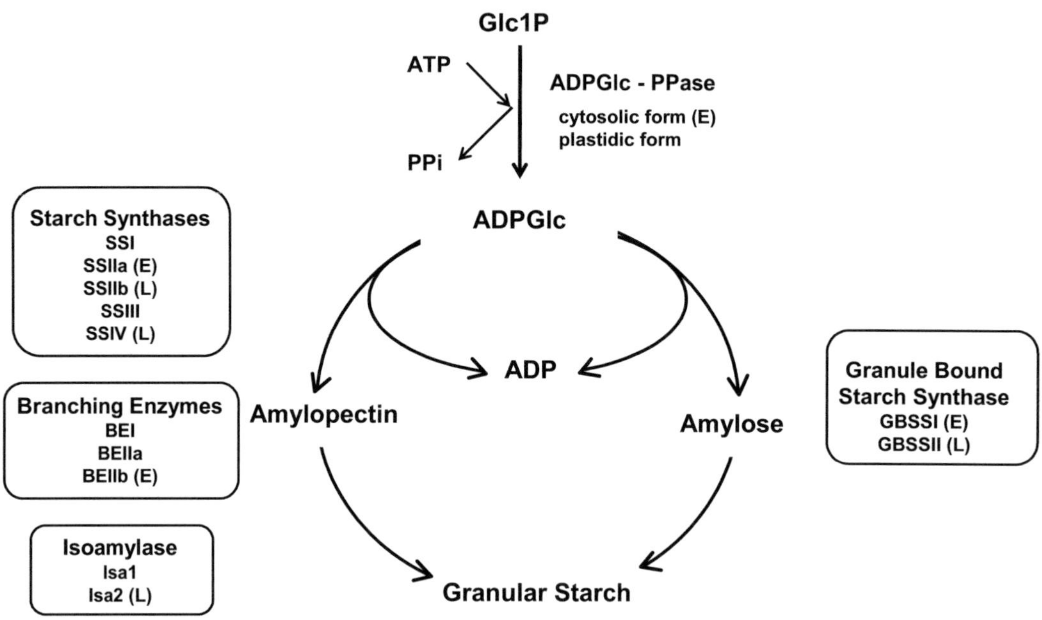

Figure 1. Starch biosynthesis pathway in plants from Glc1P. The scheme indicates the involvement of different isoforms in cereal leaf (L) or endosperm (E). When not specified, the enzymes are dual localized in both compartments (adapted from [10] and [44]).

3. Carbohydrate binding modules

Many of the enzymes involved in the pathway of polysaccharide biosynthesis present a carbohydrate binding domain in its structure. The first carbohydrate binding domain described was a cellulose-binding domain [45-47]; but later it has been found other modules in related enzymes that bind polysaccharides other than cellulose. These findings compelled to redefine the nomenclature of these domains, and now are called CBM (carbohydrate binding module). To date over 300 putative sequences in more than 50 different species have been identified, and binding domains have been classified into 64 families based on amino acid sequence, the substrate binding specificity and structure [48] (see Carbohydrate-Binding Module Family Server, http://afmb.cnrsmrs.fr/~pedro/CAZY/cbm.html).

CBMs have been found in several non-hydrolytic and hydrolytic proteins. Those with hydrolytic activity, such as cellulases, have a complex molecular structure comprised of discrete modules (one catalytic domain and one or more CBMs) that are normally linked by unstructured sequences. The CBMs increase the speed of enzymatic reactions by conducting the catalysis in a close and prolonged physical association with substrates [48] The CBMs present in non-hydrolytic proteins constitute a subunit of the catalytic domain hosts generating cohesive multienzyme complex, which lose enzymatic activity when the CBMs are removed from the structure [48]. Therefore, the CBMs have three general roles with respect to the function of their cognate catalytic modules: (i) a proximity effect, (ii) a targeting function and (iii) a disruptive function [49].

The SBD (starch binding domain) is usually a distinct sequence-structural module that improves the efficiency of an amylolytic enzyme, improving the binding to starch and its hydrolysis. Because this module was first recognised in amylases and thus revealed to cope with raw starch, it was named the raw (granular) starch binding site [50]. At the present, due to the occurrence of SBDs in a wide spectrum of non-amylolytic enzymes, it has become logical to expect a more variable function of these modules. However, there is little evidence that SBD could bind polysaccharides different to starch, although the ability of pure starch binding and degrading seems to be reserved for microorganisms [50,51].

The CBMs have been clasified in ten families based on sequence comparison: (i) CBM20, such as the C-terminal SBD from *Aspergillus niger* glucoamylase; (ii) CBM21, located at the N-terminal domain in amylase proteins; (iii) CBM25, containing one (i.e. β-amylase from *Bacillus circulans*) or two (i.e. *Bacillus sp.* α-amylase) domains; (iv) CBM26, mainly organized in tandem repeats (i.e. C-terminal domains from *Lactobacillus manihotivorans* α-amylase); (v) CBM34, present in the N-terminal domains of neopullulanase, maltogenic amylase and cyclomaltodextrinase; (vi) CBM41, N-terminal SBD, present mostly in bacterial pullulanases; (vii) CBM45, originating from eukaryotic proteins from the plant kingdom (i.e. N-terminal modules of α-glucan water dikinases and α-amylases); (viii) CBM48, which display glycogen-binding properties (including SBD from the GH13 pullulanase and regulatory modules of mammalian AMP-activated protein kinase); (ix) CBM53, SBD modules from SSIII and (x) CBM58, find in α-amylase/neopullulanase of *Bacteroides thetaiotaomicron* showing maltoheptaose binding [52-54] (http://www.cazy.org). This modules becomes important in breaking down the structure of the substrate due to the presence of two polysaccharide-binding sites [55].

Using bioinformatics techniques several SBDs and several sequences exhibiting similarities to SBDs have also been recognised in enzymes and proteins that are not necessarily amylases such as dual-specific phosphatases. These enzymes should deserve special attention because of their participation in various important physiological processes in plants and mammals. It is worth mentioning that in plants these processes concern starch metabolism, whereas in mammals they participate in the metabolism of glycogen [56,57]. The presence of an SBD motif in protein phosphatases reflects their regulatory function since they are involved in polysaccharide metabolism indirectly via modulation of activity of degradative enzymes (i.e. also amylases), such as isoamylase, β-amylase and disproportionating enzyme [85]. In particular, the initial steps of starch degradation at the granule surface are regulated mainly by phosphorylation [50,51]. Furthermore, a starch biosynthetic enzyme, the starch synthase III (SSIII) from *Arabidopsis thaliana* (AtSSIII) has been reported by our group to have a regulatory role in the synthesis of transient starch [33]. This enzyme contains 1025 amino acid residues and has an N-terminal transit peptide for chloroplast localization followed by three in tandem starch-binding domains (SBD D1, D2 and D3, residues 22-591), which bind to raw starch and its individual components, amylose or amylopectin [53,54,58,59]. The adsorption experiments show that the SBD123 region binds preferentially to amylose, and that the D1 domain is mainly responsible for this selective binding. The D2 domain contains two binding sites including amino acid residues Y394 (binding site 1) and W366 (binding site 2) which act in cooperation with the D1 domain in the binding activity while G335 and W340 have a minor role [54]. It is worth mentioning that our work was the first report on the existence of an SBD in a synthesizing enzyme (AtSSIII) and the first experimental evidence of its starch binding capacity.

4. Altering the composition and amount of starch by biotechnological manipulation of enzymes

The alteration of starch quantity and quality can be achieved through the overexpression of some enzymes involved in starch synthesis [60], by mutations or RNAi technology, such as the inhibition of potato SSII, SSIII and GBSS [61], or the decrease in the expression of wheat BEIIa and BEIIb [62,63]. In this way, by affecting the catalytic activity of enzymes involved in the synthesis of amylose or amylopectin, it could be possible to obtain starches for different purposes. Table 3 presents a summary of some of the varieties of plants (transgenic, mutant or silenced by RNAi) that exhibit altered levels of amylose.

The production of high amylose starch is of particular interest because its amount is correlated with the amount of RS in food. Foods with higher content of RS have the potential to improve human health and lower the risk of serious noninfectious diseases. As described above, the amylose content can be increased by the inactivation of the enzymes involved in amylopectin synthesis. In this way, RNAi was used to down-regulate the two different isoforms of starch-branching enzyme (BE) II (BEIIa and BEIIb) in wheat endosperm. Whereas the inhibition of BEIIb expression alone had no effect on amylose content; the decrease of both, BEIIa and BEIIb expression, resulted in the accumulation of starch containing more than 70% of amylose. When this high amylose starch was used to feed rats as a whole meal, it was observed that short-chain fatty acids such as butyrate, propionate and acetate increased with respect to controls. Short chain fatty acids are derived from the anaerobic fermentation of polysaccharides in the large intestine and are important in improving colonic health. These results indicate that this high-amylose wheat has a significant potential to improve human health through its RS content [62].

The decrease of BEIIb enzyme activity in rice is also traditionally associated with elevated amylose content, increased gelatinization temperature, and a decreased proportion of short amylopectin branches. To further elucidate the structural and functional role of this enzyme, the phenotypic effects of down-regulating BEIIb expression in rice endosperm were characterized by Buttardo and coworkers [64] by artificial microRNA (amiRNA) and hairpin RNA (hp-RNA) gene silencing. The results showed that RNA silencing of BEIIb expression in rice grains did not affect the expression of the other major isoforms of BE or SS proteins. The increase in about 2-fold of amylose content was not due to an increase in the relative proportion of amylose chains but instead was due to significantly elevated levels of long and intermediate chains of amylopectin. Rice altered by the amiRNA technique produced a more extreme starch phenotype than those modified using the hp-RNA technique, with a greater increase in the proportion of long and intermediate chains of amylopectin. The major structural modifications of starch produced in the amiRNA lines led to more severe alterations in starch granule morphology and crystallinity as well as digestibility of freshly cooked grains [64].

LINES	AMYLOSE (%)	EVENT	ENZYME INVOLVED
Standard maize	20-30	reference	NA
Sugary-1	37	mutation	Isoamylase
Dull	31	mutation	SSIII
Amylose extender	56	mutation	BEIIb
Indica rice	27	reference	NA
Japonica rice	16	mutation	SSIIa
OsSSIIIa	8	retrotransposon or mutagenesis	SSIIIa
OsSSI	60	mutation	SSI
Standard barley	20-30	reference	NA
Waxy	9	mutation	GBSS
Himalaya 292	71	mutation	SSIIa
BEIIa + BEIIb	75	RNA-silencing	BEIIa/BEIIb
Wheat	18-36	reference	NA
Sgp-1 triple null	31-38	mutation	SSIIa
BEIIa + BEIIb	70	RNAi	BEIIa/BEIIb
Potato	29	reference	NA
BE-II	38	antisense	BEII
BEI + BEII	77-87	antisense	BEI/BEII

Ref: NA, not applicable

Table 3. Amylose content of different lines (wt, mutant and/or transgenic). Adapted from [10].

The roles of BEIIa and BEIIb in defining the structure of amylose and amylopectin were also examined in barley (*Hordeum vulgare*) endosperm. Barley lines with low expression of either BEIIa, BE IIb or both isoforms were generated through RNA-mediated silencing technology. These lines enabled the study of the role of each of these proteins in determining the amylose content, the distribution of chain lengths, and the frequency of branching in both amylose and amylopectin. A high amylose phenotype (> 70%) was observed in lines expressing lower levels of BEIIa and BEIIb, while a reduction in the expression of either of these isoforms alone had minor impact on amylose content. The structure and properties of the barley high amylose starch resulting from the decrease in the expression of both BEII isoforms were found to be similar to those observed in amylose mutants of maize, which result from mutations that decrease the expression of the BEIIb gene. The analysis of amylopectin chain length distribution indicated that both BEIIa and BEIIb isoforms have distinct roles in determining the fine structure of amylopectin. A significant reduction in the frequency of branches in amylopectin was observed only when both BEIIa and BEIIb were reduced, whereas there was a significant increase in the branching frequency of amylose when BEIIb alone was reduced [61,65].

Other way of modifying amylase content is by SS expression. Amylose and amylopectin of rice mutants deficient in endosperm SS isoforms, either SSI (ΔSSI) or SSIIIa (ΔSSIIIa), were found to have an altered structure respect to to their parent (cv. *Nipponbare*, Np). The amylose content was higher in the mutants (Np, 15.5%; ΔSSI, 18.2%; ΔSSIIIa, 23.6%), and the molar ratio of branched amylose and its side chains was increased. In addition, the chain-length distribution of the β-amylase limit dextrins of amylopectin showed high regularity, which is consistent with the reported cluster structure. The mole % of the B(1)-B(3) fractions was changed slightly in ΔSSI, which is consistent with the proposed role of SSI in elongating the external part of clusters. In ΔSSIIIa, it has been observed a significant increase in the B(1) fraction and a decrease in both, the B(2) and B(3) fractions. The internal chain length of the B(2) and B(3) fractions appeared to be slightly altered, suggesting that the deficiency in SS affected the actions of branching enzyme(s) [66].

In another approach, SSIIIa null mutants of rice (*Oryza sativa*) were generated using retrotransposon insertion and chemical mutagenesis. The amylopectin B(2) to B(4) chains with degree of polymerization (DP) >/= 30 and the M(r) of amylopectin were reduced to about 60% and 70% in the mutants, suggesting that SSIIIa plays an important role in the elongation of amylopectin B(2) to B(4) chains. Chains with DP 6 to 9 and DP 16 to 19 decreased while chains with DP 10 to 15 and DP 20 to 25 increased in the amylopectin mutants. These changes in the SSIIIa mutants are almost opposite images of those of SSI-deficient rice mutant and were caused by 1.3- to 1.7-fold increase of the amount of SSI in the mutant endosperm. Furthermore, the amylose content and the extralong chains (DP >/= 500) of amylopectin were increased by 1.3- and 12-fold, respectively. These changes of starch composition of the mentioned mutants are due to the increase in about 1.7-fold of GBSSI activity. The starch granules of the mutants were found to be smaller with round shape and less crystalline. Thus, SSIIIa deficiency, the second major SS isoforrm in developing rice endosperm, affected either the structure of amylopectin, amylase content, and also the physicochemical properties of starch granules in two ways: directly by the SSIIIa deficiency itself and indirectly by up-regulation of both SSI and GBSSI mRNA [67].

By a different approach Safford et al [68] reported no effect on the amylose content of potato starch after the downregulation of the expression of the major branching enzyme isozyme (BE). However, a notable increase (50 – 100%) of the phosphorous content was detected.

Although the almost complete suppression of the branching enzyme activity (less than 5% respect to wt levels) in transgenic potato tubers, no changes in amylose content of the starches derived from these transgenic lines were detected. Differences in the gelatinization properties (an increase of up to 5°C in the peak temperature and viscosity onset temperature) are reported, suggesting that these changes correlated with the branching pattern of the starch that result in changes of the double helix length. It is also possible that the increased phosphate content observed in the transgenic starches resulted in the elevation of the gelatinization temperature [68].

Other strategy to obtain high amylose starches was carried out by Itoh et al [60]. The Waxy (Wx) gene encodes a granule-bound starch synthase (GBSS) that plays a key role in the amylose synthesis of rice and other plant species. In rice, it has been described two functional Wx alleles: Wx(a), which produces higher amounts of amylose, and Wx(b), which produces low amounts of this polymer due to a mutation in the 5' splicing site of intron 1. When the Wx(a) cDNA was introduced into null-mutant Japonica rice (wx) the amylose content were 6-11% higher than that of the original cultivar, Labelle, which carries the Wx(a) allele, although the levels of the Wx protein in the transgenic rice were equal to those of cv. Labelle [60].

Finally, using *A. thaliana* null mutant lines for the SSIII locus, it has been postulated that SSIII has a regulatory role in the starch synthesis process [33]. These mutant lines show a higher accumulation of leaf starch during the day due to an apparent increase in biosynthetic rate. Besides, starch granules show physical alterations and higher phosphate content [33]. These data suggest that SSIII might have a negative regulatory role in starch synthesis. Previously, SSIII had been associated to a starch-excess phenotype, although indirectly through its association with regulatory proteins such as 14-3-3 [69]. In addition other SS isoform, SSIV, has been described to be essential for the initiation process of starch granule synthesis since *A. thaliana* SSIV mutant plants show just one large starch granule per plastid. The role of this isoform in the formation of the starch granule could be replaced in part by the SSIII isoform since the concomitant elimination of both enzymes in Arabidopsis block the starch synthesis. These data suggests that the remaining synthase activities are unable to start the synthesis of the starch granule. Recently, SSIV has been postulated to be also involved in the regulation of starch accumulation since its overexpression increases the starch levels in Arabidopsis leaves by 30%–40%. In addition, SSIV-overexpressing lines display a higher growth rate. The increase in starch content as a consequence of enhanced SSIV expression is also observed in long-term storage starch organs such as potato tubers [70].

5. Use of carbohydrate-binding modules to change amylose - amylopectin ratio and obtaining of modified starches.

In the past few years the search for different strategies in order to produce starches with new properties was intensified. One of these strategies is to evaluate the possibility whether the microbial starch binding domains (SBDs) could be used as a universal tool for starch modification in plant biotechnology.

It has been reported that SBDs are also present in microbial starch degrading enzymes. As mentioned above, one of the functions of SBD is to attach amylolytic enzymes to the insoluble starch granule. The amino acid sequences of these modules are very well conserved among different enzymes (i.e. glucoamylase, α-amylase, β-amylase, etc.), as well as among different species such as *Clostridium thermosulfurogenes, Bacillus circulans, Aspergillus niger, Klebsiella pneumonia, Streptomyces limosus, Pseudomonas stutzeri,* etc. [50,71-73]. Several studies have shown that these enzymes lose (most of) their catalytic activity towards raw starch granules upon removal of the SBD, whereas their activity on soluble substrates remains unaltered. Besides their affinity for starch granules, SBDs can also bind maltodextrins and cyclodextrins [71]. Ji et al [72], explored the possibility of engineering artificial granule-bound proteins, which can be incorporated in the granule during biosynthesis. The SBD-encoding region of cyclodextrin glycosyltransferase from *B. circulans* was fused to the sequence encoding the transit peptide (amyloplast entry) of potato GBSSI. The synthetic gene was expressed in the tubers of two potato cultivars and one amylose-free (amf) potato mutant. The results showed that SBDs are accumulated inside starch granules, not at the granule surface and amylose-free granules contained 8 times more SBD than the amylose-containing ones. However, no consistent differences in physicochemical properties between transgenic SBD starches and their corresponding controls were found, suggesting that SBD can be used as an anchor for effector proteins without having side-effects [72].

On the other hand it was also evaluated whether is it possible to produce an amylose-free potato starch by displacing GBSSI, from the starch granule by engineering multiple-repeat CBM20 SBD (two, three, four and five). The constructs were introduced in wild type potato cultivar, and the starches of the resulting transformants were compared with those expressing amf potato clones. The amount of SBDs accumulated in starch granules was increased progressively from SBD to SBD3 and not when were used SBD4 and SBD5; however, a reduction in amylose content was not achieved in any of the transformants. It was shown that SBDn expression can affect the physical process underlying granule assembly in both potato genetic backgrounds, without altering the primary structure of the constituent starch polymers and the granule melting temperature. Granule size distribution of the starches obtained from transgenic plants was similar to untransformed controls, irrespective of the amount of SBDn accumulated. In the amf background, granule size is severely affected [74].

In the case of starches which require chemical modifications to enhance their properties, such as the improved stability in solution by acetylation, a drawback is generated when pollutant chemicals are used. A biological alternative to the derivatization process was investigated by the expression of an amyloplast-targeted *Escherichia coli* maltose acetyltransferase (MAT) in tubers of wild-type and mutant amf potato plants. MAT was expressed alone, or fused in its N- or C-terminus to a SBD to be target to the starch granule. Starch granules derived from transgenic plants contained acetyl groups in low number. In addition, MAT protein on the starch granules present catalytic activity even after post-harvesting, when supplied with glucose or maltose and acetyl-coenzyme A, but it was not able to acetylate starch polymers in vitro. Starch granules from transformants where MAT was expressed alone also showed MAT catalytic activity, indicating that MAT is accumulated in starch granules, and could bind to the polymer without the presence of any SBD. Furthermore, the fusion of MAT and SBD affects granule morphology: in potato transformants, the percentage of altered granules when the SBD was located at the C-terminal end correlated with the amount of fusion protein accumulated. When SBD was located at the N-terminus of MAT or it is absent, no differences were found respect to the untransformed controls, indicating that not only is the simultaneous presence of SBD and MAT important for altering granule morphology, but also their localization in the fusion protein [75].

Another approach to obtain modified starches involves the bacterial glucansucrases [76]. Certain bacteria possess an array of enzymes, so-called glucansucrases, which can attach (contiguous) 1,6-linked or 1,3-linked glucosyl residues to maltodextrins. This, together with the presence of sucrose inside the potato tuber amyloplast [77], suggests that glucansucrases are of great interest for diversifying starch structure. With few exceptions, glucansucrases are extracellular enzymes, which are produced by lactic acid bacteria such as *Leuconostoc mesenteroides*, oral *Streptococci* and some species of *Lactococcus* and *Lactobacillus* [78]. The glucansucrases catalyze the polymerization of glucose residues from sucrose, which leads to the production of a large variety of α-glucans with different sizes and structures, and composed of diverse linkage types. Most glucansucrases share a common structure composed of four different regions: a signal peptide, a variable region, a catalytic domain, and a glucan-binding domain (GBD) [76].

Production of water-insoluble mutan polymers in wild type potato tubers was investigated by Kok-Jacon et al (2005) after expression of full-length GTFI (mutansucrase) and a truncated version without glucan-binding domain from *Streptococcus downei.* Mutan polymers are bacterial polysaccharides that are secreted by oral microorganisms and have adhesive properties and different degrees of water-solubility [81]. They account for about 70% of the carbohydrates present in dental plaque [79] in addition to dextrans and levans [80]. When the short form of the protein was expressed, low amounts of mutan polymer attached to the starch granules has been detected. Besides, these plants exhibited severely altered tuber phenotype and starch granule morphology in comparison to those expressing the full-length GTFI gene, whereas no changes at the starch level were observed. Finally, the rheological properties of the starch obtained from plants expressing the truncated protein were also altered, showing a higher retrogradation during cooling of the starch paste [80].

Subsequently, the same group of investigators fused the truncated form of a mutansucrase (without glucan binding domain) to an N- or C- terminal SBD. The different enzymes were introduced into two genetically different potato backgrounds (wild type and amf lines), in order to attach the enzyme to the growing starch granules, and to facilitate the incorporation of mutan polymers in starch. Starches from the chimeric transformants seemed to contain less amounts of mutan than those from plants expressing the mutansucrase alone, suggesting that SBD might inhibit the catalytic activity of the enzyme. Scanning electron microscopy showed that expression of SBD-mutansucrase fusion proteins resulted in alterations of granule morphology in both genetic backgrounds. Surprisingly, the amf starches containing the chimeric form had a spongeous appearance, as the granule surface contained many small holes and grooves, indicating that this fusion protein can interfere with the lateral interactions of amylopectin sidechains. No differences in physicochemical properties of the transgenic starches were observed [82].

Finally, all the knowledge gained about the characteristics, structure, function and occurrence of SBD and GBD will support current and future experimental research. Since SBD are domains which retain their structural fold and functional properties independently of the remaining parts of the protein molecule including the catalytic domain, they can be applied in various fields of biotechnology [48,83-86]. It is important to note that most of the applications have involved only the CBM20 SBD. One of the most attractive fields is represented by starch processing in the food industry, especially the hydrolysis of starch into maltodextrins and maltooligosaccharides [87]. Since conventional processes require starch gelatinization at elevated temperature and thus use of thermostable amylolytic enzymes [88], the possibility of carrying out the processes without gelatinization, by utilizing new enzymes with attached SBD is desirable [52,89,90].

6. Conclusions

Food production in terms of quality and quantity, as well as for new plants commodities and products in developed and developing countries, cannot based only on classical agriculture [91]. The metabolic engineering of plants has yielded remarkable results by increasing the production of minor components (essential oils, vitamin A, vitamin E and flavonoids) and, as well as the composition of major components, such as starch or fatty acids [92]. The improvement in the food we eat is necessary and crucial in societies that have bad eating habits. The health benefits provided by the intake of resistant starches have been properly tested and it will be desirable that these kinds of starches could be incorporated into the human diet. Molecular tools available at the present and those likely to be developed in the near future, will enable the development of new strategies to increase the content of resistant starch in grains and other vegetables. Manipulation of the starch synthesis pathway through the modification of enzymes belonging to this route, and the use of CBM (and specifically SBD) of both microbial and plant, are alternatives that are desirable to explore in more detail.

Author details

María Victoria Busi and Diego F. Gomez-Casati
Centro de Estudios Fotosintéticos y Bioquímicos (CEFOBI-CONICET),
Suipacha 570, Rosario, Argentina
IIB-INTECH, Universidad Nacional de General San Martín (UNSAM),
San Martín, Buenos Aires, Argentina

Mariana Martín
Centro de Estudios Fotosintéticos y Bioquímicos (CEFOBI-CONICET),
Universidad Nacional de Rosario, Suipacha 570, Rosario, Argentina

Acknowledgement

Our work is supported in part by grants from the Biotechnology Program from Universidad Nacional de General San Martin (UNSAM) (PROG07F / 2-2007), Consejo Nacional de Investigaciones Científicas y Técnicas (CONICET, PIP 00237) and Agencia Nacional de Promoción Científica y Tecnológica (ANPCyT, PICT 2010 – 0543 and PICT 2010 – 0069). MVB, MM and DGC are research members from CONICET.

7. References

[1] Halpin, C. (2005). Gene stacking in transgenic plants--the challenge for 21st century plant biotechnology. Plant Biotechnol J 3, 141-55.

[2] Cassidy, A. (2004) Serials, Nuts and Pulses. In Plants: Diet and health (Goldberg, G., ed.), pp. 134-146. Blackwell Sciences, Oxford, UK.

[3] Nugent, A.P. (2005). Health properties of resistant starch. Nutrition Bull 30

[4] Zhao, F.J. and Shewry, P.R. (2010). Recent developments in modifying crops and agronomic practice to improve human health Food Policy

[5] Lopez, X. (2009). Obesity: How did resistant starches might contribute? Enfasis Alim 2, 32-38.

[6] Dominguez Vergara, A.M., Vazquez-Moreno, L. and Ramos-Clamont Mon Fort, G. (2009). Review of the role of prebiotic oligosaccharides in the prevention of gastrointestinal infections. Arch Latin Nutr 59, 358-368.

[7] Topping, D.L., Fukushima, M. and Bird, A.R. (2003). Resistant starch as a prebiotic and synbiotic: state of the art. Proc Nutr Soc 62, 171-6.

[8] Robertson, M.D. (2012). Dietary-resistant starch and glucose metabolism. Curr Opin Clin Nutr Metab Care

[9] Shamaia, K., Bianco-Peled, H. and Shimonic, E. (2003). Polymorphism of resistant starch type III Carbohydr Polym 54, 363-369.

[10] Morell, M.K., Konik-Rosse, C., Ahmed, R., Li, Z. and Rahman, S. (2004). Synthesis of resistant starch in plants. J AOAC Int 87, 740-748.

[11] Bird, A.R., Flory, C., Davies, D.A., Usher, S. and Topping, D.L. (2004). A novel barley cultivar (Himalaya 292) with a specific gene mutation in starch synthase IIa raises large bowel starch and short-chain fatty acids in rats. J Nutr 134, 831-5.

[12] Higgins, J.A., Higbee, D.R., Donahoo, W.T., Brown, I.L., Bell, M.L. and Bessesen, D.H. (2004). Resistant starch consumption promotes lipid oxidation. Nutr Metab (Lond) 1, 8.

[13] Farhat, I.A., Protzmann, J., Becker, A., Valles-Pamies, B., Neale, R. and Hill, S.E. (2001). Effect of the extent of conversion and retrogradation on the digestibility of potato starch. Starch/Stärke 53, 431-436.

[14] Sivak, M. and Preiss, J. (1998). Advances in Food Nutrition Research. Academic Press, San Diego, California. 41

[15] Alonso, A.G., Calixto, F.S. and Delcour, J.A. (1998). Influence of botanical source and proccesing on formation of resistant starch type III. Cereal Chem 75, 802-804.

[16] Thompson, D.B. (2000). Strategies for the manufacture of resistant starch. Trends Food Sci Technol 11, 245-253.
[17] Jenner, H.L. (2003). Transgenesis and yield: what are our targets? Trends Biotechnol 21, 190-2.
[18] Willmitzer, L. (1999). Plant Biotechnology: Output traits-the second generation of plant biotechnology products is gaining momentum. Curr Opin Biotechnol 10, 161-162.
[19] Morell, M.K. et al. (2003). Barley sex6 mutants lack starch synthase IIa activity and contain a starch with novel properties. Plant J 34, 173-85.
[20] Blauth, S.L., Kim, K.N., Klucinec, J., Shannon, J.C., Thompson, D. and Guiltinan, M. (2002). Identification of Mutator insertional mutants of starch-branching enzyme 1 (sbe1) in Zea mays L. Plant Mol Biol 48, 287-97.
[21] Gao, M., Fisher, D.K., Kim, K.N., Shannon, J.C. and Guiltinan, M.J. (1997). Independent genetic control of maize starch-branching enzymes IIa and IIb. Isolation and characterization of a Sbe2a cDNA. Plant Physiol 114, 69-78.
[22] Roach, P.J. (2002). Glycogen and its metabolism. Curr Mol Med 2, 101-20.
[23] Dauvillee, D. et al. (2009). Genetic dissection of floridean starch synthesis in the cytosol of the model dinoflagellate Crypthecodinium cohnii. Proc Natl Acad Sci U S A 106, 21126-30.
[24] Cavalier-Smith, T. (2009). Predation and eukaryote cell origins: a coevolutionary perspective. Int J Biochem Cell Biol 41, 307-22.
[25] Patron, M.J. and Keeling, P. (2005). Common evolutionary origin of starch biosynthetic enzymes in green and red algae. J Phycol 41, 1131-1141.
[26] Deschamps, P., Moreau, H., Worden, A.Z., Dauvillee, D. and Ball, S.G. (2008). Early gene duplication within chloroplastida and its correspondence with relocation of starch metabolism to chloroplasts. Genetics 178, 2373-87.
[27] Deschamps, P., Haferkamp, I., d'Hulst, C., Neuhaus, H.E. and Ball, S.G. (2008). The relocation of starch metabolism to chloroplasts: when, why and how. Trends Plant Sci 13, 574-82.
[28] Tetlow, I.J. (2011). Starch biosynthesis in developing seeds. Seed Sci Res 21, 5-32.
[29] Smith, A.M., Denyer, K. and Martin, C. (1997). The Synthesis of the Starch Granule. Annu Rev Plant Physiol Plant Mol Biol 48, 67-87.
[30] Ball, S.G. and Morell, M.K. (2003). From bacterial glycogen to starch: understanding the biogenesis of the plant starch granule. Annu Rev Plant Biol 54, 207-33.
[31] Delvalle, D. et al. (2005). Soluble starch synthase I: a major determinant for the synthesis of amylopectin in Arabidopsis thaliana leaves. Plant J 43, 398-412.
[32] Maddelein, M.L. et al. (1994). Toward an understanding of the biogenesis of the starch granule. Determination of granule-bound and soluble starch synthase functions in amylopectin synthesis. J Biol Chem 269, 25150-7.
[33] Zhang, X., Myers, A.M. and James, M.G. (2005). Mutations affecting starch synthase III in Arabidopsis alter leaf starch structure and increase the rate of starch synthesis. Plant Physiol 138, 663-74.

[34] Delrue, B. et al. (1992). Waxy Chlamydomonas reinhardtii: monocellular algal mutants defective in amylose biosynthesis and granule-bound starch synthase activity accumulate a structurally modified amylopectin. J Bacteriol 174, 3612-20.
[35] Wattebled, F., Planchot, V., Dong, Y., Szydlowski, N., Pontoire, B., Devin, A., Ball, S. and D'Hulst, C. (2008). Further evidence for the mandatory nature of polysaccharide debranching for the aggregation of semicrystalline starch and for overlapping functions of debranching enzymes in Arabidopsis leaves. Plant Physiol 148, 1309-23.
[36] Coppin, A. et al. (2005). Evolution of plant-like crystalline storage polysaccharide in the protozoan parasite Toxoplasma gondii argues for a red alga ancestry. J Mol Evol 60, 257-67.
[37] Ritte, G., Heydenreich, M., Mahlow, S., Haebel, S., Kotting, O. and Steup, M. (2006). Phosphorylation of C6- and C3-positions of glucosyl residues in starch is catalysed by distinct dikinases. FEBS Lett 580, 4872-6.
[38] Baunsgaard, L., Lutken, H., Mikkelsen, R., Glaring, M.A., Pham, T.T. and Blennow, A. (2005). A novel isoform of glucan, water dikinase phosphorylates pre-phosphorylated alpha-glucans and is involved in starch degradation in Arabidopsis. Plant J 41, 595-605.
[39] Kotting, O., Pusch, K., Tiessen, A., Geigenberger, P., Steup, M. and Ritte, G. (2005). Identification of a novel enzyme required for starch metabolism in Arabidopsis leaves. The phosphoglucan, water dikinase. Plant Physiol 137, 242-52.
[40] Kotting, O., Kossmann, J., Zeeman, S.C. and Lloyd, J.R. (2010). Regulation of starch metabolism: the age of enlightenment? Curr Opin Plant Biol 13, 321-9.
[41] Myers, A.M., Morell, M.K., James, M.G. and Ball, S.G. (2000). Recent progress toward understanding biosynthesis of the amylopectin crystal. Plant Physiol 122, 989-97.
[42] Blennow, A. and Svensson, B. (2010). Dynamics of starch granule biogenesis -the role of redox regulated enzymes- and low-affinity carbohydrate binding modules. Biocatal Biotransf 28, 3-9.
[43] Rahman, S. et al. (2003). The sugary-type isoamylase gene from rice and Aegilops tauschii: characterization and comparison with maize and arabidopsis. Genome 46, 496-506.
[44] Morell, M.K. and Myers, A.M. (2005). Towards the rational design of cereal starches. Curr Opin Plant Biol 8, 204-10.
[45] Gilkes, N.R., Warren, R.A., Miller, R.C., Jr. and Kilburn, D.G. (1988). Precise excision of the cellulose binding domains from two Cellulomonas fimi cellulases by a homologous protease and the effect on catalysis. J Biol Chem 263, 10401-7.
[46] Tomme, P., Driver, D.P., Amandoron, E.A., Miller, R.C., Jr., Antony, R., Warren, J. and Kilburn, D.G. (1995). Comparison of a fungal (family I) and bacterial (family II) cellulose-binding domain. J Bacteriol 177, 4356-63.
[47] Tomme, P., Warren, R.A. and Gilkes, N.R. (1995). Cellulose hydrolysis by bacteria and fungi. Adv Microb Physiol 37, 1-81.
[48] Shoseyov, O., Shani, Z. and Levy, I. (2006). Carbohydrate binding modules: biochemical properties and novel applications. Microbiol Mol Biol Rev 70, 283-95.
[49] Boraston, A.B., Bolam, D.N., Gilbert, H.J. and Davies, G.J. (2004). Carbohydrate-binding modules: fine-tuning polysaccharide recognition. Biochem J 382, 769-81.

[50] Machovic, M. and Janecek, S. (2006). Starch-binding domains in the post-genome era. Cell Mol Life Sci

[51] Machovic, M. and Janecek, S. (2006). The evolution of putative starch-binding domains. FEBS Lett 580, 6349-56.

[52] Janecek, S., Svensson, B. and MacGregor, E.A. (2011). Structural and evolutionary aspects of two families of non-catalytic domains present in starch and glycogen binding proteins from microbes, plants and animals. Enzyme Microb Technol 49, 429-440.

[53] Wayllace, N.Z., Valdez, H.A., Ugalde, R.A., Busi, M.V. and Gomez-Casati, D.F. (2010). The starch-binding capacity of the noncatalytic SBD2 region and the interaction between the N- and C-terminal domains are involved in the modulation of the activity of starch synthase III from Arabidopsis thaliana. Febs J 277, 428-40.

[54] Valdez, H.A., Peralta, D.A., Wayllace, N.Z., Grisolía, M.J., Gomez-Casati, D.F. and Busi, M.V. (2011). Preferential binding of SBD from Arabidopsis thaliana SSIII to polysaccharides. Study of amino acid residues involved. Starch/Stärke 63, 451-460.

[55] Southall, S.M., Simpson, P.J., Gilbert, H.J., Williamson, G. and Williamson, M.P. (1999). The starch-binding domain from glucoamylase disrupts the structure of starch. FEBS Lett 447, 58-60.

[56] Kerk, D., Conley, T.R., Rodriguez, F.A., Tran, H.T., Nimick, M., Muench, D.G. and Moorhead, G.B. (2006). A chloroplast-localized dual-specificity protein phosphatase in Arabidopsis contains a phylogenetically dispersed and ancient carbohydrate-binding domain, which binds the polysaccharide starch. Plant J 46, 400-13.

[57] Roma-Mateo, C. et al. (2011). Laforin, a dual specificity protein phosphatase involved in Lafora disease, is phosphorylated at Ser25 by AMP-activated protein kinase. Biochem J

[58] Palopoli, N., Busi, M.V., Fornasari, M.S., Gomez-Casati, D., Ugalde, R. and Parisi, G. (2006). Starch-synthase III family encodes a tandem of three starch-binding domains. Proteins 65, 27-31.

[59] Valdez, H.A., Busi, M.V., Wayllace, N.Z., Parisi, G., Ugalde, R.A. and Gomez-Casati, D.F. (2008). Role of the N-terminal starch-binding domains in the kinetic properties of starch synthase III from Arabidopsis thaliana. Biochemistry 47, 3026-32.

[60] Itoh, K., Ozaki, H., Okada, K., Hori, H., Takeda, Y. and Mitsui, T. (2003). Introduction of Wx transgene into rice wx mutants leads to both high- and low-amylose rice. Plant Cell Physiol 44, 473-80.

[61] Jobling, S.A., Westcott, R.J., Tayal, A., Jeffcoat, R. and Schwall, G.P. (2002). Production of a freeze-thaw-stable potato starch by antisense inhibition of three starch synthase genes. Nat Biotechnol 20, 295-9.

[62] Regina, A. et al. (2006). High-amylose wheat generated by RNA interference improves indices of large-bowel health in rats. Proc Natl Acad Sci U S A 103, 3546-51.

[63] Sestili, F., Janni, M., Doherty, A., Botticella, E., D'Ovidio, R., Masci, S., Jones, H.D. and Lafiandra, D. (2010). Increasing the amylose content of durum wheat through silencing of the SBEIIa genes. BMC Plant Biol 10, 144.

[64] Butardo, V.M. et al. (2011). Impact of down-regulation of starch branching enzyme IIb in rice by artificial microRNA- and hairpin RNA-mediated RNA silencing. J Exp Bot 62, 4927-41.

[65] Regina, A., Kosar-Hashemi, B., Ling, S., Li, Z., Rahman, S. and Morell, M. (2010). Control of starch branching in barley defined through differential RNAi suppression of starch branching enzyme IIa and IIb. J Exp Bot 61, 1469-82.
[66] Hanashiro, I., Higuchi, T., Aihara, S., Nakamura, Y. and Fujita, N. (2011). Structures of starches from rice mutants deficient in the starch synthase isozyme SSI or SSIIIa. Biomacromolecules 12, 1621-8.
[67] Fujita, N. et al. (2007). Characterization of SSIIIa-deficient mutants of rice: the function of SSIIIa and pleiotropic effects by SSIIIa deficiency in the rice endosperm. Plant Physiol 144, 2009-23.
[68] Safford, R.E. et al. (1998). Consequences of antisense RNAi of starch branching enzyme activity on properties of potato starch. Carbohydr Polym 35, 155-160.
[69] Sehnke, P.C., Chung, H.J., Wu, K. and Ferl, R.J. (2001). Regulation of starch accumulation by granule-associated plant 14-3-3 proteins. Proc Natl Acad Sci U S A 98, 765-70.
[70] Szydlowski, N. et al. (2009). Starch granule initiation in Arabidopsis requires the presence of either class IV or class III starch synthases. Plant Cell 21, 2443-57.
[71] Svensson, B., Jespersen, H., Sierks, M.R. and MacGregor, E.A. (1989). Sequence homology between putative raw-starch binding domains from different starch-degrading enzymes. Biochem J 264, 309-11.
[72] Ji, Q., Vincken, J.P., Suurs, L.C. and Visser, R.G. (2003). Microbial starch-binding domains as a tool for targeting proteins to granules during starch biosynthesis. Plant Mol Biol 51, 789-801.
[73] Janecek, S. and Sevcik, J. (1999). The evolution of starch-binding domain. FEBS Lett 456, 119-25.
[74] Firouzabadi, F.N., Vincken, J.P., Ji, Q., Suurs, L.C., Buleon, A. and Visser, R.G. (2007). Accumulation of multiple-repeat starch-binding domains (SBD2-SBD5) does not reduce amylose content of potato starch granules. Planta 225, 919-33.
[75] Nazarian Firouzabadi, F., Vincken, J.P., Ji, Q., Suurs, L.C. and Visser, R.G. (2007). Expression of an engineered granule-bound Escherichia coli maltose acetyltransferase in wild-type and amf potato plants. Plant Biotechnol J 5, 134-45.
[76] Kok-Jacon, G.A., Ji, Q., Vincken, J.P. and Visser, R.G. (2003). Towards a more versatile alpha-glucan biosynthesis in plants. J Plant Physiol 160, 765-77.
[77] Gerrits, N., Turk, S.C., van Dun, K.P., Hulleman, S.H., Visser, R.G., Weisbeek, P.J. and Smeekens, S.C. (2001). Sucrose metabolism in plastids. Plant Physiol 125, 926-34.
[78] Robyt, J.F. (1995). Mechanisms in the glucansucrase synthesis of polysaccharides and oligosaccharides from sucrose. Adv Carbohydr Chem Biochem 51, 133-68.
[79] Loesche, W.J. (1986). Role of Streptococcus mutans in human dental decay. Microbiol Rev 50, 353-80.
[80] Kok-Jacon, G.A., Vincken, J.P., Suurs, L.C. and Visser, R.G. (2005). Mutan produced in potato amyloplasts adheres to starch granules. Plant Biotechnol J 3, 341-51.
[81] Sutherland, I. (2001). Biofilm exopolysaccharides: a strong and sticky framework. Microbiology 147, 3-9.

[82] Nazarian Firouzabadi, F., Kok-Jacon, G.A., Vincken, J.P., Ji, Q., Suurs, L.C. and Visser, R.G. (2007). Fusion proteins comprising the catalytic domain of mutansucrase and a starch-binding domain can alter the morphology of amylose-free potato starch granules during biosynthesis. Transgenic Res 16, 645-56.

[83] Juge, N. et al. (2006). The activity of barley alpha-amylase on starch granules is enhanced by fusion of a starch binding domain from Aspergillus niger glucoamylase. Biochim Biophys Acta 1764, 275-84.

[84] Ohdan, K., Kuriki, T., Takata, H., Kaneko, H. and Okada, S. (2000). Introduction of raw starch-binding domains into Bacillus subtilis alpha-amylase by fusion with the starch-binding domain of Bacillus cyclomaltodextrin glucanotransferase. Appl Environ Microbiol 66, 3058-64.

[85] Latorre-Garcia, L., Adam, A.C., Manzanares, P. and Polaina, J. (2005). Improving the amylolytic activity of Saccharomyces cerevisiae glucoamylase by the addition of a starch binding domain. J Biotechnol 118, 167-76.

[86] Christiansen, C., Abou Hachem, M., Janecek, S., Vikso-Nielsen, A., Blennow, A. and Svensson, B. (2009). The carbohydrate-binding module family 20--diversity, structure, and function. Febs J 276, 5006-29.

[87] Van Geel-Schutten, G.H. et al. (1999). Biochemical and structural characterization of the glucan and fructan exopolysaccharides synthesized by the lactobacillus reuteri wild-type strain and by mutant strains. Appl Environ Microbiol 65, 3008-14.

[88] Bertoldo, C. and Antranikian, G. (2002). Starch-hydrolyzing enzymes from thermophilic archaea and bacteria. Curr Opin Chem Biol 6, 151-60.

[89] Vikso-Nielsen, A., Andersen, C., Hoff, T. and Pedersen, S. (2006). Development of new alpha-amylases for raw starch hydrolysis. Biocatal Biotransf 24, 121-127.

[90] Wang, P., Singh, V., Xue, H., Johnston, D.B., Rausch, K.D. and Tumbleson, M.E. (2007). Comparisson of raw starch hydrolysis enzyme with conventional liquetfaction and saccharification enzymes in dry-grind corn proccesing. Cereal Chem 84, 10-14.

[91] Altman, A. (1999). Plant Biotechnology in the 21st century: the challenges ahead. Electr J Biotechnol 2, 51-55.

[92] Morandini, P. (2009). Rethinking metabolic control. Plant Sci 176, 441-451.

Section 4

Applications in the Pharmaceutical Industry

Chapter 19

Bioactive Polysaccharides of American Ginseng *Panax quinquefolius* L. in Modulation of Immune Function: Phytochemical and Pharmacological Characterization

Edmund M. K. Lui, Chike G. Azike, José A. Guerrero-Analco,
Ahmad A. Romeh, Hua Pei, Sherif J. Kaldas,
John T. Arnason and Paul A. Charpentier

Additional information is available at the end of the chapter

http://dx.doi.org/10.5772/50741

1. Introduction

Ginseng has a long history of use as a traditional medicine; and it is one of the top selling medicinal herbs in the world. It is a multi-action herb with a wide range of pharmacological effects on the central nervous system, cardiovascular system and endocrine secretion, and the reproductive and immune systems [1]. Ginseng is a deciduous, perennial plant of the Araliaceae family. There are two major species of ginseng: *Panax ginseng and Panax quinquefolius;* and the roots are primarily used for medicinal benefits. Ginseng's wide range of pharmacological activities is believed to be due to the presence of a host of bioactive compounds. The primary ones are the ginsenosides, which are steroidal saponins conjugated to different sugar moieties and polysaccharides (PS) which account for 10-20% by weight of ginseng.

Polysaccharide components of ginseng have received much attention recently because of the emergence of different biological activities, such as immunomodulatory, antibacterial, anti-mutagenic, radioprotective, anti-oxidative, anti-ulcer, antidepressant, anti-septicaemic and anti-inflammatory activities [2]. Specifically, the polysaccharide fraction of ginseng has been shown to have immunomodulatory effects in both preclinical and clinical studies [3-6], although they are poorly characterized.

Several polysaccharides have been identified in *P. ginseng* and *P. notoginseng* but these compounds, including arabinogalactan, pectins, and acidic polysaccharides, have been

INTECH
open science | open minds

© 2012 The Author(s). Licensee InTech. This chapter is distributed under the terms of the Creative Commons Attribution License http://creativecommons.org/licenses/by/3.0), which permits unrestricted use, distribution, and reproduction in any medium, provided the original work is properly cited.

rarely studied in the North American species. They are made up of a complex chain of monosaccharides rich in L-arabinose, D-galactose, L-rhamnose, D-galacturonic acid, D-glucuronic acid and D-galactosyl residues [7]. The actual structural characteristics and the heterogeneity of PS components are poorly understood due to a lack of methodologies for separation as well as quantitative and qualitative analysis.

Most studies have focused on Asian ginseng PS and mostly *in vitro* experimental models. In this chapter, we will focus on *Panax quinquefolius* (American ginseng). The use of gel permeation chromatography (GPC) with multiple detectors has provided enhanced resolution for phytochemical analysis. And we have used both *in vivo* and *in vitro* models to evaluate its immunomodulatory activity.

2. Materials

Ginseng. Four-year-old American ginseng roots collected in 2007 from five different farms in Ontario, Canada were provided by the Ontario Ginseng Growers Association. Ginseng extracts from each farm were prepared individually and combined to produce composite extracts which were used for phytochemical and pharmacological studies [6]. The AQ extract has no detectable endotoxin contamination as determined by Limulus test.

2.1. Chemicals and biologicals

Sephadex G75 was purchased from GE Healthcare Bio-Sciences AB (Sweden). The Diethylaminoethyl (DEAE)-Cellulose and monosaccharide standards were purchased from Sigma (Oakville, Ontario). All other chemicals were of analytical grade and used as received. Cell culture medium and reagents were purchased from Gibco laboratories (USA). BD OptEIA ELISA kits tumour necrosis factor-α and interleukin-6 (BD Biosciences, USA). LPS from *Escherichia coli* and Griess reagent were purchased from Sigma-Aldrich (USA).

Animals. Adult male Sprague-Dawley rats (200–250g; Charles River, St. Constant, QC, Canada) were used. The Animal Ethics Review Committee of the Western University approved the study (Protocol No: 2009-070).

3. Methods

Preparation of the aqueous (AQ) and polysaccharide (PS) ginseng extracts. Dried ginseng root samples were shipped to Naturex (USA) for extraction. Samples were ground between ¼ and ½ inch and used to produce the AQ extract [6]. Briefly, 4kg ground ginseng roots were soaked three times during five hours in 16L of water solution at 40°C. After extraction, the solution was filtered at room temperature. The excess solvent was then removed by a rotary evaporator under vacuum at 45°C. The three pools were combined and concentrated again until the total solids on a dry basis were around 60%. These concentrates were lyophilized with a freeze dryer (Labconco, USA) at -50°C under reduced pressure to produce the AQ ginseng extract in powder form. Yield of the powder extracts from the concentrates was about 66%. The yields of the final extract (mean ± standard deviation of % extractive) from the initial ground root were 41.74±4.92.

A solution of AQ extract in distilled water (10g/10mL) was prepared, and the crude PS was precipitated by the addition of four volumes of 95% ethanol. The PS fraction was collected by centrifugation at 350×*g* (Beckman Model TJ-6, USA) for 10 minutes and lyophilized to produce the crude PS extract.

To prepare the water soluble polysaccharide extract (WSPE), ginseng roots (500 g) were extracted with 7.0 L of MilliQ (EMD Millipore) water at 100 ºC for 4 h and filtered through sheets of glass fiber. The solid material was extracted twice under identical conditions. The filtrates were combined, then centrifuged to remove water insoluble materials and supernatants were concentrated (1.0 L) and precipitated by the addition of 95% ethanol (4 to 1 volumes). After centrifugation, the precipitate was washed and dried by solvent exchange, first using 95% ethanol, and then absolute ethanol. Crude water soluble polysaccharide extract (WSPE) was obtained with a 20.0 % yield (relative to dry weight of plant material) (Figure 1).

3.1. Chromatography of ginseng extracts

High performance liquid chromatography (HPLC) analysis for ginsenoside determination

HPLC analysis on the composition of ginsenosides in the AQ extracts (100mg/ml methanol) was performed with a Waters 1525 HPLC System with a binary pump and UV detector [6]. A reversed-phase Inspire C18 column (100mm×4.6 mm, i.d. 5μm) purchased from Dikma Technologies (USA) was used for all chromatographic separations. Gradient elution consisted of [A] water and [B] acetonitrile at a flow of 1.3mL/min as follows: 0min, 80-20%; 0-60min, 58-42%; 60-70min, 10-90%; 70-80min, 80-20%. Absorbance of the eluates was monitored at 203nm.

Sephadex G-75 chromatography

AQ ginseng extract (500mg) was dissolved in 5mL distilled water and then fractionated by loading to a calibrated Sephadex G-75 column (47×2.5cm) equilibrated and eluted with distilled water mobile phase at 4°C with a flow rate of 1mL/min [6]. Absorbance of the eluates was monitored at 230nm. Fractions were collected and lyophilized for the study of bioactivity distribution.

Preparation of the de-proteinated water soluble polysaccharide extracts (DWSPE)

WSPE (50 g) were re-dissolved in 1.5 L of MilliQ water and partitioned five times with Sevag reagent (1:4 n-butanol:chloroform, v/v, 500 mL each) to remove proteins [8]. Polysaccharides were precipitated again by ethanol and dried by solvent exchange. This procedure yielded 46.0 g of the de-proteinated water soluble polysaccharide fraction (DWSPE). The procedure for the preparation of DWSPE from *P. quinquefolius* is shown in Figure 1.

Total fractionation of the DWSPE by ion exchange chromatography on DEAE-Cellulose

DWSPE fraction (20 g) was dissolved in MilliQ water (200 mL) and loaded on a DEAE-Cellulose column (10.0 X 20 cm, Cl^-) pre-equilibrated with MilliQ water. The column was eluted first with 4.0 L of MilliQ water at a flow rate of 10 mL/min (4 bar column pressure) to obtain the unbound or neutral fraction (DWSPE-N) and then with 4.0 L of 0.5 M NaCl to obtain the bound or acidic fraction (DWSPE-A). The fractions were concentrated, dialyzed (cut off pore size of 2 KDa) against MilliQ water and freeze dried to give 15.0 g (13.8 %) of the DWSPE-N and 0.9 g (0.83 %) of the acidic fraction DWSPE-A. DWSPE-A (0.6 g) was dissolved in 50 mL MilliQ water and loaded on a DEAE-Cellulose column (10 X 20 cm, Cl^-). The column was eluted by a stepwise gradient with 2.0 L of NaCl aqueous solutions (0.0, 0.1, 0.2, 0.3 and 0.5 M each) at a flow rate of 10 mL/min (4 bar column pressure). A total of 120 eluate fractions were collected (50 mL each), dialyzed and lyophilized (Figure 2).

High performance gel permeation chromatography-evaporative light scattering detection (HPGPC-ELSD) analysis

HPGPC was carried out at 40 ºC using a TSK-gel G-3000PWXL column (7.8 X 300 mm, TOSOH, Japan) connected to a HPLC system coupled with Diode Array and Evaporative Light Scattering Detectors (DAD-ELSD). The column was calibrated with standard dextrans (5 to 410 KDa range, Figure 7-B). Ten microliters of 20 mg/mL solutions of DWSPE, DWSPE-N and DWSPE-A were separately injected and eluted with HPLC grade water at a flow rate of 0.8 mL/min and monitored using ELSD with a temperature setting at 80 °C.

High performance gel permeation chromatography-multi-detector analysis

AQ and PS ginseng extracts were analysed at 40°C with TSK-gel PWXL G-4000PWXL column (7.8 X 300 mm, TOSOH, USA) connected to a Viscotek (Varian Instruments, USA) gel permeation chromatography system with Omnisec software (version 4.5, Viscotek, USA) for data acquisition. Solutions of AQ and PS extract (1mg/mL) were filtered with 0.2μm nylon filter and used for analysis. Each sample (100μl) was injected and eluted with 0.3M sodium chloride (NaCl) mobile phase at a flow rate of 1mL/min and monitored using a multiple detectors system for light scattering, refractive index and viscosity. Pullulan polysaccharide reference standard was analyzed as a positive control.

3.2. Analysis of the monosaccharides composition in WSPE by HPLC-ELSD

Carbohydrate analysis represents a challenge in analytical chemistry since neutral or acidic saccharides (mono, oligo and poly) have little UV activity. In our study, evaporative light scattering detection (ELSD) was used. The ELSD does not require the solutes of interest to have any optical properties; and the only requirement is that the eluent be more volatile than the solutes. WSPE (20 mg) was dissolved in 10 mL of 2N HCl solution and was boiled for 2 h. The hydrolyzed product was neutralized (pH 6-7) and centrifuged before analysis.

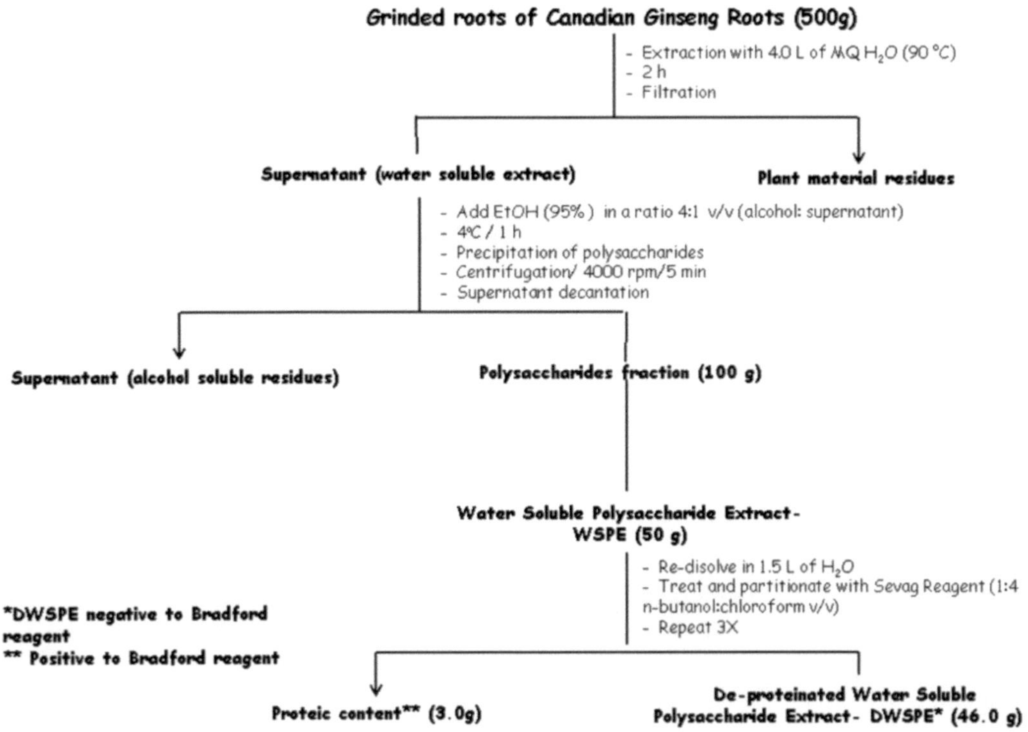

Figure 1. Preparation of the water soluble polysaccharides extract (WSPE) and the de-proteinated extract from the roots of 4-year-old Ontario-grown Panax quinquefolius.

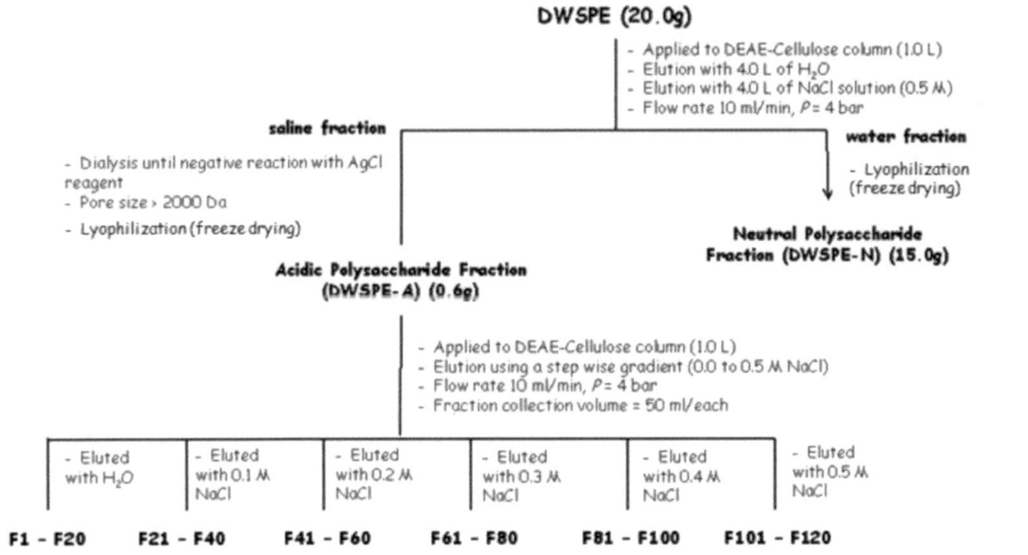

Figure 2. Step wise fractionation of the de-proteinated water soluble ginseng polysaccharide extract (DWSPE).

HPLC analysis was conducted using an 1100 series HPLC-DAD-ELSD system (Agilent Technologies Inc., Santa Clara, CA, USA). To enhance resolution of monosaccharides, analysis was performed using two separate columns (Figure 2). Glucose, galactose, arabinose, mannose and xylose were eluted using a Rezex RPM Monosaccharide PB+2 (8%) (Phenomenex, Torrance, California) column with a mobile phase of 100 % water (Chromasolv Plus, HPLC grade) isocratically at 80 °C and a flow rate of 0.6 mL/min. The ELSD temperature was set to 80 °C. Galacturonic acid and rhamnose content were examined using a Luna 5 μ NH_2 100Å column (Phenomenex, Torrance, California) and eluted with a mobile phase of acetonitrile and water (80:20) at 40 °C and a flow rate of 3 mL/min. The ELSD temperature was set to 44 °C.

3.3. Pharmacological evaluation

3.3.1. *In vivo study*

Adult male rats (250-300 gm) were treated with 125 mg/kg of AQ extract or crude PS fraction dissolved in saline by gastric garvage (10 ml/Kg bwt) once daily for 3 or 6 consecutive days, and examined 24 hr after the last dose. Animals were anesthetized with i.m. injection (80 and 5 mg/kg b.wt. ketamine and xylazine, respectively) and the trachea was cannulated for lung bronchoalveolar lavage (BAL) with Dulbecco's phosphate-buffered saline (PBS) to collect alveolar macrophages.

Blood was collected into heparinized tubes from rats by intracardiac puncture, samples were immediately centrifuged at 3000 rpm for 10 minutes and the plasma was separated, aliquoted and stored at -20 °C until use.

3.3.2. *In vitro study*

Cell culture

Rat alveolar macrophages were collected by BAL using cannulated 10-ml syringe with three 10ml washes of PBS. Fluid recovered from BAL was centrifuged at 1000 rpm for 5 minutes. Cells were cultured in RPMI-1640 medium supplemented with 10% Fetal Bovine Serum (FBS), 25mM HEPES, 2mM Glutamine, 100IU/ml penicillin and 100μg/ml streptomycin in 96-well tissue culture plates, at a density of 2×10^5 cells per well at 37°C maintained in a humidified incubator with 5% CO_2.

Cell treatment

Immuno-stimulatory effect

Experiments to evaluate dose-related stimulation of inflammatory mediators profile *in vitro* were carried out by treating and incubating rat alveolar macrophages with 0, 50, 100 and 200μg/ml of ginseng extracts for 24 hours and washed before challenging with LPS (1μg/mL) was used as positive control. The 24 hours-production of NO, TNF-α and IL-6 in culture medium was determined.

LPS-induced immuno-suppression

To examine the direct inhibitory effect of ginseng extracts on LPS-stimulated immune function, we pre-treated the macrophages with 0, 50, 100 or 200μg/ml of ginseng extracts for 24 hours and washed before challenging with LPS (1 μg/ml). The 24-hour cytokine production induced by LPS was determined by measuring NO, TNF-α and IL-6 levels in the culture medium.

3.3. Quantification of NO, TNF-α and IL-6

TNF-α and IL-6 concentrations in supernatants from cultured cells and plasma were analyzed with ELISA [6]. Samples were evaluated with rat cytokine-specific BD OptEIA ELISA kits (BD Biosciences, USA) according to the manufacturer's protocol. NO production was analyzed as accumulation of nitrite in the culture medium. Nitrite in culture supernatants was determined with Griess reagent (Sigma-Aldrich, USA) as previously described [6].

3.4. Statistical analysis

In vivo and *in vitro* experiments were performed at least three separate times. All statistical analyses were performed with GraphPad prism 4.0a Software (GraphPad Software Inc., USA). Data were presented as the mean ± standard deviation (SD) of triplicates from three independent experiments. Data sets with multiple comparisons were evaluated by one-way analysis of variance (ANOVA) with Dunnett's *post-hoc* test. $P<0.05$ was considered to be statistically significant.

4. Results

4.1. Phytochemistry

Ginsenoside Compositon. AQ extract contained a total ginsenoside content of 13.87% dry weight of extract and showed characteristics of *Panax quinquefolius* with Rb1 and Re as the predominant ginsenosides, with no detectable Rf and minimal levels of Rg1 (Fig.3).

Crude PS extract. The yield of crude PS fraction by four volumes of 95% ethanol precipitation was 10 % dry weight of root materials. A representative G-75 chromatographic profile of the crude PS extract is shown in Figure 4 [6]. The major PS peak (with a elution volume of 100 ML) had an estimated average molecular weight of 73kDa, while there were two minor, less well- resolved peaks (Fig 4).

Monosaccharide composition of the WSPE determined by HPLC-ELSD. Representative chromatograms showing individual monosaccharides in water soluble polysaccharide extract (WSPE) are shown in Figure 5 and 6. Glucose was found to be the major neutral monosaccharide present in WSPE with amounts ranging from 77 to 86 % (w/w). Galactose and arabinose were present in similar amounts with galactose being present at levels of 6.8

to 7.5 % (w/w) and arabinose being present at levels between 4.5 to 5.9 % (w/w). Galacturonic acid was also identified on WSPE at levels ranging from 8.7 to 9.5 % (w/w). Mannose and xylose were also monitored but were not detected in the sample.

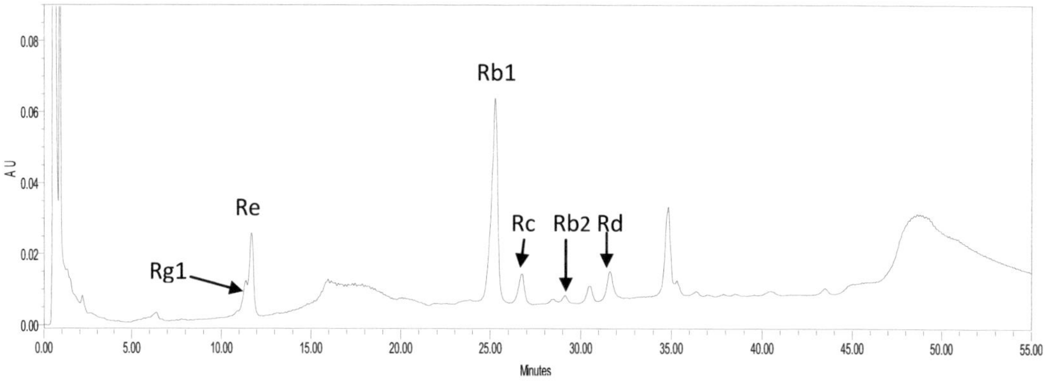

Figure 3. HPLC ginsenoside profile of Ontario-grown American ginseng aqueous extract. A reversed-phase Inspire C18 column was used with gradient elution consisted of [A] water and [B] acetonitrile: 0min, 80-20%; 0-60min, 58-42%; 60-70min, 10-90%; 70-80min, 80-20%. Absorbance of the eluates was monitored at 203nm.

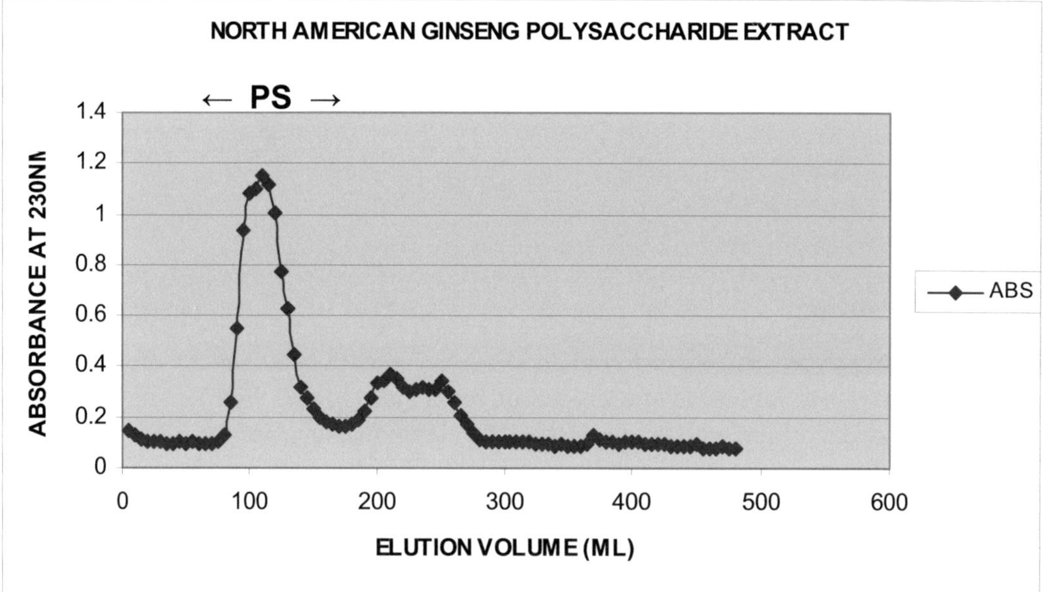

Figure 4. Sephadex G-75 (47×2.5cm) chromatographic fractionation of PS extracts of ginseng. A calibrated column was loaded with 500mg of the extract, and then eluted with distilled water at a flow rate of 1mL/min. The y-axis is the absorbance at 230nm while the x-axis represents the elution volume (mL).

Total fractionation of the DWSPE by ion exchange chromatography on DEAE-Cellulose. To further analyse the WSPE, the material was de-proteinated using the Sevag method, giving a

DWSPE with a yield of 92% and the protein content yield of 1.8% (relative to the dry weight of WSPE). Fractionation of WSPE by a combination of anion-exchange on DEAE-cellulose and gel permeation chromatographies with a procedure shown in Figure 1 revealed the elution of a neutral fraction with water (N-DWSPE, 75 % relative to the dry weight of DWSPE) and an acidic fraction (A-DWSPE, 4.5%) with a 0.5 M NaCl solution (Figure 7). Both fractions presented a wide and complex molecular weight distribution ranging from 5 to 410 KDa. The acidic fraction A-DWSPE containing most likely uronic acids was subjected to a second fractionation by DEAE-cellulose chromatography.

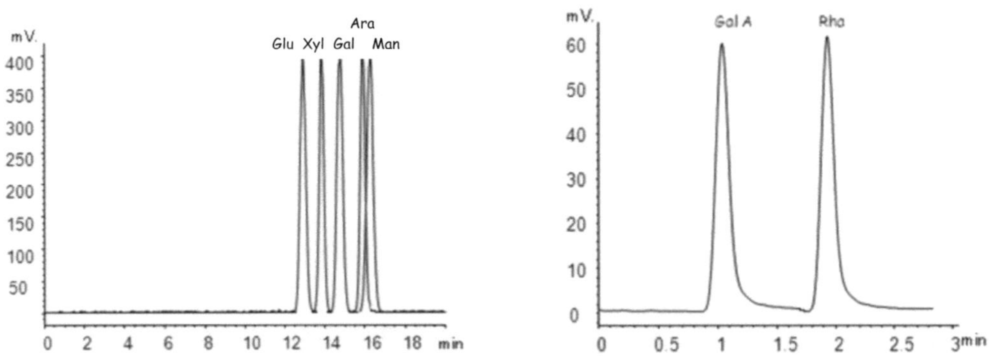

V (A) (B)

Figure 5. Representative chromatograms of the novel analytical method by HPLC-ELSD for the analysis of seven monosaccharides including glucose (Glu), galactose (Gal), arabinose (Ara), Xylose (Xyl), Mannose (Man) (A), galacturonic acid (Gal A) and rhamnose (Rha) (B).

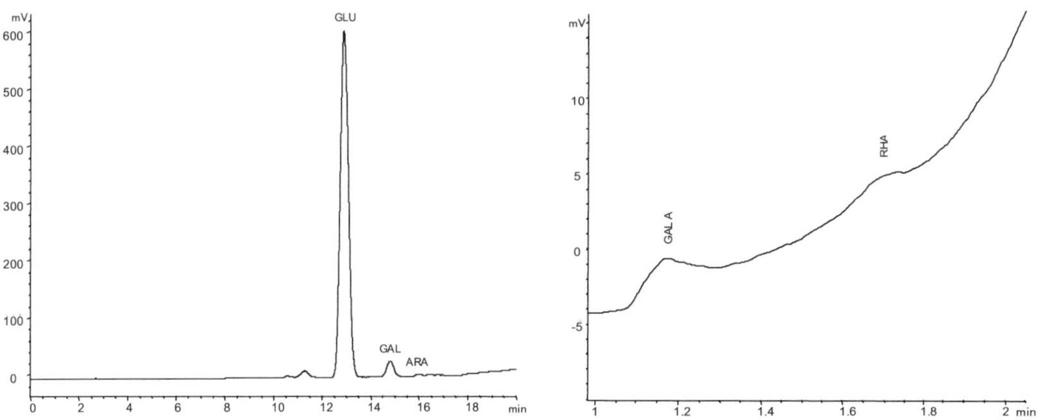

Figure 6. Representative HPLC-ELSD Chromatograms of monosaccharide detected in WSPE, including glucose (Glu), galactose (Gal), arabinose (Ara) (A), galacturonic acid (Gal A) and rhamnose (Rha) (B).

Secondary Fractionation with the A-DWSPE on DEAE-Cellulose column chromatography. The elution of 0.6 g of A-DWSPE on a DEAE-cellulose chromatography was carried out using a stepwise gradient of NaCl. With these elution steps, A-DWSPE was separated into six fractions (Figure 2): A-DWSPE-1 (F1-F20, 96 mg, 16 %), A-DWSPE-2 (F21-F40, 54mg, 9%), A-DWSPE-3 (F41-F60, 162 mg, 27%), A-DWSPE-4 (F61-F80, 120mg, 20%), A-DWSPE-5 (81-F100, 78mg, 13%) and A-DWSPE-6 (F101-F120, 60mg, 10%) corresponding to the elution with 0.0, 0.1, 0.2, 0.3, 0.4 and 0.5 M NaCl solutions, respectively.

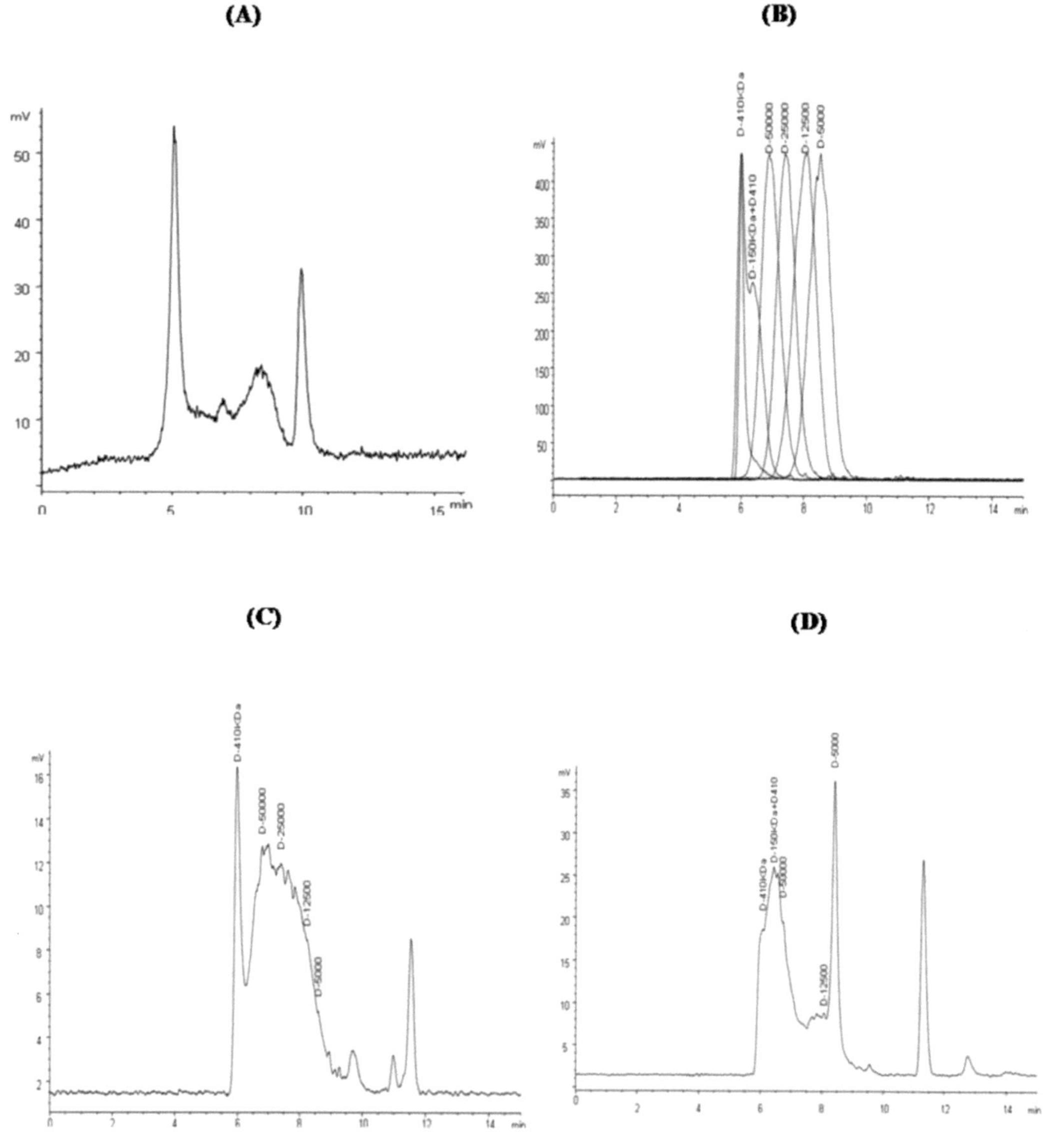

Figure 7. Representative chromatrograms by HPGPC-ELSD analysis; (**A**) DWSPE sample, (**B**) Dextran standards mixture, (**C**) DWSPE-N and (**D**) DWSPE-A. All samples were eluted under the same conditions.

In order to further analyze the unique NA ginseng polysaccharide samples, GPC with multiple detectors was utilized. Figure 8 shows the results of multiple pullulan polysaccharide standards ranging from 1800 Da to 1,050,000 Da. The utilized column gave good resolution of the individual standards over this range. Figure 8 shows the results of the multi-detector system containing 4 different detectors, i.e. refractive index (RI), right angle light scattering, low angle light scattering and viscometer. In Figure 9, the crude polysacchride extract is observed to have three major peaks with M_w values of 1092 kDa, 135 kDa and 12 kDa (Table 1). The major peak at 1092 kDa accounts for 66% of the weight fraction. Figure 9 shows the deproteinated polysaccharide fraction, in which the M_w values are largely unchanged (Table 1). However, the protein content fraction (PTF) contains only a small amount of carbohydrates with the majority of the high molecular weight fraction removed. The acid fraction contains 3 peaks, of generally lower M_w values, while the neutral fraction obtained from anionic exchange of DWSPE contains only one broad peak (Figure 9).

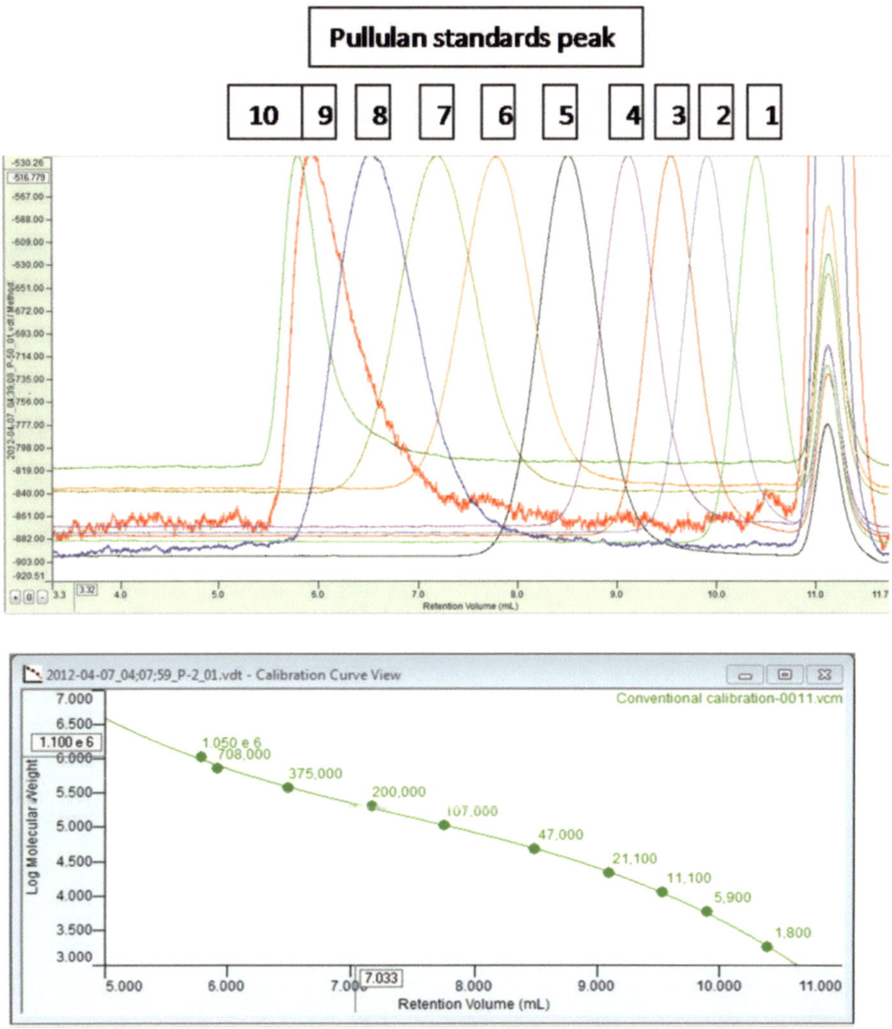

Figure 8. a) Resolution of individual pullulan standards: peak 1, 1800; peak 2, 5900; peak 3, 11,100; peak 4, 21,100; peak 5, 47,000; peak 6, 107,000; peak 8, 375,000; peak 9, 708,000; peak 10, 1,050,000 and b) calibration curve of pullulan standards through G-4000PWXL column.

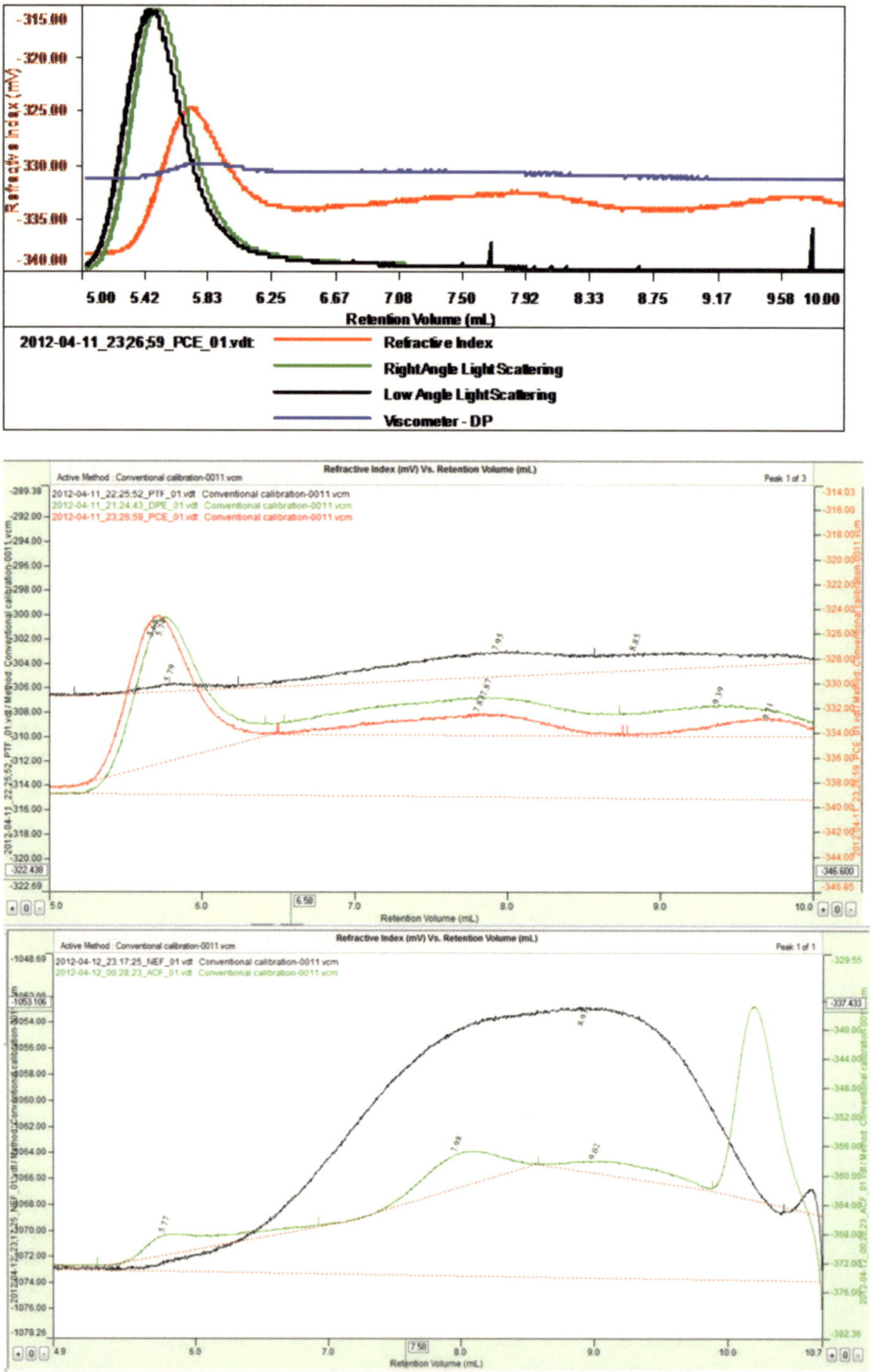

Figure 9. a) GPC traces showing traces from the individual detectors of the PCE (Crude extract of polysaccharides), b) RI plot for the PCE (red), DWSPE (green) and PTF (black) samples. The high

molecular weight peak at retention volume 5.79mL has very low intensity in the PTF sample, which shows that this sample has a very low content of polysaccharide., c) RI plot of the neutral extract (black) and the acidic extract (green) from the ion exchange column, the neutral extract shows much bigger RI area, therefore much higher sample recovery.

PCE	1	2	3
Mn (Dalton)	995,589	106,001	9,502
Mw(Dalton)	1,092,000	135,019	12,037
Mz(Dalton)	1,188,000	172,473	15,433
RI area (mVmL)	5.84	2.03	1.03
Weight fraction (% of RI area)	65.6%	22.8%	11.6%

Table 1. Molecular weight data of the PCE Extract.

5. Immunomodulatory activity

In vivo effect. Treatment with AQ extract for 3-6 days produced marked stimulation of alveolar macrophages as determined by increased production of NO and TNF- and IL-6 following culturing for 24 hrs, reaching activities that were 50-100% of the positive (LPS) control (Figure 10). This immunostimulatory effect was also reflected in the elevation of plasma TNF- and IL-6 levels of treated animals (Figure 11). However, the responsiveness of macrophages collected from ginseng treated animals to LPS stimulation *ex vivo* showed >50% to 100% reduction in NO, TNF-α and IL-6 production as compared to those non-ginseng treated controls, especially with 6 days of ginseng treatment (Figure 10). These data showed that orally administered AQ extract had both immuno-stimulatory and anti-inflammatory effect. Data presented in Figure 12 showed that this immunomodulatory activity could be extended to the PS extract on the basis of its effect on macrophage NO production and the LPS responsiveness.

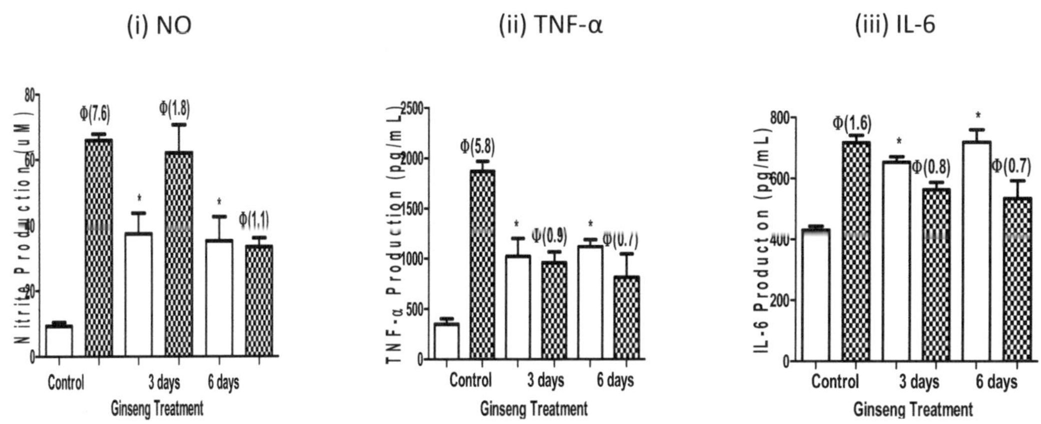

Figure 10. Figure 10. Orally administered ginseng AQ extract: elevated cytokine production and reduced LPS-stimulatd cytokine production in cultured alveolar macrophages. Alveolar macrophages of rats treated orally with 0 and 125mg/kg ginseng AQ extract for 3 and 6 Days were cultured for 24 hours to measure production of NO and cytokines (by ELISA) . To determine responsiveness to

LPS stimulation, ginseng treated macrophages were exposed to 1ug/ml LPS in culture to determine changes in 24 hr NO and cytokine production . Three independent experiments were performed and the data were shown as mean ± SD. Datasets were evaluated by ANOVA. * Values P<0.05 compared to the untreated control were statistically significant. Φ values in bracket denote fold increase in LPS-stimulated cytokine production over control.

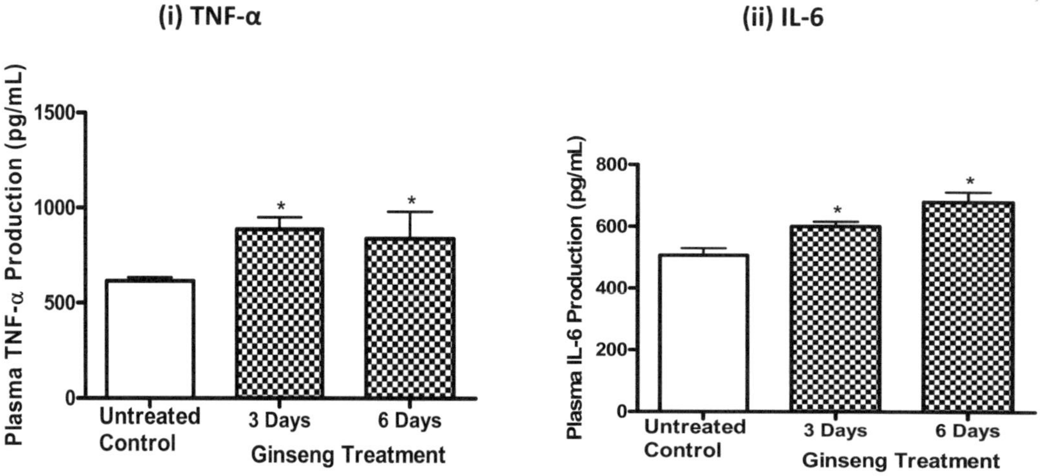

Figure 11. Ginseng AQ extract treatment elevated plasma (i) TNF-α and (ii) IL-6 levels. Rats were treated orally with 125mg/kg ginseng AQ extract for 3 and 6 days. Plasma cytokine concentrations were determined by ELISA. Three independent experiments were performed and the data were shown as mean ± SD. Datasets were evaluated by ANOVA. * Values $P<0.05$ compared to the untreated control were statistically significant.

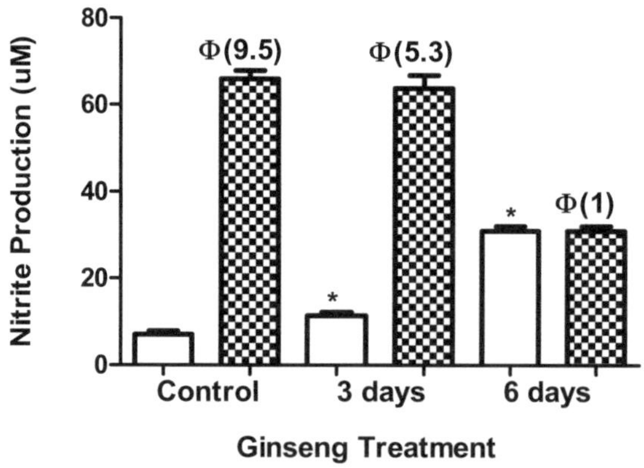

Figure 12. Orally administered ginseng PS extract (125 mg/kg) up-regulated NO production and reduced responsiveness to LPS (LPS 1 μg/ml) stimulation in cultured alveolar macrophages. Cells from untreated controls were treated with LPS 1 μg/ml as positive control for macrophage responsiveness. NO was determined by Griess reaction assay. Three independent experiments were performed and the data were shown as mean ± SD. Datasets were evaluated by ANOVA. * Values P<0.05 compared to the untreated control were statistically significant. Φ values in bracket denote fold increase in LPS-stimulated NO production over control.

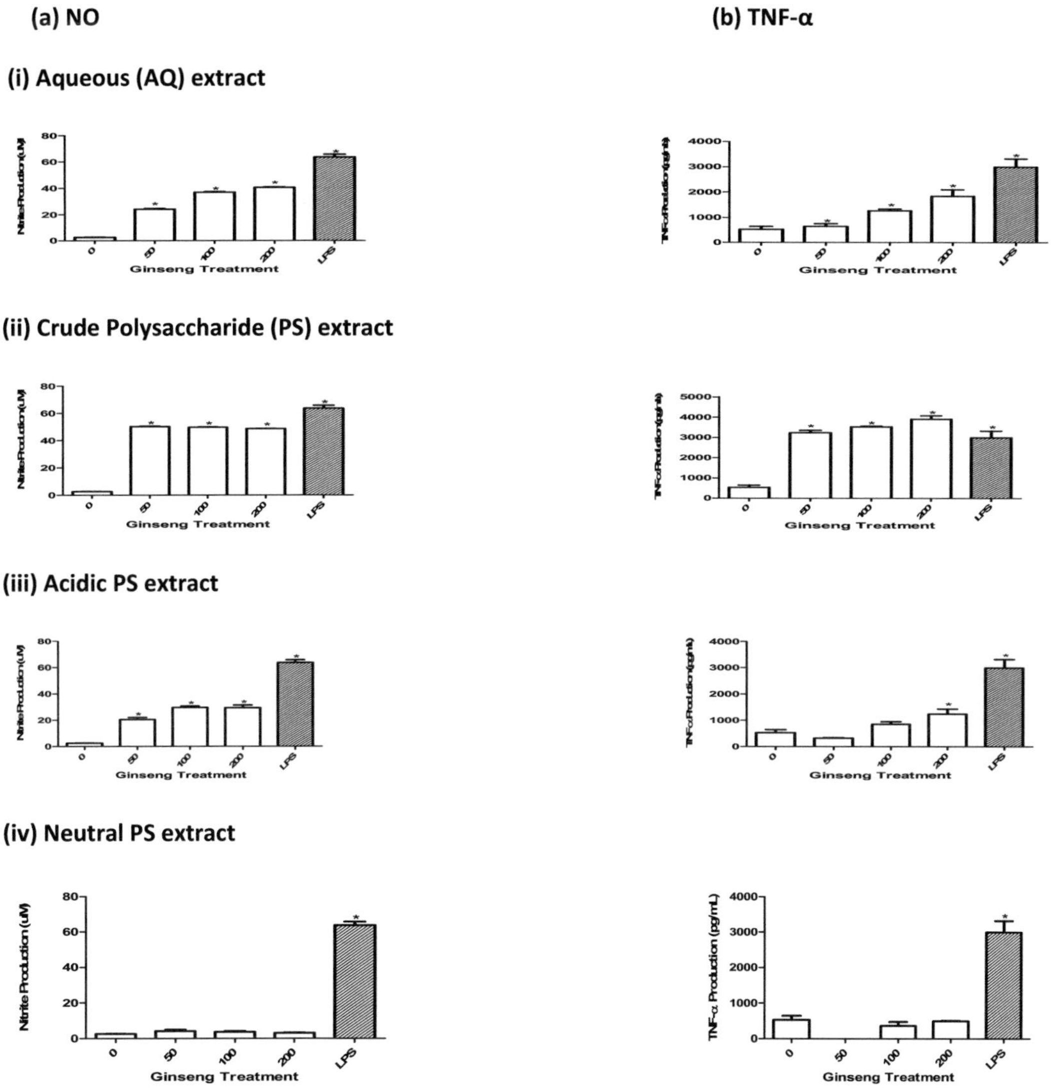

Figure 13. Immuno-stimulatory effects *in vitro* of AQ, crude PS, acidic PS and neutral PS extracts on 24 hours macrophage production of (a) NO and (b) TNF-α. Alveolar macrophages isolated from control rats were treated with 0, 50, 100 and 200μg/ml of ginseng extracts for 24 hours, and the culture supernatants were analysed for NO and TNF-α by Griess reaction assay and ELISA, respectively. Cells treated with LPS (1 μg/ml) were used as positive controls. Three independent experiments were performed and the data were shown as mean ± SD. Datasets were evaluated by ANOVA. * Values $P < 0.05$ compared to the untreated (vehicle) control were statistically significant.

In vitro effect

Both AQ, crude PS and acidic PS showed stimulation of NO and TNF-α production by alveolar macrophages *in vitro* (Figure 13). The magnitude of the response to PS was greater than those induced by acidic PS or AQ extract. Neutral PS was devoid of activity. The lack of concentration-dependent effect of PS was probably due to its high potency and inducing its maximum effect at the concentration studied. Data presented in Figure 14 showed how pretreatment with various extracts for 24 hrs altered the subsequent response to LPS challenge. Since prior LPS treatment was known to cause desensitization of subsequent response to LPS, this was used as a positive control to evaluate the immunosuppressive effect of ginseng extracts. It was apparent that PS was the most effective in reducing the NO and TNF-α response to LPS, while the AQ and acidic PS extracts were similar, and neutral PS was inactive.

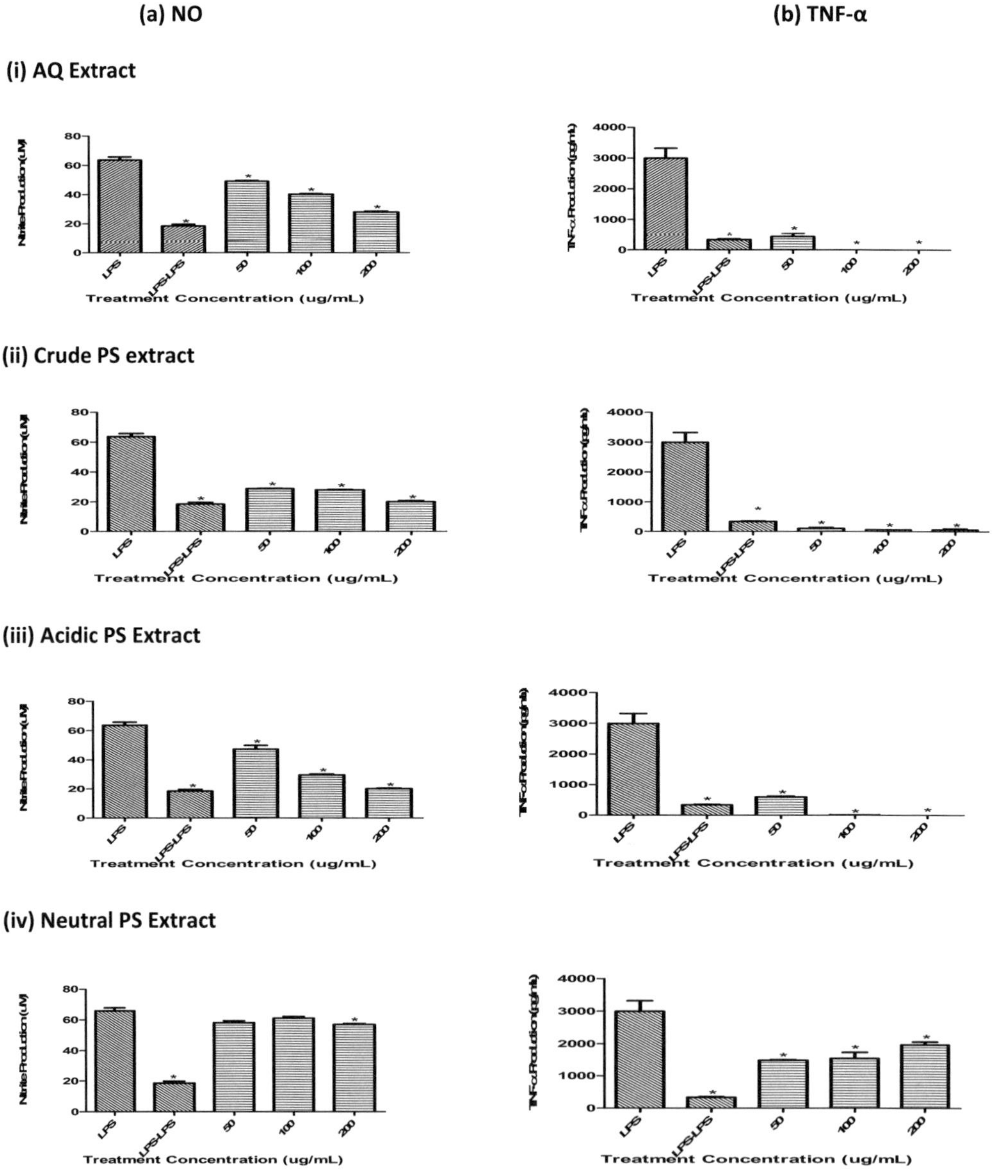

Figure 14. *In vitro* effects of AQ, crude PS, acidic PS and neutral PS extracts on LPS-stimulated 24 hours macrophage production of (a) NO and (b) TNF-α. Rat Alveolar macrophages were pre-treated with

ginseng extracts (0, 50, 200 μg/ml) for 24 hours and were washed before challenged with LPS 1 μg/ml. 24 hr-pretreatment with LPS prior to LPS stimulation was used as positive control to demonstrate desensitization of macrophage responsiveness (LPS-LPS). NO and TNF-α were determined by Griess reaction assay and ELISA, respectively. Three independent experiments were performed and the data were shown as mean ± SD. Datasets were evaluated by ANOVA. * Values $P < 0.05$ compared to the LPS positive control were statistically significant.

6. Discussion

Medicinal plants have been in use for human health for thousands of years, yet polysaccharides have only been recognized recently as a major contributor to the bioactivity of these traditional medicines. Polysaccharides from plant sources with immunomodulatory, anti-tumor, anti-viral, anti-bacterial, anti-inflammatory, anti-oxidant, and anti-diabetogenic activities have been reported [2, 9-15]. And a few polysaccharides, including lentinan, *Astragalus* polysaccharide, polyporus polysaccharide and *Achyranthes bidentata* polysaccharide have been licensed for clinical application in China [16]. In Canada, a polysaccharide-enriched American ginseng extract (Cold Fx®) has been licensed in 2007 as a natural health product to 'help reduce the frequency, severity and duration of cold and flu symptoms by boosting the immune system' with an estimated annual sale of over $48M [4, 5]. Our findings on the paradoxical effects of AQ and PS extract on macrophage function *in vivo* may have significant implication in the use of American ginseng polysaccharides in several clinical applications.

The structure and biological activities of the polysaccharides from the roots, leaves and fruits of *Panax ginseng* have recently been reviewed by Sun [2]. There is limited information on American ginseng. Our study *in vitro* has demonstrated the up-regulation of inflammatory mediators production by AQ and crude PS extracts in rat alveolar macrophages (Figure 13) which validated what we have previously reported [3 and 6]. In addition we have demonstrated specificity of PS in that acidic but not neutral species of the PS was bioactive. Following sub-acute oral administration, both PS and AQ extracts were also immuno-stimulatory based on elevation of plasma cytokine levels and increase in the function of alveolar macrophages *ex vivo*. The immuno-stimulatory dosage used in the present *in vivo* study was comparable to those proven to be effective for cardiovascular health and for protection against diabetic retinopathy, neuropathy and cardiomyopathy reported by other investigators using identical ginseng extracts and in the same animal species [17-19]. However, the magnitude of the immunostimulatory effect *in vivo* was smaller than the *in vitro* response. This may be related to the lower bioavailability of the orally-administered ginseng extract. The pharmacokinetics of oral ginseng PS is not known, but the recovery of major ginsenosides in plasma after oral administration was quite low [20]. Result of our *in vitro* and *in vivo* studies was supportive of what was reported for CVT-E002 (a patented, poly-furanosyl-pyranosyl polysaccharide-rich extract of the root of North American ginseng): stimulation of normal mouse spleen cells and immunoglobulin G production as well as activation of peritoneal exudate macrophages leading to enhanced cytokine stimulation in treated mice.

In addition to the well-recognized immuno-stimulatory activity of ginseng, an anti-inflammatory effect was shown in the present study as reflected in the reduced responsiveness of alveolar macrophages collected from ginseng-treated animals to LPS challenges *ex vivo* (Figure 10). This apparent anti-inflammatory effect of ginseng PS is different from what we have previously reported for a specific component(s) of the alcoholic extract of American ginseng as well as the anti-inflammatory effects that have been ascribed to some ginsenosides and their metabolites [6, 21-22]. This potential anti-inflammatory mechanism is being validated by evaluating changes in LPS–induced inflammatory response following polysaccharide pretreatment *in vivo* in our on-going research. The intent is to determine whether ginseng PS causes desensitization of immune cells as reported for LPS [23]. This action of ginseng, when proven, may be particularly relevant to bacterial infection and related toxicemia. An anti-inflammatory effect of ginseng polysaccharide has been reported by Zhao et al. using a model of auto-immune disease, as evidenced by the reduction in the expression of TNF-α and IFN-γ in lymphocytes in the enteric mucosal immune system of rats with collagen induced arthritis [24]. In view of the diverse immunomodulatory effects of ginseng polysaccharide, the identification of specific polysaccharides with unique property and biological action will be of great interest.

It appears that AQ and PS ginseng extracts have a paradoxical effect on macrophage function: stimulation under normal condition, but reduction when the biological system is under a pro-inflammatory state. In the context of sepsis, AQ and PS immuno-stimulatory effect will be beneficial as a first line of defense during the initial infection stage of bacterial infection by rendering macrophages to be cytotoxic [25] whereas the immune-suppressive activity may be effective in antagonizing the cytokine storm at the later stages of infection by suppressing LPS activation of macrophages.

Carbohydrates analysis represents a major challenge in analytical chemistry since neutral or acidic saccharides (mono, oligo and poly) have little UV activity. The refractive index (RI) detector, which is commonly used in HPLC analysis, has issues with baseline stability and sensitivity. The ELSD used in the present study has the advantage of its independence of any optical properties in the solutes of interest. Our HPLC-based analysis allowed the measurement of 7 mono-saccharides within a 30 minute total run time. The monosaccharide composition of *P. quinquefolius* reported in this study was similar to that described previously [3] with the addition of galacturonic acid. The monosaccharide composition provides insight into the types of polysaccharides which may be found in *P. quinquefolius*. Polysaccharides structures in *P. quinquefolius* have not previously been thoroughly studied, though several polysaccharides have been isolated from *P. ginseng* and *P. notoginseng*. Glucose and galacturonic acid were the most prominent monosaccharides detected. Previously, polysaccharide fractions from *P. ginseng* with high levels of glucose have been determined to contain starch-like glucans and arabinogalactans and fractions with high levels of galacturonic acid have been shown to contain pectins with several linked galacturonic acid domains [26]. It is possible that polysaccharides similar to these may be present in Ontario-grown American ginseng though further work will have to be done to characterize their structures.

Polysaccharides are very complex with a wide range of MW, varying monosaccharide composition and conformation (degree of branching or linearity), which contribute to their diverse structure and biological activities, these can also hamper the study of their structure–function relationships [16]. The multi-detector GPC instrument provided additional information on the polysaccharide structure. This instrument uses Triple detection with a concentration detector (refractive index detector), viscometer and light scattering detector, with each detector providing different although complementary information [20]. Pullulan standards of up to 1,000 kDa showed that the measured ginseng polysaccharide molecular weights for the crude, water soluble and deproteinated extracts are within the range of standards. The light scattering detector, which is based on fluctuations in interference between macromolecules (e.g. polysaccharides) scattering a coherent monochromatic laser beam, is considered an absolute detector for Mw values, confirming the reported values of Table 1. The viscometer is sensitive to branching effects, although the trace in Figure 9 shows no significant branching effect for the ginseng polysaccharides. The RI detector was rather sensitive to the carbohydrate fractions, showing 3 major polysaccharide components for each of the three measured extracts. The high molecular weight peak has a very low intensity for the PTF sample, indicating that the deproteination step removed the high molecular weight fraction. The neutral extract gave similar molecular weights as the acidic polysaccharide extract, although with a higher sample recovery.

7. Conclusions and future directions

Our study has revealed a paradoxical immunomodulatory effect of AQ and PS extracts isolated from Ontario-grown *Panax quinquefolius*. This activity may be relevant to clinical application involving bacterial infection and toxemia. Challenges in development of botanical polysaccharides for clinical application include lack of methodology for structure identification and characterization, isolation and purification, and product quality control, has been recognized [11]. Our research will focus on the application of new methodologies to characterize the polysaccharide structures from ginseng and to elucidate the structure-biological activity relationship.

Author details

Edmund M. K. Lui*, Chike G. Azike and Hua Pei
Ontario Ginseng Innovation and Research Consortium
Department of Physiology & Pharmacology, Schulich School of Medicine and Dentistry, Canada

José A. Guerrero-Analco and John T. Arnason
Ontario Ginseng Innovation and Research Consortium
Department of Biology, University of Ottawa, Ottawa Ontario, Canada

* Corresponding Author

Ahmad A. Romeh
Department of Chemical and Biochemical Engineering, Faculty of Engineering, University of Western Ontario, London Ontario, Canada

Sherif J. Kaldas
Department of Biology, University of Ottawa, Ottawa Ontario, Canada

Paul A. Charpentier
Ontario Ginseng Innovation and Research Consortium, Department of Chemical and Biochemical Engineering, Faculty of Engineering, University of Western Ontario, London Ontario, Canada

Acknowledgement

This research was supported by Ontario Ginseng Research & Innovation Consortium (OGRIC) funded by the Ministry of Research & Innovation, Ontario Research Funded Research Excellence program for the project 'New Technologies for Ginseng Agriculture and Product Development'(RE02-049 awarded to EMK Lui). We acknowledge the contribution of PolyAnalytik London Ontario, Canada in providing instrument for the gel permeation chromatography analysis of ginseng extracts.

8. References

[1] Attele AS, Wu JA, Yuan CS (1999) Ginseng Pharmacology; Multiple Constituents and Multiple Actions. Biochem. pharmacol. 58:1685–1693.

[2] Sun Y (2011) Structure and Biological Activities of the Polysaccharides from the Leaves, Roots and Fruits of *Panax ginseng* C.A. Meyer: An Overview. Carbohyd. polym. 85: 490-499.

[3] Assinewe VA, Arnason JT, Aubry A, Mullin J, Lemaire I (2002): Extractable Polysaccharides of *Panax quinquefolius* L. (North American ginseng) Root Stimulate TNF-α Production by Alveolar Macrophages. Phytomedicine 9:398–404.

[4] Biondo PD, Goruk S, Ruth MR, O'Connell E, Field CJ (2008) Effect of CVT-E002™ (COLD-fX®) Versus a Ginsenoside Extract on Systemic and Gut-Associated Immune Function. Int. immunopharmacol. 8:1134–1142.

[5] Predy GN, Goel V, Lovlin RE, Basu TK (2006) Immune Modulating Effects of Daily Supplementation of COLD-FX (a Proprietary Extract of North American Ginseng) in Healthy Adults. J Clin Biochem Nutr 39: 162-167.

[6] Azike CG, Charpentier PA, Hou J, Pei H, Lui EMK. (2011). The Yin and Yang actions of North American ginseng root in modulating the immune function of macrophages Chin Med. 6, 1-12.

[7] Wang M, Guilbert LJ, Li J, Wu Y, Pang P, Basu TK, Shan JJ (2004) A Proprietary Extract from North American Ginseng (*Panax quinquefolium*) Enhances IL-2 and IFN-gamma Productions in Murine Spleen Cells Induced by Con-A. Int. immunopharmacol. 4:311–315.

[8] Sevag M G, Lackman D B, Smolens J (1938). The isolation of the components of streptococcal nucleoproteins in serologically active form. Journal of Biological Chemistry, 124, 425–436.

[9] Igor A. Schepetkin, Mark T. Quinn (2006). Botanical polysaccharides: Macrophage Immunomodulation & therapeutic potential. International Immunopharmacology 6: 317–333.

[10] Meiqi Wang, Larry J. Guilbert, Lei Ling, Jie Li, Yingqi Wu, Sharon Xu, Peter Pang and Jacqueline J. Shan (2001). Immunomodulating activity of CVT-E002, a proprietary extract from North American ginseng (*Panax quinquefolium*). Journal of Pharmacy and Pharmacology 53: 1515–1523.

[11] Jie-Young Song, Seon-Kyu Han, Eun-Hwa Son, Suhk-Neung Pyo, Yeon-Sook Yun, Seh-Yoon, Induction of secretory and tumoricidal activities in peritoneal macrophages by ginsan, Yi International Immunopharmacology 2 (2002) 857–865.

[12] Yoo DG, Kim MC, Park MK, Park KM, Quan FS, Song JM, Wee JJ, Wang BZ, Cho YK, Compans RW, Kang SM (2012). Protective effect of ginseng polysaccharides on influenza viral infection. PLoS One. 2012;7(3):e33678.

[13] Ahn JY, Song JY, Yun YS, Jeong G, Choi IS (2006). Protection of *Staphylococcus aureus*-infected septic mice by suppression of early acute inflammation and enhanced antimicrobial activity by ginsan. FEMS Immunol Med Microbiol; 46:187-97.

[14] Dianhui Luo, Baishan Fang, Structural identification of ginseng polysaccharides and testing of their antioxidant activities (2008). Carbohydrate Polymers 72 376–381.

[15] Xie JT, Wu JA, Mehendale S, Aung HH, Yuan CS (2004). Anti-hyperglycemic effect of the polysaccharides fraction from American ginseng berry extract in ob/ob mice. Phytomedicine. 11:182-7.

[16] He X, Niu X, Li J, Xu S , Lu A (2012). Immunomodulatory Activities of Five Clinically Used Chinese Herbal Polysaccharides. J. exp. integr. med. 2:15-27.

[17] Wu Y, Lu X, Xiang FL, Lui EMK, Feng Q (2011) Ginseng protects the heart from ischemia and reperfusion injury via up regulation of endothelial nitric oxide synthase. Pharmacological Research. 64:195-202.

[18] Subhrojit S, Chen S, Feng B, Wu Y, Lui E, Chakrabarti S (2012) Preventive effects of North American ginseng (*Panax quinquefolius*) on diabetic retinopathy and cardiomyopathy. Phytotheraphy Res. (In Press)

[19] Subhrojit S, Chen S, Feng B, Wu Y, Lui E, Chakrabarti S (2012) Preventive effects of North American ginseng (*Panax quinquefolius*) on diabetic neuropathy. Phytomedicine. 19: 494– 505.

[20] Zhou D, Tong L, Wan M, Wang G, Ye Z, Wang Z, Lin R (2011). An LC-MS method for simultaneous determination of nine ginsenosides in rat plasma and its application in pharmacokinetic study. Biomed Chromatogr, 25::720-6.

[21] Lee DC, Yang CL, Chik SC, Li JC, Rong JH, Chan GC, Lau AS (2009) Bioactivity-Guided Identification and Cell Signalling Technology to Delineate the Immunomodulatory Effects of *Panax ginseng* on Human Promonocytic U937 cells. J. transl. med. 34:1-10.

[22] Oh GS, Pae HO, Choi BM, Seo EA, Kim DH, Shin MK, Kim JD, Kim JB, Chung HT (2004). 20(S)-Protopanaxatriol, one of ginsenoside metabolites, inhibits inducible nitric

oxide synthase and cyclooxygenase-2 expressions through inactivation of nuclear factor-kappaB in RAW 264.7 macrophages stimulated with lipopolysaccharide. Cancer Lett, 205:23-29.

[23] Leon-Ponte M, Kirchhof MG, Sun T, Stephens T, Singh B, Sandhu S, Madrenas J. (2005). Polycationic lipids inhibit the pro-inflammatory response to LPS. Immunol Lett. 96:73-83.

[24] Zhao H, Zhang W, Xiao C, Lu C, Xu S, He X, Li X, Chen S, Yang D, Chan ASC, Lu A (2011) Effect of Ginseng Polysaccharide on TNF-α and INF-γ Produced by Enteric Mucosal Lymphocytes in Collagen Induced Arthritic Rats. J. med. plants res. 5:1536-1542.

[25] Lim DS, Bae KG, Jung IS, Kim CH, Yun YS, Song JY (2002) Anti-Septicaemic Effect of Polysaccharide from *Panax ginseng* by Macrophage Activation. J. infection. 45: 32-38.

[26] Zhang X, Yu L, Bi H, Li X, Ni W, Han H, Li N, Wang B, Zhou Y, Tai G (2009) Total Fractionation and characterization of the Water-Soluble Polysaccharides Isolated from *Panax ginseng* C.A. Meyer. Carbohyd. polym. 77: 544-552.

Chapter 20

Polysaccharides from Red Algae: Genesis of a Renaissance

María Josefina Carlucci, Cecilia Gabriela Mateu, María Carolina Artuso and Luis Alberto Scolaro

Additional information is available at the end of the chapter

http://dx.doi.org/10.5772/47445

1. Introduction

The Red Queen Effect is an evolutionary hypothesis [1]. This evolutionary concept is named for the Red Queen´s comment to Alice in Through the Looking Glass that "it takes all the running you can do, to stay in the same place". It posits that multicellular organisms with long life cycles must constantly change, adaptation process driven by the changing conditions of the environment, in order to survive the onslaught of potentially lethal pathogens which have much shorter life cycles and can thus evolve orders of magnitude faster.

Viruses play a relevant role in these new evolutionary mechanisms by transferring of genes to and from the hosts they parasite [2]. Over the past three decades, it has become apparent that viruses are ubiquitous, abundant and ecologically important in the environment [3]. As phylogenetic analysis shows, nearly all organisms of all kingdoms have become infected by viruses since the beginning of life. The great impact that viruses can have on the genetic systems is well illustrated by the evolution of mitochondria. In reference [4] have shown, the existence of a strong selection pressure has pushed for the replacement of cellular enzymes by viral ones in mitochondria and chloroplasts. In both organelles, this replacement has been associated with profound modifications in the mechanism of DNA replication and chromosome structure [5]. The fact that viruses are probably very ancient allows better understanding their extraordinary diversity, explaining why most viral proteins inferred from genome sequencing have no cellular homologues [6]. Besides, the existence in the biosphere of an unlimited reservoir of viral proteins has provided opportunities at different steps of the evolutionary process, to introduce new functions into organisms.

INTECH
open science | open minds

© 2012 The Author(s). Licensee InTech. This chapter is distributed under the terms of the Creative Commons Attribution License http://creativecommons.org/licenses/by/3.0), which permits unrestricted use, distribution, and reproduction in any medium, provided the original work is properly cited.

At the present, it remains controversial the inclusion of viruses in the "tree of life". Several authors assume viruses are non-living organisms and believe their properties are driven solely by thermodynamically spontaneous reactions while others give priority to the fact that phylogenetic tree is based on the genomic content of its components, not the physical manifestations of these genomes. Moreover, the fact that viral genomes carried inside virions encode gene products that allow for adaptation and response to changing intracellular and extracellular conditions favors the inclusion of these agents in the tree of life [7,8].

The oligosaccharides chains (glycans) attached to cell surface and extracellular proteins and lipids are known to mediate many important biological roles [9,10]. However, for many glycans, there are still no evident functions that are of obvious benefit to the organism that synthesizes them. In 1949, Haldane postulated "Now every species of mammal and bird so far investigated has shown quite surprising biochemical diversity by serological tests. The antigens concerned seem to be proteins to which polysaccharides are attached. We do not know their functions in the organism, though some of them seem to be part of the structure of the cell membrane. I wish to suggest that they may play a part in disease resistance, a particular race of bacteria or virus being adapted to individuals of a certain range of biochemical constitutions, while those of other constitutions are relatively resistant" [11]. In [12], suggested that glycan diversification in complex multicellular organisms is driven by evolutionary selection pressures of both endogenous and exogenous origin. They also argued that exogenous selection pressures mediated by viral and microbial pathogens and parasites that recognize glycans have played a more prominent role, favoring intra-and interspecies diversity.

2. Red algae and carrageenans

Red algae (Division: Rhodophyta) are one of the oldest and largest groups of eukaryotic algae with more than 10000 species described (Figure 1). They are distributed worldwide but grow best in waters of near 15°C. They have the characteristic of all eukaryotes including the nuclei which in some algae are smaller than their plastids. However, their cells lack of flagella so they need the water movement to carry masculine cells to the oocyte. They also have disorganized chloroplasts lacking of external endoplasmic reticulum and containing unstacked thylakoids. Their red colour is due to the presence of the phycoerythrin pigment which reflects red light and absorbs blue/green ones. Since blue light penetrates water to a greater depth than light of longer wavelengths, red algae are able to photosynthesize and live in water of 260 m in deep which receive 0.1% of surface irradiance; this means one thousand times less light than the surface. Those rhodophytes that have small amounts of this pigment might seem green or bluish from the chlorophyll and other pigments present in them.

Over the last 2.45 billion years, algae have been diversifying [13] in order to survive in competitive ecological niches. This adaptation led to evolution of a large and diverse array of biochemical constituents.

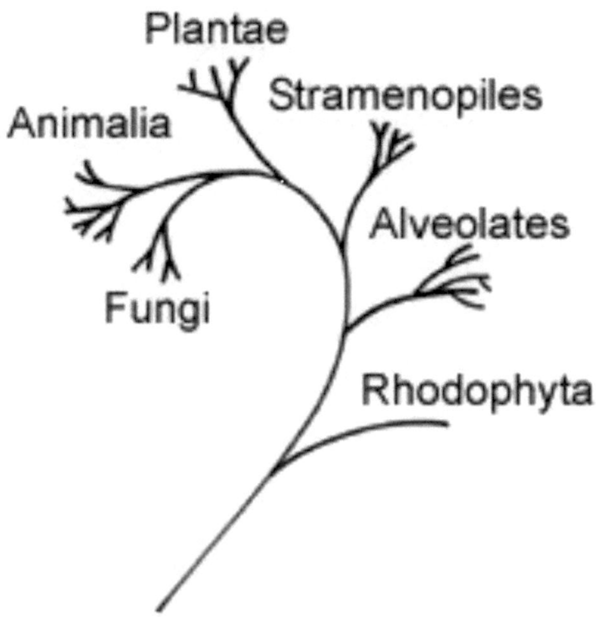

Figure 1. Rhodophyta branched off very early in the tree of life

Red algae contain large quantities of polysaccharides in the cellular wall, thereof, most are sulfated galactans. These galactans are generally constituted by alternatively repeated units of bonds 1,3-α-galactopyranose and 1,4-β-D-galactopyranose can defer in the level and pattern of sulfation, in the substitution by methoxy and/or pyruvate groups and other sugars such as mannose and xylose. They also defer in the 3,6-anhydrogalactose content and the 1,3 -α-galctopyranose residues configuration.

Among these galactans, the carrageenans (CGNs) may be mentioned, which have similar structures to the pattern observed in the galactosaminoglycans. They comprise a wide group of structures that may be divided in two families: the κ-family, defined by the presence of a sulfated C4 group in the unit β-D, and formed by CGNs-κ/ι and the carrageenans-μ/ν, and the λ-family, characterized by a sulfate-C2 group and constituted by all the varieties of λ structures. The λ- and ι-carrageenan types are more strongly sulfated than the most of the heparan sulfate (HS) derived from tissues [14]. In general, this type of carrageenans exhibits a viral inhibitory potential a little greater than the κ-carrageenans.

Polysaccharides are composed of building-blocks that although being not numerous, their almost infinite combination led to an array of polysaccharides with an important structural complexity [15]. The diversity of polysaccharides can be further increased by acetylation, methylation and, more commonly in the case of many marine algal polysaccharides, sulfation [16]. Moreover, many algal polysaccharides are metabolically active, either as a storage molecule which undergo structural changes during their life cycles or as a structural component [17].

The sulfated polysaccharides are highly abundant and accessible compounds that may be isolated from various natural sources. Micro- and macroalgae are under investigation for numerous commercial food, agri- and horticultural, pharmaceutical, cosmetic and bioenergy applications [18-21]. Polysaccharides are also known for their wide and variable physicochemical properties which make them suitable for different applications in the fields of medicine and pharmacology. They have proved to be useful tools due to their immune-modulator and antitumoral activity, their interference in the clotting system and in the inflammatory processes, in dermatology, in dietary programs and moreover by affecting the viral replication. Among the natural sources where they can be found are the cell walls from algae. Depending on the type of algae, those with similar structures to the glycosaminoglycans (GAGs) and wide antiviral activity can be isolated [22] (Table 1). Antiviral activity has been documented for retrovirus: human immunodeficiency virus type 1 and 2 (HIV-1 and HIV-2), herpesvirus: herpes simplex virus (HSV) type 1 and 2, human cytomegalovirus (HCMV); pseudorabies virus; flavivirus: dengue virus type 2; smallpox virus: variola virus; hepadnavirus: hepatitis B virus (HBV); orthomyxovirus: influenza A virus (inf A); paramyxovirus: respiratory syncytial virus (RSV) and parainfluenza virus; rhabdovirus: vesicular stomatitis virus (VSV); arenavirus: Junin virus, Tacaribe virus and togavirus: Sindbis virus, Semliki Forest virus and against some naked viruses, such as encephalomyocarditis virus, Hepatitis A virus [23] and papilloma virus (HPV) [24], of both DNA and RNA viral types (Table 1). For most of these viruses the initial bond of the virus to the cells would be mainly mediated by the interaction of virus with a GAG of the cellular surface known as HS [14]. In general, sulfated polysaccharides have a chemical structure very similar to the HS. Thus, they might block viral infection by competing against virion attachment to the cell surface.

GAGs are linear polysaccharides constituted of successive repetition of a disaccharide unit which may be sulfated. GAGs can be divided in two groups; Glucosaminoglycans, like HS and galactosaminoglycans like chondroitin sulfate. The initial incorporation of saccharide units of *N*-acetylglucosamine or *N*-acetylgalactosamine, respectively, gives their names. An important difference between these groups is that Glucosaminoglycans are attached by 1,4 union while galactosaminoglycans are attached by 1,3 and 1,4. GAGs are found mainly in the cell surface and in much of the intracellular matrix of the mesodermic tissue as is shown in Table 2 (connective, cartilage, muscle and bone). Frequently, they are linked to a core protein and one or more covalently attached glycosaminoglycan chains, known it as Proteoglycans.

GAGs are negatively charged molecules that can have a physiological significance like hyaluronic acid, dermatan sulfate, chondroitin sulfate, heparin, HS and keratan sulfate.

Algae	Compound	Virus	Reference
		Red algae	
Schizymenia pacifica	λ- carrageenan	HIV-1, AMV	[25]
Schizymenia dubyi	Sulfated galactans with uronic acid	HIV-1, HSV-1, HSV-2, VSV	[26]
Nothogenia fastigiata	(Xylo)mannans	HIV-1,HIV-2, SIV, HSV-1, HSV-2, HCMV, Inf A, RSV, Junin, Tacaribe	[27,28]
Aghardiella tenera	Sulfated Agarans	HIV-1, HIV-2, HSV-1, HSV-2, HCMV, VSV, Inf A, RSV, togavirus, parainfluenza virus, smallpox	[29]
Digenea simples	sp non-characterized	HIV	[30]
Nothogenia fastigiata	Xylogalactans	HSV-1, HSV-2	[31]
Pterocladiella capillacea	Sulfated Agarans and hybrid DL- galactans	HSV-1, HSV-2, HCMV	[32]
Gigartina skottsbergii	λ-,κ/ι- and μ/ν-carrageenans	HSV-1, HSV-2	[33,34]
Cryptopleura ramosa	Sulfated agarans	HSV-1, HSV-2	[35]
Stenogramme interrupta	Carrageenans	HSV-1, HSV-2	[36]
Asparagopsis armata	Sulfated agarans	HIV	[37]
Bostrychia montagnei	Sulfated agarans	HSV-1, HSV-2	[38]
Gymnogongrus torulosus	Hybrid DL- galactans	HSV-2, virus dengue 2	[39]
Gracilaria corticata	Sulfated agarans	HSV-1, HSV-2	[40]
		Brown algae	
Pelvetia fastigiata	Fucans	HBV	[41]
Fucus vesiculosus	Fucans	HIV-1	[42]
Sargassum horneri	Fucans	HSV-1, HCMV HIV-1	[43]
Leathessia difformis	Fucans	HSV-1, HSV-2	[44]
Adenocystis utricularis	Fucans	HSV-1, HSV-2	[45]
Microalga			
Cochlodinium polykrikoides	Extract	HIV-1, RSV, Inf A, Inf B	[46]
Porphyridium sp	Extract	HSV-1, HSV-2	[47]
		Green algae	
Monostroma latissinum	Sulfated Rhamnans	HSV-1, HCMV, HIV-1	[48]

Table 1. Antiviral activity of sulfated polysaccharides extracted from marine algae

GAG	Location	Comments
Hialuronates	Synovial fluid, vitreous humor, extracellular matrix with loss of connective tissue (vasculogenesis).	Long polymers (containing no sulfates), shock-absorbing.
Chondroitin sulfate	cartilages, bone and cardiac valves.	More abundant GAGs.
Heparan sulfate	Basal Membrane and components of the cellular surface.	
Heparin	Components of the intracellular granules of the mastocytes, coating of the lung arteries, liver and skin.	More sulfated than the Heparan sulfate.
Dermatan sulfate	Skin, cardiac valves and blood vessels.	Long polymers (no sulfates), shock-absorbing.
keratan sulfate	Cornea, bone and cartilage.	More abundant GAGs.

Table 2. Normal distribution of GAGs in the body.

3. Relationship among glycosaminoglycans, carrageenans and viruses

The discovery that viruses are highly abundant in natural waters initiated renewed research on the impact of viral infection and lysis on aquatic microorganisms [49]. It is believed that viruses influence the composition of marine communities and are a major force behind biogeochemical cycles. Each infection has the potential to introduce new genetic information into an organism or progeny virus, thereby driving the evolution of both, host and virus [50].

The eukaryotic algae represent the oldest known eukaryote for which there exist clear geological data [51] and all classes of algae have their specific DNA virus. The HSV is an ancient DNA virus which is widespread in nature and has coevolved with its hosts. Many viruses interact with their host polysaccharides present on the cell wall; HSV uses HS.

The basic structural motifs and modifications of HS glycosaminoglycans seem to have been conserved for several hundred million years of evolution [52,53].

One suggested explanation is that endogenous heparan sulfate-binding proteins may have developed different binding specificities with evolutionary time. At first glance, this may seem an exception to the suggestion made here, in view that extensive glycan diversification has accompanied species evolution. However, HS can generate numerous intrinsic structural variations, and there are currently inadequate data about the extent of speciesspecific differences in the specificities of the binding proteins and/or the expression of structural motifs in different cell lineages.

HS is highly sulfated and it is thought to be the most biologically active GAG. The sulfated monosaccharide sequences within HS determine the protein binding specificity and regulate fundamental biological functions including growth control, signal transduction, cell adhesion, homeostasis, morphogenesis, lipid metabolism and pathophysiology [14]. Numerous viruses including herpesviruses utilize cell surface HS as receptor to infect target cells.

It has been reported that in the course of an inflammation, an infection or tissue damage, the proteoglycan HS is cleaved causing fragments of soluble HS [54]. On the other hand, in healthy tissues, no significant fractions of soluble HS are found, though they can be found in the fluids of damaged tissues –at concentrations within the required ranges to stimulate dendritic cells [55] and in the infected individuals urine [56].

HSV attaches to cells by an interaction between the envelope glycoprotein C and cell surface HS. The virus-cell complex is formed by ionic interactions between the anionic (mainly sulfate) groups in the polysaccharide and basic amino acids of glycoproteins, and non-ionic ones depending on hydrophobic amino acids interspersed between the basic ones in the glycoprotein-binding zone [57]. This interaction is a decisive step in virus multiplication and may be differentiated but not dissociated from an evolutive point of view.

CGNs resemble to some extent the naturally occurring GAGs owing to their backbone composition of sulfated disaccharides are believed to be of potential therapeutic importance because they can mimic with GAGs present in cell membranes.

Natural CGNs, extracted from red seaweeds, are well known as potent and selective inhibitors of HSV-1 and HSV-2. CGNs chemical structures are similar to that of HS that serves as a primary receptor for adsorption of HSV onto cells. Mode of antiviral action is mediated by the interference with HSV attachment to cells, blocking the interaction virus-HS, a mandatory step during the multiplication cycle to achieve a productive infection that involves viral glycoproteins. In this work were used the κ and ι CGNs, their structures are present in the compound named 1C3 CGN, which is an "hybrid" κ/ι-and partially cyclized μ/ν CGN (Figure 2).

Since most pathogens and the toxins they produce bind to specific sugar sequences to initiate infection and disease, it is reasonable to assume that at least some glycan variation must have arisen from this selection pressure [12]. On this basis, pressure of selection *in vitro* with an antiviral drug like HS in the case of HSV may be employed to shorten the time necessary for attenuation. Moreover, in this last case we may speculate that if herpesviruses which are extensively spread in the environment are exposed to sulfated polysaccharide (its natural receptor), in the form of CGN, the appearance of virus variants would readily occur as a consequence of an intense virus-host interaction. Our results indicate that attenuation is a common trait of HSV obtained under selective pressure of CGN.

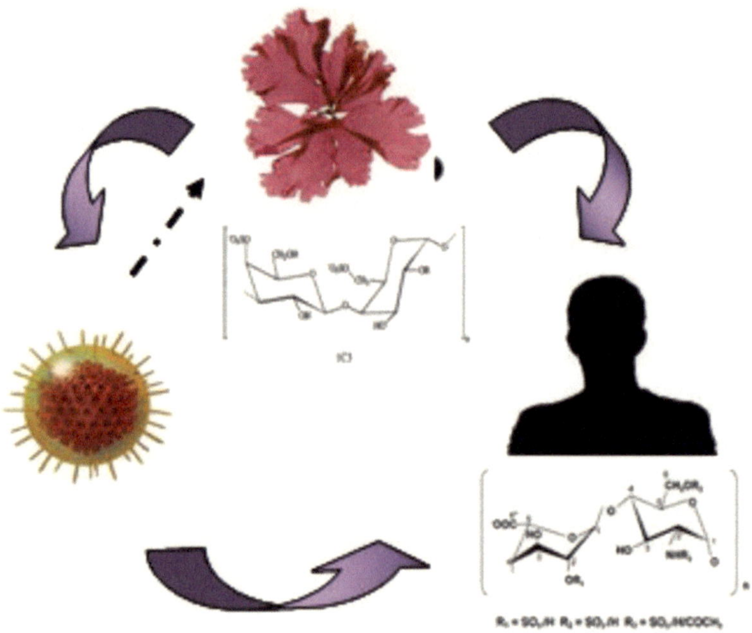

Figure 2. 1C3 CGN isolated from the red algae *Gigartina skottbergii* has a chemical structure similar to cellular HS. Attenuated HSV can be isolated from viral populations grown in the presence of increasing concentrations of CGN. This procedure may reflect an accelerated evolution process for HSV where biological modification of the viral particle can be demonstrated.

4. Viruses: Friend or enemy

It is tempting to postulate that the driving forces of evolutionary novelty are not randomly derived from chance mutations of the genetic text, but from a precise genome editing by omnipresent viruses [58]. For decades, non-coding regions of the genome have been ignored or declared as "junk"-DNA. Recently, scientists have realized that these regions incorporate decisive higher-order regulatory functions. New research has shown that these non-coding repetitive sequences originated primarily from retroviral RNA [59,60].

For a long time, viruses were interpreted as causing acute infections of susceptible organisms, using the host cellular machinery to reproduce, and achieving their lytic nature only in order to infect other cells. Although this narrative remains valid, it merely represents a case of viruses that were unable to reach persistent or chronic status of infection [61]. Most viruses, however, are stable, persistent agents that are able to establish a complex relationship with the host cell and in many cases this interaction lasts for the entire lifespan of the cell even when a competent immune system is present. The immune system is of crucial importance in defense against infection. It has to cope with a large number of different pathogens that relentlessly develop new ways to avoid recognition or elimination. Yet most infections are cleared. Immune-system genes must evolve to keep pace with increasingly sophisticated evasion by pathogens. To remain effective, defense demands creativity and competence because many pathogens have sophisticated and rapidly evolving evasion mechanisms [62]. This defense response is triggered by the presence of pathogens within the host, however there exists an immune response that comprises a set of autoreactive "natural antibodies" that do not rely on exogenous antigen stimulation to be synthesized by autoantibody-secreting B lymphocytes [63,64]. On the other hand, these autoreactive natural antibodies are reactive against components of the host´s immune system (i.e. cytokines) exacerbating ongoing infectious diseases or predisposing host to infection [65,66].

Individual resistance to pathogens depends on the combination of receptors on cells from the immune system although non-immune genes also influence resistance [67]. Signs of natural selection in a human population are especially illustrative, when a mutation in a certain gene is dangerous in normal conditions but confers resistance to infections widespread in the region. Among the better known are the mutations in hemoglobin and glucose-6-phosphate-dehydrogenase affecting red blood cells and conferring resistance to malaria [68]. Another example is the deletion at the 5´end of the CCR5 chemokine receptor conferring resistance to HIV infection. This molecule serves as the principal co-receptor, with CD4, for HIV-type 1. The allele with the deletion was intensely selected in Europe probably because it also provided resistance to plague and smallpox [69]. More subtle, but nonetheless important, relationship between cell and virus is that associated to changes in cell physiology due to viral infection that regulates cell death, transformation, secretory pathways, cell stress response, etc [70-72]. This panorama may account for a "symbiotic evolution" of cell and virus, although viruses have much shorter generation times than cells. Studies of genomic polymorphism of HSV-1 suggest that the evolution of this virus would be very slow and host-dependent [73].

5. Effect of carrageenans on the virus

In our laboratory, viral variants of HSV-1 (strain F) and HSV-2 (strain MS) were obtained by successive passages in Vero cells under selective pressure with the 1C3 CGN (is an "hybrid" κ/ι-and partially cyclized μ/ν-CGN) and ι or κ-CGNs respectively in order to test the ability of the CGN to generate resistant variants during the selection process and to study the CGN-virus modulation. Different clones were plaque purified and pretested to exclude reversion to wild type (Figure 3). 1C314-1 and 1C317-2 are viral variants derived from HSV-1 strain F (1C314-1 means, "1C3" is the type of CGN that was used for the selection of the variant, "14" number of pasagge that we chose for cloning, and "1" is the number of selected clon). κ22-12, κ22-13, ι22-9 and ι22-12 are variants derived from HSV-2 strain MS. In this case, κ and ι CGNs were used for the selection of these variants, respectively.

After the viral selection and cloning, all the variants showed a syncytial phenotype on Vero cells, this phenotype was also observed on mouse lung and genitals primary cell cultures [74,75].

Figure 3. Variants of HSV were obtained by successive passages in Vero cells under selective pressure with different types of CGNs with increasing concentration of them.

In order to characterize the obtained viral variants, the susceptibility to CGNs was assessed. All the viral variants showed low or middle levels of resistance to CGNs, heparin, Aciclovir (ACV), and Foscarnet (PFA) (Figure 4).

The results are shown as relative resistance (RR): IC50 viral variant/IC50 parental strain.(IC50 : inhibitory concentration 50%, is the concentration in μg/ml required to reduce plaque number by 50%).

Although some variants showed 15 or 17 fold of RR, those values are not significant compared with viral variants selected with ACV (ACVp6-F and ACVp6-MS), that showed higher values, > 50 RR, after only 6 passages with ACV. (unpublished data).

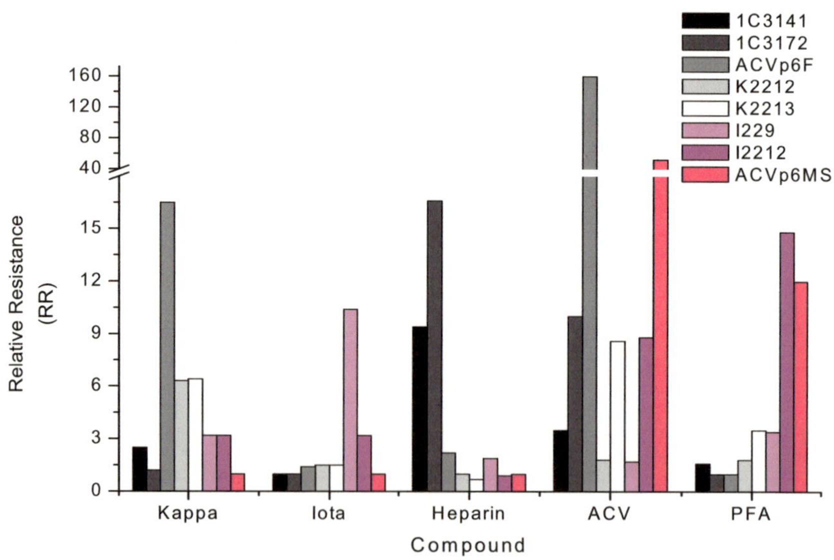

Figure 4. Susceptibility of HSV viral variants to several compounds with known antiherpetic activity. IC_{50}values were determinate on Vero cells by a reduction plaque assay; the RR values were calculated as (IC_{50}: viral variants/ IC_{50} parental strain). Kappa: κ CGN; iota: ι CGN; ACV: Aciclovir; PFA: Foscarnet.

Mucosal surface represents the primary site for replication and spread of several viruses including HSV type 1 or 2. In order to resemble natural routes of infections, virulence of the viral variants was assessed in a mouse model by two routes of infection, intravaginal or intranasal using either BALB/c or C57BL/6 mice.

The HSV-1 variants were avirulent for BALB/c mice infected by intravaginal route, although the parental strain F was highly lethal (Table 3). The attenuation of the variants correlated with low levels of pro-inflammatory cytokines (IL-6 and TNF-α) in vaginal lavages with respect to the parental strain, despite that viral titers were similar between the viral variants and the F strain. Nevertheless, the variants were highly lethal for BALB/c mice inoculated by the intranasal route, with a generalized organ spreading of virus.

On the other hand all the clones of HSV-2 were less virulent for mice intravaginally and intranasally inoculated, particularly for C57BL/6 mice, whereas MS strain produced 100% of mortality. In contrast to HSV-1 variants, the attenuation correlated with high levels of pro-inflammatory cytokines (TNF-α and IL-6) detected in vaginal lavages (Table 3).

Despite the differences in the virulence of variants, the levels of infectivity in the site of inoculation did not differ from those observed for the parental strains. These results suggested that the observed attenuation was not due by a lower viral replication and it could be explained by a differentiated immunological response.

Viral Characterization	HSV-1 (F)		HSV-2 (MS)	
	Parental strain	Viral variants	Parental strain	Viral variants
Cytopathic effect (*in vitro*)	Cell rounding	Syncytial	Cell rounding	Syncytial
Intravaginal virulence	High	Avirulent	High	Low
Intranasal virulence	High	High	High	Low
Levels of IL-6 and TNF-α	High	Low	Low	High

Table 3. Characterization of HSV variants obtained by selective pressure with CGN. Viral variants obtained under pressure of selection with carrageenans were characterized regarding their phenotype, *in vivo* virulence for BALB/c mice and immune response. The parental strains were also assessed for final comparison.

Moreover, by the same methodology, variants of HSV-2 obtained by selective pressure with nonsulfated compounds (from the condensation of mandelic acid) have been obtained. Viral variants were not syncytial and showed resistance to the same drug, heparin and the carrageenan 1C3 in the order of 2.6 to 6.7 times with respect to the control virus. However, in vivo, these variants showed no difference of pathogenicity and mortality with the parental strain.

6. The evolutive beliefs

About seven hundred and fifty million years ago, the first multicellular organisms appeared on earth. Since then, cells from multicellular organisms have found a way to become smarter. Multicellular life forms were initially isolated communities or "colonies" of unicellular organisms. However, the evolutionary advantages of living in a community soon led to communities composed of millions and billions of individual cells socially interactive. The evolutionary trend towards increasing complexity in the community is but a reflection of the biological imperative of survival. The better an organism perceives the surrounding environment, the more chances to survive. Unfortunately, we "forget" that cooperation is conveniently necessary for the evolution when Charles Darwin brought out a radically different theory about the origin of life. One hundred and fifty years ago, Darwin concluded that living organisms are involved in a constant "struggle for survival". In the final chapter of The Origin of Species by Means of Natural Selection, or The maintenance of favored races in the struggle for existence, Darwin speaks of an unstable "struggle for existence" and that evolution is conditioned by "the war of nature, from famine and death". This concept is related to the Darwinian notion that evolution occurs at random [76]. While Darwin is the most famous evolutionist, the first scientist to consider evolution as a scientific fact was the distinguished French biologist Jean Baptiste de Lamarck [77,78]. Even Ernst Mayr, the principal agent of the "neo-Darwinism," a modernization of Darwin's theory that incorporates molecular genetics of the twentieth century, admits that Lamarck was the pioneer [79]. Lamarck presented his theory not only fifty years before Darwin, but also offered a far less violent theory of evolutionary mechanisms. Lamarck's theory suggests that evolution is based on a cooperative and “instructive” interaction between organisms and the environment that allows living things to survive and evolve in a dynamic world. His idea was that organisms acquire and pass on the necessary adaptations to survive in a changing environment. One of the reasons why scientists are rethinking the theories of Lamarck is that evolutionists continue to remind the invaluable role of cooperation in maintaining life in the biosphere. In the book Darwin's Blind Spot written in 2002 by British physicist Frank Ryan has been recorded a number of these relationships [80]. Today, the understanding of cooperation in nature is much deeper than that obtained by simple observation. Biologists are increasingly aware that the animals have coevolved with different sets of microorganisms necessary for a healthy and “normal” development [81]. Recent advances in genetics have revealed an additional mechanism of cooperation between species. It has been discovered that genes are shared not only among individual members of a species (sexual reproduction) but also between members of different species. The distribution of information through gene transfer accelerates the process of evolution, in view that organisms can learn lessons "learned" by others [82-85]. Because of this distribution of genes, organisms can no longer be considered as isolated entities, there are no walls between species [86]. The distribution of genes is not a “by chance” mechanism. Is the method that nature uses to increase the survival of the biosphere. Genes are nothing more than a physical memory of the experiences learned by the organisms, the exchange of these genes between species spreads these “memories” and, consequently, influences the survival of all organisms that constitute a living community. Timothy Lenton has shown that evolution is more dependent on the interaction between species than on the interaction among individuals of the same species [87]. The evolution thus becomes a question of survival of fittest groups, not of individuals better adapted.

7. Genesis of a renaissance – Cooperative biocommunication

Darwinian theory emphasizes competition and selection of the individual as a main guiding force in evolution (not cooperation or group membership). Evidence supports the fact that viruses and other genetic parasites are key elements in the evolution of all living organisms. In "Life: The Communicative Structure", Dr. Witzany suggests that the genesis of new species, genera, and realms of organisms would not occur in any neo-Darwinistic sense via "chance mutations" and selection, but via a kind of innovation code (evolution code, creation code, text generating code), which is capable of DNA/RNA text editing [88]. It turns out that the genetic code that encodes proteins -practically the sole subject of current bioengineering- is only a kind of structuring vocabulary, and not a complete structure in itself, and is subjected to a high-order regulatory code that lies hidden in the nonprotein-coding regions of the DNA, which have been identified as RNA agents many years later [89]. There is increasing evidence that all cellular life is colonized by exogenous and/or endogenous viruses in a nonlytic but persistent lifestyle. A persistent lifestyle in cellular life-forms most often seems to derive from an equilibrium status reached by at least two competing genetic and the immune function of the host that keeps them in balance. If we imagine that humans and one of the simplest animals, *Caenorhabditis elegans*, share a nearly equal number of genes (ca. 20,000) it become obvious that the elements that create the enormous diversity are not the protein coding genes but their higher order regulatory network that is processed by the mobile genetic elements, such as transposons and retroposons and noncoding RNAs [90]. If we consider the important role of the highly structured and ordered regulatory network of noncoding RNAs as not being randomly derived, one of the most favorable models with explanatory power is the virus-first thesis [91].

For many decades it was common practice to speak about the "genetic code" with its inherent language-like features. The concept postulated by Manfred Eigen that nucleic acid sequences are comparable to and function like a real language, coherent with a (molecular) syntax, linguistic and a vocabulary, was commonly used in genetics, cell biology, and molecular biology.

In contrast with the evolutionary paradigm of random assemblies of nucleic acids that constitute the genetic text we do not know any real-life languages or codes which emerged as a randomly derived mixture of characters. Every language is based on signs, whether they are signals or symbols. In humans and other animals they are transported auditively, visually, or tactilely. In nonhuman living beings they are transported by small molecules in crystallized, fluid, and gaseous form. Additionally these signs can be combined coherently with combinatorial rules (syntax). Signs are not generated and used by themselves, but in real-life languages by living beings. These sign-generating and sign-using agents live *in vivo* in continued changing interactions and environmental circumstances (interaction HSV-cells-carrageenans). This is the context (pragmatics) in which a living being is interwoven. This context determines the meaning (semantics) of the signs in messages that are used to communicate and to coordinate single as well as group behavior (changes in HSV-glycoproteins in contact with cell surface modificate innate response). Therefore, we may understand that the same sentence, or the same syntactic sequence order, of any language or code can have different, and in extreme cases, opposite meanings and therefore transport different messages. The important consequence of this fact is that it is not possible to extract the meaning of an information content solely out of the syntactic structure, but someone has to identify the context within which the living being uses this syntactic structure. The primary agents are not the sequences of signs, nor the rules which determine sequences, but the living agents. Without living agents there are no signs, no semiotic rules, no signalling, and no communication, no living agents could coordinate growth and development. If we assume the genetic code to function language-like, knowing that no language which has been observed functions by itself, then we have to postulate living agents that are competent to use signs coherent with syntactic, pragmatic, and semantic rules. Adapted to the genetic code, this means that there must be living agents competent in generation and integration of meaningful nucleotide sequences, and meaningful nucleotide sequences are not a randomly derived mixture of nucleotides. In accord with Dr. Witzany, this view could change the construction of research projects, that is, shifting the focus from mutational (random) changes of nucleotide sequences to investigating nucleotide sequences from the perspective of viral-derived sequences that now play important roles in the regulation of cellular functions, for example, HSV asymptomatic infection where the virus have reached an equilibrium status balance by the immune response of the infected host to achieve a latent lifestyle. Their status within one of many addiction modules (genetic and genomic innovations) together with the host immune system, each of them a unique culture-dependent habitat can be changed by nonbeneficial circumstance for the cell (e.g., stress) and they may become lytic again, resulting in a variety of diseases [59].

8. Conclusion

This work invites us to think about a possible alternative to attenuate virus for basic science study, therapeutical or prophylatic applications, employing natural compounds with chemical structures already "seen" by the pathogen and present in the host as essential cellular components widely distributed in nature. Besides, selective reexpression of viral ligands (using different type of polysaccharides) in conjunction with pathogenesis experiments will allow the testing of predictions about putative protective roles played by some glycans in certain tissues. Studying the comparative glycobiology of closely and distantly related species should also help, by ascertaining the rates of glycan diversification during evolution.This strategy could be considered as a natural evolutionary process where the virus contributes with valuable "updated" information gathered from previous ancestral infections and making it available for "new" actual hosts, generating a reciprocal benefit between host and virus.

Author details

Carlucci María Josefina*, Mateu Cecilia Gabriela,
Artuso María Carolina and Scolaro Luis Alberto
Departamento de Química Biológica- IQUIBYCEN (CONICET-UBA),
Laboratorio de Virología, Facultad de Ciencias Exactas y Naturales, Ciudad Universitaria,
Universidad de Buenos Aires, Argentina

Acknowledgement

Research in the authors´ laboratory was supported by Agencia Nacional de Promoción Científica y Tecnológica y Consejo Nacional de Investigaciones Científicas y Tecnológicas (CONICET).

9. References

[1] Van Valen (1974) Two modes of evolution. Nature, 252, 298-300.
[2] Villarreal L.P, Witzany G (2010) Viruses are essential agents within the roots and stem of the tree of life. J.Theor.Biol. 262, 698-710.
[3] Breitbart M, Rohwer F (2005) Here a virus, there a virus, everywhere the same virus? Trends Microbiol. 13, 278-84.
[4] Gray M.W, Lang B.F (1998) Transcription in chloroplasts and mitochondria: a tale of two polymerases. Trends Microbiol. 6, 1-3.
[5] [5] Forterre P (2006) The origin of viruses and their possible roles in major evolutionary transitions.Virus Res. 117, 5-16.
[6] Daubin V, Ochman H (2004) Start-up entities in the origin of new genes. Curr. Opin. Genet. Dev. 14 ,616-619.
[7] Hegde N.R, Maddur M.S, Kaveri SV, Bayry, J (2009) Reasons to include viruses in the tree of life. Nat. Rev. Microbiol. 7, 615.
[8] Ludmir E.B, Enquist L.W (2009) Viral genomes are part of the phylogenetic tree of life. Nat. Rev. Microbiol. 7, 615.
[9] Gahmberg C.G, Tolvanen M (1996) Why mammalian cell surface proteins are glycoproteins. Trends Biochem. Sci. 21,308-311.
[10] Varki A, Marth J (1995) Oligosaccharides in vertebrate development. Semin. Dev. Biol. 6, 127-138.
[11] Haldane J.B (1949) Disease and evolution. In Symposium sui fattori ecologi e genetici della specilazione negli animali, Supplemento a la Ricerca Scientifica Anno 19th, 68-75.
[12] Gagneux P, Varki A (1999) Evolutionary considerations in relating oligosaccharide diversity to biological function. Glycobiology 9, 747-755.
[13] Rasmussen B, Fletcher I.R, Brocks J.J, Kilburn M.R (2008) Reassessing the first appearence of eukaryotes and cyanobateria. Nature; 255:1101–4.

* Corresponding Author

[14] Esko J. D, Selleck S. B (2002) Order out of chaos: assembly of ligand binding sites in heparan sulfate. Annu Rev Biochem 71, 435-471.

[15] Stengel D. B, Connan S, Popper Z. A (2011) Algal chemodiversity and bioactivity: sources of natural variability and implications for commercial application. Biotechnol *Adv* 29, 483-501.

[16] Wijesekara I, Pangestuti R, Kim S-K (2011) Biological activities and potential health benefits of sulphated polysaccharides derived from marine algae. Carbohydr Polymers 84:14–21.

[17] Craigie J.S (1990) Cell walls. In: Cole KM, Sheath RM, editors. Biology of the red algae. Cambridge: Cambridge University Press; pp. 221–57.

[18] Demirbas A, Demirbas M.F(2011) Importance of algae oil as a source of biodiesel. Energy Convers Manage 52:163–70.

[19] Milledge J.J (2010) Commercial application of microalgae other than as biofuels: a brief review. Rev. Environ. Sci. Biotechnol. 10:31–41.

[20] Plaza M, Herrero M, Cifuentes A, Ibanez E (2009) Innovative natural functional ingredients from microalgae. J. Agric. Food Chem. 57:7159–70.

[21] Smit A.J (2004) Medicinal and pharmaceutical uses of seaweed natural products: a review. J. Appl. Phycol. 16:245–62.

[22] Pujol C.A, Carlucci M.J, Matulewicz M.C, Damonte E.B (2007) Natural sulfated polysaccharides for the prevention and control of viral infections. Spring-Verlag Top Heterocycl. Chem. 11: 259-281.

[23] Girond S, Crance J. M, Van Cuyck-Gandre H, Renaudet J, Deloince R (1991) Antiviral activity of carrageenan on hepatitis A virus replication in cell culture. *Res Virol* 142, 261-270.

[24] Buck C. B, Thompson C. D, Roberts J. N, Muller M, Lowy D. R, Schiller J. T (2006) Carrageenan is a potent inhibitor of papillomavirus infection. *PLoS Pathog* 2, e69.

[25] Nakashima H, Kido Y, Kobayashi N, Motoki Y, Neushul M, Yamamoto N (1987) Purification and characterization of an avian myeloblastosis and human immunodeficiency virus reverse transcriptase inhibitor, sulfated polysaccharides extracted from sea algae. Antimicrob. Agents Chemother. 31, 1524-1528.

[26] Bourgougnon N, Roussakis C, Kornprobst JM, Lahaye M (1994) Effects in vitro of sulfated polysaccharide from Schizymenia dubyi (Rhodophyta, Gigartinales) on a non-small-cell bronchopulmonary carcinoma line (NSCLC-N6). Cancer Lett. 85(1):87-92.

[27] Kolender A. A, Pujol C. A, Damonte E. B, Matulewicz M. C, Cerezo A. S (1997) The system of sulfated alpha-(1-->3)-linked D-mannans from the red seaweed Nothogenia fastigiata: structures, antiherpetic and anticoagulant properties. Carbohydr. Res. 304, 53-60.

[28] Damonte E, Neyts J, Pujol CA, Snoeck R, Andrei G, Ikeda S, Witvrouw M, Reymen D, Haines H, Matulewicz MC, et al (1994) Antiviral activity of a sulphated polysaccharide from the red seaweed Nothogenia fastigiata..Biochem Pharmacol. 47(12):2187-92.

[29] De Clercq E (2000) Current lead natural products for the chemotherapy of human immunodeficiency virus (HIV) infection. Med Res Rev. 20(5):323-49.

[30] Sekine H, Ohonuki N, Sadamasu K, Monma K, Kudoh Y, Nakamura H, Okada Y, Okuyama T (1995) The inhibitory effect of the crude extract from a seaweed of Dygenea simplex C. Agardh on the in vitro cytopathic activity of HIV-1 and it's antigen production. Chem Pharm Bull (Tokyo) 43, 1580-1584.
[31] Damonte E. B, Matulewicz M. C, Cerezo A. S, Coto C. E (1996) Herpes simplex virus-inhibitory sulfated xylogalactans from the red seaweed Nothogenia fastigiata. Chemotherapy 42, 57-64.
[32] Pujol CA, Errea MI, Matulewicz MC, Damonte EB (1996) Antiherpetic activity of S1, an algal derived sulphated galactan. Phytother Res. 10(5):410-413.
[33] Carlucci M. J, Pujol C. A, Ciancia M, Noseda M. D, Matulewicz M. C, Damonte E. B, Cerezo, A. S (1997) Antiherpetic and anticoagulant properties of carrageenans from the red seaweed Gigartina skottsbergii and their cyclized derivatives: correlation between structure and biological activity. Int J Biol Macromol 20, 97-105.
[34] Carlucci M. J, Ciancia M, Matulewicz M. C, Cerezo A. S, Damonte E. B (1999) Antiherpetic activity and mode of action of natural carrageenans of diverse structural types. Antiviral Res 43, 93-102.
[35] Carlucci MJ, Scolaro LA, Matulewicz MC, Damonte EB (1997) Antiviral activity of natural sulphated galactans on herpes virus multiplication in cell culture. Planta Med. 63 :429-432.
[36] Caceres P. J, Carlucci M. J, Damonte E. B, Matsuhiro B, Zuniga E. A (2000) Carrageenans from chilean samples of Stenogramme interrupta (Phyllophoraceae): structural analysis and biological activity. Phytochemistry 53, 81-86.
[37] Haslin C, Lahaye M, Pellegrini M, Chermann J. C (2001) In vitro anti-HIV activity of sulfated cell-wall polysaccharides from gametic, carposporic and tetrasporic stages of the Mediterranean red alga Asparagopsis armata. Planta Med 67, 301-305.
[38] Duarte M. E, Noseda D. G, Noseda M. D, Tulio S, Pujol C. A, Damonte E. B (2001) Inhibitory effect of sulfated galactans from the marine alga Bostrychia montagnei on herpes simplex virus replication in vitro. Phytomedicine 8, 53-58.
[39] Pujol C. A, Estevez J. M, Carlucci M. J, Ciancia M, Cerezo A. S, Damonte E. B (2002). Novel DL-galactan hybrids from the red seaweed Gymnogongrus torulosus are potent inhibitors of herpes simplex virus and dengue virus. Antivir Chem. Chemother. 13, 83-89.
[40] Mazumder S, Ghosal P. K, Pujol C. A, Carlucci M. J, Damonte E. B, Ray B (2002) Isolation, chemical investigation and antiviral activity of polysaccharides from Gracilaria corticata (Gracilariaceae, Rhodophyta). Int. J. Biol. Macromol. 31, 87-95.
[41] Venkateswaran P. S, Millman I, Blumberg B. S (1989) Interaction of fucoidan from Pelvetia fastigiata with surface antigens of hepatitis B and woodchuck hepatitis viruses. Planta Med. 55, 265-270.
[42] Beress A, Wassermann O, Tahhan S, Bruhn T, Beress L, Kraiselburd E. N Gonzalez L. V, de Motta G. E, Chavez P. I (1993) A new procedure for the isolation of anti-HIV compounds (polysaccharides and polyphenols) from the marine alga Fucus vesiculosus. J. Nat. Prod. 56, 478-488.

[43] Hoshino T, Hayashi T, Hayashi K, Hamada J, Lee J. B, Sankawa U (1998) An antivirally active sulfated polysaccharide from Sargassum horneri (TURNER) C. AGARDH. Biol. Pharm. Bull. 21, 730-734.

[44] Feldman S. C, Reynaldi S, Stortz C. A, Cerezo A. S, Damonte E. B (1999) Antiviral properties of fucoidan fractions from Leathesia difformis. Phytomedicine 6, 335-340.

[45] Ponce N. M, Pujol C. A, Damonte E. B, Flores M. L, Stortz, C. A (2003) Fucoidans from the brown seaweed Adenocystis utricularis: extraction methods, antiviral activity and structural studies. Carbohydr. Res. 338, 153-165.

[46] Hasui M, Matsuda M, Okutani K, Shigeta S (1995) In vitro antiviral activities of sulfated polysaccharides from a marine microalga (Cochlodinium polykrikoides) against human immunodeficiency virus and other enveloped viruses. Int J Biol Macromol. 17, 293-297.

[47] Huheihel M, Ishanu V, Tal J, Arad S. M (2002) Activity of Porphyridium sp. polysaccharide against herpes simplex viruses in vitro and in vivo. J Biochem Biophys. Methods 50, 189-200.

[48] Lee J. B, Hayashi K, Hayashi T, Sankawa U, Maeda M (1999) Antiviral activities against HSV-1, HCMV, and HIV-1 of rhamnan sulfate from Monostroma latissimum. Planta Med 65, 439-441.

[49] Bergh O, Borsheim K. Y, Bratbak G, Heldal M (1989) High abundance of viruse found in aquatic environments. *Nature* 340, 467-468.

[50] Carlucci M. J, Damonte E B, Scolaro L A (2011) Virus driven evolution: a probable explanation for "Similia Similibus Curantur" philosophy. *Infect Genet Evol* 11, 798-802.

[51] Knoll A. H (1992) The early evolution of eukaryotes: a geological perspective. Science 256, 622-627.

[52] Cassaro C. M, Dietrich C. P (1977) Distribution of sulfated mucopolysaccharides in invertebrates. *J Biol Chem* 252, 2254-2261.

[53] Dietrich C. P, Nader H. B, Straus A. H (1983) Structural differences of heparan sulfates according to the tissue and species of origin. *Biochem Biophys Res Commun* 111, 865-871.

[54] Ihrcke N. S, Parker W, Reissner K. J, Platt J. L (1998) Regulation of platelet heparanase during inflammation: role of pH and proteinases. J Cell Physiol 175, 255-267.

[55] Kainulainen V, Wang H, Schick C, Bernfield M (1998) Syndecans, heparan sulfate proteoglycans, maintain the proteolytic balance of acute wound fluids. J Biol Chem 273, 11563-11569.

[56] Oragui E. E, Nadel S, Kyd P, Levin M (2000) Increased excretion of urinary glycosaminoglycans in meningococcal septicemia and their relationship to proteinuria. Crit Care Med 28, 3002-3008.

[57] Damonte E. B, Matulewicz M. C, Cerezo A. S (2004) Sulfated seaweed polysaccharides as antiviral agents. *Curr Med Chem* 11, 2399-2419.

[58] Witzany G (2006) Natural Genome-Editing competences of viruses. Acta Biotheor. 54, 235-253 .

[59] Villarreal L.P (2009) Origin of Group Identity. Springer. New York,USA

[60] Ryan F.P (2006) Genomic creativity and natural selection: A modern synthesis. Biol. J. Linn. Soc. 88, 655-672

[61] Villarreal L.P (2005) The dilemma of the transition in Evolution: the Eukaryotes, in: Viruses and the Evolution of life. Washington. American Society for Microbiol. Press, pp.101-142 .
[62] Trowsdale J, Parham P (2004) Defense strategies and immunity-related genes. Eur J Immunol 34,7-17.
[63] Kohler H, Bayry J, Nicoletti A, Kaveri S.V (2003) Natural Autoantibodies as tools to predict the outcome of immune response? Scand. J. Immunol.58, 285-9
[64] Ochsenbein A. F, Fehr T, Lutz C, Suter M, Brombacher F, Hengartner H, Zinkernagel R. M (1999) Control of early viral and bacterial distribution and disease by natural antibodies. Science 286, 2156-2159.
[65] Maddur M.S, Vani J, Desmazes-Lacroix S, Kaveri S, Bayry J (2010) Autoimmunity as a predisposition for infectious diseases. Plos Pathog. 6,11 e1001077.
[66] van de Vosse E, van Dissel J, Ottenhoff T (2009) Genetic deficiencies of innate immune signalling in human infectious disease. Lancet Infect Dis 9, 688-698.
[67] Danilova N (2006) The evolution of immune mechanisms. J. Exp. Zool. (Mol. Dev. Evol). 306:,496-520.
[68] Bamshad M, Wooding S.P (2003) Signatures of natural selection in the human genome. Nat. Rev. Genet. 4,99-111 .
[69] Galvani A.P, Slatkin M (2003) Evaluating plague and smallpox as historical selective pressures for the CCR5-Delta 32 HIV-resistance allele. Proc Natl Acad Sci. USA 100, 15276-79.
[70] Bureau J-F, Le Goff S, Thomas D, Parlow A.F, de la Torre J.C, Homann D, Brahic M, Oldstone M.B.A (2001) Disruption of differentiated functions during viral infection in vivo. V. Mapping of a locus involved in susceptibility of mice to growth hormone deficiency due to persistent lymphocytic choriomeningitis virus infection. Virology 281, 61-66.
[71] Shadan F.F, Villarreal L.P (1993) Coevolution of persistently infecting small DNA viruses and their hosts linked to host-interactive regulatory domains. Proc. Natl. Acad. Sci. USA. 1,(90) 4117-21.
[72] Villarreal L.P (2009) The source of self: genetic parasites and the origin of adaptative immunity. Ann. NY Acad. Sci. 1178, 194-232.
[73] Sakaoka H, Kurita K, Iida Y, Takada S, Umene K, Kim Y.T, Ren C.S, Nahmias A.J (1994) Quantitative analysis of genomic polymorphism of herpes simplex virus type 1 strains from six countries. studies of molecular evolution and molecular epidemiology of the virus. J. Gen. Virol. 75, 513-527.
[74] Carlucci M.J, Scolaro L.A, Damonte E.B (2002) Herpes simplex virus type 1 variants arising after selection with antiviral carrageenan: lack of correlation between drug-susceptibility and syn phenotype. J. Med. Virol. 68, 82-91.
[75] Mateu C, Perez Recalde M, Artuso M, et al (2011) Emergence of HSV-1 syncytial variants with altered virulence for mice after selection with a natural carrageenan. J Sex. Transm. Dis. 38(6), 548-554
[76] Lipton B.H (2008) The Biology of belief. Hay House Publishers India

[77] Lamarck J.B (1809) Philosophie zoologique, on exposition des considerations relatives á l'histoire naturelle des animaux. Libraire, Paris.
[78] Lamarck J.B (1963) Zoological philosophy. Hafner Publishing Co, New York. USA.
[79] Mayr E (1976) Evolution and the diversity of life: selected essays. The Belknap Press of Harvard University Press, Cambridge, Mass.
[80] Ryan F (2002) Darwin's blind spot: Evolution beyond natural selection, Houghton Mifflin, NewYork, USA.
[81] Ruby E, Henderson B, et al (2003) We get by with a little help from our (little) friends. Science 303:1305-1307.
[82] Nitz N, Gomes C, et al (2004). Heritable integration of kDNA minicircle sequences from Tripanosoma cruzi into the Avian genome: Insights into human Chagas disease. Cell 118:175-186.
[83] Pennisi E (2004) Researchers trade insights about gene swapping. Science 305:334-335.
[84] Boucher Y, Douady C.J, et al (2003) Lateral gene transfer and the origins of prokaryotic groups. Annual Review of Genetics 37:283-328.
[85] Gogarten J.P (2003) Gene Transfer: Gene swapping craze reaches eukaroytes. Current Biology, 13:R53-R54.
[86] Pennisi E (2001) Sequences reveal borrowed genes. Science 294:1634-1635.
[87] Lenton T (1998) Gaia and natural selection. Nature 394:439-447.
[88] Witzany G (2000) Life: the communicative structure: a new philosophy of biology. Norderstedt: Libri Books on Demand.
[89] Witzany G (2009) Noncoding RNAs: persistent viral agents as modular tools for cellular needs. Ann. NY Acad. Sci. 1178:244-267.
[90] Claverie JM (2005) Fewer genes, more noncoding RNA. Science 309:1529-1530.
[91] Villarreal L.P (2005) Viruses and the Evolution of life. ASM Press.Washington, USA.

Chapter 21

1,3-β-Glucans: Drug Delivery and Pharmacology

Mohit S. Verma and Frank X. Gu

Additional information is available at the end of the chapter

http://dx.doi.org/10.5772/50363

1. Introduction

Natural polysaccharides are used in a variety of applications due to their unique properties. These applications range from paper manufacturing to wound healing [1]. One interesting class of polysaccharides comprises 1,3-β-glucans, which are glucopyranose polysaccharides with (1,3) glycosidic linkages and varying degree of (1,6) branches [2]. 1,3-β-glucans can form single or triple helical structures, which can be used to synthesize resilient gels by applying heat and humidity [3,4]. The properties of these gels are governed by the structure of the polysaccharide, which is determined by the degree of branching and the molecular weight. Thus, controlling the microscopic structure allows control over the macroscopic function. This is especially advantageous in the field of drug delivery because the encapsulation of different agents can be facilitated by the use of different polysaccharides. The properties of 1,3-β-glucans can also be modified by covalently attaching functional units to the polysaccharide backbone [5].

1,3-β-glucans are derived from microbial [6] and fungal [2] sources and hence have innate immunomodulatory properties. When these 1,3-β-glucans are a component of the foreign pathogens, they can act as recognition sites for macrophages to facilitate the elimination and removal of these pathogens [7]. When extracted 1,3-β-glucans are administered to animals or humans, they recruit macrophages and stimulate the immune system through a similar mechanism [8,9]. This result has been utilized for various pharmacological applications including cancer inhibition [10-17], infection resistance [18-21] and wound healing [22-24]. Current research is focusing on combining the structural properties of 1,3-β-glucans with the pharmacological ones to further enhance the efficacy of hybrid systems thus created.

2. Crystallinity of 1,3-β-glucans

2.1. Structure

1,3-β-glucans are semi-crystalline polysaccharides comprising a combination of single helices, triple helices and random coils. The crystallinity of these polysaccharides has been

INTECH
open science | open minds

© 2012 The Author(s). Licensee InTech. This chapter is distributed under the terms of the Creative Commons Attribution License http://creativecommons.org/licenses/by/3.0), which permits unrestricted use, distribution, and reproduction in any medium, provided the original work is properly cited.

studied using X-ray diffraction (XRD). In this study, curdlan, which is a linear 1,3-β-glucan, was used as a model polysaccharide [3]. Different forms and states of curdlan demonstrate different crystallinity. One example that was studied in detail is the annealed "dry" state, where the curdlan is dissolved in dimethyl sufoxide, extracted in methanol and annealed in the presence of water at 145 °C. Curdlan is then dried *in vacuo* to obtain the sample for XRD experiments. The results from XRD measurements conclude that six-fold triple-helices are formed with an advance of 2.935 Å per monomer unit. The model of this structure confirms that the three strands of triple helices are held together by hydrogen bonding between O(2) hydroxyls while the helices are brought together by O(4) and O(6) hydrogen bonding [3].

An alternate structure of curdlan helices is presented based on semi-empirical modeling. It is proposed that hydrogen bonding of the strands occurs along the helix axis rather than perpendicular to it. The different structures are illustrated in Figure 1. It is demonstrated that this alternate structure provides a more stable structure of curdlan and hence is likely to have higher population [25].

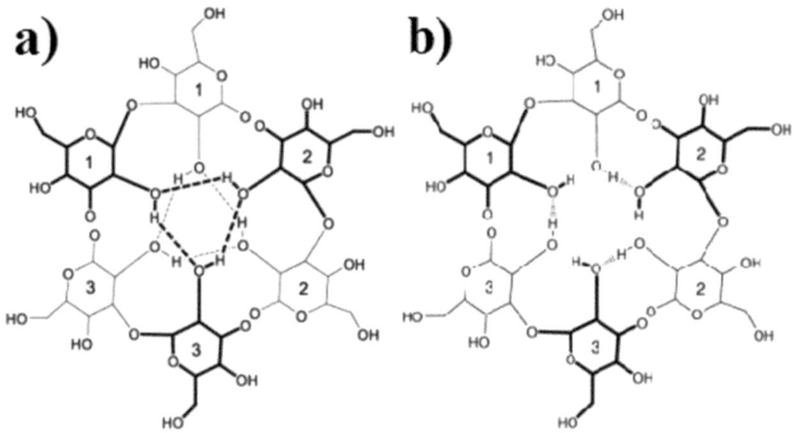

Figure 1. Illustration of possible orientations of curdlan triple helices: a) hydrogen bonding perpendicular to helix axis; b) hydrogen bonding along helix axis. Figure adapted from [25].

1,3-β-glucans have also been complexed with nucleotides to form crystalline structures. In the example of curdlan and poly(cytidylic acid) complex, semi-empirical modeling suggests that two glucose units of different curdlan chains form hydrogen bonds with one base of the nucleotide chain [26]. This property of curdlan complexing with nucleotides has been exploited in forming liquid crystalline gels with deoxyribonucleic acid (DNA). Such structures could be synthesized at varying scales ranging from nanometers to centimeters [27].

2.2. Liquid crystalline gels

Curdlan can be used to form liquid crystalline gels when it is exposed to transition metal salts [28,29]. The crystallinity of these gels depends on the molecular weight of the gels [30]. DNA has also been used to synthesize gel beads [31]. When used together, DNA and curdlan provide control over the size and morphology of the synthesized hybrid structures. Various structures can be obtained by modifying the concentration of curdlan and DNA [27].

Curdlan is insoluble in water but it dissolves in alkaline solutions. Thus, DNA and curdlan are mixed together in a basic solution and then this mixture is added to a solution of calcium chloride salt either directly or through a dialysis membrane. Direct addition leads to formation of structures at the nanometer and millimeter scales. Dialysis allows for the formation of centimeter sized gels. The macroscopic structures are assessed by using crossed nicols (Figure 2), while the nano- and micro-structures are characterized using transmission electron microscopy (TEM, Figure 4). When viewing the centimeter scale gels between two perpendicularly placed polarizers, orthogonal dark lines are observed on the gels. These lines are known as isogyres and indicate the anisotropy in liquid crystalline gels. It is observed that increasing the concentration of DNA decreases the crystallinity of the gel as the isogyres become less defined. This is illustrated in Figure 2 [27].

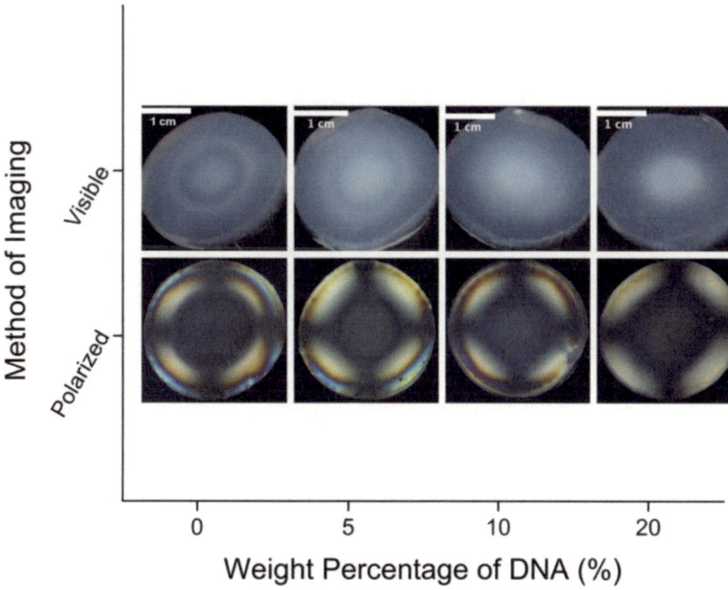

Figure 2. Liquid crystalline gels of curdlan and DNA: 100% curdlan, 5% DNA, 10% DNA and 20 % DNA as seen under visible light (top) and crossed nicols (bottom). Scale bars are 1 cm each. Figure obtained from [27].

A similar phenomenon was observed at the milimeter scale when the structure was observed under the microscope. It was seen that although DNA provided rigidity and well-defined shape to the structure, it reduced the crystallinity. This is likely because DNA forms a less crystalline structure compared to curdlan. It is possible that DNA might not be forming helices with curdlan, but instead forming a gel with microphase separation. The results from millimetre scale are highlighted in Figure 3. The opacity and lack of isogyres in DNA sample implies low crystallinity. Thus, a simple method is presented to determine degree of crystallinity of gels qualitatively [27].

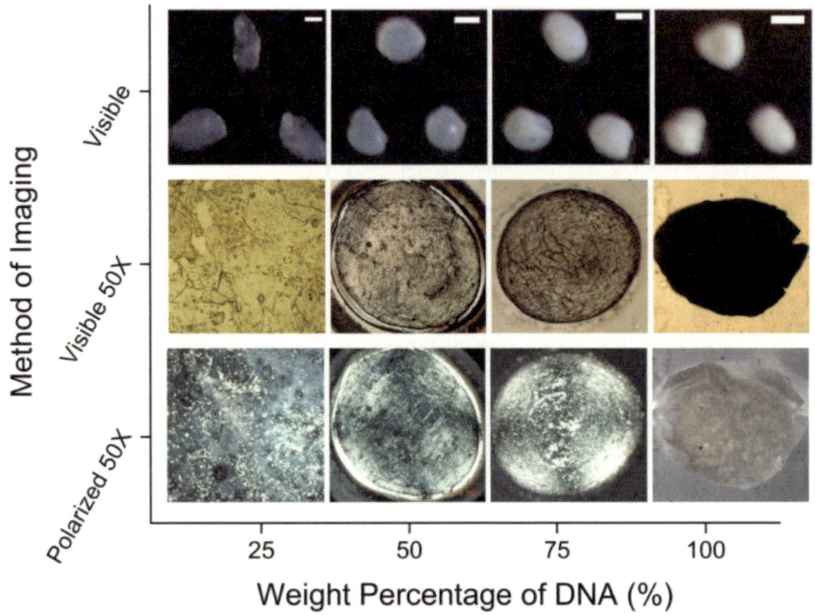

Figure 3. Spherical gels of curdlan and DNA observed under visible light (top and middle) and polarized light (bottom). Scale bars are 1 mm each. Figures obtained from [27].

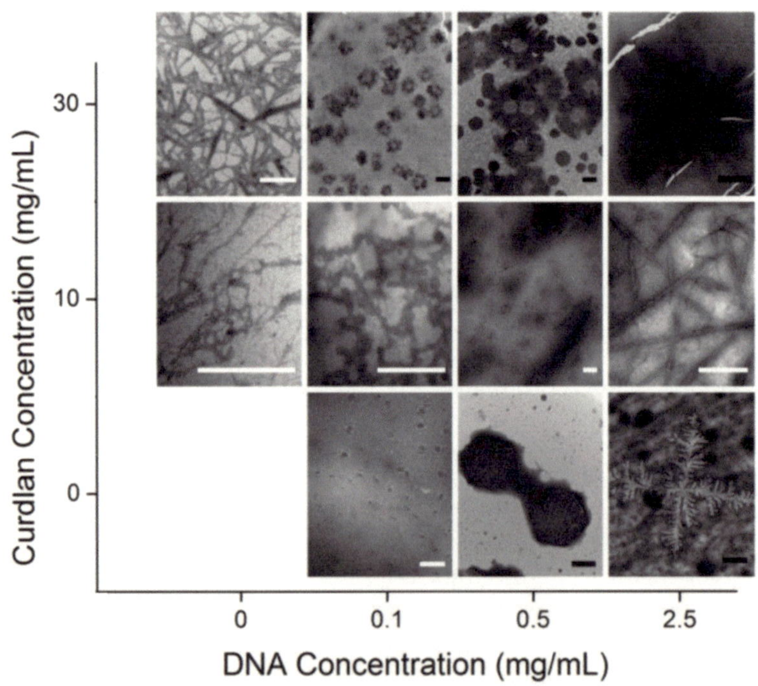

Figure 4. TEM images of micro- and nano-structures of curdlan and DNA. White scale bars are 500 nm and black scale bars are 2000 nm. Figure reproduced from [27].

At the micro- and nano-scale levels, the morphology of the gels could be changed between spheres and fibers by modulating the concentration of the DNA or curdlan. This is summarized in Figure 4. These hybrid liquid crystalline gel systems have the potential of creating advanced drug delivery vehicles where the crystalline regions of the system prevent degradation of the biomolecule and the amorphous regions maintain functionality of the encapsulated moiety. This hybrid system also serves as a tool for further studies of the molecular structure undertaken by 1,3-β-glucan and DNA in various microenvironments [27].

3. Applications in drug delivery

The ability to control the structure of 1,3-β-glucan based carriers has encouraged the application of these glucans in drug delivery. The glucans can be used as gels, nanoparticles, microparticles or complexes.

3.1. Encapsulation within gels

Curdlan is the commonly used 1,3-β-glucan for formation of gels. Curdlan gels can be prepared by heating the suspension of curdlan in aqueous solution and then cooling it down. If the suspension is heated to 60 °C, a low-set thermally reversible gel is formed whereas if the temperature is above 80 °C, a high-set thermally irreversible gel is formed. Drugs such as indomethacin, salbutamol sulfate and prednisolone have been encapsulated in curdlan gels. The gels are prepared by mixing curdlan in drug solutions at 5-10% curdlan concentration, adding the suspension to a glass test tube, heating the test tube in water bath at the desired temperature for 10 minutes and then unmolding the gel. The experiments conducted in [32] have demonstrated that a high-set gel can lower the rate of drug release. The curdlan based gels have been able to provide a sustained release of drugs for compared to commercially available formulations. These gels can be used as drug delivery suppositories for rectal administration, which bypasses hepatic first pass clearance [32].

Curdlan gels have also been used for developing protein delivery devices. Since proteins can denature at high temperatures, the temperature required for forming curdlan gels needs to be lowered. Curdlan can form aqueous gels in the presence of hydrogen bond disrupting agents such as dimethyl sulfoxide, urea and thiocyanates [33]. This property encourages the use of chaotropes for lowering the gelling temperature. It has been demonstrated that the presence of 8 M urea can decrease the gelling temperature of curdlan from 55 °C to 37 °C. This has been used for encapsulating bovine serum albumin (BSA) as a model protein. Although gels synthesized using urea are able to demonstrate sustained release of BSA over 100 hours, the toxicity of urea is a concern. Urea also has the possibility of disrupting the hydrogen bonds of BSA and hence denaturing the protein. Thus, an alternative method of reducing the gelling temperature has been developed by modifying the backbone of curdlan to form a hydroxyethyl derivative. This system was also able to form gels at 37 °C but the BSA release was sustained for a shorter time period of 75 hours [34].

3.2. Microparticles and nanoparticles

Nanoparticles are typically used in drug delivery applications because of high drug encapsulation efficiency [35], controlled drug release and incorporation of diagnostic agents [36,37]. One of the major challenges with drug delivery is achieving specificity with cellular uptake. One method of overcoming this hurdle is by attaching targeting ligands on the surface of nanoparticles to allow enhanced uptake by specific cells [38]. Another strategy is the encapsulation of these drug loaded nanoparticle in glucan based microspheres. Glucan microspheres are derived from the cell walls of *Sacchamoryces cerevisiae* (Baker's yeast) and are 2-4 μm in size. These are porous and hollow microparticles composed mainly of 1,3-β-D-glucan and small quantities of chitin [39]. The β-glucan on the microspheres serves as a specific target for uptake by immune cells such as macrophages and dendritic cells [40]. Glucan microspheres can be used for delivering various payloads such as proteins [19], DNA [40], siRNA [16,41] and small molecules [39]. Typically small molecules are not easily entrapped within these glucan microspheres and hence the use of nanoparticles is necessary. These nanoparticles can then be loaded in glucan microspheres for enhanced uptake. Recently, two types of nanoparticles have been encapsulated within glucan microspheres: fluorescent polystyrene nanoparticles and doxorubicin loaded mesoporous silica nanoparticles [39]. The polystyrene particles are encapsulated using capillary forces from the pores of microspheres, whereas the silica nanoparticles are loaded using electrostatic interactions. These particles are then used *in vitro* to demonstrate enhanced uptake by dectin-1 expressing fibroblast cells (NIH3T3-D1 cell line). It has been observed that doxorubicin loaded silica nanoparticles were more effective when encapsulated within glucan microspheres because of enhanced uptake. These results were consistent with fluorescently tagged polystyrene nanoparticle uptake as well [39].

Aside from being used as a targeting moiety, 1,3-β-glucans have also been used as structural units for encapsulating insoluble drugs. One example is the use of short chain curdlan with a molecular weight of 990 Da, derived from Vietnam medicinal mushroom *Hericium erinaceum*. This curdlan has been used to synthesize a nanoparticle formulation for encapsulating curcumin, which is a water-insoluble compound with promising anti-cancer activity. Curcumin is derived from the rhizomes of the herb *Curcuma longa* and has demonstrated cancer prevention and suppression through various signaling pathways [42]. The delivery of curcumin has been enhanced by encapsulating it within curdlan to form 50 nm nanoparticles. These nanoparticles are prepared by adding curcumin dissolved in ethanol to a solution of curdlan in water and stirring at room temperature. The solvents are then evaporated and nanoparticles are purified by centrifugation to remove excess curdlan and larger aggregates. These curcumin loaded curdlan nanoparticles have been able to inhibit tumor growth in Hep-G2 cell line *in vitro* [10].

Amphiphilic derivatives of 1,3-β-glucans have also been extensively studied for formation of micellar nanoparticles with drug encapsulating capabilities. One example of an amphiphilic system is based on cholesterol-carboxymethylcurdlan. The hydrophilicity of curdlan is increased by modifying the backbone with carboxymethyl groups. The loading of hydrophobic drugs is increased by inclusion of cholesterol moieties in the curdlan backbone [17]. This system uses a remote loading method for encapsulating epirubicin, where a pH gradient between the interior and exterior of the nanoparticle is utilized to achieve high ratio of drug to carrier [43]. Remote loading method is implemented by preparing blank cholesterol-carboxymethylcurdlan nanoparticles using probe sonication, resuspending dried nanoparticles in ammonium sulfate, performing buffer exchange in sodium chloride and finally adding the desired amount of epirubicin to the solution. This method is able to achieve up to 39.6% drug loading, which is remarkable for polymeric systems. This curdlan based delivery device has been able to enhance the cytotoxicity of epirubicin *in vitro* when assessed using human cervical carcinoma (HeLa) cell lines as compared to free epirubicin. The delivery vehicle also increases the circulation half-life *in vivo* by 4.31 times as compared to free epirubicin when tested in Wistar rats. These results provide promising opportunities for utilizing a curdlan based drug delivery vehicle for enhanced efficacy of encapsulated drugs.

Alternatively, curdlan can also be used as the hydrophobic component in an amphiphilic drug delivery vehicle. To achieve this a graft copolymer of curdlan and poly(ethylene glycol) (PEG) has been synthesized [5]. PEG is a commonly used polymer for enhanced biocompatibility [44-48]. Doxorubicin can be incorporated in graft copolymer nanoparticle by nanoprecipitation method. In this technique, the copolymer and drug are dissolved in a common water miscible solvent and then added to magnetically stirred water in a drop-wise manner. The mixing causes self-assembly of polymers where hydrophobic components are at the core of the nanoparticle and hydrophilic components are at the surface. In the case of doxorubicin and curdlan-graft-PEG, dimethyl sulfoxide is used as the common solvent [5]. A schematic of the nanoparticle is presented in Figure 5.

The structure of curdlan-graft-PEG nanoparticles has been confirmed using TEM. The samples were stained using phosphotungstic acid, which acts as a negative stain. This makes the background appear dark and the samples of interest appear bright. Additionally, since phosophotungstic acid is hydrophilic, hydrophobic components will be excluded from the acid and hence appear brighter than hydrophilic components. As observed in Figure 6, the nanoparticles are about 109 nm in size. Doxorubicin is visible as the bright center of the nanoparticles. Figure 6 (B) shows the three distinct layers of the nanoparticle, which correspond well with the schematic presented in Figure 5 [5].

The doxorubicin formulation with curdlan-graft-PEG was tested *in vitro* to determine its drug release profile and it demonstrated sustained release over 24 hours following a Fickian diffusion release model. Since PEG graft is supposed to improve the biocompatibility of the nanoparticles, a hemolysis assay was implemented to assess this property. Sheep red blood cells (RBCs) were used for this assay and nanoparticles of interest were incubated with the

RBCs for one hour at 37 °C. The amount of lysis was compared to a negative control of veronal buffered saline and positive control of deionized water. Curdlan-graft-PEG nanoparticles showed hemolysis below 5% at clinically relevant concentrations [5], which is considered biocompatible [35].

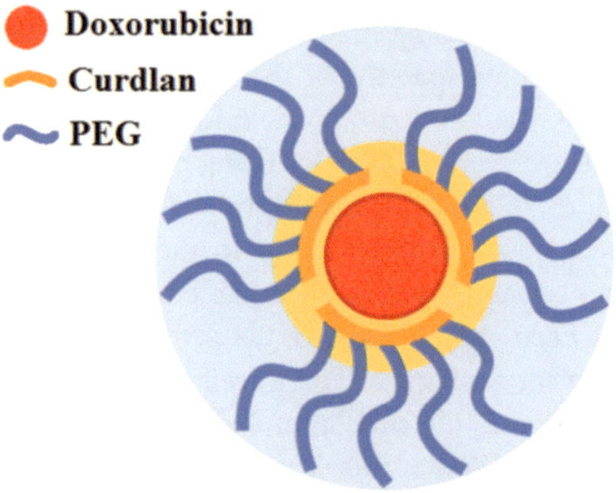

Figure 5. Schematic of nanoparticle synthesized using curdlan-graft-poly(ethylene glycol) and doxorubicin. Doxorubicin is at the core, surrounded by curdlan and PEG forms the shell. Figure adapted from [5].

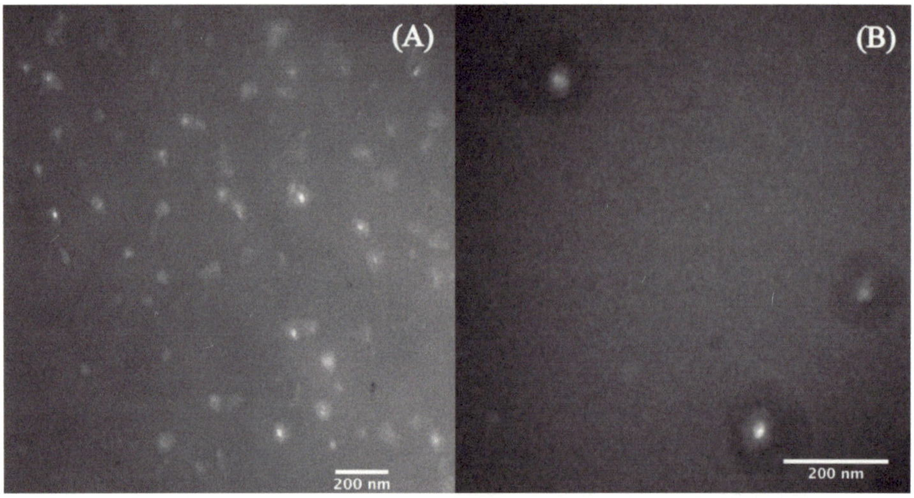

Figure 6. TEM images of doxorubicin encapsulated within curdlan-graft-PEG nanoparticles. (A) Bright center indicates doxorubicin and surroundings are darker because of higher electron density; (B) Three layers of the nanoparticle can be seen: bright doxorubicin core, dense curdlan layer in the middle and sparse PEG layer on the shell. Scale bars are 200 nm. Figure adapted from [5].

Therefore, nanoparticles and microparticles based on 1,3-β-glucans provide promising opportunities for drug delivery as structural units and as targeting ligands. The combination of these properties can be exploited for developing a drug delivery vehicle with enhanced potency.

3.3. Glucan complexes with polynucleotides

Polynucleotides have found several applications in drug delivery as active agents due to their therapeutic effects [16,41,49]. One of the challenges faced by polynucleotide delivery is their rapid degradation *in vivo*. Encapsulation of polynucleotides thus becomes necessary for maintaining their function. 1,3-β-glucans are a preferred choice for forming these complexes because of their helix forming capabilities. In the past, several polysaccharides have been tested for their ability to form complexes with the polynucleotide poly(C) and the complexation was assessed using circular dichorism (CD) where a change in the spectrum indicates complex formation. Only schizophyllan (1,3-β-glucan with one 1,6 branch every three units) and lentinan (1,3-β-glucan with two 1,6 branch every five units) have demonstrated changes in CD spectra, whereas curdlan, amylose (1,4-α-glucan), dextran (1,6-α-glucan) and pullulan (1,4-α-1,6-α-glucan) did not show any changes. This implies that only soluble 1,3-β-glucans are able to form complexes with poly(C). Commercially available curdlan is unable to form a complex because of its low solubility in water, which leads to precipitation [50]. The solubility of curdlan can be improved by reducing the molecular weight of the polymer. At lower molecular weight, curdlan is able to form complexes with poly(C) but these complexes are not as stable as the ones formed with schizophyllan [51]. Complexes have also been formed with poly(A) using schizophyllan but poly(U) showed no complexation, which suggests that polysaccharides can bind to polynucleotides in a specific manner [52]. Curdlan backbone has also been modified with carbohydrate molecules using click chemistry to improve the solubility and hence induce complex formation with poly(C) [53].

Hitherto, only homo-sequence polynucleotides have been discussed. Often therapeutic polynucleotides are composed of heterogeneous base pairs. Drug delivery vehicles for hetero-sequence oligonucleotides have also been developed by synthesizing cationic curdlan chains. CpG DNA has demonstrated immune stimulating effects but it needs to be preserved from degradation [54]. Cationic curdlan, synthesized using click chemistry, has been able to form complexes with CpG DNA and increase the cellular uptake in macrophage-like cell line J774.A1. The complex also induces an increase in cytokine (IL-12) secretion, which suggests activation of the macrophage cells [54]. Another strategy of binding hetero-sequence oligonucleotides is by modification of one terminal to attach a homo-sequence such as poly(A). A schizophyllan derivative has been utilized for binding modified antisense oligonucleotides. Schizophyllan has been modified with galactose and PEG units to enhance cellular uptake. It has been observed that the antisense effect was maximized with the use of schizophyllan derivative when the complex is administered to hepatoblastoma HepG2 and melanoma A375 cell lines [55]. Thus, 1,3-β-glucan based drug delivery devices serve as biocompatible carriers for a variety of nucleotides.

4. Applications in immunotherapy

1,3-β-glucans have been known to generate an immune response including stimulation of cytokine production, oxidative burst, increased phagocyte and lymphocyte proliferation as well as phagocytosis of opsonized tissues. Various mechanisms are responsible for this activation and this response has been utilized for varied applications including cancer resistance, disease immunity and wound healing [56].

4.1. Biological pathways of activation

The complement system is responsible for innate immunity and can be activated by either classical, alternative or lectin pathways. Using lentinan, pachyman and pachymaran polysaccharides, it has been demonstrated that 1,3-β-glucans use the alternative pathway of complement activation for generating an immune response since they show an increased consumption of C3 and C5 proteins from the complement cascade [57]. Additionally, opsonization is an important component of innate immune response. Typically complement protein C3b gets coated on pathogens and is later detected by complement receptor 1 or deactivated to form iC3b for regulation. In the presence of 1,3-β-glucans, iC3b can be detected by complement receptor 3 (CR3) or dectin-1 and this mechanism can be exploited for attacking cancer cells coated with iC3b [8]. Although the effect of CR3 has been negligible in immune activation in murine models [9,58], functional CR3 is essential for phagocytosis in human neutrophils [59].

Most of the existing studies have focused on macrophages and neutrophils but some 1,3-β-glucans such as lentinan are responsible for stimulation of T-cells and natural killer cells and hence affect the acquired immune response [60,61]. Curdlan has also been used to demonstrate increased proliferation of lymphocytes, which can in turn enhance the immune response [62]. Oxidative burst is yet another immunomodulatory effect that has been demonstrated by 1,3-β-glucans. While curdlan has shown the induction of inducible nitric oxide synthase in rat macrophages [63], other 1,3-β-glucans have shown an increase in the production of reactive oxygen species [62]. Some studies have demonstrated that immune activation is only possible by linear 1,3-β-glucans [64] while others emphasize that complex branching is important for most effective stimulation of immune response [65]. Although, several aspects of the biological pathways of activation have been discovered, further studies are necessary to gain a better understanding of the intricate interactions between 1,3-β-glucans and the immune system.

4.2. Tumor suppression

The use of 1,3-β-glucans in cancer therapy has been present in Japan since 1986, where they have been used for gastric, lung and cervical cancers [66]. Lentinan and pachymaran have demonstrated high tumor inhibition ratios of 99.6% and 96%, when tested against subcutaneous implantation of sarcoma 180 in mice [64]. It is hypothesized that the effect of lentinan and pachymaran is highly dependent on the activation of T-lymphocytes because the removal of the thymus from mice caused a suppression of antitumor effects from the 1,3-β-glucans [60]. Additionally, it is also speculated that deactivation of protein helices might be important for antitumor effects because a study has demonstrated that only polysaccharides that deactivated bovine serum albumin showed antitumor activity [67]. Besides lentinan and pachymaran, other 1,3-β-glucans have also exemplified tumor suppression. Some of the prominent examples are scleroglucan with an inhibition ratio of 90.4%, curdlan with inhibition of 99-100% [68], grifolan with inhibition of 97.9% [69] and 1,3-β-glucan from *Agaricus blazei* with inhibition of 99.3% [65]. These results have been encouraging and hence 1,3-β-glucans have been used in combinatorial therapies with antibodies in implanted human tumor xenografts from melanoma, epidermoid carcinoma, breast carcinoma, metastatic lymphoma and daudi lymphoma. The mice had higher survival rates in the presence of 1,3-β-glucans as compared to treatment with antibodies alone [12].

Subsequently, water soluble 1,3-β-glucans have been derived to improve the usability of these polysaccharides. Some examples of these polysaccharides include carboxymethylpachymaran with tumor inhibition ratio of 99.6%, hydroxymethylpachymaran and hydroxypropylpachymaran with up to 100% tumor inhibition when assessed against solid sarcoma 180 at a dose of 5 mg/kg [70]. Several derivatives of curdlan including carboxymethyl, glucosyl, sulfoethyl and sulfopropyl attachments to the backbone have also retained antitumor activity [71,72]. These examples demonstrate the versatility of 1,3-β-glucans in cancer inhibition and thus these polysaccharides can be modified to suit the desired application.

4.3. Infection prevention

Most common infections are caused by bacteria and fungi and since 1,3-β-glucans can induce inflammatory response, these glucans can be used for providing infection resistance. It has been shown that when administering 1,3-β-glucan to mice, their survival against *Staphylococcus aureus* infection increased from 70% to 97% [73]. Other 1,3-β-glucans such as glucan phosphate, laminarin and scleroglucan have been studied in detail for assessing their pharmacokinetic profile following oral administration in rats. It has been observed that these glucans are able to translocate from the gastrointestinal tract to systemic circulation. The glucans were able to increase secretion of interleukins, increase expression of dectin-1 on macrophages and increase expression of toll-like receptor 2 on dendritic cells. Thus, these effects increase the long-term survival of rats from 0% to 40% when challenged with *Candida albicans* fungal infection and from 0% to 50% when challenged with *Staphyloccous aureus* bacterial infection [74].

When considering larger mammals, infection resistance in pigs, dogs and horses has also found applications of 1,3-β-glucans. When piglets were fed with β-glucans after weaning, they demonstrated lower infection from enterotoxigenic *Escherichia coli*, which was highlighted by decreased diarrhoea and decreased content of inoculated *Escherichia coli* in the faeces as compared to control groups. These results present a significant advancement in veterinary medicine for pigs because the immunity of pigs usually suffers severely right after weaning and the current vaccines against *Escherichia coli* take a long period to be effective [20]. In the case of horses, the administration of 1,3-β-glucan to pregnant mares has been able to increase the cellular immune response in foals. This becomes very useful for preventing premature deaths of neonates [75]. While most studies have focused on the effects of 1,3-β-glucans on improving innate immunity, when these polysaccharides are administered to dogs, they showed enhancement in humoral immunity as indicated by changes in serum IgM and IgA levels [18].

1,3-β-glucans have also been used in aquatic animals as exemplified by the use of schizophyllan with 60-80% survival rate and lentinan and scleroglucan with 55-75% survival rate when tested against *Edwardsiella tarda* bacteria attack on *Cyrinus carpio L.* carp. The survival rate was also increased when infected with *Aeromonas hydrophila* with survival of 60% with schizophyllan, 70% with lentinan and 80% with scleroglucan administration. In the absence of glucans, the carp underwent complete mortality upon any infection [76]. These results have been repeatable when tilapia and grass carp is exposed to *Aeromonas hydrophila* [77]. As more studies are conducted on the use of 1,3-β-glucans, the quality of veterinary health care can be improved further and these results can eventually be transferred to human applications.

In addition to bacterial and fungal resistance, 1,3-β-glucans have also demonstrated promising results against malaria [78], herpes simplex virus [79] and human immunodeficiency virus (HIV) [80-82]. Expanding on HIV research, various complexes have been synthesized with curdlan sulfate in order to enhance the efficacy of the polysaccharide. Some prominent examples include covalent conjugation of azidothymidine to curdlan sulfate for drug delivery to the lymph nodes and bone marrow [83] and conjugation of fullerene C_{60} with curdlan for combining their anti-HIV effects [84].

4.4. Wound healing

1,3-β-glucans can have an impact on wound healing by recruiting macrophages to the wound site [85] and by increasing collagen deposition [24]. Beta glucan collagen matrix wound dressings have been used in children suffering from partial thickness burns and the dressings were able to simplify wound care by reducing analgesic requirements, improving cosmetic results and eliminating the need for repetitive dressing changes [86]. Other composites of β-glucan have been created with poly(vinyl alcohol) [23] and chitosan [22]. Poly(vinyl alcohol)/β-glucan composite was able to speed up the wound healing process when tested using rat models and hence decreased the healing time by 48% as compared to cotton gauze [23]. When using a composite of β-glucan and chitosan, a transparent dressing was obtained, which showed better results compared to commercially available chitosan based Beschitin® W. The synthesized chitosan composite did not dissolve during application period and was easy to remove because it did not adhere to wounds [22].

5. Conclusion

An assortment of 1,3-β-glucans have been explored for their structural and pharmacological capabilities. The ability of 1,3-β-glucans to form helical structures and gels has been advantageous for forming complexes with small molecules and macromolecules. The immunomodulatory effects of 1,3-β-glucans have served to fight cancer and infections and to promote wound healing. Research is moving towards combining the ability of 1,3-β-glucans to encapsulate bioactive agents with their own bioactivity for creating potent therapeutic devices against current challenges.

Author details

Mohit S. Verma and Frank X. Gu*

Chemical Engineering, University of Waterloo, Waterloo, ON, Canada

Waterloo Institute for Nanotechnology, University of Waterloo, Waterloo, ON, Canada

Acknowledgement

This work was financially supported by Natural Sciences and Engineering Research Council of Canada (NSERC) and 20/20 NSERC – Ophthalmic Materials Network. Mohit S. Verma is also financially supported by NSERC Vanier Canada Graduate Scholarship. We would also like to acknowledge Benjamin C. Lehtovaara for his contributions to the outline of this book chapter.

6. References

[1] Czaja W, Krystynowicz A, Bielecki S, Brown RM. Microbial cellulose - the natural power to heal wounds. Biomaterials 2006;27(2):145-151.

[2] Wasser SP. Medicinal mushrooms as a source of antitumor and immunomodulating polysaccharides. Applied Microbiology and Biotechnology 2002;60(3):258-274.

[3] Deslandes Y, Marchessault RH, Sarko A. Packing Analysis of Carbohydrates and Polysaccharides .13. Triple-Helical Structure of (1-]3)-Beta-D-Glucan. Macromolecules 1980;13(6):1466-1471.

[4] McIntire TM, Brant DA. Observations of the (1 -> 3)-beta-D-glucan linear triple helix to macrocycle interconversion using noncontact atomic force microscopy. Journal of the American Chemical Society 1998;120(28):6909-6919.

[5] Lehtovaara BC, Verma MS, Gu FX. Synthesis of curdlan-graft-poly(ethylene glycol) and formulation of doxorubicin-loaded core-shell nanoparticles. Journal of Bioactive and Compatible Polymers 2012;27(1):3-17.

[6] Sutherland IW. Microbial polysaccharides from Gram-negative bacteria. International Dairy Journal 2001;11(9):663-674.

[7] Steele C, Rapaka RR, Metz A, Pop SM, Williams DL, Gordon S, et al. The beta-glucan receptor dectin-1 recognizes specific morphologies of Aspergillus fumigatus. Plos Pathogens 2005;1(4):323-334.

[8] Vetvicka V, Thornton BP, Ross GD. Soluble beta glucan polysaccharide binding to the lectin site of neutrophil or natural killer cell complement receptor type 3 (CD11b/CD18) generates a primed state of the receptor capable of mediating cytotoxicity of iC3b-opsonized target cells. Journal of Clinical Investigation 1996;98(1):50-61.

[9] Brown GD, Herre J, Williams DL, Willment JA, Marshall ASJ, Gordon S. Dectin-1 mediates the biological effects of beta-glucans. Journal of Experimental Medicine 2003;197(9):1119-1124.

* Corresponding Author

[10] Le Mai Huong, Ha Phuong Thu, Nguyen Thi Bich Thuy, Tran Thi Hong Ha, Ha Thi Minh Thi, Mai Thu Trang, et al. Preparation and Antitumor-promoting Activity of Curcumin Encapsulated by 1,3-beta-Glucan Isolated from Vietnam Medicinal Mushroom Hericium erinaceum. Chemistry Letters 2011;40(8):846-848.

[11] Liu J, Gunn L, Hansen R, Yan J. Combined yeast-derived beta-glucan with anti-tumor monoclonal antibody for cancer immunotherapy. Experimental and molecular pathology 2009;86(3):208-214.

[12] Cheung NKV, Modak S, Vickers A, Knuckles B. Orally administered beta-glucans enhance anti-tumor effects of monoclonal antibodies. Cancer Immunology Immunotherapy 2002;51(10):557-564.

[13] Na K, Park KH, Kim SW, Bae YH. Self-assembled hydrogel nanoparticles from curdlan derivatives: characterization, anti-cancer drug release and interaction with a hepatoma cell line (HepG2). Journal of Controlled Release 2000;69(2):225-236.

[14] Ross GD, Vetvicka V, Yan J, Xia Y, Vetvickova J. Therapeutic intervention with complement and beta-glucan in cancer. Immunopharmacology 1999;42(1-3):61-74.

[15] Sasaki T, Abiko N, Nitta K, Takasuka N, Sugino Y. Anti-Tumor Activity of Carboxymethylglucans obtained by Carboxymethylation of (1-]3)-Beta-D-Glucan from Alcaligenes-Faecalis Var Myxogenes Ifo-13140. European Journal of Cancer 1979;15(2):211-215.

[16] Tesz GJ, Aouadi M, Prot M, Nicoloro SM, Boutet E, Amano SU, et al. Glucan particles for selective delivery of siRNA to phagocytic cells in mice. Biochemical Journal 2011;436:351-362.

[17] Li L, Gao F, Tang H, Bai Y, Li R, Li X, et al. Self-assembled nanoparticles of cholesterol-conjugated carboxymethyl curdlan as a novel carrier of epirubicin. Nanotechnology 2010;21(26):265601-265601.

[18] Stuyven E, Verdonck F, Van Hoek I, Daminet S, Duchateau L, Remon JP, et al. Oral Administration of beta-1,3/1,6-Glucan to Dogs Temporally Changes Total and Antigen-Specific IgA and IgM. Clinical and Vaccine Immunology 2010;17(2):281-285.

[19] Huang H, Ostroff GR, Lee CK, Wang JP, Specht CA, Levitz SM. Distinct Patterns of Dendritic Cell Cytokine Release Stimulated by Fungal beta-Glucans and Toll-Like Receptor Agonists. Infection and Immunity 2009;77(5):1774-1781.

[20] Stuyven E, Cox E, Vancaeneghem S, Arnouts S, Deprez P, Goddeeris BM. Effect of beta-glucans on an ETEC infection in piglets. Veterinary Immunology and Immunopathology 2009;128(1-3):60-66.

[21] Murphy EA, Davis JM, Brown AS, Carmichael MD, Carson JA, Van Rooijen N, et al. Benefits of oat beta-glucan on respiratory infection following exercise stress: role of lung macrophages. American Journal of Physiology-Regulatory Integrative and Comparative Physiology 2008;294(5):R1593-R1599.

[22] Kofuji K, Huang Y, Tsubaki K, Kokido F, Nishikawa K, Isobe T, et al. Preparation and evaluation of a novel wound dressing sheet comprised of beta-glucan-chitosan complex. Reactive & Functional Polymers 2010;70(10):784-789.

[23] Huang M, Yang M. Evaluation of glucan/poly(vinyl alcohol) blend wound dressing using rat models. International Journal of Pharmaceutics 2008;346(1-2):38-46.

[24] Portera CA, Love EJ, Memore L, Zhang LY, Muller A, Browder W, et al. Effect of macrophage stimulation on collagen biosynthesis in the healing wound. American Surgeon 1997;63(2):125-130.
[25] Miyoshi K, Uezu K, Sakurai K, Shinkai S. Proposal of a new hydrogen-bonding form to maintain curdlan triple helix. Chemistry & Biodiversity 2004;1(6):916-924.
[26] Miyoshi K, Uezu K, Sakurai K, Shinkai S. Polysaccharide-polynucleotide complexes. Part 32. Structural analysis of the Curdlan/poly(cytidylic acid) complex with semiempirical molecular orbital calculations. Biomacromolecules 2005;6(3):1540-1546.
[27] Lehtovaara BC, Verma MS, Gu FX. Multi-phase ionotropic liquid crystalline gels with controlled architecture by self-assembly of biopolymers. Carbohydrate Polymers 2012;87(2):1881-1885.
[28] Dobashi T, Nobe M, Yoshihara H, Yamamoto T, Konno A. Liquid crystalline gel with refractive index gradient of curdlan. Langmuir 2004;20(16):6530-6534.
[29] Dobashi T, Yoshihara H, Nobe M, Koike M, Yamamoto T, Konno A. Liquid crystalline gel beads of curdlan. Langmuir 2005;21(1):2-4.
[30] Nobe M, Kuroda N, Dobashi T, Yamamoto T, Konno A, Nakata M. Molecular weight effect on liquid crystalline gel formation of curdlan. Biomacromolecules 2005;6(6):3373-3379.
[31] Moran MC, Miguel MG, Lindman B. DNA gel particles: Particle preparation and release characteristics. Langmuir 2007;23(12):6478-6481.
[32] Kanke M, Tanabe E, Katayama H, Koda Y, Yoshitomi H. Application of Curdlan to Controlled Drug-Delivery .3. Drug-Release from Sustained-Release Suppositories In-Vitro. Biological & Pharmaceutical Bulletin 1995;18(8):1154-1158.
[33] Renn DW. Purified curdlan and its hydroxyalkyl derivatives: preparation, properties and applications. Carbohydrate Polymers 1997;33(4):219-225.
[34] Kim BS, Jung ID, Kim JS, Lee J, Lee IY, Lee KB. Curdlan gels as protein drug delivery vehicles. Biotechnology Letters 2000;22(14):1127-1130.
[35] Verma MS, Liu S, Chen YY, Meerasa A, Gu FX. Size-tunable nanoparticles composed of dextran-b-poly(D,L-lactide) for drug delivery applications. Nano Research 2012;5(1):49-61.
[36] Sun D. Nanotheranostics: Integration of Imaging and Targeted Drug Delivery. Molecular Pharmaceutics 2010;7(6):1879-1879.
[37] Rosen JE, Yoffe S, Meerasa A, Verma MS, Gu FX. Nanotechnology and Diagnostic Imaging: New Advances in Contrast Agent Technology. Journal of Nanomedicine & Nanotechnology 2011:1000115.
[38] Cho KJ, Wang X, Nie SM, Chen Z, Shin DM. Therapeutic nanoparticles for drug delivery in cancer. Clinical Cancer Research 2008;14(5):1310-1316.
[39] Soto ER, Caras AC, Kut LC, Castle MK, Ostroff GR. Glucan particles for macrophage targeted delivery of nanoparticles. Journal of Drug Delivery 2012;2012:143524.
[40] Soto ER, Ostroff GR. Characterization of multilayered nanoparticles encapsulated in yeast cell wall particles for DNA delivery. Bioconjugate chemistry 2008;19(4):840-848.
[41] Aouadi M, Tesz GJ, Nicoloro SM, Wang M, Chouinard M, Soto E, et al. Orally delivered siRNA targeting macrophage Map4k4 suppresses systemic inflammation. Nature 2009;458(7242):1180-1184.

[42] Sa G, Das T. Anti cancer effects of curcumin: cycle of life and death. Cell Division 2008;3:14.
[43] Lewrick F, Suss R. Remote loading of anthracyclines into liposomes. Methods in Molecular Biology 2010;605:139-45.
[44] Dhar S, Gu FX, Langer R, Farokhzad OC, Lippard SJ. Targeted delivery of cisplatin to prostate cancer cells by aptamer functionalized Pt(IV) prodrug-PLGA-PEG nanoparticles. Proceedings of the National Academy of Sciences of the United States of America 2008;105(45):17356-17361.
[45] Esmaeili F, Ghahremani MH, Ostad SN, Atyabi F, Seyedabadi M, Malekshahi MR, et al. Folate-receptor-targeted delivery of docetaxel nanoparticles prepared by PLGA-PEG-folate conjugate. Journal of Drug Targeting 2008;16(5):415-423.
[46] Cheng J, Teply BA, Sherifi I, Sung J, Luther G, Gu FX, et al. Formulation of functionalized PLGA-PEG nanoparticles for in vivo targeted drug delivery. Biomaterials 2007;28(5):869-876.
[47] Luu YK, Kim K, Hsiao BS, Chu B, Hadjiargyrou M. Development of a nanostructured DNA delivery scaffold via electrospinning of PLGA and PLA-PEG block copolymers. Journal of Controlled Release 2003;89(2):341-353.
[48] Avgoustakis K, Beletsi A, Panagi Z, Klepetsanis P, Karydas AG, Ithakissios DS. PLGA-mPEG nanoparticles of cisplatin: in vitro nanoparticle degradation, in vitro drug release and in vivo drug residence in blood properties. Journal of Controlled Release 2002;79(1-3):123-135.
[49] Davis HL, Weeratna R, Waldschmidt TJ, Tygrett L, Schorr J, Krieg AM. CpG DNA is a potent enhancer of specific immunity in mice immunized with recombinant hepatitis B surface antigen. Journal of Immunology 1998;160(2):870-876.
[50] Kimura T, Koumoto K, Sakurai K, Shinkai S. Polysaccharide-polynucleotide complexes (III): A novel interaction between the beta-1,3-glucan family and the single-stranded RNA poly(C). Chemistry Letters 2000(11):1242-1243.
[51] Koumoto K, Kimura T, Kobayashi H, Sakurai K, Shinkai S. Chemical modification of curdlan to induce an interaction with poly(C)(1). Chemistry Letters 2001(9):908-909.
[52] Sakurai K, Shinkai S. Molecular recognition of adenine, cytosine, and uracil in a single-stranded RNA by a natural polysaccharide: Schizophyllan. Journal of the American Chemical Society 2000;122(18):4520-4521.
[53] Hasegawa T, Numata M, Okumura S, Kimura T, Sakurai K, Shinkai S. Carbohydrate-appended curdlans as a new family of glycoclusters with binding properties both for a polynucleotide and lectins. Organic & Biomolecular Chemistry 2007;5(15):2404-2412.
[54] Krieg A. CpG motifs: the active ingredient in bacterial extracts? Nature medicine 2003;9(7):831-835.
[55] Karinaga R, Anada T, Minari J, Mizu M, Koumoto K, Fukuda J, et al. Galactose-PEG dual conjugation of beta-(1 -> 3)-D-glucan schizophyllan for antisense oligonucleotides delivery to enhance the cellular uptake. Biomaterials 2006;27(8):1626-1635.
[56] Lehtovaara BC, Gu FX. Pharmacological, structural, and drug delivery properties and applications of 1,3-beta-glucans. Journal of Agricultural and Food Chemistry 2011;59(13):6813-6828.

[57] Hamuro J, Hadding U, Bittersuermann D. Solid-Phase Activation of Alternative Pathway of Complement by Beta-1,3-Glucans and its Possible Role for Tumor Regressing Activity. Immunology 1978;34(4):695-705.
[58] Brown GD, Taylor PR, Reid DM, Willment JA, Williams DL, Martinez-Pomares L, et al. Dectin-1 is a major beta-glucan receptor on macrophages. Journal of Experimental Medicine 2002;196(3):407-412.
[59] van Bruggen R, Drewniak A, Jansen M, van Houdt M, Roos D, Chapel H, et al. Complement receptor 3, not Dectin-1, is the major receptor on human neutrophils for beta-glucan-bearing particles. Molecular Immunology 2009;47(2-3):575-581.
[60] Maeda YY, Chihara G. Lentinan, a New Immuno-Accelerator of Cell-Mediated Responses. Nature 1971;229(5287):634-634.
[61] Fujimoto T, Omote K, Mai M, Natsuumesakai S. Evaluation of Basic Procedures for Adoptive Immunotherapy for Gastric-Cancer. Biotherapy 1992;5(2):153-163.
[62] Sonck E, Stuyven E, Goddeeris B, Cox E. The effect of beta-glucans on porcine leukocytes. Veterinary Immunology and Immunopathology 2010;135(3-4):199-207.
[63] Ljungman AG, Leanderson P, Tagesson C. (1 -> 3)-beta-D-glucan stimulates nitric oxide generation and cytokine mRNA expression in macrophages. Environmental Toxicology and Pharmacology 1998;5(4):273-281.
[64] Chihara G, Hamuro J, Maeda Y, Arai Y, Fukuoka F. Antitumour Polysaccharide Derived Chemically from Natural Glucan (Pachyman). Nature 1970;225(5236):943-944.
[65] Ohno N, Furukawa M, Miura NN, Adachi Y, Motoi M, Yadomae T. Antitumor beta-glucan from the cultured fruit body of Agaricus blazei. Biological & Pharmaceutical Bulletin 2001;24(7):820-828.
[66] Bohn JA, BeMiller JN. (1->3)-beta-D-glucans as biological response modifiers: A review of structure-functional activity relationships. Carbohydrate Polymers 1995;28(1):3-14.
[67] Hamuro J, Chihara G. Effect of Antitumor Polysaccharides on Higher Structure of Serum-Protein. Nature 1973;245(5419):40-41.
[68] Sasaki T, Abiko N, Sugino Y, Nitta K. Dependence on Chain-Length of Anti-Tumor Activity of (1-]3)-Beta-D-Glucan from Alcaligenes-Faecalis Var Myxogenes, Ifo 13140, and its Acid-Degraded Products. Cancer Research 1978;38(2):379-383.
[69] Iino K, Ohno N, Suzuki I, Miyazaki T, Yadomae T. Structural Characterization of a Neutral Antitumour Beta-D-Glucan Extracted with Hot Sodium-Hydroxide from Cultured Fruit Bodies of Grifola-Frondosa. Carbohydrate Research 1985;141(1):111-119.
[70] Hamuro J, Yamashit.Y, Ohsaka Y, Maeda YY, Chihara G. Carboxymethylpachymaran, a New Water Soluble Polysaccharide with Marked Antitumour Activity. Nature 1971;233(5320):486-488.
[71] Demleitner S, Kraus J, Franz G. Synthesis and Antitumor-Activity of Derivatives of Curdlan and Lichenan Branched at C-6. Carbohydrate Research 1992;226(2):239-246.
[72] Demleitner S, Kraus J, Franz G. Synthesis and Antitumor-Activity of Sulfoalkyl Derivatives of Curdlan and Lichenan. Carbohydrate Research 1992;226(2):247-252.
[73] Diluzio NR, Williams DL. Protective Effect of Glucan Against Systemic Staphylococcus-Aureus Septicemia in Normal and Leukemic Mice. Infection and Immunity 1978;20(3):804-810.

[74] Rice PJ, Adams EL, Ozment-Skelton T, Gonzalez AJ, Goldman MP, Lockhart BE, et al. Oral delivery and gastrointestinal absorption of soluble glucans stimulate increased resistance to infectious challenge. Journal of Pharmacology and Experimental Therapeutics 2005;314(3):1079-1086.

[75] Krakowski L, Krzyzanowski J, Wrona Z, Siwicki AK. The effect of nonspecific immunostimulation of pregnant mares with 1,3/1,6 glucan and levamisole on the immunoglobulins levels in colostrum, selected indices of nonspecific cellular and humoral immunity in foals in neonatal and postnatal period. Veterinary Immunology and Immunopathology 1999;68(1):1-11.

[76] Yano T, Matsuyama H, Mangindaan REP. Polysaccharide-Induced Protection of Carp, Cyprinus-Carpio L, Against Bacterial-Infection. Journal of Fish Diseases 1991;14(5):577-582.

[77] Wang WS, Wang DH. Enhancement of the resistance of Tilapia and grass carp to experimental Aeromonas hydrophila and Edwardsiella tarda infections by several polysaccharides. Comparative Immunology Microbiology and Infectious Diseases 1997;20(3):261-270.

[78] Evans SG, Morrison D, Kaneko Y, Havlik I. The effect of curdlan sulphate on development in vitro of Plasmodium falciparum. Transactions of the Royal Society of Tropical Medicine and Hygiene 1998;92(1):87-89.

[79] Zhang M, Cheung PCK, Ooi VEC, Zhang L. Evaluation of sulfated fungal beta-glucans from the sclerotium of Pleurotus tuber-regium as a potential water-soluble anti-viral agent. Carbohydrate Research 2004;339(13):2297-2301.

[80] Yoshida T, Hatanaka K, Uryu T, Kaneko Y, Suzuki E, Miyano H, et al. Synthesis and Structural-Analysis of Curdlan Sulfate with a Potent Inhibitory Effect Invitro of Aids Virus-Infection. Macromolecules 1990;23(16):3717-3722.

[81] Yoshida T, Yasuda Y, Uryu T, Nakashima H, Yamamoto N, Mimura T, et al. Synthesis and In-Vitro Inhibitory Effect of L-Glycosyl-Branched Curdlan Sulfates on Aids Virus-Infection. Macromolecules 1994;27(22):6272-6276.

[82] Yoshida T, Yasuda Y, Mimura T, Kaneko Y, Nakashima H, Yamamoto N, et al. Synthesis of Curdlan Sulfates having Inhibitory Effects In-Vitro Against Aids Viruses Hiv-1 and Hiv-2. Carbohydrate Research 1995;276(2):425-436.

[83] Gao Y, Katsuraya K, Kaneko Y, Mimura T, Nakashima H, Uryu T. Synthesis, enzymatic hydrolysis, and anti-HIV activity of AZT-spacer-curdlan sulfates. Macromolecules 1999;32(25):8319-8324.

[84] Ungurenasu C, Pinteala M. Syntheses and characterization of water-soluble C-60-curdlan sulfates for biological applications. Journal of Polymer Science Part A-Polymer Chemistry 2007;45(14):3124-3128.

[85] Browder W, Williams D, Lucore P, Pretus H, Jones E, Mcnamee R. Effect of Enhanced Macrophage Function on Early Wound-Healing. Surgery 1988;104(2):224-230.

[86] Delatte SJ, Evans J, Hebra A, Adamson W, Othersen HB, Tagge EP. Effectiveness of beta-glucan collagen for treatment of partial-thickness burns in children. Journal of Pediatric Surgery 2001;36(1):113-118.

Chapter 22

Complexes of Polysaccharides and Glycyrrhizic Acid with Drug Molecules – Mechanochemical Synthesis and Pharmacological Activity

A. V. Dushkin, T. G. Tolstikova, M. V. Khvostov and G. A. Tolstikov

Additional information is available at the end of the chapter

http://dx.doi.org/10.5772/48095

1. Introduction

1.1. Using poly- and oligo-saccharides for drug delivery; Possibility for mechanochemical synthesis of supramolecular systems

Providing efficient ways of delivering active drug molecules to their destinations in target organisms, the so-called drug delivery, is among major challenges in today's pharmacy. An important relevant issue is to enhance the efficacy and safety of pharmaceutical compounds by correcting their solubility [1-3]. Polysaccharides (e.g., derivatives of cellulose, chitosan, and alginic and hyaluronic acids) make part of compositions with controlled or retarded drug release [4-6], while oligosaccharides (alpha-, beta-, and gamma-cyclodextrins and their derivatives) are broadly used to increase solubility and dissolution rates as they can form guest-host supramolecular complexes with poorly soluble drugs [7, 8]. Until recently little was known whether complexes of this kind may result from the activity of natural plant-derived or synthetic water-soluble polysaccharides though these are common elements in dietary supplements or drugs. Polysaccharides have aroused no interest in this respect, possibly because the technology for producing supramolecular complexes requires liquid phases (solutions or melts): The complexes form by molecular interaction in the liquid, and the solid phase is extracted then on drying (solutions) or cooling (melts). However, being easily soluble in water, polysaccharides are almost insoluble in other solvents and, moreover, decompose on heating rather than melt. The target drugs, instead, often dissolve rapidly in non-aqueous solvents but are poorly soluble or insoluble in water. Therefore, the liquid-phase synthesis of polysaccharide-drug complexes has been impeded by the lack of co-solubility.

INTECH
open science | open minds

© 2012 The Author(s). Licensee InTech. This chapter is distributed under the terms of the Creative Commons Attribution License http://creativecommons.org/licenses/by/3.0), which permits unrestricted use, distribution, and reproduction in any medium, provided the original work is properly cited.

This difficulty may be surmountable with solid-state chemistry approaches, specifically, with mechanochemical transformations in mixtures of solids [9-11]. Unlike the liquid-phase synthesis, mechanochemical treatment is a simpler single-stage process going without solvents or melts and respective additional procedures. The flow chart in Fig. 1 shows a simplified sequence of transformations the powder mixtures experience during dry milling in various mills.

There may be three types of main products relevant to our study, depending on the properties of starting materials:

1. "molecular dispersions", or solid solutions of drugs in excess filling (dispersion medium);
2. supramolecular complexes or products of chemical reactions between the components;
3. composite materials: aggregates of powdered particles.

In fact, they all are solid dispersions that form supramolecular structures (complexes or micelles) that enclose drug molecules and provide their solubility.

Generally, solid-phase processes have a number of advantages in laboratory and technological uses as they yield, in a shorter time, materials which the classical liquid-phase technology can never provide and allow avoiding problems associated with melts or solvents and side reactions. The high potentiality of mechanical activation was proven in our previous studies [12-14], e.g., on quick-dissolving pharmaceutical compositions [15-18] and synthesis of polyfluorinated aromatic compounds [19, 20].

In this synopsis we present techniques for synthesizing supramolecular complexes of poorly soluble drugs with water-soluble polysaccharides or with glycyrrhizic acid (a plant-derived glycoside), describe physicochemical properties of their solid forms and solutions, and report the results of pharmacological testing.

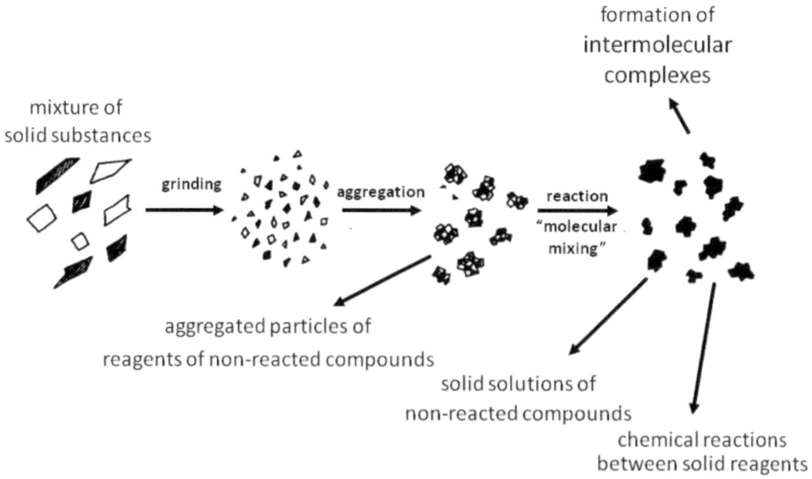

Figure 1. Mechanochemical transformations in mixtures of solids organic substances.

2. Physicochemical description of supramolecular systems including polysaccharides and glycyrrhizic acid with drug molecules

2.1. Synthesis of supramolecular systems of water-soluble polysaccharides and their stability in aqueous solutions

In the course of this study, different conditions of mechanochemical synthesis and various complexing agents have been tried and compared in terms of efficacy. The complexing agents were: arabinogalactan (AG), a water-soluble larch polysaccharide derived from *Larix sibirica* Ledeb. and *Larix gmelinii* (Rupr.), fibregum (FG), a glycoprotein of acacia gum, fruit pectin (PC), hydroxyethyl starch (HES200/0,5), dextrans (D) 10. 40. 70, and ß-cyclodextrin (CD), the latter chosen as standard for being widely used in pharmaceutics. The mixtures of powdered components (polysaccharides/complexing agents and drugs) were dispersed in ball mills at greater or lower intensities in laboratory planetary- and rotary-type mills, respectively. Milder rotary milling was predominantly applied because the molecular-level mixing in a planetary mill, common to laboratory studies of mechanochemical modification of drugs [9], may partly destroy the material and pose scaling problems. The materials processed by nondestructive rotary milling [10, 11, 21], instead, interact to produce solid dispersed systems of components (composite aggregates of superdispersed particles), and the process is easily scaled onto industrial flow mills.

The obtained compositions were checked for drug contents to avoid unwanted chemical reactions. Formation of supramolecular complexes was identified from changes in solubility of drugs in the water solution of the compositions [22].

Dissolution and complexing of poorly water-soluble drugs can be illustrated by such simplified equations:

$$\mathrm{Drug}_{\mathrm{solid}} \leftrightarrow \mathrm{Drug}_{\mathrm{solution}} \quad (1)$$

$$\mathrm{Drug}_{\mathrm{solution}} + \mathrm{CA}_{\mathrm{solution}} \leftrightarrow (\mathrm{Drug} \cdot \mathrm{CA})_{\mathrm{solution}} \quad (2)$$

Equilibrium, according to (2), is given by

$$K_{DCA} = [(\mathrm{Drug} \cdot \mathrm{CA})_{\mathrm{solution}}] / [\mathrm{Drug}_{\mathrm{solution}}] \times [\mathrm{CA}_{\mathrm{solution}}], \quad (3)$$

where $\mathrm{Drug}_{\mathrm{solid}}$ is the drug in a crystalline solid phase, in equilibrium with the solution; $\mathrm{Drug}_{\mathrm{solution}}$ is the drug existing in the free form in the solution; $\mathrm{CA}_{\mathrm{solution}}$ is the free complexing agent in the solution; $(\mathrm{Drug} \cdot \mathrm{CA})_{\mathrm{solution}}$ is the complexing agent-drug complex in the solution; K_{DCA} is the constant of supramolecular complexing.

The value $\mathrm{Drug}_{\mathrm{solution}}$ corresponds to the thermodynamic equilibrium solubility in the absence of complexing agents. In the case of complexing, the total concentration of the dissolved drug C_{drug} equals the sum its free and bound forms.

$$C_{Drug} = [\mathrm{Drug}_{\mathrm{solution}}] + [(\mathrm{Drug}\ \mathrm{CA})_{\mathrm{solution}}] \quad (4)$$

Thus, the solubility increase of a drug in the solution (X) in the presence of a complexing agent is

$$X = C_{Drug} / [Drug_{solution}] = 1 + K_{DCA} \cdot [CA_{solution}] \quad (5)$$

In our view, X is a good proxy of binding strength in the supramolecular complexes drugs may form with various water-soluble polymers.

All poorly soluble drugs we studied have shown a notable solubility increase when became incorporated into compositions with complexing agents. Table 1 below shows solubility data reported in [22-32] as far as published for the first time in this review.

The binding strength in the complexes grows in the series "dextran 70 < dextrans 40 and 10 ~ < HES < ß-cyclodextrin, fibregum < pectin < arabinogalactan". Complexing of pectin with mezapam and clozapine most probably occurs by acid-base reactions, which accounts for quite a high binding strength. However, other complexing agents lack acid-base groups and the interaction mechanism is most likely "hydrophobic", as in the case of cyclodextrin complexes. Thus, the mechanochemical treatment strengthens considerably the drug binding in compositions. The solubility of drugs increases, depending on the way of mixing, in the series "mixing without milling < high-rate milling < low-rate milling.

The obtained compositions were analyzed by X-ray powder diffraction and thermal methods. All non-processed mixtures showed X-ray and thermal features typical of crystalline drugs, which disappeared or decreased markedly after milling. Therefore, drugs in the ground mixtures partly or fully loose their crystallinity, possibly, as their solid phase becomes disordered and their molecules are dispersed into the excess solid phase of complexing agents, with formation of solid solutions or supramolecular complexes. In the latter case, the solubility changes evidence that the analyzed compositions form more strongly bound complexes when form in the solid phase than in the solution.

2.2. Molecular dynamics and structure of arabonigalactan complexes

AG-drug systems were investigated by ^1H NMR spectroscopy [22] for the molecular dynamics of complexes and the mobility of arabinogalactan (AG) molecule fragments. NMR relaxometry is applicable to molecular complexes as the spin-lattice and spin-spin relaxation times (T_1 and T_2, respectively) are highly sensitive to interactions and diffusion mobility of molecules. As a molecule becomes bound in a complex, its diffusion mobility slows down, and the proton relaxation times decrease notably. In the case of rapid complex-solution molecular exchange, the NMR signal decays according to the mono-exponential law. Otherwise, if the exchange is slower than the relaxation time, the kinetics is biexponential:

$$A(t) = P_1 \cdot \exp(-t / T_{21}) + P_2 \cdot \exp(-t / T_{22}) \quad (6)$$

API	Complexing agent/drug mass ratios mass	Solubility pure drug [Drugsolution} g/l / Solubility by complexation CDrug g/l	Solubility increase, X[1]	Reference
Diazepam	Arabinogalactan (1/10)[2]	0.048/0.058	1.2	[22,24]
	Arabinogalactan (1/10)[3]	0.048/0.115	2.4	[22,24]
	Arabinogalactan (1/10)[4]	0.048/2.31	48.2	[23]
	Pectin (1/10)[3]	0.048/0.67	14.2	[23]
	Hydroxyethylstarch (1/10)[4]	0.048/0.075	1.53	[23]
	Beta-cyclodextrin (1/10)[3]	0.048/0.086	1.8	[23]
	Glycyrrhizic acid (1/10)[3]	0.048/0.16	3.4	[29]
	Dextran 10[4]	0.048/0.09	1.9	[23]
	Dextran 40[4]	0.048/0.092	1.9	[23]
	Dextran 70[4]	0.048/0.057	1.2	[23]
Indomethacin	Arabinogalactan (1/10)[2]	0.04/0.044	1.1	[22,24]
	Arabinogalactan (1/10)[3]	0.04/0.396	9.9	[22,24]
	Arabinogalactan (1/10)[4]	0.04/1.59	39.7	[23]
	Hydroxyethylstarch (1/10)[4]	0.04/0.54	13.5	[23]
	Beta-cyclodextrin (1/10)[3]	0.04/0.096	2.4	[23]
Mezapam	Arabinogalactan (1/10)[2]	0.02/0.98	4.9	[22,24]
	Arabinogalactan (1/10)[3]	0.02/0.382	19.1	[22,24]
	Arabinogalactan (1/10)[4]	0.02/2.81	140.6	[23]
	Pectin (1/10)[3]	0.02/1.54	77.1	[23]
	Hydroxyethylstarch(1/10)[4]	0.02/0.04	2.0	[23]
Clozapine	Arabinogalactan (1/10)[2]	0.04/0.176	4.4	[22,24]
	Arabinogalactan (1/10)[3]	0.04/0.82	20.5	[22,24]
	Arabinogalactan (1/10)[4]	0.04/4.32	107.9	[23]
	Pectin (1/10)[3]	0.04/1.63	40.8	[23]

	Hydroxyethylstarch(1/10)⁴	0.04/0.222	5.5	[23]
	Beta-cyclodextrin (1/10)³	0.04/0.60	15.1	[23]
	Glycyrrhizic acid (1/10)³	0.04/0.088	2.2	[29]
Nifedipine	Arabinogalactan (1/10)³	0.18/1.24	6.9	[26]
	Arabinogalactan (1/20)³	0.18/2.46	13.7	[26]
	Glycyrrhizic acid (1/10)³	0.18/0.92	5.1	[30]
Dihydro-quercitin	Arabinogalactan (1/10)⁴	0.65/3.75	5.9	[23,27]
	Hydroxyethylstarch(1/10)⁴	0.65/1.97	3.0	[23]
	Fibregum(1/10)⁴	0.65/5.72	8.8	[23,27]
Quercitin	Arabinogalactan (1/10)³	0.019/0.21	11.6	[28]
	Arabinogalactan (1/20)³	0.019/1.28	71.0	[28]
Ibuprofen	Arabinogalactan (1/10)²	0.03/0.036	1.2	[29]
	Arabinogalactan (1/10)⁴	0.03/0.85	28.4	[29]
	Hydroxyethylstarch (1/10)⁴	0.03/0.08	2.6	[29]
	Glycyrrhizic acid (1/10)³	0.03/0.441	14.7	[29]
Beta-Carotene	Arabinogalactan (1/40)³	< 0.001/2.65	> 2000	[25], This article
Warfarin	Arabinogalactan (1/40)³	0.021/0.111	5.3	[31]
Contaxantine	Arabinogalactan (1/40)³	< 0.001/2.64	> 2000	[25], This article
Albendazol	Arabinogalactan (1/10)⁴	0.003/0.174	58.0	[32]
	Hydroxyethylstarch(1/10)⁴	0.003/0.094	31.3	[32]
Carbenazim	Arabinogalactan (1/10)⁴	0.009/0.146	16.2	[32]
	Hydroxyethylstarch(1/10)⁴	0.009/0.020	2.1	[32]
Simvastatin	Glycyrrhizic acid (1/10)³	0.0012/0.314	260	This article
	Arabinogalactan (1/10)³	0.0012/0.044	36,7	This article

1 – To determine the solubility of the drug, machined mixture of complexing agent/drug, in amounts of 0.4 grams, as well as the linkage of individual substances which are equivalent to their content in the above mixture was dissolved in 5 ml of water while stirring with a magnetic stirrer at +25 ° C till reaching constant concentration. The concentration of drug in the solution was analyzed by HPLC.

2 – mixing without mechanical treatment;

3 – treatment in a planetary mill, acceleration 40 g;

4 – treatment in a rotary ball mill, acceleration 1 g;

Table 1. Increase in water solubility of some drugs as a result of complexing.

The fast component P_1 and the slow component P_2 correspond, respectively, to the shares of molecules in the complex and in the solution. Typical T_2 values are 0.5-1 s for molecules in the solution and 0.03-0.09 s for those bound in the complex. Shorter T_{21} times mean lower mobility of drug molecules in the latter case.

Similar considerations apply to the mobility within polymers when parts of a macromolecule differ in mobility, possibly, controlled by their spin and conformations.

2.2.1. T_2 measurements

Arabinogalactan shows biexponential relaxation patterns. The calculated parameters for arabinogalactan and AG-drug complexes are listed in Table 2.

Sample	P_1 %	T_{21} msec	P_2 %	T_{22} msec
Arabinogalactan, native[2]	80	17	22	250
Arabinogalactan, treated in planetary mill[2]	65	25	35	250
Clozapine/Arabinogalactan 1/20 w/w No treatment[3]	88	90	12	1000
Clozapine/Arabinogalactan 1/20 w/w Mixture treated in planetary mill[3]	90	40	10	1000
Mezapam/Arabinogalactan 1/20 w/w No treatment[3]	55	50	45	250
Mezapam/Arabinogalactan 1/20 w/w Mixture treated in planetary mill[3]	90	30	10	250
Diazepam/Arabinogalactan 1/20 w/w No treatment[3]	mono	150	-	-
Diazepam/Arabinogalactan 1/20 w/w Mixture treated in planetary mill[3]	20	60	80	800
Indomethacin/Arabinogalactan 1/20 w/w No treatment[3]	58	50	42	900
Indomethacin/Arabinogalactan 1/20 w/w Mixture treated in planetary mill[3]	67	40	33	900

1 - T2 measurements were performed for the aromatic protons of drug molecules, to an accuracy of ± 10%
2 – solvent D_2O;
3 - solvent 70% D_2O + 30% CD_3OD;

Table 2. Spin-spin relaxation times of protons for arabinogalactan and drug molecules in solutions[1]

The short relaxation times may correspond to the interior protons and the long times may represent the exterior protons of the polymer compound. Mechanical activation in a planetary mill increases the molecular mobility of the interior fragments but decreases their percentage. A relatively narrow ~ 6 kHz band in the ^1H NMR spectra of AG powder, which stands out against a broad line associated with dipole-dipole interaction non-averagable in solids, represents a mobile phase with its integral intensity up to ~ 15% of the number of hydrogen nuclei in the sample. The mobile phase may correspond to fragments of AG macromolecules, possibly, side chains, as one may reasonably hypothesize given that water content in AG never exceeds 2 wt.%. This very fact appears to facilitate AG-Drug molecular complexing on mechanical activation of solids.

AG-Drug systems most often exhibit distinct biexponential kinetics as evidence that the drug molecules are either free or bound in complexes with AG. The bound molecules are more abundant and less mobile in milled samples, while the free ones keep almost invariable NMR relaxation times. The characteristic ^{1}H NMR bands of clozapine and mezapam move to low field on complexing, possibly because the molecules become protonated at the account of minor remnant uronic acid present in AG, the shift being likewise greater in the milled samples. However, no complexing-related shifting appears in the cases of indomethacin and diazepam. The life time of drug molecules in complexes with AG must to be $\sim\geq 100$ ms, judging from the conditions of slow exchange.

The system AG-diazepam offers an illustrative example. Solutions of these mixtures not subjected to mechanical treatment show mono-exponential relaxation behavior, but with shorter times than in free diazepam, likely as a result of rapid solution-complex molecular exchange. The milled mixtures, on the contrary, have biexponential kinetics corresponding to slower exchange of molecules and stronger binding.

Thus, dynamic NMR spectroscopy of all Drug-AG solutions indicates formation of supramolecular drug-polysaccharide complexes, like the data on solubility increase. Most likely, the complexing sites are at side chain spaces in the branching macromolecules. Unlike cyclodextrins, ensembles of polysaccharide molecules (including arabainogalactan) are micro-heterogeneous in mass and structure. As a result, molecular modeling of the complex is very difficult. The binding mechanism appears to lie mainly with hydrophobic interactions [33, 34] which are typical of guest-host cyclodextrin complexes. A certain support to this hypothesis comes from stronger binding of highly lipophilic drugs which are almost insoluble in water. In this case, the branched structure of AG macromolecules [35, 36] is especially favorable for complexing. However, Coulomb interactions may contribute as well in the presence of acid-base groups in polysaccharides and drugs [37].

2.3. Transformations of polysaccharides in solid state and in solutions

Macromolecules characteristically have broadly varying molecular weights, from $\sim 10^3$ to $\sim 10^7$ Da. Macromolecules in polymers involved in technological production of various materials may experience mechanical action and partial destruction (breakdown of chains) whereby their molecular weight becomes ever more heterogeneous and diminishes on average [38]. The destructive change may be especially prominent in "dry" technological processes, such as pulverization, pelleting, or mixing, e.g., in mechanochemical solid-state complexing of drugs with water-soluble polymers (polysaccharides). Partial destruction of polymers may change their toxicological properties which have to be cautiously monitored when making new drugs and food products.

Molecular weight patterns were studied [39] in polysaccharides (dextrans 10, 40, and 70, HES 200/0.5, and larch AG and acacia FG gum) by gel permeation chromatography (GPC) [40] of samples treated in rotary and planetary mills; the obtained materials were tested for their toxicity.

See Fig. 2 for example chromatograms of AG and Table 3 for calculated molecular weights of the analyzed polysaccharides before and after mechanical treatment.

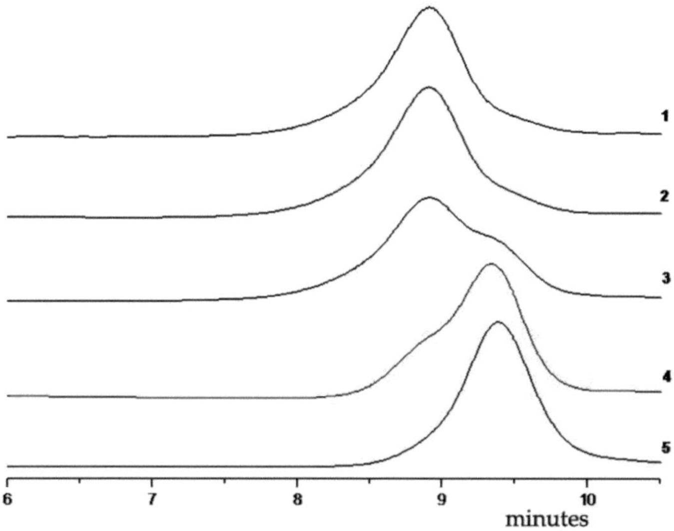

Figure 2. GPC chromatograms of 0.02 wt.% arabinogalactan water solution. 1 = native; 2 – 5 = subjected to mechanical treatment: ball mill, 2 hours (2), ball mill, 24 hours (3), planetary mill, 10 min (4), extremely intense treatment, mixed ball loading (5); Eluent: H_2O/0.1 N $LiNO_3$

Polysaccharides ground in a high-rate planetary mill diminish markedly in molecular weight and change slightly their polydispersity index Mw/Mn. The mechanical destruction is stronger in polymers with larger molecular weights, which agrees with published evidence [39]. Note that highly branching macromolecules (HES and AG) break down into roughly equal fragments. The Mw/Mn ratios in polysaccharides do not grow much, possibly, because destruction mostly affects their high-molecular fractions. Destruction is apparently controlled by the structure of polysaccharide molecules and physicochemical chain breakdown mechanisms. According to a model for linear synthetic polymers [41], the chains that occur in the middle of macromolecules are especially prone to failure. Destruction of HES and AG is qualitatively similar to that model, though dextrans and fibregum may deform by a different mechanism.

The results for larch arabinogalactan are worth of special consideration. High-rate treatment in a planetary mill, especially with mixed ball loading reduces strongly the AG molecular weight. According to chromatograms (Fig. 2), its Mw 17.3 kDa macromolecules split quantitatively into two almost equal parts of Mw = 8.3 kDa, while their Mw/Mn ratio decreases to 1.08 [23]. Therefore, the native AG molecules may consist of two relatively weakly bonded fragments of equal molecular weights and easily break down on milling

[23,39]. Note that AG macromolecules with MM (Molecular Mass) ~9 kDa are likewise the main product of chemical destruction of Canadian larch AG [42].

Furthermore, the analyzed polysaccharides experience almost no mechanical failure on low-rate grinding in a rotary mill (Table 3). Thus, ball rotary milling appears to be most often preferable, as molecular mass changes in technologically produced polymers are commonly unwanted in view of their further use in dietary supplements and drugs, otherwise additional tests and standardization may be required.

Sample	Treatment	Mn, kDa	Mw, kDa	Mw/ Mn	Weight shares of macromolecules, kDa	
					10%	90%
Fibregum	Native	146.6	256.7	1.8	<75.9	<528.2
	Planetary mill, 20g, 20min	31.4	55.2	1.8	<16.3	<113.5
	Ball mill, 1g, 4hours	120.3	231.6	1.9	<60.3	<478.4
Arabinogalactan	Native	13.9	17.3	1.2	<9.0	<27.9
	Planetary mill, 20g, 20min	9.3	11.2	1.2	<6.1	<18.4
	Ball mill, 1g, 4hours	13.1	16.3	1.2	<8.1	<26.2
Hydroxyethyl starch 200/0,5	Native	47.9	116.9	2.4	<20.9	<265.3
	Planetary mill, 20g, 20min	26.6	55.2	2.1	<12.7	<118.9
	Ball mill, 1g, 4hours	45.6	105.5	2.3	<20.0	<237.6
Dextran 70	Native	30.9	76.4	2.5	<14.0	<174.7
	Planetary mill, 20g, 20min	22.7	54.8	2.4	<10.4	<123.5
	Ball mill, 1g, 4hours	29.6	73.5	2.5	<13.4	<169.2
Dextran 40	Native	24.6	38.0	1.5	<13.3	<72.3
	Planetary mill, 20g, 20min	19.5	31.9	1.6	<10.4	<61.3
	Ball mill, 1g, 4hours	24.3	37.4	1.5	<13.0	<71.2
Dextran 10	Native	8.3	13.4	1.6	<4.2	<26.4
	Planetary mill, 20g, 20min	8.0	12.1	1.5	<41.9	<22.7
	Ball mill, 1g, 4hours	8.3	13.4	1.6	<4.2	<26.3

Table 3. Molecular mass distribution of polysaccharides

Toxicological tests of the milled polysaccharides show that a single intragastric injection administered at doses from 500 to 6000 mg/kg body weight caused no death in experimental animals. Their appearance, behavior, and state were within the background over the whole dose range; no statistically significant changes in body temperature relative to the control was observed, and body weight growth was uniform in all groups. Injections of the tested polysaccharides neither induced any considerable effect on the central nervous system of the mice. Patomorphological postmortem examination of mice in 14 days after polysaccharide administration revealed no pathology in thoracic and abdominal cavities. The median lethal dose LD_{50} for all polysaccharides was over 6000 mg/kg body weight on single intragastric injection.

2.4. Supramolecular structures of glycyrrhizic acid (GA) and poorly soluble drugs water solutions: Synthesis and properties

Biosynthetic and natural plant-derived carbohydrate-bearing metabolic agents have been increasingly used for obtaining complexes (clathrates) with drugs in drug delivery research. The mechanism of GA-pharmacon interaction in solutions may consist in involving drug molecules into self-associates (micelles) that exist in a wide range of concentrations in GA solutions. Until recently however, the existence of micelles in GA solutions had no direct proof but was either inferred from measured concentration dependences of solution viscosity [43] or was studied by dynamic NMR spectroscopy in water-methanol solutions [44]; the latter (30% concentration) were at the same time used as solvent for technical reasons. Thus, the molecular mechanism of drug complexing remained unclear, whether it was incorporation into micelles or supramolecular complexes with GA in water solutions, without organic solvents which change notably the GA-pharmacon reactions.

GA water solutions, with and without the presence of poorly soluble drugs, were investigated in [29] using gel permeation chromatography, which allows detecting self-associates/micelles and estimating their sizes and concentrations. On the other hand, solid GA-drug dispersions were obtained with the mechanochemical approach developed earlier [10, 11]. The binding strengths in GA-pharmacon supramolecular complexes or GA micelles in water solutions were compared using the criterion of solubility increase in poorly soluble drugs [22] and studied in terms of pharmacological activity.

Chromatograms of GA water solutions (Fig. 3) show peaks of high-molecular (~ 46-67 kDa) forms over all studied concentration ranges, while the GA molecular weight is 836.96 Da (Table 4).

The peak areas, being proportional to the solution concentrations and calculated relative to the known amounts of standard dextrans, show that almost all GA is stored in the solution. This, in our view, is evidence for the existence of GA self-associates (micelles). The critical concentration of micelle formation (CMC) was estimated earlier [43] from viscosity change in GA solutions to be 0.004 wt. % (0.05 mM). In our case estimating exact CMC is difficult for the limited sensitivity of refractometric detector and for dilution in the chromatographic column. However, it may be inferred from the time to the chromatographic peak (~0,5 min) and the elution rate. The solution we studied underwent about 10-fold dilution, and the derived CMC is ~≤ 0.0001 wt.%, (0.001mM), or far less than in water-methanol solution (0.04-0.08 wt.% or 0.5-1.0 mM) [44]. In diluted 0.01-0.001 wt. % solutions, there is only one type of micelles (~ 66 kDa) with a very low Mw/Mn ratio of 1.08-1.06. As the GA solutions reach concentrations of 0.5 wt.%, micelles decrease in weight to form ~ 46 kDa bodies and increase in Mw/Mn ratios. Therefore, almost all glycyrrhizic acid in water solutions from 0.0001 to 0.5 wt.% exists in the self-associated form of micelles, out of which the ones with MM= ~ 66 kDa consisting of about 80 GA molecules are most stable.

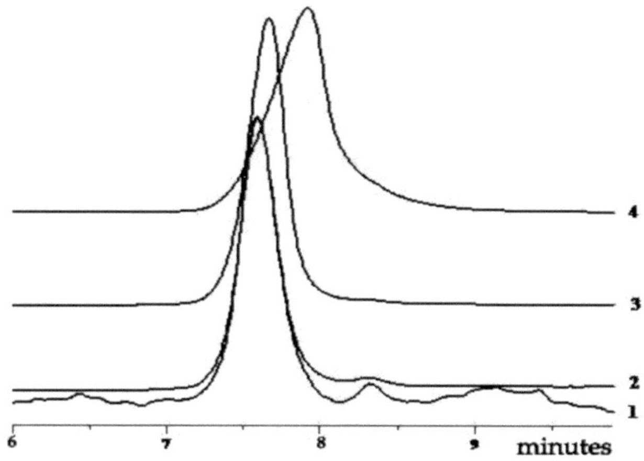

Figure 3. GPS chromatograms of glycyrrhizic acid water solution. 1 - concentration - 0.001, 2 - 0.01, 3 - 0.1, 4 - 0.5 wt.%.

COMPOSITION	Molecular masses	Solution concentration, wt. %			
		0.001	0.01	0.1	0.5
GLYCYRRIZIC ACID	Mw/Mn kDa	65.93/61.0	66.2/62.5	60.7/57.3	45.3/36.5
GLYCYRRIZIC ACID/ /IBUPROFEN 10/1 w/w Treated in planetary mill 3 min	Mw/Mn kDa	69.0/67.2	69.4/65.3	65.2/61.7	48.6/39.0

Table 4. Molecular-mass characteristics of micelles in solutions of glycyrrhizic acid and its composition with ibuprofen

2.4.1. GA solid dispersions with poorly water-soluble drugs

Solid GA dispersions with ibuprofen, phenylbutazone, clozapine, and diazepam were obtained by milling with GA (10:1 mass, or 2.5/1 – 4/1 molar ratios). Their thermal analysis data indicate that the crystalline phase of the drugs becomes disordered, until complete loss of crystallinity. In our view, the molecules of drugs may disperse into the excess solid GA with formation of solid solutions. Other investigated systems behave in a similar way.

As the dispersions dissolve, the drugs become more soluble in water (Table 1), this being evidence of the efficiency of GA as a solubilizing agent and the mechanochemical treatment as a tool for synthesis of water-soluble solid dispersions. GA has a nearly intermediate solubilizing effect higher than HES but lower than AG.

2.4.2. *GPC of dissolved GA-drug solid dispersions*

The GPC data for GA-ibuprofen dispersions in water are shown in Table 4. Similar results of MM increase in micelles were obtained for the GA-diazepam, GA- phenylbutazone, and GA-clozapine systems. The peak areas, proportional to the concentrations of the analyzed solutions, indicate that they bear almost the entire mass of GA-drug samples. Thus, the dissolved drugs in the GA-drug complexes are likewise self-associated in micelles which are stable in a broad range of concentrations, as well as in the solutions of native GA. Therefore, we suggest that poorly soluble drugs increase their solubility by incorporating into GA micelles/self-associates. GA molecules contain a hydrophilic (two glucuronide residues) and a hydrophobic (triterpene) components (Fig. 4).

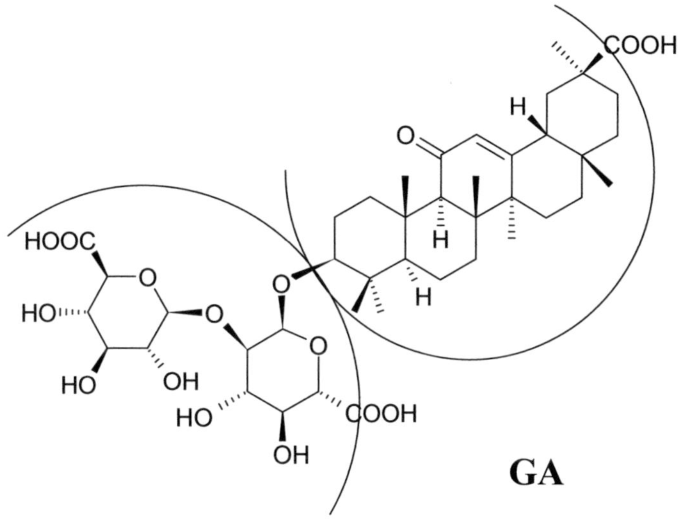

Figure 4. The structure of glycyrrhizic acid

In micelles, the latter are most likely oriented inward and the former toward the outer surface of the self-associate, while the drug molecules may occur either in the interior hydrophobic part or complex with the exterior hydrophilic part of the micelles. Unfortunately, the experimental evidence is insufficient to judge about these subtle GA-drug interaction mechanisms. Generally, the MM of GA-pharmacon micelles are 5-7% higher than those in GA water solutions, all over the studied range of concentrations. Another possibility is that drug molecules may substitute for some GA molecules while the micelles generally grow in size. As the solution concentration increases, the size difference of micelles grows correspondingly. Note that the region of high GA concentrations is proximal to the conditions in which the solubility of drugs was measured, this being additional evidence for the suggested mechanism of water solubility increase. An explanation for the decreasing MM differences on dilution may be that drug molecules can escape from GA micelles during GPC process to diluted solutions and appear in chromatograms as individual substances eluted at different rates [45]. Anyway, further investigation is needed to gain more insights into the GA-drug interaction.

3. Pharmacological activity of supramolecular complexes

The pharamacological activity of the complexes was investigated *in vivo* on females and males of outbred white mice and on Wistar rats, obtained from the SPF vivarium of the Institute of Cytology and Genetics, Novosibirsk. All animal procedures and experiments followed the 1986 Convention on Humane Care and Use of Laboratory Animals. The activity was determined using standard pharmacological tests [46]. All studied complexes, with some exceptions, are AG-Drug or GA-Drug compositions of 10:1 weight ratio, which was found out to be of greatest efficacy [22].

3.1. Arabinogalactan-drug complexes

As a special study has shown, AG complexing with nonsteroidal anti-inflammatory drugs and non-narcotic analgesics of various action mechanisms can reduce the required dose for 5-100 times and, hence, avoid the side effects typical of the drugs.

For instance, complexing with indomethacin allows 10 to 20 times dose reduction relative to the standard and induces twice fewer cases of lesion to gastric mucous membrane, with the same high anti-inflammatory activity [22, 24].

Administration of phenylbutazone in a complex with AG, at ten times lower pharmacon content, stimulated analgesic activity, both in the model of chemical effect and on thermal action, which may expand its applicability scope (Table 5).

Compounds	Hot plate, s	Acetic acid writhing model, amount
Control	28.73±2.34	4.76±0.26
Phenylbutazone:AG 1:10. 120 mg/kg per os	42.20±5.20*	2.00±0.09*
Phenylbutazone, 12 mg/kg per os	11.40±0.80	4.25±0.50
Ibuprofen:AG 1:10. 200 mg/kg per os	23.50±2.43	0.75±0.01*
Ibuprofen, 200 mg/kg per os	19.5±2.45	1.50±0.03*
Metamizole sodium: AG 1:10. 50 mg/kg per os	32.50±3.70	1.75±0.08
Metamizole sodium, 50 mg/kg per os	30.40±3.20	0.63±0.10
Metamizole sodium, 5 mg/kg per os	17.70±2.30	4.25±0.75
*p <0.05 relative to control		

Table 5. Analgesic activity of AG complexes with non-narcotic analgesics

AG complexes with 10 times smaller doses of ibuprofen and metamizole sodium showed a high analgesic activity in visceral pain models (the former) and in two models of test pain.

Studies in this line were continued, with the above approach and proceeding from the obtained results, on AG complexes with drugs that involve the central nervous system. Specifically, AG complexing with diazepam allowed reducing the dose for ten times and enhanced the anxiolytic effect. AG complexes with mezapam acted as standard anxiolytics at 20 times lower doses of the phamacon. The AG-clozapine complexing provides a two-fold dose decrease at a higher sedative action.

Another objective was to study drugs involving the blood coagulation and cardiovascular systems. The AG-warfarin (WF) complex was tested on intragastric injection in females of Wistar rats, with prothrombin time (PT, in seconds) as the principal criterion of the action. PT is the classical laboratory test of the exterior blood coagulation pathway used to evaluate the system of hemostasis in general and the efficacy of warfarin therapy in particular. The complex was administered once, in a dose of 20 mg/kg body weight, which is equivalent to 2 mg/kg warfarin. In 24 hours after the injection, PT increased considerably (to 30 s against the 11.63 s for the intact control). With free warfarin, this time was 42 s, or 28.5% longer than with the AG-warfarin complex, but it equalized (21 s) for both agents in 48 hours after a single injection (Fig. 5,6).

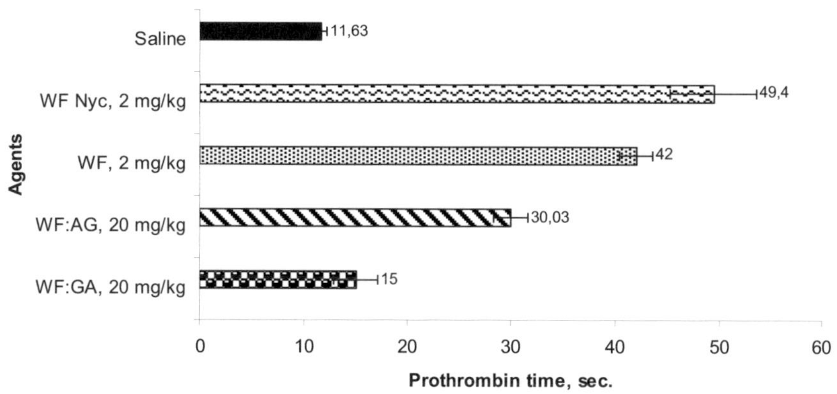

Figure 5. Prothrombin time, 24 hours after single dosing of WF:AG and WF:GA.

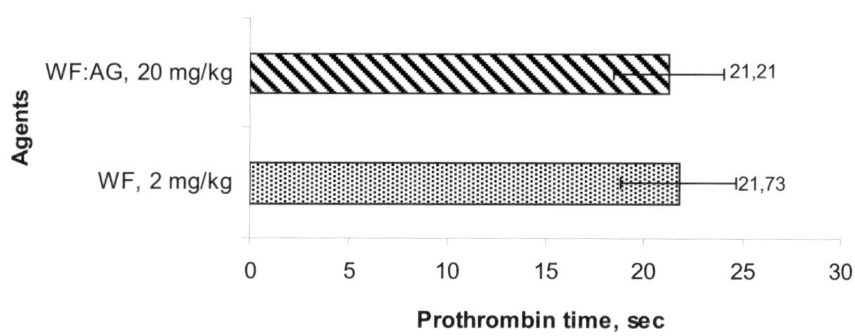

Figure 6. Prothrombin time, 48 hours after single dosing of WF:AG and WF.

The difference in pharmacokinetics between the AG:WF complex and free warfarin was further explored after a single administration of 20 and 2 mg/kg body weight, respectively. Blood was sampled in 1, 8, 10, 12, 24, 48, and 72 hours after injection on decapitation. Fig. 7 shows average plasma warfarin contents, and pharmacokinetic parameters are listed in Table 6. The concentrations of the compounds increase in a similar way but free warfarin reaches C_{max} seven hours sooner (T_{max}) than in complexes with AG, and the concentrations become equal in 24 hours after single dosing. Excretion, on the contrary, is slower in pure warfarin than in the complex, which is consistent with 27 % higher clearance (CL) in the complex than in the free drug. Thus, warfarin increases more smoothly when bound with AG than in the free form and thus poses lower bleeding risks associated with its abrupt rise during dosage adjustment. Furthermore, shorter mean retention times (MRT) for the AG:WF complex may secure the following injection and accelerate warfarin excretion in the case of drug withdrawal.

Thus, complexing of warfarin with AG increases its safety and reduces unwanted bleeding risks in the case of anti-coagulant treatment.

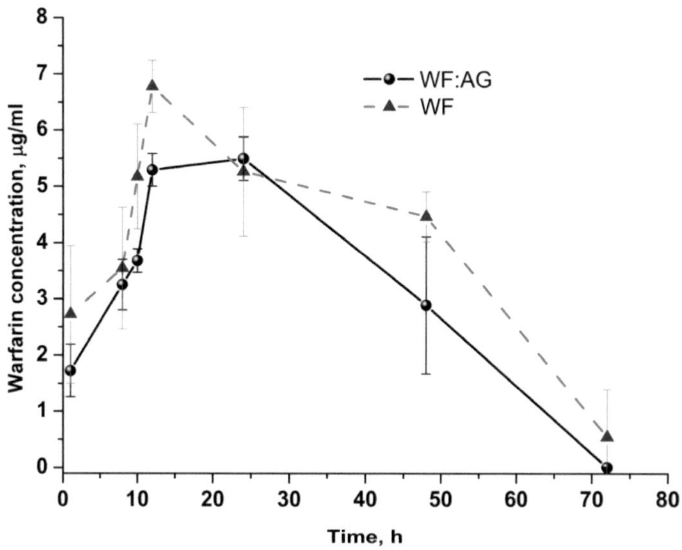

Figure 7. Mean plasma concentration–time profile of WF:AG and blank WF after single oral administration at a dose of 20 mg/kg (dose of WF is equal to 2 mg/kg) and 2 mg/kg, respectively.

Compounds	WF, 2 mg/kg	WF:AG, 20 mg/kg
CL, ml/h	1.52±0.03	1.93±0.18*
MRT, h	31.39±1.82	21.81±2.38*
Terminal half life, $T_{1/2}$, h	5.11±0.24	6.38±2.55
T_{max}, h	11.00±1.41	18.00±8.49
C_{max}, µg/ml	6.47±0.91	5.64±0.19
AUC, µg h/ml	263.01±0.02	208.34±20.03*
*p<0.05 against WF		

Table 6. Pharmacokinetic parameters of WF and WF:AG

Another drug in which we studied the pharmacological effect of complexing with AG was nifedipine (NF). NF is a dihydropyridine blocker of slow calcium channels that dilates coronary and peripheral vessels and reduces the oxygen demand in myocardium.

Nifedipine exerts a minor negative inotropic effect and a very weak antiarrhythmic action. Intravenous injection of 3.5 mg/kg AG:NF complex (0.35 mg/kg NF) caused 26 % drop of blood pressure, measured via a carotid cannula, while 0.35 mg/kg NF can provide only a 9% decrease.

Being aware that nifedipine has a pleiotropic antiarrhythmic effect, besides the basic hypotensive action, the NF:AG complex was investigated in this respect on intravenous injection in a model of arrhythmia induced with 250 mg/kg calcium chloride. The complex administered in a dose of 0.175 mg/kg body weight (0.0175 mg/kg NF) arrested lethal heart rate disorder in 100% and 65% of cases, respectively, when applied prior to and after exposure to the arrhythmogen. Pure NF at 0.0175 mg/kg had no antiarrhythmic effect in the model of calcium chloride arrhythmia.

Thus, the NF:AG complex has demonstrated a stronger hypotensive and antiarrhythmic action than pure NF on intravenous administration, while its effective hypotensive dose is ten times smaller. Arabinogalactan itself does not induce any statistically significant decrease in blood pressure and cardiac rates. Furthermore, it is important that the new method adds another water-soluble form of NF as there is the only soluble nifedipine (adalate) available in the market.

A similar hypotensive effect was obtained with nisoldipine, another dihydropyridine.

Complexing of hypoglycemic drugs (metformin, rosiglitazone, insulin) with AG increased their solubility but allowed no dose reduction, though it improved notably the state of animals exposed to a toxic dose of alloxan, an agent simulating trial hypergclycemia.

The reported pharmacological data on the AG complexes with these pharmacons agree with the results on complexing with terpenoids (glycirrhizic acid, stevioside, and rebaudioside). In both cases, complexing increases the basic activity of drugs, allows dose reduction and forms new properties. We suggest to call this effect complexing or clathration of pharmacons with plant-derived carbohydrate-bearing metabolic agents.

3.2. Glycyrrhizic acid (GA)-drug complexes

3.2.1. Nonsteroidal anti-inflammatory drugs (NSAID)

GA:NSAID complexes with acetylsalicylic acid (ASA), diclofenac (OF), phenylbutazone (BD), and indomethacin (IM) were synthesized in solutions [24] and in solid state [23]. Complexing was confirmed by spectrometry. In IR spectra, the bands of hydroxyl and carbonyl groups of the glycoside were shifted to short wave numbers.

All mentioned complexes show anti-inflammatory action in smaller doses than the primary drug, and have 3-11 times larger therapeutic index (LD_{50} /ED_{50}) [47,48] (Table 7).

GA complexes with aspirin and diclofenac (GA:ASA, GA:OF) exerted a prominent anti-inflammatory action in six models of acute inflammation induced by carrageenan, formalin, histamine, serotonin, Difko's agar, and trypsin, as well as in the cases of chronic inflammation (cotton and pocket granulomas) in intact and adrenalectomized animals [48].

Compounds	Dose range of complex*	Dose range of free NSAID
GA : ASA (1:1)	4500/82=54.8	1900/98=19.4
GA: OF (1:1)	1750/12.5=140	310/8=33.7
GA : BD (1:1)	3150/62=50.8	880/56=15.7
GA : AN (1:1)	8000/68=117.6	570/55=10.3

Table 7. Anti-inflammatory effects of GA complexes and free NSAID.*LD_{50}/ED_{50}; ED_{50} – effective dose

The anti-inflammatory action of the GA:IM complex is stronger than of the free drug at equal dosing (10 mg/kg body weight). GA:NSAID complexing also potentiates other (analgesic, antipyretic) biological activities [29]. The GA:OF complex exerted a more prominent anesthetic action than diclofenac in electric and thermal stimulation (57.5±2.0 and 43.2±2.6) and exceeded the amidopyrine effect in the case of thermoalgesic stimulation (23.4±1.1 and 18.5±1.4). The anesthetic activity of the GA: metamisole sodium (GA:AN) complex is 11.4 times higher relative to metamisole sodium (AN) alone [23] while and that of the GA:ASA complex exceeds the effect of aspirin in animals exposed to thermoalgesic stimuli. The GA complexes with ASA and OF are 3 and 2.3 times more potent pain relievers than the respective NSAID in acetyl choline writhing model. The GA:ASA complex demonstrates a 4 times higher therapeutic index than aspirin in the acetic acid writhing model. The GA:ASA and GA:OF complexes show high antipyretic activity, twice larger than in the pure pharmacons [47,48].

Thus, water-soluble GA:ASA and GA:OF complexes evoke prominent anti-inflammatory and antipyretic effects, their spectrum of pharmacological activities and therapeutic ratio being larger than in the respective NSAID. The complexes also induce a marked membrane-stabilizing effect and reduce accumulation of primary and secondary products of lipid peroxidation in animals with chronic inflammation.

Complexing of glycyrrhizic acid with ibuprofen increases the analgesic action of the latter at twice lower doses.

Furthermore, GA complexes irritate less strongly the gastric mucosal membrane than their NSAID counterparts. For instance, the GA:ASA complex promotes reparation of ulcers though the ulcerogenic activity of GA:OF is minor. Both complexes diminish E1 and E2 blood prostaglandins in animals with chronic inflammation. The complexes can be recommended for clinical trials as anti-inflammatory agents, including for patients that suffer from ulcer of stomach and duodenum. The acute toxic effect of GA:NSAID complexes is 2 to 14 times as low as in the respective pharmacons (Table 8) [47, 48].

GA:NSAID complexing obviously exerts a synergetic effect of higher biological activity along with lower toxicity and weaker ulcerogenic action on the gastrointestinal tract, and has a higher water solubility. It is evident on comparing the pharmacological activities of the ASA:GA and OF:GA supramolecular complexes obtained by dissolution and solid-state mehcanochemical synthesis that the latter opens a promising perspective of a technologically preferable and saving way of producing highly active NSAI drugs [23].

Complexes	LD_{50}, mg/kg	Pharmacon LD_{50}, mg/kg
GA : ASA (1:1)	4500	1900
GA: OF (1:1)	1750	310
GA : BD (1:1)	3150	880
GA : AN (1:1)	8000	570

Table 8. Acute toxicity of GA:NSAID complexes in mice (per os)

The GA:Rofecoxib complex at doses 50 and 100 mg/kg body weight shows no active anti-inflammatory and analgesic effects but is more highly soluble.

3.2.2. Prostaglandins

Prostaglandins have been of broad use in human and veterinary medicine for their ability, in small doses, of stimulating womb muscles.

Veterinary uses of prostaglandins have been especially important: prostaglandin-based drugs are employed in swine breeding for farrow synchronization and are highly potent in preparing cattle and horse females for artificial insemination; these drugs allowed solving many problems with puerperal complications in cows and horses.

Kloprostenol, one main veterinary drug, is made by multistage synthesis, like other prostaglandins, which imposes its high price. It is urgent to reduce the effective dose and at the same time to increase the stability of labile prostaglandins in finished drugs. Both solutions have been found through complexing E and F prostaglandins (PGE1, PGE2, PGF2α), sulprostone (SP) and kloprostenol with GA. The complexes were synthesized and tested for uterotonic activity (Table 9).

In experiments on rats and guinea pigs, the GA:PGE1 (1:1) and GA:PGF2α (1:1) complexes changed the amplitude of uterine actions twice more strongly than the same concentration of PGE1 sodium (10-8 g/ml). SP and PGE2 as complexes with GA (1:1) induced three times greater amplitudes and increased uteric tonicity [49].

An efficient veterinary drug of klatraprostin has been developed on the basis of GA complexing with kloprostenol, a well known synthetic luteolitic prostaglandin, at doses five times as low as in the world practice [48, 49]. The drug is cheaper than its imported analogs while its action is more physiological than in the best foreign counterparts.

Klatiram, which contains an aminoacid (tyrosine) besides GA and kloprostenol [48,49], has a still greater potency. It is more effective than estrofan though bears 100 times less prostaglandin (kloprostenol).

Compound	Change of uterine contraction amplitude, %	P	Change in uterus tonus, %	P
GA: PGE1	53.4±5.0	<0.002	49.4±1.2	<0.002
PGE1	24.3±1.5	<0.05	30.7±2.2	<0.05
GA : SP	150.0±11.0	<0.001	135.0±10.0	<.001
SP	50.0±5.0	<0.001	115.0±9.5	<0.001
GA : PGE2	63.5±6.0	<0.001	40.7±4.0	<0.002
PGE2	20.0±2.8	<0.05	33.5±2.4	<0.05
GA: PGF2α	55.6±5.0	<0.001	61.0±5.6	<0.001
PGF2α	27.8±1.5	<0.05	39.4±5.3	<0.02

Table 9. Uterotonic activity of prostaglandins and their GA complexes (1:1) in rats *ex vivo*, phosphate buffer C = 10-8 g/ml. PGE1 , PGE2, SP, PGF2α.

3.2.3. Cardiovascular drugs and anticoagulant warfarin

Pharmacological activity was also investigated in GA complexes with antiarrhythmic Lappaconitine hydrobromide (LA) and antihypertensive nifedipine (NF) drugs.

Lappaconitine hydrobromide belongs to the group of clinically used antiarrhythmic drugs and is administered to patients with various rhythm disorders, especially, ventricular arrhythmia, paroxysmal ciliary arrhythmia, and monofocal atrial tachycardia, but it has a drawback of high toxicity.

In special experiments on antiarrhythmic action of LA:GA complexes, the one patented as alaglizin [50] showed the highest efficacy. Alaglizin, being ten times less toxic than LA, induced antiarrhythmic effects in models of calcium chloride and aconitine arrhythmia and had the highest antiarrhythmic therapeutic index (LD_{50}/ED_{50}) among all available drugs of this kind. When administered at 0.125 mg/kg and 0.250 mg/kg body weight, alaglizin causes no influence on electrocardiogram parameters, according to an extended study in models of calcium chloride and adrenal arrhythmia. Intravenous injection of 0.125 mg/kg alaglizin prior to exposure to a lethal dose of calcium chloride blocked arrhythmia in 80% of rats; 0.250 mg/kg of alaglizin applied after arrhythmogenic $CaCl_2$ stopped the already developed arrhythmia in 50% of animals. A single injection of 0.125 mg/kg and 0.250 mg/kg alaglizin in a model of adrenal arrhythmia prevented full development of arrhythmia; 50% and 100 % of animals recovered normal ECG parameters on receiving 0.125 mg/kg and 0.250 mg/kg alaglizin, respectively. Alaglizin in the model had an ED_{50} of 0.125 mg/kg body weight against 0.290 mg/kg in LA and thus contained 14 times lower lappaconitine.

The NF:GA complex exerted hypotensive action on intravenous injection of its water solution in rats at a dose with ten times lower NF [51, 52]. The hypotensive effect of GA complexing with nisoldipine (another dihydropyridine) was similar to that of complexes with AG.

Complexing can have an important consequence of amplifying the pleiotropic effect of pharmacons (see above), such as NF. The wanted antiarrhythmic effect of NF can be only achieved with a dose that causes almost critical blood pressure drop, but the NF:GA complex induces the same effect with 29 times lower NF than the hypotensive dose.

Thus, the NF:GA complex is a promising parenteral drug with universal activity against hypertensive crises attendant with arrhythmia.

A presumable action mechanism of the NF:GA complex was studied *in vitro* on neurons isolated from peripharyngeal ganglia of *Lymnacea stagnalis* molluscs. The provoked responses are arrested completely with an NF concentration as high as 3.0 mM but with 30 times as low concentration (0.1 mM) of NF:GA. The responses to NF are blocked faster and recover sooner after neuron washing than the responses induced by the complex, which indicates stronger binding of the latter with receptors and its more prolonged action. The higher receptor affinity of the complex is corroborated by comparing the NF and NF:GA effects on calcium channels [51] (Fig. 8)

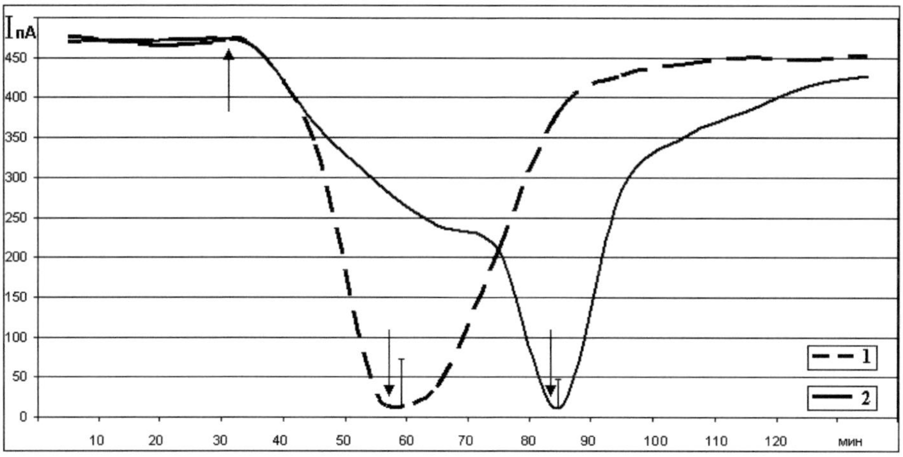

1. amplitude of calcium currents induced by Nifedipine.
2. amplitude of calcium currents induced by Nifedipine clathrate with glycyrrhizic acid.
Upward arrows show start point of the blocker action. Downward arrows show start point of blockers washout.
X axis set as time in minutes.
Y axis set as amplitude of incoming current in pA.

Figure 8. Averaged changes in calcium current amplitude induced by action and washout of blockers within a group of neurons (12 cells in each group).

The propranolol-GA complex studied in terms of hypotensive activity ensured 14% blood pressure drop in normotensive animals at a dose of 0.2 mg/kg body weight (minimum dose 0.0025 mg/kg gave a 11% decrease). Pure propranolol decreases blood pressure for 11% in a dose of 0.2 mg/kg; a similar pressure drop may be achieved by the minimum dose 0.0025 mg/kg, but it is statistically unconfident. Therefore, complexing enhances and stabilizes the hypotensive action of propranolol and allows 12.5-fold reduction of its dose.

Furthermore, complexing amplifies the antiarrhythmic action (pleiotropic effect) of propranolol. Namely, 100% of animals exposed to 0.3 mg/kg arrhythomogenic 0.1 % adrenalin survived on administration of 0.0025 mg/kg GA:Propranolol, while only 40 %

survived among those who received 0.0002 mg/kg propranolol (this being the dose contained in the complex). Control animals responded to adrenalin by fatal disrhythmia grading into ventricular fibrillation, and 100% of them died [53].

The complex of GA with anticoagulant warfarin was investigated in a dose of 20 mg/kg body weight, which corresponds to 2 mg/kg WF. Prothrombin time was first measured six hours after a single intragastric injection. The interval was chosen proceeding from pharmacokinetics of warfarin: its plasma metabolites reach the maximum in 6-12 hours in rats [54]. In our experiments, PT has shown a confident change only in the positive control group with 2 mg/kg WF, but it is not clinically significant as it fails to ensure the required increase in coagulation time. A significant PT increase was observed on single intragastric administration of WF, while the WF:GA complex induced a smooth PT rise as late as in 30 hours (two injections) after the beginning of the experiment and the values corresponding to the positive control (WF) were reached only in 54 hours (three injections). Thus, WF:GA complexing made WF more soluble in water but slowed down its wanted anticoagulant action, possibly because GA molecules screened its active centers (Fig. 9).

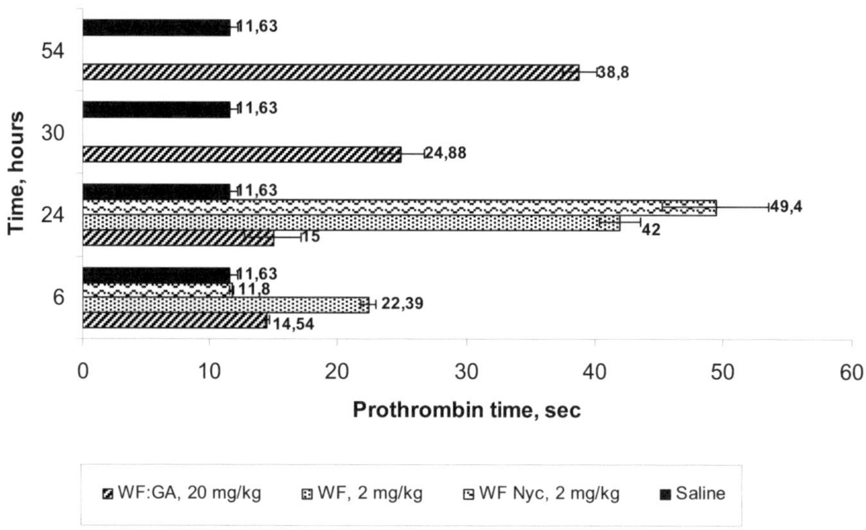

Figure 9. Prothrombin time of WF:GA

3.2.4. *Psychotropic drugs*

The complexation effect was discovered in a pharmacological study of GA complexes with antidepressant fluoxetine and with anxiolytic gamma-amino-beta-phenylbutirate hydrochloride. Fluoxetine (FL), N-methyl-3-phenyl-3-[4-(trifluoromethyl)phenoxy]propan-1-amine acts as a depression antagonist by inhibiting serotonin neuronal uptake in CNS. Antidepressants are known to have many drawbacks, such as large doses, a narrow activity spectrum, high toxicity and prolonged elimination that involves liver cells and exerts deleterious effects on the renal function.

The study of fluoxetine complexes with GA (hereafter fluaglizin) aimed primarily at checking the possibility to alleviate the side effects of the pharmacon. In preliminary testing, the 1:10 FL:GA complex showed the highest activity. The complex, with 0.072 pharmacon parts in its one weight part, was patented under the name of fluaglizin (FG) [55,56]. Note that its LD_{50} exceeds 5000 mg/kg body weight against LD_{50} =248 mg/kg in fluoxetine.

Fluaglizin exerts a more prominent antidepressant effect than fluoxetine on single-dose administration in the Porsolt test and is a stronger serotonin uptake inhibitor. For instance, it suppressed the action of chloral hydrate more effectively than fluoxetine in a test with 5-hydroxytryptophan. Like fluoxetine, fluaglizin lacks anxiolytic activity.

The antidepressive effect of fluaglizin was identical to that of fluoxetine in a model of social depression in mice, namely, the animals became twice more communicative toward familiar and unfamiliar partners (had twice greater frequency and time of contacts). The dose of fluoxetine in the FL:GA complex is 1.08 mg/kg body weight against the standard 15 mg/kg. Fluaglizin, like fluoxetine, prevents blood glucose drop and attenuates peroxidation normalizing the antioxidant status of depressed individuals [55, 56].

In order to understand the mechanism of FL:GA action and compare it with fluoxetine, we studied its effect on contents of catecholamines and their precursors in different brain parts on single and therapeutic 25 mg/kg administration. The 17 times lower dose of fluoxetine in FL:GA complexes induced a weaker effect on serotonin uptake and triggered dophamine exchange in brain [57]. The nootropic activity of fluoxetine, first mentioned in [48], shows up also in the FL:GA complex.

Fluoxetine (30 μM) is known to suppress epileptiform activity which is evident in 50% lower-amplitude oscillations of electric potential (reflecting the activity of nerve cells) in response to pulse stimulation of hippocampus on the background of picrotoxin action.

Like fluoxetine, fluaglizin inhibits bikukulin-induced epileptiform activity in hippocampus sections in rats and acts as epilepsy antagonist [48].

Phenybut (FB), 4-amino-3-phenyl-butyric acid, is a nootropic and tranquilizing drug which relieves stress and anxiety and improves sleep. It is used in clinical practice against asthenia, neurotic anxiety, and sleep disorders, prior to surgery, and for preventing naupathia. It, however, has drawbacks of inducing somnolence and allergic reactions.

The FB:GA complex is twice less toxic than FB and stimulates cognitive activity in the same way as the pharmacon and GABA but, unlike the two latter, it provides a 20% increase in memorizing abilities in animals and attenuates sedative effects [48].

3.2.5. Anticancer drugs

GA complexes (1:1) with 5-fluorouracil, tegafur, and daunorubicin hydrochloride were synthesized in solution [58]. Complexing decreases the toxicity of the drugs and increases their solubility in water. The GA complex with fluorouracil exerts antineoplastic action with respect to Pliss lymphosarcoma, B-16 melanoma, and Geren carcinoma. The effective indices are, respectively, 3.05; 2.11, and 1.76. The tumor growth inhibition is 67.2%; 53.4 %; 87.1 %.

3.3. Antimicrobial drugs

GA complexes with antibiotics (chloramphenicol) and sulfanilamides (sulfamethoxypyridazine, saladimetoxine, sulfamonomethoxine, sulfadimidine, and sulfaguanidine), as well as with isoniazid and nitrofural, 1:1, were likewise synthesized in solution and compared in terms of their microbial activity. The highest survival rate (90%) was observed on the 10th day after exposure to infection in animals with staphylococcosis that received 50 mg/kg GA-chloramphenicol, while only 30% survived when pure chloramphenicol was given. The percentage of survived animals with *Pseudomonas aeruginosa, Proteus vulgaris,* and *E. coli* infections reached 80% on GA: chloramphenicol administration but it was only 20-50% on treatment with free chloramphenicol. The complex was also shown to stimulate humoral and cell immune responses [59].

3.4. Simvastatin (hypocholesterolemic drug)

Inhibitors of 3-hydroxy-3-methylglutaryl coenzyme A (3HMG-CoA) reductase, the so-called statins, are known to successfully reduce low-density lipoproteins, and are used for this in atherosclerosis therapy. However, most statins cause side effects and have to be replaced by safer drugs, with a more prolonged action. NMR spectroscopy of the behavior of simvastatin (SMS) in solutions with GA indicates formation of stable complexes. The synthesized SMS:GA complex is stable in water solutions at GA above 0.2 mM [60]. The complex patented as simvaglysine (SMG) [61] demonstrates uncompetitive inhibition of 3HMG-CoA reductase. SMG, in doses with three times lower statin, turns out to be more potent and safer than SMS.

Thus, simvaglysine acts as an uncompetitive inhibitor/proinhibitor of the 3HMG-CoA reductase reaction by inducing inhibition of cholesterol synthesis in liver microsomal fraction of rats *in vitro,* being no less potent than simvastatin. Within the inhibition constants from 100 to 300 nM, SMS inhibits 37.7- 42.0% mevalonate formation, while SMG provides a 31-33% total blood cholesterol decrease after 14 days of administration in rats with hypercholesterolemia, at doses with 3 times lower SMS, which is as effective as the therapeutic dose of simvastatin. The greater safety of SMG is confirmed by a lower blood CPK (creatine phosphokinase) increase after 14 days of treatment than in the case of SMS: 2.3 lower than with 2-5 times larger SMS doses [60, 61].

4. Conclusion

Thus, water-soluble molecular complexes of various polysaccharides and glycyrrhizic acid with drugs that normally dissolve poorly in water have been synthesized and tested for binding strength and pharmacological activity in comparison with the constituent drugs.

The suggested mechanochemical synthesis of solid dispersed systems "drug-complexing agent" ensures high strength of the complexes, on condition of low-energy nondestructive treatment of polysaccharide macromolecules.

Arabinogalactan, a water soluble polysaccharide of *Larix sibirica* Ledeb. and *Larix gmelinii* (Rupr.), when used as a complexing agent, provides the highest solubility among other studied poly- and oligosaccharides. Another advantage of arabinogalactan is its exceptional, almost infinite, raw material source in the Northern Hemisphere (in Russia and Canada), as well as the availability of extraction and purification technologies [62].

The molecules of poorly water soluble drugs can be carried by micelles that form in water solutions of glycyrrhizic acid. Complexing with polysaccharides, in the same way as with GA, increases the solubility of pharmacons and allows reducing significantly (up to ten times) the effective dose avoiding, at the same time, some unwanted side effects.

Thus, a new perspective is opening of obtaining highly effective and safe drugs and new ways of drug delivery.

Author details

A. V. Dushkin
Institute of Solid State Chemistry and Mechanochemistry, Siberian Branch,
Russian Academy of Sciences, Russia

T. G. Tolstikova, M. V. Khvostov and G. A. Tolstikov
N. N. Vorozhtsov Novosibirsk Institute of Organic Chemistry, Siberian Branch,
Russian Academy of Sciences, Russia

Acknowledgement

This work is supported in part by the program of partnership fundamental research of the SB RAS, Grant 19.

5. References

[1] Anil J Shinde (2007) Solubilization of Poorly Soluble Drugs: A Review. Pharmainfo.net. Available: http://www.pharmainfo.net/reviews/solubilization-poorly-soluble-drugs-review. Submitted 2007 Nov 13.

[2] Payghan S, Bhat M, Savla A, Toppo E, Purohit S (2008) Potential Of Solubility In Drug Discovery And Development. Pharmainfo.net. Available:

http://www.pharmainfo.net/ reviews/potential-solubility-drug-discovery-and-development. Submitted 2008 Oct 21.
[3] Yellela S.R. Krishnaiah (2010) Pharmaceutical Technologies for Enhancing Oral Bioavailability of Poorly Soluble Drugs. Journal of Bioequivalence & Bioavailability. 2(2): 028-036.
[4] Robert H. Marchessault, François Ravenelle, Xiao Xia Zhu editors (2006) Polysaccharides for Drug Delivery and Pharmaceutical Applications. V.934. ACS Symposium Series. 365p.
[5] Gordon A. Morris, M. Samil Kok, Stephen E. Harding and Gary G. Adam (2010) Polysaccharide drug delivery systems based on pectin and chitosan. Biotechnology and Genetic Engineering Reviews. 27: 257-284.
[6] Sabyasachi Maiti, Somdipta Ranjit, Biswanath Sa (2010) Polysaccharide-Based Graft Copolymers in Controlled Drug Delivery. International Journal of PharmTech Research. 2(2):1350-1358.
[7] Mark E.Davis, Marcus E. Brewster (2004) Cyclodextrin-based pharmaceutics: past, present and future. Nature reviews I Drug discovery. 3: 1023-1035.
[8] Erem Bilensoy, editor (2011) Cyclodextrins in Pharmaceutics, Cosmetics, and Biomedicine: Current and Future Industrial Applications. WILEY. 440p.
[9] A.M.Dubinskaya (1999) Transformations of organic compounds under the action of mechanicalal stress. Russian Chemical Reviews. 68(8): 637- 652.
[10] Dushkin A.V. (2004) Potential of Mechanochemical Technology in Organic Synthesis and Synthesis of New Materials. Chemistry for Sustainable Development. 12(3): 251 – 273.
[11] Dushkin A.V. (2010) Mechanochemical synthesis of organic compounds and rapidly soluble materials : in M.Sopicka-Lizer, editor, High-energy ball milling. Mechanochemical processing of nanopowders. Woodhead Publishing Limited. Oxford, pp. 249-273.
[12] A.V. Dushkin, E.V.Nagovitsina, V.V.Boldyrev, A.G.Druganov (1991) Mechanochemical reactions of solid organic compounds. Siberian Chemical J. 5: 75 -81.
[13] A.V.Dushkin,_I.B.Troitskaya,V.V.Boldyrev and I.A.Grigorjev (2005) Solid-phase mechanochemical incorporation of a spin label into cellulose. Russian Chemical Bulletin, International Edition, 54(5): 1155—1159.
[14] E.S.Meteleva, A.V.Dushkin, V.V.Boldyrev (2007) Mechanochemical obtaining of chitosan's derivatives. Chemistry for Sustainable Development. 15: 127-133 (rus).
[15] A. V. Dushkin, L. M. Karnatovskaya and others. (2001) Synthesis and evaluation of ulcerogenic activity of instant disperse solid systems based on acetylsalicylic acid and biologically active licorice components. Pharmaceutical Chemistry Journal. 35(11): p.605-607.
[16] A,V.Dusjkin, N.V.Timofeeva (2001) Method of producing quick soluble tabletted form of acetyl-salicylic acid, Russia Patent 2170582.
[17] A.V.Dushkin, S.A.Guskov, V.N.Bugreev (2005), Method of preparing powder-like water-soluble sparking composition, Russia patent 2288594.

[18] Gus'kov S, Dushkin A and Boldyrev V (2007) Physical-chemical bases of mechanochemical obtaining of rapidly soluble disperse systems. Chemistry for Sustainable Development. (15): 35-43 (rus).

[19] Dushkin A, Karnatovskaia L and others (2000), Solid-state mechanochemical synthesis of fluoroaromatic compounds, Doklady Chemistry, 371, 632-635.

[20] Dushkin A, Karnatovskaia L and others. (2001b), Solid-phase mechanochemical synthesis of fluoroaromatic compounds, Synth. Commun., 31, 1041-1045.

[21] A.V.Dushkin, Z.Yu.Rykova, V.V.Boldyrev, T.P.Shaktshneider (1994) Aggregation processes in the reactivity of mechanicalally activated organic solids. Int. J. Mechanochem. Mechanochem. Alloying. 1: 1 – 10.

[22] A.V.Dushkin, E.S.Meteleva, T.G.Tolstikova and others (2008) Mechanochemical preparation and pharmacological activities of water-soluble intermolecular complexes of arabinogalactan with medicinal agents. Rus. Chem. Bull. 6: 1299-1307.

[23] A.V.Dushkin, E.S.Meteleva, T.G.Tolstikova, M.V.Khvostov, G.A.Tolstikov (2010) Mechanochemical Preparation and Properties of Water-Soluble Intermolecular Complexes of Polysaccharides and β-Cyclodextrin with Pharmaceutical Substances. Chemistry for Sustainable Development. 18(6):719-728.

[24] A.V.Dushkin, T.G.Tolstokova and others (2008) Water-soluble pharmaceutical composition and method for its preparation. Russia patent 2337710.

[25] N.E.Polyakov, T.V.Leshina, E.S.Meteleva, A.V.Dushkin and others (2009) Water Soluble Complexes of Carotenoids with Arabinogalactan. J. Phys. Chem. B. 113 (1): 275-282.

[26] T.G.Tolstikova, G.A.Tolstikov and others (2008) Drugs for treatment of hypertension and heart rhythm disorders. Russia patent 2391980.

[27] A.V.Dushkin, E.S.Meteleva and others (2011) The composition of high pharmacological activity on the basis of dihydroquercetin and plant polysaccharides (options). Russia patent 2421215.

[28] L.N.Pribytkova, S.A.Gus'kov, A.V.Dushkin, S.I.Pisareva (2011) Mechanochemical Obtaining of Water-soluble Composition on the Base of Quercetin. Chemistry of Natural Compounds. 3: 333-336.

[29] A. V. Dushkin, E. S. Meteleva, T. G. Tolstikova and others (2010) Complexation of Pharmacons with Glycyrrhizic Acid as a Route to the Development of the Preparations with Enhanced Efficiency. Chemistry for Sustainable Development 18(4): 517-525.

[30] T.G.Tolstikova, M.V.Khvostov and others (2010) Improvement of pharmacological values of the nifedipine by means of mechanochemical complexation with glycyrrhizic acid. Biomedical chemistry. 56(2): 187-194 (rus).

[31] Tolstikova T. G., Khvostov M. V. And others (2011) Alteration of Warfarin's Pharmacologic Properties in Clathrates with Glycyrrhizic Acid and Arabinogalactan. Letters in Drug Design & Discovery. 8(3): 201-204.

[32] S.S.Khalikov, A.V.Dushkin, M.S. Khalikov and others (2011) Mechanochemical modification of properties of anthelmintic preparations. Chemistry for Sustainable Development. 19(6): 699-703 (rus).

[33] T. Loftsson, M.E. Brewster (1996) Pharmaceutical Applications of Cyclodextrins. 1. Drug Solubilization and Stabilization. J.Pharm.Sci. 85(10): 1017-1025.

[34] Ramnik Singh, Nitin Bharti, Jyotsana Madan, S. N. Hiremath. (2010) Characterization of Cyclodextrin Inclusion Complexes – A Review. Journal of Pharmaceutical Science and Technology. 2 (3): 171-183.

[35] Clarcke A.E., Anderson R.L., Stone B.A. (1979) Form and function of arabinogalactans and arabinogalactan-proteins. Phytochemistry. 18: 521–540.

[36] Marta Izydorczyk, Steve W. Cui, and Qi Wang (2005) Polysaccharide Gums: Structures, Functional Properties, and Applications. Food Carbohydrates Chemistry, Physical properties and Applications. Available: http://uqu.edu.sa/files2/tiny_mce/plugins/filemanager/files/4300270/1/2/1574_C006.pdf.

[37] Philippe H. Hünenberger, Ulf Börjesson, Roberto D. Lins (2001) Electrostatic Interactions in Biomolecular Systems. Chimia. 55: 861–866.

[38] C Oprea, F Dan (2006) Macromolecular Mechanochemistry Polymer Mechanochemistry. Cambridge International Science Publishi. (1)1: 495p.

[39] A.V.Dushkin, E.S.Meteleva, T.G.Tolstikova, M.V.Khvostov (2012) GPC and toxicology investigation of mechanochemical transformation of water-soluble polysaccharides. Pharmaceutical Chemistry Journal. 2012(7) In press.

[40] A.M.Striegel, J.J.Kirkland, W.W.Yau, D.D.Bly (2009) Modern Size Exclusion Liquid Chromatography, Practice of Gel Permeation and Gel Filtration Chromatography, 2nd ed.; Wiley: NY. 494p.

[41] N.Grassie, G.Scott (1985) Polymer degradation and stabilization. Cambridge University Press.(1985: 201-207.

[42] J.H.Prescott, P.Enriquez, Chu Jung, E.Menz and others (1995) Larch arabinogalactan for hepatic drug delivery: isolation and characterization of a 9 kDa arabinogalactan fragment. Carbohydrate Research. 278: 113-128.

[43] T.V.Romanko, Yu.V.Murinov (2001) Some features of the flow of dilute solutions of glycyrrhizic acid. Rus.J.Phys.Chem A. 75(9): 1459-1462.

[44] N.E.Polyakov, V.K.Khan, M.B.Taraban, T.V.Leshina (2008) Calcium Receptor Blocker Nifedipine with Glycyrrhizic Acid. J. Phys. Chem. B. 112: 4435-4440.

[45] H.Determan. (1969) Gel Chromatography. Springer Verlag, New York. 252p.

[46] R.U.Khabriev (2005) Guidance for experimental (preclinical) study of new pharmacological compounds, Medicine, Moscow, 832p. (rus).

[47] T.G.Tolstikova, M.V.Khvostov, A.O.Bryzgalov (2009) The Complexes of Drugs with Carbohydrate-Containing Plant Metabolites as Pharmacologically Promising Agents. Mini-Reviews in Medicinal Chemistry. 9: 1317-1328.

[48] G.A.Tolstikov, L.A.Baltina, V.P.Grankina, R.M.Kondratenko, T.G. Tolstikova (2007) Licorice: biodiversity, chemistry and application in medicine, Academic Publishing House "GEO", Novosibirsk, 311p. (rus).
[49] G.A.Tolstikov, Yu.I.Murinov and others (1991) Complexes of β-glycyrrhizic acid with prostaglandins – new class of uterotonic agents. Pharm. Chem. J. 3: 42-44 (rus).
[50] A.O.Bryzgalov, T.G.Tolstikova, I.V.Sorokina and others (2005) Antiarrhythmic activity of alaglizin. J. Expert Clin. Pharm. 68(4): 24-28 (rus).
[51] T.G.Tolstikova, A.O.Bryzgalov, I.V.Sorokina, M.V.Hvostov and others (2007) Increase in Pharmacological Activity of Drugs in their Clathrates with Plant Glycosides. Letters in Drug Design & Discovery. 4: 168-170.
[52] T.G.Tolstikova, M.V.Khvostov, A.O.Bryzgalov, A.V.Dushkin, E.S.Meteleva (2009) Complex of Nifedipine with Glycyrrhizic Acid as a Novel Water-Soluble Antihypertensive and Antiarrhythmic Agent. Letters in Drug Design & Discovery. 6: 155-158.
[53] T.G.Tolstikova, G.I.Lifshits, M.V.Khvostov, A.O.Bryzgalov, I.V.Sorokina. (2007) Complexing in the development of low-dose beta-adrenoceptor antagonists (experimental study). Herald of the NSU. Series: biology, clinical medicine. 2(5): 70-73 (rus).
[54] X.Zhu,. W.G.Shin (2005) Gender Differences in Pharmacokinetics of Oral Warfarin in Rats. Biopharm. Drug Dispos. 26: 147-150.
[55] D.F.Avgustinovich, I.L.Kovalenko, I.V.Sorokina, T.G.Tolstikova, A.G.Tolstikov (2004) Ethological study of antidepressant effect of fluaglisine in conditions of chronic social stress in mice. Bul. Exp. Biol. and Med. 137(1):99-103.
[56] T.G.Tolstikova, I.V.Sorokina, I.L.Kovalenko and others (2004) Influence of clathrate formation to the pharmacon activity in complexes with glycyrrhizic acid. Doklady Akademii Nauk. 394(2):707-709.
[57] G.T.Shishkina, N.N.Dygalo, A.M.Yudina and others (2005) Effects of Fluoxetine and Its Complexes with Glycyrrizhinic Acid on Behavior and Brain Monoamine Levels in Rats. I. P. Pavlov Journal of Higher Nervous Activity. 2: 207-212 (rus).
[58] L.A.Baltina, Yu.I.Murinov, A.F.Ismagilova, G.A.Tolstikov and others (2003) Production and antineoplastic action of complex compounds of β-glycerrhizic acid with some antineoplastic drugs. Pharm. Chem. J. 37: 3-4 (rus).
[59] R.M.Kondratenko, L.A.Baltina, G.A.Tolstikov and others (2003) Complex compounds of glycerrhizic acid with antimicrobial drugs. Pharm. Chem. J. 37: 32-35 (rus).
[60] V.A.Vavilin, N.F.Salakhutdinov, Yu.I.Ragino and others (2008) Hypocholesteremic properties of complex compound of simvastatin with glycerrhizic acid (simvaglizin) in experimental models. Biomed. Chem. 54: 301-313.
[61] Yu.I.Ragino, V.A.Vavilin, N.F.Salakhutdinov and others (2008) Examination of the hypocholesteremic effect and safety of simvaglizin on the model of hypercholesterolemia in rabbits. Bull. Exp. Biol. Med. 145: 285-287.

[62] E. N. Medvedeva, V. A. Babkin, L. A. Ostroukhova (2003) Larch arabinogalactan - features and prospects, Review Chemistry of Vegetable Raw Materials [Khimiya rastitel'nogo syr´y], 1: 27-37 (rus).

Chapter 23

The Chitosan as Dietary Fiber: An *in vitro* Comparative Study of Interactions with Drug and Nutritional Substances

Máira Regina Rodrigues, Alexandre de Souza e Silva and Fábio Vieira Lacerda

Additional information is available at the end of the chapter

http://dx.doi.org/10.5772/48311

1. Introduction

Obesity is considered a disease that has grown significantly in the last two decades worldwide. The concern with this increase is justified because obesity develops and remains due to different factors, causing several other diseases and sometimes, can lead to death. Several strategies have been searched to control its progress usually through therapies and, in critical cases, surgeries. Despite all efforts, statistical studies show that obesity is still growing consequently generates the necessity of further studies on the subject (Dacome, 2005).

Epidemiological researches have studied the impact of overweight and obesity on the risk of chronic disease, as coronary heart disease, type 2 diabetes mellitus, hypertension, stroke, dyslipidemia, insulin-resistance, glucose intolerance, metabolic syndrome, and cancers of the breast, endometrium, prostate and colon (Aslander-van Vliet et al., 2007). Health consequences and compromised quality of life associated with obesity provide incentives to abate the obesity epidemic. However, despite recognition of these effects, the epidemic of obesity and overweight is not reversed (Johnson et al., 2007).

In general, treatments for obesity are based on regular exercise, nutritional reeducation, pharmacological treatment, behavioral therapy and use of dietary fibers that promote the reduction of fat absorption (Aslander-van Vliet et al., 2007). Much research has been conducted on the dietary supplements that promote the reduction of body weight and fat mass (Saper et al., 2004). These ingredients reportedly act as a fiber to increase satiety and also to decrease the absorption of fat by binding to it (Kumirska et al., 2010).

INTECH
open science | open minds

© 2012 The Author(s). Licensee InTech. This chapter is distributed under the terms of the Creative Commons Attribution License http://creativecommons.org/licenses/by/3.0), which permits unrestricted use, distribution, and reproduction in any medium, provided the original work is properly cited.

A natural substance that helps in these anti-obesity treatments that has been highly recommended to control obesity is chitosan (Hennen, 2005). Chemically speaking, chitosan (Figure 1) is a linear polysaccharide of β(1→4)-linked-2-amino-2-deoxy-D-glucopyranose obtained by deacetylation of chitin, the main component of the exoskeleton of insects and crustaceans (Kumirska et al., 2010). It has many important properties, such as non-toxicity, biocompatibility, biodegradability, antimicrobial activity, chemical reactivity (Cummings et al., 2010), industrial applications (Hennen, 2005), as well as carrier for body fat (Ni Mhurchu et al., 2004; Ni Mhurchu et al., 2005; Jull et al., 2008; Lois & Kumar, 2008), cholesterol and triglyceride (Razdan & Petterson, 1994; Liu et al., 2008; Zhang et al., 2008). Many mechanisms (Tapola et al., 2008; Prajapati, 2009) to explain the carriers and absorptive properties of microenvironment produced by chitosan in solution have been proposed.

However, the use of chitosan is still controversial, and studies in favor and against the use of chitosan have been constantly reported. Many studies have confirmed the hypocholesterolemic activity of chitosan (Sugano et al., 1978; Liao et al., 2007; Yao et al., 2008; Liu et al., 2008; Zhang et al., 2008). The same way, works have reported that the triglyceride and cholesterol absorption have been inhibited and the cholesterol concentration of mice fed with a high fat diet plus chitosan has been decreased (Razdan & Petterson, 1994; Liu et al., 2008; Zhang et al., 2008). Other studies reported that chitosan is efficacious in facilitating the reducing body fat and weight loss in obese individuals (Schiller et al., 2001; Kaats et al., 2006).

On the other hand, studies have shown that oral administration of chitosan has weak action on the reduction of triglyceride and plasma cholesterol in rabbits (Hirano & Akiyama, 1995). Other works have reported that the effect of chitosan on body weight is minimal and unlikely to be of clinical significance (Ni Mhurchu et al., 2004; Ni Mhurchu et al., 2005; Lois & Kumar, 2008, Jull et al., 2008), as well as that the fat trapped was clinically insignificant in studies with overweight adults treated with chitosan capsules before each meal (Pittler et al., 1999; Pittler & Ernst, 2004; Gades & Stern, 2005).

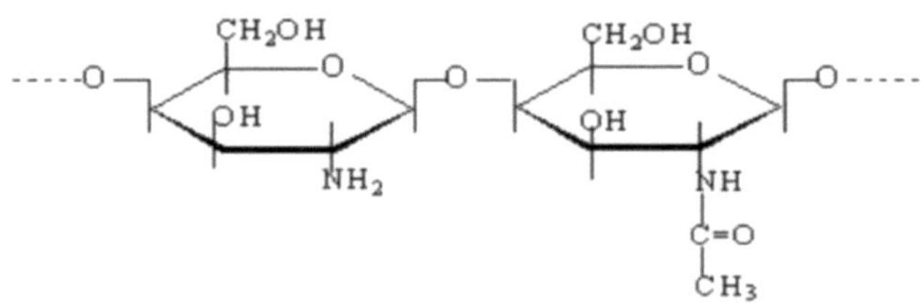

Figure 1. Chemical structure of chitosan.

Is well known that chitosan produces microenvironments with carriers and absorptive properties in acidic aqueous solution. These begin to form above a certain concentration, critical aggregate concentration, CAC (Rodrigues, 2005). The mechanism of solubilization of molecules is well known (Rodrigues, 2005; Rodrigues et al., 2008) however, the process by which chitosan acts as a carrier of fat is not yet fully understood and two mechanisms have been suggested (Prajapati, 2009; Tapola et al., 2008). One of these mechanisms describe the effect of chitosan fiber network, were chitosan also binds neutral lipids like cholesterol and triglycerides through hydrophobic bonds (Tapola et al., 2008; Prajapati, 2009). In other mechanism, the positive charges (NH_3^+ group generated by stomach acids) on chitosan attract and binds to fatty and bile acids (both negatively charged). This complex is indigestible by the body and excreted in the feces (Tapola et al., 2008; Prajapati, 2009).

Regardless of the solubilization mechanism, nutrients can also be solubilized in chitosan microenvironments, as reported in some studies. Works demonstrated that chitosan causes significant decrease in protein digestibility (Deuchi et al., 1994) and its effect on nutrient digestibility (Ho et al. 2001). Nevertheless, studies on the interaction of chitosan with nutrients are still rare and inconclusive (Gades & Stern, 2005; Hennen, 2005; Kaats et al., 2006; Barbosa et al., 2007; Tapola et al., 2008).

In this context, we present a comparative study of interactions of the chitosan with molecules of two vitamins and one drug. To each molecule, the study was conducted in acid aqueous solution, condition similar to the stomach environment, where occurs formation of chitosan gel responsible for solubilizing molecules.

Drug fluoxetine was chosen for this study. The need for anti-depressive drugs with few side effects, as anticholinergic activity and cardiovascular accidents, boosted the development of new anti-depressant compounds (Böer et al., 2010), as fluoxetine, which inhibits the uptake of serotonin by the neurons in the brain, enhances serotonin neurotransmission and had the longest half-life that other selective serotonine reuptake inhibitors (SSRIs) (Rizo et al., 2011). The precise mechanism of action is not clear but it has less cardiovascular, sedative and anticholinergic effects than the tricyclic antidepressant drugs (Shah et al., 2008). The main indications for the prescription of fluoxetine are for obsessive-compulsive disorder, depression therapy, bulimia nervosa, alimentary disorders and obesity (Suarez et al., 2009). Besides drug, the nutritional reeducation and intake of dietary fibers as chitosan has been recommended in treatments for obesity (Aslander-van Vliet et al., 2007). Based on the possible concurrent use of fluoxetine and chitosan, it is important to evaluate the interactions between both substances.

Vitamins chosen for this study were the B2 and B12. Vitamin B2 or riboflavin is a vitamin B complex that participates in numerous metabolic reactions and physiological functions (United States Pharmacopeia, 2007). Vitamin B12 or cyanocobalamin is an essential component in human diet, plays a key role in cell nucleus, enzymatic processes in the mitochondria, and cytoplasm; it is necessary for the synthesis of red blood cells, for the maintenance of the nervous system, and for the growth and development in children (Wang et al., 2007). Both vitamins are not produced by the body and are consumed only in small quantities (Sommer, 2008); deficiency can cause many diseases (Sun et al., 2007).

The interactions between chitosan-vitamin and chitosan-drug have been verified by monitoring the photophysical properties of these components. For this, fluorescence and UV-Vis absorption measurements were initially evaluated in acid aqueous solution and after in weakly acidic solution of chitosan given information about the interactions between this chemical component in conditions that approaches the stomach chemical environment.

2. Experimental

2.1. Chemicals

Chitosan and fluoxetine hydrochloride were purchased from Aldrich Chemical Co. (St. Louis, MO, USA), and vitamin B2 (riboflavin, 96%) and vitamin B12 (cyanocobalamin,, USP Grade) from Vetec Co. (Duque de Caxias, RJ, Brazil).and Merck Co. (Darmstadt, Hessen, Federal Republic of *German), respectively.* Other chemicals were ultraviolet/high-performance liquid chromatography grade and used without further purification; ultrapure water was supplied by a Milli-Q system.

2.2. Spectroscopic measurements

Previous studies have shown that the best conditions to solubilize chitosan are: chitosan 1% (w/v) dissolved in aqueous solution of glacial acetic acid 1% (v/v) under stirring (Signini & Campana Filho, 1999; Rodrigues, 2005; Rodrigues et al., 2008), so measurements in presence of chitosan were conducted in these conditions.

Chitosan has no fluorescence emission or absorption in the experimental conditions. The absorption spectra of chemicals (fluoxetine, B2 and B12) were measured using quartz cuvettes with 1 cm of optical pathway. The fluorescence measurements were performed at λexcit= 275 nm and λemis= 305 nm for vitamin B12 (Li and Chen, 2000); at λexcit= 440 nm and λemis= 305 nm for vitamin B2 (United States Pharmacopeia, 2007; Association of Official Analytical Chemists, 2005); and at λexcit= 230 nm and λemis= 290 nm for fluoxetine (United States Pharmacopeia, 2007; Association of Official Analytical Chemists, 2005).

Absorbance measurements were taken at maximum of the absorption spectra and performed at room temperature.

Initially, variations of both fluorescence and absorption spectra of the chemicals (fluoxetine, B12 and B2) were taken as a function of their concentration in acid aqueous solution and after in different concentrations of chitosan in aqueous acid solution, at the same range of chemicals concentration. The variation in the spectra of the chemicals (fluoxetine, vitamins B2 and B12) was also studied by keeping fixed the concentration of vitamin and varying the concentration of chitosan in acid aqueous solution).

3. Results and discussion

Absorption spectra and chemical structures of chemicals are shown in Figure 2. Spectral data in Figure 2.a shows that vitamin B12 absorb significantly within 425–600 nm range, as described in the literature (Zheng & Lu, 1997; British Pharmacopoeia, 1998) and the present work chose the absorption peak at ~550 nm to assess the spectral behavior. Similar graphs have been obtained for vitamin B2 and fluoxetine. Figure 2.b shows that the absorption maximum for the vitamin B2 occurs at ~440 nm, data consistent with the literature ((United States Pharmacopeia, 2007). Figure 2.c. shows the absorption spectrum of fluoxetine consistent with the literature (Fregonezi-Nery et al., 2008) that exhibits two absorption maxima at 270 and 275 nm. The last one maximum was chosen to monitor the spectral behavior of fluoxetine.

The chitosan-chemicals (fluoxetine, vitamins B2 and B12) interaction have been studied in aqueous acid solution by the monitoring the fluorescence and UV-visible spectra of chemicals, each monitored separately.

In three cases, the increase in concentration of chemical causes an increase in both absorption and fluorescence intensities due to the increase of species that absorb and emit light and this increases is linear profile always indicating that self-aggregation processes are not occurring in this concentration range (data not shown).

Subsequently, chemicals (fluoxetine, vitamins B2 and B12) were studied in the absence and the presence of chitosan, at concentrations 0.050 $g.L^{-1}$, 0.60 $g.L^{-1}$ and 1.0 $g.L^{-1}$ of polysaccharide, keeping fixed the chemicals concentration ($8.5x10^{-5}$ $mol.L^{-1}$). With the increase of chitosan concentration both fluorescence and absorption intensities of chemicals are increased.

Figures 3, 4 and 5 show the behavior of fluorescence intensities to fluoxetine, vitamin B12 and B2, respectively. In all graphics, fluorescence intensities significantly increase when chitosan concentration goes from zero to 1.0 $g.L^{-1}$. This is a common behavior of fluorescent molecules when they migrate from the solution for environment of different polarity (Kalyanasundaram, 1987) and is due to the influence of microenvironment formed by chitosan on the photophysics of the chemical that is changed due to spatial hindrance that it suffers and due to loss of part of rotational freedom of substituent groups, (Kalyanasundaram & Thomas, 1977; Valeur, 2001). Then, with increase concentration of chitosan, the microenvironment becomes more rigid and the lifetime of the chemicals (fluoxetine, vitamins B2 and B12) in the excited states are living longer (Kalyanasundaram, 1987). However, fluorescence intensities of vitamin B12 and fluoxetine show similar increase rate while vitamin B2 is markedly lower. The increase of fluorescence intensities with the polysaccharide concentration has been observed also to vitamin in pharmaceutical formulations containing dextran (Alda et al., 1996).

Absorption intensities of chemicals (fluoxetine, vitamins B2 and B12) increase with the chitosan concentration similarly to the of fluorescence intensities, Figure 6. The reason for this behavior is the increased stiffness of environment generated by chitosan chains. However, in this case, intensities show the following increasing order: vitamin B2, fluoxetine and vitamin B12.

In chemical structure of all chemicals (fluoxetine, vitamins B2 and B12) there are rings with double bonds and polar groups that can interact strongly with the similar groups of chitosan. There are also OH groups in the molecular structure of chitosan favor hydrogen-bonding type interactions with polar groups of chemicals. These interactions can influence the absorption and the emission processes of radiation of molecules reducing the rotational degrees of freedom of the molecule (Ramamurthy, 1991).

In general way, results demonstrated that three chemicals (fluoxetine, vitamins B2 and B12) are transferred to microenvironment generated by weakly acidic solution of chitosan but in different proportions, due to the structural feature and solubility of each. Table 1 describes the relative increase of the fluorescence and absorption intensities of the chemicals when chitosan concentration ranges from zero to 1.0 g.L^{-1}.

Both vitamins belong to the class of hydro soluble vitamins (Sun et al., 2007), while the fluoxetine drug is slightly soluble in water (Darwish, 2005). The low solubility promotes some molecules of fluoxetine to migrate from aqueous environment to the more rigid environment generated by chitosan (the higher the concentration) causing a proportionately greater increase of absorbance and fluorescence intensities. However, the fluorescence intensities of vitamin B12 also increase in the same proportion and the absorbance intensities in proportion even higher, despite the hydro soluble nature of this vitamin.

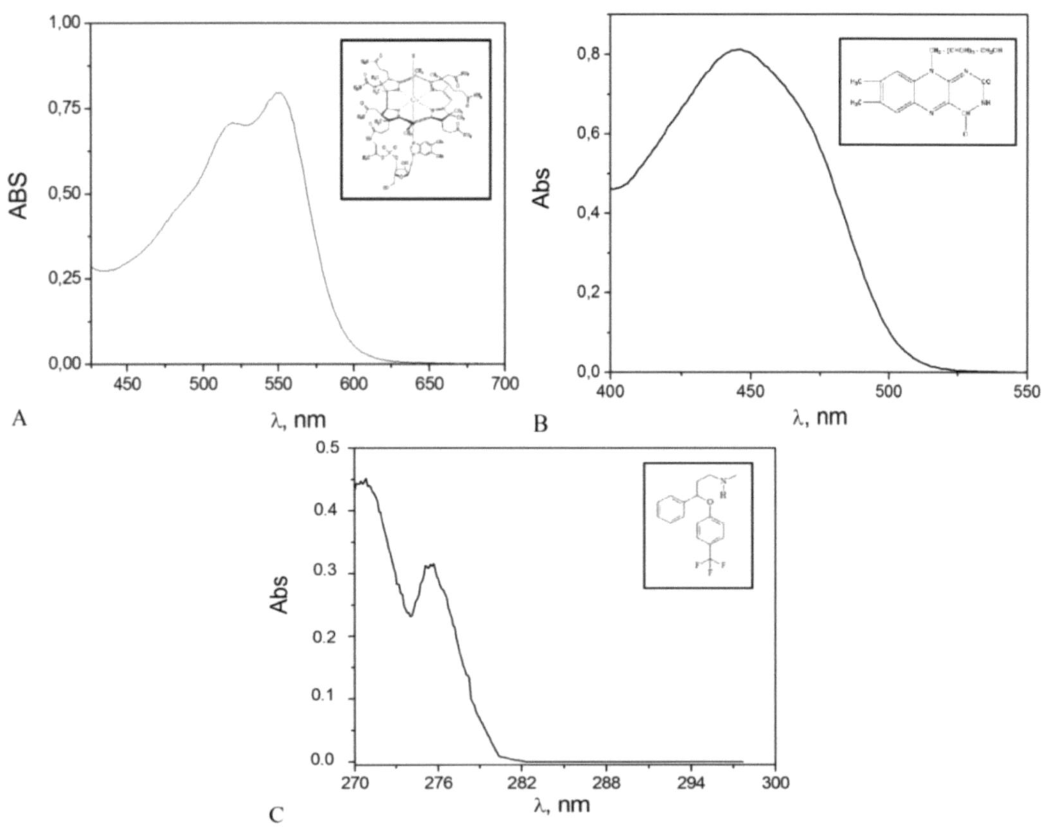

Figure 2. Absorption spectra and chemical structures of vitamin B12 (A), vitamin B2 (B) and fluoxetine (C).

These results demonstrate that the transfer process of chemicals from aqueous environment to the chitosan aggregates is influenced by solubility of the molecules in water and/or by the molecular structure. Particularly, the structure molecular of vitamin B12 seems to have an interesting effect in this case. Vitamin B12 belong to the cobalamins, a class of octahedral Co(III) complexes which contain a planar framework called a corrin, with the metal center coordinated in the equatorial position by the four corrin nitrogens. Her energies of excited states are sensitive to the nature of the ligands of center coordinated and are influenced by water content of the surrounding environment (Solheim et al., 2011). These characteristics may be the reason for the more significant increase of the spectral properties of the vitamin B12, with increasing concentration of chitosan (lower water content), compared with the vitamin B2.

In fact, some molecules of fluoxetine or vitamins B2 or B12, are transferred to microenvironment generated by weakly acidic solution of chitosan. Among the three, the vitamin B2 is transferred in a smaller proportion. However, for all of them is expected that its loss in the diet, caused by administration of chitosan, is not so significant.

From our observations, possible risks to the patient should be considered when prolonged treatment with chitosan is prescribed and perhaps extra care should be taken when chitosan and fluoxetine are prescribed together in slimming diets. In the case of vitamins, essential for many physiological functions, there must be some precautions to minimize the impacts generated by this therapeutic, as the replacement of nutrients in the diet of patient.

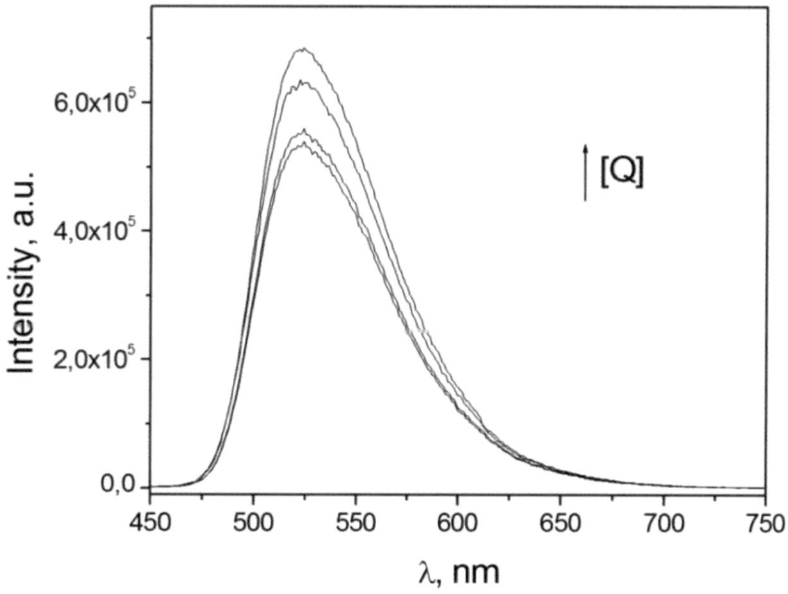

Figure 3. Fluorescence spectra of fluoxetine in acid aqueous solution of chitosan. Chitosan concentrations: 0.00; 0.050; 0.60 and 1.0 (from the base to the top).

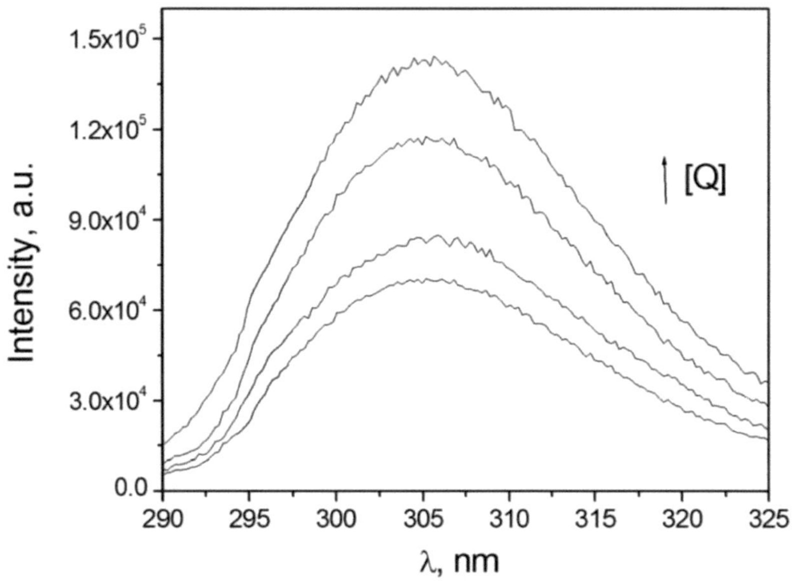

Figure 4. Fluorescence spectra of vitamin B12 in acid aqueous solution of chitosan. Chitosan concentrations: 0.00; 0.050; 0.60 and 1.0 (from the base to the top).

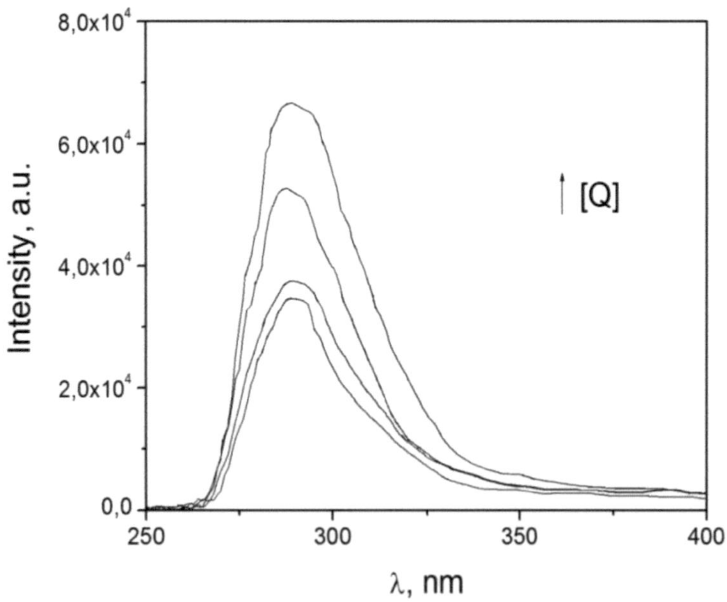

Figure 5. Fluorescence spectra of vitamin B2 in acid aqueous solution of chitosan. Chitosan concentrations: 0.00; 0.050; 0.60 and 1.0 (from the base to the top).

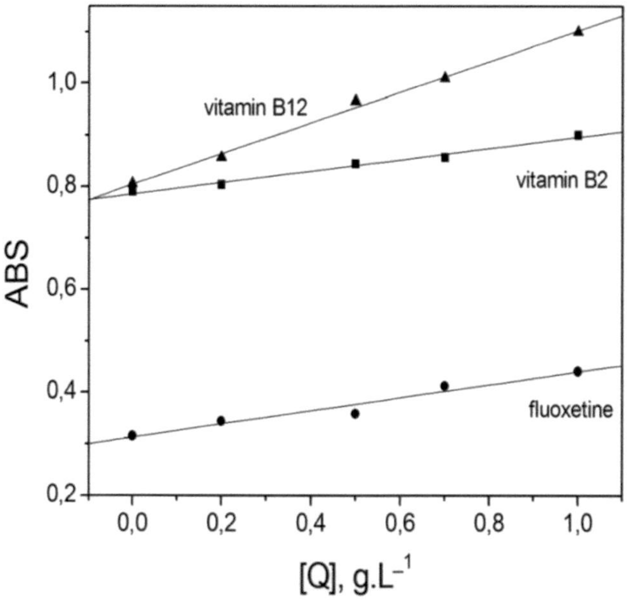

Figure 6. Behavior of absorption intensities of fluoxetine, vitamin B12 and B2.

Chemicals	I	ABS
Vitamin B12	107 %	80 %
Vitamin B2	30 %	14 %
fluoxetine	109 %	40 %

Table 1. Relative increase on the fluorescence (I) and absorption (ABS) intensities of the fluoxetine, vitamins B2 and B12 when chitosan concentration ranges from zero to 1.0 $g.L^{-1}$.

This paper seeks to warn to possible problems connected with the excessive loss of vitamins and other nutrients by the body during prolonged treatment with chitosan, as well as due the concomitant use of chitosan and fluoxetine.

4. Conclusions

Innumerous studies have described the formation of aggregates in naturals (Kim et al., 2000; Pelletier et al., 2000; Zhbankov et al., 2003), as chitosan ((Rodrigues, 2005, Hennen, 2005; Rodrigues et al., 2008) and synthetic (Kalyanasundaram, 1987; Neumann & Rodrigues, 1994; Neumann et al., 1995; Gomes et al., 2006; Gomes et al., 2007; Sur, 2010) polymers solutions and these molecular structures occur due to intra- and intermolecular interactions.

Chitosan is a polysaccharide precursor of materials suitable to release and/or dissolve drugs into the human body (Hennen, 2005), among other uses. However, the study of spectral properties of chemicals, fluoxetine, vitamins B2 and B12, demonstrated that the microenvironment generated by weakly acidic solution of chitosan also is able to sequester some B2, B12 and fluoxetine molecules, despite the hydro soluble nature of vitamins.

The results described in the current work demonstrated the wide range of possibilities in the studies of interaction between chitosan, used in diets as anti-obesity supplement, and molecules that are present in the human body, as well as with other drugs. Moreover, our work demonstrates the need for more studies on the subject as a means of providing information on the use of chitosan in diets as anti-obesity supplement.

Author details

Máira Regina Rodrigues*
Universidade Federal Fluminense, Polo Universitário de Rio das Ostras, Rio das Ostras, RJ, Brazil

Alexandre de Souza e Silva and Fábio Vieira Lacerda
Centro Universitário de Itajubá, Instituto de Ciências Biológicas, BairroVarginha, Itajubá, MG, Brazil

Acknowledgement

The authors thank CNPq and FAPESP for the financial support.

5. References

Alda, M.J.L.; Parceiro, P.; Goicoechea, A.G. (1996). Rapid zero-order and second-derivative UV spectrophotometric determination of cyanocobalamin in pharmaceutical formulations containing dextran and preservatives. *Journal of Pharmaceutical and Biomedical Analysis,* 14:363-365.

Aslander-van Vliet, E.; Smart, C.; Waldron, S. (2007). Nutritional management in childhood and adolescent diabetes. *Pediatric Diabetes,* 8:323-339.

Association of Official Analytical Chemists. (2005). Official Methods of Analysis of the Association Analytical Chemists. 18th ed. Gaithersburg, MD: AOAC.

Barbosa, L.S.; Pazdziora, A.Z.; Kojima, A.S.G.; Martins, M.S.F.; Ferreira, C.L.P.; Macêdo, G.S.; Latorraca, M.Q.; Gomes-da-Silva, M.H.G. (2007). Effects of chitosan over some nutritional and biochemical parameters in rats. *Journal of Brazilian Society of Food and Nutrition,* 32:1-13.

Böer, U.; Noll, C.; Cierny, I.; Krause, D.; Hiemke, C.; Knepel, W. (2010). A common mechanism of action of the selective serotonin reuptake inhibitors citalopram and fluoxetine: Reversal of chronic psychosocial stress-induced increase in CRE/CREB-directed gene transcription in transgenic reporter gene mice. *European Journal of Pharmacology,* 633:33-38.

British Pharmacopoeia. (1998). Her Majesty's *Stationery* Office, London, p.413.

Cummings, J.H.; Roberfroid, M.B.; Andersson, H.; Barth, C.; Ferro-Luzzi, A.; Ghoos, Y.; Gibney, M.; Hermansen, K.; James, W.P.T.; Korver, O.; Lairon, D.; Pascal, G.; Voragen,

* Corresponding Author

A.G.S. (2010). A new look at dietary carbohydrate: chemistry, physiology and health. *European Journal of Clinical Nutrition,* 64:334-339.

Dacome, L. (2005). Useless and pernicious matter: corpulence in eighteenth-century britain. In: *Cultures of the abdomen: diet, digestion and fat in the modern world.* Eds: A. Carden-Coyne and C. Forth. Palgrave, New York.

Darwish, I.A. (2005). Development and validation of spectrophotometric methods for determination of fluoxetine, sertraline, and paroxetine in pharmaceutical dosage forms. *Journal of AOAC International,* 88:38-45.

Deuchi, K.; Kanauchi, O.; Imasato, Y.; Kobayashi, E. (1994). Decreasing effect of chitosan on the apparent fat digestibility by rats fed on a high-fat diet. *Bioscience Biotechnology & Biochemistry,* 58:1613-1616.

Fregonezi-Nery, M.M.; Baracat, M.M.; Casagrande, R.; Machado, H.T.; Miglioranza, B.; Gianotto, E.A.S. (2008). Validação de métodos para determinação de fluoxetina em cápsulas. *Química Nova,* 31:1665-1669.

Gades, M.D.; Stern, J.S. (2005). Chitosan supplementation and fat absorption in men and women. *Journal of the American Dietetic Association,* 105:72-77.

Gomes, A.J.; Faustino, A.S.; Machado, A.E.H.; Zaniquelli, M.E.; Rigoletto, T.P.; Lunardi, C.N. (2006). Characterization of PLGA microparticle as a drug carrier for the compound 3-ethoxycarbonyl-2H-benzofuro[3,2-d]-1-benzopyran-2-one. Ultra structural study of cellular up take and intracellular distribution. *Drug Delivery,* 13:447–454.

Gomes, A.J.; Assunção, R.M.N.; Rodrigues Filho, G.; Espreafico, E.M.; Machado, A.E.H. (2007). Preparation and characterization of poly(D,L-lactic-co-glycolicacid) nanoparticles containing 3-(benzoxazol-2-yl)-7-(N,N-diethylamino)cromen-2-one. *Journal of Applied Polymer Science,* 105:964–972.

Hennen, W.J. (2005). *Chitosan.* Woodland Publishing Inc., Pleasant Grove, Utah.

Hirano, S.; Akiyama, Y. (1995). Absence of a hypocholesterolaemic action of chitosan in high-serum-cholesterol rabbits. *Journal of the Science of Food and Agriculture,* 69:91-94.

Ho, S.C.; Tai, E.S.; Eng, P.H.; Tan, C.E.; Fok, A.C. (2001). In the absence of dietary surveillance, chitosan does not reduce plasma lipids or obesity in hypercholesterolaemic obese Asian subjects. *The Singapore Medical Journal,* 42:6–10.

Johnson, R.J.; Segal, M.S.; Sautin, Y.; Nakagawa, T.; Feig, D.I.; Kang, D.H.; Gersch, M.S.; Benner, S.; Sánchez-Lozada, L.G. (2007). Potential role of sugar (fructose) in the epidemic of hypertension, obesity and the metabolic syndrome, diabetes, kidney disease, and cardiovascular disease. *The American Journal of Clinical Nutrition,* 86:899-906.

Jull, A.B.; Ni Mhurchu, C.; Bennett, D.A.; Dunshea-Mooij, C.A.; Rodgers, A. (2008). Chitosan for overweight or obesity. *Cochrane Database System Review,* 16:CD003892.

Kaats, G.R.; Michalek, J.E.; Preuss, H.G. (2006). Evaluating eficacy of a chitosan product using a double-blinded, placebo-controlled protocol. *Journal of the American College of Nutrition,* 25:389-394.

Kalyanasundaram. K.; Thomas, J.K. (1977). Environment effects on vibronic band intensities in pyrene monomer fluorescence and their application in studies of micellar systems. *Journal of the American Chemical Society,* 99:2039–2044.

Kalyanasundaram, K. (1987). *Photochemistry in microheterogeneous systems*. Academic, New York.

Kim, I.S.; Jeong, Y.L.; Kim, S.H. (2000). Core-shell type polymeric nanoparticles composed of poly(l-lacticacid) and poly(N-isopropylacrylamide). *The International Journal of Pharmaceutics,* 205:109-115.

Kumirska, J.; Czerwicka, M.; Kaczyński, Z.; Bychowska, A.; Brzozowski, K.; Thöming, J.; Stepnowski, P. (2010). Application of spectroscopic methods for structural analysis of chitin and chitosan. *Marine Drugs,* 8:1567–1636.

Li, H.B.; Chen, F. (2000). Determination of vitamin B12 in pharmaceutical preparations by a highly sensitive fluorimetric method. *Fresenius' Journal of Analytical Chemistry,* 368:836–838.

Liao, F.H.; Shieh, M.J.; Chang, N.C.; Chien, Y.W. (2007). Chitosan supplementation lowers serum lipids and maintains normal calcium, magnesium, and iron status in hyperlipidemic patients. *Nutrition Research*, 27: 146-151.

Liu, J.; Zhang, J.; Xia, W. (2008). Hypocholesterolaemic effects of different chitosan samples in vitro and in vivo. *Food Chemistry*, 107:419-425.

Lois, K.; Kumar, S. (2008). Pharmacotherapy of obesity. *Therapy,* 5:223-235.

Neumann, M.G.; Rodrigues, M.R. (1994). Photochemical determination of aggregation in polyelectrolytes and the binding constants of quencher molecule. *Journal of Photochemistry and Photobiology A: Chemistry*, 83:161-164.

Neumann, M.G.; Tiberti, L.; Rodrigues, M.R. (1995). Binding of metallic cations to poly(styrenesulphonate)polymers and copolymers. *Journal of the Brazilian Chemical Society,* 6:179-195.

Ni Mhurchu, C.; Poppitt, S.D.; McGill, A.T.; Leahy, F.E.; Bennett, D.A.; Lin, R.B.; Ormrod, D.; Ward, L.; Strik, C.; Rodgers, A. (2004). The effect of the dietary supplement, chitosan, on body weight: a randomised controlled trial in 250 overweight and obese adults. *International Journal of Obesity,* 28:1149–1156.

Ni Mhurchu, C.; Dunshea-Mooij, C.; Bennett, D.; Rodgers, A. (2005). Effect of chitosan on weight loss in overweight and obese individuals: a systematic review of randomized controlled trials. *Obesity Reviews*, 6:35–42.

Pelletier, S.; Hubert, P.; Lapicque, F.; Payan, E.; Dellacherie, E. (2000). Amphiphilic derivatives of sodium alginate and hyaluronate: synthesis and physic-chemical properties of aqueous dilute solutions. *Carbohydrate Polymers,* 43:343-349.

Pittler, M.H.; Abbot, N.C.; Harkness, E.F.; Ernst, E. (1999). Randomized, double-blind trial of chitosan for body weight reduction. *European Journal of Clinical Nutrition,* 53:379-381.

Pittler, M.H.; Ernst, E. (2004). Dietary supplements for body-weight reduction: a systematic review. *American Journal of Clinical Nutrition*, 79:529-536.

Prajapati, B.G. (2009). Chitosan a marine medical polymer and its lipid lowering capacity. *The internet Journal of Health*, 9:1-7.

Ramamurthy, V. (1991). Use of photophysical techniques in the study of organized assemblies. In: *Photochemistry in organized and constrained media*. Eds: C. Bohne, RW Redmond, JC Scaiano. VCH, New York.

Razdan, A.; Petterson, D. (1994). Effect of chitin and chitosan on nutrient digestibility and plasma lipid concentration in broiler chickens. *British Journal of Nutrition,* 72:277-288.

Rizo, C.; Deshpande, A.; Ing, A.; Seeman, N. (2011). A rapid, web-based method for obtaining patient views on effects and side-effects of antidepressants. *Journal of Affective Disorders,* 130:290-293.

Rodrigues, M.R. (2005). Synthesis and investigations of chitosan derivatives formed by reaction with acyl chlorides. *Journal of Carbohydrate Chemistry,* 24:41-54.

Rodrigues, M.R.; Lima, A.; Codognoto, C.; Villaverde, A.B.; Pacheco, M.T.T.; de Oliveira, H.P.M. (2008). Detection of polymolecular associations in hydrophobized chitosan derivatives using fluorescent probes. *Journal of Fluorescence,* 8:973–977.

Saper, R.B.; Eisenberg, D.M.; Phillips, R.S. (2004). Common dietary supplements for weight loss. *American Family Physician,* 70:1731-1738.

Schiller, R.N.; Barrager, E.; Schauss, A.G.; Nichols, E.J. (2001). A randomized, double-blind, placebo-controlled study examining the effects of a rapidly soluble chitosan dietary supplement on weight loss and body composition in overweight and mildly obese individuals. *Journal of the American Nutraceutical Association,* 4:42-49.

Shah, J.; Jan, R.; Rehman, F. (2008). Flow injection spectrophotometric determination of fluoxetine in bulk and in pharmaceutical preparations. *Journal of the Chilean Chemical Society,* 3:53-59.

Signini, R.; Campana Filho, S.P. (1999). On the preparation and characterization of chitosan hydrochloride. *Polymer Bulletin,* 42:159–163.

Solheim, H.; Kornobis, K.; Ruud, K.; Kozlowski, P.M. (2011). Electronically Excited States of Vitamin B12 and Methylcobalamin: Theoretical Analysis of Absorption, CD, and MCD Data. *The Journal of Physical Chemistry B,* 115:737–748.

Sommer, A. (2008). History of nutrition - Vitamin A deficiency and clinical disease: An historical overview. *Journal of Nutrition,* 138:1835-1839.

Suarez, W.T.; Sartori, E.R.; Batista, É.F.; Fatibello-Filho, O. (2009). Flow-injection turbidimetric determination of fluoxetine hydrochloride in pharmaceutical formulations. *Química Nova,* 32:2396-2400.

Sugano, M.; Fujikawa, T.; Hiratsuji, Y.; Hasegawa, Y. (1978). Hypocholesterolemic effects of chitosan in colesterol-fed rats. *Nutrition Reports International,* 18:531-537.

Sun, J.; Zhu, X.; Wu, M. (2007). Hydroxypropyl-β-Cyclodextrin Enhanced Determination for the Vitamin B12 by Fluorescence Quenching Method. *Journal of Fluorescence,* 17:265–270.

Sur, U.K. (2010). Stimuli responsive bio-inspired synthetic polymer nanocomposites. *Current Science,* 98:1562-1563.

Tapola, N.S.; Lyyra, M.L.; Kolehmainen, R.M.; Sarkkinen, E.S.; Schauss, A.G. (2008). Safety aspects and cholesterol-lowering efficacy of chitosan tablets. *Journal of* the American College of Nutrition, 27:22–30.

United States Pharmacopeia (USP). (2007). United States Pharmacopeial Convention. 30 th ed. Rockville.

Valeur, B. (2001). *Molecular fluorescence: principles and applications.* Wiley-VCH, Verlag.

Wang, X.; Wei, L.; Kotra, L.P. (2007). Cyanocobalamin (vitamin B12) conjugates with enhanced solubility. *Bioorganic & Medicinal Chemistry,* 15:1780–1787.

Yao, H.T.; Huang, S.Y.; Chiang, M.T. (2008). A comparative study on hypoglycemic and hypocholesterolemic effects of high and low molecular weight chitosan in streptozotocin-induced diabetic rats. *Food and Chemical Toxicology,* 46:1525-1534.

Zhang, J.; Zhang, J.; Liu, J.; Li, L.; Xia, W. (2008). Dietary chitosan improves hypercholesterolemia in rats fed high-fat diets. *Nutrition Research,* 28:383-390.

Zhbankov, R.G.; Firsov, S.P.; Grinshpan, D.D.; Baran, J.M.K.; Marchewkac, Ratajczak, H. (2003). Vibrational spectra and noncovalent interactions of carbohydrates molecules. *Journal of Molecular Structure,* 645:9-16.

Zheng, D.; Lu, T. (1997). Electrochemical reactions of cyanocobalamin in acidic media. *Journal of Electroanalytical Chemistry,* 429:61-65.

Chapter 24

The Future of Synthetic Carbohydrate Vaccines: Immunological Studies on *Streptococcus pneumoniae* Type 14

Dodi Safari, Ger Rijkers and Harm Snippe

Additional information is available at the end of the chapter

http://dx.doi.org/10.5772/48326

1. Introduction

Studies on synthetic carbohydrates to be used as potential vaccine candidates for polysaccharide encapsulated bacteria were started in the mid-1970s. They were the logical follow-up to studies being performed at that time on the immunogenicity of antigens composed of carrier proteins and synthetic hapten groups. Hapten-carrier complexes were first introduced in immunology by Karl Landsteiner in the early 1900s [1]. He discovered that (i) small organic molecules with a simple structure, such as phenyl arsonates and nitrophenyls, do not provoke antibodies by themselves, but (ii) if those molecules are attached covalently, by simple chemical reactions, to a protein carrier, then antibodies against those small organic molecules are evoked. Since their introduction, these hapten-carrier complexes have become excellent tools to elucidate the role of different antigen-reactive cells in the immune response [2]. The key players in this immunological process are thymus-derived T cells and bone marrow-derived B cells. The former group of lymphoid cells is responsible for various phenomena of cell-mediated immunity, e.g. delayed hypersensitivity, allograft-, and graft-versus-host reactions, and reacts with specific determinants on the carrier protein (T cell epitopes). The latter group of lymphoid cells (B cells) give rise to the precursors of antibody-secreting cells, and reacts with both the carrier protein and the synthetic haptenic determinants. This results in antibody formation to both the carrier and the hapten.

The reason to apply the above concepts and techniques to carbohydrate antigens was to address an immunological problem: polysaccharide molecules are classified as so-called thymus-independent (TI) antigens, because they do not require T cells to induce an immune response of B cells. As a result, the antibodies formed are mainly of the IgM class and have a

INTECH
open science | open minds

© 2012 The Author(s). Licensee InTech. This chapter is distributed under the terms of the Creative Commons Attribution License http://creativecommons.org/licenses/by/3.0), which permits unrestricted use, distribution, and reproduction in any medium, provided the original work is properly cited.

low avidity. Moreover, no immunological memory is generated and the antigens are poorly immunogenic in infants. Latter characteristic has major implications for development of vaccines against polysaccharide encapsulated bacteria. It was hypothesized that by linking small carbohydrates (oligosaccharides) to a carrier protein, the immunogenic behavior would change to that of a thymus-dependent (TD) antigen. Therefore, the studies of both Goebel [3, 4] and Campbell and Pappenheimer [5], who first isolated the antigenic determinant of *Streptococcus pneumoniae* type 3, were combined and extended. The hapten-inhibition studies by Mage and Kabat [6] demonstrated that the antibody-combining site of type 3 pneumococcal polysaccharide consists of two to three cellobiuronic acid units. In the dextran-anti-dextran system extensively studied by Kabat and colleagues [7] the upper size limit of the antibody-combining site appeared to be a hexa- or heptasaccharide and the lower limit was estimated to be somewhat larger than a monosaccharide. Snippe and colleagues [8] proved in 1983 that small synthetic oligosaccharides (tetra- and hexasaccharides) of *S. pneumoniae* type 3 could be transformed into TD antigens by conjugating them to a protein carrier. This opened the way to explore the synthesis and immunogenicity of numerous oligosaccharide-carrier protein conjugates of different pneumococcal serotypes. Those studies culminated in 2004 in the large-scale synthesis and introduction of a synthetic oligosaccharide vaccine for *Haemophilus influenzae* type b for use in humans in Cuba [9]. The recent exploration of gold nanoclusters coated with synthetic oligosaccharides and peptides as a vaccine are a promising platform towards the development of fully synthetic carbohydrate-based vaccines [10].

2. Streptococcus pneumoniae

Streptococcus pneumoniae (*S. pneumoniae* or pneumococcus) is a leading cause of bacterial pneumonia, meningitis, and sepsis in children worldwide. It is estimated that 1.6 million people die from these infections each year, of whom one million are children [11, 12]. *S. pneumoniae* are lancet-shaped, gram-positive, and alpha-hemolytic bacteria that colonize the mucosal surfaces of the upper respiratory tract [13]. Three major surface layers can be distinguished from the inside to the outside: the plasma membrane, the cell wall, and the capsule (Fig. 1) [14]. The cell wall consists of a triple-layered peptidoglycan backbone that anchors the capsular polysaccharide, the cell wall polysaccharide, and also various proteins such as pneumococcal surface protein A (pspA) and hyluronate lyase (Hyl) (Fig. 1). The capsule is the thickest layer, completely concealing the inner structures of exponentially growing *S. pneumoniae* bacteria.

3. Capsular polysaccharide

Capsular polysaccharides are well known as the major virulence factors of *S. pneumoniae*. Today more than 92 serotypes have been identified based on the different chemical structures of these polysaccharides [16, 17]. This diversity determines the ability of the serotypes to survive in the bloodstream and very likely the ability to cause invasive disease, especially in the respiratory tract [14, 16]. Recently, new *S. pneumoniae* serotypes have been identified, e.g. serotype 6C [17], 6D [18, 19], and 11E [20]. Capsular polysaccharides are large polymers (0.5-$2x10^6$ Da), composed of multiple repeating units of up to eight sugar residues [14]. The capsular polysaccharides are generally synthesized by the Wzx/Wzy-dependent pathway, except for type 3 and 37 which are synthesized by the synthase pathway [21, 22] (Fig. 2). In the synthase pathway capsule is produced through processive transferase activity [23, 24].

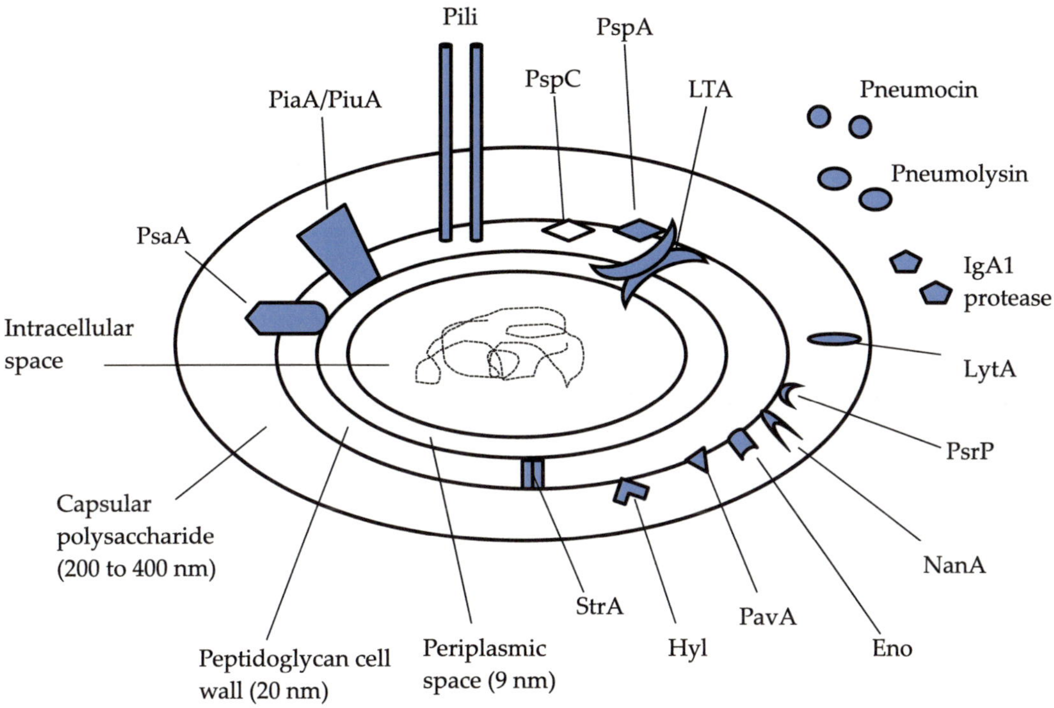

Figure 1. Schematic structure of *S. pneumoniae*. StrA=sortase A. Hyl=hyluronate lyase. PavA=pneumococcal adhesion and virulence. Eno=enolase. NanA=neuraminidase. PsrP=pneumococcal serine-rich repeat protein. LytA=autolysin. LTA=lipoteichoic acid. PspA=pneumococcal surface protein A. PspC=pneumococcal surface protein C. PiaA/PiuA=pneumococcal iron acquisition and uptake. PsaA=pneumococcal surface antigen A. (Adopted from van der Poll, T. and Opal, S.M. [15] and de Velasco, E.A. et al [14])

Many studies have demonstrated that antibodies directed against the capsular polysaccharide are essential for protection against pneumococcal disease [25-27]. However, the native capsular polysaccharides are well-known thymus-independent type-2 (TI-2) antigens that lack T-helper epitopes and therefore mainly induce IgM antibodies, and to a lesser degree IgG [28]. The TI-2 characteristics of polysaccharides can be altered by conjugation of polysaccharide to a protein carrier (glycoconjugate) resulting in a switch to an anti-polysaccharide antibody response with characteristics of a T-cell-dependent response. This is reflected by the generation of memory B and T cells and the induction of high titers of anti-polysccharide IgG antibodies after booster immunization [29].

It should be noted that not all polysaccharides behave as TI-2 antigens. Zwitterionic polysaccharides such as *S. pneumoniea* type 1 polysaccharide: [→3)-α-AATGal-(1→4)-α-D-Gal*p*A-(1→3)-α-D-Gal*p*A-(1→]$_n$ with a right-handed helix with repeated zwitterionically charged grooves elicit potent T cell responses *in vivo* and *in vitro* [30, 31].

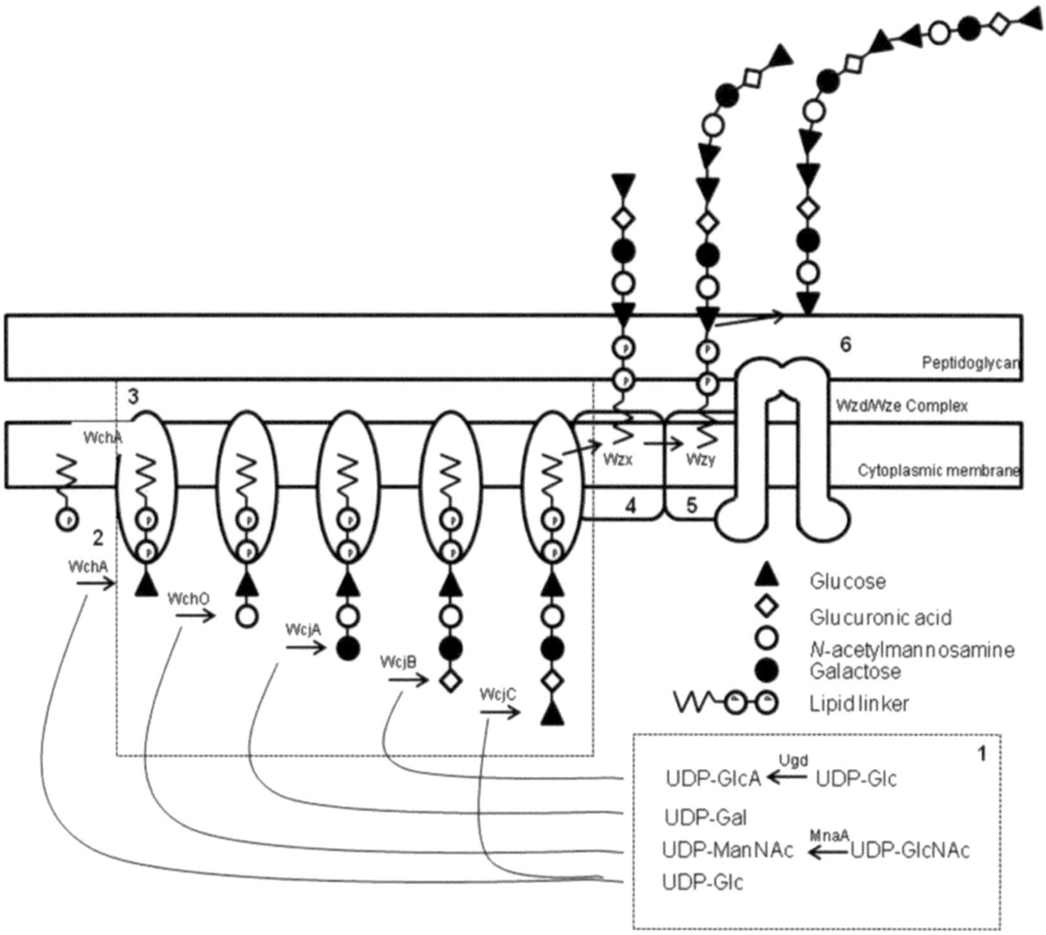

Figure 2. Representation of the Wzx/Wzy-dependent pathway for biosynthesis of CPS 9A (Adopted from Bentley. S.D. et al [21]). Representation of the Wzx/Wzy-Dependent Pathway Pictured is a hypothetical model for capsule biosynthesis in *S. pneumoniae* based on a mixture of experimental evidence and speculation.

1. Non-housekeeping nucleotide sugar biosynthesis.
2. The initial transferase (WchA in this case) links the initial sugar as a sugar phosphate (Glc-P) to a membrane-associated lipid carrier (widely assumed to be undecaprenyl phosphate).
3. Glycosyl transferases sequentially link further sugars to generate repeat unit.
4. Wzx flippase transports the repeat unit across the cytoplasmic membrane.
5. Wzy polymerase links individual repeat units to form lipid-linked CPS.

6. Wzd/Wze complex translocates mature CPS to the cell surface and may be responsible for the attachment to peptidoglycan.

4. Development of pneumococcal vaccines

Although the first pneumococcal vaccines, including the application of the principle of conjugate vaccination, were already initiated in the beginning of the previous century, most of these developments stopped when antibiotics were introduced. Existing vaccines were even withdrawn from the market. By now, in many parts of the world, the antibiotic resistance of *S. pneumoniae* bacteria has increased: America [32, 33], Africa [34], Europe [35, 36], Asia [37-39], and Australia [40]. This makes treatment of pneumococcal infections more difficult and stresses the importance of the development of effective vaccines as a strategy to reduce morbidity and mortality caused by *S. pneumoniae* infection worldwide.

4.1. Pneumococcal polysaccharide-based vaccines.

Currently two vaccine types against *S. pneumoniae* are commercially available: a pneumococcal polysaccharide vaccine (PPV) and a pneumococcal conjugate vaccine (PCV) [41]. The first multivalent pneumococcal polysaccharide vaccine (PPV) contains 23 purified capsular polysaccharides (25 μg of each capsule type; Pneumovax®, PPV23: 1, 2, 3, 4, 5, 6B, 7, 8, 9N, 9V, 10A, 11A, 12F, 14, 15B, 15F, 18C, 19A, 19F, 20, 22F, 23F, 33F) which is licensed for use in adults and children older than 2 years of age [42]. This vaccine was shown to be moderately effective in young adults [43] but not in young children [44] and elderly [45] and also not in immunocompromised patients, e.g HIV infected people [46, 47].

In early 2000, a polysaccharide-protein conjugate vaccine targeting seven pneumococcal serotypes was licensed in the United States for use in young children (Prevnar®, PCV7: 4, 6B, 9V, 14, 18C, 19F, 23F). The polysaccharides are conjugated to the non-toxic cross reactive material from diphtheria toxin, CRM_{197} and each dose contains 2μg of each capsule type, except for 6B, for which 4 μg is included in every vaccine dose[48]. The PCV7 vaccine produces a significant effect regarding prevention of invasive pneumococcal disease in children younger than 24 months (based on a meta-analysis of published data from trials on pneumococcal vaccine) [49]. Large scale introduction of PCV7 has resulted in an overall decline in infectious pneumococcal disease (IPD). However, IPD caused by the non-vaccine serotypes serotypes 1, 19A, 3, 6A, and 7F has increased (replacement disease), highlighting the need for inclusion of these serotypes in future improved vaccine formulations [50]. Apart from the CRM_{197} based PCV7, several new candidate pneumococcal conjugate vaccines have been developed to cover more serotypes with different protein carriers and most of them are in clinical trials, such as PCV10 vaccine (1, 4, 5, 6B, 7F, 9V, 14, 18C, 19F, 23F) [51, 52] and PCV13 vaccine (1, 3, 4, 5, 6A, 6B, 7F, 9V, 14, 18C, 19A, 19F, 23F) [53].

4.2. Pneumococcal protein-based vaccines

An alternative vaccine strategy focuses on the use of pneumococcal surface-associated proteins which are to be assumed to elicit protection in all age groups against all, or nearly all, pneumococcal serotypes (Fig. 1). Protection induced by the proteins should be serotype-independent and possibly cheaper and thus within reach of developing countries [54]. Currently, several surface pneumococcal proteins are investigated as a candidate vaccine against *S. pneumoniae* infection with single or combination of recombinant proteins, such as pspA family fusion protein [55]; pneumolysis and pspA1/pspA2 combined [56]. Recently new candidate protein antigens were discussed at the 8th International Symposium on Pneumococci and Pneumococcal Diseases at Iguaçu Falls, Brazil (2012), phtD (pneumococcal histidin triad protein D) and PcpA (pneumococcal choline binding protein A) [57].

4.3. Pneumococcal synthetic oligosaccharide-based vaccines

The current polysaccharide conjugate vaccines are based on natural polysaccharides, purified form bacterial cultures. Synthetic oligosaccharide–protein conjugates (neoglycoconjugate), involving functional mimics of the natural polysaccharide antigens have emerged as an attractive option [58]. The advantages of neoglycoconjugates are well-defined chemical structures (chain length, epitope conformation, and carbohydrate/protein ratio) as well as a lack of the impurities present in polysaccharides obtained from bacterial cultures [59, 60].

The chemical synthesis of oligosaccharide fragments however is complex. According to the sequence in the natural polysaccharide, monosaccharide residues have to be linked in such a way that they form an oligosaccharide with the required stereospecificity (epitope). Various methodologies and strategies for synthesis of carbohydrates have successfully been used for production of experimental neoglycoconjugates, as reviewed by Kamerling [16]. In 2001, the first automated synthesis of oligosaccharides was reported by Plante, O.J. et al [61].

Neoglycoconjugates have been prepared for saccharides of different microorganisms. In 2004, Verez Bencomo et al., reported the large-scale synthesis and the introduction of a synthetic oligosaccharide vaccine for *Haemophilus influenzae* type b for use in humans in Cuba [9]. The immunogenicity of the synthetic oligosaccharide fragment of the O-specific polysaccharide (O-PS) of *Vibrio cholera* O1, serotype Ogawa, conjugated to bovine serum albumin has been investigated in a mouse model [62, 63]. A multimeric bivalent synthetic hexasaccharide fragment of the O-specific polysaccharide of *Vibrio cholera* O1, serotype Ogawa, in combination with Inaba:1 or a synthetic disaccharide tetrapeptide peptidoglycan fragment as adjuvant were prepared and conjugated to recombinant tetanus toxin H(C) fragment as protein carrier [64]. The immunogenicity of synthetic oligosaccharides mimicking the O-antigen of the *Shigella flexneri* 2a lipopolysaccharide (LPS) was also investigated in mice [65, 66]. Immunization of mice with synthetic hexasaccharide of glycosylphosphatidylinositol malarial toxin conjugated to a protein carrier was reported to protect the mice from an otherwise lethal dose of malaria parasites [67]. A fully synthetic carbohydrate-based antitumor candidate vaccine for the common T-synthase was recently reported [68].

Meanwhile we and other groups have been working on improving the immunogenicity of neoglycoconjugates against different *S. pneumoniae* serotypes in animal models: Di-, tri-, and tetrasaccharides related to polysaccharide type 17F conjugated to keyhole limpet hemocyanin (KLH) protein[69, 70] and tri- and tetrasaccharides related to type 23 conjugated to KLH protein [71]; Di-, tri-, and tetrasaccharides related to type 6B conjugated to KLH protein [72]; Di-, tri-, and tetrasaccharide related to type 3 conjugated to the cross-reactive material of diphteria toxin (CRM_{197}) protein [60] and most recently overlapping oligosaccharide varying from tri- to dodecasaccharides related to polysaccharide type 14 conjugated to CRM_{197} protein [73, 74].

5. Immunogenicity of synthetic oligosaccharide based vaccines

This review focuses on the *S. pneumoniae* type 14 capsular polysaccharide (Pn14PS) which consists of biosynthetic repeating units of the tetrasaccharide {6)-[β-D-Galp-(1→4)-]β-D-GlcpNAc-(1→3)-β-D-Galp-(1→4)-β-D-Glcp-(1→}n [75] (Fig. 3).

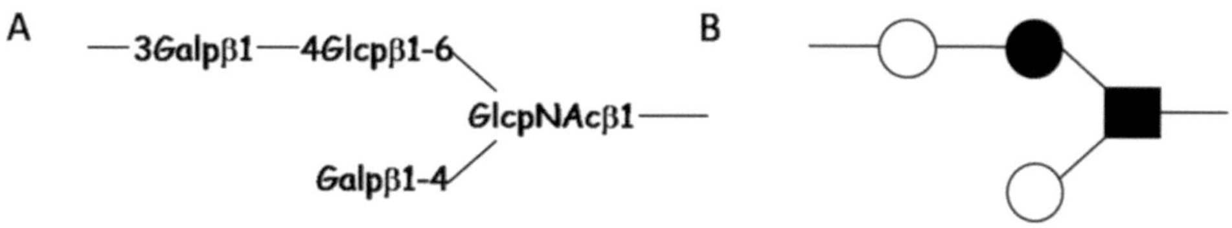

Figure 3. A branched tetrasaccharide repeating unit of *S. pneumoniae* type 14 capsular polysaccharide (A) and its nomenclature symbol (B): filled circle = glucose (Glc); open circle = galactose (Gal), and filled square = N-acetylglucosamine (GlcNAc)

5.1. Identification of the minimal structure of oligosaccharide capable in evoking anti-Pn14PS antibodies.

It was reported that a synthetic branched tetrasaccharide, corresponding to a single structural repeating unit of Pn14PS conjugated to the cross-reactive material of diptheria toxin (CRM_{197}), was found to induce anti-polysaccharide type 14 antibodies by Mawas, F. et al [74]. We continued to investigate further how small the minimal structure in Pn14PS can be and still produce specific antibodies against native polysaccharide type 14 [73]. 16 overlapping oligosaccharide fragments of Pn14PS were synthesized as described previously [76-79] and were conjugated to the protein carrier CRM_{197}. The mice immunization studies were performed to investigate the immunogenicity of the neoglycoconjugates. We found that the fragments with a linear and/or incomplete branched structure did not elicit specific antibodies against native Pn14PS (Fig. 4: JJ118, JJ42, JJ141, DM65, JJ153, JJ9, JJ6 and DM35) [73]. High titer of anti-Pn14PS IgG antibodies was observed when the complete branched structure fragments, conjugated to the protein carrier were used in the mouse model (Fig. 4: JJ1, DM66, DM36, ML1, ML2, and CRM_{197}-Pn14PS as a positive control), excepted for JJ5 and JJ10 which elicited low titer of anti-Pn14PS antibodies.

We also tested the phagocytic capacity of mice sera by human polymorph nuclear cells and a mouse macrophage cell line. We found that the sera containing antibodies against Pn14PS were also capable of promoting the phagocytosis of *S. pneumoniae* type 14. Conjugates that did not evoke specific antibodies against polysaccharide type 14 also did not display phagocytic capacity [73].

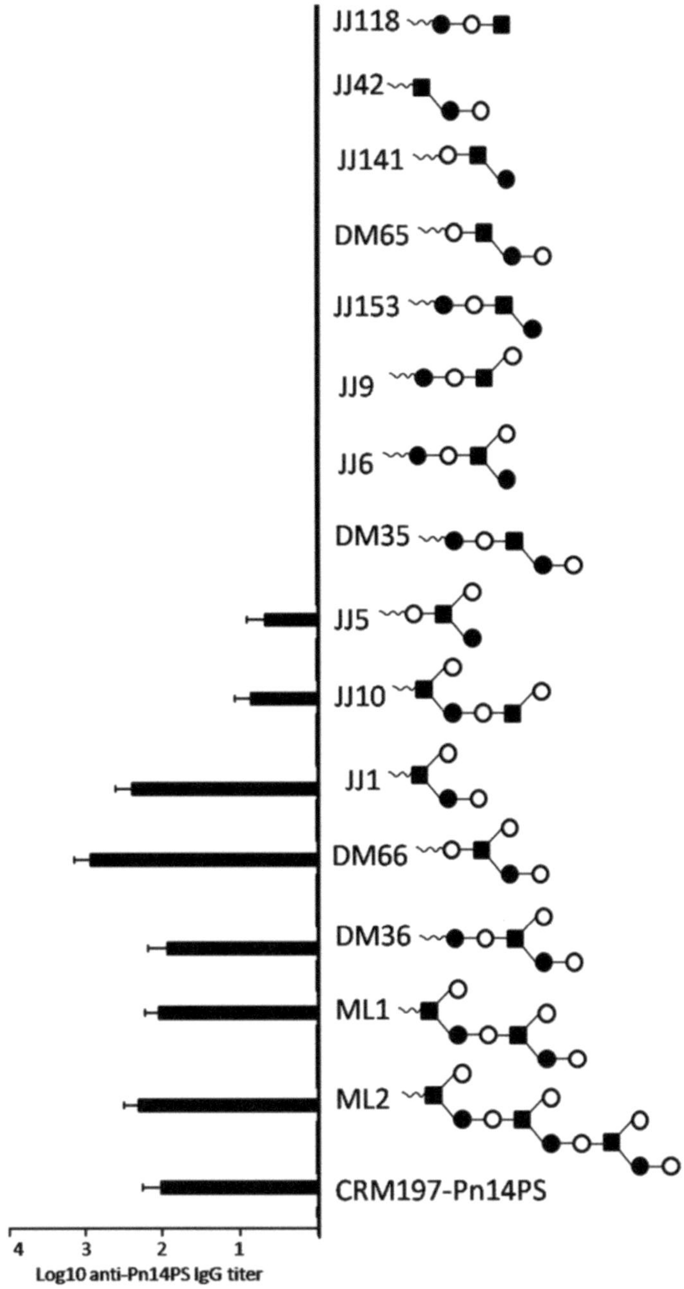

Figure 4. Level of anti-Pn14PS antibodies and schematic structure of overlapping synthetic oligosaccharide fragments of Pn14PS (Adopted from Safari et al 2008 [73]). The oligosaccharides were conjugated to CRM_{197} protein and the immunogenicity of those conjugates were studies in a mouse model. Mice were immunized with polysaccharide type 14 conjugated to CRM_{197} (CRM_{197}-Pn14PS) as a positive control. Enzyme-linked immunosorbent assay was employed to measure specific anti-Pn14PS IgG antibodies after the booster immunization. Antibody titers were expressed as the log10 of the dilution Filled circle = glucose (Glc); open circle = galactose (Gal), and filled square = N-acetylglucosamine (GlcNAc).

In conclusion, the present study has shown that the branched trisaccharide Glc-(Gal-)GlcNAc is the core structure inducing Pn14PS-specific antibodies and that the neighboring galactose at the non-reducing end significantly contributes to the induction of phagocytosis-promoting antibodies [73]. Our study provides evidence that the branched tetrasaccharide Gal-Glc-(Gal-)GlcNAc is a prime candidate for a synthetic oligosaccharide conjugate vaccine against infections caused by *S. pneumoniae* type 14 [73].

5.2. Relationship between polysaccharide of Pn14PS and GBSIII

We also determined the minimal epitope in group B streptococcus type III polysaccharide (GBSIIIPS), using both a panel of anti-Pn14PS mouse sera and sera of humans vaccinated with either Pn14PS or GBSIIIPS as reported by Safari et al [80]. Native Pn14PS is structurally related to and has cross-reactivity with GBSIIIPS [81]. The branched structures of Pn14PS and GBSIIIPS differ only in the absence (in Pn14PS) or presence (in GBSIIIPS) of the (α2→3)-linked sialic acid N-acetylneuraminic acid (Neu5Ac) in their side chains: {→4)-β-D-Glcp-(1→6)-[±α-Neu5Ac-(2→3)-β-D-Galp-(1→4)-]β-D-GlcpNAc-(1→3)-β-D-Galp-(1→}n [82]. We reported that type-specific Pn14PS antibodies which recognize the branched structure of Pn14PS have a low affinity for the native GBSIIIPS and do not promote opsonophagocytosis of GBSIII, however desialylation of GBSIIIPS, however, resulted in dramatically higher affinity of anti-Pn14PS antibodies in mice when GBSIIIP was treated by nurimindase (desialylation) [80]. These results revealed that GBSIII bacteria are protected from binding of antibodies against Pn14PS by a residue of (α2→3)-linked sialic acid, as described previously [83, 84].

5.3. Booster immunization either with either neoglycoconjugate or native polysaccharide

We investigated further the immune response to a neoglycoconjugate of Pn14PS (GC) on the outcome of sustained immunity to *S. pneumoniae* type 14 in a mouse model after the booster injection with either (GC) or native Pn14PS (PS) [85]. We found, as we expected, that the amount of specific IgG antibodies against Pn14PS increased substantially when a GC booster was given to mice previously primed with the same GC [85]. The induced antibodies were capable to opsonise *S. pneumoniae* type 14. Boosting with PS following a primary conjugate vaccine injection did not result in IgG antibody formation to Pn14PS (Table 1).

In order to explain these phenomena we investigated how a booster immunization with a GC or PS affects the cell-mediated immune response by measuring the production profile of a panel of cytokines [85]. We observed a high level of IL-5 in serum after a booster injection with GC (GC-GC or GC-GC-GC). Boosting with PS did not result in the induction of IL-5 nor of any of the other tested cytokines (Table 1; GC-PS and GC-PS-PS). We conclude that induction of the cytokine IL-5 in serum is an early sign of a successful booster immunization and is a prerequisite for the production of specific anti-polysaccharide IgG antibodies [85]. In-vitro spleen cell cultures were also used to investigate the effect of a booster injection on activation of memory T cells. IL-5 which well known Th2 cytokines, were evoked by the GC in spleen cell cultures of mice previously primed and boosted with the same GC [85]. In conclusion, the inability of polysaccharide to boost primed mice might be due to the incapability to induce the cytokines.

Immunization[1]	IgG titer (Log_{10})[2]	Level of Cytokine IL-5 (pg/ml)	
		In serum[3]	After stimulation[4]
GC-GC	2.18±0.22	1022.3±275.2	571.2±20.0
GC-PS	0.34±0.47	0.3±0.5	66.1±0.4
GC-GC-GC	3.02±0.17	2700.4±112.3	1172.8±7.1
GC-PS-PS	0.0	0.0	664.9±221.
Saline	0.0	6.9±1.1	0.0

[1]Five mice per group were immunized with a CRM-neoglycoconjugate (GC), a synthetic branched tetrasaccharide of Pn14PS that is conjugated to a CRM_{197} protein. Booster doses containing either a GC (GC-GC and GC-GC-GC) or a native polysaccharide of Pn14PS (PS) (GC-PS, GCGC-PS, and GC-PS-PS) were injected at Weeks 5 and 10.
[2]ELISA was employed to measure specific anti-Pn14PS IgG antibodies, and expressed as the log10 of the sera dilution
[3]Cytokine levels in sera from mice receiving booster injection. Sera were collected on Day 1 after the primary immunization
[4]Splenocytes were isolated 7 days after the first booster injection. Spleen cells were cultured in vitro and stimulated with CRM-neoglycoconjugate and supernatants were collected 72 h after culture initiation.

Table 1. Effect of booster immunization either with with either the same neoglycoconjugate or a native polysaccharide (Adopted from Safari, D. el at [85] with permission)

5.4. Improvement of anti-Pn14PS antibodies level by coadjuvant administration

The immunogenicity of neoglycoconjugate was increased with adjuvant coadministration [73, 86]. We set out to investigate in a mouse model the effect of adjuvant coadministration i.e. Quil-A, MPL, DDA, CpG and Alum on both the antibody- and cell-mediated immune response against a neoglycoconjugate as reported by Safari et al [87]. In the absence of adjuvant, immunization with neoglycoconjugate leads after a booster merely to IgG1 antibodies against PnP14PS. Coadministration of adjuvant had multiple effects: a diversified anti-Pn14PS IgG antibody response (also other IgG subclasses than IgG1 were evoked), an enhanced avidity and increased opsonic activity of these antibodies [87]. We found that next to Quil-A also DDA as a single dose or in combination with CpG had similar effects on the diversification of eliciting a broader variety of anti-Pn14PS IgG antibody subclasses. Meanwhile, CpG or alum on their own showed in majority IgG1 antibodies after booster immunization in a same pattern as in non adjuvant groups [87]. Compared to other adjuvants, codelivered Quil-A strongly improved the antibody avidity and enhanced the phagocytosis of *S. pneumoniae* type 14 [87].

6. Future researches

In this review, synthetic oligosaccharide-protein conjugates are proven to be effective vaccines in mice model. A logical next step would be a feasibility and immunogenicity study in human volunteers. Before that, a study should be started with synthetic oligosaccharide-protein conjugates for at least the pneumococcal serotypes 1, 4, 5, 9V and 18C and should even have been completed, because the minimal epitopes for these polysaccharides are still unknown.

To improve the immunogenicity of oligosaccharide-protein conjugates co-delivery of adjuvants are required. As an alternative to the addition of adjuvants, studies should be initiated to direct oligosaccharide-protein conjugates to dendritic cells by incorporation of specific ligands. Targeting to and activation of dendritic cells by TLR5 is a possibility to be explored.

Author details

Dodi Safari*
Eijkman Institute for Molecular Biology, Jakarta, Indonesia

Ger Rijkers
Department of Sciences, Roosevelt Academy, Middelburg and Department of Medical Microbiology and Immunology, St. Antonius Hospital, Nieuwegein, The Netherlands

Harm Snippe
Department of Medical Microbiology, University Medical Center Utrecht, Utrecht, The Netherlands

Acknowledgement

This review is a part of the PhD thesis of Dodi Safari with the title: Development of new synthetic oligosaccharide vaccines: the immunogenicity of oligosaccharide- CRM_{197} neoconjugates and oligosaccharide/peptide hybrid gold nanoparticles based on the capsular polysaccharide structure of *Streptococcus pneumoniae* type. 14. Utrecht University. 2010. This work was supported by a grant from the European Commission under Contract No: MRTN-CT-2004-005645.

7. References

[1] Landsteiner K, Jacobs J. Studies on the sensitization of animals with simple chemical compounds. J.Exp.Med. 1935;61:643-56.
[2] Mitchison NA; Rajewsky K; Taylor RB. Sterzl J and Riha I, editors.Prague symposium on developmental aspects of antibody formation and structure. Praha: Academia Publishing House; 1970.
[3] Goebel WF. Studies on antibacterial immunity induced by artificial antigens. I. Immunity to experimental pneumococcal infection with an antigen containing cellobiuronic acid. J.Exp.Med. 1939;69:353-64.
[4] Goebel WF. Studies on antibacterial immunity induced by artificial antigens. II. Immunity to experimental pneumococcal infection with antigens containing saccharides of synthetic origin. J.Exp.Med. 1940;72:33-48.
[5] Campbell JH, Pappenheimer AM. Quantitative studies of the specificity of anti-pneumococcal polysaccharide antibodies, type III and VIII. I. Isolation of

* Corresponding Author

oligosaccharides from acid and from enzymatic hydrolysates. Immunochemistry 1966;3:195-212.

[6] Mage RG, Kabat EA. Immunochemical studies on dextrans: III. The specificities of rabbit antidextrans. Further findings on antidextrans with 1,2- and 1,6-specificities. J.Immunol. 1963;(91):633-40.

[7] Kabat EA; Mayer MM. Experimental immunochemistry. IL, USA: Springfield; 1948.

[8] Snippe H, van Dam JEG, van Houte AJ, Willers JMN, Kamerling JP, Vliegenthart JFG. Preparation of a semisynthetic vaccine to *Streptococcus pneumoniae* type 3. Infect.Immun. 1983;42(2):842-4.

[9] Verez-Bencomo V, Fernandez-Santana V, Hardy E, Toledo ME, Rodriguez A, Baly A, Harrera L, Izquierdo M, Villar A, Valdes Y, Cosme K, Deler ML, Montane M, Gracia E, Ramos A, Aguilar A, Medina E, Torano G, Sosa I, Hernandez I, Martinez R, Muzachio A, Carmenates A, Costa L, Cardoso F, Campa C, Diaz V, Roy R. A synthetic conjugate polysaccharide vaccine against *Haemophilus influenzae* type b. Science 2004;305:522-5.

[10] Safari D, Maradi M, Chiodo F, Dekker HA, Shan Y, Adamo R, Oscarson S, Rijkers GT, Lahmann M, Kamerling JP, Penades S, Snippe H. Gold nanoparticles as carriers for a synthetic *Streptococcus pneumoniae* type 14 conjugate vaccine. Nanomedicine 2012;doi:10.2217/NNM.11.151

[11] Bryce J, Boschi-Pinto C, Shibuya K, Black RE. WHO estimates of the causes of death in children. Lancet 2005;365(9465):1147-52.

[12] World Health Organization. Pneumococcal conjugate vaccine for childhood immunization--WHO position paper. Wkly.Epidemiol.Rec. 2007;82(12):93-104.

[13] Kadioglu A, Weiser JN, Paton JC, Andrew PW. The role of *Streptococcus pneumoniae* virulence factors in host respiratory colonization and disease. Nat.Rev.Microbiol 2008;6(4):288-301.

[14] de Velasco EA, Verheul AF, Verhoef J, Snippe H. *Streptococcus pneumoniae*: virulence factors, pathogenesis, and vaccines. Microbiol.Rev. 1995;59(4):591-603.

[15] van der PT, Opal SM. Pathogenesis, treatment, and prevention of pneumococcal pneumonia. Lancet 2009;374(9700):1543-56.

[16] Kamerling JP. Tomasz A, editors. *Streptococcus pneumoniae,* molecular biology & mechanisms of disease. New York: Mary Ann Liebert; 1999;Pneumococcal polysaccharides: A chemical view. p. 81-114.

[17] Park IH, Pritchard DG, Cartee R, Brandao A, Brandileone MC, Nahm MH. Discovery of a new capsular serotype (6C) within serogroup 6 of *Streptococcus pneumoniae.* J Clin.Microbiol 2007;45(4):1225-33.

[18] Bratcher PE, Kim KH, Kang JH, Hong JY, Nahm MH. Identification of natural pneumococcal isolates expressing serotype 6D by genetic, biochemical and serological characterization. Microbiology 2010;156:555-60.

[19] Oftadeh S, Satzke C, Gilbert GL. Identification of the newly described *Streptococcus pneumoniae* serotype 6D using Quellung reaction and PCR. J Clin.Microbiol 2010;

[20] Calix JJ, Nahm MH. A new pneumococcal serotype, 11E, has a variably inactivated wcjE gene. J Infect.Dis. 2010;202(1):29-38.

[21] Bentley SD, Aanensen DM, Mavroidi A, Saunders D, Rabbinowitsch E, Collins M, Donohoe K, Harris D, Murphy L, Quail MA, Samuel G, Skovsted IC, Kaltoft MS, Barrell B, Reeves PR, Parkhill J, Spratt BG. Genetic analysis of the capsular biosynthetic locus from all 90 pneumococcal serotypes. PLoS.Genet. 2006;2(3):e31

[22] Mavroidi A, Aanensen DM, Godoy D, Skovsted IC, Kaltoft MS, Reeves PR, Bentley SD, Spratt BG. Genetic relatedness of the *Streptococcus pneumoniae* capsular biosynthetic loci. J.Bacteriol. 2007;189(21):7841-55.

[23] Llull D, Garcia E, Lopez R. Tts, a processive beta-glucosyltransferase of *Streptococcus pneumoniae*, directs the synthesis of the branched type 37 capsular polysaccharide in Pneumococcus and other gram-positive species. J Biol.Chem 2001;276(24):21053-61.

[24] Arrecubieta C, Lopez R, Garcia E. Type 3-specific synthase of *Streptococcus pneumoniae* (Cap3B) directs type 3 polysaccharide biosynthesis in Escherichia coli and in pneumococcal strains of different serotypes. J Exp.Med 1996;184(2):449-55.

[25] Breukels MA, Rijkers GT, Voorhorst-Ogink MM, Zegers BJ, Sanders LA. Pneumococcal conjugate vaccine primes for polysaccharide-inducible IgG2 antibody response in children with recurrent otitis media acuta. J.Infect.Dis. 1999;179(5):1152-6.

[26] Rennels MB, Edwards KM, Keyserling HL, Reisinger KS, Hogerman DA, Madore DV, Chang I, Paradiso PR, Malinoski FJ, Kimura A. Safety and immunogenicity of heptavalent pneumococcal vaccine conjugated to CRM_{197} in United States infants. Pediatrics 1998;101(4 Pt 1):604-11.

[27] Prymula R, Kriz P, Kaliskova E, Pascal T, Poolman J, Schuerman L. Effect of vaccination with pneumococcal capsular polysaccharides conjugated to Haemophilus influenzae-derived protein D on nasopharyngeal carriage of *Streptococcus pneumoniae* and *H. influenzae* in children under 2 years of age. Vaccine 2009;28(1):71-8.

[28] French N. Use of pneumococcal polysaccharide vaccines: no simple answers. J.Infect. 2003;46(2):78-86.

[29] Ada G, Isaacs D. Carbohydrate-protein conjugate vaccines. Clin.Microbiol.Infect. 2003;9(2):79-85.

[30] Kalka-Moll WM, Tzianabos AO, Bryant PW, Niemeyer M, Ploegh HL, Kasper DL. Zwitterionic polysaccharides stimulate T cells by MHC class II-dependent interactions. J Immunol. 2002;169(11):6149-53.

[31] Choi YH, Roehrl MH, Kasper DL, Wang JY. A unique structural pattern shared by T-cell-activating and abscess-regulating zwitterionic polysaccharides. Biochemistry 2002;41(51):15144-51.

[32] Karnezis TT, Smith A, Whittier S, Haddad J, Saiman L. Antimicrobial resistance among isolates causing invasive pneumococcal disease before and after licensure of heptavalent conjugate pneumococcal vaccine. PLoS One. 2009;4(6):e5965

[33] Valenzuela MT, de Quadros C. Antibiotic resistance in Latin America: a cause for alarm. Vaccine 2009;27 Suppl 3:C25-C28

[34] Vlieghe E, Phoba MF, Tamfun JJ, Jacobs J. Antibiotic resistance among bacterial pathogens in Central Africa: a review of the published literature between 1955 and 2008. Int.J Antimicrob.Agents 2009;34(4):295-303.

[35] Siira L, Rantala M, Jalava J, Hakanen AJ, Huovinen P, Kaijalainen T, Lyytikainen O, Virolainen A. Temporal trends of antimicrobial resistance and clonality of invasive *Streptococcus pneumoniae* isolates in Finland, 2002 to 2006. Antimicrob.Agents Chemother. 2009;53(5):2066-73.

[36] Vila-Corcoles A, Bejarano-Romero F, Salsench E, Ochoa-Gondar O, de DC, Gomez-Bertomeu F, Raga-Luria X, Cliville-Guasch X, Arija V. Drug-resistance in *Streptococcus pneumoniae* isolates among Spanish middle aged and older adults with community-acquired pneumonia. BMC.Infect.Dis. 2009;9:36

[37] Imai S, Ito Y, Ishida T, Hirai T, Ito I, Maekawa K, Takakura S, Iinuma Y, Ichiyama S, Mishima M. High prevalence of multidrug-resistant Pneumococcal molecular epidemiology network clones among *Streptococcus pneumoniae* isolates from adult patients with community-acquired pneumonia in Japan. Clin.Microbiol.Infect. 2009;

[38] Shibl AM. Distribution of serotypes and antibiotic resistance of invasive pneumococcal disease isolates among children aged 5 years and under in Saudi Arabia (2000-2004). Clin.Microbiol.Infect. 2008;14(9):876-9.

[39] Yang F, Xu XG, Yang MJ, Zhang YY, Klugman KP, McGee L. Antimicrobial susceptibility and molecular epidemiology of *Streptococcus pneumoniae* isolated from Shanghai, China. Int.J.Antimicrob.Agents 2008;32(5):386-91.

[40] Gottlieb T, Collignon PJ, Robson JM, Pearson JC, Bell JM. Prevalence of antimicrobial resistances in *Streptococcus pneumoniae* in Australia, 2005: report from the Australian Group on Antimicrobial Resistance. Commun.Dis.Intell. 2008;32(2):242-9.

[41] Barocchi MA, Censini S, Rappuoli R. Vaccines in the era of genomics: the pneumococcal challenge. Vaccine 2007;25(16):2963-73.

[42] O'Brien KL, Hochman M, Goldblatt D. Combined schedules of pneumococcal conjugate and polysaccharide vaccines: is hyporesponsiveness an issue? Lancet Infect.Dis. 2007;7(9):597-606.

[43] Singleton RJ, Hennessy TW, Bulkow LR, Hammitt LL, Zulz T, Hurlburt DA, Butler JC, Rudolph K, Parkinson A. Invasive pneumococcal disease caused by nonvaccine serotypes among alaska native children with high levels of 7-valent pneumococcal conjugate vaccine coverage. JAMA 2007;297(16):1784-92.

[44] Douglas RM, Paton JC, Duncan SJ, Hansman DJ. Antibody response to pneumococcal vaccination in children younger than five years of age. J Infect.Dis. 1983;148(1):131-7.

[45] Ortqvist A, Hedlund J, Burman LA, Elbel E, Hofer M, Leinonen M, Lindblad I, Sundelof B, Kalin M. Randomised trial of 23-valent pneumococcal capsular polysaccharide vaccine in prevention of pneumonia in middle-aged and elderly people. Swedish Pneumococcal Vaccination Study Group. Lancet 1998;351(9100):399-403.

[46] Teshale EH, Hanson D, Flannery B, Phares C, Wolfe M, Schuchat A, Sullivan P. Effectiveness of 23-valent polysaccharide pneumococcal vaccine on pneumonia in HIV-infected adults in the United States, 1998--2003. Vaccine 2008;26(46):5830-4.

[47] Veras MA, Enanoria WT, Castilho EA, Reingold AL. Effectiveness of the polysaccharide pneumococcal vaccine among HIV-infected persons in Brazil: a case control study. BMC.Infect.Dis. 2007;7:119

[48] Whitney CG, Farley MM, Hadler J, Harrison LH, Bennett NM, Lynfield R, Reingold A, Cieslak PR, Pilishvili T, Jackson D, Facklam RR, Jorgensen JH, Schuchat A. Decline in invasive pneumococcal disease after the introduction of protein-polysaccharide conjugate vaccine. N.Engl.J.Med. 2003;348(18):1737-46.

[49] Pavia M, Bianco A, Nobile CG, Marinelli P, Angelillo IF. Efficacy of pneumococcal vaccination in children younger than 24 months: a meta-analysis. Pediatrics 2009;123(6):e1103-e1110

[50] Isaacman DJ, McIntosh ED, Reinert RR. Burden of invasive pneumococcal disease and serotype distribution among *Streptococcus pneumoniae* isolates in young children in Europe: impact of the 7-valent pneumococcal conjugate vaccine and considerations for future conjugate vaccines. Int.J.Infect.Dis. 2009;14(3):e197-209.

[51] Knuf M, Szenborn L, Moro M, Petit C, Bermal N, Bernard L, Dieussaert I, Schuerman L. Immunogenicity of routinely used childhood vaccines when coadministered with the 10-valent pneumococcal non-typeable Haemophilus influenzae protein D conjugate vaccine (PHiD-CV). Pediatr.Infect.Dis.J. 2009;28(4 Suppl):S97-S108

[52] Bermal N, Szenborn L, Chrobot A, Alberto E, Lommel P, Gatchalian S, Dieussaert I, Schuerman L. The 10-valent pneumococcal non-typeable Haemophilus influenzae protein D conjugate vaccine (PHiD-CV) coadministered with DTPw-HBV/Hib and poliovirus vaccines: assessment of immunogenicity. Pediatr.Infect.Dis.J. 2009;28(4 Suppl):S89-S96

[53] Meng C, Lin H, Huang J, Wang H, Cai Q, Fang L, Guo Y. Development of 5-valent conjugate pneumococcal protein A - Capsular polysaccharide pneumococcal vaccine against invasive pneumococcal disease. Microb.Pathog. 2009;47(3):151-6.

[54] Bogaert D, de Groot R, Hermans PWM. *Streptococcus pneumoniae* colonisation: the key to pneumococcal disease . Lancet.Infect.Dis. 2004;4:144-54.

[55] Xin W, Li Y, Mo H, Roland KL, Curtiss R, III. PspA family fusion proteins delivered by attenuated Salmonella enterica serovar Typhimurium extend and enhance protection against *Streptococcus pneumoniae*. Infect.Immun. 2009;77(10):4518-28.

[56] Francis JP, Richmond PC, Pomat WS, Michael A, Keno H, Phuanukoonnon S, Nelson JB, Whinnen M, Heinrich T, Smith WA, Prescott SL, Holt PG, Siba PM, Lehmann D, van den Biggelaar AH. Maternal antibodies to pneumolysin, but not pneumococcal surface protein A, delay early pneumococcal carriage in high-risk Papua New Guinean infants. Clin.Vaccine Immunol. 2009;16(11):1633-8.

[57] Ljutic B, Ochs M, Messham B, Ming M, Dookie A, Harper K, Ausar SF. Formulation, stability and immunogenicity of a trivalent pneumococcal protein vaccine formulated with aluminum salt adjuvants. Vaccine 2012;Epub ahead of print

[58] Pozsgay V. Recent developments in synthetic oligosaccharide-based bacterial vaccines. Curr.Top.Med Chem 2008;8(2):126-40.

[59] Jansen WT, Snippe H. Short-chain oligosaccharide protein conjugates as experimental pneumococcal vaccines. Indian.J.Med.Res. 2004;119:7-12.

[60] Benaissa-Trouw B, Lefeber DJ, Kamerling JP, Vliegenthart JF, Kraaijeveld K, Snippe H. Synthetic polysaccharide type 3-related di-, tri-, and tetrasaccharide- CRM_{197} conjugates

induce protection against *Streptococcus pneumoniae* type 3 in mice. Infect.Immun. 2001;69(7):4698-701.

[61] Plante OJ, Palmacci ER, Seeberger PH. Automated solid-phase synthesis of oligosaccharides. Science 2001;291(5508):1523-7.

[62] Saksena R, Ma X, Wade TK, Kovác P, Wade WF. Length of the linker and the interval between immunizations influences the efficacy of *Vibrio cholerae* O1,Ogawa hexasaccharide neoglycoconjugates. FEMS Immunol.Med.Microbiol. 2006;47:116-28.

[63] Saksena R, Ma X, Wade TK, Kovac P, Wade WF. Effect of saccharide length on the immunogenicity of neoglycoconjugates from synthetic fragments of the O-SP of *Vibrio cholerae* O1, serotype Ogawa. Carbohydr.Res. 2005;340:2256-69.

[64] Bongat AF, Saksena R, Adamo R, Fujimoto Y, Shiokawa Z, Peterson DC, Fukase K, Vann WF, Kovac P. Multimeric bivalent immunogens from recombinant tetanus toxin H(C) fragment, synthetic hexasaccharides, and a glycopeptide adjuvant. Glycoconj.J. 2009;27(1):69-77.

[65] Phalipon A, Tanguy M, Grandjean C, Guerreiro C, Belot F, Cohen D, Sansonetti PJ, Mulard LA. A synthetic carbohydrate-protein conjugate vaccine candidate against Shigella flexneri 2a infection. J.Immunol. 2009;182(4):2241-7.

[66] Said HF, Phalipon A, Tanguy M, Guerreiro C, Belot F, Frisch B, Mulard LA, Schuber F. Rational design and immunogenicity of liposome-based diepitope constructs: application to synthetic oligosaccharides mimicking the Shigella flexneri 2a O-antigen. Vaccine 2009;27(39):5419-26.

[67] Schofield L, Hewitt MC, Evans K, Siomos MA, Seeberger PH. Synthetic GPI as a candidate anti-toxic vaccine in a model of malaria. Nature 2002;418(6899):785-9.

[68] Jeon I, Lee D, Krauss IJ, Danishefsky SJ. A new model for the presentation of tumor-associated antigens and the quest for an anticancer vaccine: a solution to the synthesis challenge via ring-closing metathesis. J.Am.Chem.Soc. 2009;131(40):14337-44.

[69] de Velasco EA, Verheul AF, Veeneman GH, Gomes LJ, van Boom JH, Verhoef J, Snippe H. Protein-conjugated synthetic di- and trisaccharides of pneumococcal type 17F exhibit a different immunogenicity and antigenicity than tetrasaccharide. Vaccine 1993;11(14):1429-36.

[70] Jansen WT, Verheul AFM, Veeneman GH, van Boom JH, Snippe H. Revised interpretation of the immunological results obtained with pneumococcal polysaccharide 17F derived synthetic di-, tri- and tetrasaccharide conjugates in mice and rabbits. Vaccine 2002;20:19-21.

[71] de Velasco EA, Verheul AF, van Steijn AM, Dekker HA, Feldman RG, Fernandez IM, Kamerling JP, Vliegenthart JF, Verhoef J, Snippe H. Epitope specificity of rabbit immunoglobulin G (IgG) elicited by pneumococcal type 23F synthetic oligosaccharide- and native polysaccharide-protein conjugate vaccines: comparison with human anti-polysaccharide 23F IgG. Infect.Immun. 1994;62(3):799-808.

[72] Jansen WT, Hogenboom S, Thijssen MJL, Kamerling JP, Vliegenthart JFG, Verhoef J, Snippe H, Verheul AFM. Synthetic 6B di-, tri-, and tetrasaccharide-protein conjugates contain pneumococcal type 6A and 6B common and 6B-specific epitopes that elicit protective antibodies in mice. Infect.Immun. 2001;69(2):787-93.

[73] Safari D, Dekker HA, Joosten JA, Michalik D, de Souza AC, Adamo R, Lahmann M, Sundgren A, Oscarson S, Kamerling JP, Snippe H. Identification of the smallest structure capable of evoking opsonophagocytic antibodies against *Streptococcus pneumoniae* type 14. Infect.Immun. 2008;76(10):4615-23.

[74] Mawas F, Niggemann J, Jones C, Corbel MJ, Kamerling JP, Vliegenthart JFG. Immunogenicity in a mouse model of a conjugate vaccine made with a synthetic single repeating unit of type 14 pneumococcal polysaccharide coupled to CRM_{197}. Infect.Immun. 2002;70(9):5107-14.

[75] Lindberg B, Lonngren J, Powel DA. Structural studies on the specific type 14 pneumococcal polysaccharide. Carbohydr.Res. 1977;58:177-86.

[76] Joosten JA, Lazet BJ, Kamerling JP, Vliegenthart JF. Chemo-enzymatic synthesis of tetra-, penta-, and hexasaccharide fragments of the capsular polysaccharide of *Streptococcus pneumoniae* type 14. Carbohydr.Res 2003;338(23):2629-51.

[77] Joosten JA, Kamerling JP, Vliegenthart JF. Chemo-enzymatic synthesis of a tetra- and octasaccharide fragment of the capsular polysaccharide of *Streptococcus pneumoniae* type 14. Carbohydr.Res 2003;338(23):2611-27.

[78] Michalik D, Vliegenthart JFG, Kamerling JP. Chemoenzymic synthesis of oligosaccharide fragments of the capsular polysaccharide of *Streptococcus pneumoniae* type 14. J.Chem.Soc., Perkin Trans.1 2002;1973-81.

[79] Sundgren A, Lahmann M, Oscarson S. Block synthesis of *Streptococcus pneumoniae* type 14 capsular polysaccharide structures. J.Carbohydr.Chem. 2005;24(4):379-91.

[80] Safari D, Dekker HA, Rijkers G, van der Ende A, Kamerling JP, Snippe H. The immune response to group B streptococcus type III capsular polysaccharide is directed to the -Glc-GlcNAc-Gal- backbone epitope. Glycoconj.J. 2011;28:557-67.

[81] Guttormsen HK, Baker CJ, Nahm MH, Paoletti LC, Zughaier SM, Edwards MS, Kasper DL. Type III group B streptococcal polysaccharide induces antibodies that cross-react with *Streptococcus pneumoniae* type 14. Infect.Immun. 2002;70(4):1724-38.

[82] Kadirvelraj R, Gonzalez-Quteirino J, Foley BL, Beckham ML, Jennings HJ, Foote S, Ford MG, Woods RJ. Understanding the bacterial polysaccharide antigenicity of *Streptococcus agalactiae* versus *Streptococcus pneumoniae*. Proc.Natl.Acad.Sci.USA 2007;103(21):8149-54.

[83] Jennings HJ, Lugowski C, Kasper DL. Conformational aspects critical to the immunospecificity of the type III group B streptococcal polysaccharide. Biochemistry 1981;20(16):4511-8.

[84] Miernyk KM, Butler JC, Bulkow LR, Singleton RJ, Hennessy TW, Dentinger CM, Peters HV, Knutsen B, Hickel J, Parkinson AJ. Immunogenicity and reactogenicity of pneumococcal polysaccharide and conjugate vaccines in alaska native adults 55-70 years of age. Clin.Infect.Dis. 2009;49(2):241-8.

[85] Safari D, Dekker HA, de Jong B, Rijkers G, Kamerling JP, Snippe H. Antibody- and cell-mediated immune responses to a synthetic oligosaccharide conjugate vaccine after booster immunization. Vaccine 2011;29(38):6498-504.

[86] Lefeber DJ, Benaissa-Trouw B, Vliegenthart JFG, Kamerling JP, Jansen WTM, Kraaijeveld K, Snippe H. Th1-directing adjuvants increase the immunogenicity of

oligosaccharide-protein conjugate vaccines related to *Streptococcus pneumoniae* type 3. Infect.Immun. 2003;71(12):6915-20.

[87] Safari D, Dekker HA, Rijkers G, Snippe H. Codelivery of adjuvants at the primary immunization site is essential for evoking a robust immune response to neoglycoconjugates. Vaccine 2011;29(4):849-54.